Magnetic Resonance
of Carbonaceous Solids

ADVANCES IN CHEMISTRY SERIES **229**

Magnetic Resonance of Carbonaceous Solids

Robert E. Botto, EDITOR
Argonne National Laboratory

Yuzo Sanada, EDITOR
Hokkaido University

Developed from a symposium sponsored
by the International Chemical Congress
of the Pacific Basin Societies

American Chemical Society, Washington, DC 1993

Library of Congress Cataloging-in-Publication Data

Magnetic resonance of carbonaceous solids: developed by a symposium sponsored by the 1989 International Chemical Congress of Pacific Basin Societies, Honolulu, Hawaii, December 17–22, 1989 / Robert E. Botto, editor, Yūzō Sanada, editor.

p. cm. — (Advances in chemistry series, ISSN 0065–2393 ; 229)

Includes bibliograhical references and index.

ISBN 0–8412–1866–8

1. Coal—Analysis—Congresses. 2. Carbon—Analysis—Congresses. 3. Nuclear magnetic resonance spectroscopy—Congresses.

I. Botto, Robert E., 1946– . II. Sanada, Yūzō, 1932–
III. International Chemical Congress of Pacific Basin Societies (1989: Honolulu, Hawaii) IV. Series.

QD1.A355 no. 229
[TP325]
540 s—dc20   92–36495
[662.6'22]   CIP

The paper used in this publication meets the minimum requirements of American National Standard for Information Sciences—Permanence of Paper for Printed Library Materials, ANSI Z39.48–1984.

Copyright © 1993

American Chemical Society

All Rights Reserved. The appearance of the code at the bottom of the first page of each chapter in this volume indicates the copyright owner's consent that reprographic copies of the chapter may be made for personal or internal use or for the personal or internal use of specific clients. This consent is given on the condition, however, that the copier pay the stated per-copy fee through the Copyright Clearance Center, Inc., 27 Congress Street, Salem, MA 01970, for copying beyond that permitted by Sections 107 or 108 of the U.S. Copyright Law. This consent does not extend to copying or transmission by any means—graphic or electronic—for any other purpose, such as for general distribution, for advertising or promotional purposes, for creating a new collective work, for resale, or for information storage and retrieval systems. The copying fee for each chapter is indicated in the code at the bottom of the first page of the chapter.

The citation of trade names and/or names of manufacturers in this publication is not to be construed as an endorsement or as approval by ACS of the commercial products or services referenced herein; nor should the mere reference herein to any drawing, specification, chemical process, or other data be regarded as a license or as a conveyance of any right or permission to the holder, reader, or any other person or corporation, to manufacture, reproduce, use, or sell any patented invention or copyrighted work that may in any way be related thereto. Registered names, trademarks, etc., used in this publication, even without specific indication thereof, are not to be considered unprotected by law.

PRINTED IN THE UNITED STATES OF AMERICA

1993 Advisory Board
Advances in Chemistry Series
M. Joan Comstock, *Series Editor*

V. Dean Adams
Tennessee Technological University

Robert J. Alaimo
Procter & Gamble Pharmaceuticals, Inc.

Mark Arnold
University of Iowa

David Baker
University of Tennessee

Arindam Bose
Pfizer Central Research

Robert F. Brady, Jr.
Naval Research Laboratory

Margaret A. Cavanaugh
National Science Foundation

Dennis W. Hess
Lehigh University

Hiroshi Ito
IBM Almaden Research Center

Madeleine M. Joullie
University of Pennsylvania

Gretchen S. Kohl
Dow-Corning Corporation

Bonnie Lawlor
Institute for Scientific Information

Douglas R. Lloyd
The University of Texas at Austin

Robert McGorrin
Kraft General Foods

Julius J. Menn
Plant Sciences Institute,
U.S. Department of Agriculture

Vincent Pecoraro
University of Michigan

Marshall Phillips
Delmont Laboratories

George W. Roberts
North Carolina State University

A. Truman Schwartz
Macalaster College

John R. Shapley
University of Illinois
at Urbana–Champaign

Peter Willett
University of Sheffield (England)

FOREWORD

The ADVANCES IN CHEMISTRY SERIES was founded in 1949 by the American Chemical Society as an outlet for symposia and collections of data in special areas of topical interest that could not be accommodated in the Society's journals. It provides a medium for symposia that would otherwise be fragmented because their papers would be distributed among several journals or not published at all.

Papers are reviewed critically according to ACS editorial standards and receive the careful attention and processing characteristic of ACS publications. Volumes in the ADVANCES IN CHEMISTRY SERIES maintain the integrity of the symposia on which they are based; however, verbatim reproductions of previously published papers are not accepted. Papers may include reports of research as well as reviews, because symposia my embrace both types of presentation.

ABOUT THE EDITORS

ROBERT E. BOTTO is a research chemist in the Chemistry Division at Argonne National Laboratory in Argonne, Illinois. His research involves the characterization of fossil fuels, polymers, and heterogeneous catalysts by solid-state NMR spectroscopy and imaging. He currently holds the position of Director of the Chemistry Division NMR Laboratory.

In addition to authoring more than 50 scientific publications, he has presented as many papers at national and international meetings and has organized numerous national and international symposia. He is a member of the Sigma Xi Honorary Research Society and the American Chemical Society, where he served as Secretary of the Fuel Chemistry Division from 1987 to 1989 and Newsletter Editor in 1990.

He received his A.B. degree in chemistry in 1968 from Rutgers College in New Brunswick, New Jersey. He received his M.S. degree in chemistry from Michigan State University in East Lansing, Michigan, in 1970. Following two years of service in the U.S. Army Science and Engineering Corps, he returned to Michigan State University where he received his Ph.D. degree in 1975.

YUZO SANADA is a professor with the Faculty of Engineering of the Metals Research Institute at Hokkaido University. After receiving his B.Eng. degree from Toyama University and his Ph.D. from Tohoku University, he performed his postdoctoral work with the Research Council of Alberta in Edmonton.

Before joining the faculty at Hokkaido University, he performed research for the Resources Research Institute (1954–1965) and was a senior researcher at the National Research Institute for Pollution and Resources (1965– 1975). His research expertise is in coal chemistry, heavy hydrocarbons, and carbon material science, and he specializes in the constitution and structural analysis of coal and related carbonaceous materials, the chemistry of coal liquefaction, and the pyrolysis and carbonization of coal and pitch. He received the Japan Petroleum Institute Medal in 1975 and 1990 and the Fuel Society of Japan Medal in 1976.

He has written several books on coal and carbonization engineering. He has also authored more than 200 papers. In addition, he is a member of the Editorial Advisory Board of *Energy and Fuels* and a board member of both the Fuel Society of Japan and the Carbon Society of Japan.

CONTENTS

Preface xiii

OVERVIEWS

1. Quantitation in ^{13}C NMR Spectroscopy of Carbonaceous
   Solids 3
   Robert A. Wind, Gary E. Maciel, and Robert E. Botto

2. ^{1}H NMR Spectroscopy: Approaches for Carbonaceous
   Solids 27
   Gary E. Maciel, Charles E. Bronnimann,
   and Cynthia F. Ridenour

3. Solid Materials Research with NMR and Dynamic Nuclear
   Polarization Spectroscopy 45
   Robert A. Wind, Russ Lewis, Herman Lock,
   and Gary E. Maciel

4. Advances in Electron Nuclear Double Resonance
   Spectroscopy 65
   Hans Thomann and Marcelino Bernardo

5. Strategies for Measurement of Electron Spin Relaxation 91
   Michael K. Bowman

6. Multifrequency Electron Paramagnetic Resonance
   Spectroscopy 107
   R. L. Belford and R. B. Clarkson

7. High-Temperature Electron Spin Resonance and NMR
   Methods Applied to Coal 139
   Yuzo Sanada and Leo J. Lynch

NMR SPECTROSCOPY

8. Diffusion-Coupled Relaxation and the Submicroscopic
   Structure of Bituminous Coals 175
   Wesley A. Barton, Leo J. Lynch, David S. Webster,
   and Greg Simms

9. ^{13}C NMR Spectroscopy of Pyridine and Alkylpyridines Sorbed onto Coal 201
   Anthony M. Vassallo

10. Distortion-Free ^{13}C NMR Spectroscopy in Coal: ^{1}H Rotating-Frame Dynamic Nuclear Polarization and ^{1}H–^{13}C Cross-Polarization 217
    Robert A. Wind

11. Proton Magnetic Resonance Thermal Analysis of Argonne Premium Coals 229
    Richard Sakurovs, Leo J. Lynch, and Wesley A. Barton

12. Multivariate Statistical Analysis and Pattern Recognition: Principal-Component Analysis of Coal Solid-State NMR Relaxation Times 253
    David E. Axelson

13. Structural Parameters of Argonne Coal Samples: Solid-State ^{13}C NMR Spectroscopy 269
    Yoshio Adachi and Minoru Nakamizo

14. Characterization of Coal Liquefaction Residues by Solid-State ^{13}C NMR Spectroscopy 281
    Natsuko Cyr and Nosa O. Egiebor

15. ^{1}H NMR Spectroscopy and Spin–Lattice Relaxation of Argonne Premium Coals 295
    Kikuko Hayamizu, Shigenobu Hayashi, Kunio Kamiya, and Mitsutaka Kawamura

16. High-Field NMR Studies of Argonne Premium Coals 311
    Jiangzhi Hu, Liyun Li, and Chaohui Ye

17. Solid-State ^{13}C NMR Studies on Coal and Coal Oxidation 323
    J. Anthony MacPhee, Hiroyuki Kawashima, Yasumasa Yamashita, and Yoshio Yamada

18. Measurement of Spin–Lattice Relaxation in Argonne Premium Coal Samples 341
    Chihji Tsiao and Robert E. Botto

19. Quantitation of Protons in the Argonne Premium Coals by Solid-State ^{1}H NMR Spectroscopy 359
    Luisita dela Rosa, Marek Pruski, Bernard Gerstein

20. Bloch-Decay and Cross-Polarization–Magic-Angle Spinning ^{13}C NMR Study of the Argonne Premium Coals: Effects of High-Speed Spinning 377
    James A. Franz and John C. Linehan

21. High-Resolution ^{1}H NMR Studies of Argonne Premium Coals 401
    Antoni Jurkiewicz, Charles E. Bronnimann, and Gary E. Maciel

22. Measurement of ^{13}C Chemical-Shift Anisotropy in Coal 419
    Anita M. Orendt, Mark S. Solum, Naresh K. Sethi, Craig D. Hughes, Ronald J. Pugmire, and David M. Grant

ELECTRON PARAMAGNETIC RESONANCE SPECTROSCOPY

23. 2-mm Band and X-Band Electron Spin Resonance and Electron Spin-Echo Investigations of Some Carbonaceous Materials 443
    Yuri D. Tsvetkov, Sergei A. Dzuba, and Victor I. Gulin

24. Electron Spin Resonance, Electron Nuclear Double Resonance, and Electron Spin-Echo Spectroscopic Studies of Argonne Premium Coals 451
    Xinhua Chen, Hugh McManus, and Larry Kevan

25. Electron Paramagnetic Resonance Spin-Probe Studies of Porosity in Solvent-Swelled Coal 467
    Ross Spears, Janina Goslar, and Lowell D. Kispert

26. Dynamic In Situ 9-GHz Electron Paramagnetic Resonance Studies of Argonne Premium Coal Samples 483
    H. A. Buckmaster and Jadwiga Kudynska

27. Electron Magnetic Resonance of Standard Coal Samples at Multiple Microwave Frequencies 507
    R. B. Clarkson, Wei Wang, D. R. Brown, H. C. Crookham, and R. L. Belford

28. Novel Characterization of Argonne Premium Coals with Electron Donors or Acceptors by Electron Paramagnetic Resonance Spectroscopy 529
    T. Keneko, M. Sasaki, T. Yokono, and Yuzo Sanada

29. Electron Paramagnetic Resonance and Electron Spin-Echo Spectroscopy of Argonne Premium Coals 539
Bernard G. Silbernagel, L. A. Gebhard, Marcelino Bernardo, and H. Thomann

30. Pulsed Electron Nuclear Double Resonance Spectroscopy of Argonne Premium Coals 561
Hans Thomann, Marcelino Bernardo, and Bernard G. Silbernagel

31. Temperature Dependence of the Electron Paramagnetic Resonance Intensity of Whole Coals: The Search for Triplet States 581
Kurt S. Rothenberger, Richard F. Sprecher, Salvatore M. Castellano, and Herbert L. Retcofsky

32. Measurement of Electron Dipolar Fields and Dynamics in Solids 605
Michael K. Bowman

CONCLUSION

33. Advanced Magnetic Resonance Techniques Applied to Argonne Premium Coals 629
Bernard G. Silbernagel and Robert E. Botto

INDEXES

Author Index 647

Affiliation Index 647

Subject Index 648

PREFACE

THE GREATLY EXPANDING ROLE of magnetic resonance techniques in the characterization of carbonaceous solids over the past two decades has significantly enhanced our fundamental understanding of the structure of these materials. NMR spectroscopy of solids has matured to the point that the experimental parameters obtained are fairly well understood, and thus the usefulness of the method as an analytical structural tool is being pursued vigorously. Developments in the field of electron paramagnetic resonance (EPR) spectroscopy have led to novel pulse experiments that are providing new, quantitative information on nuclei in the vicinity of free electrons and on the properties of the free electrons themselves.

The symposium upon which this book is based was designed to foster research on new magnetic resonance techniques, and the focus was on a common, homogeneous set of carbonaceous samples, the Argonne Premium coals. The symposium provided a timely forum on the fundamental structural implications of magnetic resonance studies on this unique suite of samples and presented the opportunity for evaluating and comparing different techniques for the analysis of complex carbonaceous fuels.

This volume is divided into several sections: The first section is composed of seven invited chapters on experimental techniques in magnetic resonance. This overview section is followed by two sections containing 25 contributed papers from the symposium in the areas of NMR and EPR spectroscopy. A conclusion chapter summarizes the results and offers a good overview of the relevant issues regarding analysis of carbonaceous solids by magnetic resonance methods.

The seven overview chapters on magnetic resonance techniques provide comprehensive reviews and useful detailed information on experimental methods in NMR and EPR spectroscopy of solids. The topics include quantitative aspects of carbon and multipulse proton NMR spectroscopy, spectral enhancement with dynamic nuclear polarization, pulsed electron–nuclear double resonance (ENDOR) and electron spin-relaxation studies, and experiments performed at high frequencies and at very high temperatures.

The chapters in this volume that were selected from the symposium present new measurements and results together with a critical evaluation of other relevant work on magnetic resonance spectroscopy of carbonaceous solids. In many instances, the experiments are quite novel.

Contributions in the area of NMR spectroscopy explore quantitative issues in carbon and proton spectroscopy, new signal-enhancement techniques, ultrafast magic-angle spinning methods, and the use of relaxation measurements to probe the physical structure of coal at the microscopic level.

New techniques in EPR spectroscopy, particularly electron spin-echo (ESE) and pulsed ENDOR spectroscopy, give new insights into the carbon radical environment. Several chapters explore the nature of the carbon radicals by using analyses of EPR line shapes and ESE relaxation times. Other contributions deal with structure changes induced by oxidation, thermolysis, or solvent swelling, and the use of the anisotropic chemical-shift tensors and the implementation of multivariate statistical analyses of relaxation data to expand our understanding of coal structure.

We express our gratitude to the sponsors of the symposium for financial support: The Petroleum Research Fund, Amoco Oil Company, Exxon Engineering and Research Company, Varian Instruments, Bruker Instruments, and Merck Isotopes, Inc. Only with their help was it possible to gather such a distinguished field of scientists form four different continents to participate in this timely forum on the fundamental aspects of magnetic resonance to carbonaceous solids. Special appreciation is due to Karl Vorres for distributing Argonne Premium coals to the symposium participants, Debbie Vervack for much help in organizing the conference and for editorial assistance, and to Julie Poudrier Skinner, Cheryl Shanks, and M. Joan Comstock at ACS books who were actively involved in many of the details and decisions that have made this volume become a reality.

ROBERT E. BOTTO
Argonne National Laboratory
Argonne, IL 60439

YUZO SANADA
Hokkaido University
Sapporo 060 Japan

October 28, 1991

OVERVIEWS

Quantitation in ^{13}C NMR Spectroscopy of Carbonaceous Solids

Robert A. Wind[1], Gary E. Maciel[2], and Robert E. Botto[3]

[1]Chemagnetics, Inc., 2555 Midpoint Drive, Fort Collins, CO 80525
[2]Department of Chemistry, Colorado State University, Fort Collins, CO 80523
[3]Chemistry Division, Argonne National Laboratory, Argonne, IL 60439

> *This chapter provides an overview of the fundamental issues concerning quantitation in ^{13}C NMR spectroscopy of carbonaceous solids. General factors governing quantitation in solid-state ^{13}C NMR spectroscopy (such as sample heterogeneity, the presence of unpaired electron spins, interference of proton decoupling by molecular motion, magic-angle spinning (MAS) effects, and implementation of the proper recycle-delay time) are discussed together with those factors that play a major role in cross-polarization (CP) experiments (Hartmann–Hahn match, proton spin-locking, cross-polarization spin dynamics, and interference of cross-polarization from MAS). Technical aspects and requirements of the solid-state ^{13}C NMR experiment are outlined, and effective strategies to obtain the most reliable results are presented.*

THE POTENTIAL USE OF SOLID-STATE ^{13}C NMR spectroscopy for the characterization of carbonaceous solids such as coal, the determination of the various functional groups, and the study of fossil-fuel conversion processes has been recognized for more than three decades. Numerous investigations have been reported in the literature, and several reviews have appeared (*1–6*). With the advent of ^{1}H decoupling (*7*), cross-polarization (CP) (*8*), and magic-angle spinning (MAS) (*9*), which have made it possible to obtain high-resolution ^{13}C NMR spectra in solids in a relatively short measuring time, CP–MAS techniques have been applied extensively in coal research (*2–6, 10–16*). In fact, ^{13}C CP–MAS NMR spectroscopy has become one of the most widely used methods to investigate coal. However, concern has been growing over the past decade about the quantitativeness of CP–MAS spectroscopy for coal analyses. In the very first ^{13}C CP NMR study of coal, VanderHart and Retcofsky (*10*) estimated that only ~50% of the total carbon spins could be detected via this technique. Since then the issue of quantitation in NMR analyses of coal and related materials has been the topic of much debate (*10, 12, 15, 17–25*). Although the quantity of observable carbons for coals and their individual organic constituents (macerals) have been shown to vary widely, the general consensus is that for reasons that can be related to both specific coal properties and the applied NMR techniques, a substantial fraction of the carbons is not observed.

In principle, two NMR techniques can be used to measure ^{13}C spectra in coal; these are the CP experiment mentioned and the conventional single-pulse (SP) experiment, consisting of the simple 90° pulse-acquisition–recycle-delay sequence. Figure 1 shows the radio frequency (rf) pulse sequences employed in SP and CP, as well as the time constants and time delays involved in both experiments. In this chapter, sources that can limit the quantitative information in ^{13}C NMR spectroscopy in coal will be reviewed, and possible remedies will be given to improve the situation. We shall discriminate between limiting factors that play a role in solid-state ^{13}C NMR spectroscopy generally, and limiting factors that occur when CP is employed.

General Factors Governing Quantitation in Solid-State ^{13}C NMR Spectroscopy

In both SP and CP it is assumed that, during the ^{13}C signal acquisition, the protons are decoupled from the carbons via irradiation with a strong rf field at the proton Larmor frequency, and the MAS is applied to remove the chemical-shift anisotropies. The following factors can influence the quantitativeness of the ^{13}C spectra:

1. sample heterogeneity
2. unpaired electrons
3. interference from molecular motions on ^{1}H decoupling
4. magic-angle spinning
5. recycle delay

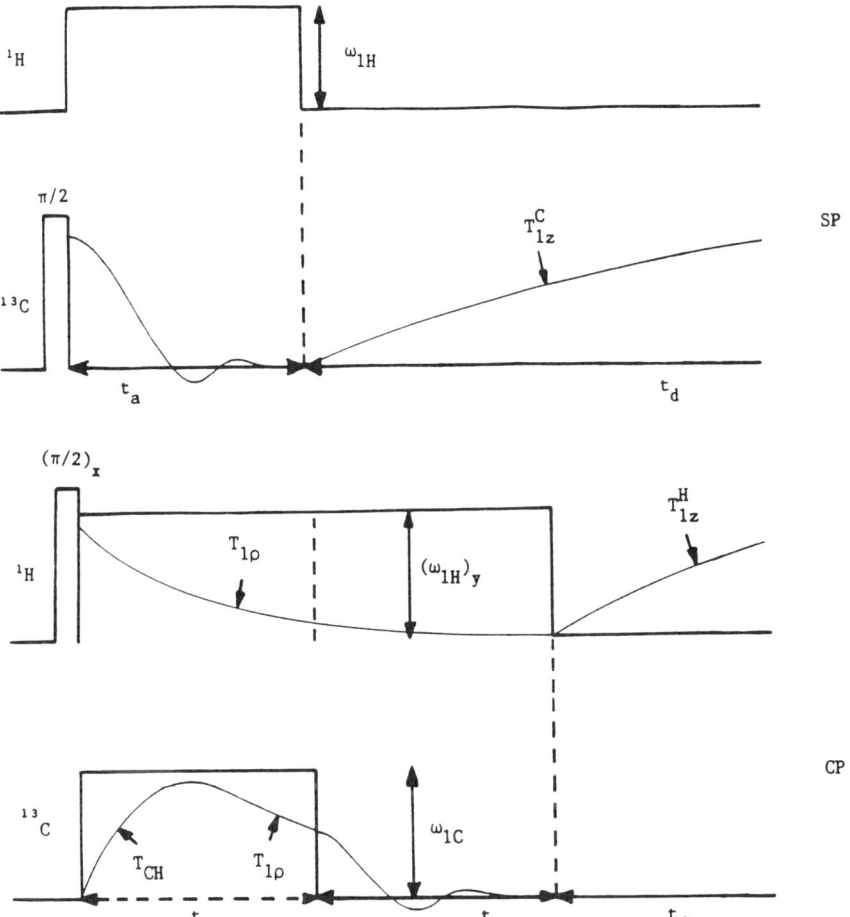

Figure 1. The single-pulse (SP) and cross-polarization (CP) experiments, where T_{1z}^C is the ^{13}C Zeeman relaxation time, T_{1z}^H is the ^1H Zeeman relaxation time, $T_{1\rho}$ is the ^1H rotating-frame relaxation time, T_{CH} is the ^1H–^{13}C polarization-transfer time, t_{CP} is the matching time, t_a is the acquisition time, t_d is the recycle delay time, and ω_{1H} and ω_{1C} are the amplitudes of the rf fields applied at the ^1H and the ^{13}C Larmor frequencies, respectively.

Sample Heterogeneity. Coal is one of the most complex carbonaceous materials known. It is by nature an extremely heterogeneous solid composed of a number of distinct organic phases termed macerals and, to a lesser extent, an inorganic mineral phase. Each maceral and mineral phase exhibits a unique set of physical and chemical properties that contribute to the overall behavior of coal. The macerals typically range in size from micrometers upward, and pure bands up to meters thick commonly can be found in coal seams. Consequently, maceral dimensions are sufficiently large to suggest that nuclear spin diffusion across maceral boundaries should be incomplete on the NMR time scale, typically on the order of 10–20 and 100–200 Å in the limits of proton relaxation in the rotating frame and laboratory frame, respectively. This incomplete spin diffusion means that nuclei within different maceral phases can act as isolated spin reservoirs with very different magnetic properties.

Additional factors must be considered as well. Individual macerals may be in intimate association with inorganic minerals, some of which may contain paramagnetic ions. Moreover, the macerals themselves may contain different concentrations of organic free radicals and thus cause relaxation time differences that ultimately govern the evolution of magnetization within these isolated domains. The presence of paramagnetic centers allows the possibility that significant portions of the carbons are sufficiently broadened or shifted outside the spectral range to render them invisible in the NMR experiment. Moreover, variations in free radical content among these isolated phases can lead to serious distortions in carbon signal measured for the entire coal sample.

Thus, complexities in the organic, inorganic, and physical structures of coal limit the quantitative reliability of solid-state ^{13}C NMR measurements. In a previous investigation (24) on a well-characterized set of maceral concentrates, we had found that individual macerals exhibit unique carbon polarization profiles in CP contact-time experiments. Carbon spin-counting experiments established that the percentage of invisible carbons varies greatly among different macerals and can be correlated to the free radical concentration directly. This experimental bias in the NMR signal response of macerals implied that carbon aromaticities of whole coals are underestimated regardless of the pulse sequence employed, SP or CP.

The inherent limitations associated with measuring a complex solid such as coal are unavoidable. However, a fundamental understanding of these complexities is necessary to establish confidence limits of the measurements so that meaningful parameters can be obtained from the data.

Unpaired Electrons. By nature coal contains a large amount of unpaired electrons that are present either in the form of paramagnetic inorganic ions associated with detrital or precipitated mineral matter or as organic free radicals (1). For ^{13}C nuclei located in the vicinity of these electrons, the electron–^{13}C interactions, which are in general either dipolar or scalar, can become so large that the corresponding resonance lines become very broad or are even shifted completely out of the ^{13}C spectral region. This event will give rise to carbons in a sample that are not observed and whose number will increase with increasing unpaired electron concentration, N_e. The percentages of undetected carbons are expected to be especially large in higher rank coals, where N_e values are high (1, 2). Moreover, these effects may also be large in low-rank coals that contain reasonably high levels of paramagnetic metal ions associated with the organic structures (26).

To estimate the percentage of undetected carbons, we first consider solids containing fixed paramagnetic centers. For such a system, a radius R around each electron can be defined as the average radius of influence of an electron. Radius R can be approximated by $4/3(\pi N_e R^3) = 1$ (27). It follows that if carbons inside some radius b around the electrons are not detected, then the percentage of carbons not observed, $\%C_{loss}$, is given by

$$\%C_{loss} = \frac{4/3(\pi b^3)}{4/3(\pi R^3)} \times 100 = 4/3(\pi b^3 N_e) \times 100 \quad (1)$$

From a study of a resinite sample that had been doped with variable amounts of the stable radical 1,3-bis(diphenylene)-2-phenylallyl (BDPA) (25), it can be deduced that $\%C_{loss}$ is approximately given by

$$\%C_{loss} \approx 5 \times 10^{-19} N_e \quad (2)$$

It follows from equations 1 and 2 that carbons inside a radius of ~10 Å around an electron are not detected.

The question arises whether eq 2 can be used also to estimate the carbon percentage not observed in coal. The unpaired electrons are present mainly in the aromatic regions of coal (15). This feature suggests that the signal losses should occur mainly for the aromatic carbons. In fact, some work (28) supports this viewpoint. Selective reduction of the free radicals in a subbituminous coal with SmI_2 brought about a significant increase in the percentage of carbons observable by NMR spectroscopy. The spectrum of the treated coal exhibited about 15% higher aromatic carbon content than the untreated sample and showed more intense absorptions for nonprotonated carbons in the aromatic region.

The observed broadening of a ^{13}C line depends on both the electron–^{13}C interactions and the electron–electron interactions. This relationship can be understood as follows: The electron–electron interactions, which can be either electron–electron spin-exchange interactions (29), electron–electron dipolar interactions, or both, contain terms proportional to $K(S^+_i S^-_j + S^-_i S^+_j)$, where K is a measure of the strength of these interactions and S_i and S_j are spin operators to two electrons i and j. As a result, electron–electron flip-flop transitions occur at a rate of the order of K, which reduces the lifetime, τ, of an electron spin in a given state ($\tau \sim K^{-1}$). Hence the electron spin operator fluctuates in time, and this fluctuation can partly or completely average out the local field arising from the electron–^{13}C interactions, if K is comparable to or larger than this local field in absence of electron flip-flops. If we define the local field arising from electron–^{13}C interactions as ω_L, then under the condition $K^2 \gg \omega_L^2$, the observed broadening, ω_m, is given by the approximation (30):

$$\omega_m \approx \omega_L^2 \tau \approx \omega_L^2 K^{-1} \ll \omega_L \qquad K^2 \gg \omega_L^2 \qquad (3)$$

It follows that if the electrons are "self-decoupling" from the ^{13}C nuclei via the electron flip-flop mechanisms described, a reduction occurs in the broadenings of the ^{13}C lines, hence in the percentage of undetected carbons.

Moreover, the electron flip-flop transition rate depends on the unpaired electron spin concentration, N_e. In solids containing a large amount of unpaired electrons (typically $N_e > 6 \times 10^{19}$ cm^{-3}), the electron spin-exchange interactions become so large that exchange-narrowing effects are observed in the electron spin resonance (ESR) lines (15, 31, 32). Given that the ESR line width in absence of exchange effects is of the order of 30 MHz (15), τ must be much smaller than 5×10^{-9} s. In this case, it is expected that even carbons located at a distance considerably smaller than 10 Å from the electrons can be observed, and the result is a percentage of nonobservable carbon that is smaller than that predicted by eq 2. This result is in accordance with the results obtained for a fusinite sample (24), where $N_e = 2.3 \times 10^{20}$ cm^{-3}. Equation 2 would predict a loss of 100%, whereas the observed value was "only" 57%. It also explains the nonlinearity seen in the plot of %C$_{loss}$ versus the electron spin concentration for a resinite sample that had been doped with increasing amounts of the stable organic free radical BDPA (25). Figure 2 shows that the linear relationship between %C$_{loss}$ and N_e breaks down at concentrations higher than ~6×10^{19} spins/g. Thereafter, increasing the spin concentration has a less pronounced effect on loss of signal. In fact, extrapolating the high concentration line to N_e of the fusinite sample yields a predicted loss of about 60%, which is very close to the value observed.

For coals with N_e values smaller than ~6×10^{19} cm^{-3}, the unpaired electrons behave more or less as fixed paramagnetic centers (15). Consequently, the electron spin-exchange interactions can be neglected, and only the electron–electron dipolar couplings have to be taken into account.

The flip-flop transition rates arising from these interactions depend on the electron concentration and can be estimated from the following formula (33):

$$K \approx 0.9\gamma_e^2 \hbar N_e \approx 3 \times 10^{-13} N_e \qquad (4)$$

where N_e is equal to the spin concentration per cubic centimeter, γ_e is the electron magnetogyric ratio, and $\hbar$ is the Planck constant divided by 2π.

As an example, we consider the case that $N_e = 2 \times 10^{19}$ cm^{-3}. Then, according to eq 4, $K \approx 6 \times 10^6$ s^{-1}. Obviously this rate is not sufficiently large to average out local fields of tens of megahertz, as experienced by ^{13}C nuclei in the immediate vicinity of the unpaired electrons, but it is capable of reducing the smaller local fields at the carbon sites more remote from the electrons. In fact, this local field reduction explains why ^{13}C nuclei at a distance of 10 Å from an electron can be observed at all. It can easily be calculated that the local field these nuclei are experiencing from the electron in absence of electron flip-flops, ω_L, is about 6×10^4 s^{-1}. Hence the width of the corresponding ^{13}C line would be about 10 kHz, which would make these carbons almost undetectable. However, as

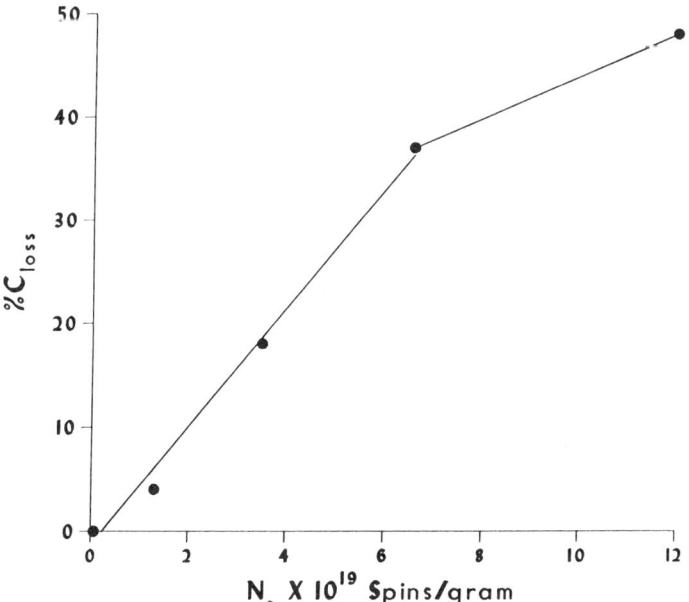

Figure 2. Variation in %C_{loss} in a SP experiment on a resinite sample with increasing BDPA radical concentration.

K becomes much larger than ω_L, line-narrowing occurs. In our example, according to eq 3 the observed broadening, ω_L, is 6×10^2 s^{-1}, or 100 Hz only, a value that indeed makes these carbons observable. Hence, for coals with $N_e \leq 6 \times 10^{19}$ cm^{-3}, eq 2 provides a reasonable estimate of the percentage of undetected carbons. For an anthracite coal, where $N_e = 6 \times 10^{19}$ cm^{-3}, eq 2 predicts a loss of 30%, comparable to the observed value of 20% (15).

The signal losses due to the presence of unpaired electrons in coal can be investigated by the use of an NMR standard. If the coal is mixed with a known amount of an internal reference material, the ^{13}C intensities can be compared and the amount of undetected carbons in the coal can be determined. A suitable standard developed for this purpose is tetrakis-(trimethylsilyl)silane (TKS) (34). This compound meets several important criteria for its use as a reference standard. It exhibits a resonance that is quantitative, narrow (~2 Hz), and distinct. It is also grindable, nonvolatile, and chemically inert, with a high solubility in a wide range of organic solvents so it can be removed easily from the sample under investigation. In addition, it displays a chemical shift at 3.5 ppm, which is well outside the typical chemical-shift range of most organic molecular solids and which is independent of magnetic field strength, and it has favorable relaxation properties and cross-polarization dynamics for use in CP and SP experiments.

Another very promising standard has been developed (35), namely, bicyclo[3.2.1]-4-pyrrolidino-N-methyl-8-octanone triflate, in which the carbonyl group is 97% ^{13}C enriched. The isotopically enriched carbonyl carbon produces a fairly narrow (1.5-ppm) line at 217 ppm, which is usually outside of the spectrum of a coal.

Of course, the use of a standard provides information only about the total amount of undetected carbons in coal, and not whether these are aromatic or aliphatic carbons. In view of previous free radical quenching experiments (28), the best estimate is to assign most of the undetected carbons as being aromatic.

A second approach to investigate signal loss involves direct measurement of N_e by electron paramagnetic resonance (EPR) spectroscopy. Using this information, one can solve eq 2 to estimate the loss in carbon signal for coals with $N_e \leq 6 \times 10^{19}$ spins/cm^3.

Perhaps the best strategy to deal with the adverse effects of free radicals is to remove them by special chemical treatment. For instance, selective chemical quenching of free radicals in a subbituminous coal with samarium iodide in tetrahydrofuran greatly reduced %C_{loss} from 42% to 15% (28). Preliminary experiments using this reducing agent suggest that the judicious application of SP methods combined with pretreatment of the coal can be used effectively for improving the quantitative reliability of solid-state ^{13}C NMR measurements.

A second intriguing possibility to remove the effects of radicals involves decoupling of the electron spins. The experiment would consist of decoupling electron spins from the carbons via irradiation with strong field at the electron Larmor frequency. However, experimentally this method is difficult to realize, and so far no attempt has been reported in the literature.

Interference from Molecular Motions on ^{1}H Decoupling.
Irradiation with a strong rf field with an amplitude ω_{1H} at the proton Larmor frequency during ^{13}C-signal acquisition removes the broadenings of the ^{13}C lines arising from proton–carbon dipolar interactions. The dipolar term responsible for the broadening is equal to dI_zS_z, where d is an amplitude depending on the magnitude and orientation of the proton–carbon vector and I_z and S_z are carbon and proton spin operators, respectively. The rf irradiation generates a time dependence in S_z, which oscillates with a frequency ω_{1H}. Hence, if ω_{1H} is large compared to d, the broadening is averaged out. In practice, ω_{1H} must also be large compared to the strength of the proton–proton dipolar coupling in order to get efficient decoupling (7). If a solid has molecular motions with frequencies comparable to $\omega_{1H}/2\pi$ (typically on the order of 40–80 kHz), then d becomes time-dependent as well. In this case, interferences between $d(t)$ and $S_z(t)$ (where t is time) can result in time-independent carbon–proton interactions and thereby cause severe broadening of the ^{13}C lines and hence apparent signal losses (36, 37). Whether this phenomenon occurs in coal has not yet been investigated extensively. For a bituminous coal, no change in the ^{13}C intensity was found when $\omega_{1H}/2\pi$ was varied from 40 to 82 kHz (24), but obviously more work is needed before more definite conclusions can be drawn.

Magic-Angle Spinning. Although magic-angle spinning has been widely used to remove the chemical-shift anisotropies (CSAs), the application of MAS can give rise to two problems that can affect the quantitative character of the ^{13}C spectra: (1) spinning sidebands and (2) interference between MAS and molecular motions.

Spinning Sidebands. If the MAS frequency, ω_r, is small compared to the CSAs, spinning sidebands (SSB) will occur in the spectra. This result is especially the case for the aromatic ^{13}C nuclei, which typically have CSA values on the order of ~150 ppm. This condition can make quantitative interpretation of the spectra difficult, because of an overlap of spinning sidebands corresponding to aromatic resonances with resonances originating from aliphatic carbons. Although methods have been developed to suppress spinning sidebands (*38–40*), the quantitative character of the spectra, corrected in this way, is uncertain (*25*). Therefore, it is advisable to measure the ^{13}C spectra under conditions where spinning sidebands are avoided altogether. To achieve this condition, the spinning frequency has to be chosen equal to or greater than the largest CSA. For instance, for a ^{13}C Larmor frequency of 25 MHz, the spinning frequency should be ≥3.8 kHz. Another possibility of avoiding spinning sidebands is the application of the magic-angle hopping (MAH) technique, in which instead of a continuous rotation, a three-step rotation "hopping" of the sample is applied over an angle of 120°, around an axis aligned along the magic angle (*41, 42*). The disadvantages of MAH are the rather poor sensitivity and the fact that the ^{13}C relaxation times need to be rather long to obtain quantitative results. This condition is not always fulfilled in coal, especially in lower rank coals with an oxygen percentage larger than ~10% (dry, mineral-matter free or dmmf) (*15, 26*).

Interference Between MAS and Molecular Motions. If molecular motions in a coal make the CSA patterns time-dependent, and if these motions occur at frequencies comparable to the MAS frequency, then broadening effects arise as a result of an interference between the two time-dependencies (*36, 43*). This effect is analogous to the interference between molecular motions and the ^{1}H decoupling field (vide supra). Because aromatic carbons possess the largest CSA values, interference between CSA and MAS will primarily influence the aromatic intensities in the ^{13}C coal spectrum. Indeed, a decrease in the apparent aromaticity has been observed at higher MAS frequencies, and this observation indicates a loss in the aromatic signal at larger spinning speeds (*15, 25, 44*). Whether this effect is due to the interference between MAS and molecular motions is not certain, because these results have been obtained via ^{1}H–^{13}C CP measurements. In CP, signal losses can occur also because of a modulation of the ^{1}H–^{13}C coupling, induced by MAS (*see* the later section MAS Modulation of T_{CH}). Nevertheless, it seems advisable to measure spectra at different spinning frequencies (preferably while avoiding spinning sidebands at all frequencies) in order to investigate possible interference effects, even when performing a SP experiment. Also, MAH might be useful in this case, keeping in mind the constraints of this method already mentioned.

Recycle Delay. The value chosen for the time delay between successive experiments, t_d (*see* Figure 1), depends on the type of experiment that is used. In CP, t_d has to be chosen large compared to the ^{1}H spin–lattice relaxation time, T_1^H, which is easy to realize, as T_1^H in coal is relatively short [hundreds of milliseconds or less (*15*)]. However, in SP t_d

must be chosen much larger than the ^{13}C spin–lattice relaxation time, T_1^C, which in higher rank coals can be very long [tens of seconds or more in degassed coals (15)]. This condition means that t_d must be made very long in order for the carbon magnetization to recover to its full thermal equilibrium value during the recycle delay. In this respect, due to the lack of spin diffusion among the ^{13}C nuclei, often a distribution of T_1^C values is observed (45); this possibility imposes a requirement that an extremely long value of t_d must be employed.

To illustrate this need for a long value of t_d, we consider the case of single-exponential relaxation and the case of a distribution of relaxation times, in which the overall relaxation is governed by $\exp[-(AT)^{1/2}]$ (28) where AT is "apparent" relaxation time defined as $T_1 = t_{1/2}/\ln 2$, where $t_{1/2}$ is the time it takes for the carbon magnetization to relax to one-half its thermal equilibrium value. Then, in order for the carbon magnetization to attain 99% of this thermal equilibrium value after the recycle delay, t_d must be $4.6T_1$ in an exponential relaxation, whereas $t_d = 31T_1$ is required in a distribution of relaxation times. Given that ^{13}C T_1 values in coal have been measured as long as 45 s (15), t_d must be chosen to be at least 1400 s! Thus, the SP experiment is very time-consuming. A possibility to decrease t_d is to expose the coal to air. Such exposure reduces the ^{1}H relaxation time by a factor of 2–4 (15, 46), and it was recently shown that ^{13}C relaxation times are also reduced by an appreciable amount (see Chapter 18). However, the usefulness of this approach has still to be established.

Factors Governing Quantitation in Cross-Polarization Experiments

As already mentioned, almost all ^{13}C studies in coal have been carried out with the CP technique. The advantages of CP over the traditional SP method are the substantial gain in sensitivity (a theoretical maximum enhancement factor of 4), which is possible by transfer of magnetization from the abundant proton spins to the rare carbon spins, and the ability to repeat the experiment on the time frame of the proton spin–lattice relaxation times, which are generally much shorter than those of carbons. However, the CP method poses some potential hazards with regard to its quantitative reliability owing to the complex, time-dependent evolution of the carbon magnetization. Indeed, many different CP experiments are needed to investigate and establish the quantitative character of the ^{13}C spectra. In fact, this need for many experiments can make the total experiment as time-consuming as, or even more time-consuming than, the standard SP experiment. The reason is that several factors can influence the quantitativeness of ^{13}C spectra obtained via CP; factors that do not play a role in a SP experiment.

Hartmann–Hahn Matching Condition. In a CP experiment without MAS or combined with low-speed MAS, the polarization-transfer time, T_{CH}, which determines the build-up of the ^{13}C magnetization during CP (Figure 1), becomes minimal when the Hartmann–Hahn condition is fulfilled, that is, when $\omega_{1C} = \omega_{1H}$ (*8*). Even a slight misadjustment of this condition can result in a decrease of the apparent carbon aromaticity (*24, 46*). The reason is that a large fraction of the aromatic ^{13}C nuclei have T_{CH} values that are considerably larger than those corresponding to the aliphatic carbons (*15, 16, 24*), and signal intensity losses due to a "mismatch" are larger for ^{13}C nuclei with larger T_{CH} values because of the nonhomogeneous spin temperature of the rare spins (*47*). Therefore, a precise adjustment of the Hartmann–Hahn condition is required to avoid signal losses, especially those of aromatic carbons.

MAS Modulation of T_{CH}. Magic-angle spinning causes a time-modulation in the $^{13}C-^1H$ interactions. As a result, when the MAS frequency, ω_r, is comparable to or larger than the local fields that the ^{13}C nuclei experience from the protons, this modulation causes an increase from the minimum value of T_{CH} at the Hartmann–Hahn condition, $\omega_{1C} = \omega_{1H}$. New minima occur in a plot of T_{CH} versus ω_{1C} (for fixed ω_{1H}); these minima occur at $\omega_{1C} = \omega_{1H} \pm n\omega$ ($n = 1, 2, ...$) (*48, 49*). Moreover, the changes in the "matching curve" under the influence of MAS depend on the T_{CH} value (*48*). In this instance, the matching curve is defined as the ^{13}C signal observed after a certain matching time, t_{CP}, as a function of the off-set frequency, $\omega_{1C} - \omega_{1H}$. In general, a decrease in ^{13}C intensity for increasing ω_r values is observed, with a larger decrease for ^{13}C nuclei with larger T_{CH} values. For instance, in Torlon, a poly(amide–imide) manufactured by Amoco, spinning at a frequency of 6 kHz reduced the ^{13}C intensities of carbons located at one, two, and three bonds away from the nearest proton by 3, 20, and 40%, respectively (*48*). Hence, it follows that even moderate spinning frequencies can reduce carbon intensities considerably. In coal, the aromatic ^{13}C nuclei have, on the average, larger T_{CH} values than the aliphatic carbons. This difference might explain the substantial decreases in the aromatic carbon intensities that have been observed when spinning was increased from about 4 to 8 kHz (*15, 25, 43*). Another possible explanation of this signal loss is interference effects between MAS and molecular motions, as discussed previously. At present this issue has not been fully resolved.

The time modulation of the C–H coupling during CP may be avoided by applying MAH in the place of MAS (*41, 43*), by changing the spinning axis during CP (*50*), or by applying a so-called stop-and-go (STAG) experiment, where the spinning is stopped during CP (*51*).

Spin-Locking ^{1}H Magnetization. To avoid losses in ^{1}H magnetization, hence subsequent losses in the ^{13}C signal after CP, the entire proton magnetization must be spin-locked along the rf field, ω_{1H}. For this reason, ω_{1H} has to be large compared to the local fields that the protons experience from each other. If the spin-locking field is insufficient, signal losses occur as a result of two factors: (1) protons that experience local fields larger than ω_{1H} dephase during the spin-locking period, and (2) a single spin temperature is established between the system characterized by the proton polarization along the spin-lock field and the proton dipolar system (*52*). The latter phenomenon results in a net ^{1}H magnetization given by $M_H = M_{H0}\omega^2_{1H}(\omega^2_{1H} + D^2)^{-1}$, where M_{H0} is the magnetization at thermal equilibrium and D is the local field. Hence, M_H becomes smaller than M_{H0} as ω_{1H} approaches D.

Figure 3 shows the decay of the ^{1}H magnetization along a spin-lock field, $\omega_{1H}/2\pi$, of 43 kHz, measured via ^{1}H NMR spectroscopy. The rapid

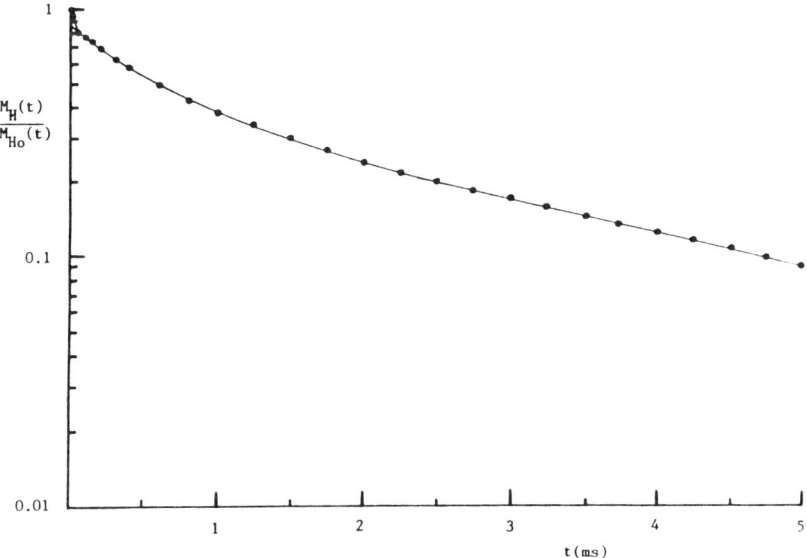

Figure 3. ^{1}H rotating-frame relaxation curve of a low-volatile bituminous coal measured directly via ^{1}H NMR spectroscopy. In this experiment $\omega_{1H}/2\pi = 43$ kHz.

decay of the magnetization during the first 100 µs can be ascribed to the effects just described. For this spin-lock field, the loss in the proton magnetization is about 15%. This result means that a similar loss can be expected in the ^{13}C signal after CP.

The loss in proton magnetization can be minimized by using a larger lock field. Although systematic investigations (19, 24) have demonstrated that varying the proton decoupling field between 40 and 82 kHz had negligible effects on the signal-to-noise (S/N) ratios of coal spectra or on the derived aromaticity values, preliminary results (53) indicate that using $\omega_{1H}/2\pi$ in excess of 80 kHz increases the signal, due to coal carbons, significantly.

Cross-Polarization Spin Dynamics. The quantitative aspects in CP associated with the nonideal behavior of T_{CH} and $T_{1\rho}$ (spin–lattice relaxation time in the rotating reference frame) has been a topic of major concern in coal investigations employing this technique. In this regard, problems arise because the matching time, t_{CP}, should be large compared to the ^{13}C–^{1}H polarization-transfer time, T_{CH}, in order to obtain the full ^{13}C magnetization. On the other hand, t_{CP} should be small in order to prevent serious signal losses due to the ^{1}H relaxation in the rotating frame, characterized by $T_{1\rho}$ (see Figure 1). In a typical coal, a distribution of T_{CH} values results from the different ^{13}C–^{1}H distances and motions present in the coal. $T_{1\rho}$ can be very short for some coals, so that t_{CP} cannot be made longer than about 1–2 ms. The interplay between these two competing relaxation pathways ultimately governs the success of a CP experiment for quantitative analysis and requires that the condition $T_{CH} \ll t_{CP} \ll T_{1\rho}$ must be fulfilled fairly rigorously. Experimentally determined values of T_{CH} are 25–50 µs for rigid, protonated carbons and on the order of 1 ms or more for nonprotonated carbons (54). Furthermore, aliphatic and aromatic carbons in coal often exhibit different $T_{1\rho}$ constants (19, 20, 55). Estimated values of $T_{1\rho}$ in coals are found in the range of 0.5–15 ms (10, 15, 19, 20, 55–57).

Therefore, a major problem is that the condition $T_{CH} \ll T_{1\rho}$ is only marginally satisfied for certain carbons with the results that these components achieve only partial polarization or none at all. This possibility is usually investigated by means of a variable contact-time experiment, in which the intensity of the ^{13}C signal is observed as a function of t_{CP}. Figure 4 shows the result of a typical experiment for aliphatic and aromatic carbons in a low-volatile bituminous coal, with volatile material (VM) = 19.9%, C = 90.2%, and H = 4.6% (dmmf). MAS is avoided in this case in order to avoid the problems mentioned in previous sections. In Figure 4, the aliphatic carbon intensity has been multiplied by 9.6, so that the aliphatic and aromatic curves coincide for $t_{CP} \geq 3$ ms. The increase in the

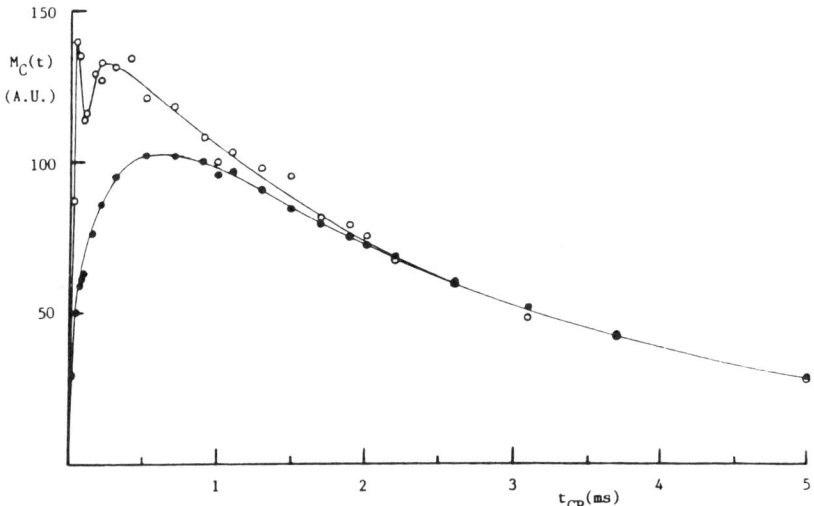

Figure 4. Aromatic (●) and aliphatic (○) ^{13}C intensities of a low-volatile bituminous coal as a function of the matching time, t_{CP}. The aliphatic signal intensity has been multiplied by a factor of 9.6. In this experiment $\omega_{1H}/2\pi = \omega_{1C}/2\pi = 43$ kHz.

^{13}C intensities for short t_{CP} values is due to T_{CH} effects, whereas the decrease in these intensities for larger values of t_{CP} is due to $T_{1\rho}$ effects. Similar variable-t_{CP} curves have been observed in many other coals (2–6, 15, 24, 25). Usually, the experimental data have been analyzed with the assumption that the time dependence in the ^{13}C magnetization, $M_C(t)$, is given by (58)

$$M_C(t) = \frac{4M_0}{1 - T_{CH}/T_{1\rho}} \left[\exp\left(-\frac{t_{CP}}{T_{1\rho}}\right) - \exp\left(-\frac{t_{CP}}{T_{CH}}\right) \right] \quad (5)$$

In eq 5, M_0 is the thermal equilibrium value of the ^{13}C magnetization. The effect of the ^{13}C spin–lattice relaxation in the rotating frame has been neglected, an approach that is usually justified in coal. By using the experimental variable-t_{CP} curves, $T_{1\rho}$ is determined from the long-term behavior of $M_C(t)$ and T_{CH} from the short-term behavior. Then, using eq 5, M_0 can be calculated and compared with the value obtained from a standard sample. Finally, the weight percentage of the carbon detected via CP can be determined and compared with the weight percentage obtained via ultimate analysis.

By using this type of an analysis, the following results have been found:

1. The aliphatic and aromatic ^{13}C nuclei manifest single, but different, T_{CH} values.
2. The ^{1}H rotating-frame relaxation is exponential; $T_{1\rho}$ of aromatic protons is generally greater than $T_{1\rho}$ of aliphatic protons.
3. Only a fraction of the carbons is observed in CP, considerably less than is found in a SP experiment.

The third result has been explained by inefficient carbon polarization of more slowly polarizing carbons in samples that contain fairly high concentrations of free radicals, in which $T_{1\rho}$ relaxation times are shortened through their interaction with paramagnetic centers via spin diffusion. An additional 30–40% of the carbons were not observed with CP (24). The same interpretation was invoked to explain results from dynamic nuclear polarization (DNP) "back-match" experiments, where 40% of the aromatic carbons in a mid-volatile bituminous (MVB) coal were found to have long T_{CH} and were not observed in the usual CP experiments (59). Direct measurements of $T_{1\rho}$ in coals also indicate that 23–50% of the protons relax too quickly to allow CP of all carbon nuclei (59, 60). In point of fact, the foregoing conclusions are not completely accurate because the analytical approaches provide an oversimplified picture of the relaxation phenomena in these systems.

To illustrate this situation, we first consider the proton rotating-frame relaxation results given in Figure 3. Even neglecting the initial drop-off in magnetization during the first 100 μs, the relaxation is indeed nonexponential. During the first few milliseconds, the decrease in the ^{1}H magnetization is much larger than is predicted by the long-term relaxation behavior, which approaches an exponential decay. This large decrease means that the loss in ^{13}C magnetization during the first few milliseconds would be much larger than the loss predicted by eq 5. In other words, if the correction of magnetization for $T_{1\rho}$ effects as illustrated by Figure 4 had been analyzed directly via ^{1}H NMR spectroscopy, much larger values of M_0 would have been obtained from the calculation. This situation is illustrated in Figure 5, in which the solid curves 1, 2, and 3 have been calculated for different values of T_{CH}, assuming that the protons interacting with the ^{13}C nuclei are relaxing according to Figure 3 (again neglecting the initial drop-off) and assuming an exponential CP factor of $1 - \exp(-t_{CP}/T_{CH})$. For simplicity, M_0 has been taken as unity. The dashed curves 4, 5, and 6 are obtained by substituting the parameters indicated by the solid curves into eq 5. Incorrect values of T_{CH}, $T_{1\rho}$, and M_0 are obtained, although for longer values of T_{CH} the overall behavior of the dashed curves is close to that of the "exact" solid curves. The apparent

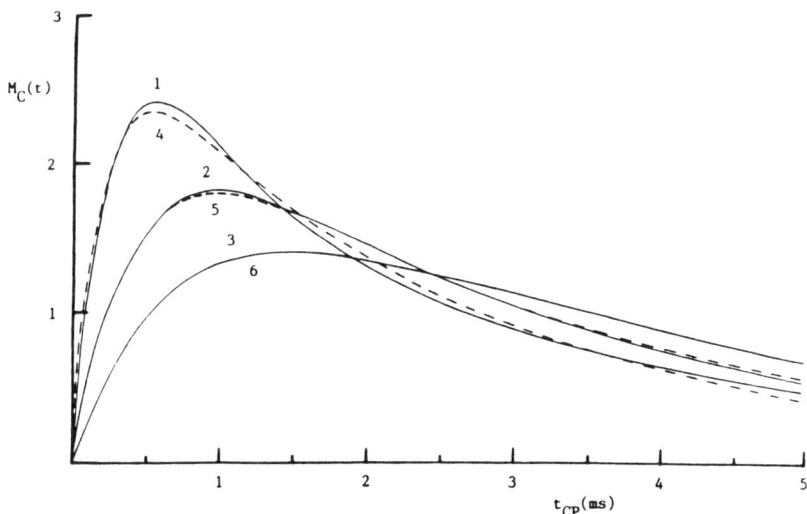

Figure 5. ^{13}C *variable-*t_{CP} *experiment. The solid curves (1, 2, and 3) were calculated from an assumed exponential CP factor* $(1 - \exp(-t_{CP}/T_{CH}))$ *using the* ^1H *rotating-frame relaxation data given in Figure 3. Curve 1:* $T_{CH} = 0.3$ *ms; curve 2:* $T_{CH} = 0.7$ *ms; curve 3:* $T_{CH} = 1.3$ *ms,* $M_{C0} = 1$. *Dashed curves (4, 5, and 6): calculated with eq 5. Curve 4:* $T_{CH} = 0.18$ *ms,* $T_{1\rho} = 2.5$ *ms,* $M_{C0} = 0.72$; *curve 5:* $T_{CH} = 0.40$ *ms,* $T_{1\rho} = 3.1$ *ms,* $M_{C0} = 0.61$; *curve 6:* $T_{CH} = 0.70$ *ms,* $T_{1\rho} = 3.7$ *ms,* $M_{C0} = 0.52$ *(curve 6 coincides completely with curve 3).*

values of M_0 are considerably lower than unity, and therefore the analysis of the data by eq 5 results in an underestimation of the fraction of observed carbons.

The foregoing results suggest that the correct procedure for analyzing the variable-t_{CP} experiment might be to use the ^1H rotating-frame relaxation results obtained directly via ^1H NMR spectroscopy. However, even this procedure is not entirely suitable because the aromatic and aliphatic protons of the coal exhibit different overall $T_{1\rho}$ relaxation behavior (*19, 20, 24*). This issue can be investigated by again measuring $T_{1\rho}$ via CP with a fixed value of t_{CP} and a variable delay period, t_1, prior to CP; during t_1 only the ω_{1H} field is applied (*see* Figure 6a). The disadvantages of this procedure are that proton relaxation also occurs during t_{CP}, and that the relaxation behavior is not measured during this time. However, the loss of magnetization can be minimized by using a small t_{CP} value. The ^1H rotating-frame relaxation curve obtained in this manner is shown for the aliphatic and aromatic protons in the low-volatile bituminous coal in Figure 6b, where $t_{CP} = 60$ s. The intensities of the aromatic and aliphatic carbons are adjusted so that they appear equal for $t_1 \approx 3$ ms. The aromatic protons relax faster than the aliphatic protons, and this observation is in accordance with the results of a previous investigation *(61)* that indicated that the ^{1}H rotating-frame relaxation is governed by coal organic radicals that are located mainly in the aromatic regions *(15)*.

On the other hand, this finding is in complete disagreement with previous observations *(19, 20, 24)* that variable-t_{CP} experiments have shown $(T_{1\rho})$ aromatic protons > $(T_{1\rho})$ aliphatic protons. The reason for this dis-

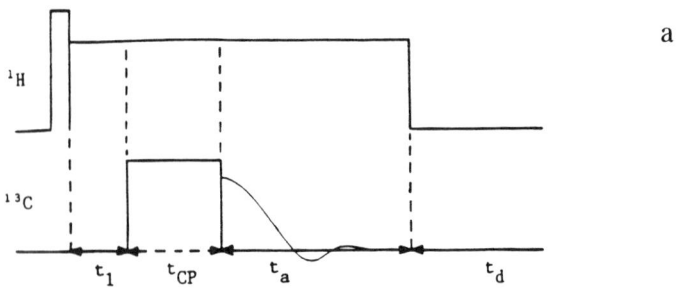

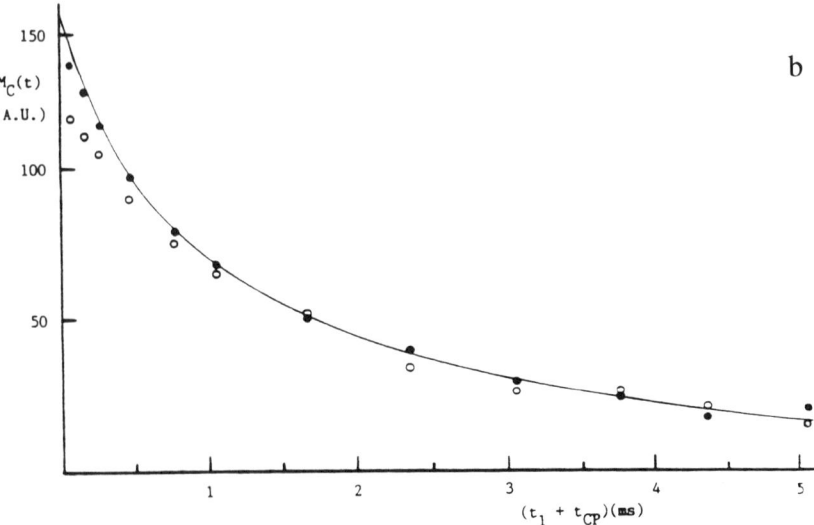

Figure 6. a: The pulse sequence used in the variable-t_1 experiment. b: Aromatic (●) and aliphatic (○) ^{13}C intensities of a low-volatile bituminous coal as a function of $(t_1 + t_{CP})$ for $t_{CP} = 60$ μs. Solid curve: ^1H rotating-frame relaxation measured via ^1H NMR spectroscopy. The aliphatic signal has been multiplied by a factor of 3.2. In this experiment $\omega_{1H}/2\pi = \omega_{1C}/2\pi = 43$ kHz.

crepancy is the occurrence of nonprotonated carbons with large T_{CH} values ($\approx$1.3 ms) that comprise a major fraction of the coal aromatic carbons. Using $t_{CP} = 60$ μs selectively attenuates their contribution to the time-dependent behavior of $M_C(t)$ for t_{CP} values <2.5 ms. This attenuation causes an apparent decrease in the overall relaxation rate and more accurately reflects proton relaxation in the rotating frame.

A complete analysis of all results shown in the Figures 3, 4, and 6b demonstrated that the CP dynamics of the aliphatic carbons can be characterized by two T_{CH} time constants, and that of the aromatic carbons by as many as three T_{CH} time constants. Thus, the interplay between T_{CH} and $T_{1\rho}$ leads to complicated results, and many different experiments (^{1}H rotating-frame relaxation measured via ^{1}H NMR spectroscopy, variable-t_{CP} experiments, and variable-t_1 experiments) are needed to extract the proper quantitative information.

The analyses of both the variable-t_{CP} and variable-t_1 experiments indicated that the percentage of detected carbons is much larger than that estimated from an analysis of only the variable-t_{CP} data by means of eq 5. In fact, using this comprehensive analysis method to characterize CP spin dynamics, we conclude that the observable percentage of carbons measured via CP and SP are comparable. This conclusion means that the general consensus that a larger fraction of the carbons is too remote from protons to be polarized via CP is incorrect. Rather, loss of signal observed in conventional CP experiments is simply a matter of relaxation effects.

The influence of $T_{1\rho}$ can be decreased by applying larger rf fields. With larger rf fields, the ^{1}H rotating-frame relaxation becomes longer and has more exponential character; thus the quantitative reliability of the measurements is improved.

Summary and Conclusions

Table I summarizes the various factors that can influence the quantitative character of ^{13}C spectra of coal as well as the remedies to improve the analysis. On the basis of experience from previous work, the following experimental conditions are recommended for the most reliable results.

1. The ^{1}H lock field, ω_{1H}, should be chosen as large as possible (e.g., $\omega_{1H}/2\pi \geq 60$ kHz). The use of large ω_{1H} is especially advantageous in CP experiments where signal losses from ^{1}H dephasing effects are minimized as a result of longer ^{1}H rotating-frame relaxation times. Moreover, slight increases in the resolution of the ^{13}C spectra of coal have been observed when the decoupling field was increased from 45 to 90 kHz or higher. Preferably, ^{13}C spectra should be

Table I. Factors Limiting Quantitation in ^{13}C Single-Pulse and Cross-Polarization Experiments and Possible Remedies

Limiting Factor	Possible Remedies
Both types of experiment	
Unpaired electrons	1. Remove electrons by chemical treatment
	2. Measure N_e via ESR spectroscopy and estimate loss in aromatic carbons from eq 2
Interference between molecular motions and the ^1H decoupling field, ω_{1H}	Measure spectra at different ω_{1H} fields, e.g., $60 < \omega_{1H}/2\pi < 100$ kHz
Spinning sidebands due to insufficiently fast MAS	1. Use low external magnetic field ($B_0 \leq 2.3$ T)
	2. Use magic-angle hopping
Interference between molecular motions and MAS	1. Measure spectra at different MAS speeds
	2. Use magic-angle hopping
Single-pulse experiments	
^{13}C spin–lattice relaxation time, T_{1z}^{C}	1. Use long recycle delays
	2. Reduce T_{1z}^{C} by exposing coal to air
Cross-polarization experiments	
Hartmann–Hahn condition	Precisely adjust the condition $\omega_{1H} = \omega_{1C}$
Modulation of C–H coupling due to MAS	1. Use low MAS frequencies (< 4 kHz)
	2. Use stop-and-go experiment
	3. Use magic-angle hopping
Dephasing of ^1H magnetization during CP	Use large ^1H rf field during CP ($\omega_{1H}/2\pi > 60$ kHz)
Interplay between $T_{1\rho}$ and T_{CH}	1. Measure $T_{1\rho}$ via ^1H NMR spectroscopy
	2. Measure $T_{1\rho}$ via variable t_{CP}, t_1 experiments
	3. Use ^1H–^{13}C nuclear solid effect
	4. Use ^1H–^{13}C RF DNP–CP

NOTE: In all experiments, an NMR intensity standard should be used.

obtained with a few values of ω_{1H} in order to investigate possible interference effects.

2. The MAS frequency should be taken as low as possible, preferably 3–4 kHz, which is sufficiently large to minimize the occurrence of spinning sidebands in the ^{13}C spectra. Consequently, it is preferable to perform the ^{13}C experiments in a small external field, that is, 1.4–2.3 T, where correspondingly smaller spinning speeds and large-sample volume can be used. Measuring spectra at lower external field strengths does not result in a decreased spectral resolution; its only disadvantage is the lower sensitivity. Recently developed large-volume spinners are capable of spinning volumes up to 6 cm^3 at 2.5 kHz (62); hence the NMR sensitivity is increased dramatically. In addition, the influence of MAS during CP matching may be minimized by either changing the spinning axis during t_{CP} or applying a STAG experiment. Furthermore, the experiment should be performed at a few MAS frequencies to investigate possible interference effects.

3. A large spectral width should be used for the detection of the ^{13}C spectra (50 kHz or more). This approach facilitates measurement of ^{13}C resonances that are broadened by the presence of unpaired electrons or from interference effects.

4. A standard sample should be mixed with the coal sample to determine the percentage of observable carbons. An excellent reference standard is tetrakis(trimethylsilyl)silane (TKS) (34) or the triflate compound already mentioned (35), although a simple standard like hexamethylbenzene has proven to be very useful (24). In CP experiments, observed carbon signal intensities for these standards must be corrected for T_{CH} and $T_{1\rho}$ effects.

A major concern is the selection of the appropriate pulse sequence, either SP or CP, to obtain the most quantitative ^{13}C spectra. Comparing the two methods from previous work found in the literature, we conclude that the SP technique is by far preferable for the quantitative characterization of carbonaceous solids, including coals. In fact, the time has come to abandon CP altogether for quantitative studies. However, this technique may be used for a qualitative survey when sensitivity is an issue or when more detailed knowledge of the coal structure is desired, for example, to confirm the presence of different molecular structures that are in low abundance or to determine the distribution of carbon–proton distances and relaxation times for various carbons and protons. Useful CP results can be achieved by applying the variable-t_{CP} and variable-t_1 experiments described herein. However, no rigorous analysis can describe in detail all processes occurring during CP, so that qualitative information can be expected at best.

The main disadvantage of the SP technique is the long recycle delay required between successive pulses. However, for lower rank coals (typically with O > 10%) the ^{13}C Zeeman relaxation time, $T_{1z}{}^C$, is rather short (on the order of several seconds) (26) and $T_{1z}{}^C$ in higher-rank coals can be decreased by exposing the coal to air. This step, together with the possibility of using large-volume rotors, makes the SP experiment very feasible. The remaining factor limiting the quantitativeness in SP experiments is the broadening effects due to the unpaired electrons. However, especially for lower rank coals in which the electron concentrations are typically lower than about 3×10^{19} cm^{-3}, the maximum intensity loss to be expected is probably not more than 10–15%. In samples in which the presence of free radicals poses a problem, the judicious application of SP methods combined with pretreatment of the coal with an appropriate radical quenching agent, such as samarium iodide (28), can be used effectively for improving the quantitative reliability of the measurements.

Finally, in other ^{1}H–^{13}C polarization-transfer techniques, part of the problems arising in CP can be circumvented. These other approaches are polarization transfer via the nuclear solid effect (NSE) (63, 64) and polarization transfer via rotating-frame (RF) DNP–CP (65, 66). In both instances, dephasing of the protons via $T_{1\rho}$ processes is avoided. These experiments are discussed in greater detail in Chapter 10.

Acknowledgments

R. E. Botto acknowledges support of the Office of Basic Energy Sciences, Division of Chemical Sciences, U.S. Department of Energy, under Contract No. W–31–109–ENG–38. R. A. Wind and G. E. Maciel acknowledge support of Fossil Energy, U.S. Department of Energy, under Contract No. DE–AC22–89PC8840.

References

1. Van Krevelen, D. W. *Coal;* Elsevier: Amsterdam, Netherlands, 1969.
2. *Analytical Methods for Coal;* Karr, C., Jr., Ed.; Academic: New York, 1976.
3. *Coal Science;* Gorbaty, M. L.; Larsen, J. W.; Wender, I., Eds.; Academic: New York, 1982.
4. *Magnetic Resonance. Introduction. Advanced Topics and Applications to Fossil Energy;* Petrakis, L.; Fraissard, J. P., Eds.; D. Reidel: Dordrecht, Netherlands, 1984.

5. Axelson, D. E. *Solid State Nuclear Magnetic Resonance of Fossil Fuels: An Experimental Approach;* Multiscience: Montreal, Canada, 1985.
6. Davidson, R. M. *NMR Studies of Coal;* Report No. ICTIS/TR 32, IEA Coal Research, 1986.
7. Pines, A.; Gibby, M. G.; Waugh, J. S. *J. Chem. Phys.* **1972,** *56,* 1776.
8. Pines, A.; Gibby, M. G.; Waugh, J. S. *J. Chem. Phys.* **1973,** *59,* 596.
9. Stejskal, E. O.; Schaefer, J.; McKay, R. A. *J. Magn. Reson.* **1977,** *25,* 569.
10. VanderHart, D. L.; Retcofsky, J. L. *Fuel* **1976,** *55,* 202.
11. Maciel, G. E.; Bartuska, V. J.; Miknis, F. P. *Fuel* **1979,** *58,* 391.
12. Miknis, F. P.; Sullivan, M. J.; Bartuska, V. J.; Maciel, G. E. *Org. Geochem.* **1981,** *3,* 19.
13. Zilm, K. W.; Pugmire, R. J.; Grant, D. M.; Larter, S. R.; Allen, J. *Fuel* **1981,** *60,* 17.
14. Botto, R. E.; Winans, R. E. *Fuel* **1983,** *62,* 271.
15. Wind, R. A.; Duijvestijn, M. J.; Lugt, C. van der; Smidt, J.; Vriend, J. *Fuel* **1987,** *66,* 876.
16. Solum, M. S.; Pugmire, R. J.; Grant, D. M. *Energy Fuels* **1989,** *3,* 187.
17. Wemmer, D. E.; Pines, A.; Whitehurst, D. D. *Philos. Trans. R. Soc. London* **1981,** *A300,* 15.
18. Hagaman, E. W.; Woody, M. C. *Proc. Int. Conf. Coal Sci.;* Verlag Glueckauf GmbH, 1981; p 807.
19. Sullivan, M. J.; Maciel, G. E. *Anal. Chem.* **1982,** *54,* 1615.
20. Dudley, R. L.; Fyfe, C. A. *Fuel* **1982,** *61,* 651.
21. Murphy, P. D.; Cassady, T. J.; Gerstein, B. C. *Fuel* **1981,** *61,* 1233.
22. Wilson, M. A.; Pugmire, R. J.; Karas, J.; Alemany, L. B.; Woolfenden, W. R.; Grant, D. M.; Given, P. H. *Anal. Chem.* **1984,** *56,* 933.
23. Dereppe, J. M. *Magnetic Resonance. Introduction. Advanced Topics and Applications to Fossil Energy;* Petrakis, L.; Fraissard, J. P., Eds.; D. Reidel: Dordrecht, Netherlands, 1984; p 535.
24. Botto, R. E.; Wilson, R.; Winans, R. E. *Energy Fuels,* **1987,** *1,* 173.
25. Snape, C. E.; Axelson, D. E.; Botto, R. E.; Delpuech, J. J.; Tekely, P.; Gerstein, B. C.; Pruski, M.; Maciel, G. E.; Wilson, M. A. *Fuel* **1989,** *68,* 547.
26. Botto, R. E.; Axelson, D. E. *Prepr. Div. Fuel Chem. ACS* **1988,** *33,* 50.
27. Lowe, I. J.; Tse, D. *Phys. Rev.* **1968,** *166,* 279.
28. Muntean, J. V.; Stock, L. M.; Botto, R. E. *Energy Fuels* **1988,** *2,* 108.
29. Wertz, J. E.; Bolton, J. R. *Electron Spin Resonance;* Chapman and Hall: London, 1986.
30. Abragam, A. *The Principles of Nuclear Magnetism;* Clarendon: Oxford, England, 1961; p 195.
31. Smidt, J.; van Krevelen, D. W. *Fuel* **1958,** *38,* 355.
32. Smidt, J., thesis, Delft University of Technology, 1960.
33. Atsarkin, V. A.; Denidov, V. V.; *Sov. Phys. JETP* **1980,** *52,* 726.
34. Muntean, J. V.; Stock, L. M.; Botto, R. E. *J. Magn. Reson.* **1988,** *76,* 540.
35. Hall, R. A.; Maciel, G. E.; unpublished.
36. VanderHart, D. L.; Earl, W. L.; Garroway, A. N. *J. Magn. Reson.* **1981,** *44,* 361.
37. Rothwell, W. P.; Waugh, J. S. *J. Chem. Phys.* **1981,** *74,* 2721.
38. Dixon, W. T. *J. Magn. Reson.* **1981,** *44,* 220.

39. Dixon, W. T.; Schaefer, J.; Sefcik, M. D.; Stejskal, E. O.; McKay, R. A. *J. Magn. Reson.* **1982**, *49*, 341.
40. Hemminga, M. A.; de Jager, P. A. *J. Magn. Reson.* **1983**, *51*, 339.
41. Bax, A.; Szeverenyi, N. M.; Maciel, G. E. *J. Magn. Reson.* **1983**, *52*, 147.
42. Szeverenyi, N. M.; Bax, A.; Maciel, G. E. *J. Magn. Reson.* **1985**, *61*, 440.
43. Suwelack, D.; Rothwell, W. P.; Waugh, J. S. *J. Chem. Phys.* **1980**, *73*, 2559.
44. Pruski, M.; dela Rosa, L.; Gerstein, B. C. *Energy Fuels* **1990**, *4*, 160.
45. van der Lugt, C.; Smidt, J.; Wind, R. A. *Investigation of Coal and Coal Products by Means of Magnetic Resonance. Part I. Investigation by Means of ^1H and ^{13}C NMR in Combination with Dynamic Nuclear Polarization and ESR;* final report of research project No. 4351, supported by a grant from the Project Office for Energy Research of the Netherlands Energy Research Foundation ECN, within the framework of the Dutch Coal Research Program, 1985.
46. Wind, R. A.; Jurkiewicz, A.; Maciel, G. E. *Fuel* **1990**, *69*, 830.
47. Wu, X.-L.; Zilm, K. W. *J. Magn. Reson.* **1991**, *93*, 265.
48. Wind, R. A.; Dec, S. F.; Lock, H.; Maciel, G. E. *J. Magn. Reson.* **1988**, *79*, 136.
49. Stejskal, E. O.; Schaefer, J.; Waugh, J. S. *J. Magn. Reson.* **1977**, *28*, 105.
50. Bax, A.; Szeverenyi, N. M.; Maciel, G. E. *J. Magn. Reson* **1983**, *55*, 494.
51. Zeigler, R. C.; Wind, R. A.; Maciel, G. E. *J. Magn. Reson.* **1988**, *79*, 299.
52. Goldman, M. *Spin Temperature and Nuclear Magnetic Resonance in Solids;* Oxford University Press: London, 1970; Chapter 3.
53. Zilm, K., private communication, Yale University, 1989.
54. VanderHart, D. L.; Retcofsky, H. L. *Prepr. Coal Chem. Workshop* 1976, p 202.
55. Hagaman, E. W.; Chambers, R. R.; Woody, M. C. *Anal. Chem.* **1986**, *58*, 387.
56. Axelson, D. E.; Parkash, S. *Fuel Sci. Technol. Int.* **1986**, *4*, 45.
57. Packer, K. J.; Harris, R. K.; Kenwright, A. M.; Snape, C. E. *Fuel* **1983**, *62*, 999.
58. Mehring, M. *High Resolution NMR in Solids;* Springer Verlag: Berlin, Germany, 1983.
59. Wind, R. A. *Prepr. Div. Fuel Chem. ACS* **1986**, *31*, 223.
60. Gerstein, B. C. *Analytical Methods for Coal and Coal Products;* Karr, C., Ed.; Academic: New York, 1978; Vol. III, Chapter 51.
61. Jurkiewicz, A.; Wind, R. A.; Maciel, G. E. *Fuel* **1990**, *69*, 830.
62. Zhang, M.; Maciel, G. E. *J. Magn. Reson.* **1989**, *85*, 176.
63. Wind, R. A.; Yannoni, C. S. *J. Magn Reson.* **1986**, *68*, 373.
64. Wind, R. A.; Yannoni, C. S. *J. Magn. Reson.* **1987**, *72*, 108.
65. Wind, R. A.; Li, L.; Lock, H.; Maciel, G. E. *J. Magn. Reson.* **1988**, *79*, 577.
66. Wind, R. A.; Lock, H. *Adv. Magn. Reson.* **1990**, *15*, 51.

RECEIVED for review February 25, 1991. ACCEPTED revised manuscript September 17, 1991.

2

^{1}H NMR Spectroscopy

Approaches for Carbonaceous Solids

Gary E. Maciel[1], Charles E. Bronnimann[2], and Cynthia F. Ridenour

Department of Chemistry, Colorado State University, Fort Collins, CO 80523

> *This chapter provides an introduction and overview of the ^{1}H CRAMPS (combined rotation and multiple-pulse spectroscopy) technique as applied to carbonaceous solids, primarily coal. The basic nature and characteristics of the CRAMPS experiment are explained. Its applications to coal, oil shale, kerogen, humic acid, and fulvic acid are described. Some of the technical characteristics and requirements of the ^{1}H CRAMPS experiment are outlined. Directions for future ^{1}H CRAMPS applications to carbonaceous solids are discussed.*

AS AN APPROACH FOR CHARACTERIZING CARBONACEOUS samples, including solids, NMR spectroscopy has a history of more than 30 years. Even before the implementation of magic-angle spinning (MAS) (*1–3*) in the study of carbonaceous solids, ^{13}C techniques showed promise, and ^{13}C MAS techniques of various types have assumed a very important role in the study of a wide range of carbonaceous solids (*see* Chapter 1 in this volume).

Most carbonaceous solids contain substantial quantities of hydrogen, so ^{1}H NMR techniques also have been of interest for many years. How-

---

[1] Corresponding author
[2] Current address: Chemagnetics, Inc., 2555 Midpoint Drive, Fort Collins, CO 80525

ever, because of the combination of a large nuclear magnetic moment and a large natural abundance (nearly 100%), ^{1}H NMR techniques for solids must overcome the effects of the strong ^{1}H–^{1}H dipolar interactions (4–6), for which the Hamiltonian (the zeroth-order term) is shown in equation 1 for a collection of n protons.

$$H_D^{(0)} = \sum_{i<j}^{n} \sum_{j=1}^{n} b_{ij}(\mathbf{I}_i \cdot \mathbf{I}_j - 3I_{zi}I_{zj}) \tag{1}$$

where

$$b_{ij} = \left(\frac{\gamma^2 \hbar^2}{2r_{ij}^3}\right)(3\cos^2\Theta_{ij} - 1)$$

and $\mathbf{I}_i$ and $\mathbf{I}_j$ are the ith and jth protons, r_{ij} is the magnitude of the internuclear vector ($\mathbf{r}_{ij}$), Θ_{ij} is the angle between $\mathbf{r}_{ij}$ and the static magnetic field ($\mathbf{B}_0$), and γ is the magnetogyric ratio of a proton. Resulting from these strong ^{1}H–^{1}H dipolar interactions are ^{1}H–^{1}H spin–spin flip-flops, which have important consequences in the overall nuclear spin behavior of protons. Analogous complications are avoided in ^{13}C NMR spectroscopy because of ^{13}C's low natural abundance.

In the absence of suitable techniques for eliminating the effects of ^{1}H–^{1}H dipolar interactions, ^{1}H NMR spectra of typical carbonaceous solids consist of very broad bands. Sufficiently rapid MAS can average the ^{1}H–^{1}H dipolar interactions, but MAS speeds in excess of what is currently available (~27 kHz) are required to adequately narrow the peaks due to protons in proton-rich parts of an organic structure, especially in methylene groups (7). Consequently, until a few years ago, ^{1}H NMR studies of carbonaceous solids were largely limited to wide-line studies or low-resolution relaxation experiments. Such studies have provided valuable insights into the mobilities and phases involved in coal structure (8–10).

For many carbonaceous solids, like typical coals, the b_{ij} values in eq 1 can be very large, for example, 50 kHz for a geminal hydrogen pair (as in a CH_2 group). Hence, magic-angle spinning would be useful as a routine high-resolution ^{1}H technique only if very high MAS speeds, for example, 40–50 kHz, could be obtained. In 1968, Haeberlen and Waugh (11) reoriented the attention of the NMR community away from thoughts of mechanically averaging ^{1}H–^{1}H dipole–dipole effects to averaging in spin space via a multiple-pulse approach. In this approach (12), the influence of the homonuclear dipole–dipole interaction on the evolution of the nuclear spin magnetization over an entire cycle of pulses is compensated by the effects of the short, strong pulses (with precisely adjusted durations, spacings, and phases). If one simultaneously spins the sample about the magic angle, then chemical-shift anisotropy (CSA) and inhomogeneous heteronuclear dipolar interactions are averaged, so sharp peaks can be obtained. This combination of multiple-pulse and MAS line narrowing was introduced by Schnabel et al. (13) and by Gerstein and co-workers (14, 15), who coined the acronym CRAMPS (for combined rotation and multiple-pulse spectroscopy).

Figure 1 shows the strategy of the ^{1}H CRAMPS approach (5). The approach was applied first to coals by Gerstein (16), whose early 60-MHz results were viewed in most quarters with unwarranted disappointment. Later, Rosenberger and colleagues (17, 18) obtained what seemed to be somewhat more promising results at 270 MHz on European coals, which often consist of less complex structural mixtures and, hence, display more attractive fine structure. In the earlier work from the Gerstein laboratory, the apparent "resolution" was limited by the nature of the sample and not by the technique or its implementation.

The ^{1}H CRAMPS Experiment for Carbonaceous Solids: An Overview

The ancestor of all the multiple-pulse homonuclear dipolar line-narrowing techniques on which current ^{1}H CRAMPS experiments are based is the WAHUHA method (12), shown in Figure 2. (WAHUHA, MREV, and BR are acronyms created from the surnames of the originators of the methods.) The shorthand notation for this cycle of four 90° pulses of different phases is $(\bar{X}Y)$ $(\bar{Y}X)$, where the letters indicate phases ($\bar{X}$ representing a $90°_{-x}$ pulse) and the parentheses enclose symbols for pulses separated by the time 2τ (with the cycle time, τ_c, equal to 6τ). Other multiple-pulse sequences used frequently in ^{1}H CRAMPS experiments are the MREV–8 sequence, $(X\bar{Y})$ $(Y\bar{X})$ $(\bar{X}Y)$ (YX) (19–21), and the BR–24 sequence, (XY) $(\bar{Y}\bar{X})$ $(\bar{X}Y)$ $(\bar{Y}X)$ (YX) $(\bar{X}\bar{Y})$ $(\bar{Y}X)$ (YX) $(\bar{X}\bar{Y})$ $(\bar{Y}X)$ $(\bar{X}Y)$ $(\bar{X}Y)$ (22).

Figure 3 shows the effects of the two line-narrowing techniques, MAS and multiple-pulse (BR–24), separately and together in ^{1}H NMR experiments on monoethyl fumarate (5). Figure 3b shows that relatively little line-narrowing is achieved by only MAS at the relatively low MAS speed of the CRAMPS experiment (2–3 kHz in the cases shown here). The application of only BR–24 in the absence of MAS (Figure 3c) eliminates the ^{1}H–^{1}H dipolar broadening and leaves patterns made up of inhomogeneous broadening effects [principally CSA, with contributions likely

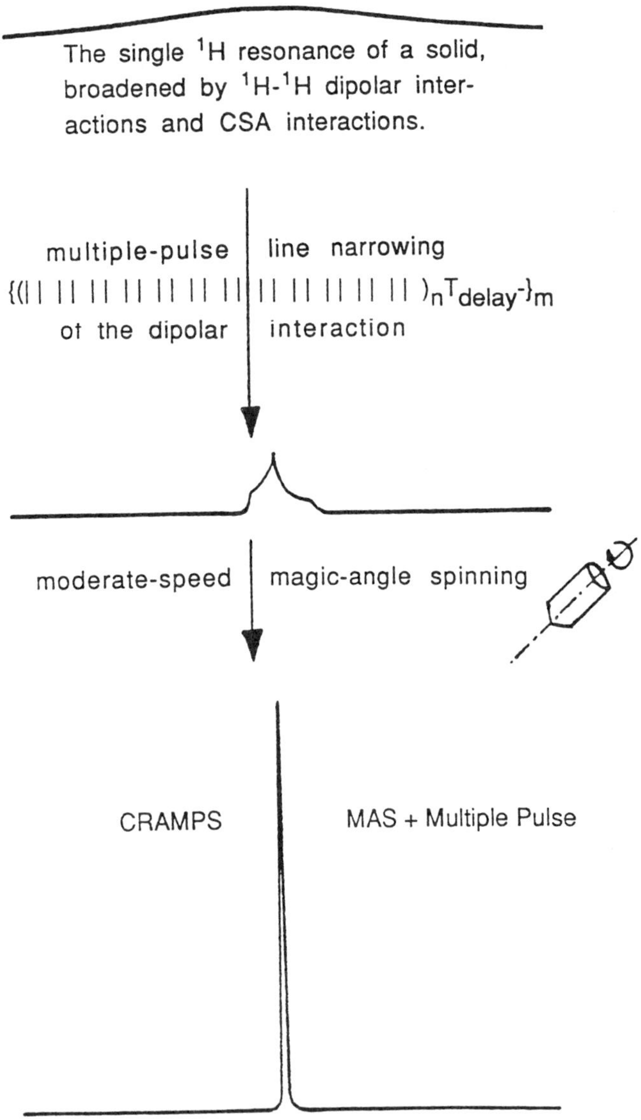

Figure 1. Diagram of the ^1H CRAMPS strategy, where n is the number of BR–24 cycles in each spectral acquisition and m is the number of repetitions of spectral acquisition.

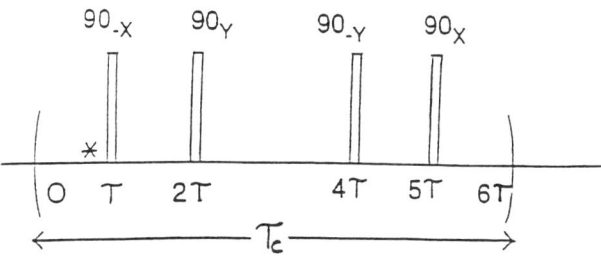

Figure 2. The WAHUHA multiple-pulse homonuclear dipolar line-narrowing sequence, showing the total cycle time $\tau_c = 6\tau$. In our experiments the pulse width is approximately 1.2 μs, τ is 2–4 μs, and the asterisk is a data-acquisition point.

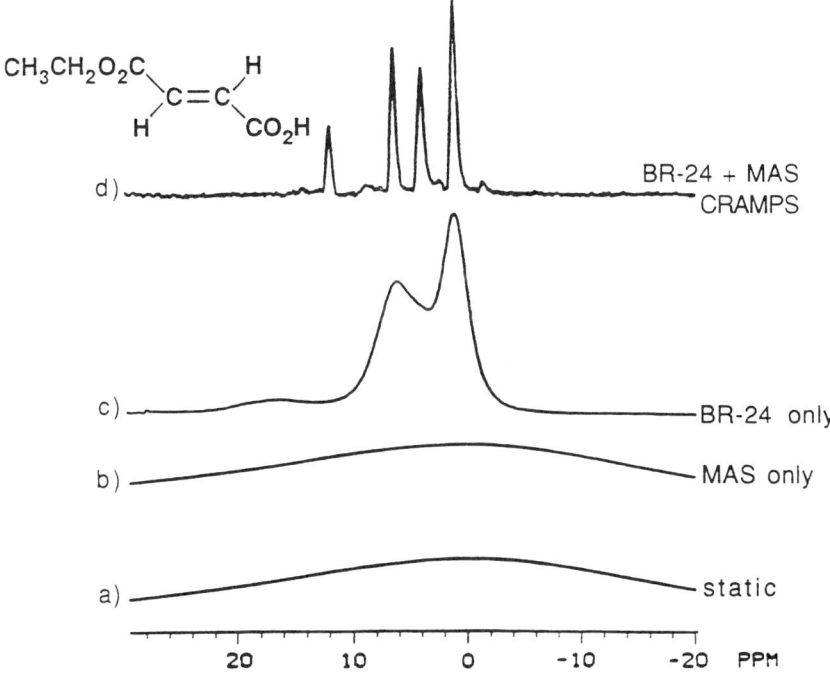

Figure 3. 187-MHz ^{1}H NMR spectra of monoethyl fumarate: (a) single pulse, no MAS; (b) single pulse with 2-kHz MAS; (c) BR–24 multiple pulse with no MAS; (d) combination of BR–24 and MAS, that is, CRAMPS. (Reproduced with permission from reference 6. Copyright 1990 Academic Press.)

from bulk susceptibility effects (6)]; such effects can largely be eliminated by MAS. Thus, in the CRAMPS spectrum (Figure 3d), a level of resolution that is useful in chemical applications is achieved.

Figure 4a shows the ^{1}H CRAMPS spectrum of citric acid as a powdered crystalline material. The level of resolution in that sample is representative of what is achievable for powdered, crystalline samples of a

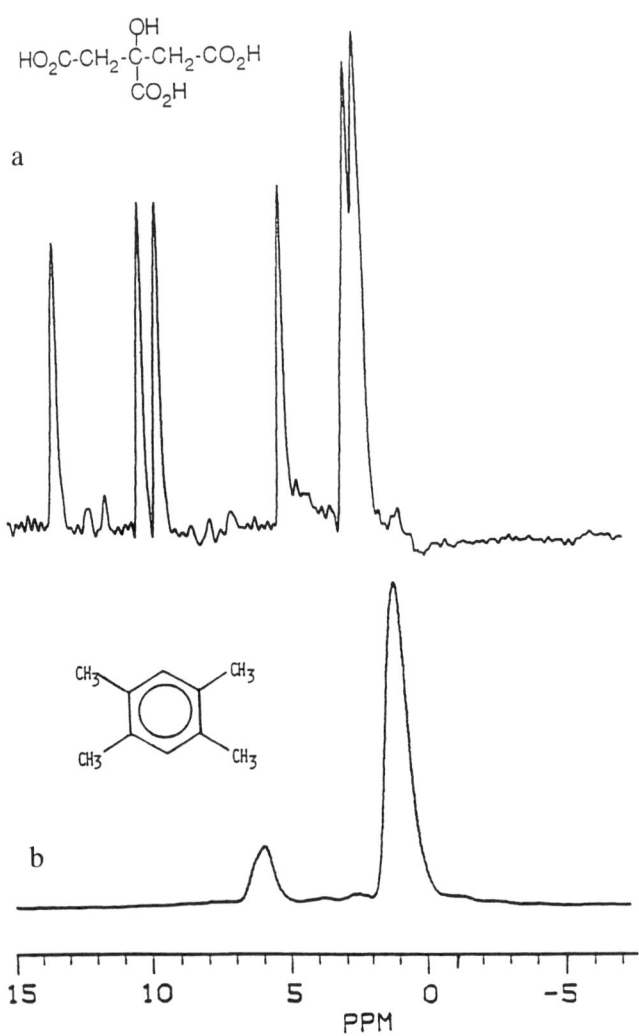

Figure 4. 187-MHz ^{1}H CRAMPS spectra of powdered (a) citric acid and (b) durene. (Reproduced with permission from reference 6. Copyright 1990 Academic Press.)

pure organic substance, although the line widths usually obtained on aromatic compounds are typically somewhat larger than for aliphatic compounds (e.g., Figure 4b); these line widths primarily reflect the effects of large anisotropies of magnetic susceptibility in the aromatic compounds.

The apparent resolution of ^{1}H CRAMPS spectra of typical coals, oil shales, and humic or fulvic acids is much lower. Figure 5 shows the ^{1}H CRAMPS spectra of three Argonne Premium coals at two different static magnetic field strengths corresponding to 187- and 360-MHz proton resonance frequencies. One general characteristic that is obvious from the ^{1}H CRAMPS spectra of these coals is that they each consist primarily of two broad contributions, a band associated with protons attached to aromatic carbons (centered at about 7–8 ppm) and a band due to protons attached to aliphatic carbons (centered at about 2–3 ppm). This situation is analogous to the ^{13}C MAS spectra of coals (*see* Chapter 1), which consist primarily of a band due to sp^2 carbons (roughly 150–110 ppm) and a band due to sp^3 carbons (roughly 0–80 ppm), although the separation between the two bands is cleaner (essentially base-line separation) in the ^{13}C case than in the ^{1}H case. In both cases, the breadth of the lines is due to inhomogeneous effects, primarily a dispersion of isotropic chemical shifts, with a contribution due to anisotropic bulk susceptibility effects. The magnitude of the latter effect should be the same for the ^{1}H and ^{13}C cases, but it has a more dramatic effect on the overall aromatic-versus-aliphatic separation in the ^{1}H case because of the smaller range of ^{1}H chemical shifts than for ^{13}C.

Another pattern is seen in Figure 5 by comparing the 187- with the 360-MHz spectra for the same coal: The ^{1}H CRAMPS spectra are not substantially improved by increasing the static magnetic field strength. This result is what one expects for the situation in which the peak widths

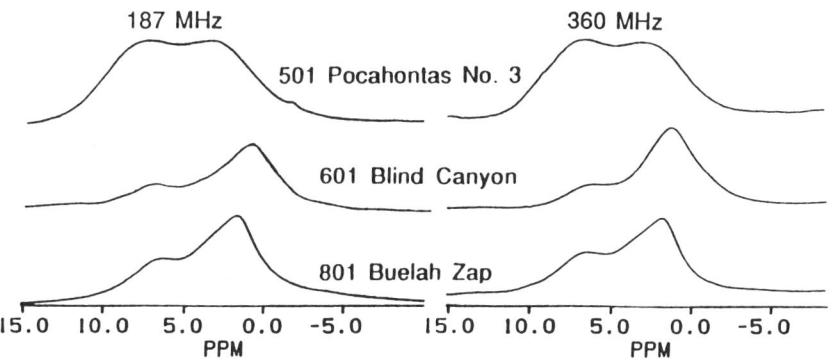

Figure 5. ^{1}H *CRAMPS spectra of Argonne Premium coals: 187 MHz (left) (29) and 360 MHz (right).*

are determined primarily by inhomogeneous effects, rather than homogeneous effects such as residual ^{1}H–^{1}H dipolar broadening. The kind of ^{1}H CRAMPS performance demonstrated on model systems, as in Figures 2 and 3, demonstrates that residual ^{1}H–^{1}H dipolar effects should not be significant contributions to the peak widths observed in the spectra shown in Figure 5.

Figure 6 shows representative ^{1}H CRAMPS spectra of a humic acid (23), a fulvic acid (23), an oil shale (24), and its kerogen concentrate (24). The humic and fulvic acids show somewhat more fine structure than for a coal, with relatively distinct contributions from various OH groups. The

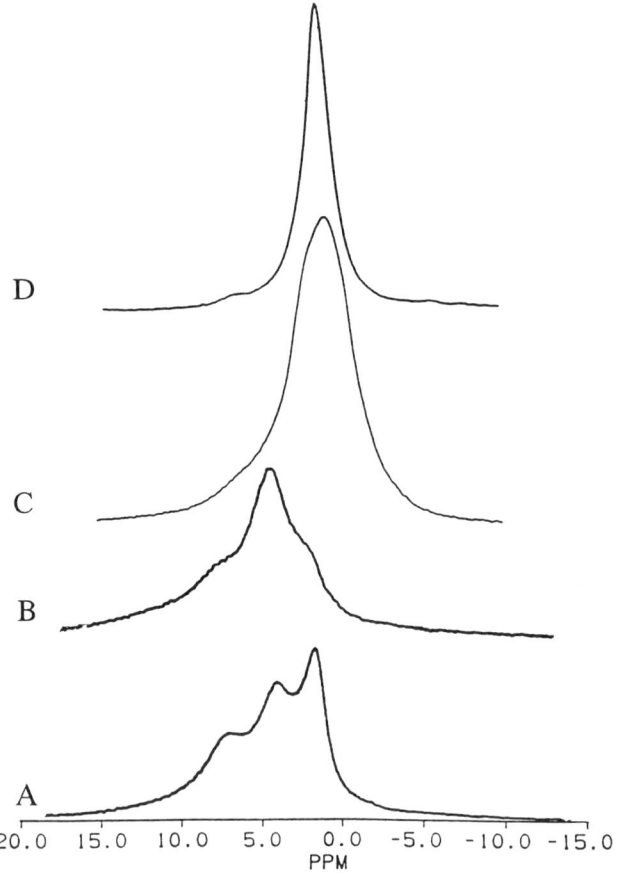

Figure 6. 187-MHz ^{1}H CRAMPS spectra of (A) Minnesota peat (Rice Lake) humic acid (23), (B) Mattole soil fulvic acid (23), (C) Colorado oil shale, and (D) kerogen concentrate of Colorado oil shale (24).

kerogen concentrate spectrum, although showing a much greater aliphatic proton content than that of a typical coal, displays roughly comparable resolution characteristics. The spectrum of the corresponding raw oil shale is somewhat "inferior" in terms of apparent overall line width. This property may be due to the very large quantity of inorganic "matrix" in oil shale, which contains substantial quantities of paramagnetic (and even ferromagnetic) centers, and these can yield large line-broadening effects.

Another interesting pattern that is often observed in the ^{1}H CRAMPS spectrum of coal (25, 26) is the effect of saturation with perdeuteropyridine (27–29). Figure 7 shows this effect. Comparing the spectra of unadulterated (dry) coals with the spectrum of the corresponding coal saturated with C_5D_5N illustrates the improved level of resolution typically obtained. This effect may be partially due to some degree of susceptibility averaging, for example, when pyridine molecules fill voids in the coal structure and partially mobilize locally anisotropic moieties. There should be a reduction in line width due to partial mobilization of certain structural moieties in coal caused by pyridine saturation; although this postulated partial mobilization should not significantly reduce residual ^{1}H–^{1}H dipolar interactions (which we believe are minimal even without C_5D_5N saturation), it could reduce the spread of isotropic chemical shifts by partially averaging disparate chemical shifts that are frozen into the system in the absence of mobilization by C_5D_5N and average anisotropic magnetic susceptibilities. Saturation with C_5D_5N also has a substantial effect on oil shale or kerogen spectra (24).

The issue of molecular-level mobility in carbonaceous solids is amenable to study via time-domain ^{1}H CRAMPS experiments. A variety of relaxation techniques that are direct carryovers from liquid-sample NMR

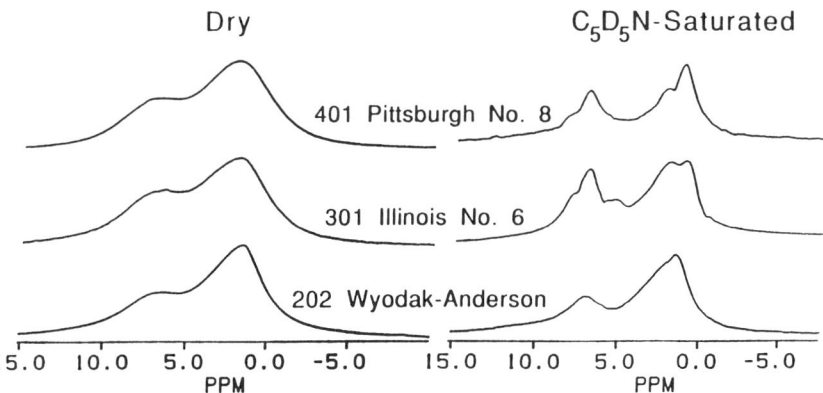

Figure 7. 187-MHz ^{1}H CRAMPS spectra of dry and C_5D_5N-saturated Argonne Premium coals (29).

spectroscopy or from ^{13}C cross-polarization (CP)–MAS NMR spectroscopy have been employed (*30*). One of the most useful of these, especially for studying issues of molecular mobility, is a ^{1}H–^{1}H dipolar-dephasing ^{1}H CRAMPS technique that is directly analogous to the well-known ^{1}H–^{13}C dipolar-dephasing technique in ^{13}C CP–MAS NMR spectroscopy. In the ^{1}H CRAMPS version (*30*), for which the pulse sequence is shown in Figure 8A, a dephasing period (*t*, with a "refocusing" π pulse in the middle) is inserted between the initial $\pi/2$ pulse that places ^{1}H magnetization in the transverse plane and initiation of the BR–24 sequence that is used for high-resolution (CRAMPS) detection. During this *t* period, the magnetization due to those protons that experience strong resultant ^{1}H–^{1}H dipolar interactions is strongly attenuated for *t* greater than about 20–30 μs. The ^{1}H–^{1}H dipolar interactions of protons in those structural moieties that experience rapid motion are strongly attenuated by the motion, and this attenuation is reflected in slower dephasing in the ^{1}H CRAMPS dipolar-dephasing experiment. The dipolar-dephasing behavior of each peak in the ^{1}H CRAMPS spectrum (especially for C_5D_5N-saturated samples) can often be analyzed in terms of eq 2 (*27–29*)

$$M(t) = \sum_i a_i \exp(-t^2/2T_{2ai}^2) + \sum_j b_j \exp(-t/T_{2bj}) \qquad (2)$$

which includes both Gaussian and Lorentzian terms. In eq 2, $M(t)$ represents the transverse ^{1}H magnetization after a dephasing period *t*, a_i is the fraction of proton magnetization dephasing according to the Gaussian time constant T_{2ai}, and b_j is the fraction of proton magnetization dephasing according to the Lorentzian parameter T_{2bj}. The Gaussian terms describe the more rapid dephasing, which is associated with the more rigid macromolecular component(s) of the sample, and the Lorentzian terms describe less rapid dephasing, which is associated with the more mobile structural components in the sample. This technique has been applied successfully to examine the relationship between chemical structure and mobility in Premium coals and their pyridine extracts and residues (*27–29*). Figure 8B shows typical ^{1}H CRAMPS dipolar-dephasing behavior of a subbituminous coal. This technique is also useful in studies of oil shales and kerogens (*24*).

Details of the ^{1}H CRAMPS Experiment

Choice of Pulse Sequence. The pulse sequences most often used in CRAMPS are derived from the solid-echo (*1*), which is essentially the first half of the WAHUHA sequence shown in Figure 2. The most

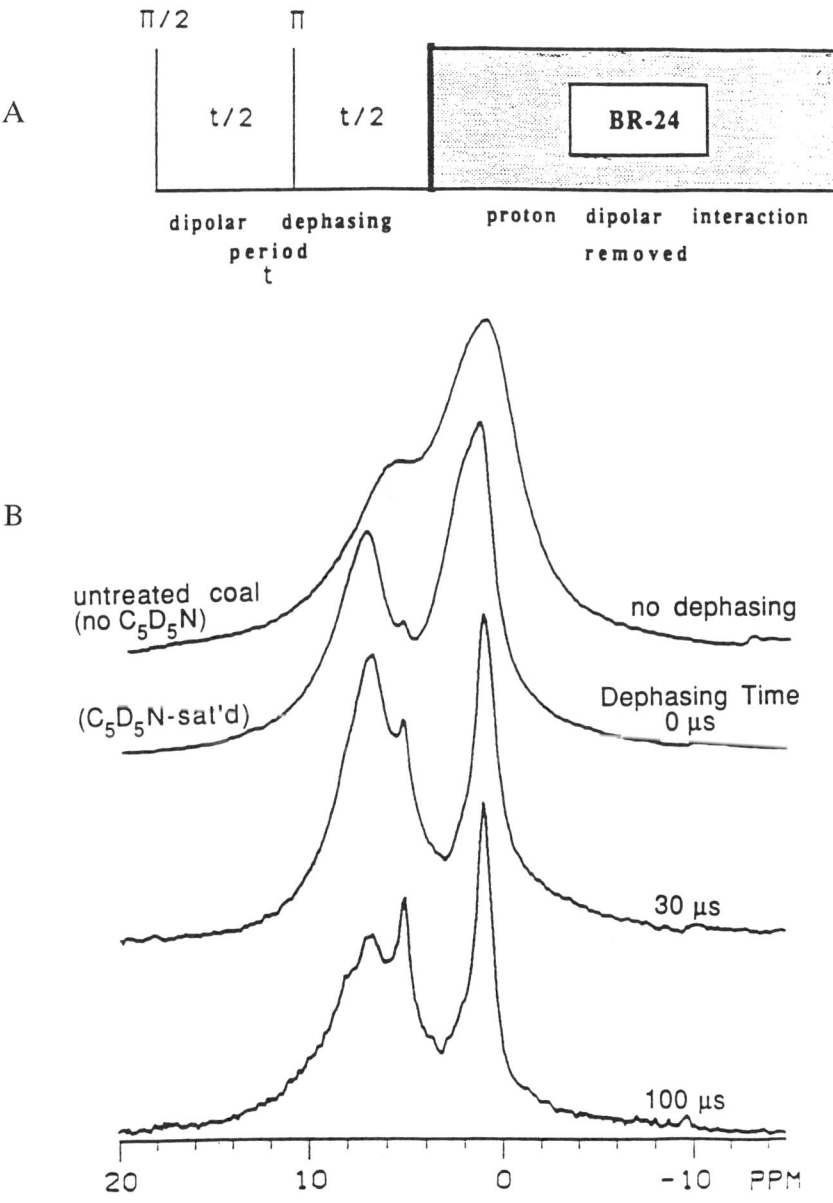

Figure 8. A: ^1H–^1H dipolar-dephasing sequence with CRAMPS detection. B: Application of the ^1H CRAMPS dipolar-dephasing sequence to a subbituminous Polish coal (27), showing the effect of dipolar-dephasing time (τ) on the spectrum of a C_5D_5N-saturated sample, with the untreated coal spectrum for comparison (27).

popular of these sequences are, in order of complexity, the WAHUHA sequence *(12)*, MREV–8 *(19–21)*, and BR–24 *(22)*. Windowless sequences *(31)* and those based on time reversal *(32, 33)* are of more limited use in obtaining high-resolution ^{1}H spectra and are not discussed here.

Pulse sequences can be evaluated according to (1) the ease with which one can obtain efficient homonuclear dipolar decoupling, (2) their chemical-shift scaling factor, and (3) the bandwidth achieved. The more complex pulse sequences are generally easier to implement and achieve more complete decoupling. However, these improvements are achieved at the cost of larger cycle times and smaller scaling factors (i.e., a larger degree of scaling).

Factors influencing the ease of implementation of a given pulse sequence are the radio frequency (rf) power and pulse-switching speed requirements and sensitivity to pulse imperfection. The size of the chemical-shift scaling factor determines the extent to which homogeneous line-broadening mechanisms, such as molecular motion, will degrade resolution. The overall effectiveness of a pulse sequence with a smaller scaling factor will be more severely affected by homogeneous broadening than that of a sequence with less severe scaling. The bandwidth of a pulse sequence is determined by its cycle time. Longer cycles have correspondingly smaller bandwidths and should therefore be limited to spectra with small frequency widths. Furthermore, the relatively low data-acquisition rate of longer cycles seriously reduces their effective sensitivity (signal-to-noise ratio, S/N) relative to shorter cycles. This reduction occurs because one cannot use an audio filter width matched to the bandwidth of the pulse sequence. Filters must instead be set to a frequency width of approximately τ^{-1} in order to avoid saturation by the pulse train. Compensating for poor sensitivity by simply averaging for very long times may not be a good idea because spectral resolution can be degraded if the spectrometer is unstable over the long term.

The WAHUHA sequence contains two solid-echo pulse groups. The chemical-shift scaling factor of this sequence is 0.575, the maximum possible. The WAHUHA cycle time (6τ) is the shortest cycle time possible for a sequence based on the solid echo; thus WAHUHA has the largest data-acquisition rate and bandwidth of the three common sequences. On the other hand, WAHUHA is the least efficient in removing homonuclear dipolar coupling. It is frequently stated that the WAHUHA sequence fails because it does not properly average the dipolar interaction (H_D) even to zeroth order in the case of finite pulse width. In fact, the pulse width can be adjusted according to the duty factor of the cycle to remove $H_D^{(0)}$ *(34)*. With the proper pulse width, WAHUHA will null both $H_D^{(0)}$ and $H_D^{(1)}$ [the latter by virtue of reflection symmetry *(4)*].

The main difficulty of WAHUHA is in its intolerance of pulse imperfections, especially rf inhomogeneity and the antisymmetric phase transient. These pulse errors result in a residual homonuclear dipolar line width that is proportional to τ. The phase transient can be reduced by reducing probe and transmitter Qs, which increases power demands. The rf homogeneity can be improved by reducing the sample volume, but at the cost of lower sensitivity. The line-broadening effects of both types of error term can be minimized by minimizing τ. In practice, the ultimate resolution available from WAHUHA is on the order of 2–4 ppm.

MREV–8 combines two WAHUHA cycles to reduce sensitivity to both rf inhomogeneity and phase transient errors. MREV–8 also removes $H_D^{(0)}$ in the case of finite 90° pulses and thus makes setup slightly easier. The MREV–8 cycle time is 12τ, yielding half the data-acquisition rate and bandwidth of WAHUHA. The data-acquisition rate of MREV–8 can easily be doubled by sampling each WAHUHA subcycle once. Oversampling does not increase the spectral width because sampling more than once per cycle introduces artifacts that will appear at the cycle frequency (35). The scaling factor of MREV–8 is 0.475, slightly worse than that of WAHUHA.

In a well-tuned spectrometer, the principal source of broadening for MREV–8 is the second-order dipole term, $H_D^{(2)}$. The line-broadening contribution due to $H_D^{(2)}$ is proportional to τ and can thus be reduced by decreasing τ. Experimentally we find that τ values of 2.0–2.5 μs are sufficient to eliminate the effects of $H_D^{(2)}$ (6). Operation at $\tau = 2$ μs places great demands on receiver recovery, transmitter power, and on the probe.

BR–24 was derived from MREV–8 by using the principle of pulse-cycle decoupling (22), which, loosely stated, is as follows: If pulse cycles are combined in the right way, their zeroth-order and first-order average Hamiltonians add to each other. In BR–24, three MREV–8 cycles are combined to remove $H_D^{(2)}$, while the compound cycle retains the favorable error-compensation characteristics of MREV–8. BR–24 should thus achieve efficient dipolar narrowing at longer τ values than does MREV–8. Experimentally, τ values of 3.0–3.5 μs are found to be adequate (6). This increased dipolar line-narrowing efficiency comes at the expense of a long cycle time (36τ) and a scaling factor of 0.375. It is easy to increase the data-acquisition rate of BR–24 by sampling twice per BR–24 cycle.

Choice of Static Magnetic Field Strength. Many considerations are involved in selecting the optimal static magnetic field strength for ^{1}H CRAMPS experiments. Apparently, in principle, the experiment can only improve by going to higher magnetic fields, but a number of technical problems make high-field CRAMPS very difficult at this time.

The main advantages of increasing the magnetic field strength are that the sensitivity and resolution of CRAMPS spectra should improve. Homogeneous contributions to the line width, including contributions from pulse imperfections and dipolar broadening, both homonuclear and heteronuclear, are typically independent of the magnetic field strength or are actually reduced at increased fields. Thus, the effect of dipolar broadening on spectral resolution should become smaller relative to the chemical-shift dispersion at higher magnetic fields. The benefits of increased sensitivity are also important, both in terms of lowering the limit of detection of ^{1}H CRAMPS, which is currently on the order to 10^{-2}–10^{-3} wt% hydrogen, and in terms of reducing overall experiment times.

CRAMPS experiments are currently being performed successfully at proton Larmor frequencies up to 400 MHz. Several challenges are involved in going beyond this frequency. With solenoid coils and traditional circuit geometries, it is difficult to achieve intense and highly homogeneous rf fields at higher frequencies. Up to now, short pulse widths have been obtained by reducing the coil volume. The homogeneous volume of the coil is thus also reduced, and this lower volume requires the use of smaller sample sizes. Most likely, future progress will involve alternate-coil geometries and matching methods.

Higher magnetic fields require larger spectral bandwidths and MAS rates. On the other hand, multiple-pulse sequences have limited bandwidths and perform best at low MAS rates. Multiple-pulse sequences efficiently decouple over only a limited frequency width (*6, 36*). For BR–24 the usable frequency width is usually greater than the bandwidth, but WAHUHA and MREV–8 do not effect homonuclear decoupling efficiently over their full bandwidths (*6, 20–22*). In our hands, BR–24 is the most sensitive of these three pulse sequences to interference from MAS. Because of its small bandwidth and sensitivity to MAS speed, BR–24 will probably be inferior to MREV-8 at 400 MHz and higher, in spite of the relatively poor response of MREV–8 to large offsets.

Another likely disadvantage of increasing the static magnetic field without limit is in the area of spin–lattice relaxation. The proton spin–lattice relaxation time, T_1^H, usually increases as the field strength increases, a result requiring larger repetition delays and canceling (at least partially) the higher-field sensitivity advantage.

Choice of Operating Parameters. The CRAMPS experiment involves a great deal of parameter variation. Preparation pulses, receiver phase, pulse width, cycle time, offset (magnitude and sign), and MAS rate should all be treated as parameters that can be varied for optimization of the CRAMPS spectrum. Coals and oil shales provide an excellent opportunity to fully exercise this flexibility.

The proper selection of receiver phase and the phase and width of the preparation pulse is important in avoiding the pedestal signal that arises from magnetization that is prepared parallel to the effective field of the multiple-pulse train (4, 6). If the pedestal is broad, as it will be if there is significant homogeneous broadening, it can interfere with the spectrum and destroy quantitation. The aromatic signals of coals and shales are particularly prone to distortion by the pedestal. If the pedestal cannot be removed, then the offset must be increased to the point where the pedestal does not overlap with the spectrum, often at the cost of degraded performance in the aliphatic region because of the enhanced offset for this region of the coal or oil shale spectrum (assuming the carrier frequency is on the low-shielding side of the spectrum). Prescriptions exist for removing the pedestal (4, 6), but for geochemical samples our experience is that these prescriptions just serve as a starting point for trial-and-error optimization.

The time τ (Figure 2) should be set to the largest value for which homonuclear decoupling is still adequate in order to allow maximum receiver recovery before sampling and to reduce the effects of pulse imperfections. For the same reasons, pulse widths should be kept to a minimum.

The optimum offset and the MAS rate are strongly coupled parameters. The averaging process of MAS couples with pulse errors to produce "rotor frequency lines" at multiples of the spinning rate (37). If spectral intensity overlaps a rotor frequency line, then severe distortion can result. Often the distortion is not immediately obvious, so one must vary the spinning rate and verify that the spectrum does not depend on the spinning rate. If spectral intensity is essentially continuous across the ^{1}H spectral width of interest, as it normally is with geochemical samples, then it is usually advisable, especially at lower fields, to spin fast enough to set the first rotor frequency line outside the spectrum and accept a slight loss of resolution due to rapid MAS. Also, one must be alert to folded-in spinning sidebands. These are often not easy to characterize by using only BR−24, but a quick measurement with MREV−8 (for which the sidebands are less likely to be folded in) will usually identify them. Many shales and high-rank coals cannot be run with BR−24 because of their large spinning-sideband patterns.

The Water Problem. The presence of water in coals is a problem in the sense that the unequivocal characterization of the water resonance in the ^{1}H CRAMPS spectrum of an untreated coal remains elusive. Water, as defined by other chemical analyses, must be present in coals, sometimes at relatively high concentrations. Until we are better able to identify water, the quantitation of structural types in coal will be uncertain in proportion to the amount of water present. If water in coal is either very tightly bound or extremely mobile, then it should make a distinct contribution to the ^{1}H CRAMPS spectrum, as a sharp peak if it is extremely mobile, or possibly, as a broad peak if water is tightly bound and if a wide range of static environments (e.g., hydrogen-bonding situations) is present. A sharp, liquidlike peak is not observed in the ^{1}H CRAMPS spectrum, even after a short dephasing interval. If water in coal manifests mobility with a correlation time comparable to the period of the multiple-pulse cycle, then a broad, perhaps undetected, resonance may result. MAS-only and low-temperature experiments may help to characterize this issue. ^{1}H CRAMPS spectra of Premium coals that have been saturated with pyridine-d_5 indeed reveal a small liquidlike water peak that corresponds to 1–5% of the total hydrogen content (*27, 28*). Measurements made on treated samples give only lower limits for the water content of coals.

Directions for Further Progress

Perhaps the most exciting recent development in CRAMPS spectroscopy is the ready availability of variable-temperature capabilities. One can now readily reach temperatures of −160 to +250 °C in the ^{1}H CRAMPS experiment. When coupled with dephasing or with measurements of T_{1mp} (representing relaxation under the influence of a multiple-pulse train) (*38*), variable-temperature CRAMPS should be a powerful new tool in the study of structure and dynamics in coals and other material. It appears that temperatures as high as 500 °C will soon be practical.

Significant advances in ^{1}H imaging have occurred recently with the use of multiple-pulse methods (*39–41*). A spatial resolution of better than 100 μm for a 1-cm^3 sample has been achieved. The potential resolution of solids ^{1}H imaging is much higher, being ultimately limited by molecular motion, which can presumably be reduced at low temperature or by other line-broadening influences (e.g., unpaired electrons). Applications of multiple-pulse imaging methods to coals are already under way (*42*).

Acknowledgments

The authors gratefully acknowledge support of this work under Department of Energy Grant No. DE–FG22–90PC90290.

References

1. Lowe, I. J. *Phys. Rev. Lett.* **1959**, *2*, 285.
2. Kessemeier, H.; Norberg, R. E. *Phys. Rev* **1967**, *155*, 321.

3. Andrew, E. R. *Philos. Trans. R. Soc. London* **1981**, *A299*, 505.
4. Haeberlen, U. *High Resolution NMR in Solids: Selective Averaging;* Academic: New York, 1976.
5. Bronnimann, C. E.; Hawkins, B. L.; Zhang, M.; Maciel, G. E. *Anal. Chem.* **1988**, *60*, 1743.
6. Maciel, G. E.; Bronnimann, C. E.; Hawkins, B. L. In *Advances in Magnetic Resonance: The Waugh Symposium;* Warren, W. S., Ed.; Academic: San Diego, CA, 1990; Vol. 14, p 125.
7. Dec, S. F.; Bronnimann, C. E.; Wind, R. A.; Maciel, G. E. *J. Magn. Reson.* **1989**, *82*, 454.
8. Jurkiewicz, A.; Murec, A.; Pislewski, N. *Fuel* **1982**, *61*, 647.
9. Barton, W. A.; Lynch, L. J.; Webster, D. S. *Fuel* **1984**, *63*, 1262.
10. Kamienski, B.; Pruski, M.; Gerstein, B. C.; Given, P. H. *Energy Fuels* **1987**, *1*, 45.
11. Haeberlen, U; Waugh, J. S. *Phys. Rev.* **1968**, *175*, 453.
12. Waugh, J. S.; Huber, L. M.; Haeberlen, U. *Phys. Rev. Lett.* **1968**, *20*, 180.
13. Schnabel, B.; Haubenreisser, U.; Scheler, G.; Müller, R. *Proc. 19th Congr. Ampere* (Heidelberg) **1976**, 441.
14. Gerstein, B. C.; Pembleton, R. G.; Wilson, R. D.; Ryan, L. J. *J. Chem. Phys.* **1977**, *66*, 361.
15. Gerstein, B. C.; Chou, C.; Pembleton, R. G.; Wilson, R. C. *J. Phys. Chem.* **1977**, *81*, 565.
16. Gerstein, B. C. *Philos. Trans. R. Soc. London* **1981**, *A299*, 521.
17. Rosenberger, H.; Scheler, G. *Z. Chem.* **1983**, *23*, 34.
18. Schmiers, H.; Rosenberger, H.; Scheler, G. *Forschungsergebnisse* **1982**, *1*, 1.
19. Mansfield, P.; Orchard, M. J.; Stalker, D. C.; Richards, K. H. B. *Phys. Rev. B Solid State* **1973**, *7*, 90.
20. Rhim, W. K.; Elleman, D. D.; Vaughan, R. W. *J. Chem. Phys.* **1973**, *58*, 1772.
21. Rhim, W. K.; Elleman, D. D.; Vaughan, R. W. *J. Chem. Phys.* **1973**, *59*, 3740.
22. Burum, D. P.; Rhim, W. K. *J. Chem. Phys.* **1979**, *71*, 944.
23. Frye, J. S.; Bronnimann, C. E.; Maciel, G. E. In *NMR of Humic Substances and Coal;* Wershaw, R. L.; Mikita, M. A., Eds.; Lewis Publishers: Chelsea, MI, 1987; p 33.
24. Jurkiewicz, A.; Ridenour, C. F.; Maciel, G. E., unpublished.
25. Bronnimann, C. E.; Maciel, G. E. *Org. Geochem.* **1989**, *14*, 189.
26. Davis, M. F.; Quinting, G. R.; Bronnimann, C. E.; Maciel, G. E. *Fuel* **1989**, *68*, 763.
27. Jurkiewicz, A.; Bronnimann, C. E.; Maciel, G. E. *Fuel* **1989**, *68*, 872.
28. Jurkiewicz, A.; Bronnimann, C. E.; Maciel, G. E. *Fuel* **1990**, *69*, 804.
29. Jurkiewicz, A.; Bronnimann, C. E.; Maciel, G. E. In *Magnetic Resonance of Carbonaceous Solids;* Botto, R. E.; Sanada, Y., Eds.; Advances in Chemistry 229; American Chemical Society: Washington, DC, 1993.
30. Bronnimann, C. E.; Zeigler R. C.; Maciel, G. E. *J. Am. Chem. Soc.* **1988**, *110*, 2023.
31. Paoles, J. G.; Mansfield, P. *Phys. Lett.* **1962**, *2*, 58.
32. Burum, D. P.; Linder, M.; Ernst, R. R. *J. Magn. Reson.* **1981**, *44*, 173.
33. Takegoshi, K.; McDowell, C. A. *Chem. Phys. Lett.* **1985**, *116*, 100.
34. Wu, X. *Chem. Phys. Lett.* **1989**, *156*, 82.

35. Mehring, M. *Z. Naturforsch* **1972,** *27A,* 1634.
36. Rhim, W. K.; Burum, D. P.; Vaughan, R. W. *Rev. Sci. Instrum.* **1976,** *47,* 720.
37. Haeberlen, U.; Ellet, J. D.; Waugh, J. S. *J. Chem. Phys.* **1971,** *55,* 53.
38. Vega, S.; Olejniczak, E. T.; Griffin, R. G. *J. Chem. Phys.* **1984,** *80,* 4832.
39. Gründer, W.; Schmiedel, H.; Freude, D. *Ann. Phys. (Leipzig)* **1971,** *27,* 409.
40. Cory, D. G.; de Boer, J. C.; Veeman, W. S. *Macromolecules* **1989,** *22,* 1618.
41. Cory, D. G.; Miller, J. B.; Turner, R.; Garroway, A. N. *Mol. Phys.* **1990,** *70,* 331.
42. Cory, D. G.; Miller, J. B.; Garroway, A. N. *J. Magn. Reson.* **1990,** *90,* 205.
43. Dieckman, S. L.; Gopalsami, N.; Botto, R. E. *Energy Fuels* **1990,** *4,* 417.

RECEIVED for review January 30, 1991. ACCEPTED revised manuscript July 1, 1991.

3

Solid Materials Research with NMR and Dynamic Nuclear Polarization Spectroscopy

Robert A. Wind[1], Russ Lewis[2], Herman Lock[2], and Gary E. Maciel[2]

[1]Chemagnetics, Inc., 2555 Midpoint Drive, Fort Collins, CO 80525
[2]Department of Chemistry, Colorado State University, Fort Collins, CO 80523

> *In solids containing both magnetic nuclei and unpaired electrons, the nuclear NMR signal can be enhanced via irradiation at or near the electron Larmor frequency to yield the dynamic nuclear polarization (DNP) effect. DNP combined with modern solid-state NMR spectroscopy can be used to study properties of materials that could not be investigated by NMR spectroscopy alone, such as the dynamics of the unpaired electrons, molecular structures in the vicinity of the unpaired electrons, and the presence of small amounts of nuclei. In this chapter a review of the various mechanisms that can determine the DNP enhancement is given, and applications of the DNP NMR technique are shown in coal, undoped trans-polyacetylene, (fluoranthenyl)$_2$PF$_6$, a ceramic fiber, and a vapor-deposited diamond.*

POLARIZATION TRANSFER between the electron and nuclear spin systems can be obtained in solids containing both magnetic nuclei and unpaired electrons via irradiation at or near the electron Larmor frequency. The result is an enhanced nuclear polarization and, henceforth, an enhanced NMR signal. This effect is called dynamic nuclear polarization (DNP) (*1–4*). DNP is an old technique, and the first DNP experiment dates as early as 1953 (5). In the earlier days of magnetic resonance, DNP was applied mainly at low temperatures in high-energy physics, where it was used for the production of polarized targets or to study magnetic ordering at extremely low spin temperatures. During the past decade, however, other applications have been developed in which DNP was combined with modern solid-state NMR techniques such as cross-polarization–magic-angle spinning (CP–MAS) and two-dimensional (2D) NMR spectroscopy (6–8).

DNP NMR spectroscopy can be applied successfully in a large variety of solid materials in which unpaired electrons occur naturally, in materials that contain paramagnetic impurities, or in materials in which unpaired electrons have been embedded by doping the solid with a suitable agent. Examples of materials in which unpaired electrons occur naturally are fossil fuels and carbonaceous chars containing π-type organic radicals (9, 10); amorphous materials in which unpaired electrons occur in the form of fixed dangling bonds (11); and undoped *trans*-polyacetylene containing mobile dangling bonds, the so-called "solitons" (12, 13). Examples of materials containing paramagnetic impurities are natural and artificial diamonds (14), and examples of doped materials are doped polymers (15) and organic conductors (16).

In this chapter a review of applications of DNP NMR spectroscopy in solid materials research will be given. First, a brief description of the DNP phenomenon will be given. The conditions under which a noticeable DNP enhancement can be expected will be discussed, and the loss in NMR signal due to the presence of the unpaired electrons will be addressed. Applications of DNP NMR spectroscopy will be given, and representative illustrations of these applications will be shown. Finally, future developments of DNP NMR spectroscopy will be discussed.

Theory

Overview of Dynamic Nuclear Polarization.

Extensive reviews of the different mechanisms that can dominate in a DNP experiment are given in references 1–4, 7, and 15. Here we confine ourselves to a treatment of the main features of these mechanisms. Moreover, we restrict ourselves to DNP of nuclear spin-½ systems, and we assume that the spin diffusion among the nuclei is fast enough to average out possible differences in enhancement factors. The kind of DNP experiment that can be applied, the magnitude of the nuclear polarization enhancement, and the circumstances under which the maximal enhancement occurs depend on the nature and time dependence of the electron–nuclear interaction term H_{en}. If H_{en} contains time-independent terms, the nuclear polarization enhancement can arise from three DNP mechanisms: the solid-state effect, the direct thermal mixing effect, and the indirect thermal mixing effect. For these effects the enhancement of the nuclear polarization minus unity, $E - 1$, as a function of the frequency of the microwave field, ω, is antisymmetrical about the electron Larmor frequency ω_e. We assume that the nuclear Larmor frequency, ω_n, is larger than the electron spin resonance (ESR) line width, $\omega^e_{\frac{1}{2}}$. Then the solid-state effect gives a maximum when $\omega \approx \omega_e \pm \omega_n$, the direct thermal mixing effect becomes maximal when $\omega \approx \omega_e \pm \omega^e_{\frac{1}{2}}$, and the nuclear polarization enhancement due to the indirect thermal mixing effect becomes optimal when $\omega = \omega \pm \omega_1$, with $\omega^e_{\frac{1}{2}} < \omega_1 < \omega_n$.

If H_{en} contains time-dependent terms with frequencies comparable to ω_e, an Overhauser effect is observed (17). Then the enhancement curve, $(E - 1)$ as a function of the microwave offset frequency, $\omega - \omega_e$, which reflects the (saturated) ESR line shape, is often symmetrical about ω_e, and $(E - 1)$ becomes maximal when $\omega = \omega_e$. In Figure 1 characteristic DNP enhancement curves are given for the four DNP mechanisms mentioned.

In theory the DNP enhancement can be very large, of the order of ω_e/ω_n or more, but in practice the observed enhancement is often considerably less. To investigate this issue, in the following sections theoretical expressions of the DNP enhancement factors will be given for static and time-dependent electron–nuclear interactions.

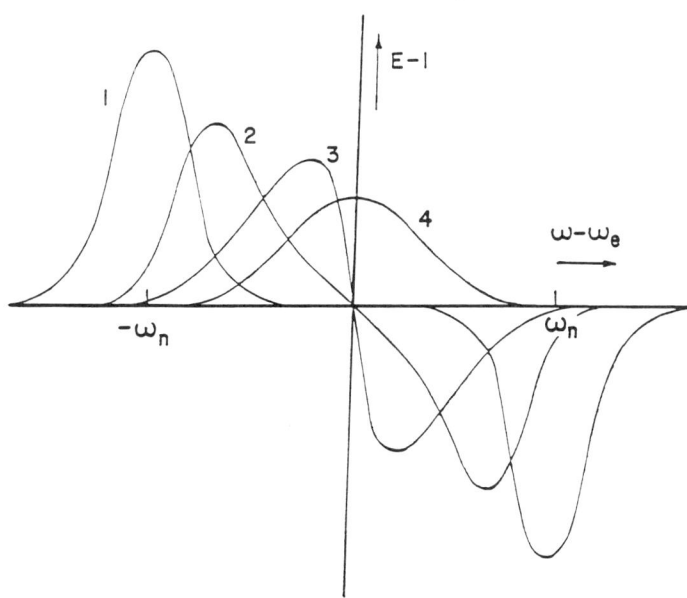

Figure 1. Frequency dependence of the various DNP effects: curve 1, solid-state effect; curve 2, indirect thermal mixing effect; curve 3, direct thermal mixing effect; and curve 4, Overhauser effect.

DNP Due to Static Electron–Nuclear Interactions. Usually the overall enhancement due to both the solid-state and thermal mixing effects becomes maximal for $\omega \approx \omega_e \pm \omega_n$, and the maximal enhancement factor is the same as that found for the solid-state effect alone (7). Therefore we shall confine ourselves to the the solid-state effect, which arises because the static nonsecular electron–nuclear dipolar interactions produce small admixtures of the unperturbed electron–nuclear product wave functions. As a result, irradiation with a frequency near $\omega_e \pm \omega_n$ leads to nonzero transition probabilities, the so-called "forbidden" transitions, $W^{\pm}(\omega)$, where electron and nuclear spins flip simultaneously. For electron–nuclear pairs with an interelectron–nuclear distance r, irradiation with a frequency ω equal to, for example, $\omega_e - \omega_n$, results in a nuclear polarization enhancement, $(E - 1)_s$, given by (7)

$$(E - 1)_s = \frac{W^-(0)}{W_z^n} \cdot \frac{|\gamma_e|}{\gamma_n} = \frac{3\pi}{20} \cdot \frac{d^2(r)}{\omega_n^2} \cdot \frac{\omega_{1e}^2 g(0)}{W_z^n} \cdot \frac{|\gamma_e|}{\gamma_n} \qquad (1)$$

In eq 1, W_z^n is the nuclear Zeeman relaxation rate; γ_e and γ_n are the electron and nuclear magnetogyric ratios, respectively; $\omega_{1e} = \gamma_e H_{1e}$; H_{1e} is the amplitude of the microwave field; $g(\omega)$ is the normalized ESR line-shape function; $d(r) = \gamma_e \gamma_n \hbar r^{-3}$ is a measure of the local field an electron and a nucleus are experiencing from each other; and $W^-(0)$ and $g(0)$ are the maximum values of $W^-(\omega)$ and $g(\omega)$, respectively. For simplicity, in eq 1 the angular dependence in the electron–nuclear interactions has been replaced by its averaged value.

In fact an electron is surrounded by many nuclei at different distances r, a condition that, according to eq 1, generally results in a distribution of enhancement factors. For nuclei undergoing rapid spin diffusion, the differences in enhancement factors are averaged out, and the result is a single enhancement factor for all nuclei. To calculate this enhancement, we define two radii around each electron: a radius R that is related to the average distance between the electrons via $(4/3)(\pi R^3 N_e) = 1$, where N_e is the number of electrons per unit volume, and a radius b such that for nuclei at distances $r < b$, the local fields these nuclei are experiencing from the electrons are so large that these nuclei do not contribute to the bulk NMR signal. Then for a homogeneous distribution of the nuclei, the average enhancement can be calculated by integrating eq 1 over the volume between the spheres with radii b and R. The result is

$$(E - 1)_s = \frac{\pi^2}{5} \cdot \frac{\gamma_e^2 \gamma_n^2 \hbar^2}{\omega_n^2} \cdot \frac{N_e}{b^3} \cdot \frac{\omega_{1e}^2 g(0)}{W_z^n} \cdot \frac{|\gamma_e|}{\gamma_n} \qquad (2)$$

It follows that the radius b is an important parameter, because not only it is one of the key parameters determining the enhancement, but it also determines the percentage of nuclei not observed as a result of the presence of the unpaired electrons. (The latter issue is discussed in more detail in Chapter 1.) To investigate the value of b, we first assume that H_{en} is completely static. Then the local field that a nucleus is experiencing from an electron, $\omega_L(r)$, is of the order of $\frac{1}{2}(\gamma_e\gamma_n\hbar r^{-3})$. Then the value of b can be estimated by requiring that $\omega_L(b)$ is approximately equal to the line width of the NMR line in the absence of the electrons, $\omega^n_{1/2}$:

$$\frac{1}{2}\gamma_e\gamma_n\hbar b^{-3} \cong \omega^n_{1/2} \qquad (3)$$

In reality H_{en} is time-dependent, even in the absence of molecular motions. This property is mainly due to fluctuations in the electron spin operator, which occur as a result of electron–electron dipolar or spin-exchange interactions (*1*). Both interactions give rise to electron–electron flip-flop transitions, resulting in local fields at the nuclear sites that are fluctuating with the inverse electron–electron flip-flop time, T_{ff}^{-1}. Usually the solid-state effect is observed in solids containing a relatively small amount of unpaired electrons (typically $<10^{20}$ cm^{-3}). Then the average distance between the unpaired electrons is so large that the spin-exchange interactions, which rapidly decrease for increasing distances (*18*), can be ignored. Then T_{ff} is caused mainly by electronic dipolar interactions. In this case T_{ff} is approximately given by (*1*)

$$T_{ff}^{-1} \cong 0.9\gamma_e^2\hbar N_e \cong 3 \times 10^{-13} N_e \qquad (4)$$

where N_e is in reciprocal cubic centimeters. It follows from the motional-averaging theory of Anderson and Weiss (*19*) that if T_{ff} is so small that $\omega_L^2 T_{ff}^2 << 1$, the local field becomes partially averaged out, and that the resulting time-averaged field, ω_a, is given by

$$\omega_a(r) \cong \omega_L^2(r) T_{ff} \qquad (5)$$

Hence for nuclei close to the electrons, where $\omega_L^2 T_{ff}^2 >> 1$, the local field remains unaltered, but for more remote nuclei, for which ω_L is decreased to such an extent that $\omega_L^2 T_{ff}^2 << 1$, the static field is reduced to ω_a, given by eq 5. As a consequence, if ω_a becomes comparable to $\omega^n_{1/2}$ for nuclei at such a distance b from an electron that the condition $\omega_L^2(b) T_{ff} << 1$ is valid, the radius b can be estimated from

$$\omega_a(b) = \omega_L^2(b)T_{ff} = \frac{1}{4}(\gamma_e\gamma_n\hbar)^2 b^{-6} T_{ff} \cong \omega^n_{1/2} \quad (6)$$

Compared with the static case, eq 4, estimation with eq 6 results in a reduced value of b and, consequently, in an increased value of the DNP enhancement factor (*see* eq 2). An example is ^{1}H DNP in polystyrene, doped with the stable radical 1,3-bis(diphenylene)-2-phenylallyl (BDPA). The following values for the different parameters occurring in equations 2, 4, and 6 were obtained (*15*): $\omega^n_{1/2} \approx 10^5$ s^{-1}, $N_e = 1.6 \times 10^{19}$ cm^{-3}, $W_z^H = 1.4$ s^{-1}, $g(0) = 5.7 \times 10^{-9}$ s, $H_{1e} = \omega_{1e}/\gamma_e = 0.7$ G, $(\omega_H/2\pi) = 60$ MHz. Under these conditions a maximum enhancement of 59 was observed. By using the expressions just derived, equations 2, 4, and 6 yield the following values:

1. If the influence of T_{ff} on the value of the local field is ignored, then $b = 13.6 \times 10^{-8}$ cm, and $(E_H - 1)_s = 8.9$.

2. Equation 4 predicts a value for T_{ff} of 2×10^{-7} s. Then according to eq 6, $b = 7.1 \times 10^{-18}$ cm, resulting in an enhancement factor, $(E_H - 1)_s$, of 62.

We conclude that incorporation of electron–electron flip-flop transitions on the time-averaged values of the electron–nuclear interactions leads to an enhancement factor that is much closer to the experimental results than that obtained without taking this effect into account.

By combining equations 2, 4, and 6, the following expression for the DNP enhancement due to the static electron–nuclear interactions is obtained for a microwave frequency $\omega = \omega_e - \omega_n$:

$$(E - 1)_s = 3.74 \frac{|\gamma_e|^3 (\hbar)^{3/2}}{\omega_n^2} \cdot \frac{\omega_{1e}^2 (\omega^n_{1/2})^{1/2} (N_e)^{3/2} g(0)}{W_z^n} \quad (7)$$

Equation 7 can be used to calculate the enhancement factors expected in practice, depending on the specific values of the various parameters. The best results can be expected in a small external field, for a large microwave power, a broad NMR line, a narrow ESR line, a large electron concentration, and a small nuclear Zeeman relaxation rate (even in the presence of the unpaired electrons). For protons, typical values for which a noticeable enhancement factor can be obtained are $H_0 \leq 1.4 \times 10^4$ G, $\omega_{1e}/\gamma_e \geq 0.3$ G, $\omega^n_{1/2} \geq 10^5$ s^{-1}, $N_e \geq 10^{18}$ cm^{-3}, $g(0) \geq 3 \times 10^{-9}$ s, and $W_z^n \leq 1$ s^{-1}. However, the combination of these values, rather than the individual values, determines the enhancement factor, and a large enhancement factor can be obtained under completely different conditions.

The Overhauser Enhancement. As already mentioned, the Overhauser effect is caused by electron–nuclear hyperfine interactions that are time-dependent on a time scale comparable to w_e^{-1}, which is typically of the order of 10^{-11} to 10^{-12} s. This time dependence can occur in the amplitude of the interaction term (as occurs, e.g., in liquids and materials containing conducting electrons), or it can arise from fluctuations in the electron spin operator as a result of electron–electron interactions (*see* the previous section). For electron–electron interactions, the electron–electron flip-flop time due to the electronic dipolar interactions can be ignored, as T_{ff} is much larger than W_e^{-1} for realistic values of the electron concentration N_e (*see* eq 4).

However, the situation is different for the electron–electron spin-exchange interactions, because strong exchange couplings can result in flip-flop times as short as 10^{-13} to 10^{-14} s (*1, 18*). Besides being necessary for the Overhauser effect, this fast time dependence in the electron–nuclear interactions as an important consequence, which is that a much larger fraction of the nuclei is observed via NMR spectroscopy than in the static interactions just discussed. In fact, in many cases the flip-flop time is so short that even for nuclei directly bonded to the molecules containing the unpaired electrons, for which the amplitude of the hyperfine field can be several megahertz or more (*20*), the time-averaged hyperfine interaction is equal to or less than $w^n_{½}$. Then 100% of the nuclei are observed. In the following we shall assume this condition to be true.

For electron–nuclear pairs, the maximum Overhauser enhancement, $(E - 1)_{ov}$, obtained under the condition $\omega = \omega_e$, is given by (*21*)

$$(E - 1)_{ov} = \frac{1}{4}[a^2(r) - d^2(r)] \cdot \frac{J(\omega_e, \tau)}{W_z^n} \cdot S \cdot \frac{|\gamma_e|}{\gamma_n} \qquad (8)$$

In eq 8, $a(r)$ is the magnitude of the scalar interaction; $J(\omega, \tau)$ is the spectral density function characterizing the time dependence in the electron–nuclear interactions; and S is the saturation factor given by $S = W(W + W_e)^{-1}$, where $W = \pi\omega_{1e}^2 g(0)$ is the allowed transition probability induced by the microwave irradiation and W_e is the electron Zeeman relaxation rate. The other parameters in eq 8 have the same meaning as those in eq 1.

When the electrons are surrounded by many nuclei at different distances, the procedure to calculate the average Overhauser enhancement is different from that used for the calculation of the solid-state enhancement. The reason is that, unlike for the solid-state effect, for the Overhauser effect the nuclei directly attached to the molecules containing the unpaired electrons are also observed, and these nuclei experience by far the largest hyperfine fields [this is especially true for the scalar interactions, which decrease rapidly with increasing distance (*18*)]. Therefore we assume that the Overhauser enhancement is caused mainly by these hyperfine interactions, and that polarization of the other nuclei in the material is enhanced via spin diffusion among the nuclei. Then the average Overhauser enhancement is given by

$$(E - 1)_{ov} = \frac{1}{4} \frac{xN_e \rho_e^2}{N_n} \cdot [a^2(r_1) - d^2(r_1)] \cdot \frac{J(\omega_e, \tau)}{W_z^n} \cdot S \cdot \frac{|\gamma_e|}{\gamma_n} \qquad (9)$$

In eq 9, x represents the amount of nuclei interacting directly with an electron, N_n is the nuclear concentration, and r_1 is the shortest distance between an electron and a nucleus. The parameter ρ_e denotes the electron density near the molecular sites and has been introduced to account for the possibility that the electrons are delocalized over the molecules, a condition resulting in a value of ρ_e less than unity.

Equation 9 can be used to predict the Overhauser enhancement in a solid. An example is the ^{1}H Overhauser enhancement in a cellulosic char (*22*). The char was prepared by pyrolyzing cellulose at 600 °C under an inert atmosphere for several hours. During the pyrolysis, radicals are formed as a result of bond ruptures. ^{13}C NMR spectroscopy revealed that the char is highly aromatic, which means that π-type organic radicals are created, where the π electrons are stabilized on aromatic macromolecules of sufficient size. Hyperfine interactions between unpaired electrons present in π-orbitals and in-plane aromatic protons have been observed of the order of 10 MHz (*23, 24*). This observation has been explained theoretically by McConnell (*25*) by considering the hyperfine interaction between a proton and a unpaired electron delocalized in the π-orbital of a benzene ring. Consistent results were found by assuming that the electron spin density at the carbon site to which the proton is bonded, ρ_e, is 0.1. We shall adopt these results and take the scalar interaction, $\rho_e a(r_1)$, as $2\pi 10^7 \approx 6 \times 10^7$ s^{-1}. By using $\rho_e = 0.1$, the dipolar interaction strength, $\rho_e d(r_1)$, is found to be $\sim 3 \times 10^7$ s$^{-1} \approx \frac{1}{2}\rho_e a(r_1)$, in good agreement with other results obtained in systems where the unpaired electron is occupying a p_z-orbital (*26*). The other parameters occurring in eq 9 have been determined via continuous-wave ESR, ^{1}H NMR, and ^{13}C NMR spectroscopy. The results are as follows: $x = 1.5$, $N_e = 1.4 \times 10^{20}$ cm^{-3}, $N_H = 1.2 \times 10^{22}$ cm^{-3}, $\tau = 6 \times 10^{-11}$ s, $W_z^H = 45$ s^{-1}, $S = 0.9$. Then, assuming a Lorentzian spectral density function,

$$J(\omega_e, \tau) = 2\tau(1 + \omega_e^2 \tau^2)^{-1}$$

eq 9 predicts an Overhauser enhancement of 81, whereas experimentally an enhancement of 50 has been measured. In view of all the uncertainties the agreement is satisfactory.

Applications of DNP NMR Spectroscopy

Both the DNP enhancement curves and the NMR spectra enhanced via DNP can be used to obtain detailed and often unique information about the chemical and physical properties of solid materials, such as

- the structures of the bulk of the materials and the vicinity of the unpaired electrons
- the location and dynamics of the unpaired electrons.

In this section representative illustrations of these various applications will be given. The experiments that will be described have been performed in an external field of 14 kG, corresponding to an electron Larmor frequency of 40 GHz and nuclear Larmor frequencies of 60 MHz for protons, 15 MHz for ^{13}C nuclei, and 12 MHz for ^{29}Si nuclei. An elaborate description of the experimental setup is given in reference 6. The microwave power was provided by a 10-W Varian klystron (VKA-7010D). In the DNP NMR probe, the microwave irradiation is obtained by means of a horn antenna positioned in front of the NMR coil. The amplitude of the microwave field, H_{1e}, is estimated to be 0.3–0.5 G. All measurements were performed at room temperature.

The DNP Enhancement Curves. The symmetrical or antisymmetrical character of the DNP curve provides information about the time-dependence in the electron–nuclear interactions of Figure 1. This difference in symmetry makes it possible to determine whether both types of interactions are present, and if so, to separate the respective enhancement curves. This feature is especially true for nuclei that have large magnetogyric ratios (γ) like protons, for which the maximum enhancement due to the static interactions is shifted farther away from the centrum of the enhancement curve, than for nuclei with small γ.

This ability to separate the enhancement curves is illustrated in Figure 2, which shows the ^{1}H and ^{13}C DNP enhancement curves of a low-volatile bituminous coal. The ^{1}H DNP curve, Figure 2a, consists of both a symmetric and an antisymmetric part, a result indicating that both static and rapidly fluctuating electron–proton interactions occur, despite the fact that the aromatic free radicals in coal are long-lived with a low mobility. This result can be explained by assuming that a distribution of electron spin-exchange exists in the coal as a result of a distribution in distances

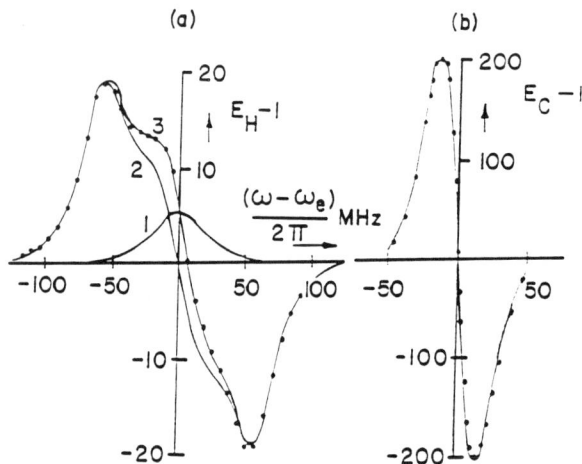

Figure 2. DNP enhancement curves obtained in a low-volatile bituminous coal. (a) ^1H DNP curves: curve 1, symmetric part; curve 2, antisymmetric part; curve 3, overall enhancement; dots, experimental. (b) ^{13}C DNP curve.

between the unpaired electrons and the delocalizations of the electrons (27). For the major part of the unpaired electrons, the exchange frequencies are so small that the electron–proton interactions can be considered to be static, whereas a fraction of the unpaired electrons is submitted to strong spin-exchange interactions, a condition resulting in time-dependent electron–proton interactions. Even if the fraction of electron–proton interactions is small, 1% or less, it gives rise to a noticeable Overhauser effect, so that DNP is a powerful tool to detect small amounts of rapidly spin-exchanging unpaired electrons.

The ^{13}C DNP curve, Figure 2b, is antisymmetric within measuring error. This property is probably due to both the much larger enhancement resulting from the static electron–carbon carbon interactions (7), which is overshadowing the relatively small Overhauser enhancement, and the relatively small ^{13}C Larmor frequency, which renders the separation between the symmetrical and antisymmetrical enhancement curves more difficult.

DNP and NMR Spectroscopy. The nuclear polarization enhancement via DNP will increase the NMR sensitivity. This property can be used to reduce the measuring time of an experiment or to measure small amounts of a certain nuclear species. This ability is especially important for rare-spin NMR spectroscopy, where the sensitivity is usually low, even after application of other polarization-transfer techniques like abundant-spin-to-rare-spin cross polarization (CP) (28). An example of such a sensitivity increase via DNP is given in Figure 3, in which ^{13}C spectra of the low-volatile bituminous coal are shown. Figure 3a shows the ^{13}C coal spectrum obtained via ^{1}H–^{13}C CP, and Figure 3b shows the so-called ^{13}C DNP–CP spectrum. In DNP–CP, the ^{13}C polarization is indirectly enhanced via DNP: First the ^{1}H polarization is increased via DNP, and then the enhanced ^{1}H polarization is transferred to the ^{13}C nuclei via CP. The advantage of this method is that the proton system is enhanced uniformly as a result of the fast spin diffusion among these nuclei, and this enhancement, after CP, results in a uniform enhancement of the ^{13}C polarization as well. Figure 3 illustrates that for this particular coal the use of DNP reduces the measuring time of a ^{13}C spectrum from 3.3 h to 1 min, a result that indicates the possibility of using DNP NMR spectroscopy to control, for example, coal burning processes.

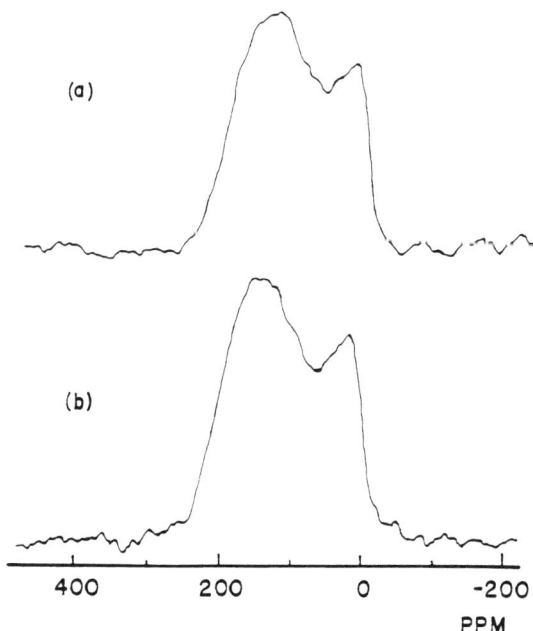

Figure 3. ^{13}C spectrum of a low-volatile bituminous coal, obtained via different methods: (a) ^1H–^{13}C CP; number of acquisitions, 20,000; recycle delay, 0.6 s; measuring time, 3.3 h; and (b) ^1H–^{13}C DNP–CP; number of acquisitions, 100; recycle delay, 0.6 s; measuring time, 1 min. (Reproduced with permission from reference 8. Copyright 1991 Marcel Dekker.)

DNP can be used also to enhance the rare-spin polarization directly. This enhancement can be achieved by applying the microwave irradiation under the condition for which the rare-spin polarization enhancement becomes maximal, and observing the rare-spin NMR signal via a simple $\pi/2$ pulse. This approach has been called the DNP–single-pulse (DNP–SP) experiment. The difference between DNP of abundant and rare spins is that, if a distribution of enhancement factors exists, for rare spins this distribution is in general not averaged out because of the slow spin diffusion among these nuclei. As a consequence, in a DNP–SP experiment the various peaks in the rare-spin spectrum can be enhanced differently, depending on the strengths of the various electron–nuclear interactions.

Compared to DNP–CP the disadvantage of the DNP–SP method is that the spectral distortions that may occur render a quantitative interpretation of the results difficult. Still, DNP–SP spectroscopy is very useful for obtaining additional information about the sample:

1. As in general the DNP enhancement factors of the rare spins close to the unpaired electrons are larger than those more remote, DNP–SP spectroscopy provides information about the molecular structure in the vicinity of the unpaired electrons. This feature is illustrated in Figure 4, which shows the ^{13}C spectra of the low-volatile bituminous coal obtained via DNP–CP (Figure 4a) and DNP–SP (Figure 4b) spectroscopy. It follows that in the DNP–SP experiment mainly the aromatic carbons are observed, a result confirming the generally accepted picture that the unpaired electrons are located in the aromatic structures of the coal.

2. DNP–SP spectroscopy can be used to determine electron-density distributions. An example of this possibility is the organic conductor (fluoranthenyl)$_2$PF$_6$. At room temperature this material behaves as a metal, where the conduction electrons are moving more or less freely parallel to the molecular stacks (*29*). The electron mobility is so large that in this sample only an Overhauser effect is observed (*16*). The average ^{1}H DNP enhancement is positive, which means that the positive enhancement due to the electron–proton scalar interactions are dominating the negative enhancement due to the electron–proton dipolar interactions, (cf. eq 9). As a consequence, the overall enhancement of the ^{13}C DNP–CP–MAS spectrum is positive as well, and Figure 5a shows the result of this experiment. Completely different results are obtained in the ^{13}C DNP–SP–MAS experiment (Figure 5b). Here some lines are enhanced positively and some negatively, whereas for other lines the DNP enhancement is so small that they are not observed at all. This result is due to the distribution of the electron density over the fluoranthenyl molecules,

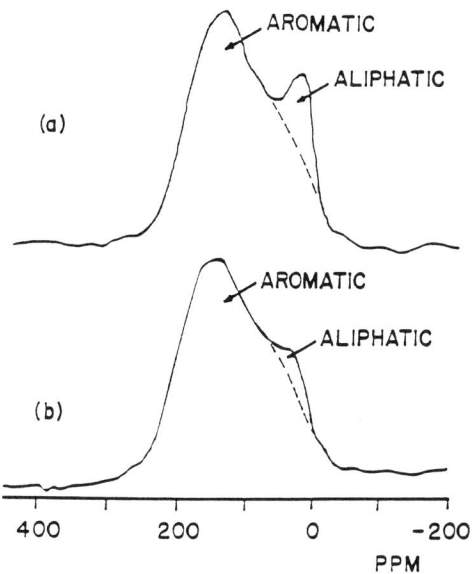

Figure 4. ^{13}C *spectrum of a low-volatile bituminous coal, obtained via different methods: (a) DNP–CP and (b) DNP–SP. (Reproduced with permission from reference 8. Copyright 1991 Marcel Dekker.)*

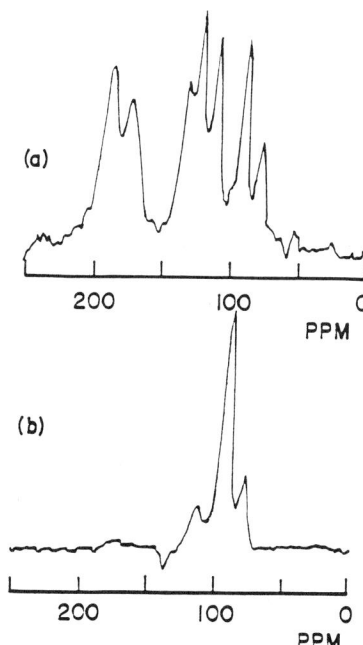

Figure 5. ^{13}C *spectra of (fluoranthenyl)$_2$PF$_6$, obtained via different methods: (a) ^1H–^{13}C DNP–CP–MAS and (b) ^{13}C DNP–SP–MAS. (Reproduced with permission from reference 8. Copyright 1991 Marcel Dekker.)*

which is different at the different sites of the carbons, and which results in different ratios of the electron–carbon scalar and dipolar interactions. Hence, DNP–SP–MAS experiments can be used to determine the electron density at the sites of the chemically different carbons.

In rare-spin DNP–SP experiments in solids containing both abundant and rare-spin species, an interesting effect, called the three-spin effect, may occur (30). During the microwave irradiation, both the abundant-spin and rare-spin polarizations are enhanced via DNP, and the enhanced polarization of the abundant spins can be transferred to that of the rare spins via the nuclear Overhauser effect (NOE) if the relaxation of the rare spins is caused by interactions between the rare and abundant spins. If the three-spin effect occurs, the rare-spin spectrum obtained via DNP–SP will be determined by a combination of direct DNP and indirect DNP–NOE. These two effects can be separated by keeping the abundant-spin polarization equal to its thermal equilibrium value during the microwave irradiation. This condition can be achieved by a partial saturation of the abundant spins. No three-spin effect has been observed in the two DNP–SP examples given here, a finding that indicates that other mechanisms, presumably electron–nuclear interactions, are dominating in the rare-spin relaxation.

Finally, in materials containing rare-spin nuclei only, DNP–SP–MAS is obviously the only method that can be applied. Examples of this class of solids are ceramics and vapor-deposited diamonds. Ceramics can be prepared by pyrolysis of various organosilicon polymers; this technique has certain advantages over conventional techniques, such as the ability to prepare specific ceramic shapes, lower processing temperatures, and larger purities (31). During the pyrolysis, unpaired electrons occur as a result of band ruptures, which can be used for DNP.

Figure 6 shows the ^{29}Si and ^{13}C DNP–SP–MAS spectra of a Si–N–C fiber derived form hydridopolysilazane (HPZ). A detailed discussion of the results will be published elsewhere; here we confine ourselves to remarking that for both types of nuclei, large DNP enhancements were obtained (several hundred), and this result makes it possible to measure rare-spin spectra with a good signal-to-noise ratio in a relatively short time. This ability is especially important for the ^{13}C nuclei, as the carbon weight percentage in this fiber is small, only 10%.

The development of diamonds by chemical vapor deposition has attracted much attention because of the unique properties of these materials (32). Still, the mechanisms determining the diamond nucleation and growth are not well understood. DNP–NMR spectroscopy might be one of the tools that can be applied to improve this situation. During the diamond formation, unpaired electrons are generated, presumably in the

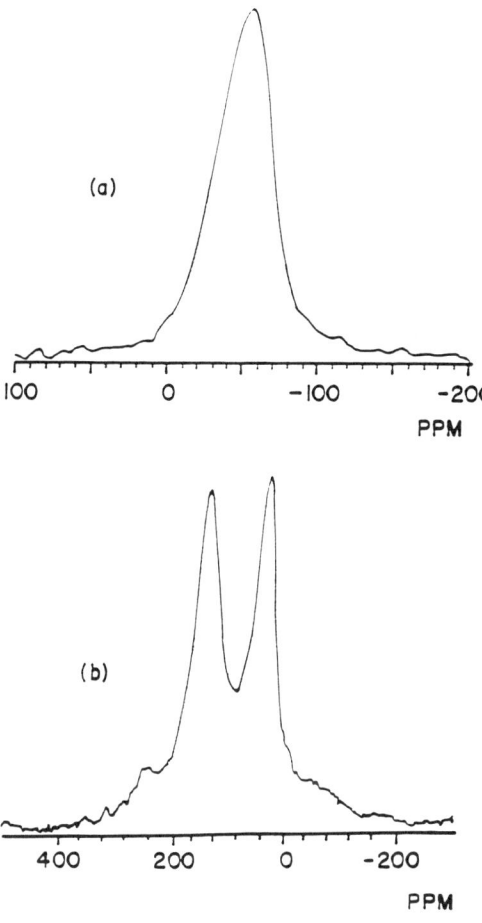

Figure 6. Rare-spin spectra of a Si–N–C ceramic fiber derived from hydridopolysilazane measured via DNP–SP–MAS: (a) ^{29}Si spectrum; number of acquisitions, 16; recycle delay, 80 s; and (b) ^{13}C spectrum; number of acquisitions, 4000; recycle delay, 5 s.

form of dangling bonds, and this situation makes ^{13}C DNP–NMR spectroscopy feasible. Indeed large DNP enhancement factors have been observed, of the order of 300 in a chemically vapor-deposited diamond.

Figure 7 shows the ^{13}C DNP–SP–MAS spectrum of this diamond. A detailed study of this and other results obtained in this material will be published elsewhere; here we restrict ourselves to remarking that after only 200 scans a spectrum with an excellent signal-to-noise ratio is obtained, and this spectrum makes it possible to distinguish two types of carbon, the main diamond peak at 35 ppm, and a weak shoulder at about

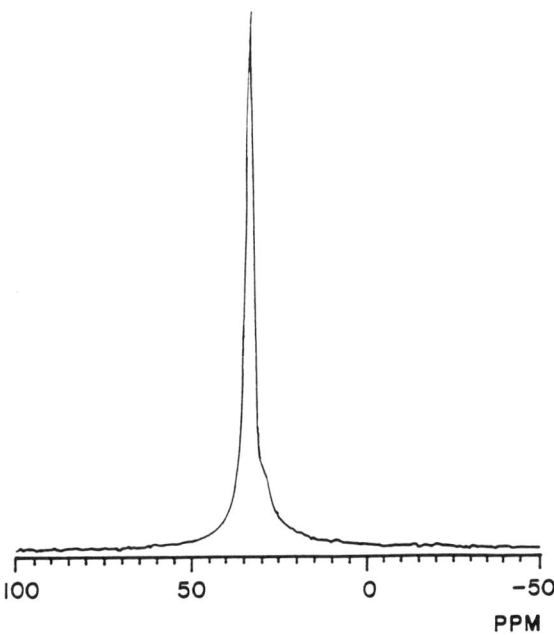

Figure 7. ^{13}C *spectrum of a vapor-deposited diamond obtained via DNP–SP–MAS; number of scans, 200; recycle delay, 60 s.*

28 ppm, attributed to carbon–proton impurities. The exact nature of these impurities and the role they play in the diamond properties are currently under investigation.

Conclusions

The use of DNP NMR spectroscopy in solid materials research is manifold:

1. DNP can significantly enhance the NMR signals, and this enhancement makes it possible to detect small concentrations of nuclei, to shorten data-measuring times by several orders of magnitude, and to examine solids containing rare spins only, in which the relaxation times can be very long.
2. The mechanism of DNP can provide valuable information about the microscopics of a system, such as the mobility of the electrons and the electron spin-exchange rates.

3. Selective DNP enhancement can be a very powerful tool in investigating molecular structures in the vicinity of the unpaired electrons and electron density distributions.

4. DNP NMR spectroscopy can be applied to study a rich variety of solids, including fossil fuels, charred materials, and solid materials that are in the focus of modern materials science, like organic conductors, ceramics, and vapor-deposited diamonds.

Furthermore, still other applications of DNP NMR spectroscopy are possible. Examples are the determination of Knight shifts ([16](16)); spin-imaging of DNP-enhanced regions ([33](33)); the study of interfaces ([30, 34](30,34)), doping effects, and radiation damage; and DNP NMR spectroscopy at variable temperatures and/or pressures. Also, recently, new electron–nuclear polarization techniques have been developed, such as DNP in the nuclear rotating frame ([35, 36](35,36)) and the NOVEL technique, the electron–nuclear analog of cross polarization ([37](37)).

Finally, under certain conditions much larger DNP enhancement factors have been observed than discussed in this chapter. For instance, using an oversized cavity instead of a horn antenna, a ^{1}H DNP enhancement factor of 260 was measured in a doped polymer ([15](15)), and recently a ^{1}H DNP enhancement of 10^4 was reported ([38](38)) in the photoexcited triplet state of pentacene. All these developments will enhance the viability of DNP NMR spectroscopy considerably, and it is to be expected that in the near future exciting new applications of this technique in the area of materials science will be reported.

Acknowledgments

The authors are grateful for financial support from U.S. Department of Energy Grant DE–FG22–85PC80–506, Alcoa Foundation, Amoco, Eastman Kodak, Philip Morris, and the Standard Oil Company of Ohio.

References

1. Abragam, A. *The Principles of Nuclear Magnetism;* Clarendon: Oxford, England, 1961.
2. Jeffries, C. D. *Dynamic Nuclear Orientation;* Wiley-Interscience: New York, 1963.
3. Goldman, M. *Spin Temperature and Nuclear Magnetic Resonance in Solids;* Oxford University Press: London, 1970.
4. Abragam, A.; Goldman, M. *Nuclear Magnetism: Order and Disorder;* Oxford University Press: London, 1982.

5. Carver, T. R.; Slichter, C. P. *Phys. Rev.* **1953**, *92*, 212.
6. Wind, R. A.; Anthonio, F. E.; Duijvestijn, M. J.; Smidt, J.; Trommel, J.; de Vette, G. M. C. *J. Magn. Reson.* **1983**, *52*, 424.
7. Wind, R. A.; Duijvestijn, M. J.; van der Lugt, C.; Manenschijn, A.; Vriend, J. *Progr. NMR Spectrosc.* **1985**, *17*, 33.
8. Wind, R. A. In *Modern NMR Techniques and Their Applications in Chemistry;* Popov, A. I.; Hallenga, K., Eds.; Marcel Dekker: New York, 1991.
9. Wind, R. A.; Duijvestijn, M. J.; van der Lugt, C.; Smidt, J.; Vriend, J. In *Magnetic Resonance: Introduction, Advanced Topics, and Applications to Fossil Energy;* Petrakis, L.; Fraissard, J. P., Eds; D. Reidel: Dordrecht, Netherlands, 1984; p 461.
10. Wind, R. A.; Duijvestijn, M. J.; van der Lugt, C.; Smidt, J. *Fuel* **1987**, *66*, 876.
11. Lock, H.; Wind, R. A.; Maciel, G. E.; Zumbulyadis, N. *Solid State Commun.* **1987**, *64*, 41.
12. Duijvestijn, M. J.; Manenschijn, A.; Smidt, J.; Wind, R. A. *J. Magn. Reson.* **1985**, *64*, 461.
13. Wind, R. A.; Duijvestijn, M. J.; Vriend, J. *Solid State Commun.* **1985**, *56*, 713.
14. Duijvestijn, M. J.; van der Lugt, C.; Smidt, J.; Wind, R. A.; Zilm, K. W.; Staplin D. C. *Chem. Phys. Lett.* **1983**, *102*, 25.
15. Duijvestijn, M. J.; Wind, R. A.; Smidt, J. *Physica* **1986**, *138B*, 147.
16. Wind, R. A.; Lock, H.; Mehring, M. *Chem. Phys. Lett.* **1987**, *141*, 283.
17. Overhauser, A. W. *Phys. Rev.* **1953**, *91*, 476.
18. Molin, Y. N.; Salikhov, K. M.; Zamaraev, K. I. *Spin Exchange;* Springer Verlag: Berlin, Heidelberg, and New York, 1980.
19. Anderson, P. W.; Weiss, P. R. *Rev. Mod. Phys.* **1953**, *25*, 269.
20. Wertz, J. E.; Bolton, J. R. *Electron Spin Resonance;* Chapman and Hall: New York and London, 1986.
21. Hausser, K. H.; Stehlik, D. In *Advances in Magnetic Resonance;* Waugh, J. S., Ed.; Academic: New York, 1968; Vol. 3, p 79.
22. Wooten, J. B.; Wind, R. A.; Maciel, G. E., Colorado State University, unpublished results, 1989.
23. Weissman, S. I. *J. Chem. Phys.* **1954**, *22*, 1135.
24. Venkataramand, B; Fraenkel, G. K. *J. Am. Chem. Soc.* **1955**, *77*, 2707.
25. McConnell, H. M. *J. Chem. Phys.* **1956**, *24*, 764.
26. Nechtschein, M.; Devreux, F.; Greene, R. L.; Clarke, T. C.; Street, G. B. *Phys. Rev. Lett.* **1980**, *44*, 356.
27. Blair, J.; Maciel, G. E.; Wind, R. A., Colorado State University, unpublished results, 1989.
28. Pines, A.; Gibby, M. G.; Waugh, J. S. *J. Chem. Phys.* **1973**, *59*, 569.
29. Mehring, M.; Spengler, J. *Phys. Rev. Lett.* **1984**, *53*, 2441.
30. Maresch, G. G.; Kendrick, R. D.; Yannoni, C. S.; Galvin, M. E. *J. Magn. Reson.* **1989**, *82*, 41.
31. Lipowitz, J.; Freeman, H. A.; Chen, R. T.; Prack, E. R. *Adv. Ceram. Mater.* **1987**, *2*, 121.

32. Angus, J. C.; Hayman, C. C. *Science (Washington, D.C.)* **1988,** *241,* 913.
33. Maciel, G. E.; Davis, M. F. *J. Magn. Reson.* **1985,** *64,* 356.
34. Schaefer, J.; Afework, M., Washington University, 1990, unpublished.
35. Wind, R. A.; Li, L.; Lock, H.; Maciel, G. E. *J. Magn. Reson.* **1988,** *79,* 577.
36. Wind, R. A.; Lock, H. *Adv. Magn. Opt. Reson.* **1990,** *15,* 51.
37. Henstra, A.; Dirksen, P.; Schmidt, J.; Wenckebach, W. T. *J. Magn. Reson.* **1988,** *77,* 389.
38. Wenckebach, W. T., 1990, unpublished.

RECEIVED for review January 30, 1991. ACCEPTED revised manuscript December 3, 1991.

4

Advances in Electron Nuclear Double Resonance Spectroscopy

Hans Thomann and Marcelino Bernardo

Corporate Research Laboratory, Exxon Research and Engineering Company, Annandale, NJ 08801

The introduction of pulsed techniques has significantly enhanced the versatility of electron nuclear double resonance (ENDOR) spectroscopy. In pulsed ENDOR experiments, short, intense microwave and radio frequency (rf) pulses are used to induce electron paramagnetic resonance (EPR) and NMR transitions on time scales short with respect to the electron and nuclear spin relaxation times. EPR detection of both spin-population transfer and coherence transfer among the nuclear hyperfine sublevels is then possible. In contrast to the conventional experiments using continuous-wave (CW) irradiation, ENDOR spectra recorded by pulsed population-transfer techniques are not sensitive to the ratio of the electron and nuclear spin relaxation rates. Consequently, pulsed ENDOR experiments are expected to be more widely applicable than CW experiments. Pulsed techniques further offer the possibility for greater sensitivity, enhanced spectral resolution, and for completely new types of double resonance experiments.

ELECTRON NUCLEAR DOUBLE RESONANCE (ENDOR) spectroscopy (*1*) is a well-established technique in which NMR transitions are detected via electron paramagnetic resonance (EPR) transitions. The primary application of ENDOR experiments is the measurement of hyperfine and quadrupole couplings of paramagnetically shifted nuclei (*2, 3*). The EN-

DOR experiment offers roughly 3 orders of magnitude greater sensitivity over direct NMR detection. Compared to the EPR experiment, both greater spectral resolution and spectral simplification are obtained in the ENDOR experiment (4). The enhanced spectral information is a direct consequence of the quantum selection rules in the ENDOR experiment.

In the ENDOR spectrum, a one-to-one correspondence exists between the number of coupled inequivalent nuclei and the number of ENDOR transitions observed. This feature is in contrast to the EPR spectrum in which the number of transitions observed is a multiplicative function of the number of inequivalent nuclei. For N equivalent protons, $(N + 1)$ lines appear in the EPR spectrum, but $(M + 1)(N + 1)$ lines are expected for M sets of N equivalent protons. In contrast, only M pairs of lines are expected in the ENDOR spectrum.

This spectral simplification is particularly significant in the study of conjugated organic radicals in which the delocalization of the paramagnetic electron over the molecule results in hyperfine coupling with many nuclei, especially protons (5). A striking example of this property is evident in the spectra of the relatively simple organic doublet radical, the triphenylmethyl radical. With one set of three equivalent protons and two sets of six equivalent protons, 196 lines are expected in the EPR spectrum, and only three pairs of lines are expected (and observed) in the ENDOR spectrum. In EPR spectra of organic π-radical molecules in the solid state, the individual hyperfine transitions are usually no longer resolved, and only the envelope of many transitions is observed. Such an EPR line is referred to as "inhomogeneously broadened". Inhomogeneously broadened EPR lines offer perhaps the most pronounced examples of the enhanced information content obtained in the ENDOR experiment (1–4).

The conventional continuous-wave ENDOR experiment, hereafter referred to as CW ENDOR, is a double resonance experiment in which NMR and EPR transitions are continuously irradiated. The success of the *CW* ENDOR experiment has been well-documented. The development of *pulsed* ENDOR methodology requires some motivation, especially in light of the increased cost and complexity of the instrumentation required for the pulsed experiments. The CW ENDOR signal intensity is a complex function of the induced electron and nuclear spin transition rates and the intrinsic spin relaxation rates. In fact, the CW ENDOR experiment is often only successful over a narrow temperature range or may not be possible at all because of unfavorable relaxation rates (2, 3, 5, 6). In pulsed ENDOR experiments, the NMR and EPR transitions are excited by microwave and radio frequency (rf) pulses in which the pulse widths are short compared to the intrinsic spin relaxation times. Consequently, the pulsed ENDOR signal is not sensitive to the detailed balance of the electron and nuclear spin relaxation rates. This characteristic is one of the major advantages of the pulsed ENDOR experiment.

The second important advantage of the pulsed ENDOR experiment is that completely new types of experiments become possible using pulsed excitation. These include the EPR detection of nuclear hyperfine sublevel coherence phenomena as well as multiple-resonance experiments in which two EPR and/or two NMR transitions are excited. Although the multiple experiments are also possible using CW excitation (3), greater versatility is possible with pulsed technology.

A single crystal of irradiated malonic acid serves as a convenient model sample we will use to demonstrate many of the pulsed ENDOR techniques. The electron spin-echo (ESE) spectrum of the irradiated malonic acid crystal at an arbitrary orientation is shown in Figure 1. This spectrum is recorded by plotting the electron spin-echo intensity with a fixed refocusing delay time as a function of the applied magnetic field. The doublet splitting arises from the hyperfine coupling of the unpaired electron in the carbon p_π-orbital with the α-proton of the CH fragment of the malonic acid radical molecule, $\cdot$CH(COOH)$_2$. The strongly coupled α-proton can to a very good approximation be treated as an isolated $S = ½$, $I = ½$ spin system (7). Many additional lines arising from the hyperfine coupling with the carboxyl protons and with protons on neighboring mol-

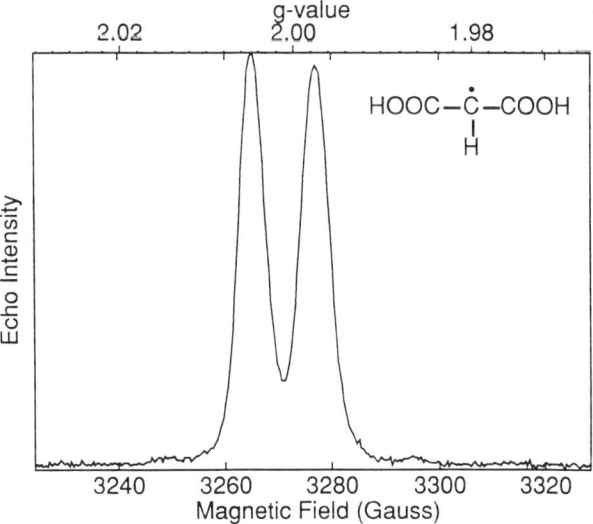

Figure 1. ESE-detected EPR absorption spectrum for the irradiated malonic acid radical at an arbitrary crystal orientation. Experimental conditions: 9.162 GHz; microwave pulse widths, 0.50 and 0.10 μs for $\pi/2$ and π pulses; 100 echo samples per magnetic field value; magnetic field resolution, 256 points/104.3 G; pulse sequence repetition rate, 500 Hz; T, 80 K.

ecules in the crystal are resolved in the ENDOR spectrum. These weaker hyperfine splittings are not resolved in the EPR spectrum but do contribute to the inhomogeneous broadening of the doublet.

Basic Concepts

The ENDOR signal is the EPR signal observed with a finite rf field. The relative ENDOR enhancement, $\Delta\chi_{ENDOR}$, or simply $\Delta\chi$, is defined from the change in the EPR signal due to the NMR transitions induced by the rf excitation:

$$\Delta\chi = \Delta\chi_{ENDOR} = \chi_{ESR}\,(\text{rf} > 0) - \chi_{ESR}\,(\text{rf} = 0) \qquad (1)$$

where χ_{ESR} identifies the paramagnetic susceptibility detected in the EPR experiment. This susceptibility is known as the ENDOR enhancement because usually $\chi_{ENDOR} > 0$. The absolute ENDOR enhancement factor can be determined from $\Delta\chi/\chi$ (rf = 0).

Most ENDOR experiments can be understood by considering only the first-order interactions describing the magnetic coupling of an $S = \tfrac{1}{2}$ electron to an $I = \tfrac{1}{2}$ nucleus. Additional simplifications assumed include an isotropic g-value and an isotropic hyperfine interaction, A, also known as the Fermi contact coupling. We further assume that the hyperfine interaction is small, so that second-order hyperfine terms (of order $h^2 A^2/g_e\beta_e B_0$) are negligible. (Here g_e is the Landé g factor, β_e is the Bohr magneton, B_0 is the externally applied magnetic field, known as the Zeeman field, and h is the Planck constant.) These assumptions lead to the simplified Hamiltonian (*8*):

$$\frac{1}{h}H_0 = \nu_e S_z - \nu_n I_z + A S_z I_z \qquad (2)$$

where $\nu_e = g_e\beta_e B_0/h$ and $\nu_n = g_n\beta_n B_0/h$ are the electron and nuclear Larmor frequencies, respectively; S_z is the electron spin polarization operator; and I_z is the nuclear spin polarization operator. The negative sign for the nuclear Zeeman term (i.e., $\nu_n I_z$) accounts for the fact that the electron and proton are of opposite charge. The Hamiltonian (eq 2) is applicable in the high-field limit in which the applied magnetic field is sufficiently large so that the electron, S, and nucleus, I, are separately quantized along the applied magnetic field direction. This condition is equivalent to stipulating that M_I and M_S are good quantum numbers. Equation 2 will not quantitatively describe the ENDOR spectrum of the α-proton hyperfine coupling of the irradiated malonic acid radical, but it will nevertheless be adequate to illustrate most pulse schemes described in this chapter.

In the high-temperature limit, $g_e\beta_e B_0/kT \ll 1$, and the spin eigenvalues, $E/h(M_S,M_I)$, in hertz, are given by

$$\frac{E}{h}(M_S,M_I) = \nu_e M_S - \nu_n M_I + A M_S M_I \qquad (3)$$

E is the energy of the spin eigenstates as defined by eq 3. The ordering of the spin energy levels depends on the relative magnitudes of hA and $h\nu_n$ and on the sign of A. For an α-proton on a C–H fragment, $A < 0$ (7, 8). This value corresponds to a negative spin density on the proton, which is in turn a consequence of spin polarization. The spin energy level diagram for $A < 0$, $|A|/2 < \nu_n$, and $g > 0$ is shown in Figure 2. For easy reference, the spin eigenstates are labeled in order of increasing energy as $E1$, $E2$, $E3$, and $E4$. The splitting of the electron Zeeman interaction by the nuclear Zeeman and hyperfine interactions produces the "sublevels" $E1$ and $E2$ in the lower electron spin manifold and the sublevels $E3$ and $E4$ in the upper manifold. In the high-temperature, high-field limit, the equilibrium spin populations are $P_1 \simeq P_2 \simeq (1 + \delta/2)$ and $P_3 \simeq P_4 \simeq (1 - \delta/2)$, where $\delta = g_e\beta_e B_0/kT$. The sublevel population, P_n, corresponds to the nuclear spin polarization in sublevel E_n. The polarization due to the nuclear Zeeman and hyperfine interactions is negligible compared to the electronic part.

EPR and NMR transitions are represented by the Hamiltonian term: $\omega_1 S_x$ and $\omega_2 I_x$, where $\omega_1 = g_e\beta_e B_1/h$, $\omega_2 = g_n\beta_n B_2/h$, and B_1 and B_2 are the magnitudes of the microwave and rf magnetic fields, respectively. The allowed EPR transition frequencies are $\nu_\pm^{EPR} = \nu_e \pm A/2$, and the al-

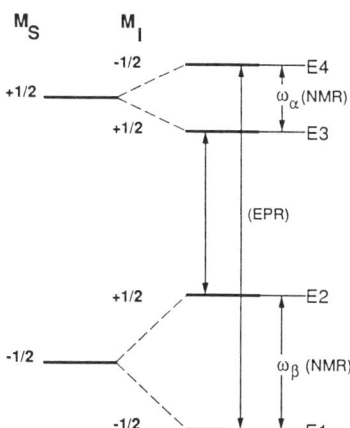

Figure 2. Energy level diagram for a four-level system.

lowed NMR transition frequencies are $\nu_{\pm}^{NMR} = \nu_n \pm A/2$. NMR transitions that are detected via the EPR transition are referred to as ENDOR transitions. In the present example with $\nu_n > A/2$, peaks in the ENDOR spectrum occur at $\nu_{\pm}^{ENDOR} = \nu_n \pm A/2$. When $\nu_n < A/2$, the ENDOR peaks occur at $\nu_{\pm}^{ENDOR} = A/2 \pm \nu_n$.

In the ENDOR experiment, microwave and rf magnetic fields are applied to induce EPR and NMR transitions. In the CW ENDOR experiment, the intensity of a partially saturated EPR transition is recorded as a function of an applied rf magnetic field while sweeping the rf frequency. The magnitude of the ENDOR enhancement depends on the ability to saturate the EPR and NMR transitions and on the relation between the electron and nuclear relaxation rates. The EPR transition must be at least partially saturated, but the NMR transition, that is, transitions between the nuclear hyperfine sublevels, must be strongly saturated. From the Bloch equations, it is readily shown that the condition for partially saturating the EPR transition is $\omega_1^2 T_{1e} T_{2e} \simeq 1$, and the condition for strongly saturating the NMR transition is $\omega_2^2 T_{1n} T_{2n} >> 1$ (9). (T_{1e} is the electron spin–lattice relaxation time, T_{2e} is the electron phase memory (or spin–spin) relaxation time, T_{1n} is the nuclear spin–lattice relaxation time, and T_{2n} is the nuclear spin–spin relaxation time.) Because $T_{1e}^{-1} >> T_{1n}^{-1}$, strong ENDOR signals requires the use of strong rf fields to drive the NMR transitions. This criterion naturally suggests the use of pulsed excitation because $\omega_1^{-2} << T_{1e} T_{2e}$ and $\omega_2^{-2} << T_{1n} T_{2n}$ for coherent pulsed excitation of the EPR and NMR transitions, respectively.

Pulse Schemes

Pulse sequences varying in the pulses applied to both the EPR and the NMR transitions have been proposed and demonstrated for pulsed ENDOR experiments (10–12). It is instructive to distinguish between pulsed ENDOR experiments in which the rf pulse(s) are applied when the electron spin magnetization in the rotating frame, $M_{x,y}(t)$, is finite in contrast to experiments in which $M_{x,y}(t) = 0$. In general experiments in which $M_{x,y}(t) = 0$ are more difficult because $M_{x,y}(t)$ decays with a time constant T_m, and the longitudinal magnetization, $M_z(t)$, decays with a time constant, T_{1e}, which is generally much longer than T_m. Here we will focus exclusively on pulsed ENDOR experiments in which the rf pulse(s) are applied when $M_{x,y}(t) = 0$ because these have been applied to carbonaceous solids.

To simplify the discussion in the following sections, the microwave and rf pulses will generally be considered selective; that is, only one allowed transition is excited. A microwave, ω_1, or rf, ω_2, pulse of duration

t_p results in a rotation angle $\theta_i = \omega_i t_p$ for $i = 1$ or 2. When $\theta = \pi$, the two spin eigenstates involved have interchanged their spin populations. A π_{ij} pulse will denote a selective π pulse between spin eigenstates i and j. In practice, usually more than one transition is excited, and pulses are not necessarily selective.

Adopting the nomenclature in the NMR literature, pulse sequences for ENDOR experiments can be generically divided into preparation, sublevel mixing, evolution, and detection periods. A sublevel polarization is created in the preparation period. With the exception of the stimulated-echo ENDOR pulse sequence, this sublevel polarization is accomplished by a selective π pulse on an EPR transition, such as the π_{14} pulse in Figure 2. This pulse creates a nonequilibrium spin population between these eigenstates: $P_1 \simeq (1 - \delta/2)$ and $P_4 \simeq (1 + \delta/2)$. Prior to the pulse, $P_{14} = +\delta$, and after the pulse, $P_{14} = -\delta$. Although no *net* electron or nuclear polarization is created, it is convenient to describe the action of the π_{14} pulse as creating a *sublevel* polarization.

The action of the sublevel mixing pulses is to transfer the polarization created in the preparation period. The details of this transfer depend on the type of experiment performed. During this time period, the longitudinal electron spin magnetization, $M_z(t)$, is finite, but $M_{x,y}(t) = 0$. The rf pulses modulate $M_z(t)$ by directly manipulating the nuclear sublevel populations or indirectly by manipulating nuclear sublevel coherence. The mixing period may or may not include an evolution period in which free precession occurs again, depending on the type of experiment performed. In practice, evolution times are never zero because a finite time delay is required between the sublevel mixing and detection pulses.

The detection period comprises microwave pulses excite one or several EPR transitions. The ENDOR effect is usually detected by creating an $M_{x,y}(t)$ signal as a readout of $M_z(t)$. With the exception of the stimulated-echo ENDOR scheme, a two-pulse electron spin echo is used in the detection period.

EPR-Detected Sublevel Population Transfer

The selective irradiation of an EPR transition in which only a subset of all possible EPR transitions is excited necessarily creates a polarization among the nuclear spin sublevels in the two electron spin manifolds. This polarization can be manipulated by pulsed rf irradiation. The most simple type of manipulation consists of the transfer of sublevel polarization among the nuclear hyperfine sublevels. Pulse sequences for creating and detecting this sublevel population transfer are described in this section.

Rabi Oscillations. The nutation of the sublevel magnetization about the rf magnetic field driving the NMR transition described by Rabi (13) can be detected in an ENDOR experiment using the pulse sequence shown in Figure 3 (14). Referring to Figure 2, suppose a microwave preparation pulse, π_{14}, is applied to create a sublevel polarization among sublevels $E3$ and $E4$. In the sublevel mixing period, an rf pulse of magnitude B_2 resonant with the $E3$–$E4$ sublevels is applied for duration t_{pl}. This rf pulse induces a nutation of the magnetization arising from these sublevels by an angle $\theta_2 = \omega_2 t_{pl}$, where $\omega_2 = \epsilon \gamma_n B_2$. ($\epsilon$ takes into account the enhancement of the magnitude of the rf field due to the hyperfine interaction. For a Hamiltonian considering only isotropic hyperfine interactions, $\epsilon = |1 + m_s A/\omega_n|$, where $m_s = \pm\frac{1}{2}$ for s = ½. γ_n is the nuclear gyromagnetic ratio: $\gamma_n = g_n \beta_n$.) The nutation of this sublevel magnetization can be detected as an amplitude modulation of an electron spin echo formed in the detection period. The echo amplitude will oscillate with a periodicity proportional to $\cos(\omega_2 t_{pl})$. The nutation can be induced by either varying the magnitude of B_2 or the pulse length, t_{pl}.

Rabi oscillations are a direct measure of the ENDOR enhancement factor. Rabi oscillations recorded by varying the pulse length at constant rf power for two lines in the ENDOR spectrum of irradiated malonic acid are shown in Figure 4. To separate the sublevel Rabi oscillations from the loss of the sublevel polarization due to the electron spin–lattice relaxation, the pulse sequence is repeated with no rf power applied on alternate pulse-sequence iterations. With approximately 1 kW of transmitter power, the rf π pulse width was 6.0 and 3.4 μs at 7.32 and 31.29 MHz, respectively. These values are representative. The particular values obtained will depend on the details of the experimental setup.

Davies ENDOR. In 1974 Davies (15) proposed a pulsed ENDOR experiment that is the pulsed analog of the conventional CW ENDOR experiment. This pulsed ENDOR is accomplished by using the pulse sequence of Figure 3 with t_p held to a constant value. Conceptually, both the CW and Davies experiments are similar: The NMR signal is detected indirectly from the difference in the EPR signal intensity observed when the applied rf magnetic field is resonant with an NMR transition. In the typical CW ENDOR experiment, an EPR transition is continuously excited by using CW irradiation of sufficient intensity to drive the transition into saturation. In the Davies ENDOR experiment, this saturation is accomplished with pulsed excitation. The essential concept is that the microwave pulse width must be short compared to the time required for the resonant spins to dissipate the absorbed energy. Under ideal conditions, this process corresponds to an "inversion" of the equilibrium spin population.

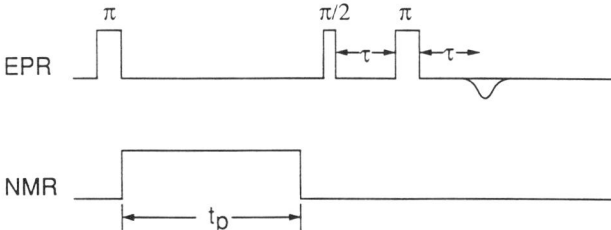

Figure 3. Pulse sequence for detecting sublevel Rabi oscillations and Davies ENDOR.

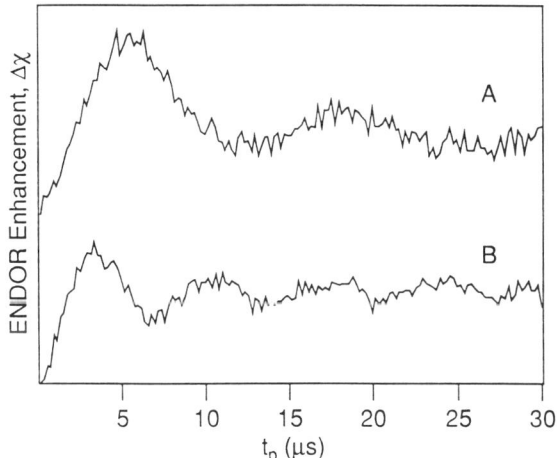

Figure 4. Sublevel Rabi oscillations detected for the irradiated malonic acid radical at rf frequencies of 7.32 (A) and 21.29 (B) MHz.

In the CW ENDOR experiment, the ENDOR signal is detected as the change in the saturated EPR signal intensity. For an ENDOR effect to be observed, the rate of rf-induced NMR transitions must therefore compete with the rate of microwave-induced EPR transitions. If these transition rates are not competitive, no ENDOR signal will be observed. This condition is in fact the reason CW ENDOR experiments often fail: The required rf magnetic field intensities in the CW ENDOR experiment are difficult to achieve. Often the temperature is used as an experimental parameter and is chosen so that the ratios of EPR and NMR transition rates are optimized to obtain the maximum CW ENDOR signal. This approach takes advantage of the fact that the electron and nuclear spin relaxation rates usually have different temperature dependences.

The large rf magnetic field intensities are much more easily achieved by using pulsed excitation, as in the Davies ENDOR experiment. The essential condition is that the rf pulse width must be short compared to the time required for either the resonant electron or nuclear spins to dissipate their absorbed energy. When this condition is satisfied, the ENDOR signal is (to first order) independent of the details of either the electron or nuclear spin dynamics. This independence has the important consequence that the pulsed ENDOR signal is no longer a sensitive function of temperature. This feature is in contrast to the CW ENDOR experiment, which is often successful only over a narrow temperature range.

The ENDOR signal is detected as a change in the EPR transition intensity. In principle, any method of detecting the EPR transition intensity could be used, including CW detection or detection of the free induction decay (FID) following single pulse excitation. In practice it is convenient to monitor the EPR transition intensity by using an electron spin echo formed by two microwave pulses. The Davies ENDOR spectrum is then recorded by plotting the intensity of the electron spin-echo signal created by the second and third microwave pulses as a function of the rf frequency. The rf frequency is incremented (or decremented) on successive pulse-sequence iterations.

An ENDOR spectrum for a single crystal of irradiated malonic acid at an arbitrary orientation recorded by using the Davies pulse sequence is shown in Figure 5. The transitions near 6 and 30 MHz are the two NMR transitions arising from the hyperfine coupling to the α-proton in the two electron spin manifolds. The transitions near 12 and 16 MHz arise from the hyperfine coupling to the carboxyl protons. The transitions centered about the proton Larmor frequency at approximately 14 MHz arise from the weakly coupled protons most likely on neighboring malonic acid molecules.

The ENDOR susceptibility in the pulsed ENDOR experiments can be readily measured. In the example in Figure 5, the ENDOR enhancement is approximately 25% of the ESE intensity. As is obvious from Figure 5, the enhancement varies across the ENDOR spectrum.

The relative intensities of the ENDOR transitions are a function of pulse conditions, particularly the microwave-inversion pulse bandwidth in the preparation period and the rf pulse in the mixing period. Using a microwave pulse in the preparation period with a large bandwidth, that is, a short, intense pulse, will suppress the intensities of ENDOR transitions from weakly coupled nuclei. Conversely, a preparation microwave pulse with a narrow bandwidth enhances the sensitivity toward the observation of weak hyperfine couplings. This enhancement arises because in an inhomogeneously broadened EPR line, the preparation pulse "burns a hole" with an approximate width of $\Delta\omega_1 = 2\pi/t_p$. This hole is sometimes referred to as a spin-alignment hole because no differential nuclear sub-

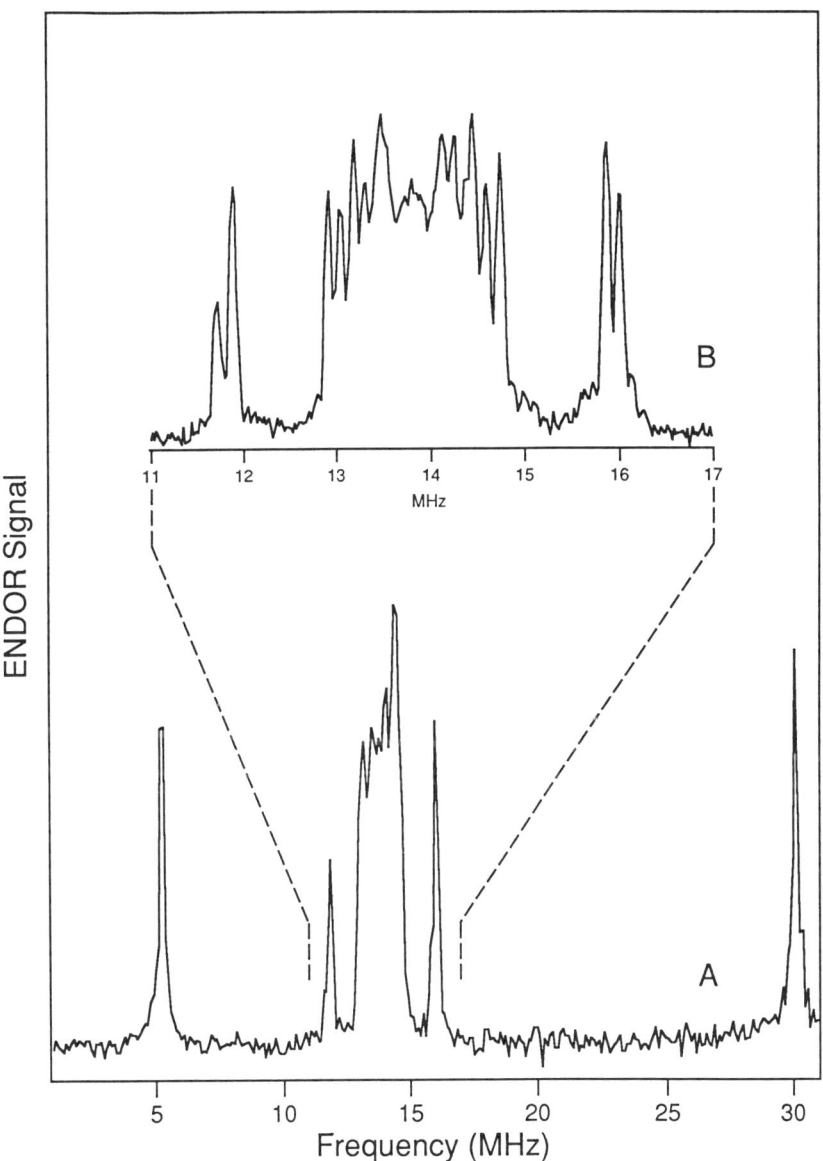

Figure 5. Davies ENDOR spectrum for an arbitrary orientation of an irradiated single crystal of malonic acid. Experimental conditions: pulse repetition rate, 100 Hz; rf sweep widths, (A) 1–31 MHz, (B) 11–17 MHz; sweep resolutions, (A) 0.1 MHz, (B) 0.023 MHz; rf pulse widths, (A) 5.0 μs, (B) 10.0 μs; microwave frequency, 9.125 GHz; microwave pulse width, 0.20 μs; magnetic field, 3248.5 G; T, 80 K.

level spin polarization is produced for NMR sublevels with hyperfine couplings $A_j < \Delta\omega_1$.

The mixing pulse bandwidth ultimately limits the resolution obtained in the ENDOR spectrum. No ENDOR lines of width $<2/t_p$, where t_p is the mixing pulse width, can be resolved in the spectrum. Greater spectral resolution can readily be obtained by using softer (longer and lower power) rf pulses and by concomitantly increasing the resolution of rf sweep. This effect is demonstrated in Figure 5B where a portion of the ENDOR spectrum was recorded under conditions to give higher spectral resolution. However, the increased spectral resolution is obtained at the expense of a lower sensitivity, which translates into a longer data-acquisition time.

Mims ENDOR. The spin-alignment hole created in the Davies or CW ENDOR experiment can be avoided by using a pulsed ENDOR scheme, shown in Figure 6, based on the three pulse-stimulated electron spin-echo sequence as first proposed and demonstrated by Mims in 1965 (*16*). An rf pulse is applied during the time t_1 between the second and third microwave pulses of the stimulated-echo pulse sequence. The ENDOR spectrum is recorded by monitoring the stimulated-echo intensity at time following the third microwave pulse as a function of the rf frequency. When the rf frequency is resonant with an NMR transition, the nuclear spin flip-shifts the local magnetic field at the electron. This flip-shift causes a shift in the precession frequency of the electron, which is observed as a reduction in the intensity of the stimulated echo.

Mims (*16, 17*) has given a simple physical description of the stimulated-echo ENDOR effect under the condition that the bandwidth of the microwave pulse is smaller than the EPR line width. This condition is generally the case in pulsed EPR spectroscopy of inhomogeneously broadened EPR lines. The first two microwave pulses burn a hole in the EPR line on which a serrated pattern of longitudinal magnetization is imposed. The formation of this magnetization pattern can be understood

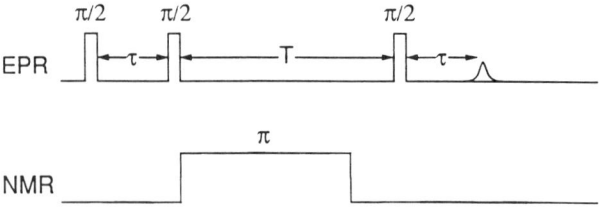

Figure 6. Stimulated-echo ENDOR, also known as Mims ENDOR, pulse sequence.

by considering the Fourier transform for a pair of rectangular pulses of finite width, t_p separated by time τ. The overall width of the hole burned into the EPR line is determined by the inverse of the microwave-pulse width, and the minima in the serrated pattern of $M_z(\omega)$ are separated by $2\pi n/\tau$, where n is a positive integer. The nuclear spin flips induced by the rf pulse effectively transfer magnetization within this serrated pattern of magnetization. This transferred magnetization tends to bleach out the serrated pattern imposed on $M_z(\omega)$, which in turn results in a reduction of the stimulated-echo amplitude. "Blind spots" in which no reduction in the stimulated-echo amplitude, that is, ENDOR effect, are obtained for values of the hyperfine coupling $A = 2\pi n/\tau$.

The stimulated-echo ENDOR spectra recorded for the irradiated malonic acid radical is shown in Figure 7A. Mims ENDOR is particularly

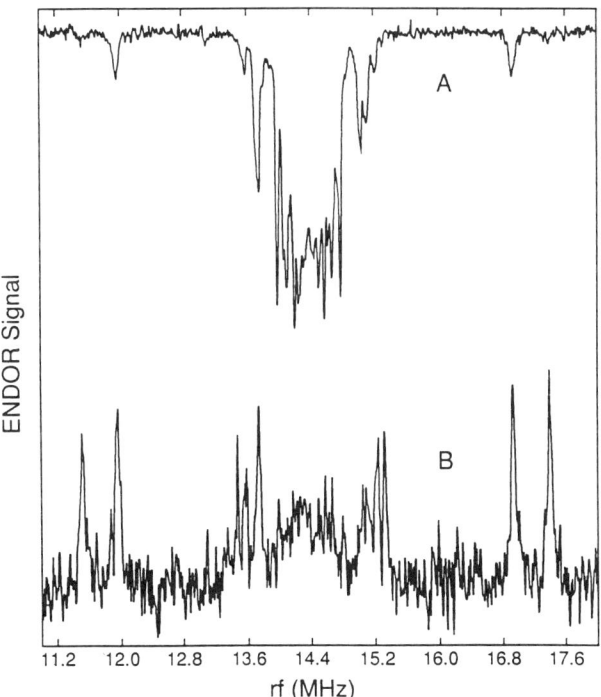

Figure 7. A: *Mims ENDOR spectrum of irradiated malonic acid radical at an arbitrary crystal orientation. Experimental conditions: microwave frequency, 9.465 GHz; microwave pulse width, 0.10 μs; τ, 0.50 μs; rf sweep width, 11–18 MHz; rf pulse width, 50 μs.* B: *Davies ENDOR under the same experimental conditions as in part A.*

sensitive to small hyperfine couplings when longer, lower power rf pulses are used. The spectrum in Figure 7A was recorded by using the same experimental conditions as the Davies ENDOR spectrum in Figure 7B. ENDOR transitions from weakly coupled nuclei are significantly enhanced in the stimulated-echo ENDOR spectrum relative to the Davies spectrum. On the other hand, the Davies spectrum is not complicated with the blind spots introduced by the τ suppression effect.

Electron Nuclear Electron Triple Resonance. Electron nuclear electron triple resonance is an extension of the electron nuclear double resonance technique in which the preparation and detection pulses are resonant with different EPR transitions (18, 19). This condition imposes an additional selection rule on transitions in the triple resonance spectrum. The electron nuclear electron triple resonance experiment is quite distinct from the electron nuclear nuclear triple resonance experiment (20). In the electron nuclear nuclear triple resonance experiment, which is also referred to as double ENDOR (21), the same EPR transition is involved in the preparation and detection periods, but two separate nuclear hyperfine transitions are pumped by using two rf irradiation fields. The double ENDOR experiment has been implemented in the pulsed mode but will not be discussed here (12, 22).

The pulse sequence for the pulsed electron nuclear electron experiment is shown in Figure 8 (19). Because two EPR transitions are involved as well as one NMR transition, this experiment can also be referred to as an ELDOR–ENDOR experiment by analogy to the electron electron double resonance (ELDOR) experiment. Two types of triple resonance experiments are possible with this pulse sequence. The first is a hyperfine selective (HS)-ENDOR, which has the advantage of generating selective ENDOR spectra for nuclei with a specified hyperfine coupling. In HS-

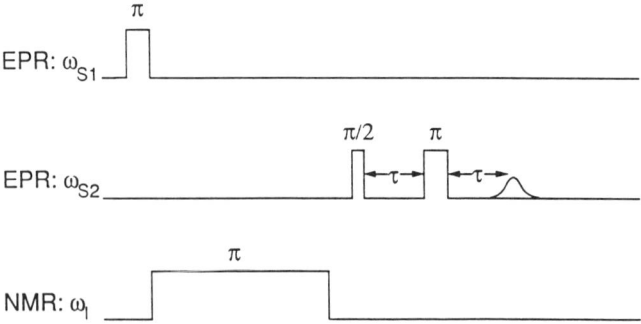

Figure 8. Pulse sequence for the electron nuclear electron triple resonance experiment.

ENDOR spectra, overlapping ENDOR lines arising from inequivalent nuclei are resolved directly without recourse to data-manipulation techniques. The second type of experiment is a method for generating EPR spectra composed of selected hyperfine sublevel transitions. This experiment effectively produces a subspectrum of the EPR spectrum and is therefore known as sublevel EPR spectroscopy. Sublevel EPR spectroscopy has the advantage of unambiguously assigning the hyperfine coupling from a single ENDOR line. An alternative method for implementing the HS-ENDOR and sublevel EPR experiments is to rapidly jump the Zeeman magnetic field between the microwave preparation and readout pulses in pulsed ENDOR experiments.

Hyperfine Selective (HS) ENDOR. In the HS-ENDOR experiment, the preparation, ω_{S1}, and detection, ω_{S2}, EPR frequencies are selected so that their difference, denoted as Δ, is equal to the hyperfine splitting, A, that separates these EPR transitions. The HS-ENDOR spectrum is generated by plotting the intensity of the EPR signal in the detection period while stepping the frequency, ω_I, of the rf pulse on successive pulse-sequence iterations. This method is analogous to the procedure for the Davies ENDOR experiment. However, in contrast to that experiment, in the HS-ENDOR experiment, a sublevel polarization transfer is detected only for those nuclei for which $A = \Delta$. When A does not equal Δ, no sublevel polarization transfer is observed. Consequently, in the HS-ENDOR spectrum, the electron spin-echo intensity detects a decrease in the sublevel spin population. This result is in contrast to the ENDOR experiment, in which the echo intensity detects the enhancement of the sublevel spin population. Transitions in the HS-ENDOR spectrum therefore have the appearance of an "emission" line rather than an "absorption" line as observed in the ENDOR spectrum.

HS-ENDOR spectra for the irradiated malonic acid radical are shown in Figure 9. The full ENDOR spectrum observed at this arbitrary orientation of the malonic acid single crystal is shown in Figure 9A. The spectrum in Figure 9B is obtained when $\Delta = 38.6$ MHz. This frequency difference corresponds to the separation of the strong doublet observed in the EPR (or ESE-detected EPR) spectrum. The doublet arises from the hyperfine coupling of the carbon π-radical with the α-proton of the CH fragment of the malonic acid radical, $\cdot$CH(COOH)$_2$. No other proton hyperfine lines are observed in this HS-ENDOR spectrum. The HS-ENDOR spectrum in Figure 9C is obtained when $\Delta = 5.17$ MHz. This frequency difference corresponds to the hyperfine coupling with the carboxyl protons. The transitions from the α-protons and the more weakly coupled protons near the nuclear Larmor frequency arise from finite pulse bandwidth effects and the excitation of forbidden transitions.

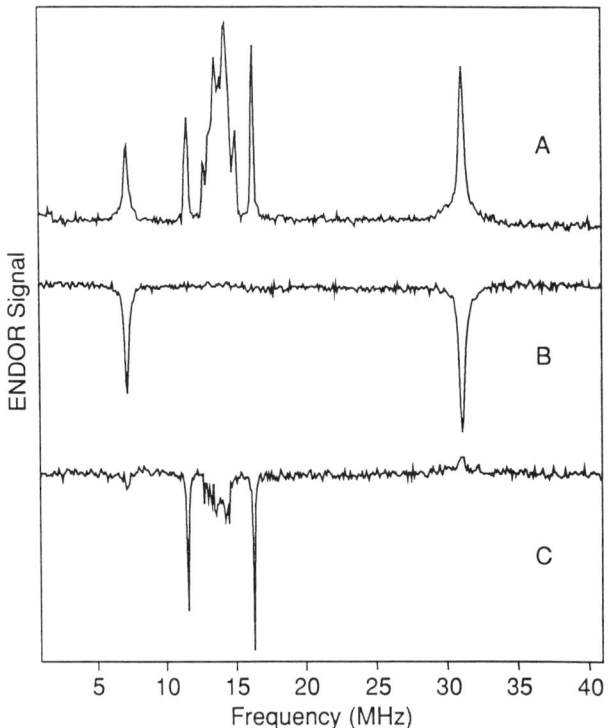

Figure 9. Davies ENDOR (A) and HS-ENDOR (B–C) spectra of irradiated malonic acid crystal. Experimental conditions: preparation frequency ν_p, 9.1600 GHz; $t_p(\pi)$, 0.40 μs; detection frequency ν_0, 9.1600 GHz; B_0, 3259.3 G; spin echo τ, 0.35 μs; $t_p(rf)$ = 5.00 μs; pulse-sequence iteration delay, 5 ms; rf increment, 0.1 MHz; 30 echoes sampled per point; 10 sweeps; T, 85 K. B: Hyperfine selective ENDOR with ν_p = 9.1214 GHz. All other conditions were the same as in part A except that $t_p(rf)$ was 4.00 μs. C: Hyperfine selective ENDOR with ν_p = 9.1547 GHz. All other conditions were the same as in part A except that $t_p(rf)$ was 8.70 μs. (Reproduced with permission from reference 12. Copyright 1990.)

Sublevel EPR Spectroscopy. Each HS-ENDOR spectrum preselects ENDOR transitions from those nuclei with $A = \Delta$. When more than one hyperfine transition is possible in each electron spin manifold, each HS-ENDOR spectrum is composed of sublevel (hyperfine) transitions from those nuclei with a preselected hyperfine coupling determined by $\Delta = A_i$, where A_i is the hyperfine coupling for the ith nucleus. Each HS-ENDOR spectrum is therefore a slice of a two-dimensional (2D) ENDOR spectrum in which one axis corresponds to the sublevel mixing frequency, as in the standard ENDOR experiment, and the other axis corresponds to the difference frequency, Δ. In this 2D-ENDOR plot, the conventional ENDOR spectrum corresponds to the spectrum along the $\Delta = 0$ axis. It is the "inverted" image of the projection on to the $\Delta = 0$ axis of all HS-ENDOR spectra recorded for $\Delta > 0$.

With the pulse sequence shown in Figure 8, it is possible to generate an EPR spectrum arising only from EPR transitions connected to the hyperfine sublevels resonant with the rf pulse. This step is accomplished by setting the rf pulse to irradiate a preselected ENDOR line and sweeping the difference frequency Δ between the excitation frequencies in the preparation and detection periods on successive pulse iterations. The spectrum recorded in this manner corresponds to a spectral slice parallel to the Δ-axis in the 2D-ENDOR plot described. A transition along the Δ-axis will be observed when $\Delta j = A_j$. The transition j is a direct measure of the hyperfine coupling associated with the ENDOR transition irradiated by the rf pulse. Additional transitions along Δ may be observed from the excitation of forbidden transitions (discussed later).

Experimentally, it is convenient to isolate the sublevel polarization transfer by taking the difference of the EPR susceptibility obtained by switching the rf pulse (in Figure 8) on and off on alternate pulse iterations. The plot of the susceptibility difference against Δ generates a plot of the sublevel saturation transfer.

Sublevel EPR spectra for the irradiated malonic acid single crystal are shown in Figure 10. The spectra in Figures 10A–10C were obtained with rf irradiation at 7.32, 31.33, and 16.33 MHz, respectively. The strong side-hole peak at −38.6 MHz identifies this value as the hyperfine coupling for the ENDOR lines at 7.32 and 31.33 MHz. This hyperfine coupling is immediately evident from a single sublevel EPR spectrum, either Figure 10A or 10B. The assignment of the hyperfine coupling does not depend on a rigorous analysis of the ENDOR spectrum such as whether the ENDOR lines are shifted by second-order hyperfine interactions. The other two side-hole peaks in Figures 10A and 10B arise from the excitation of forbidden EPR transitions. In Figure 10C, the microwave excitation pulse in the preparation period overlaps the side-hole peak at 5.17 MHz. This overlap determines the intensity of the weakly coupled protons in the HS-ENDOR spectrum.

EPR-Detected Sublevel Coherence

When the excitation pulses are short compared to the electron and nuclear spin relaxation times, sublevel coherence phenomena can be detected in pulsed ENDOR experiments. As is well-known in NMR spectroscopy, a transient nuclear spin magnetization response $M_{xy}(t)$, known

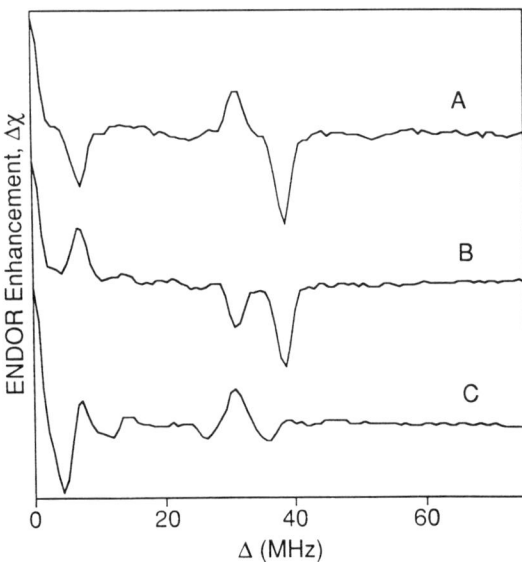

Figure 10. EPR subspectra of irradiated malonic acid. A: ν_{rf}, 7.32 MHz; $t_p(rf)$, 5.60 μs. B: ν_{rf}, 31.3 MHz; $t_p(rf)$, 3.40 μs. C: ν_{rf}, 16.31 MHz; $t_p(rf)$, 8.70 μs. All other experimental conditions were the same as for Figure 9 except that the step increment was 0.75 MHz, and 100 data points. (Reproduced with permission from reference 12. Copyright 1990.)

as the free induction decay (FID), can be observed following the application of an rf pulse resonant with the NMR transition. In NMR spectroscopy, the FID is detected and Fourier transformed to give the NMR spectrum. A wide spectrum of rf pulse sequences have been developed for manipulating the evolution of the magnetization. Many of these methods for creating and detecting nuclear spin coherence are also possible in ENDOR spectroscopy. Several are illustrated in the next section.

Sublevel Free Induction Decay. FID signals from hyperfine sublevels can be created simply by applying an rf pulse that is resonant with the NMR transition of the sublevel. However, the free precession signal cannot be directly observed in the EPR spectrometer because it is nominally at the rf frequency while the spectrometer receiver is tuned to the microwave frequency range typical of EPR transitions. The FID signal must be encoded on to an EPR transition in order to be detected. This encoding can be accomplished by converting the sublevel FID signal into a population difference by using a second rf pulse as proposed by Hofer et

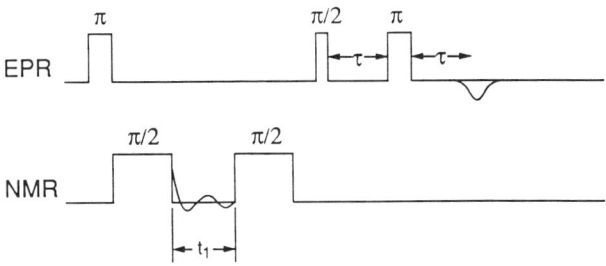

Figure 11. Pulse sequence for EPR detection of sublevel FID.

al. (23). In the ideal case, both rf pulses are $\pi/2$ pulses. The pulse sequence is shown in Figure 11. The sublevel FID is detected point by point during the evolution period, t_1, by incrementing t_1 on successive iterations of the pulse sequence. The sublevel population difference is detected as an amplitude modulation of the electron spin echo in the detection period.

EPR-detected FID waveforms recorded for the 14.42-MHz line in the ENDOR spectrum of an arbitrary orientation of a single crystal of irradiated malonic acid are shown in Figure 12A. (The ENDOR spectrum for this orientation of the crystal is shown as an inset in Figure 16.) Both the in-phase and phase-quadrature sublevel FID signals can be detected with appropriate phase-cycling schemes. The in-phase sublevel FID signal (upper trace in Figure 12A) can be detected by using rf pulses with the same phase. The phase-quadrature sublevel FID signal (lower trace in Figure 11A) is detected by phase shifting the second rf pulse by $\pi/2$ with respect to the first rf pulse. Additional phase cycling is used to cancel the overall electron spin–lattice magnetization recovery during the t_1 evolution period. This canceling is achieved by phase cycling the second rf pulse by π on alternate pulse-sequence iterations. Thus, rf pulses with relative phases of 0, $\pi/2$, π, and $3\pi/2$ (or $-\pi/2$) are required for the EPR detection of the complete quadrature sublevel FID waveforms. Additional phase cycling of the microwave pulses is used to eliminate any unwanted microwave or rf interference signals other than the electron spin echo in the detection period.

The complex Fourier transform (FT) spectrum corresponding to the FID waveforms shown in Figure 12A is shown in Figure 12B. The ENDOR line at 14.42 MHz is a single relatively narrow line that does not exhibit any fine structure (as shown in the inset in Figure 16). However, the sublevel FID ENDOR spectrum in Figure 12B reveals that it is actually composed of two and possibly three ENDOR transitions. The structure on the weaker line near +50 kHz could conceivably arise from hyperfine anisotropy effects but more likely arises from the overlap of two partially resolved sublevel transitions. The observation of a three-quantum

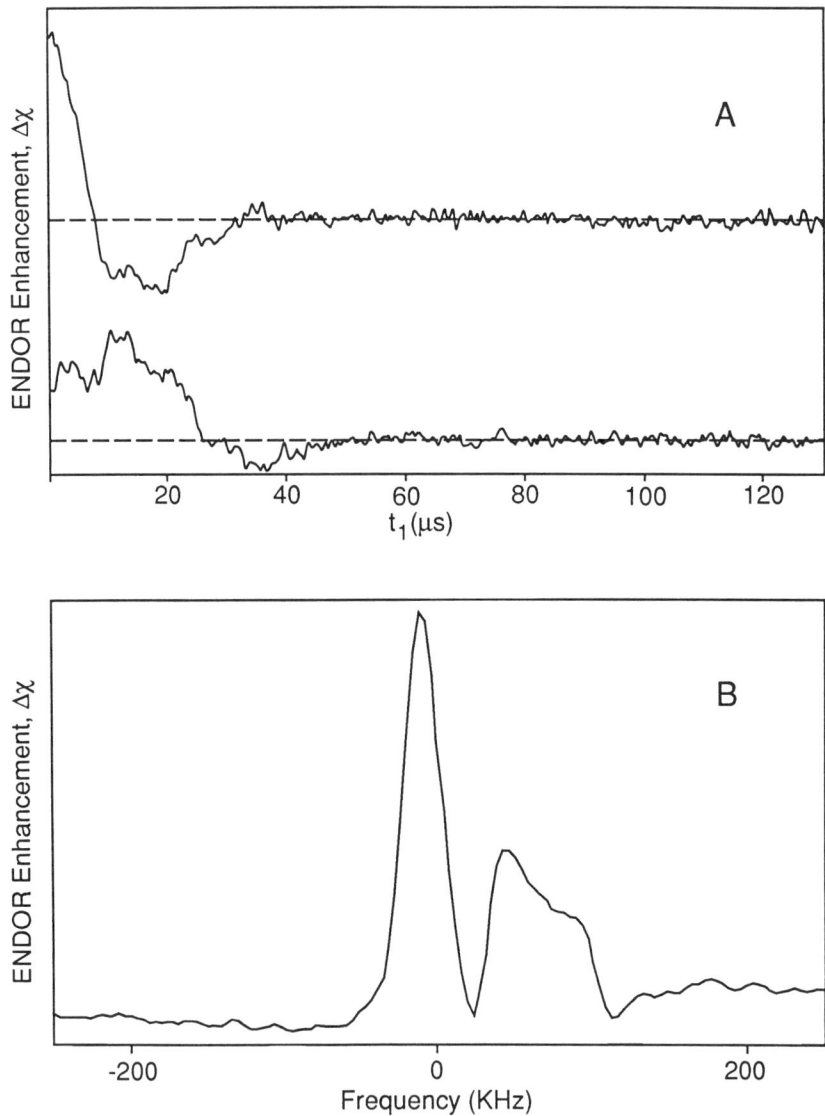

Figure 12. A: In-phase and phase-quadrature sublevel FID signals observed with the rf frequency irradiating the ENDOR line at 14.42 MHz. Experimental conditions were the same as in Figure 5 except the following: rf pulse width, 2.50 μs; t_1 increments, 0.20 μs. B: Complex FT spectrum obtained from the FID signals in part A. (Reproduced with permission from reference 12. Copyright 1990.)

transition in the multiple-quantum ENDOR spectrum for the 14.42-MHz ENDOR line (discussed in the next section) suggests that the occurrence of overlapped sublevel transitions is the correct interpretation.

Sublevel Spin Echoes. The pulse sequence in Figure 11 can be easily amended to form sublevel spin echoes, as shown in Figure 13 (*23*). Spin echoes from the hyperfine sublevels can be created by adding a refocusing pulse between the two $\pi/2$ rf pulses to form a simple Hahn spin-echo sequence, $\pi/2 - \tau_n - \pi - \tau_n - S(2\tau_n)$, where τ_n is the free precession evolution period during which nuclear coherence will evolve. The sublevel coherence is again converted to a population difference by the third rf pulse, which is applied coincident with the sublevel spin echo. The population difference is then observed as an amplitude modulation of the electron spin echo in the detection period.

NMR spin echoes from the nuclear hyperfine sublevels can be observed by sweeping the evolution time, t_1, through τ_n starting from $t_1 < \tau_n$. Two sublevel spin echoes for the ENDOR line at 30.19 MHz observed for $\tau_n = 5.0$ and 20 μs are shown in the inset in Figure 14. Phase cycling is used to eliminate unwanted signals arising from nonideal pulse conditions and to cancel the overall loss of sublevel polarization arising from the loss of electron spin magnetization during the t_1 period. The open circles in the inset figure indicate the correspondence of the amplitude of these two spin echoes to the echo envelope-decay function. The entire sublevel spin-echo envelope decay, recorded as a function of τ_n with $t_1 = \tau_n$, is shown in Figure 14. The small dots correspond to the echo amplitudes measured for a given τ. The least-squares fit to a single exponential and the residuals of the fit are also shown. The decay of the nuclear spin echo intensity, described by the spin–spin relaxation time, T_{2n}, has the value of nominally 60 μs, indicating a homogeneous line width of order 16 kHz.

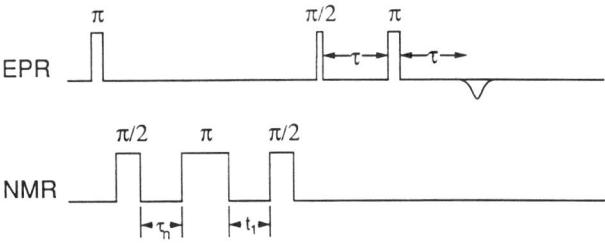

Figure 13. Pulse sequence for EPR detection of the sublevel spin echo.

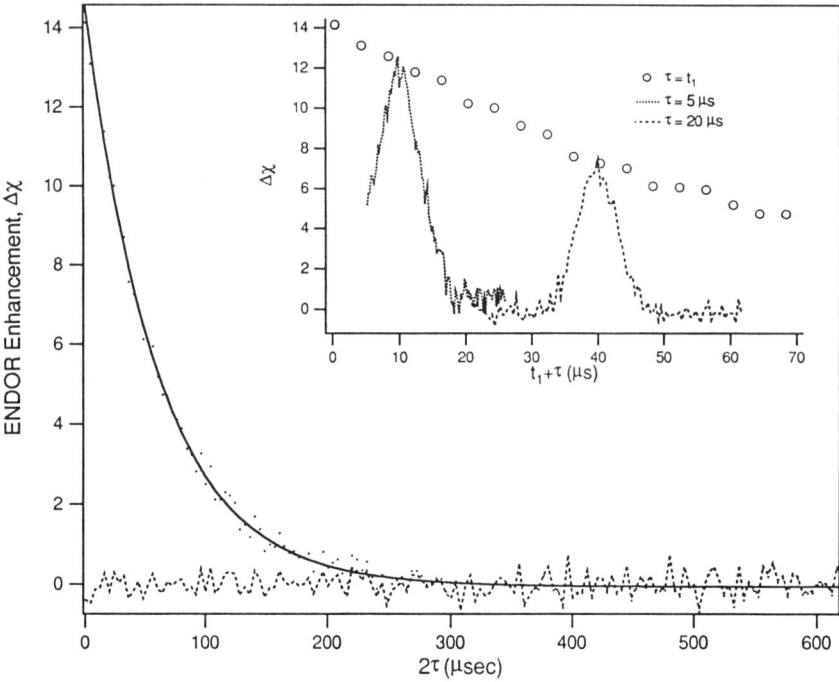

Figure 14. The sublevel spin-echo envelope decay recorded with $t_1 = \tau_n$ along with the fit to a single exponential and the residuals of the fit. Experimental conditions were the same as in Figure 5 except the following: rf frequency, 30.19 MHz; rf pulse width, 2.50 μs for a π/2 pulse. Two sublevel spin echoes observed for $\tau_n = 5.0$ and 20.0 μs are shown in the inset.

Multiple-Quantum ENDOR. One limitation of ENDOR spectroscopy is that the ENDOR line intensities are generally not proportional to the number of nuclei contributing to the ENDOR transition (24, 25). Thus it is not possible to determine from the relative line intensities alone the number of equivalent nuclei corresponding to each type of inequivalent nucleus. In CW ENDOR spectroscopy, it is possible to partially restore the proportionality by using nonsaturating NMR rf fields (24, 25). However, this approach often fails because of the low ENDOR susceptibility obtained when using nonsaturating rf fields. The sensitivity can be improved by using an extension of the ENDOR experiment, the electron nuclear nuclear triple resonance experiment (26). This experiment is a more successful approach, but it is experimentally complex (27), and it is not generally applicable in solids where electron–nuclear cross-relaxation mechanisms can rapidly transfer energy among lines in the ENDOR spectrum (*20*). The situation is improved but not completely eliminated in the pulsed versions of the ENDOR and electron nuclear nuclear triple resonance experiments (*28*).

Multiple-quantum (MQ) ENDOR spectroscopy provides another approach to counting the number of nuclei that contribute to an ENDOR line (*29*). Multiple-quantum coherences from sublevel transitions in ENDOR spectroscopy can be detected by the phase-incrementing method used in solid-state NMR spectroscopy to observe multiple-quantum NMR transitions (*30, 31*). The implementation of this pulse sequence in pulsed ENDOR spectroscopy is shown in Figure 15. Multiple-quantum orders are separated on the basis of their phase accumulation during the t_1 precession period following excitation of a sublevel transition by an rf pulse. An nth-order coherence will accumulate phase $\theta_n = n\omega t_1$ for an evolution time t_1. For example, a double-quantum coherence will accumulate twice the phase of a single-quantum coherence in a given evolution time t_1. The higher order coherences are detected by incrementing the phase τ of the second rf pulse in steps between $0 < \tau < 2\pi$ on successive pulse-sequence iterations. Analogous to the detection of the sublevel FIDs or spin echoes, an rf pulse converts the sublevel free precession signal into a sublevel population difference at time t_1. The nth-order sublevel coherence induces an nth-order modulation in the population difference over a periodicity of 2 radians. The modulation of the sublevel population difference is detected as an amplitude of the electron spin echo formed in the detection period of the pulse sequence. Thus the MQ-ENDOR waveform is recorded by monitoring the intensity of the electron spin echo as a function of the relative phase τ of the two rf pulses. The MQ-ENDOR spectrum is then obtained by cosine Fourier transformation of this waveform. Multiple-quantum coherences occurring within a spectral bandwidth of approximately $2\omega_1$ will be observed in the MQ-ENDOR spectrum.

The MQ-ENDOR spectrum obtained for the 14.42-MHz line in the ENDOR spectrum of an arbitrary orientation of an irradiated malonic acid crystal is shown in Figure 16. A portion of the ENDOR spectrum

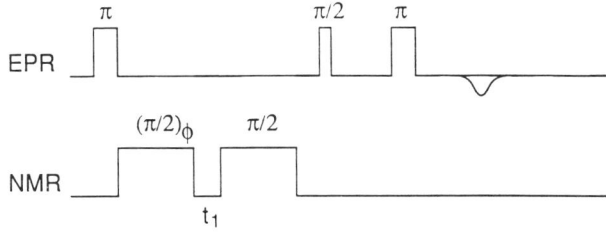

Figure 15. Pulse sequence for multiple-quantum ENDOR experiment.

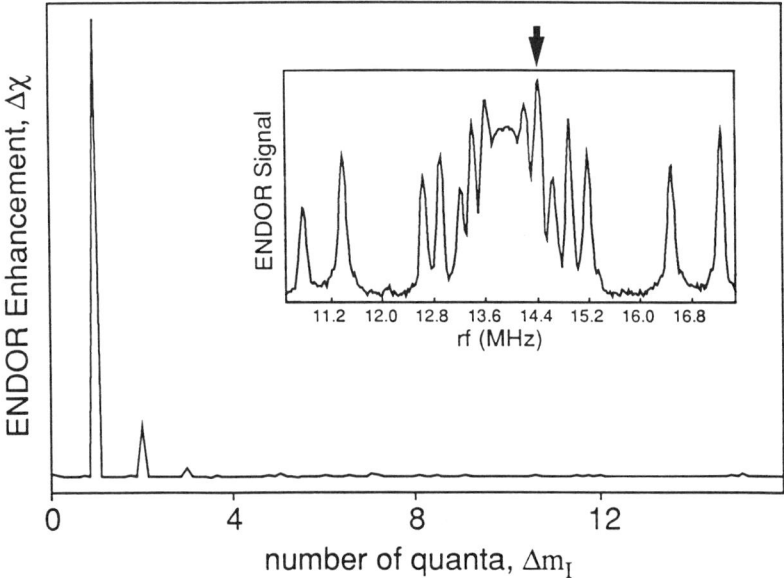

Figure 16. MQ-ENDOR spectrum for obtained for the irradiation of the 14.42-MHz line in the ENDOR spectrum of the single crystal of the irradiated malonic acid radical. The Davies ENDOR spectrum is shown in the inset. Experimental conditions were the same as in Figure 5 except the following: rf pulse width, 2.50 µs; resolution, $\pi/\Delta\tau = 16$; data collected over $\tau = 16\pi$. (Reproduced with permission from reference 12. Copyright 1990.)

with the 14.42-MHz line denoted by an arrow is shown as an insert. The 14.42-MHz line appears to be composed of only a single transition in the ENDOR spectrum. However, the three-quantum coherence observed in the MQ-ENDOR spectrum indicates that it is actually composed of at least three transitions. Three ENDOR lines are also observed in the sublevel FID ENDOR spectrum shown in Figure 12, which was recorded at the same crystal orientation as for Figure 16. These combined results indicate that the three-quantum coherence arises from three slightly inequivalent NMR transitions that give rise to the inhomogeneous broadening of the ENDOR line at 14.42 MHz. The 33-kHz line width of the lines in the sublevel FID ENDOR spectrum suggests that these lines are also inhomogeneously broadened because the homogeneous line width determined directly from the value of T_{2n} measured by sublevel spin echoes is on the order of 16 to 20 kHz.

Future of Pulsed Techniques

Pulsed technology has significantly extended the original scope of both EPR and ENDOR spectroscopies and is presently in a period of rapid evolution. New areas of applications will emerge as the versatility of this spectroscopic technique is expanded. Developments in the field of pulsed EPR spectroscopy will also likely affect future developments in pulsed ENDOR spectroscopy.

References

1. Feher, G. *Phys. Rev.* **1956**, *103*, 834.
2. *Electron Spin Double Resonance Spectroscopy;* Kevan, L.; Kispert, L. D., Eds.; Wiley-Interscience: New York, 1979.
3. *Multiple Electron Resonance Spectroscopy;* Dorio, M. M.; Freed, J. H., Eds.; Plenum: New York, 1979.
4. Hyde, J. S. In *Magnetic Resonance in Biological Systems;* Ehrenberg, A.; Malmstrom, B. G.; Vanngard, T., Eds.; Pergamon: Oxford, England, 1967.
5. Kurreck, H.; Kirste, B.; Lubitz, W. *Angew. Chem. Int. Ed.* **1984**, *23*, 173.
6. Freed, J. H. In *Multiple Electron Resonance Spectroscopy;* Dorio, M. M.; Freed, J. H., Eds.; Plenum: New York, 1979.
7. McConnell, H. M.; Heller, C.; Cole, T.; Fessenden, R. W. *J. Am. Chem. Soc.* **1960**, *82*, 766.
8. Wertz, J. E.; Bolton, J. R. *Electron Spin Resonance;* Chapman and Hall: London, 1986.
9. Allendorfer, R. D.; Maki, A. H. *J. Magn. Reson.* **1970**, *3*, 396.
10. Grupp, A.; Mehring, M. In *Modern Pulsed and Continuous Wave ESR;* Kevan, L.; Bowman, M. K., Eds.; Wiley: New York, 1990.
11. Schweiger, A. In *Modern Pulsed and Continuous Wave ESR;* Kevan, L.; Bowman, M. K., Eds.; Wiley: New York, 1990.
12. Thomann, H.; Bernardo, M. *Spectrosc. Int. J.* **1990**, *8*, 117.
13. Rabi, I. I. *Phys. Rev.* **1957**, *51*, 652.
14. Mehring, M.; Hofer, P.; Grupp, A. *Phys. Rev.* **1986**, *A33*, 3523.
15. Davies, E. R. *Phys. Lett.* **1974**, *47A*, 1.
16. Mims, W. B. *Proc. R. Soc. London* **1965**, *283*, 452.
17. Mims, W. B. In *Electron Paramagnetic Resonance;* Geschwind, S., Ed.; Plenum: New York, 1972; p 263.
18. Buhlmann, C.; Schweiger, A.; Ernst, R. R. *Chem. Phys. Lett.* **1989**, *154*, 285.
19. Thomann, H.; Bernardo, M. *Chem. Phys. Lett.* **1990**, *169*, 5.
20. Mobius, K.; Biehl, R. In *Multiple Electron Resonance Spectroscopy;* Dorio, M. M.; Freed, J. H., Eds.; Plenum: New York, 1979; p 475.
21. Cook, R. J.; Whiffen, D. H. *Proc. Roy. Soc.* **1964**, *84*, 845.
22. Mehring, M.; Hofer, P.; Grupp, A. *Ber. Bunsen-Ges. Phys. Chem.* **1987**, *91*, 1132.

23. Hofer, P.; Grupp, A.; Mehring, M. *Phys. Rev.* **1986,** *A33,* 3519.
24. Gerson, F.; Jachimowicz, J.; Mobius, K.; Biehl, R.; Hyde, J. S.; Leniart, D. S. *J. Magn. Reson.* **1975,** *18,* 471.
25. Leniart, D. S.; Connor, H. D.; Freed, J. H. *J. Chem. Phys.* **1975,** *63,* 165.
26. Freed, J. H. *J. Chem. Phys.* **1969,** *50,* 2271.
27. Dinse, K. P.; Biehl, R.; Mobius, K. *J. Chem. Phys.* **1975,** *61,* 4335.
28. Thomann, H.; Bernardo, M., unpublished.
29. Mehring, M.; Hofer, P.; Kass, H.; Grupp, A. *Europhys. Lett.* **1988,** *6,* 463.
30. Baum, J.; Munowtiz, M.; Garroway, A. N.; Pines, A. *J. Chem. Phys.* **1985,** *83,* 2015.
31. Shykind, D. N.; Baum, J.; Liu, S.-B.; Pines, A.; Garroway, A. N. *J. Magn. Reson.* **1988,** *76,* 149.

RECEIVED for review January 30, 1991. ACCEPTED revised manuscript September 5, 1991.

5

Strategies for Measurement of Electron Spin Relaxation

Michael K. Bowman

Chemistry Division, Argonne National Laboratory, Argonne, IL 60439

> *Pulsed electron paramagnetic resonance spectroscopy presents an opportunity for measuring both the relaxation and the spectra of electron spin systems. Spectra and relaxation probe two very different aspects of the spin system. However, the lack of recognizable features on most relaxation curves makes the analysis and interpretation of relaxation experiments difficult. Some of the strategies for design and interpretation of relaxation experiments are discussed.*

SPIN SYSTEMS can be viewed in two very different ways. The electron paramagnetic resonance (EPR) field-swept (or Fourier transform) spectrum is the most common view, showing absorption or dispersion as a function of the resonance condition. The less common view is that of the relaxation or the dynamics of the EPR system. These are not equivalent ways of viewing a spin system. They probe very different aspects of the spin Hamiltonian and the physical sample.

As a rule, the static parts of the spin Hamiltonian determine EPR spectra. Thus, the structure of the free radical dictates the appearance of the EPR spectrum. Consequently, spectra are useful for identifying and quantifying free radicals even if the spin Hamiltonian is not completely understood. Relaxation, on the other hand, results from dynamic elements of the spin Hamiltonian and depends on both the free radical and its environment. Although knowledge of the relaxation rates of a spin system has some utility for setting the optimal conditions in EPR or other magnetic resonance experiments, detailed interpretation of the relaxation rates requires additional information.

Relaxation of a spin system is the change from an initial nonequilibrium state to a state closer to equilibrium. An EPR signal that changes monotonically as relaxation proceeds monitors the relaxation. An individual relaxation curve is much less distinctive than an individual spectrum. Spin relaxation is often a first-order kinetic decay with an exponential relaxation curve, so that one relaxation curve looks very much like any other aside from the time scales involved. Determining the physical cause of that relaxation from a single relaxation curve is almost impossible. In contrast, the features in a spectrum often indicate the presence of hyperfine splittings, different radicals, spectroscopic g-factor anisotropy, etc.

This chapter points out some of the issues involved in a relaxation study using pulsed EPR spectroscopy. The experiment, the sample, and the data analysis affect the relaxation measurement, and their role in the experimental design is discussed. Next, the discussion considers how experimental relaxation curves can be parameterized for analysis. Finally, some major types of relaxation experiments in pulsed EPR spectroscopy are mentioned.

The aim of this discussion is not to prescribe a methodology to be followed in making relaxation measurements. Experimental electron spin relaxation is such a poorly explored field that such a global prescription is impossible. Rather, I present some of the issues that should be faced before and during relaxation experiments. The experimenter's task is to resolve the problems in the context of each experiment. Satisfactory answers cannot always be made to all the questions before measurements start. Some of the critical information simply is not available then. A few trial runs and perhaps a redesign of the experiment may be required before an interpretable set of measurements can be taken.

The discussion is in the context of relaxation measurements on dilute electron-spin systems in solids. "Spins" and "radicals" are used interchangeably to refer to paramagnetic centers. The scope of the discussion is intended to illustrate rather than be comprehensive.

Background

Before proceeding, two distinctions are needed to make the discussion easier. The first distinction is between what are called, for want of better names, "true" and "apparent" relaxation. True relaxation is the continuous, irreversible evolution of a nonequilibrium spin system toward equilibrium. It occurs spontaneously from the initially prepared state and is not driven by the experimental measurement. Examples of true relaxation are spin–spin (T_2) and spin–lattice (T_1) relaxation (as defined in the Bloch equations) (*1, 2*). In contrast, apparent relaxation is anything that is not true relaxation but has the appearance of a relaxation curve.

Consider an experimental signal that monotonically approaches zero (or some constant level) during an experiment. The signal can have the shape of a relaxation curve without the spin system spontaneously approaching equilibrium. The free induction signal from a strongly inhomogeneously broadened line appears to decay with a characteristic time T_2^*. The apparent T_2^* relaxation is reversible, and the free induction signal is recovered by applying a refocusing pulse to generate a spin echo.

The measurement process itself can even drive the signal decay. Such is the case when the Carr–Purcell experiment is performed with improperly adjusted pulses. The echo amplitudes rapidly decrease during the experiment. However, this apparent relaxation does not occur without the pulses and, at least in principle, could be reversed by carefully crafted pulses.

The second distinction is between the relaxation mechanism and the relaxation process. "Mechanism" refers to the coupling that leads to transfer of energy between the spin and its environment. "Process" characterizes the type of motion that exchanges energy with the spin system. These definitions are intended to parallel the usage of these terms in spin–lattice relaxation in solids (3). For instance, g-factor anisotropy (dependence of resonant frequency of the free radical on its orientation with respect to the applied magnetic field) can be the mechanism producing relaxation during the process of rotational diffusion. These are two distinct elements of relaxation that should be regarded separately.

In general, a relaxation measurement can be sensitive to several mechanisms and several processes and may involve mixtures of both true and apparent relaxation. The separation of these factors is challenging, but successful unweaving of the relaxation provides a very detailed view of the spin system and the sample. Failure to untangle all the contributions to relaxation can give results that are not merely incorrect, but also dangerously misleading.

Design of a Relaxation Study

Three aspects of a relaxation experiment need to be considered very carefully to have interpretable results:

1. the pulse sequence and experimental conditions in the measurement, which determine what properties of the spin system contribute to the relaxation curves

2. the relation between the detected signal and the spectrum of the radicals in the sample

3. the relaxation mechanisms and processes involved

These considerations can be posed as a series of three questions:

1. What property of the spin system is measured?
2. From what species in the sample does the signal arise?
3. How can the different mechanisms and processes be separated?

Pulse Sequence. The observable quantity in pulsed EPR experiments is usually the expectation value (denoted by $<>$) of the x component of magnetization in the rotating frame ($<S_x>$) or more generally $<S_+>$ (where the step-up operator $S_+ = S_x + iS_y$), although $<S_z>$ has been used (4, 5). This limitation is not great because almost any other spin quantity, $S_?$, can be converted into S_+ by one or more microwave pulses. The time evolution of $S_?$ can be measured in a stroboscopic fashion by allowing it to evolve or relax for a time t as $S_?$, then converting it to an observable quantity for measurement, and then repeating the process for the next value of t.

This strategy becomes complicated when, for a given pulse sequence, more than one $S_?$ contributes to $<S_+>$. For example, two microwave pulses applied to the simplest spin system described by the Bloch equations generate three different signals. Part of the S_z present at the second pulse becomes an inversion-recovery signal and reflects the spin–lattice relaxation that took place between the two pulses. Part of the S_+ magnetization created by the first pulse is refocused and generates a spin echo, sensitive to spin–spin relaxation. Some of the S_+ magnetization from the first pulse is unaffected by the second pulse and continues the original free induction decay (FID), affected by spin–spin relaxation and by inhomogeneous broadening.

Even in this simple example, three superimposed signals appear, sensitive to combinations of T_1, T_2, and T_2^*. Only T_1 and T_2 represent true relaxation; T_2^* is an apparent relaxation. With spin systems that do not obey the simple Bloch equations, the situation is even more complicated. The unwanted signals can be avoided only if it is realized that several signals are potentially present. It is necessary to be aware of every signal that can be generated in the experiment.

Both the inversion-recovery and the FID signals can be suppressed in the example by making the second pulse have a turning angle of exactly π. Unfortunately, turning angle depends on frequency offset in the EPR spectrum and on the position within the sample if the microwave magnetic field is inhomogeneous. A better strategy is to use the dependence of the phase of the individual signals on the phase of the microwave pulses. For the example, the phase of the FID depends on the phase of the first microwave pulse as ϕ_1. The spin-echo signal depends on the pulse phases of both pulses as $-\phi_1 + 2\phi_2$, and the inversion-recovery signal goes as ϕ_2.

A cycle of phases that causes any two of these signals to sum to zero can be generated. Phase cycling is a powerful means of selecting the one desired signal in an experiment (5–7).

Another useful means of separating signals is to make the measurement in a time window where the unwanted signals have negligible intensity. Apparent relaxation can be manipulated to make some signals vanish when the measurement is taken. For example, an external magnetic field gradient enhances the apparent decay of the FID and decreases the amplitude of the inversion-recovery signal, leaving only the spin echo. If these strategies fail to isolate the desired signal, a different pulse sequence might be better suited for the measurement.

Detection and the Spectrum. Every EPR spectrum is a distorted representation of the spin system. The normal first-derivative spectrum from a continuous-wave (CW) spectrometer emphasizes the sides of spectra where the absorption is changing most rapidly. The position in the EPR spectrum with the strongest absorption has zero intensity in the first-derivative spectrum. The turning points in an anisotropic spin system stand out at the expense of other orientations, and the narrowest lines appear stronger than they are. The intensity of the EPR signal depends on line width, ΔB, as ΔB^{-2} for the first derivative and ΔB^{-3} for the second derivative, etc. Although these properties are well known, it is not uncommon for a very broad signal to go unnoticed in the presence of a very narrow signal containing an extremely small fraction of the spins. Consequently, the narrow signal is assumed to characterize all the spins in the sample rather than a minor fraction.

Pulsed EPR spectroscopy has similar biases that affect relaxation measurements. A pulsed EPR signal comes from a limited spectral window. The width and shape of that window are determined by the interaction of three things: the portion of the spectrum excited, the spectral response function of the spectrometer, and the processing of the signal. Relaxation measurements can be affected, in addition, by spins lying outside this window or making a minor contribution to the signal within the window.

The portion of the spectrum excited in a pulse sequence depends on the width and turning angle of the pulses and on the pulse sequence itself (6). Generally, the shorter the pulses, the wider the spectral range excited. The greater the number of pulses, the narrower the spectral range. The calculation of the spectral range excited in a simple spin system is a straightforward if somewhat tedious exercise using the formalism of Jaynes (8) and Bloom (9). As a rule, the half-width at half-height for spectral excitation is $0.1 - 1 \times T_p^{-1}$ (in frequency), where T_p is the microwave pulse width (in time) for a square pulse (6).

The microwave resonator used in almost all experiments acts as a tuned filter, accepting EPR frequencies within its bandwidth and rejecting those outside (*10*). Normally, the portion of the EPR spectrum excited is much smaller than the bandwidth of the resonator, because either the pulses are wide or they are stretched by the quality factor, Q, of the resonator. This filtering effect is negligible unless the frequency of the microwave pulses is offset from the resonator frequency.

The largest factor in determining what is observed in a pulsed EPR experiment is the signal processing. Spin echoes and FIDs contain spectral information that appears upon Fourier transformation (*6*). If an experiment records the FID or the echo shape, then the Fourier transform spectrum is a good means to visualize what was excited by the pulses and passed by the resonator. From such a spectrum, a particular line or spectral feature can be measured and its relaxation followed cleanly. Two drawbacks to this approach are the following:

1. A single point in the Fourier transform spectrum comprises a fraction of the total signal intensity, and unless the line is integrated, the signal-to-noise ratio (S/N) suffers.

2. Recording the full echo or FID shape for each time delay for each relaxation curve generates much data to be stored and processed.

For these reasons, the echo or FID is often sampled by using a boxcar integrator or similar device. The boxcar gate is a filter with a response function. In a spectrometer with coherent detection (most recent spectrometers), the boxcar response function can be obtained by Fourier transformation of the boxcar gate with respect to the center of the echo or the origin of the FID. When a very narrow boxcar gate sits at the center of the echo, the boxcar response is flat and the output is proportional to the integral of the EPR spectrum excited. On the other hand, a gate of finite duration, T_b, has a frequency (ω) response function $(1/\omega) \sin(\omega T_b)$ and is sensitive to signals near the microwave-pulse frequency. In the limit of a very broad gate, this response function becomes a delta function at the microwave frequency, and only radicals exactly in resonance are recorded (*11*).

Careful adjustment of the EPR resonance condition, microwave pulse width, and position of the boxcar gate provides great control over what is measured in a pulsed EPR experiment while simultaneously reducing the volume of data to be stored and processed. Interpretation of relaxation experiments requires understanding both the response function and the EPR spectrum of the sample.

A pulsed experiment gives a response proportional to the EPR absorption spectrum at that frequency in the absence of any relaxation. In a pulse sequence designed to measure relaxation of some spin operator, for example, $\mathbf{S}_x$, $\mathbf{S}_z$, or $\mathbf{S}_+$, which we shall denote $\mathbf{S}_?$, the relaxation is followed as a time interval is incremented. Yet, during the remainder of the experiment, spin relaxation still takes place, attenuating the pulsed EPR signal. If several species are present, each is attenuated differently if their relaxation properties differ. As in CW EPR spectroscopy, not all spins in the sample are observable under the same sample and spectrometer conditions. As the relaxation properties of the spins change, for example, with temperature, the signal amplitudes change. It is not unusual in frozen aqueous samples for an intense, very broad spectrum (often blamed on Fe or Mn ions) to grow in and dominate pulsed EPR measurements as the sample temperature drops below 20 K. If the change in species detected goes unnoticed, spurious relaxation behavior can be reported.

Interpretation of Relaxation. Understanding the relaxation is the most difficult part of a relaxation study. Even when the measurements are perfect and relaxation of $<S_?>$ for a single spin species is measured with high accuracy, assignment of relaxation processes and mechanisms is problematic. The relaxation is occurring along several independent, parallel pathways with different rates and is producing a single relaxation curve usually with only two parameters: the decay rate and the amplitude. Unfortunately, the single decay rate can be matched by almost any relaxation process and mechanism provided the sample parameters that enter the relaxation model are not accurately known. The problem is not "garbage in, garbage out", because a good measurement and a valid model can still yield "garbage" if the model is inappropriate for that measurement. Some models can easily be eliminated from consideration. For instance, a relaxation process involving unrestrained motion of radicals in a liquid is inappropriate for the relaxation of immobilized radicals in a solid. Still, that condition leaves many plausible models to explain the relaxation in that solid.

Fortunately, the different relaxation processes have a different dependence of the relaxation rate on, for example, temperature and EPR frequency, but the different mechanisms depend differently on isotopic composition, hyperfine line, concentration, viscosity, etc. Varying these parameters can eliminate some relaxation models from consideration. Rarely can a relaxation model be proven correct, but a model should be reasonable and consistent with the data and the sample.

The Sample

Most solids studied by pulsed EPR spectroscopy are magnetically dilute single crystals, polycrystalline powders, or disordered glasses. In all three types of samples, the spins are never identical or homogeneous and can exhibit a distribution of relaxation behaviors. In powders and glasses, the radicals have many different orientations with respect to the applied magnetic field. Because many relaxation mechanisms involve anisotropic couplings that depend on the orientation of the spin system in the magnetic field, the relaxation curves are often sums of many exponentials in powders and glasses or show an orientation dependence in single crystals.

Free radical sites in a glass do not have identical potential fields like the sites in a crystal. A distribution of radical conformations or structures may occur in glasses, effectively a set of similar but different spin systems. Although it is tempting to think that the radical and the glassy matrix around it adopt a unique configuration, a dispersion of structures and interactions is likely.

More subtle differences make spins nonidentical even if they have the same physical structure and orientation. Nuclear spins with significant hyperfine interaction can affect electron relaxation and produce populations of nonequivalent radicals (at least on the time scale of the nuclear T_1). Also, all dilute spin systems have different populations distinguished by the local dipolar fields. Those fields arise from nearby nuclear spins or the more distant electron spins of the radicals. Furthermore, even a completely random distribution of radicals has regions with high local concentration and regions of low local concentration. Thus there is always some inequivalence between radicals in the sample that may manifest itself as a distribution of relaxation rates and result in nonexponential relaxation.

The nonexponential shape of the relaxation curve may aid in discovering the mechanisms and processes responsible for spin relaxation. For example, cross-relaxation to a rapidly relaxing center has a characteristic $\exp(-kT^{1/2})$ dependence (k characterizes the rate and T is time) (3, 12, 13). Also, the dependence (or independence) of spin relaxation on orientation can serve as another criterion for choosing between possible relaxation models.

One danger of having several inequivalent radical species in the same sample is that they lead to the subtle and unintentional selection of a nonrepresentative subpopulation of spins. A relaxation measurement can involve long delays between pulses during which extensive relaxation occurs. For instance, an electron dipolar field measurement using a stimulated echo sequence with a large interpulse spacing can allow extensive spin–lattice relaxation during the sequence. If that sample has a distribution of T_1 values, radicals with long T_1 values are preferentially selected. If, as one might expect, the dipolar field is correlated with T_1, the measured dipolar field will not represent most radicals in the sample. Ichikawa and Yoshida (14) showed that selection of radicals based on T_1 can produce strong resolution enhancement in field-swept spectra from the alkyl radical in γ-irradiated n-hexane single crystals. The implication is that the radicals with large inhomogeneous broadening have shorter T_1

values. In heterogeneous samples such as coal, careful consideration must be given to the possibility that relaxation measurements select non-representative populations of radicals.

Data Analysis

The raw data from a relaxation study consist of sets of relaxation curves recorded under different experimental conditions. These must be reduced to a set of rate parameters for comparison to predictions from various relaxation models. If the relaxation curves are pure exponential decays or the sum of a few exponentials, many methods can be used to extract the rate parameters from the relaxation curves. The task becomes more difficult if the relaxation curves are nonexponential because of a continuous distribution of relaxation rates. We want to characterize the decay by an easily obtained parameter, but we also want to compare that parameter with the predictions of relaxation models. In this respect, the time it takes for the signal to fall to 1/e of its initial value (or the half-life) is a poor choice if the initial value can not be observed because of spectrometer dead time.

Direct calculation of the theoretical relaxation curve as $<\exp[-k(...)t]>_{(...)}$ (where $<>_{(...)}$ denotes the appropriate average over orientation, spectral position, radical, etc.) for the distribution of relaxation rates $k(...)$ and fitting it to the experimental relaxation curve is time consuming. Sometimes the averaged theoretical decay function has a simple analytical form as in the cross-relaxation example mentioned, so that the analytical form can be fit to the data, but usually it does not. A power series or polynomial expansion of the averaged theoretical decay may not converge to a constant at long times like the experimental curves. However, the cumulant expansion, a mathematical technique popularized by Kubo (15), does provide a simple means of parameterizing the averaged, theoretical relaxation curve for the purpose of rapid fitting. The cumulant expansion has the desirable properties of excellent convergence from the start of relaxation out to some well-defined point in the decay. The expansion also can be made to converge to zero at long times.

The cumulant expansion in the present example is

$$<\exp[-k(...)t]>_{(...)} = \exp\left(\sum_{n=1}^{\infty} \frac{\kappa_n(k)t^n}{n!}\right) \qquad (1)$$

where $\kappa_n(k)$ is the nth cumulant of $k(...)$. The cumulants are closely related to the moments, $\mu_i(k) = <k^i(k)>_{(...)}$, of the distribution of rates

$k(...)$. The first three cumulants are (16)

$$\kappa_1 = \mu_1 \tag{2a}$$
$$\kappa_2 = \mu_2 + \mu_1^2 \tag{2b}$$
$$\kappa_3 = \mu_3 - \mu_2\mu_1 + 2\mu_1^3 \tag{2c}$$

Often $k(...)$ can be broken into the product of two factors, $\nu(.) \cdot f(..)$. The $\nu(.)$ is an overall relaxation velocity or speed factor depending, for example, on temperature and varying from one experimental curve to another. The $f(..)$ is a geometric factor that depends on, for example, radical orientation and local spin concentration, but is the same for a set of relaxation curves. Then equation 1 can be written as

$$<\exp[-k(...)t]>_{(...)} = <\exp[-\nu(.)f(..)t]>_{(.)} = \exp\left[\sum_{n=1}^{\infty} \frac{\kappa_n(f)\nu^n t^n}{n!}\right] \tag{3}$$

where the $\kappa(f)$ are the cumulants of $f(..)$. Convergence of this cumulant expansion to a constant is ensured if the power series in the exponential is truncated after a term with $\kappa_n(f) < 0$. As more terms are included before truncation, the approximation is valid to progressively longer times.

The cumulant expansion for a single mechanism and process contains only one adjustable parameter, ν, for fitting an experimental curve. After an initial calculation of the cumulants, curve fitting is very rapid. The variation of the ν values with temperature, isotopic composition, etc. provides a criterion for discriminating between the possible relaxation mechanisms and processes.

Relaxation Experiments

We now turn to a discussion of relaxation measurements that can be made in solids. The concepts will be kept simple so that the subtle points can be delineated clearly. Three experiments—inversion recovery, two-pulse spin echo, and the "2 + 1" experiment of Kurshev et al. (17)—are examined. These are by no means the best experiments for any particular measurement, but they illustrate the considerations involved in assigning relaxation processes and mechanisms. It is assumed that precautions are made to isolate the one response from the pulse sequence that is desired.

Inversion Recovery. Inversion-recovery or saturation-recovery experiments were one of the first to be tried in pulsed EPR spectroscopy (18, 19). The concept is quite simple. The spin system is inverted or saturated, or at least the populations of the spin levels are perturbed from Boltzmann equilibrium values. The EPR signal is then monitored by using the FID from a second pulse as the populations relax to equilibrium by transferring energy to and from the lattice. Hence the desired process is spin–lattice relaxation or the T_1 in the Bloch equations.

But what actually happens in the experiment? Saturation or inversion is experimentally achievable only over a finite frequency range. This feature is known as "burning a hole" in the spectrum and is an apt description. The hole may be much wider or much narrower than the EPR spectrum. An EPR signal is used to monitor relaxation. That EPR signal is from a finite observation window, which can be wider or narrower than the hole burned in the spectrum. The relaxation curve is the reappearance of the EPR signal within that window, ideally by spin–lattice relaxation processes. But the experiment is sensitive to anything that changes the EPR signal in the observation window. It can include cross-relaxation of a spin in the window with an unperturbed spin outside of the window. Spectral diffusion, the diffusion in frequency space of the EPR frequency of an unsaturated spin outside the window, carries signal into the observation window so that the hole gets shallower but wider. Many processes can increase the signal within the observation window at the expense of signal outside the window, none of which are spin–lattice relaxation. If two spin species (even exchange-coupled pairs) have different spin–lattice relaxation rates, the slower relaxing species can recover by cross-relaxation with the faster relaxing one.

Fortunately, some processes are experimentally distinguishable. If the hole can be broadened so the entire EPR spectrum is inverted or saturated, spectral diffusion and cross-relaxation are reduced or eliminated because there are no longer any unperturbed spins outside the observation window.

Another strategy is to enlarge the observation window and note any changes inside or outside the hole. If the hole gets wider, spectral diffusion is taking place. If the entire spectrum outside the hole decreases, cross-relaxation may be occurring and the integrated intensity of the entire spectrum will recover with T_1 (if there is only one radical species).

When we understand how these other processes interfere with the inversion-recovery experiment, diagnostic tests for their presence can be designed. For instance, long and short saturation pulses can be used to burn holes with different shapes into the spectrum. The T_1 will not change, but the influence of competing relaxation processes on the measured relaxation curves may change. No spin–lattice relaxation study is complete without a thorough check for interfering processes.

Two-Pulse Spin Echo. The two-pulse, or Hahn, spin echo uses one pulse to convert S_z into S_+, which then precesses about the applied magnetic field with an amplitude (or phase) of exp $(-i\omega_{EPR}\tau)$ within the plane normal to the magnetic field. After a time τ, a second pulse is applied that, in principle, "refocuses" the magnetization. The pulse is supposed to perform a phase conjugation on the spins so that a particular spin with phase of exp $(-i\omega_{EPR}\tau)$ before the pulse is left with a phase of exp $(+i\omega_{EPR}\tau)$. The spin precesses with the same EPR frequency and at time 2τ has a phase of exp $(+i\omega_{EPR}\tau)$ exp $(-i\omega_{EPR}\tau) = 1$. Because all spins have the same phase independent of ω_{EPR}, the sample has a nonzero $<S_+>$, and the echo appears. Any process that destroys the phase of the spin, a T_2 process, decreases the echo and thus produces a decay curve.

Other ways of decreasing the echo amplitude cannot be called T_2 in the sense of the Bloch equations. We have assumed that ω_{EPR} remains constant during the measurement. If it is not constant, then a phase error builds up and prevents the perfect refocusing of the spins. The spectral diffusion discussed earlier is one process that attenuates the echo. Depending on the speed and the amplitude of the changes in EPR frequency caused by spectral diffusion, the two-pulse echo can exhibit a variety of nonexponential decays *(20, 21)*.

The second microwave pulse itself may also be responsible for a change in ω_{EPR}. The microwave pulse can act on all electron spins in the sample. If it flips two spins located near each other in the sample, then it also changes the dipolar interaction that each one feels and thereby changes the EPR frequency. Because this type of spectral diffusion occurs instantly at the second microwave pulse, it is called instantaneous diffusion. The effect of instantaneous diffusion can be modulated by changing the turning angle (strength) of the second microwave pulse. If the pulse flips only a few spins in the sample, there is still an echo (from those spins that are flipped), but the total dipolar field does not change much, and instantaneous diffusion is suppressed *(20, 21)*. This is an apparent relaxation driven by the second microwave pulse.

The microwave pulse can also excite the "forbidden" spin-flip transitions in solids whereby an electron spin and a nuclear spin coupled to it are flipped. When the second pulse excites a spin-flip transition, the EPR frequency changes by an amount related to the hyperfine coupling. This change produces a periodic interference with the refocusing of the echo and results in electron spin-echo envelope modulation (ESEEM) *(22–25)*. Depending on the distribution of hyperfine couplings involved, the ESEEM can look like an apparent relaxation decay or a periodic modulation of the echo amplitude. The ESEEM does depend on the amplitude or width of the second microwave pulse and can be partially suppressed by using long pulses.

The "2 + 1" Sequence. The "2 + 1" sequence *(17)* consists of three microwave pulses. Two of the pulses generate a spin echo that is detected. The time between those two pulses is not varied, in contrast to the two-pulse echo experiment just discussed. This means that the echo always has a constant degree of attenuation from the T_2 decay and from the ESEEM. A relaxation curve is generated by sweeping in time a third microwave pulse (the "+1" pulse) between the other two. The important point in this experiment is that except for very weak, long pulses, none of the spins detected in the echo were flipped by the "+1" pulse. Consequently, the echo from the detected spins can change only as a result of other spins flipped by the "+1" pulse. Normally, this sequence measures only changes in the dipolar field seen by the detected spins and caused by the "+1" pulse acting on the nearby spins. The echo is used only to report on what happens to the other spins that do not contribute to the echo. This point is subtle, but it is the key to understanding the experiment. The spins that are detected in the echo are excited only by the two echo-forming pulses, which are not varied during a measurement. Therefore, the echo must remain constant during the experiment unless there is communication (a dipolar interaction) between the echo-forming spins and the spins pumped by the "+1" pulse.

The change in position of the "+1" pulse means that the change in dipolar field (or instantaneous diffusion) occurs at a different point in the dephasing and rephasing of the echo. Consequently, the echo amplitude changes as the "+1" pulse is moved in the interval between the two echo-forming pulses. Because the amplitude changes or relaxation is driven by the "+1" pulse, this experiment is an example of apparent relaxation. Although this experiment has not been used extensively, it finally offers a very clean way of measuring dipolar fields in solids without complications from T_2, spectral diffusion, or ESEEM.

One more point should be considered in its application. The echo is detected after a considerable amount of decay has taken place (in order to maximize the "2 + 1" relaxation amplitude). If the sample contains spins with a distribution of T_2 values or spectral diffusion rates, the tendency will be to select the spins with the slowest relaxation. This select population may not be representative of the dipolar fields in the rest of the spin system.

Conclusion

Relaxation studies can provide a unique view of spin systems and the solids that contain them. The growing availability of pulsed EPR spectrometers places relaxation measurements within the realm of possibility for more and more researchers. Compared to CW EPR spectroscopy, relaxation measurements are subject to a new set of considerations, limitations, and pitfalls that must be addressed. The same critical evaluation of experimental data and procedures that is made for other experimental techniques must also be made for EPR relaxation measurements. This discussion has tried to present some of the issues to be encountered.

Acknowledgment

The ideas in this paper have been honed over the years by interactions with many people, particularly with J. R. Norris. Without these interactions, I would not have learned or been able to articulate much of this. I would also like to thank both reviewers of this manuscript for their detailed and thorough comments.

This work was performed under the auspices of the Office of Basic Energy Sciences, Division of Chemical Sciences, U.S. Department of Energy, under Contract No. W–31–109–Eng–38.

References

1. Carrington, A.; McLachlan, A. D. *Introduction to Magnetic Resonance;* Harper and Row: New York, 1967.
2. Wertz, J. E.; Bolton, J. R. *Electron Spin Resonance: Elementary Theory and Practical Applications;* McGraw-Hill: New York, 1972.
3. Bowman, M. K. *Time Domain Electron Spin Resonance;* Kevan, L.; Schwartz, R. N., Eds.; Wiley: New York, 1979; pp 67–105.
4. Bloembergen, N.; Wang, S. *Phys. Rev.* **1954,** *93,* 72–83.
5. Schweiger, A. In *Modern Pulsed and Continuous-Wave Electron Spin Resonance;* Kevan, L.; Bowman, M. K., Eds.; Wiley: New York, 1990; pp 43–118.
6. Bowman, M. K. In *Modern Pulsed and Continuous-Wave Electron Spin Resonance;* Kevan, L.; Bowman, M. K., Eds.; Wiley: New York, 1990; pp 1–42.
7. Gorcester, J.; Millhauser, G. L.; Freed, J. H. In *Modern Pulsed and Continuous-Wave Electron Spin Resonance;* Kevan, L.; Bowman, M. K., Eds.; Wiley: New York, 1990; pp 119–194.
8. Jaynes, E. T. *Phys. Rev.* **1955,** *98,* 1099–1105.
9. Bloom, A. L. *Phys. Rev.* **1955,** *98,* 1105–1111.
10. Mims, W. B. In *Electron Paramagnetic Resonance;* Geschwind, S., Ed.; Plenum: New York, 1972; pp 263–351.
11. Schwartz, R. N.; Jones, L. L.; Bowman, M. K. *J. Phys. Chem.* **1979,** *83,* 3429–3434.
12. Förster, T. *Z. Naturforsch.* **1949,** *4a,* 321.
13. Salikhov, K. M.; Semenov, A. G.; Tsvetkov, Yu. D. *Electron Spin Echoes and Their Applications;* Nauk: Novosibirsk, U.S.S.R., 1977.
14. Ichikawa, T.; Yoshida, H. *J. Phys. Chem.* **1988,** *92,* 5684–5688.
15. Kubo, R. *J. Phys. Soc. Jpn.* **1962,** *17,* 1100–1120.
16. Kendall, M. G. *The Advanced Theory of Statistics;* Charles Griffin: London, 1947.

17. Kurshev, V. K.; Raitsimring, A. M.; Tsvetkov, Yu. D. *J. Magn. Reson.* **1989**, *81*, 441–454.
18. Blume, R. J. *Phys. Rev.* **1958**, *109*, 1867.
19. Kaplan, D. E.; Browne, M. E.; Cowen, J. A. *Rev. Sci. Instrum.* **1961**, *32*, 1182.
20. Brown, I. M. *Time Domain Electron Spin Resonance;* Kevan, L.; Schwartz, R. N., Eds.; Wiley: New York, 1979; pp 195–229.
21. Salikhov, K. M.; Tsvetkov, Yu. D. *Time Domain Electron Spin Resonance;* Kevan, L.; Schwartz, R. N., Eds.; Wiley: New York, 1979; pp 231–277.
22. Rowan, L. G.; Hahn, E. L.; Mims, W. B. *Phys. Rev.* **1965**, *137*, A61.
23. Mims, W. B. *Phys. Rev. B.* **1972**, *5*, 2409.
24. Mims, W. B. *Phys. Rev. B.* **1972**, *6*, 3543
25. Bowman, M. K.; Massoth, R. J. In *Electronic Magnetic Resonance of the Solid State;* Weil, J. A.; Bowman, M. K.; Morton, J. R.; Preston, K. F., Eds.; Canadian Society of Chemistry: Ottawa, Canada, 1987; pp 99–110.

RECEIVED for review December 26, 1990. ACCEPTED revised manuscript July 29, 1991.

6

Multifrequency Electron Paramagnetic Resonance Spectroscopy

R. L. Belford[1] and R. B. Clarkson[2]

[1]School of Chemical Sciences, and [2]College of Veterinary Medicine, Illinois EPR Research Center, University of Illinois at Urbana–Champaign, Urbana, IL 61801

> *Over the past decade, the advantages of performing electron paramagnetic resonance (EPR) experiments outside of the standard 9- to 35-GHz range have become increasingly exploited. For example, low-frequency (<9 GHz) EPR experiments may lead to increased spectral resolution or enhance features due to "forbidden" transitions. High-frequency (>35 GHz) experiments also may result in enhanced resolution or allow for more accurate determinations of zero-field splittings in high-spin systems or details of motion not observable at lower frequencies. The selective review presented here emphasizes applications of low- and high-frequency EPR spectroscopy with special emphasis on multifrequency applications in coal research.*

A RESONANCE PHENOMENON such as electron paramagnetic resonance (EPR) is not restricted to a particular frequency, but can be accomplished at any field–frequency combination that satisfies the resonance condition. However, for a great many years this flexibility largely went unexploited. During the development of EPR spectroscopy, the early literature contained many descriptions of spectrometer systems operating outside the 9- to 35-GHz range (*1–8*). However, practical considerations, including good sensitivity, the convenience of using 3-cm wavelength radiation and magnetic fields of a few thousand gauss, and the availability of equipment as a result of wartime technology, quickly made X-band (9–10 GHz) the frequency of choice. With the advent of high-quality commercial EPR spectrometers, practitioners of the technique virtually ignored the existence of frequencies outside the X-band (and to a far lesser extent, Q-band [34–36 GHz]) range.

But recently, stimulated by new technology and new motivations, the development of spectrometer systems outside this range, both high and low, is undergoing a renaissance. This chapter is a selective review of these developments, especially those of the past decade. Multifrequency EPR spectroscopy can be very useful for samples ranging from liquids and frozen solutions to single crystals. However, we focus our attention here on studies of polycrystalline and amorphous samples, including organic and inorganic species often found in coal, which demonstrate some unique advantages of utilizing EPR spectroscopy at very low and very high frequencies.

Low-Frequency EPR Spectroscopy

In this section, we discuss the utility of low-frequency (sub-X-band) EPR spectroscopy in the solid state as a part of the multifrequency approach. Although the discussion of factors influencing the spectra will be general, copper complexes will be most commonly employed as examples, because these complexes have accounted for many of the applications of low-frequency EPR spectroscopy to date.

The consequences and benefits of using low-frequency EPR spectroscopy will be discussed in five sections. First, we investigate how frequency plays a part in determining the line width and line shape of a given sample. Second, we show the influence of state mixing as a cause of breakdown in the high-field selection rules and how this leads to the appearance of "forbidden" ($\Delta M_S = 1$, $\Delta M_I \neq 0$, where M_S and M_I are the magnetic quantum numbers for electronic and nuclear spins, respectively) transitions. This discussion also will include examples of low-frequency pulsed EPR spectroscopy, where more extensive state mixing at lower frequencies gives rise to deeper electron spin-echo envelope modulation (ESEEM) from coal. Third, low frequency will be discussed in the context of a multifrequency approach for the determination of spin Hamiltonian parameters from spectral simulation via computer. A fourth section will describe developments in instrumentation and analysis that have made sub-X-band frequencies a practical extension of the multifrequency domain. Finally, we call attention to an important low-frequency special technique, zero-field (frequency-swept) EPR spectroscopy, in the fifth section.

Line-Shape and Line-Width Considerations. As the frequency decreases, the electron Zeeman splitting diminishes, and anisotropic *g*-matrix components collapse toward a common center.

This effect is observed in Figure 1, where a powder spectrum of bis(2,4-pentanedionato)palladium(II) [^{63}Cu–Pd(acac)$_2$ (acac is acetylace-

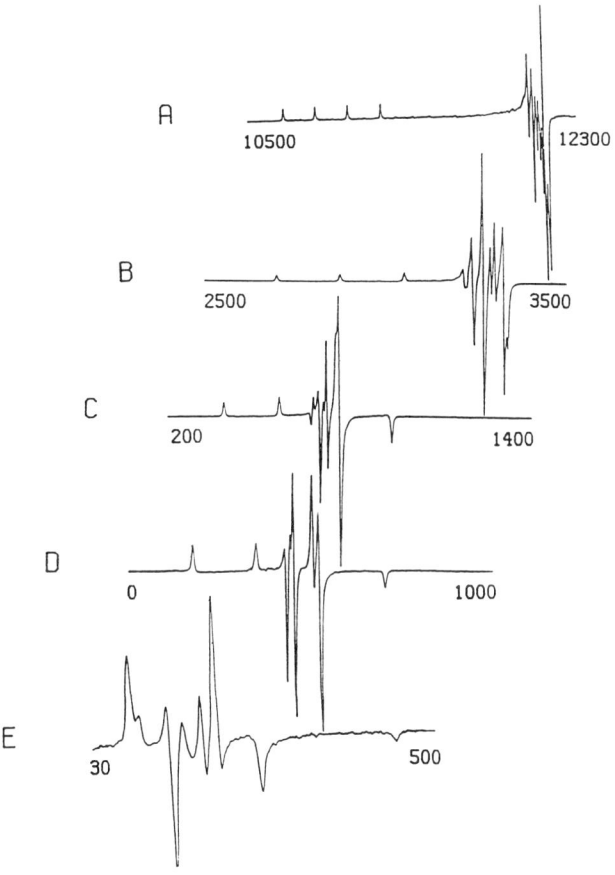

Figure 1. EPR spectra of ^{63}Cu doped into ^{63}Cu–Pd(acac)$_2$ powder at various frequencies. Field values are indicated for the beginning and end of the scan in gauss (1 T = 10^4 G) on each spectrum. All spectra were run at approximately 100 K with appropriate settings of power and modulation amplitude. Spectrum E required time averaging. Frequencies: A, 34.7803 GHz; B, 9.3760 GHz; C, 2.3899 GHz; D, 1.3931 GHz; and E, 560.4 MHz. Spectra A, B, and D were run at the Illinois ESR Research Center. Spectra C and E were run at the National Biomedical ESR Center in Milwaukee, WI.

tonate)] is shown at frequencies ranging from 600 MHz to Q-band. A system with a small g-anisotropy may be simplified and appear nearly isotropic when the frequency becomes low. This result may make it easier to determine field-independent effects such as the hyperfine couplings. When the spectrometer frequency becomes so low that the hyperfine splittings at zero field exceed the spectrometer frequency, features will be lost, just as the large electronic zero-field splittings from $S > 1/2$ systems may preclude the observation of certain transitions. This condition occurs in Figure 2, where a VO(acac)$_2$ solution spectrum is compared at X-band and at 600 MHz. A simple, isotropic eight-line pattern is expected from the $S = 1/2$, $I = 7/2$ motionally averaged system. At X-band this pattern is observed, but at 600 MHz only three lines appear. This result at 600 MHz is a consequence of the large zero-field hyperfine splitting in this compound. At zero field, the system consists of two levels, $F = 4$ and $F = 3$ (where $F = S + I, S + I - 1, \ldots$; S and I are electronic and nuclear spin quantum numbers, respectively), split by $4A_{ISO}/h$ (A_{ISO} is the isotropic hyperfine coupling factor and h is the Planck constant), or about 1200 MHz, which is twice the spectrometer frequency!

Small variations in local-site geometry may cause inhomogeneous line broadening associated with a distribution of g-factors, commonly called g strain. A similar effect is observed with the A values (A strain, where A is the hyperfine coupling factor). The g strain causes an increase in line width with increasing frequency. Therefore, resolution, as defined by the number of spectral features that can actually be distinguished, may actually improve with a decrease in frequency. Abdrakhmanov and Ivanova (9) have taken advantage of this line narrowing at 1.2 GHz for the determination of nuclear quadrupole couplings in an alkali silicate glass containing 1% copper oxide.

Froncisz and Hyde (10) have found that line widths of the $M_I = -3/2$ and $-1/2$ hyperfine lines in the g parallel region of square-planar copper complexes are frequency-dependent and go through a minimum at frequency ranges of 6–8 and 1–3 GHz, respectively. Froncisz et al. (11) exploited this phenomenon, which is caused by a correlation between A and g strains, to determine the number of nitrogen donor atoms bound to copper in complexes such as Cu transferrin. An S-band (2–4 GHz) spectrum of this compound, displayed in Figure 3, clearly shows a triplet of peaks on the $M_I = -1/2$ transition in the g parallel region, indicating the attachment of a single nitrogen to copper. This statement could not be made unambiguously on the basis of X-band data alone.

Line widths in coal show a frequency dependence that may be useful in understanding coal structure. In a recent multifrequency EPR study (9, 35, and 140 GHz), Bresgunov et al. (12) showed that the line width of an Argonne Premium coal sample increased in a nonlinear fashion, as predicted by a model that included both unresolved nuclear hyperfine couplings and random *g* dispersion. In another study, Clarkson et al. *(13)* showed that the line width of evacuated samples of fusinite, measured at 1, 9, 35, 96, and 250 GHz, displayed no change between 1 and 9 GHz, followed by a linear increase in ΔB_{pp} (the peak-to-peak line width in the first-derivative EPR spectrum), indicative of the spin-exchange narrowing condition (eq 1) that dominates the line shape in this maceral *(13)*.

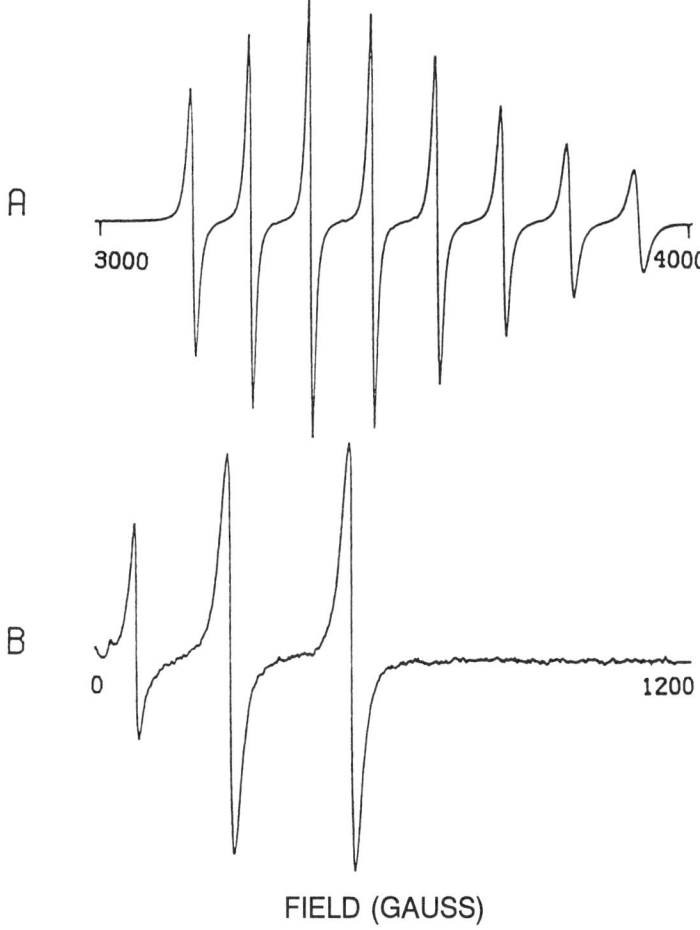

Figure 2. Room-temperature, solution EPR spectra of VO(acac)$_2$ dissolved in a 1:1 mixture of toluene and chloroform showing loss of hyperfine transitions at very low frequency. Field values are indicated for the beginning and end of the scan in gauss on each spectrum. Frequencies: A, 9.7645 GHz; and B, 594.8 MHz. Spectrum A was run at the Illinois EPR Research Center; B was run at the National Biomedical ESR Center.

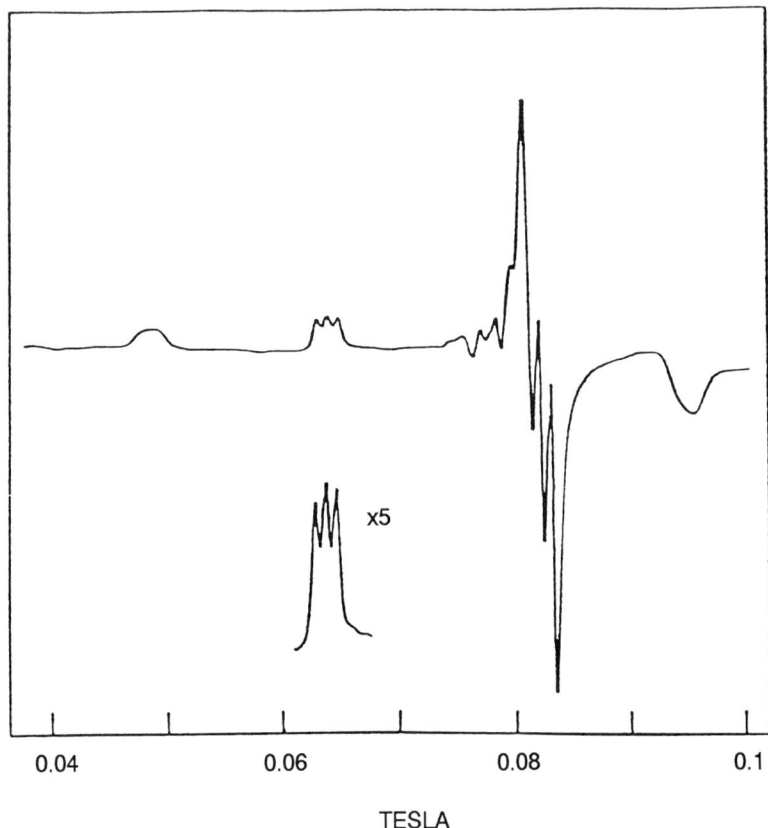

Figure 3. S-band spectrum (2.295 GHz) of dicupric transferrin enriched in ^{63}Cu, recorded in frozen solution at 100 K. (Reproduced with permission from reference 11. Copyright 1982.)

$$1/\tau_{ij} \gg |g_i - g_j| \beta_e B_o \qquad (1)$$

where $1/\tau_{ij}$ is the rate of exchange between species i and j; g_i and g_j are the g-factors of species i and j, respectively; β_e is the Bohr magneton; and B_o is the static magnetic field intensity. When the inequality is obeyed, a single, narrow resonance line at the mean g value $1/2(g_i + g_j)$ is observed, and it reduces the line-width effect of variations in the g values of species contributing to the EPR resonance. Above 9 GHz, the exchange interaction is not strong enough to fully satisfy the inequality in eq 1, and g dispersion begins to manifest itself in an increase in line width that is linear in B_o, as seen in Figure 4.

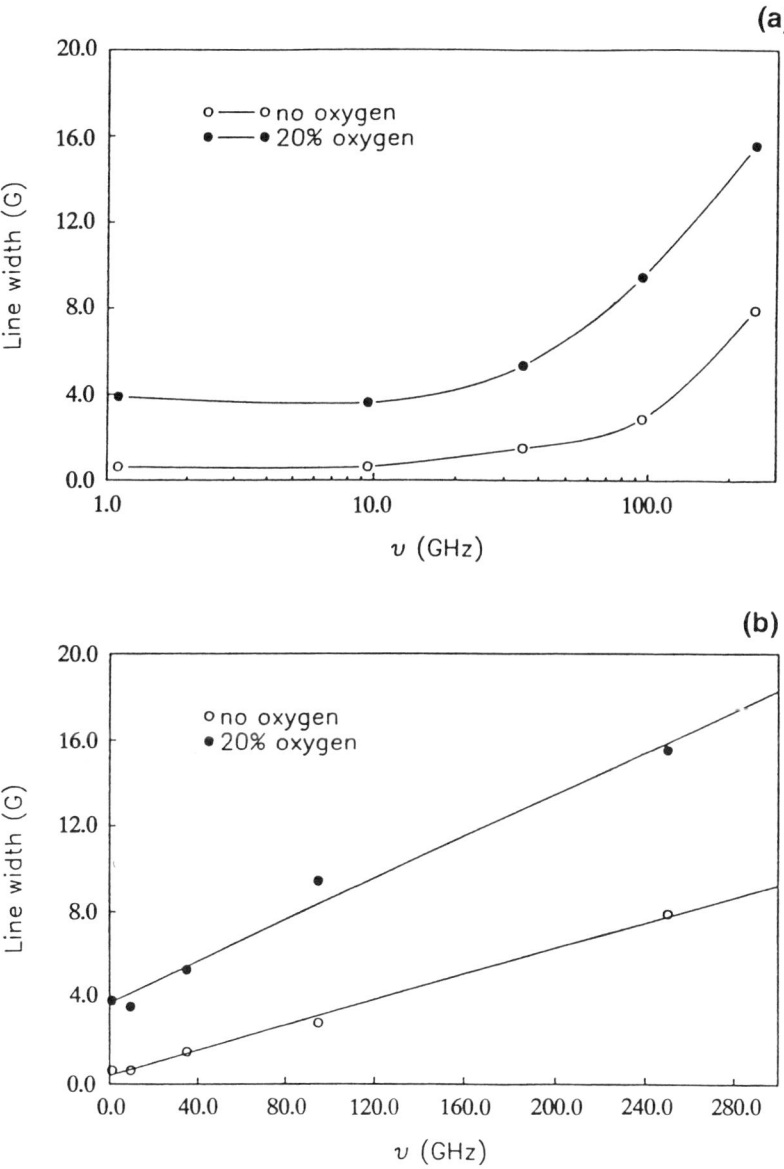

Figure 4. EPR line width (B_{pp}) of fusinite as a function of frequency from 1 to 250 GHz. (a) Semilogarithmic plot illustrating the onset of line broadening above 9 GHz. (b) Conventional plot demonstrating linear dependence of B_{pp} on frequency (or B_o). 250-GHz data were obtained at Cornell University, courtesy of Jack Freed.

State Mixing. The spin Hamiltonian for an EPR active system with a single spin ($S = 1/2$) and a nonzero nuclear moment ($I \neq 0$) is given in eq 2:

$$H = \beta_e \mathbf{S} \cdot \mathbf{g} \cdot \mathbf{B}_o + \mathbf{S} \cdot \mathbf{A} \cdot \mathbf{I} - g_n \beta_n \mathbf{I} \cdot \mathbf{B}_o + \mathrm{QD}\left[I_z^2 - \frac{I(I+1)}{3} \right] \quad (2)$$

where **S** is the electronic spin angular momentum vector (matrix); **g** is the matrix of g factors; $\mathbf{B}_o$ is the static magnetic field vector; **A** is the matrix of A values, or hyperfine coupling matrix; **I** is the nuclear spin angular momentum vector; g_n is the nuclear g factor (gyromagnetic ratio); β_n is the nuclear magneton; and successive terms represent the electronic Zeeman interaction, the electron–nuclear hyperfine interaction, the nuclear Zeeman interaction, and the nuclear quadrupole interaction (QD = $3P_{zz}/2$, where P_{zz} is the largest element of the diagonalized nuclear quadrupole coupling tensor), which is nonzero only for those systems in which $I > 1/2$.

State mixing can become significant when the interactions represented by two or more of these terms become of similar magnitude (*14*). The zero-field terms (electron–nuclear hyperfine and nuclear quadrupole) are invariant for a given system, so the mixing may be adjusted through variation of the field-dependent terms (electron and nuclear Zeeman) by obtaining spectra at different field or frequency values.

Table I compares the magnitudes of the spin Hamiltonian terms at selected field values, using parameters typical of a square-planar copper complex. At fields corresponding to low frequencies, the electron Zeeman interaction becomes comparable to the hyperfine interaction along the parallel axis of the crystallite. In this case, there is mixing of the electronic states; that is, each state with $M_S = +1/2$ also contains a small component that has $M_S = -1/2$ and vice versa. Although electron–nuclear hyperfine interactions in coal are not this strong [typical proton hyperfine couplings are in the range $A/h = 0.1$–30 MHz (*15*)], state mixing is much more pronounced at lower field–frequency values.

Altman (*16*) has done spectral simulations of a powder EPR spectrum using parameters for a typical square-planar copper complex in which the total intensity is compared to the forbidden intensity as a function of frequency. The results are compared at 8.0 and 1.0 GHz in Figure 5. The spectrum at 8.0 GHz is essentially that of the primary transitions only, whereas at 1.0 GHz some features are solely due to nonprimary intensity. In this example, the nuclear quadrupole coupling has been set to

Table I. Spin Hamiltonian Terms at Various Fields

Field (T)	$g_\parallel \beta_e B_o/h$	$g_\perp \beta_e B_o/h$	$g_n \beta_n B/h$
0.02	630	570	0.2
0.05	600	1400	0.6
0.10	3100	2900	1.2
0.32	10,100	9200	3.6
1.25	39,400	35,900	10.4

NOTE: All values are given in megahertz. In all cases, $A_\parallel/h = 500$ MHz, $A_\perp/h = 70$ MHz, and $QD/h = 15$ MHz. The data were generated from the following parameters, typical of a square-planar copper(II) complex: $g_\parallel = 2.25$, $g_\perp = 2.05$, and $g_n = 1.4804$.

zero, which is not normally a condition typical of copper complexes. Had the quadrupole coupling been nonzero, it would have increased the intensity of the forbidden spectrum at both frequencies.

Electron state mixing is derived from the perpendicular component of the ($\hat{z}$) hyperfine matrix in an anisotropic system as follows. Consider the spin Hamiltonian with B_o along the principal axis in an approximately uniaxial system. Without writing the nuclear Zeeman and quadrupole terms, we have

$$H = g_z \beta_e S_z B_o + A_\parallel S_z I_z + A_\perp (S_x I_x + S_y I_y) \qquad (3)$$

Consider the final term to be a perturbation on the rest of the system. Rewriting it in terms of the raising and lowering operators, S_+, S_-, I_+, and I_-, then applying perturbation theory, we find that the first-order correction to a state $|-1/2, M_I\rangle$ is given by

$$\frac{I(I+1) - M_I(M_I - 1)}{E(\tfrac{1}{2}, M_I - 1) - E(-\tfrac{1}{2}, M_I)} \; |\tfrac{1}{2}, M_I - 1\rangle \qquad (4)$$

This correction increases as the difference in energy between the two states (i.e., the field–frequency) decreases. The effect has been used by Rothenberger, et al. (17) in a novel experiment with $B_o \parallel B_1$ ($_1$ is the oscillating magnetic field vector) at L-band (1–2 GHz) as a means to determine nuclear quadrupole coupling in the spectra of a copper powder, ^{63}Cu–Pd(acac)$_2$. The low field–frequency serves, as shown, to mix into

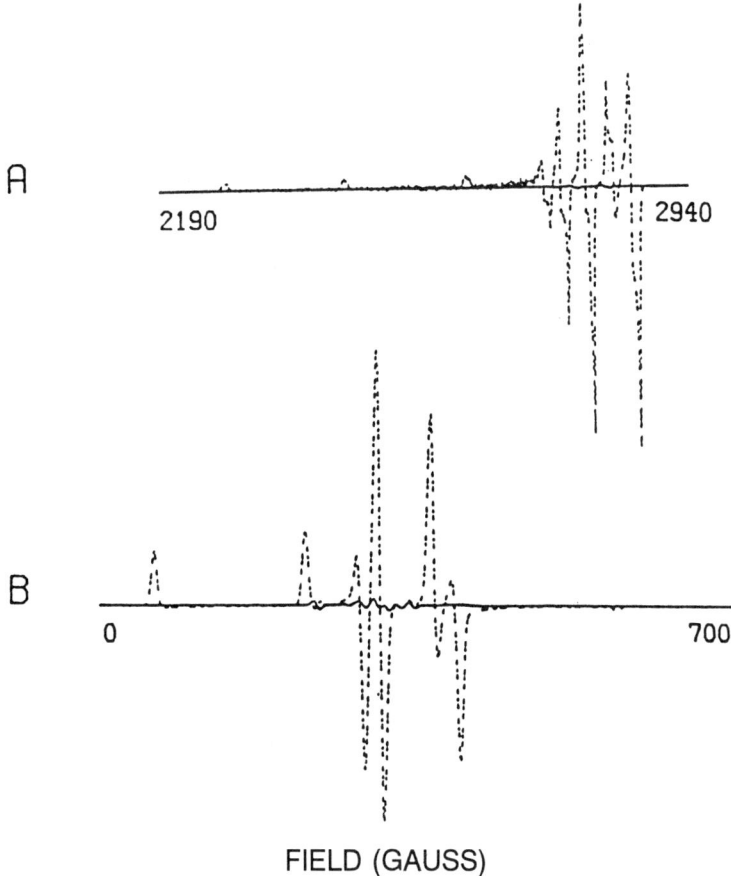

Figure 5. Comparison of simulated EPR spectra at (A) 8.0 and (B) 1.0 GHz. Each spectrum shows both the total intensity (dashed line) and the forbidden intensity (solid line) on the same scale. The spin Hamiltonian parameters were those of an idealized uniaxial system with no nuclear quadrupole interaction and $g_{\parallel} = 2.25$, $g_{\perp} = 2.05$, $A_{\parallel}/h = -540$ MHz, and $A_{\perp}/h = -84$ MHz. At 8.0 GHz, none of the features are due mostly to nonprimary intensity, but at 1.0 GHz one feature is due solely to forbidden intensity.

each state a portion of the state in the other electron-spin manifold in which the nuclear state differs by one unit of angular momentum. The parallel oscillating field ($\mathbf{B}_o \parallel \mathbf{B}_1$) alters the selection rules such that, along the z axis, the only allowed transitions are those in which no net change in angular momentum results. In this way, it connects exactly those states that have been mixed by electron-state mixing at low field.

Under these conditions, the secondary transitions, those with $\Delta M_S = \pm 1$, $\Delta M_I = \mp 1$, are rigorously allowed and can be observed directly. The nuclear quadrupole coupling information can be extracted directly from the resulting spectrum.

Enhanced state mixing at low field–frequency values also creates deeper electron spin-echo envelope modulation (ESEEM) in electron spin-echo experiments (18). The ESEEM effect results when pulsed microwave fields create coherences between one spin state and two others that differ in M_I, for example, coherences between $|-1/2, 1/2>$ and states $|1/2, 1/2>$, and $|1/2, -1/2>$ in an $S = 1/2$, $I = 1/2$ system. These two "branching transitions" lead to a modulation of the spin-echo amplitude as a function of interpulse spacing, and an analysis of the ESEEM pattern can reveal details of weaker electron–nuclear coupling. State mixing increases the probability that forbidden transitions involving $\Delta M_S = \pm 1$ and $\Delta M_I = \mp 1$ can occur. Enhancement of ESEEM depth at lower field–frequency values has been reported (18) in coals; this enhancement allows a more detailed examination of very weak hyperfine interactions, as illustrated in Figure 6.

Spectral Simulation. Computer simulation is one of the most powerful and useful methods for extracting spin Hamiltonian parameters from EPR spectra. For complicated systems, such as transition metal complexes in disordered (powder or frozen solution) states, simulation via computer may be the only practical means of analyzing the data. The EPR spectra of organic radicals in coal never exhibit the magnitude of g and A anisotropies seen in transition metal systems, but the need for spectral simulation is just as great, particularly in light of the usually heterogeneous nature of the samples. In the simulation process, parameters of the spin Hamiltonian are extracted from the line shape by varying trial parameters until the simulation matches the actual spectrum. Unfortunately, the process does not always guarantee that the parameters that provide the match are necessarily correct. A higher level of confidence may be achieved by taking spectra, and simulating, at more than one frequency. The basic effects of frequency on EPR spectra are those that might be expected. The field-dependent terms should dominate at high frequency, and the field-independent terms should become more important at low frequency. That is, usually the g-matrix will be best determined from simulation at as high a frequency as is practical (e.g., 35 GHz or higher), and the hyperfine and quadrupole coupling will be better revealed at lower frequency.

In addition, there are less obvious and predictable effects. As the frequency is varied, certain features may reinforce or cancel each other. For-

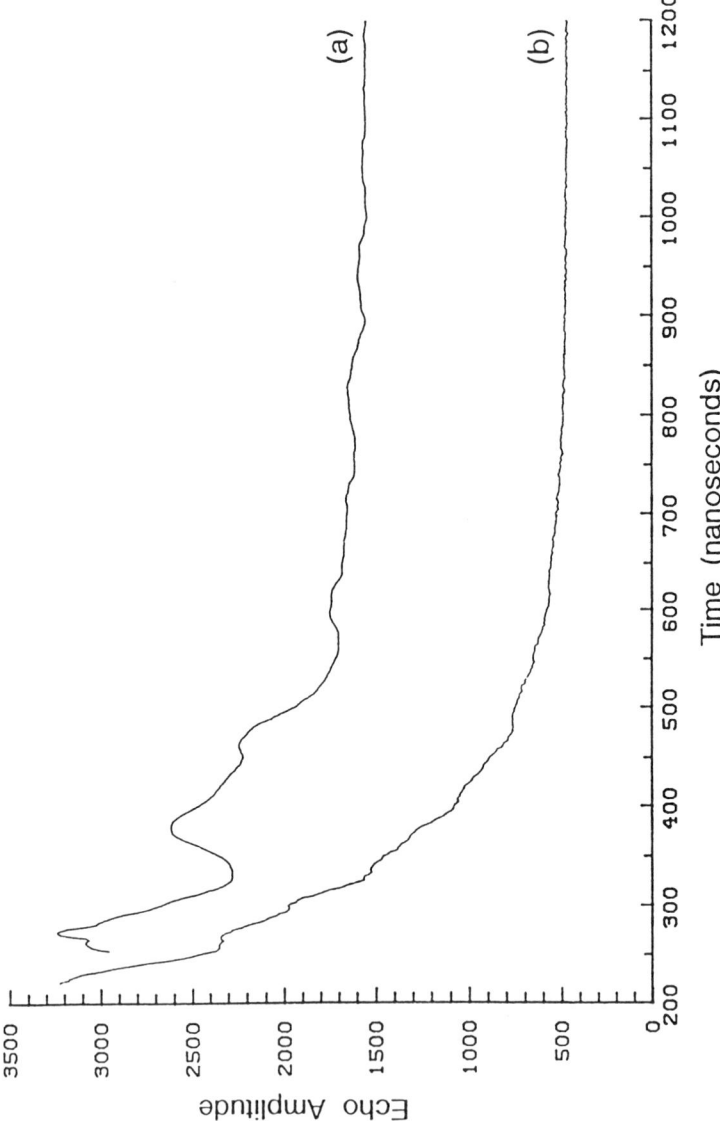

Figure 6. Comparison of vitrinite two-pulse (Hahn echo) ESEEM patterns at (a) S-band (3 GHz) and (b) X-band (9 GHz) frequencies. Experiments were performed at 298 K. Echo amplitude units are arbitrary. (Reproduced with permission from reference 18. Copyright 1989.)

bidden transitions may take on or lose intensity, with consequences as minor as a subtle change in shape or intensity of a feature or as dramatic as the appearance or loss of a peak. Off-axis extrema may develop and add "extra" lines to the spectrum (19). Slight x–y anisotropy in the g- or A-matrices or noncoincidence between the major axes of the g- and A-matrices may severely complicate the perpendicular region. Multiple species with slightly different g- and A-matrices may superimpose spectral features. In these circumstances a multifrequency analysis will often be the most useful. Ideally, the best final set of experimental parameters results from selecting each variable from a simulation that is most sensitive to that variable.

Instrumentation and Methodology. Although the instrumentation requirements at low frequency differ little from those at X-band, and in some respects are simpler, developments during the past decade have made low frequency a highly practical part of a multifrequency EPR analysis. Perhaps the most significant is novel microwave resonator design, particularly the development of the loop-gap resonator as a replacement for the more conventional cavity resonator. Cavity size is a primary concern at low frequencies: The dimensions of conventional cavities are on the order of the source wavelength. Such dimensions are perfectly acceptable at X-band, where the wavelength is a convenient 3 cm. However, setting up a rectangular TE_{102} mode cavity at a frequency of 1 GHz would require dimensions of nearly 30 cm! (TE denotes transverse electric, and 102 refers to a particular standing wave pattern, as defined, e.g., by Poole in reference 20.) Abdrakhmanov and Ivanova (9) addressed this problem and proposed several possible solutions.

However, the most attractive alternative to date has been the loop-gap resonator, such as that shown in Figure 7 (21). It has been described as a simple inductive–capacitive resonant circuit consisting of a conductive cylindrical loop cut by one or more longitudinal slots (22). The dimensions of these devices can be small compared to the wavelength, and they can be adjusted by using the proper materials. Hyde and Froncisz at the National Biomedical ESR Center in Milwaukee, Wisconsin, developed a series of loop-gap resonators covering the range from <1 to 4 GHz, each with a sample access hole that will accommodate a standard 4-mm diameter X-band EPR tube. Thus, the same sample easily can be examined over a range of frequencies spanning a factor of 20.

A second factor in going to low frequencies is the need for improved signal-to-noise (S/N) ratios, because sensitivity is scaled as (frequency)^n, where n is between 1/2 and 9/2, depending on the conditions (20). Because of its small size, the loop-gap resonator is characterized by a high filling factor and high energy density, both of which improve sensitivity.

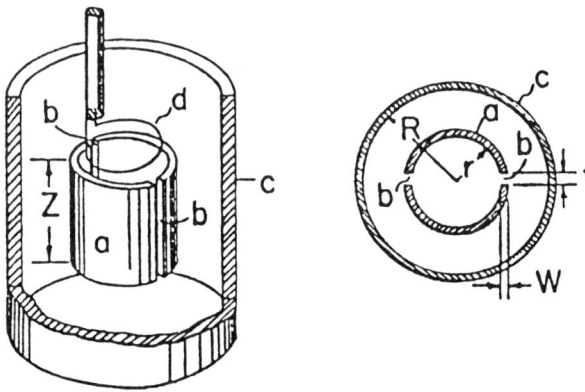

Figure 7. The loop-gap resonator showing the principal components (a, loop; b, gap; c, shield; and d, inductive coupler) and the critical dimensions (Z, resonator length; r, resonator radius; R, total radius; t, gap separation; and W, wall thickness). The microwave (B_1) field in the loop is parallel to the axis of the loop. (Reproduced with permission from reference 20. Copyright 1982.)

With the use of these resonators, Hyde and Froncisz (23) constructed an S-band spectrometer that has sensitivity comparable to an X-band instrument, and Clarkson et al. (18) built a pulsed S-band instrument with good performance characteristics. Other means of increasing sensitivity have included the use of GaAs FET (field-effect transistor) amplification of the microwave signal and the use of computers in data collection for time-averaging purposes.

Just as there exist special considerations in collecting data at low frequencies, so also do there exist considerations in analyzing the data, in particular, via computer simulation. In constructing the simulation, the standard practice is to calculate the energy levels for a system at a given magnetic field. In contrast, what is wanted for comparison to experimental data is a simulation calculating the magnetic field values corresponding to a particular energy (i.e., spectrometer frequency). At X-band or above, this conversion poses little problem, because the relationship between field and frequency is essentially linear. At low frequency, however, this assumption is no longer valid. Problems arise when field-independent terms, such as the hyperfine, become on the same order as the electron Zeeman interaction. (The same problem results at high frequency in high-spin compounds with large electronic zero-field splittings.)

Altman (16), using computer simulations with parameters from typical square-planar copper complexes, showed that minor distortions begin to occur in simulated features farthest from the field position at which the spin Hamiltonian was diagonalized, when working at a frequency of about 4.0 GHz. These distortions become significant at about 2.0 GHz, particularly in the perpendicular region. At 1.0 GHz, the entire simulation, except for those areas close to the value at which the diagonalization took place, bears little resemblance to that theoretically predicted.

Belford et al. (24) proposed two solutions to this problem. One solution, the eigenfield approach, is an exact method for obtaining the desired transition fields directly at any value of frequency. It involves diagonalization of an $n^2 \times n^2$ matrix where n is the number of basis states of the system. Its only disadvantage is the large amount of computation that is required to obtain powder-type spectra. An alternative, which takes less computer time by 1 or more orders of magnitude, is a frequency-shift perturbation technique (24). This technique has been shown to be adequate for the simulation of spectra with hyperfine couplings spanning a spectral width of >500 G at a spectrometer frequency of 0.9 GHz. This technique is illustrated in Figure 8, where a spectrum simulated with the frequency-shift perturbation taken to fourth order is compared to the same simulation taken only to first order.

Zero-Field EPR Spectroscopy. A technique related to the low-frequency EPR methods just discussed is zero-field EPR spectroscopy. In this technique, spectra are obtained not by scanning the applied magnetic field at a fixed frequency, but by scanning the frequency at zero applied field. (In practice, a small Zeeman-modulation field is typically used for detection purposes.) As noted in a review of zero-field EPR spectroscopy (25), frequencies have been scanned over octave bandwidths up to 8 GHz and over waveguide bandwidths up to 40 GHz and higher. The technique is sensitive only to the field-independent nuclear and electron fine structure terms in the spin Hamiltonian, so it is a valuable method for measuring these zero-field interactions. The frequency sweep range must, of course, correspond to energies appropriate to the zero-field interactions of interest. For example, low-frequency scan ranges (0.8–2 GHz) have been used to measure nuclear hyperfine and quadrupole spin Hamiltonian parameters in polycrystalline samples of the vanadyl ion (VO^{2+}) doped into a series of Tutton salts (26, 27); higher frequencies (1–8 GHz) were used to measure electron and nuclear zero-field interactions in Mn^{2+}-doped Tutton salts ($S = 5/2, I = 5/2$) (28). The comprehensive review by Bramley and Strach (25) presents a detailed discussion of zero-field EPR applications and methods.

High-Frequency EPR Spectroscopy

As described in the preceding section, radio frequency, L-band (1–2 GHz), and S-band (2–4 GHz) EPR spectroscopy can often provide an attractive

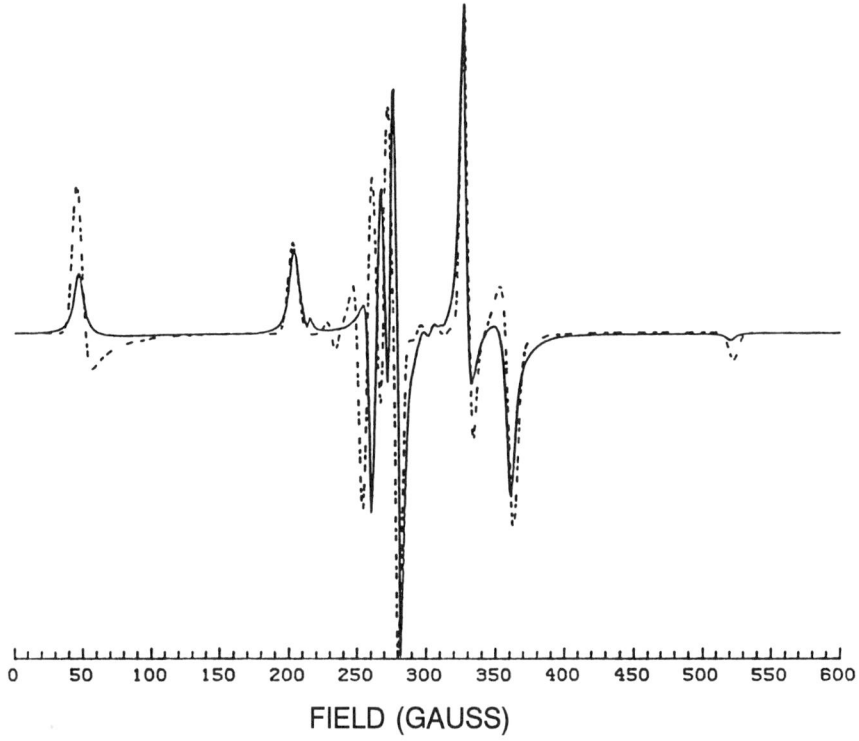

Figure 8. Computer simulations for a powder system at 0.9 GHz using the typical square-planar copper(II) parameters from Table I. The solid-line simulation results from taking the frequency-shift-perturbation routine to first order only. In the dashed-line simulation, the routine is taken to fourth order. Although the line shapes in the fourth-order simulation may be unfamiliar, their veracity has been demonstrated by exploring the effects of choice of field at which the Hamiltonian is diagonalized. The line shape near the field of diagonalization agrees with that generated by the fourth-order correction.

complement to the standard "middle-frequency" X- and Q-band experiments. In this section we discuss applications of high-frequency EPR spectroscopy, where high frequencies are considered to be those well above 35 GHz. At these frequencies, loosely referred to as "millimeter waves", the field-dependent Zeeman terms in the $S = 1/2$ spin Hamiltonian described earlier (eq 2) will dominate the field-independent terms. As a consequence, the increased Zeeman splittings will result in field-expanded spectra and can, in many instances, lead to better spectral resolution. The spectra will often be simple and interpretable without the aid of computer simulations.

High-frequency EPR spectroscopy can also be used to good advantage for the study of high-spin ($S > 1/2$) systems, where single-ion zero-field splitting (ZFS) energies are often the dominant terms in the spin Hamiltonian. By increasing the frequency to the point where the electronic Zeeman term is a substantial perturbation of the ZFS term, or even to the point where the terms are similar in magnitude, the ZFS energies can be determined more accurately.

High-frequency EPR spectroscopy is especially useful in studying coal and other fossil fuels. In these materials, spin-orbit coupling between an unpaired electron and heteroatoms such as oxygen and sulfur contributes to a small anisotropy in the g-matrix, which manifests itself only as line broadening at lower frequencies, but which sometimes can be resolved at higher frequencies to provide a direct, nondestructive route for the study of organic sulfur and oxygen species. Our discussion of some advantages and limitations of high-frequency EPR spectroscopy will be divided into three sections: (1) applications due to enhanced spectral resolution, (2) determinations of high-spin zero field splittings, and (3) high-frequency instrumentation.

Increased Resolution. Perhaps the most obvious reason to use high-frequency EPR spectroscopy is to take advantage of the potential for increased spectral resolution. Assuming that the resonance line widths do not increase significantly at higher frequencies, then for simple systems we expect a linear resolution enhancement with field of features whose positions are determined by electronic or nuclear Zeeman terms. Similar reasoning has led NMR spectroscopists to the limits of superconducting magnet technology (11.4 T and higher) in search of increased resolution and the simplified spectral interpretation that results. Lebedev and co-workers, working at 150 GHz, led the way in the application of high-frequency techniques to chemical systems, and most of their work has been included in comprehensive reviews (*29, 30*).

We will provide here brief, representative examples of the use of high-field EPR spectroscopy for the more accurate determination of g-matrix components, for spectral separation of free radicals with very similar g values, and for a more direct and informative analysis of molecular motion effects on free radical spectra.

Numerous demonstrations of the increased resolution found at higher frequencies have been provided by the groups of Lebedev and Tsvetkov (*29–33*). For this work, they used a home-built EPR spectrometer of fairly conventional design; it employs a reflection cavity, field modulation, a reference arm, and homodyne detection. One of their examples, a comparison of the 10- and 150-GHz spectra for a nitroxide radical in frozen solution, is shown in Figure 9. Each of the canonical g-matrix components is resolved, and this resolution provides a straightforward and direct measurement of the spin Hamiltonian values (g_x = 2.0089(1), g_y = 2.0062(1), g_z = 2.0024(1), $A_z/(g_e\beta_e)$, (^{14}N) = 3.1 mT, where the g- and A-matrices are assumed to be coincident, the x-axis is directed along the N–O bond vector, and the z-axis is in the plane of the N–O pi-bond). The lowest frequency at which such a direct measurement could be made is approximately 35 GHz. Although at lower frequencies the g-matrix components can be determined from single-crystal experiments, the high-frequency approach is more generally applicable.

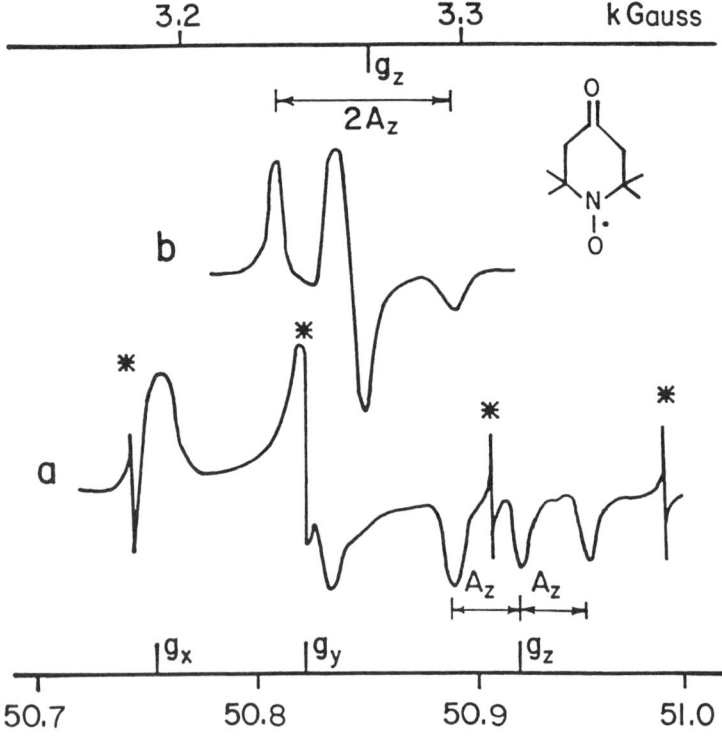

Figure 9. Comparison of (a) 150- and (b) 10-GHz EPR spectra for a nitroxide free radical in frozen toluene. In part a, resonances due to a spectral standard (Mn^{2+} in MgO) are indicated by asterisks. In part b, resonances due to the perpendicular g-matrix components are unresolved in the central feature. (1 T = 10 kG). (Reproduced with permission from reference 30. Copyright 1985.)

High-frequency resolution enhancement is possible only if the line widths do not increase dramatically with frequency, that is, when the contribution of g-strain broadening to the line widths is small (*10*). The groups of Lebedev (*29*) and of Möbius (*34*) demonstrated that for organic free radicals, g-strain broadening is not necessarily a limitation. Paramagnetic transition metal centers, with larger g-matrix anisotropies and general sensitivity to environment, are, however, more likely to exhibit g-strain broadening. Certainly if such effects are apparent at X- or Q-band frequencies, little or no resolution enhancement would be expected at higher frequencies.

The use of the increased resolution at high frequency to separate the spectra of free radicals with very similar g values has been demonstrated by several groups (*29, 34–40*). Figure 10 illustrates spectra at 10 and 150 GHz for a mixture of two similar nitroxide radicals (*30*). In this example, the isotropic g-value difference between the two radicals is 2.40×10^{-4}. Similar resolution gains have been seen in solid mixtures; the technique is particularly useful for separating weak spectral features that would otherwise be buried in the complicated patterns often seen for solid-state mixtures.

The work of Retcofsky (*41*) was the first to show that valuable information about the structure of organic radicals in coal could be obtained by a careful study of the g values calculated from EPR spectra. However, spectra from bituminous coals and anthracite usually displayed only one inhomogeneously broadened resonance; thus, data on g values were limited to a single value for each sample. Retcofsky showed that correlations existed between measured g values and heteroatom concentrations in different samples, and thereby established the link between measured g values and maceral structure. Retcofsky et al. (*42*) also demonstrated the importance of performing the EPR measurements at different frequencies in a study that considered the magnetic-field dependence of EPR line widths in a Pittsburgh (Bruceton) coal. They showed that the line width in this coal was not due in any appreciable way to a distribution of g values from many different radical species, but rather it was the result of unresolved hyperfine structure. In other studies at X- and Q-bands on asphaltenes from North American tar sands (*43*) and on oil shales (*44*), distributions of g values leading to field-dependent line widths were observed, and these observations provided evidence both for the differences in radical content and distribution in coals and other fossil-fuel sources, and for the importance of performing EPR spectroscopy at several frequencies as a means to study spectral features with magnetic-field (Zeeman) dependencies.

Recently, instruments operating at 96 and 140 GHz have been employed to investigate resolution enhancement in coals. Clarkson and co-workers (*45, 46*) developed a W-band (96 GHz) instrument to study the

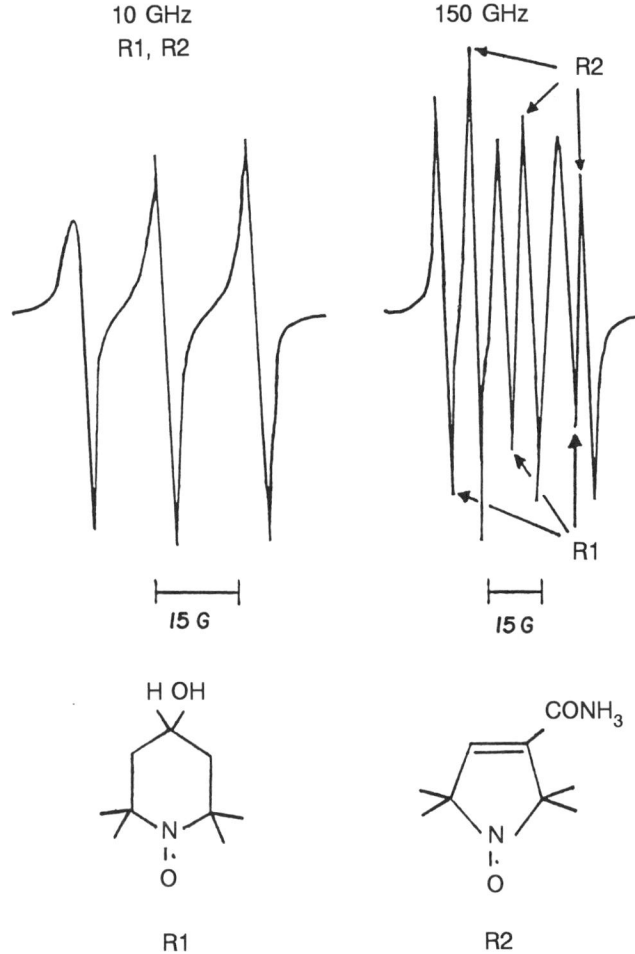

Figure 10. Comparison of 10- and 150-GHz EPR spectra for a solution mixture of radicals R1 and R2. (Reproduced with permission from reference 30. Copyright 1985.)

effects of heteroatoms (O and S) on the EPR spectra of Illinois and Argonne Premium coals. They studied macerals separated from a whole coal by density-gradient centrifugation (DGC) and showed a significant variation in EPR line shape with maceral type and organic sulfur content, as shown in Figure 11. Good correlation between sulfur content and line shape was achieved by means of a simple, two-species model. These studies also demonstrated the great sensitivity to oxygen exhibited by inertinites, which contribute spin-exchange narrowed signals to the overall spectra of many coals.

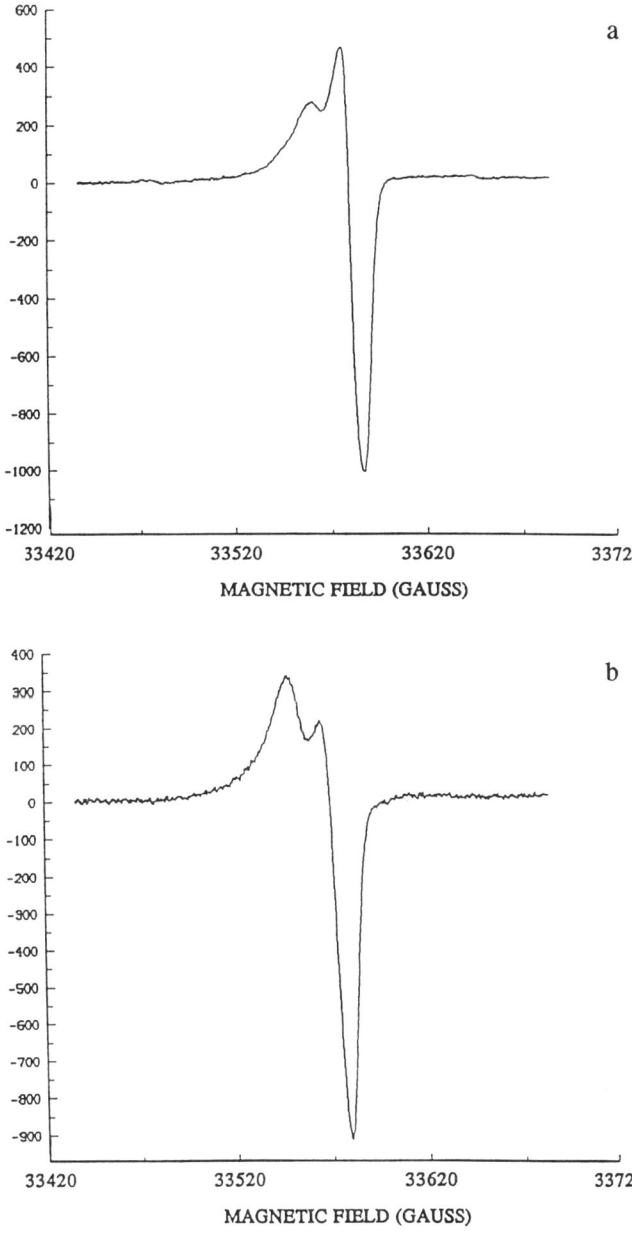

Figure 11. 96-GHz (W-band) EPR spectra of two macerals separated by DGC from an Illinois No. 6 coal: a, vitrinite (2.6% S); and b, sporinite (3.9% S).

Lebedev and co-workers (*12*) examined Argonne Premium coal samples at 140 GHz and observed excellent sensitivity of the spectra to passage effects. They also saw effects of oxygen on the EPR line shapes. These two sets of studies suggest that in many coals, g-strain broadening does increase EPR line widths as a function of magnetic field, but not in a simple linear fashion, and not at a rate fast enough to seriously detract from resolution improvements brought about by higher fields. Recently, experiments comparing 95- and 250-GHz spectra of a separated vitrinite indicate that even greater spectral resolution can be achieved by yet higher field strengths, as seen in Figure 12 (*47*).

Lebedev and co-workers (*29, 48–50*) have provided another interesting application of 150-GHz EPR spectroscopy in their studies of molecular motions. For instance, they examined several nitroxides in liquid paraffin over a temperature range of 220 to 340 K (*48*). Because of the ability to easily distinguish the three g-matrix components at this frequency, they were able to investigate anisotropic rotations of the free rad-

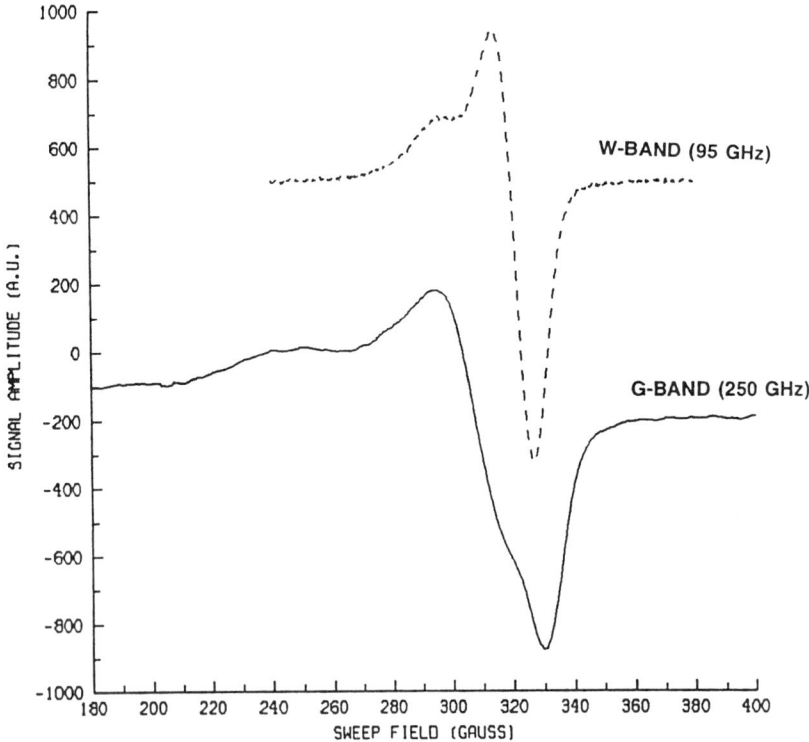

Figure 12. Comparison of EPR spectra of vitrinite separated by DGC from an Illinois No. 6 coal taken at 95 and 250 GHz.

icals by observing anisotropic line broadening as a function of temperature. Ultimately, they were able to correlate these anisotropies to molecular geometries. More examples of molecular motion studies are provided in other articles of Lebedev and co-workers (*29, 49, 50*).

Zero-Field Splitting. For high-spin systems ($S > 1/2$) the dipolar interaction between unpaired electrons results in an additional term (H_{ss}) that must be added to the spin Hamiltonian given earlier (eq 1). This zero-field splitting (ZFS) term,

$$H_{ss} = \mathbf{S}^T \cdot \mathbf{D} \cdot \mathbf{S} \qquad (5)$$

(where **D** is the ZFS matrix) can be fairly large in comparison to the other terms in the spin Hamiltonian. In eq 5 the trace of **D** is zero. The diagonal elements are usually designated in terms of two independent parameters D and E, in which case eq 5 can be written as:

$$H_{ss} = D(S_z^2 - S^2/3) + E(S_x^2 - S_y^2) \qquad (6)$$

The actual zero-field splitting is twice the value of D. In high-spin Fe(III) and Mn(III) compounds this splitting may be as great as 10–25 cm^{-1}. A whole class of compounds of biological interest has ZFS parameters much larger than the customary X- and Q-band frequencies. They include derivatives of hemoproteins (e.g., myoglobin and hemoglobin) and transition metal porphyrins, in addition to many inorganic high-spin transition metal compounds. When the probing microwave frequency is less than the ZFS, D and E cannot be measured directly. For very large zero-field cases only the transition between $M_S = +1/2$ and $M_S = -1/2$ is allowed. The zero-field parameters can be estimated only from second-order effects. One way to do this estimation is to measure the shift in the effective g-factor (g_e) as a function of the magnetic field. The following perturbation approximation formula can be used to determine $g_\perp$ and D:

$$g_\perp^e = 3g_\perp[1 - 2(g_\perp \beta_e B_o/2D)^2] \qquad (7)$$

The observed "g shift" will be greater at higher magnetic fields (i.e., higher frequencies), and will thus allow for a more accurate determination of the spin Hamiltonian parameters. Alpert et al. (*51*) used this method to measure D for methemoglobin. They used a transmission EPR spectrometer and employed a 70-GHz klystron and a series of backward wave oscillators (carcinotrons) to cover a very broad frequency range (70–400 GHz). The fitting of their data to eq 7 over this range led to a value $D/hc = 10.7$ (2) cm^{-1}. Even using the 400-GHz frequency, they were unable to excite a transition to directly measure the zero-field splitting $2D$.

Higher-frequency experiments can provide direct measurements of the ZFS spin Hamiltonian parameters. Brackett et al. (*52*) excited zero-field transitions in transmission spectra of a series of Fe(III) and Mn(III) compounds by using a Fourier transform infrared technique. These measurements were made by employing a Michelson interferometer to record a Fourier transform spectrum while stepping the field of a superconducting magnet. The source in the experiments was the Rayleigh–Jeans region (0–900 GHz) of the black-body spectrum of a high-pressure mercury arc lamp. Because of the low power of the source, a liquid-helium-cooled bolometer detector was necessary. Because of the nature of the experiment, essentially a continuous range of frequency and magnetic fields was available. With this spectrometer, zero-field splittings ranging from $D/hc = 1$ cm^{-1} through 16 cm^{-1} were probed. Champion and Sievers (*53*) used a similar spectrometer to prove the low-lying electronic states of high-spin ($S = 2$) ferrous iron in deoxymyoglobin and deoxyhemoglobin.

Lasers provide another means for obtaining the high frequencies required to cause direct transitions between the zero-field split levels. In collaboration with Robert Wagner at the Naval Research Laboratory, we have used several far-IR gas lasers (158 and 337 GHz) in a transmission EPR experiment to examine the zero-field splittings in a number of high-spin iron compounds. The spectrometer, which makes use of a sweeping superconducting magnet (14 T), is outlined in Figure 13; it is similar to a previously described system used by Wagner and White (*54*) to study the EPR of metal centers in semiconductors. Our experiments, performed at ~4.2 K, made use of neither a sample cavity nor field modulation methods. Even so, the combination of low temperature and high frequency allowed us to record spectra in single scans. A sample spectrum of cytochrome c oxidase is illustrated in Figure 14.

Instrumentation. As noted in the preceding sections, high-frequency EPR spectra have been obtained on a variety of experimental setups, ranging from broad-band, transmission spectrometers to high-sensitivity, reflection-cavity instruments that resemble in design the standard commercial X-band EPR spectrometers. In the following paragraphs, we focus on a few of the technical problems that are encountered when constructing a high-sensitivity, high-frequency EPR spectrometer.

Owing to the increased use of millimeter waves in military radar and communications, it is presently possible to purchase as catalog items near-

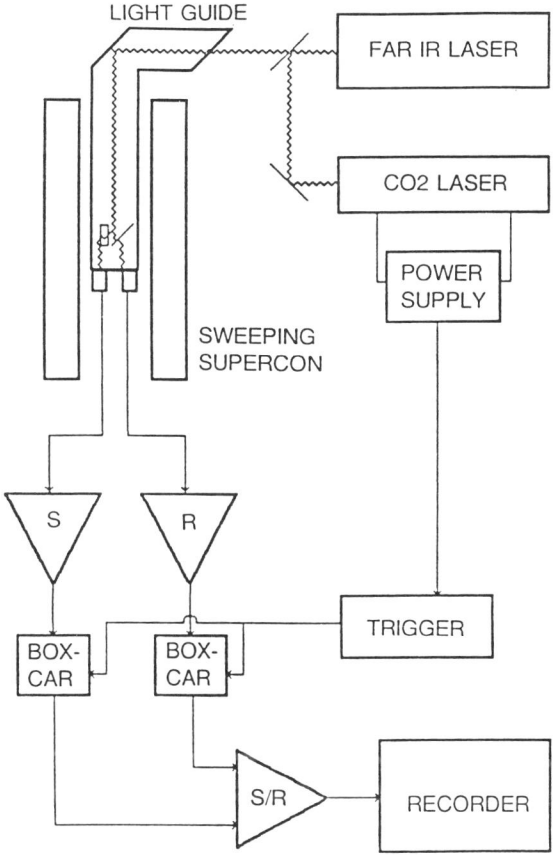

Figure 13. Block schematic for the Wagner far-IR spectrometer (Naval Research Laboratory, Washington, DC).

ly all of the waveguide components necessary to build a high-frequency EPR bridge, particularly up to 140 GHz; some components are available up to 220 GHz. However, significant differences exist between low-frequency and high-frequency waveguides and components. The most important practical difference is the increased attenuation and insertion loss encountered at higher frequencies. This problem is so severe that Möbius, and co-workers (*34*) estimated an overall 90% power loss (source-to-cavity) in their 94-GHz (W-band) spectrometer. A substantial portion of the overall loss occurs in long, straight waveguide sections (−4 dB/m for the commercial WR-10 waveguide at 94 GHz). Möbius made use of oversized Q-band guide (−0.6 dB/m at 94 GHz) to minimize these attenuation losses.

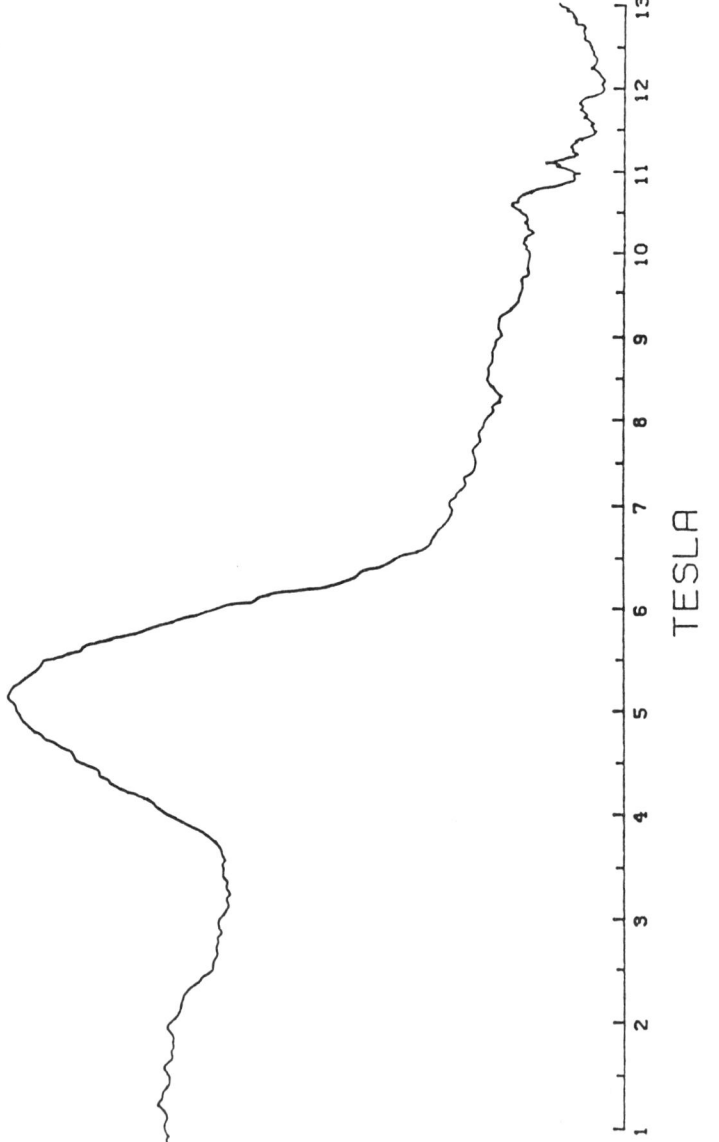

Figure 14. Cytochrome c oxidase transmission EPR spectrum taken with 337-GHz far-IR laser source; the temperature was 4.2 K. The ratio (sample signal to reference signal) is plotted as a function of magnetic field.

One solution to overcome the large losses is to use a relatively high-power source; klystrons and backward wave oscillators have been the standard power sources above 90 GHz. Lower noise solid-state Gunn oscillator sources (50–80 mW) have, however, been used effectively in ~70-GHz (V-band) spectrometers where attenuation losses are somewhat less significant (*55*). Above 90 GHz, commercially available GaAs Gunn oscillators are presently limited in power (e.g., <60 mW at 94 GHz), although advances in InP oscillator technology will undoubtedly increase this value. Solid-state sources may yet be practical above 90 GHz; commercial IMPATT oscillators and amplifiers can presently deliver >200 mW of power at 94 GHz. Our experience working at 96 GHz is that a solid-state source is excellent for studying most coal samples, but it has somewhat less than optimum power for certain special cases, for which klystron sources are available.

The absolute sensitivity of an EPR spectrometer is proportional to the frequency raised to some power ranging between 1/2 and 9/2, depending upon the experimental conditions (*22*). For instance, an improvement in sensitivity from 2×10^{10} spins/G at X-band to 5×10^9 spins/G at Q-band has been reported (*29*). (The sensitivity is specified in terms of the number of spins required to give a S/N ratio of unity when a microwave power of 1 mW is incident on the cavity and the recorder integrating time is 1 s. Dividing this number by the line width in gauss yields the units of spins per gauss.) One indication of the sensitivity of higher frequency spectrometers is that much of the experimental data has been obtained without the use of a sample cavity. Peter (*6*), for example, reported a sensitivity of 2×10^{14} spins/G at 75 GHz using a transmission spectrometer and backward wave oscillator source. This number is fairly typical for spectrometers without a cavity.

High-sensitivity work at high frequencies, however, requires the use of a sample cavity. For example, absolute sensitivities of 5×10^{11} spins/G at 70 GHz (*56*), 1×10^9 spins/G at 94 GHz (*34*), and 4×10^7 spins/G at 150 GHz (*29*) have been reported for spectrometers employing cavities. These sensitivities are limited largely by sample size and shape considerations; the difficulty is in designing a high-Q cavity that does not overly degrade upon sample loading. Among the designs that have been attempted so far are standard TE_{011} cavities (*29, 45, 56*) and Fabry–Perot resonators (*34, 57, 58*). (TE denotes transverse electric, and 011 refers to a particular standing wave pattern) For all of these resonators, sample volumes are necessarily small. Lebedev's group, for example, uses 0.7-mm diameter tubes for solids, glasses, and nonaqueous solutions, and 0.2-mm diameter tubes for aqueous solutions, all in conjunction with a TE_{011} cylindrical cavity (*29*).

Room-temperature aqueous solutions can be studied at high frequencies. In fact, the imaginary (absorptive) component of the dielectric constant of water at room temperature goes through a broad maximum in the range of 10–35 GHz and drops off by a factor of 2.5 at 100 GHz (*59*). Notably, Lebedev and co-workers (*29*), after having tried a number of resonator types, apparently favor the standard, single-mode cylindrical cavity. An attractive alternative to these cavities may be the loop-gap resonator that has been put to such good use at very low frequencies. Froncisz et al. (*60*) described a 35-GHz loop-gap resonator. Conceivably, one might extend extend some version of this design to even higher frequencies. Table II lists the reported sensitivities of many of the high-frequency, high-sensitivity instruments currently in operation.

The applied magnetic field is another important design consideration. The field-sweep linearity, homogeneity, and reproducibility all play clear roles in the final resolution and accuracy of any high-field spectrometer. Up to about 75 GHz, large but fairly conventional iron-core electromagnets can be used. At higher frequencies, however, a field-swept superconducting magnet is probably the best alternative. Although superconducting solenoids are typically thought of as operating in the constant-field (persistent) mode, several groups demonstrated (*58, 61, 62*) that they can be used quite effectively in the field-sweep mode for high-resolution EPR spectroscopy. When operating in this mode, if the magnetic-field resolution of a 100-GHz spectrometer is to equal that of a 10-GHz spectrometer, then the relative field regulation at the higher frequency (higher field) must be an order of magnitude better than the regulation at the lower frequency. In short, the importance of the magnet and its field sweep are not to be underestimated. The field is typically swept by ramping the current in the main superconducting solenoid. Lebedev and co-workers use a Hall effect probe for field regulation and internal spectral standards for calibration. Möbius and co-workers utilize a pick-up coil to obtain linear field sweeps and current regulation for setting the field. Commercial field-swept superconducting magnets are also available. Oxford Instruments, for instance, has constructed such magnets for broad-band NMR applications.

Conclusions

We have provided here a number of examples of uses of low- and high-frequency EPR experiments. Despite the clear utility of the multifrequency approach, practical obstacles stand in the way of making such an approach universally routine. Among these are the technical difficulties associated with assembling several spectrometers that operate over a wide range of frequencies. Economic obstacles also exist. In particular, the cost of equipment (sources, waveguide components, superconducting mag-

Table II. Reported Sensitivities for EPR Spectrometers at Several Frequencies for Nonlossy, Nonsaturating Samples

Band (Frequency, GHz)	Instrument	N_{min} (spins/G)	C_{min} (spins/G cm^3)
X (9.5)	Varian, TE$_{102}$ rectangular resonator, klystron	2×10^{10}	2×10^{11}
Q (35)	Varian E110, TE$_{011}$ cylindrical resonator	5×10^{9}	1×10^{12}
W (94)	Berlin (Möbius), Fabry–Perot resonator, klystron	1×10^{9}	~8.0×10^{11}
W (94)	IERC Mark I, TE$_{013}$ cylindrical resonator, Gunn oscillator at low power (2.4 mW)	~2×10^{9}	~1.3×10^{12}
W (94)	IERC Mark I, TE$_{013}$ cylindrical resonator; Varian klystron (12 mW)	~1.8×10^{8}	~1.2×10^{11}
D (140)	Moscow (Lebedev, Grinberg, et al.), TE$_{011}$ cylindrical resonator, klystron (20 mW)	4.7×10^{7}	~2×10^{11}
D (140)	Moscow (Lebedev et al.), Fabry–Perot resonator	2×10^{9}	~3×10^{11}
G (250)	Cornell (Freed), frequency-multiplied solid-state source, Fabry–Perot resonator	3×10^{11}	5×10^{12}

NOTE: Nonlossy means "not lossy", or not causing microwave losses. N_{min} is the minimum detectable number of spins. C_{min} is the minimum detectable concentration.

net, etc.) required to set up a high-frequency EPR spectrometer is no small matter.

One response to these practical difficulties has been to establish national–international EPR research centers, such as the National Biomedical ESR Center at the Medical College of Wisconsin (Milwaukee), where there is access to both pulsed and continuous-wave (CW) spectrometers operating at several frequencies. The Illinois EPR Research Center at the University of Illinois, funded by the National Institutes of Health Division of Research Resources, is another such center and has CW spectrometers operating between 1 and 96 GHz and a 2–4-GHz pulsed instrument. However, certain individual research groups, for example, Freed's at Cornell [250-GHz spectrometer (63)], Robert's at Grenoble (64), and others cited by Eaton and Eaton in a review (as yet unpublished), have successfully produced these special instruments.

Acknowledgment

Support was provided by the National Institutes of Health, Division of Research Resources, Grant No. RR01811; the U.S. Department of Energy (University Coal Research Program); the Illinois Department of Energy and Natural Resources through the Center for Research on Sulfur in Coal; and the Petroleum Research Fund, administered by the American Chemical Society.

References

1. Feher, G.; Kip, A. F. *Phys. Rev.* **1955**, *98*, 337.
2. Charru, A. *Acad. Sci. Paris C. R.* **1956**, *243*, 652.
3. Strandberg, M. W. P.; Tinkham, M.; Solt, I. H., Jr.; Davis, C. F., Jr. *Rev. Sci. Instrum.* **1956**, *27*, 596.
4. El'sting, O. G. *Instrum. Exp. Technol. Engl. Trans.* **1960**, 753.
5. Marcley, R. G. *Am. J. Phys.* **1961**, *29*, 492.
6. Peter, M. *Phys. Rev.* **1959**, *113*, 801.
7. Mock, J. B. *Rev. Sci. Instrum.* **1960**, *31*, 551.
8. Savzade, M.; Pontnau, S.; Girard, B. *Acad. Sci. Paris C. R.* **1964**, *258*, 4458.
9. Abdrakhmanov, R. S.; Ivanova, T. A. *J. Mol. Struct.* **1973**, *19*, 683.
10. Froncisz, W.; Hyde, J. S. *J. Chem. Phys.* **1980**, *73*, 3123.
11. Froncisz, W.; Aisen, P. *Biochim. Biophys. Acta* **1982**, *700*, 55.
12. Bresgunov, A. Y.; Dubinskii, A. A.; Poluektov, O. G.; Vorob'eva, G. A.; Lebedev, Ya. S. *J. Chem. Soc. Faraday Trans.* **1990**, *86*, 3185.
13. Clarkson, R. B.; Vahidi, N.; Boyer, S.; Liu, K. J.; Swartz, H. M., unpublished.
14. Hyde, J. S.; Froncisz, W. *Ann. Rev. Biophys. Bioeng.* **1982**, *11*, 391.

15. Retcofsky, H. L.; Hough, M. R.; Maguire, M. M.; Clarkson, R. B. In *Coal Structure;* Gorbaty, M. L.; Ouchi, K., Eds.; Advances in Chemistry 192; American Chemical Society: Washington, DC, 1981; pp 37–58.
16. Altman, T. E. Ph.D. Thesis, University of Illinois at Urbana—Champaign, 1981.
17. Rothenberger, K. S.; Nilges, M. J.; Altman, T. E.; Glab, K.; Belford, R. L.; Froncisz, W.; Hyde, J. S. *Chem. Phys. Lett.* **1986**, *124*, 295.
18. Clarkson, R. B.; Timken, M. D.; Brown, D. R.; Crookham, H. C.; Belford, R. L. *Chem. Phys. Lett.* **1989**, *163*, 277.
19. Ovchinnikov, L. V.; Konstantinov, V. N. *J. Magn. Reson.* **1978**, *32*, 179.
20. Poole, C. P., Jr. *Electron Spin Resonance*, 2nd ed.; Wiley and Sons: New York, 1983; p 404.
21. Froncisz, W.; Hyde, J. S. *J. Magn. Reson.* **1982**, *47*, 515.
22. Mehdizadeh, M.; Ishii, T. K.; Hyde, J. S.; Froncisz, W. *IEEE Trans. Microwave Theory Tech.* **1983**, *MTT–31*, 1059.
23. Hyde, J. S.; Froncisz, W. *Proc. Natl. Electron. Conf.* **1981**, *35*, 602.
24. Belford, R. L.; Davis, P. H.; Belford, G. G.; Lenhardt, T. M. In *Extended Interactions between Metal Ions in Transition Metal Complexes;* Interrante, L. V., Ed.; ACS Symposium Series No. 5; American Chemical Society: Washington, DC, 1974; pp 40–50.
25. Bramley, R.; Strach, S. J. *Chem. Rev.* **1983**, *83*, 49.
26. Strach, S. J.; Bramley, R. *Chem. Phys. Lett.* **1984**, *109*, 363.
27. Bramley, R.; Strach, S. J. *J. Magn. Reson.* **1985**, *61*, 245.
28. Strach, S. J.; Bramley, R. *J. Magn. Reson.* **1984**, *56*, 10.
29. Grinberg, O. Y.; Dubinskii, A. A.; Lebedev, Ya. S. *Russ. Chem. Rev. Engl. Transl.* **1983**, *52*, 850.
30. Grinberg, O. Y.; Dubinskii, A. A.; Shuvalov, V. F.; Oranskii, L. G.; Kurochkin, V. I.; Lebedev, Ya. S. *Izvest. Akad. Nauk SSSR Chem. Engl. Transl.* **1985**, *34*, 425.
31. Lebedev, Ya. S. In *Modern Pulsed and CW Electron Spin Resonance;* Kevan, L.; Bowman, M. K., Eds.; Wiley-Interscience: New York, 1990, Chapter 8.
32. Dikanov, S. A.; Gulin, V. I.; Tsvetkov, Y. D.; Gregor'ev, I. A. *J. Chem. Soc. Faraday Trans.* **1990**, *86*, 3201.
33. Gulin, V. I.; Dikanov, S. A.; Tsvetkov, Y. D. *Chem. Phys. Lett.* **1990**, *170*, 211.
34. Haindl, E.; Möbius, K.; Oloff, H. *Z. Naturforsch.* **1985**, *40a*, 169.
35. Krinichnyi, V. I.; Grinberg, O. Y.; Nazarova, I. B.; Kozub, G. I.; Tkachenko, L. I.; Khidenkl', M. L.; Lebedev, Ya. S. *Izvest. Akad. Nauk SSSR Chem. Engl. Transl.* **1985**, *34*, 425.
36. Box, H. C.; Freund, H. G.; Lilga, K. T.; Budzinski, E. E. *J. Phys. Chem.* **1970**, *74*, 40.
37. Box, H. C.; Freund, H. G. *Appl. Spectrosc.* **1980**, *34*, 293.
38. Box, H. C.; Freund, H. G.; Budzinski, E. E. *J. Chem. Phys.* **1981**, *74*, 2667.
39. Box, H. C.; Freund, H. G.; Budzinski, E. E.; Potienko, G. *J. Chem. Phys.* **1980**, *73*, 2052.
40. Box, H. C.; Budzinski, E. E.; Freund, H. G.; Potter, W. R. *J. Chem. Phys.* **1979**, *70*, 1320.
41. Retcofsky, H. L. In *Coal Science;* Gorbaty, M. L.; Larsen, J. W.; Wender, I., Eds.; Academic: New York, 1982, Chapter 3; Vol. 1, pp 43–82.

42. Retcofsky, H. L.; Stark, J. M.; Friedel, R. A. *Anal. Chem.* **1968**, *40*, 1699.
43. Malhotra, V. M.; Buckmaster, H. A. *Org. Geochem.* **1985**, *8*, 235.
44. Eaton, G. R.; Eaton, S. S. *Fuel* **1981**, *60*, 67.
45. Clarkson, R. B.; Wang, W.; Nilges, M. J.; Belford, R. L. In *Processing and Utilization of High-Sulfur Coals III;* Markuszewski, R.; Wheelock, T. D., Eds.; Elsevier: Amsterdam, Netherlands, 1990; pp 67–68.
46. Clarkson, R. B.; Wang, W.; Brown, D. R.; Crookham, H. C.; Belford, R. L. *Fuel* **1990**, *69*, 1405.
47. Clarkson, R. B.; Earle, K.; Budil, D.; Freed, J.; Belford, R. L., unpublished.
48. Dubinskii, A. A.; Grinberg, O. Ya.; Kurochkin, V. I.; Oranskii, L. G.; Poluetkov, O. G.; Lebedev, Ya. S. *Theor. Exp. Chem. Engl. Transl.* **1981**, *17*, 180.
49. Poluektov, O. G.; Grinberg, O. Ya.; Dubinskii, A. A.; Sidorov, O. Y.; Lebedev, Ya. S. *Teor. Eksp. Khim.* **1989**, *4*, 459.
50. Panferov, Y. F.; Dubinskii, A. A.; Grinberg, O. Y.; Aldoshina, M. Z.; Lyubovskaya, R. N.; Khidekel, M. L.; Lebedev, Ya. S. *Sov. J. Chem.* **1987**, *4*, 831.
51. Alpert, Y.; Couder, Y.; Tuchendler, J.; Thome, H. *Biochim. Biophys. Acta.* **1973**, *322*, 34.
52. Brackett, G. C.; Richards, P. L.; Caughey, W. S. *J. Chem. Phys.* **1971**, *54*, 4383.
53. Champion, P. M.; Sievers, A. J. *J. Chem. Phys.* **1980**, *72*, 1569.
54. Wagner, R. J.; White, A. M. *Solid State Commun.* **1979**, *32*, 399.
55. Box, H. C.; Freund, H. G.; Kreilick; R. W., private communications.
56. Van den Boom, H. *Rev. Sci. Instrum.* **1969**, *40*, 550.
57. Amity, I. *Rev. Sci. Instrum.* **1970**, *41*, 1492.
58. Galkin, A. A.; Grinev, G. G.; Kurochkin, V. I.; Nemchenko, E. D. *Sov. J. Nondestruct. Test. Engl. Transl.* **1977**, *12*, 443.
59. Hasted, J. B. In *Water: A Comprehensive Treatise;* Franks, F., Ed.; Plenum: New York, 1972, Chapter 7; Vol. 1.
60. Froncisz, W.; Oles, T.; Hyde, J. S. *Rev. Sci. Instrum.* **1986**, *57*, 1095.
61. Galkin, A. A.; Grinberg, O. Y.; Dubinskii, A. A.; Kabdin, N. N.; Krymov, V. N.; Kurochkin, V. I.; Lebedev, Ya. S.; Oranskii, L. G.; Shuvalov, V. F. *Instrum. Exp. Tech. Engl. Transl.* **1985**, *18*, 294.
62. Burghaus, O.; Haindl, E.; Plato, M.; Mobius, K. *J. Phys. E Sci. Instrum.* **1985**, *18*, 294.
63. Lynch, B.; Earle, K. A.; Freed, J. *Rev. Sci. Instrum.* **1988**, *59*, 1345.
64. Barra, A. L.; Brunel, L. C.; Robert, J. B. *Chem. Phys. Lett.* **1990**, *165*, 107.

RECEIVED for review January 8, 1991. ACCEPTED revised manuscript August 9, 1991.

7

High-Temperature Electron Spin Resonance and NMR Methods Applied to Coal

Yuzo Sanada[1] and Leo J. Lynch[2]

[1]Faculty of Engineering, Hokkaido University, Sapporo, 060, Japan
[2]Commonwealth Scientific and Industrial Research Organisation (CSIRO), Division of Coal and Energy Technology, P.O. Box 136, North Ryde, New South Wales 2113, Australia

The applications of in situ electron spin resonance (ESR) and NMR spectroscopic methods to study thermally induced transformations of coals and related materials are addressed. Brief background statements are given on the underlying principles, and the focus is on the particular methodologies developed at Hokkaido University and CSIRO during the past decade. Designs of both high-temperature and high-temperature–high-pressure ESR and NMR probes and spectrometer configurations for their use are outlined. The high-temperature in situ ESR technique is illustrated by its application to Argonne Premium coals. Applications of NMR techniques to derive chemical-shift information on coal and pitch materials during pyrolysis and liquefaction processes, and techniques that monitor molecular dynamics during pyrolysis are presented. Monitoring molecular dynamics is demonstrated to be an effective method of thermal analysis.

THERMALLY INDUCED CHANGES in coals and related materials can be studied in situ by adapting electron spin resonance (ESR) and nuclear magnetic resonance (NMR) spectroscopic methods. The thermal transformation of coal and other organic solids by pyrolysis–carbonization, combustion, or liquefaction can be investigated by equilibrium methods whereby the properties of the original and product materials at different stages of the process are assessed. An advantage of this approach is that measurement time is not a constraint to obtaining precise and detailed data. However, in situ or "reaction time" methods of observation are required to measure transient, nonequilibrium intermediate states that occur in these processes. Phenomena of particular significance for coal include thermally induced free radical reactions (1–4) and the so-called "thermoplastic" state obtained in the early stages of pyrolytic decomposition of some coals of bituminous rank (5). An in situ measurement technique is an effective method of thermal analysis if it has adequate time resolution to detect transient phenomena and thereby to monitor physical transitions and chemical changes in real time.

A range of established thermal analysis techniques has been used to study coal materials and their reactions. These techniques are based on a variety of measurements, including some that sense heat flow [differential scanning calorimetry (DSC) and differential thermal analysis (DTA)], changes in mass [thermogravimetric analysis (TGA)], and changes in dimension [thermomechanical analysis (TMA)]. Other in situ measurement techniques involve analysis of evolved products and include such methods as evolved gas analysis (EGA), combined gas chromatography–mass spectrometry (GCMS), and emission and adsorption Fourier transform infrared spectroscopy.

The adaptation of magnetic resonance spectroscopy techniques for in situ or reaction time studies at elevated temperatures is made difficult by the complexity of the measurement procedures and the uncertain time resolution that it is possible to achieve. An effective method of NMR thermal analysis requires that the detected parameters that reflect the physical and chemical properties of the substance be recorded and monitored as a function of temperature or time as the substance is subjected to a controlled heating program. Ideally the measurement should be instantaneous in order to capture the information. The extent to which this measurement can be adequately approximated in practice depends on the rate of change associated with the phenomenon of interest and the time necessary for the magnetic resonance measurements to be completed. The rate of measurement can be limited by the inherent and therefore changing nature of the specimen; therefore, close attention must be given to the procedures of measurement.

High-Temperature ESR Methods

The Spectral Parameters of ESR Spectroscopy. ESR spectroscopy, sometimes called electron paramagnetic resonance (EPR) spectroscopy, is a means of detecting direct transitions between electron Zeeman levels. The phenomenon of electron spin resonance is observed only for atomic or molecular systems having net electron spin angular momentum, that is, materials containing one or more unpaired electrons. Materials that meet this criterion include organic free radicals, molecules in their triplet states, some types of charge-transfer complexes, transition metal ions, semiconductors, metals, inorganic molecules such as oxygen that contain partially filled molecular orbitals, and crystals having certain point defects.

The initiation of chemical changes in organic substances by thermal energy, gamma irradiation, and mechanical forces usually involves the generation of free radicals. The formation of free radicals in such reactions is due to homolytic scission of a single covalent bond between two different atoms as follows:

$$A - B \rightarrow A\cdot + B\cdot$$

Detection of an unpaired electron, as in a free radical, can provide important information about the pathways of reactions that have already occurred or are in progress. Typical free radical reactions are chain reactions that occur in three steps: (1) formation; (2) propagation including atom transfer, addition, rearrangement, and fragmentation reactions; and (3) termination including combination of two radicals and disproportionation.

π-radicals are long-lived, having lifetimes of several minutes or greater, in contrast to the transient, short-lived (i.e., highly reactive) nature of σ-radicals. π-radicals occur typically in aromatic or conjugated molecules in which all the electrons occupy sp-type orbitals. The π-electron orbitals in such molecules overlap to form molecular orbitals with a discrete band of energy levels that can be described as linear combinations of the aromatic $2p$ orbitals. This delocalization of the unpaired electron results in the stabilization of the π-radical. Hence π-radicals are readily detected in conventional (equilibrium) ESR spectroscopy, and σ-radicals are generally difficult to detect. Therefore, the much greater capability of in situ ESR spectroscopy to detect short-lived σ-radicals is of great advantage in studies of coal reactivity.

The three most useful parameters that can be extracted from ESR spectra are the spectral intensity, g value, and spectral line width; these parameters provide information on spin concentration, spin types, and the molecular environment. Other parameters, such as the electron relaxation times, can sometimes be measured or estimated, but have had only limited use in coal research (3, 6).

Since the first successful ESR experiments in 1945, many investigations on coal structure and reactivity have been made with commercially available ESR spectrometers. The aspects of ESR spectrometry necessary for a basic understanding of its applications in coal research are discussed in several excellent treatises (*1–3*).

High-Temperature ESR Spectroscopy. Ingram et al. (*4*) and Ubersfeld et al. (*7*) were the first to detect stable free radicals in coal by ESR measurements. Since then, many ESR studies on carbonaceous materials such as kerogen, coal, pitch, and their derivatives have been conducted, and the signal parameters of the naturally formed free radicals have been used for coal rank determination. The ESR parameters of spin concentration, spectral line width, and g value have provided details of coal structure and of coal utilization reactions such as extraction with solvent, liquefaction, and gasification (*3, 8*).

The detection of thermally formed free radicals in heated coal was reported earlier (*2–4*). However, these studies were of samples heat-treated outside the ESR cavity prior to measurements at room temperature. The information obtained in such measurements is limited because transient radicals quenched during the course of sample preparation are not observed.

More recently the usefulness and importance of ESR in in situ high-temperature spectroscopy have been recognized and adopted by several investigators (*9–11*) because of its capability of monitoring thermally generated free radicals as they are formed.

Extensive studies at the Faculty of Engineering Laboratory, Hokkaido University, using in situ high-temperature ESR spectroscopy are directed at a better understanding of the chemistry of free radicals in coal materials (*12–15*).

High-Temperature–High-Pressure ESR Methods. Realization of the potential of in situ ESR spectroscopy to study coal pyrolysis and liquefaction reactions requires the availability of high-temperature or pressure-regulated ESR instrumentation. However, commercially available ESR cavities have a limited temperature range (<773 K) and no pressure-regulation capability. Petrakis and Grandy (*8*) developed high-temperature–high-pressure apparatus to meet this need.

The design described by Petrakis and Grandy (*8*) to allow combined high-temperature and high-pressure operations is largely a combination of aspects of previous designs for separate high-pressure and high-temperature apparatus. This apparatus is basically a water-cooled Cu–Be alloy pressure vessel with an X-band cylindrical TE_{011} brass cavity inside; details of its design are described in the figures and photographs of reference 8. (TE means transverse electric wave. TE_{lmn} shows the vibration modes of the microwave.) The design pressure of 73.8 MPa and the temperature capability of 753 K were selected because they are typical of coal liquefaction processes. Although built primarily for the study of free radicals in coal liquefaction, this cavity system could be used to investigate a variety of problems such as reactions on catalyst surfaces.

A range of apparatus for high-pressure–high-temperature in situ ESR spectroscopy has been developed successfully at Hokkaido (13). A sample pressure vessel consisting of a quartz capillary tube, open at one end and connected to a high-pressure line, is shown in Figure 1. Figure 2 illustrates a water-cooled high-temperature ESR probe with a rectangular

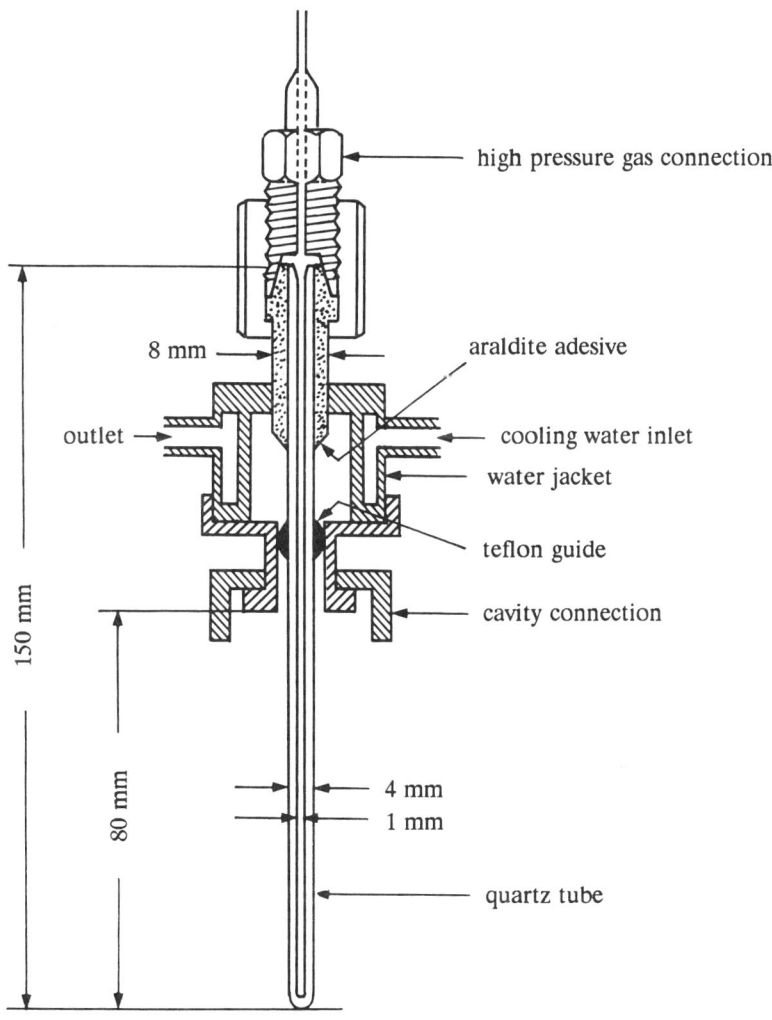

Figure 1. High-temperature–high-pressure ESR tube.

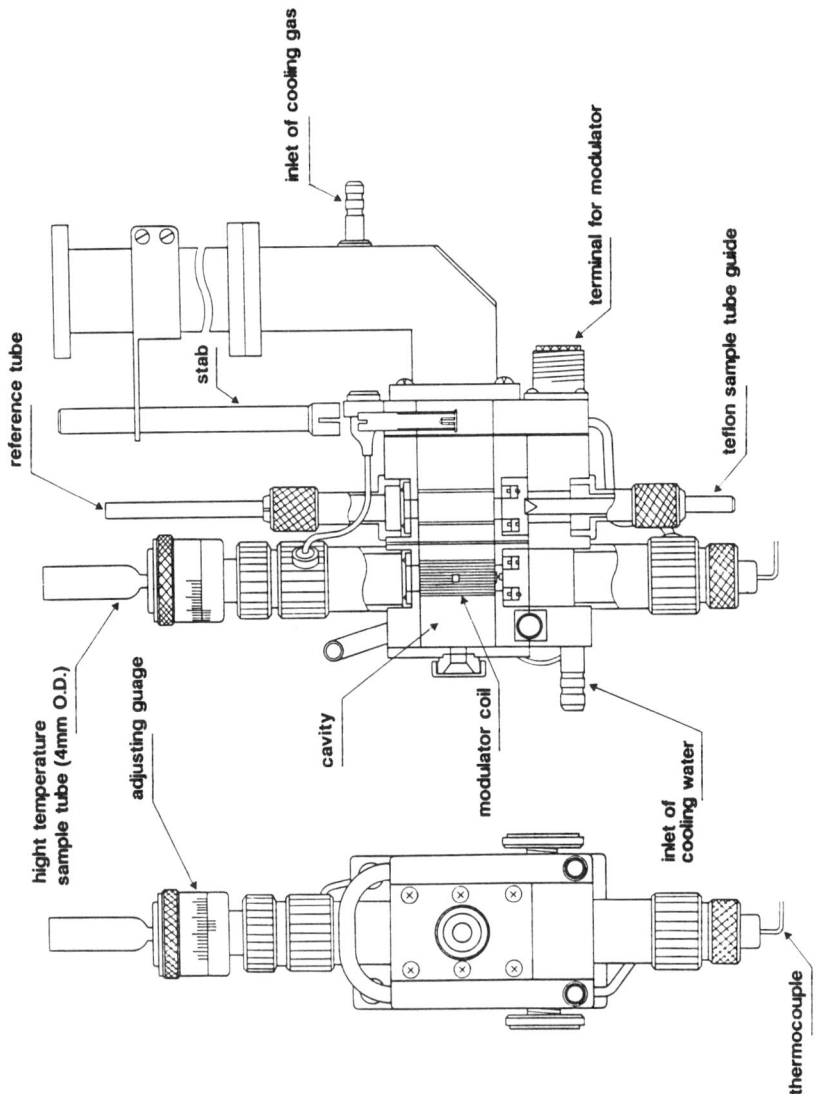

Figure 2. Water-cooled high-temperature ESR cavity.

X-band TE_{103} cavity, and Figure 3 presents an integrated high-temperature–high-pressure ESR assembly. ESR spectra are recorded by using these probes with a Varian E109 X-band spectrometer equipped with an Echo Electronics cylindrical or rectangular high-temperature cavity. Experiments are conducted at a frequency of 9.35 GHz and a modulation of 100 kHz provided by the modulation unit supplied with the Varian spectrometer. A microwave power of 30 db is used throughout most measurements. About 20 mg of sample is mounted in a 4-mm o.d. quartz tube and inserted directly into the high-temperature cavity. For atmospheric measurements all runs are carried out in a flow of nitrogen gas (kept at a constant flow rate) and at heating rates of 2–10 K/min.

The sample heater made of a Pt-plated silica tube is located outside of the sample tube, as shown in Figure 3. For measurements at high pressures the pressure vessel is inserted into the heater. The measurement procedures are similar to those at atmospheric conditions.

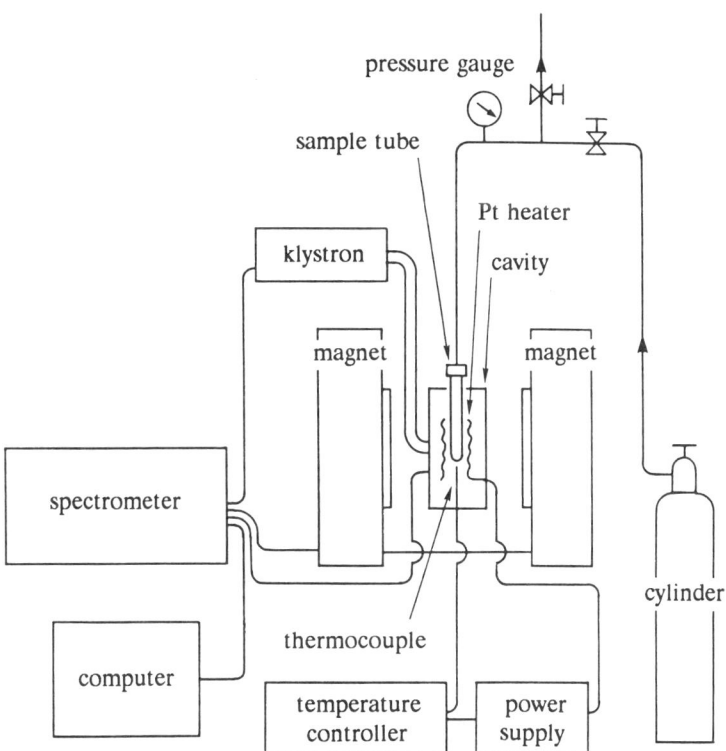

Figure 3. Assembly of high-temperature–high-pressure ESR apparatus.

Various heating rates in the range of 2–15 K/min were investigated, and a rate of 10 K/min was selected for routine measurements satisfying the objectives of our coal studies. In these experiments ESR spectra are recorded at 25-K intervals up to a maximum temperature of 973 K. Thus the number of readings taken over any period of time depends on the heating rate selected. The spectra are recorded in a scan time of 7.5 s, and the spin concentration is computed on-line with a Hewlett–Packard computer system.

Spin concentrations measured at room temperature are calculated by reference to two secondary standards (an evacuated and a sealed coal sample) and a pitch sample. The two standards are in turn calibrated against diphenylpicrylhydrazyl (DPPH). For measurements of samples at elevated temperatures, corrections are made for the Boltzmann effect and for weight loss.

These in situ methods to study thermal reactions of coals not only allow transient effects to be detected but can also provide information on the kinetics of the reactions from the dependence of the measured spin concentrations on heating rate. They also enable the study of smaller coal specimens because of the many measurements that are made on a given sample.

Pyrolysis Studies on Pristine and Oxidized Coals. The usefulness of in situ high-temperature ESR spectroscopy can be illustrated with reference to a recent study of the pyrolysis of the Argonne Premium coal samples (*16*).

Pristine Coal. Figure 4 shows the general profile of spin concentration, N_s, as a function of temperature for seven Argonne Premium coals whose carbon rank ranges from 75% C (No. 1) to 87% C (No. 7). These temperature dependences for the five lowest rank coals, from a lignite (No. 1) to a high-volatile bituminous (HVB, No. 5), behave similarly and show a single peak value in N_s.

The two highest rank coals, a HVB (No. 6) and a medium-volatile bituminous (MVB) (No. 7), exhibit a second peak in N_s at lower temperatures. This result has been reported in earlier studies (*17, 18*); however, the temperatures of 623 and 648 K of the first peak for the two coals in this study are higher. The value of the second peak in N_s and the temperature at which this peak is observed are referred to here as R_3 and T_3, respectively. Figure 5 plots the relation between T_3 and the carbon content of these pristine coals. As coal rank increases, T_3 increases, except that a plateau value of ~823 K is maintained for all the HVB coals. In

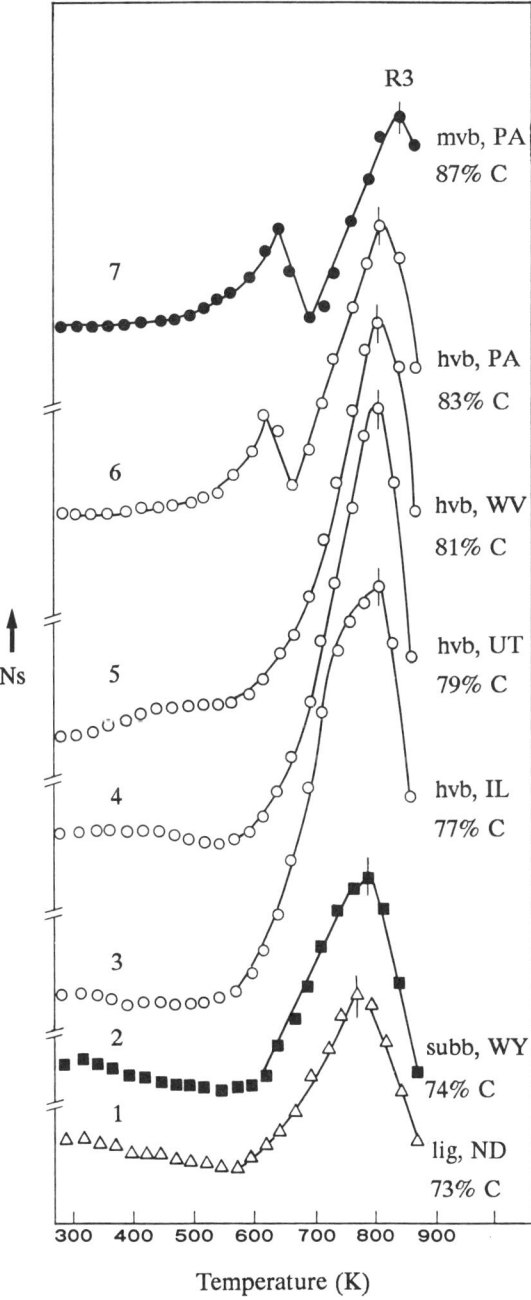

Figure 4. Temperature dependence of radical concentration (N_s) of pristine Argonne Premium coals.

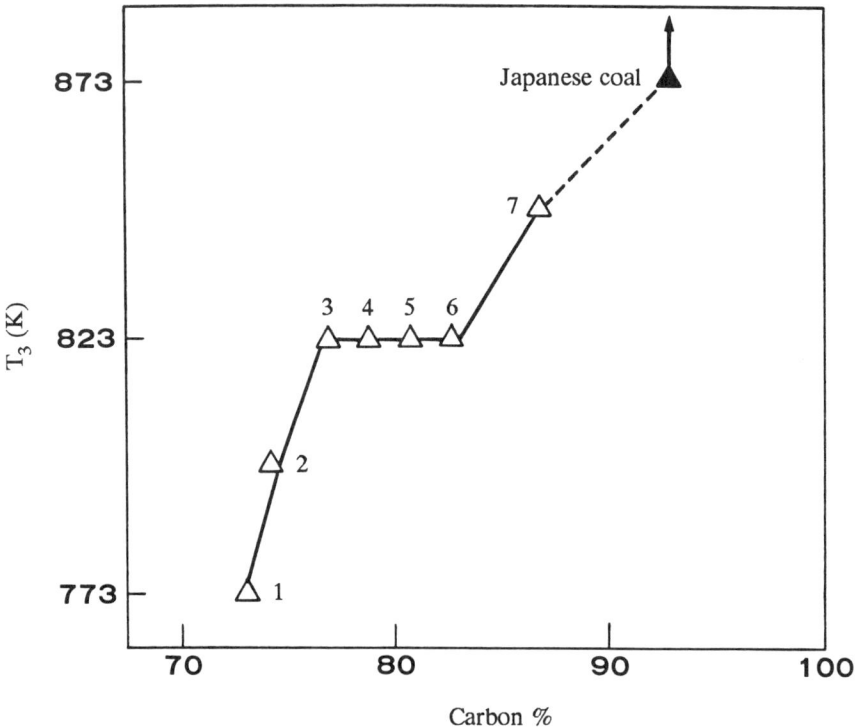

Figure 5. T_3 *of Argonne coals as function of rank (numbers are defined in Figure 4).*

Figure 6 R_3 is seen to correlate closely with the hydrogen index [the remaining hydrocarbon generation potential (*19*)], a result suggesting that hydrocarbon generation from coal proceeds via free radical reactions.

Oxidized Coal. In Figure 7, the effects of 10 days of weathering on the lignite (No. 1) and 2 months of weathering on the MVB No. 7 coal are illustrated by comparison of the N_s-versus-temperature plots of the weathered and pristine coals. A further peak value of N_s is observed at 450 K for the lignite and at 525 K for the MVB No. 7 coal; this peak can be attributed to the weathering.

The sensitivity of radical formation to the presence of oxygen during heating of coal is illustrated in Figure 8. Here the dependences of N_s versus temperature of samples of the weathered MVB No. 7 coal when heated in the presence of air and of nitrogen are compared. Three major effects are present: (1) the peak value of N_s initiated by weathering be-

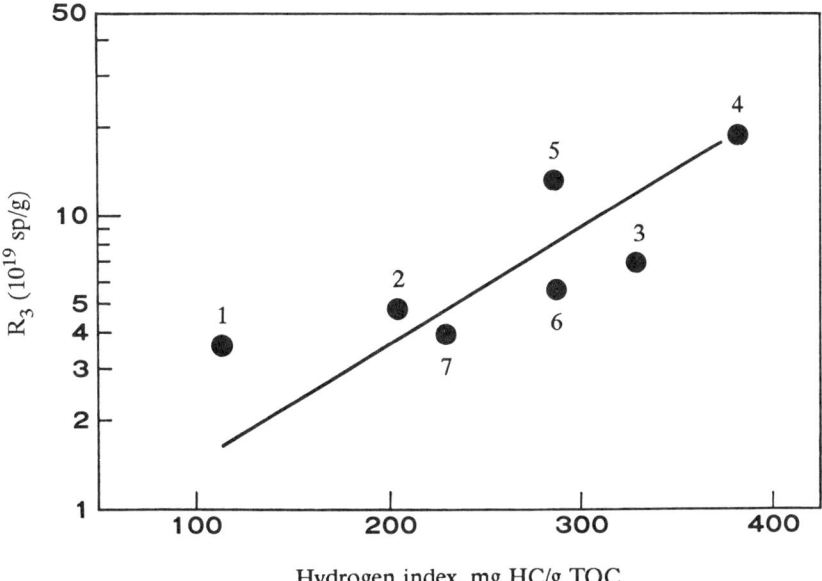

Figure 6. The second peak of free radical concentration (R_3) of Argonne coals versus hydrogen index (hydrogen index was quoted from reference 19).

comes more pronounced, (2) the two peak values in N_s that characterize the pristine coal cannot be resolved, and (3) the presence of oxygen enhances the onset of radical generation and increases the free radical production. These results suggest that weathering may be responsible for the existence of the lower temperature peak values of N_s of the higher rank (Nos. 6 and 7) pristine coals. In the detailed study of the temperature dependence of N_s of coal No. 7 oxidized at 473 K over 8 days, the lower temperature peak value of free radical concentration in the pristine coal was quenched after 1 h of oxidation. Also, as the time of oxidation proceeds, radical enhancement increases, and the process of radical generation is shifted to lower temperatures.

High-Temperature NMR Methods

Background. High-temperature and high-temperature–high-pressure NMR applications have embraced both inorganic and organic materials. The method of probe heating to attain the high experimental temperatures has usually been electric resistive, but recently methods of inductive (20) and infrared irradiation (21) heating have been used, and tem-

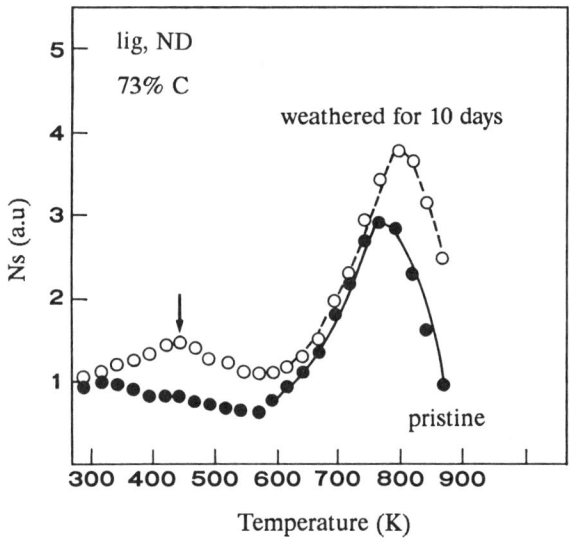
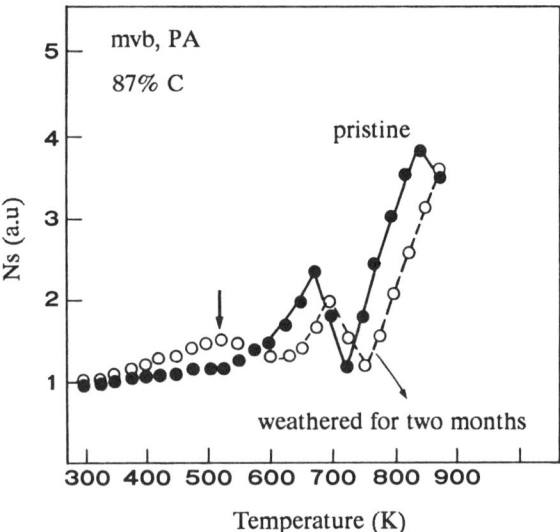

Figure 7. Temperature dependence of N_s of pristine and weathered lignite and MVB coal.

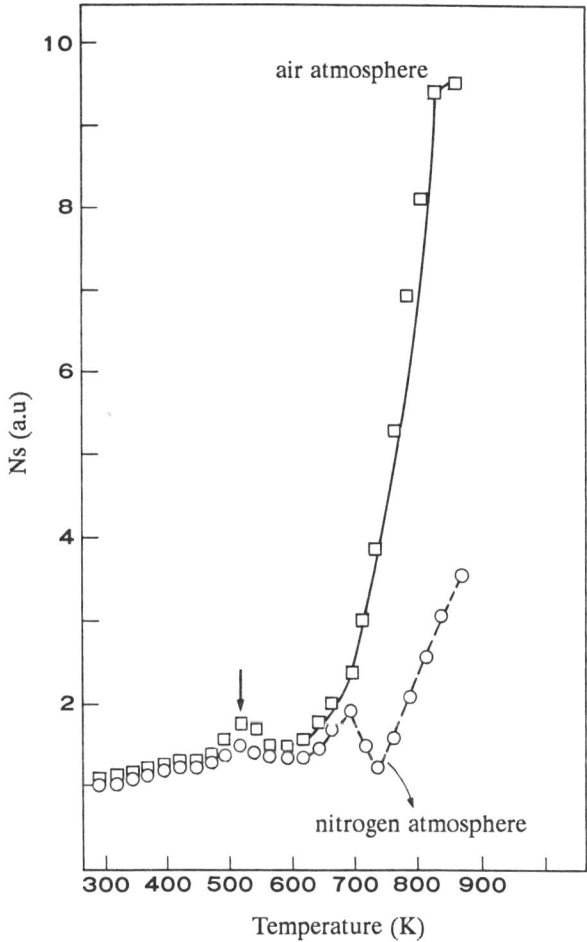

Figure 8. Temperature dependence of N_s of MVB (No. 7) weathered coal in nitrogen and air atmospheres.

peratures as high as 1400 K have been achieved (*22*). These high temperatures have been required to study mostly inorganic substances, including solid and liquid metals, minerals, glasses, and other semiconductor materials. In these applications many resonances have been probed, for example, ^{1}H (*23, 24*), ^{77}Se (*25*), ^{51}V and ^{59}Co (*22, 26*), ^{27}Al (*27*), ^{29}Si (*28–30*), ^{9}Be (*31*), ^{23}Na (*21, 29*), ^{17}O (*32*), and ^{39}K (*33*). These high-temperature studies of inorganic material have been concerned invariably with elucidation of structures and molecular dynamics and not thermal degradation processes.

Organic materials, for which ^{1}H and ^{13}C NMRs are probed, are much less thermally stable than inorganic substances, and temperatures as high as ~800 K are of main interest. This temperature range allows for the important thermophysical and thermochemical transformations of organic materials, including pyrolytic decomposition and formation of carbonized residues, to be investigated.

Studies of Carbonaceous Materials at Hokkaido University. High-temperature NMR studies of the thermal transformation of carbonaceous materials at the Faculty of Engineering, Hokkaido University, in Japan have been concerned with polymer (*34–37*), pitch (*38–46*), and coal (*38, 47–49*) materials. The work has extended to the development and use of both atmospheric high-temperature (*38*) and high-pressure–high-temperature (*35, 50*) techniques in both ^{1}H and ^{13}C (*35, 43, 50*) NMR investigations.

An outline of the design and operation of a high-temperature probe for ^{1}H NMR spectroscopy is shown in Figure 9 (*38*). This probe has a

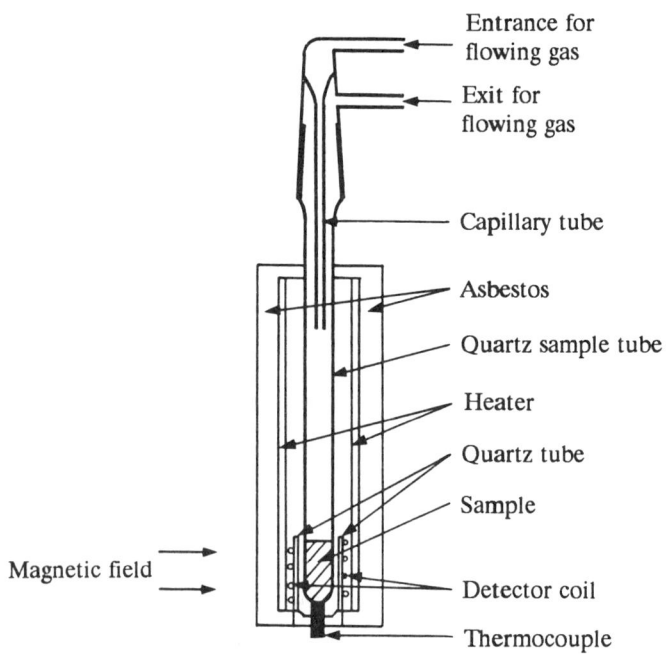

Figure 9. Outline schematic of a high-temperature ^{1}H NMR probe and flowing gas system. (Reproduced with permission from reference 38. Copyright 1979 Pergamon Press PLC.)

temperature capability to 823 K and operates in the pulsed mode at 36.4 MHz. It allows for specimen heating at different rates and the simultaneous recording of NMR spectra.

The design of a water-cooled high-pressure–high-temperature probe for high-resolution ^{13}C NMR operation in the ranges of 0.1 to 100 MPa of pressure and 293 to 823 K of temperature is shown in Figure 10 (*50*). Re-

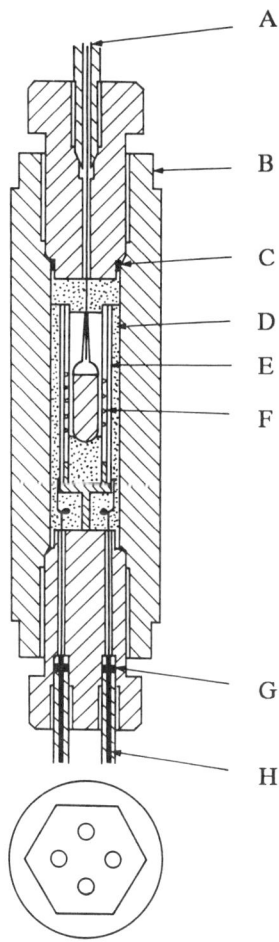

Figure 10. High-temperature–high-pressure ^{13}C NMR probe incorporating a heater internal to the pressure vessel. A, thermocouple; B, titanium alloy vessel; C, copper sealing washer; D, quartz wool; E, heater; F, rf coil; G, Teflon scaling cone; and H, Teflon tube. (Reproduced with permission from reference 50. Copyright 1983 Academic Press.)

sistive heating that is internal to the pressure vessel is used for the probe. To minimize the loss of spectral resolution through field inhomogeneity and other effects caused by the heater current, the resistance heater is noninductively wound, and a method of digital switching for alternating applications of the radio frequency (rf) pulses and the heater current is used.

Figure 11 is the block diagram of an NMR spectrometer system for high-pressure–high-temperature ^{13}C NMR experiments that incorporates this probe (50). This system is based on a Bruker SXP-100 high-power pulsed spectrometer and a 60-mm gap JEOL 3H high-resolution electromagnet and operates at a resonance frequency of 9.6 MHz for ^{13}C NMR measurements. The design of a similar spectrometer system for ^{1}H NMR operation that has also been described (36) includes an externally heated pressure vessel.

An emphasis has been to obtain chemical resolution from the NMR spectra. In studies of pitch, ^{1}H NMR spectra that have been obtained quantitatively resolve the aliphatic and aromatic moieties. The tempera-

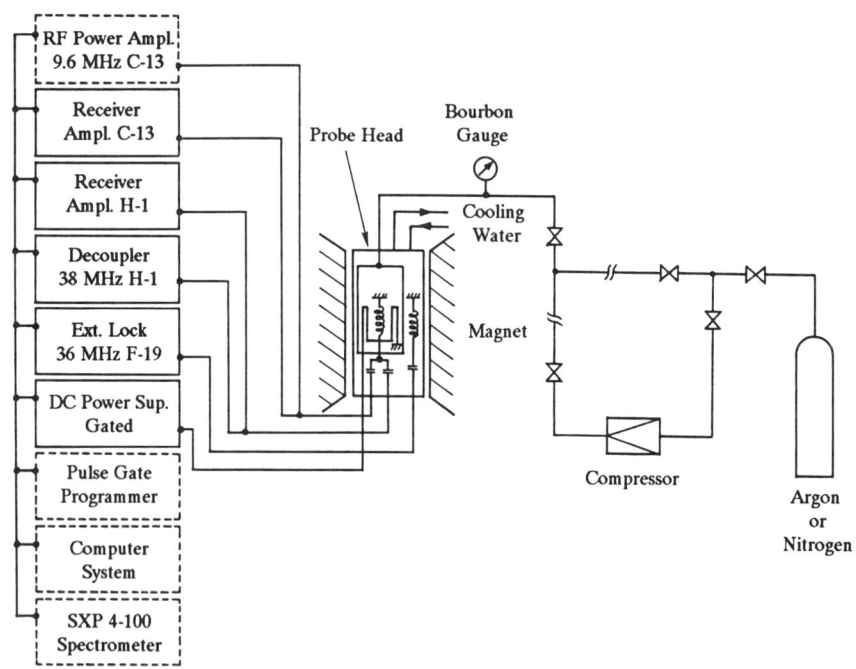

Figure 11. Block diagram of a high-temperature–high-pressure ^{13}C NMR system. (Reproduced with permission from reference 50. Copyright 1983 Academic Press.)

ture (*38, 41, 44–46*) and time (*39, 40*) dependences of aromatization processes in the course of pitch pyrolysis thus obtained have been related to mesophase formation (*39, 40, 42*). The more recent demonstration of in situ high-temperature ^{13}C NMR spectra of mesophase pitch (*43*) has revealed much greater detail in the chemical-shift patterns that possibly distinguish aromatic structures distributed between ordered mesophase and isotropic pitch phases. In another study (*47*), hydrogen-transfer reactions in heated mixtures of donor oils and pitch precursors such as acenaphthylene were investigated by observing shifts in the ^{1}H NMR spectra.

Miyazawa et al. (*38*) recorded ^{1}H NMR spectra of several coals and coal extracts during heating under flowing nitrogen at 5 K/min to temperatures above 673 K. The temperature dependence of the line width, ΔH, of these unresolved spectra was obtained. Some of these earlier results obtained during heating of a subbituminous Taiheiyo coal (77% C) are shown in Figure 12. For a brown coal ΔH decreased to a minimum near 600 K and then rapidly increased, whereas a similar decrease in ΔH for a bituminous coal occurred at higher temperatures, reaching a minimum near 670 K.

This procedure was used by Yokono et al. (*47*) to investigate a wider range of pristine and laboratory oxidized coals. They showed that a well-defined minimum in the temperature dependence of ΔH occurred for

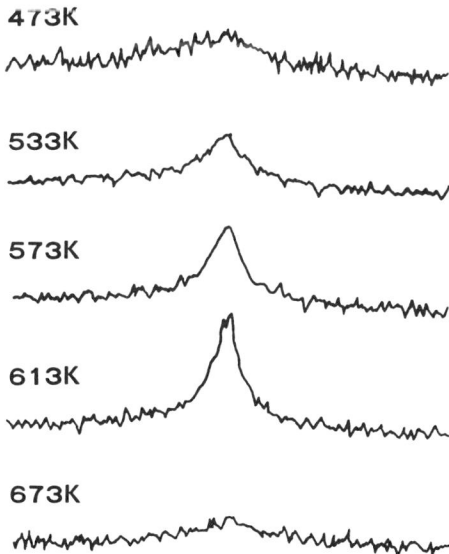

Figure 12. ^{1}*H NMR spectra obtained during heating of a subbituminous Taiheiyo coal (77% C). (Reproduced with permission from reference 38. Copyright 1979 Pergamon Press PLC.)*

bituminous coals and related it to the thermoplastic phenomenon as measured by Gieseler plastometry. They also were able to show that the molecular mobility responsible for the initial line narrowing is reduced by extensive oxidation of the coal. Yokono et al. (48) demonstrated the utility of high-pressure–high-temperature ^{1}H NMR spectroscopy to study coal liquefaction under autogenous conditions. This work illustrated how changes in the resolution of the aliphatic and aromatic bands of the ^{1}H NMR spectra relate to the nature of the liquefaction slurry and hence can be used to evaluate the efficacy of solvents and catalysts (Figure 13).

Studies of Carbonaceous Materials at CSIRO. High-temperature NMR studies of carbonaceous materials at CSIRO, North Ryde, Australia, have been of ^{1}H NMR applications at atmospheric pressure. The emphasis has been to develop this high-temperature NMR methodology as an effective technique of thermal analysis.

In that time resolution of measurement is of paramount importance for effective thermal analysis, the NMR procedure most adaptable for ap-

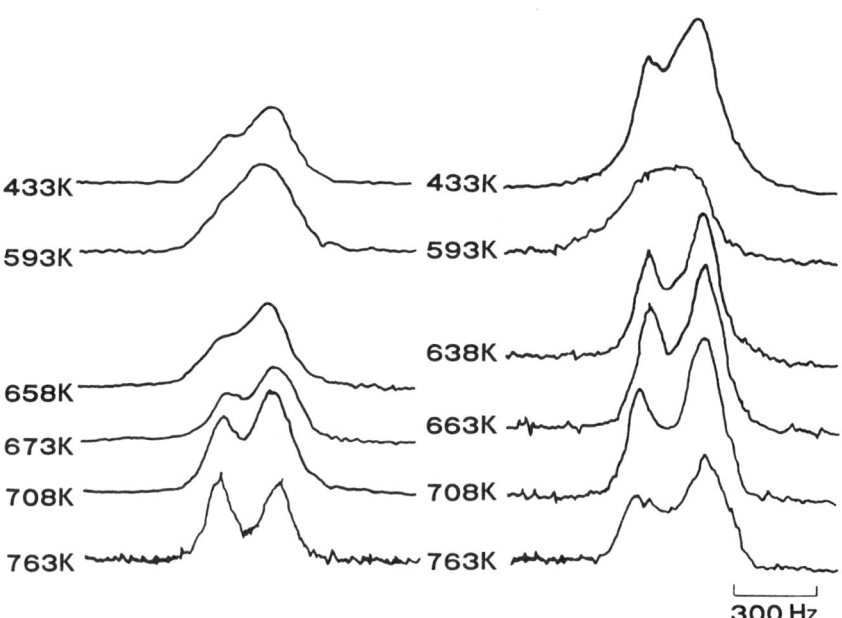

Figure 13. ^{1}H NMR spectra of coal–tetrahydroquinoline with (right) and without (left) $ZnCl_2$ catalyst under autogenous heating conditions. (Reproduced with permission from reference 49. Copyright 1986 Butterworth–Heinemann Ltd.)

plication to organic solid materials is that of simple "undecoupled" pulsed ^{1}H NMR spectroscopy. The abundance of protons in organic materials and the high relative sensitivity of the proton resonance provide an adequate signal-to-noise ratio to minimize the signal averaging, and hence the time, necessary for a given measurement. Furthermore, this form of NMR spectroscopy is highly appropriate for studies of the molecular structure and molecular dynamics of condensed organic substances, as distinct from studies of chemical composition and structure.

Whereas a number of designs of NMR probes and apparatus for high-temperature NMR spectroscopy of various forms have been demonstrated successfully, few efforts have been made to refine procedures for dynamic operation, that is, the ability to accurately control measurement in the course of changing conditions of probe and specimen. Measurement difficulties arise because of both the continually changing nature of the substance (which is in fact a component of the resonant circuit) under study and also the properties of the NMR probe circuit. Factors that influence the probe circuit are (1) changes in the dielectric properties of the specimen and (2) the direct temperature effects on the rf coil resistance and other electrical properties of the resonant probe. Thus close attention must be given to the mechanical, electrical, and structural design of the probe.

To obtain adequate and reproducible NMR measurements under thermal analysis conditions it is necessary to control and regulate the following:

1. the temperature and uniformity of temperature over the sample volume as it is heated according to a predetermined temperature regime

2. the rf pulse energy and pulse repetition rate necessary to achieve the required level(s) of resonance saturation

3. the applied magnetic field to ensure the resonance condition

4. adjustments of the power matching and resonance tuning of the probe circuit

5. the interval between individual NMR measurements and their duration and therefore the extent of signal averaging

The first requirement is met in the design of the high-temperature probe. Several successful designs based on feedback-controlled direct current (dc) heated furnaces have been described (51–53). The construction of one of these furnaces is illustrated in Figure 14. In these designs care has been taken to minimize coupling between the tuned rf and furnace heating coils and to achieve accurate temperature regulation and

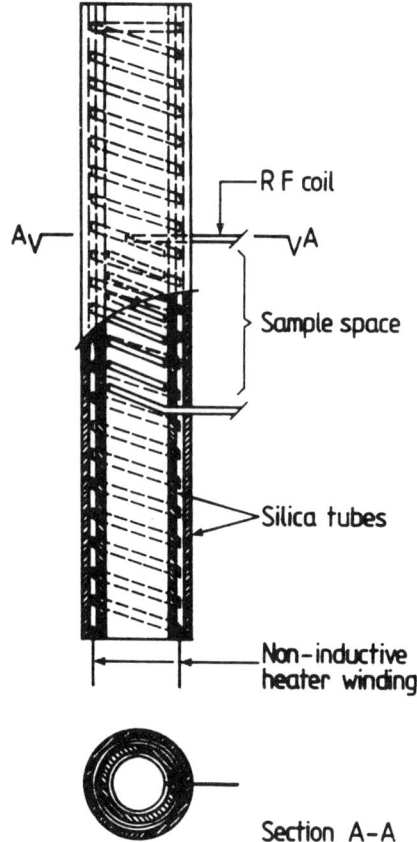

Figure 14. The mechanical design of the furnace for a ^1H NMR high-temperature probe. (Reproduced with permission from reference 52. Copyright 1988 Academic Press.)

close uniformity of temperature over the sample volume. The furnace probe is enclosed in a brass water-cooled jacket to maintain ambient temperatures adjacent to the magnet probe caps (Figure 15). Undecoupled ^1H NMR signals of solid materials decay rapidly; therefore, minimum "dead time" of the rf tuned circuit after rf pulsing is desirable. Dead-time periods of less than 6 μs are achieved with circuit Q (quality factor) values as high as 20 by using a tapped series-tuned circuit (51) and duplexing techniques [introduced by Lowe and Tarr (54)] effected by crossed diodes and quarter-wave transmission lines.

Requirements 2–5 can be achieved adequately by manual spectrome-

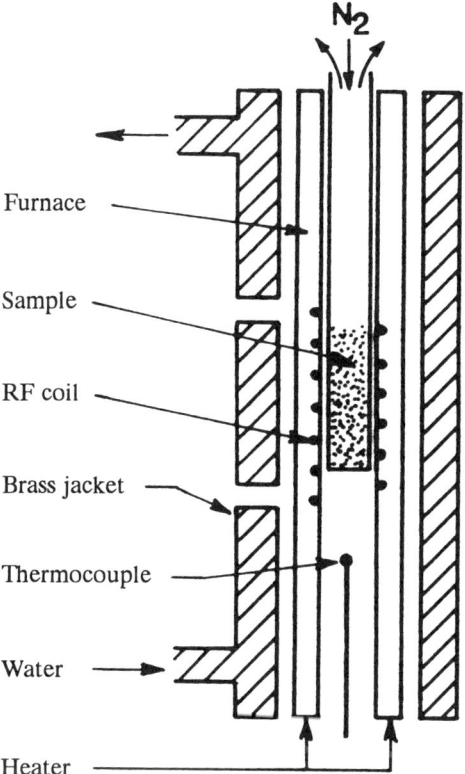

Figure 15. ^1H *NMR high-temperature probe incorporating high-temperature furnace and water-cooled jacket.*

ter adjustments performed by an experienced operator of an appropriately appointed spectrometer system. Studies (52) of the pyrolysis behavior of coals and other carbonaceous materials have been made by using such a system centered on a Bruker SXP-100 high-power spectrometer operating at 60 MHz and interfaced to a data-logging computer. However, manual operation in thermal analysis experiments is tedious and also limits the attainable time resolution. Full computer control of the features listed enables automated operation and an effective method of proton magnetic resonance thermal analysis (PMRTA). This goal has been achieved and adapted to low-field magnetic resonance spectrometers such as the Bruker Minispec. Such an instrument is described in a patent specification (53). An operational description of this PMRTA instrument is given in Figure 16, and a schematic of its operations in a pyrolysis experiment is given in Figure 17.

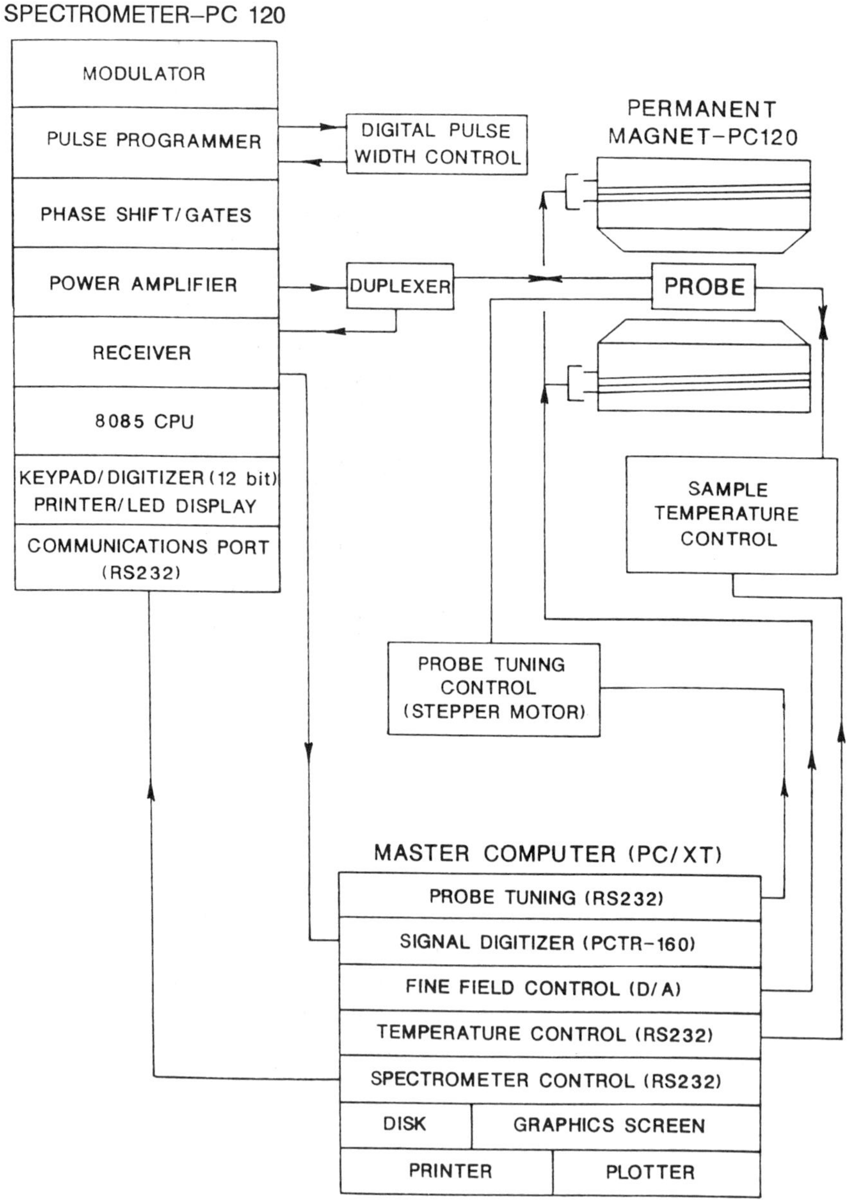

Figure 16. Block diagram of a PMRTA instrument adapted from a Bruker Minispec 120 spectrometer.

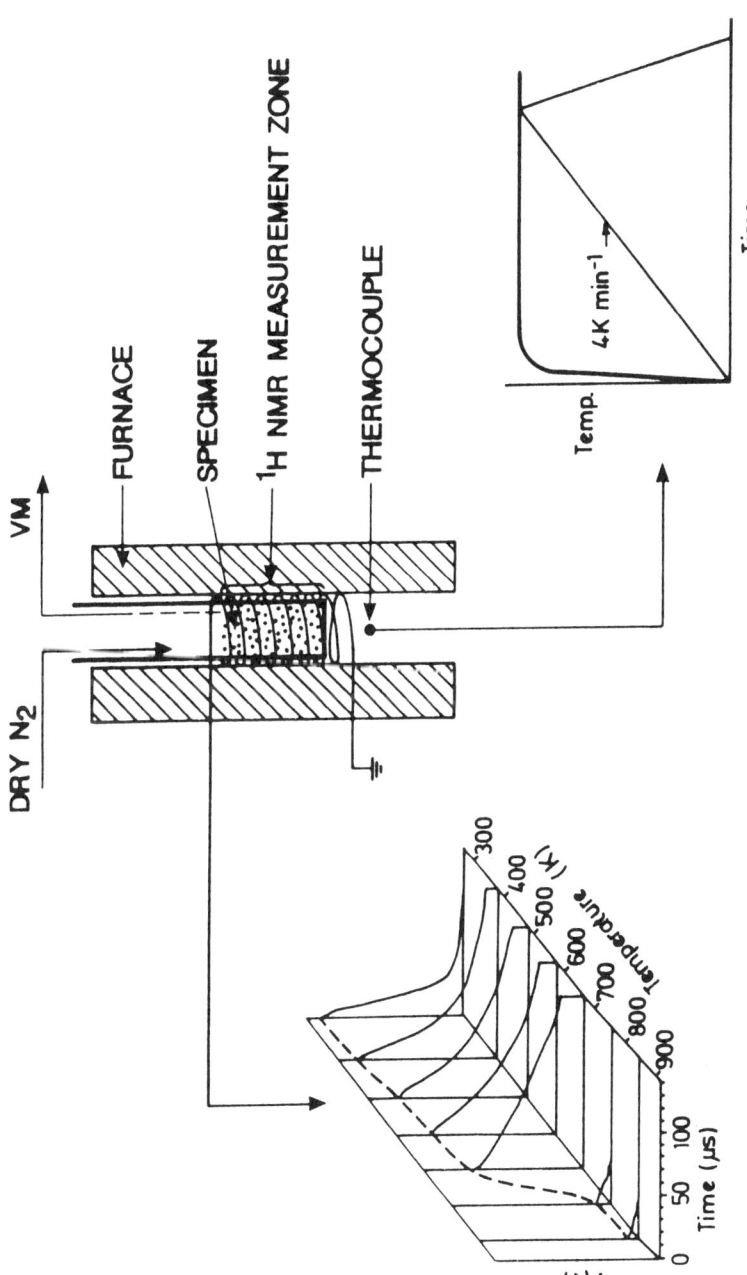

Figure 17. *Schematic of the operation of a PMRTA pyrolysis experiment.*

The block diagram in Figure 16 outlines the main components of the spectrometer (Bruker Minispec 120) interconnections to the additional hardware and functions required for thermal analysis operation. These additions include the high-temperature probe–duplexer assembly and associated sample temperature-control electronics as well as a "master" computer (IBM PC type) for data acquisition and processing and for control of the hardware functions of digital rf pulse-width adjustment, rf probe tuning, magnetic field adjustment, temperature control, and NMR signal digitization. Master computer control of the Bruker Minispec settings is through an RS232 interface. Different pulse sequences necessary for PMRTA setting up and measurement are encoded in an experiment definition module (EDM) laboratory ROM (read-only memory) set and are loaded on master computer command to the Minispec console.

A detailed outline of PMRTA methodology is given in reference 52. For a typical PMRTA experiment (Figure 17), ~500 mg of powdered and predried specimen is contained in an open glass tube under flowing dry nitrogen gas. Experiments are conducted either during uniform heating at rates up to 20 K/min or under isothermal conditions after the specimen has been bought to temperature by uniform or "shock" heating. The two-pulse $90°_x$–τ–$90°_y$ solid-echo sequence (55) is used preferably to generate the ^{1}H NMR transverse relaxation signal $I(t)$ because of its advantage over a single-pulse stimulation in reducing the loss of initial signal information due to the dead-time effect, which is significant for rigid solid materials.

The ^{1}H NMR signals so generated are recorded by signal averaging at an appropriate rate within a period of 30–60 s at regular intervals during the course of the experiment. The appropriate rate is set to avoid saturation effects and is assessed in situ by estimation of the spin–lattice relaxation rate, which will change with the thermally induced changes in the molecular properties of the specimen. The extent of signal averaging and hence interval of measurement are influenced by this changing spin–lattice rate and the deterioration in signal-to-noise (S/N) ratio as the temperature is increased. This deterioration in S/N of the NMR measurement is caused by the temperature dependence of both the macroscopic proton magnetization (Curie's law) and the response (Q) of the NMR detection circuit. Typically the S/N of ~30:1 at room temperature of a 500-mg coal specimen containing $\sim 2 \times 20^{22}$ protons will reduce to ~2:1 at 800 K because of this temperature sensitivity together with pyrolysis loss from the sample, whereas the spin–lattice limited signal averaging rate will vary by a factor of perhaps 5–10.

"Correction" for those temperature sensitivity effects that are external to the molecular properties of the specimen is achieved by calibration against the temperature dependence of the ^{1}H NMR signal amplitude of a thermally stable char specimen. This calibration allows direct comparison of the amplitudes of the ^{1}H NMR signals obtained over a range of temperatures between 290 and 875 K.

Stacked plots of ^{1}H NMR signals recorded in PMRTA pyrolysis experiments under uniform and isothermal heating conditions are shown in Figures 18 and 19, respectively.

The data shown in Figure 18 are of a highly thermoplastic bituminous coal heated under pyrolysis conditions at 4 K/min to 875 K. Much qualitative information is apparent from these raw data. In particular, the pyrolytic decomposition evidenced by the decrease in initial signal amplitude $I(0)$ above ~700 K and the thermoplastic event shown by the rapid increase and decrease in the lifetimes of the ^{1}H NMR signals between ~620 and 750 K are well-defined. Parameters that can be derived from these data, when plotted as functions of temperature, are pyrograms descriptive of the thermal-transformation properties of the specimen.

The initial signal amplitude corrected for external temperature effects $I(0)$ is in most instances (56) closely proportional to the hydrogen content of the specimen. Therefore the $I(0)$ or hydrogen-loss pyrogram plotted in Figure 20 is akin to that of a weight-loss pyrogram in thermogravimetric

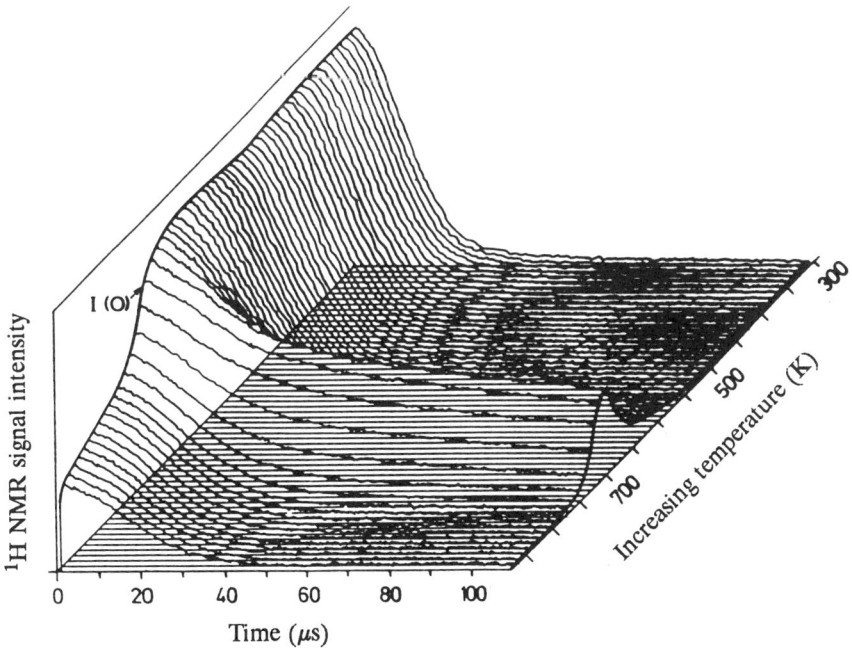

Figure 18. A stacked plot of ^{1}H NMR signals obtained during a pyrolysis thermal analysis experiment of a thermoplastic bituminous coal at a uniform heating rate of 4 K/min.

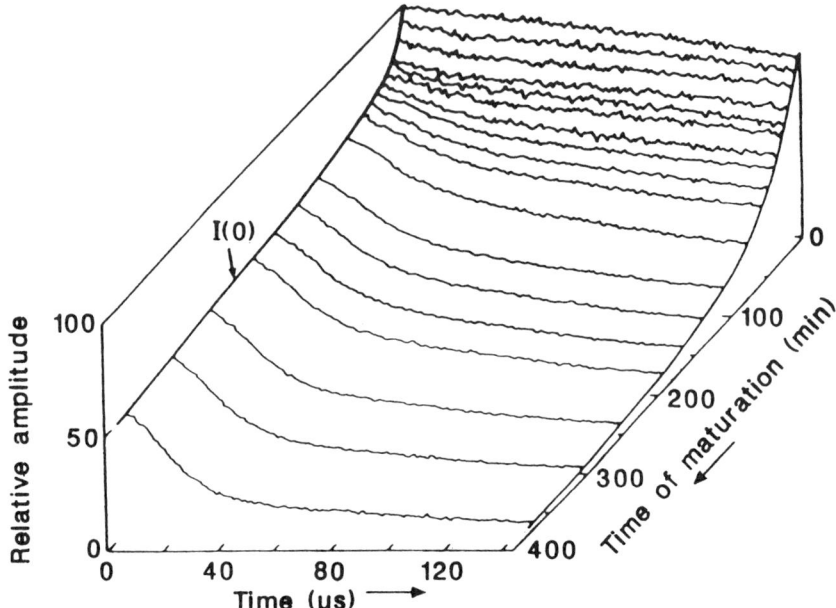

Figure 19. A stacked plot of ^1H NMR signals obtained during a maturation thermal analysis experiment of a pitch at 673 K. (Reproduced with permission from reference 52. Copyright 1988 Academic Press.)

analysis (TGA). The differential of this pyrogram, also plotted in Figure 20, gives the temperature rate of pyrolysis loss from the specimen that is seen to be a maximum for this coal at ~725 K.

In many applications the relatively slowly relaxing tail of the ^1H NMR signals can be closely approximated by exponential functions $i_m(t)$ with apparent transverse relaxation time constants typical of mobile molecular structures, and the difference signal, $i_r(t) = I(t) - i_m(t)$, is rapidly decaying and Gaussian-like, typical of rigid-like molecular structures (57). This separation of the ^1H NMR signals into components permits hydrogen-weighted "mobile" and "rigid" fractions of the molecular structure to be defined. Such a parameter, the "mobile" remaining hydrogen as a percentage fraction of the initial hydrogen content of the specimen, m_r, is plotted in Figure 21. This pyrogram clearly delineates the temperature region of thermoplasticity of this bituminous coal and shows that at ~710 K a maximum of ~60% of the hydrogen initially contained in the coal sample exists in a mobile or "thermoplastic" molecular lattice.

The variations apparent in the ^1H NMR signals during heating (Figure 18) are manifestations of their sensitivity to the thermally induced

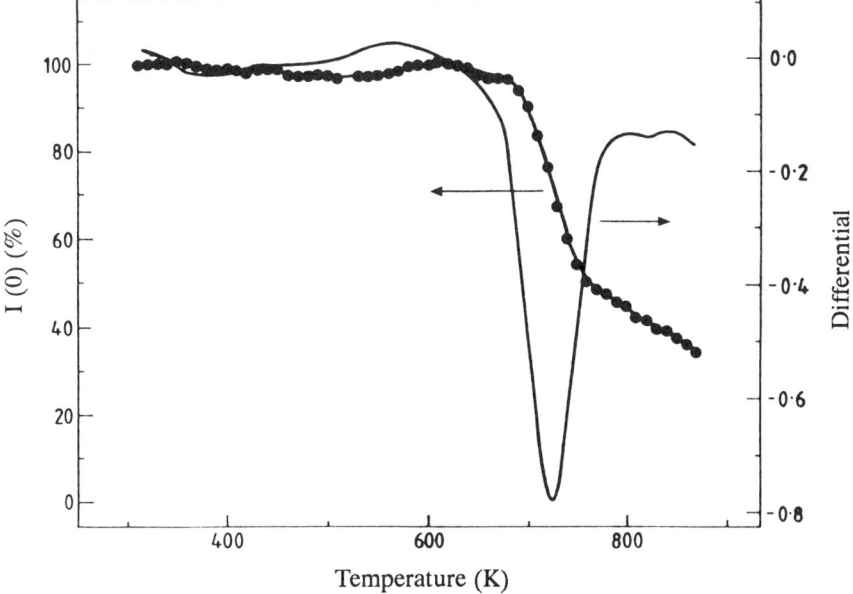

Figure 20. I(0) or hydrogen-loss pyrogram derived from the PMRTA data for a thermoplastic bituminous coal shown in Figure 18. The differential of this pyrogram is also shown.

changes in the molecular dynamics of the coal sample. A type of parameter highly effective in "capturing" this sensitivity is that calculated as an empirical (and truncated) second moment $M_{2T}\nu$ of the frequency domain spectrum $G(\nu)$ obtained by Fourier transformation of $I(t)$:

$$M_{2T}\nu = M_{2T}(\nu_T) = \frac{\int_0^{\nu_T} \nu^2 G(\nu)\, d\nu}{\int_0^{\nu_T} G(\nu)\, d\nu} \quad (1)$$

where ν and ν_T are frequencies specified relative to the resonance frequency. The value of $M_{2T}\nu$ is very sensitive to the molecular dynamics and also sensitive to the concentration and distribution of protons and other magnetic species (i.e., unpaired electrons) in the molecular lattice (58). For a material of stable composition $M_{2T}\nu$ (for $\nu \geq 16$ kHz) is a maximum when the molecular lattice is rigid and decreases rapidly with onset of molecular motions with reorientation rates greater than $\sim 10^5$ Hz—rates sufficient to effectively "motionally narrow" proton dipolar magnetic interactions.

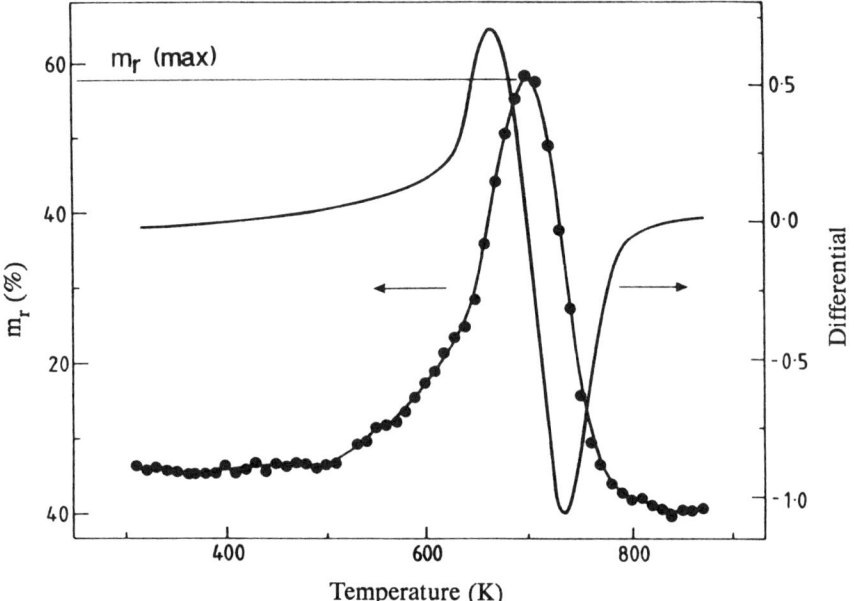

Figure 21. m_r or "rigid-remaining hydrogen" pyrogram derived from the PMRTA data for a thermoplastic bituminous coal shown in Figure 18. The differential form of this pyrogram is also shown.

Because $M_{2T}\nu$ is computed as an integral function of the measured signal, it is representative of the total specimen and also subject to lesser experimental scatter than "curve-fitting" parameters such as m_r defined previously. The significance of equation 1 and the truncation frequency used to define $M_{2T}\nu$ are discussed by Sakurovs et al. in Chapter 11. We stress here the valuable information content inherent in this $M_{2T}\nu$ parameter: It characterizes a material according to features of it structural (hydrogen distribution) and molecular dynamic properties.

The $M_{2T}\nu$ values calculated for a truncation frequency of 16 kHz ($M_{2T}16$) are plotted versus temperature in Figure 22. Again the thermoplastic nature of the coal is clearly delineated by this $M_{2T}16$ pyrogram, which can be used to characterize the coal accordingly (59). Also of significance is the information content of the secondary parameters that can be read from this pyrogram and its temperature derivative. A number of the secondary parameters are illustrated in Figure 22. These are

- $T_{M2T}(\text{min})$, the temperature of the minimum value of $M_{2T}16$;
- $T_{M2T}16(\text{max}-)$, the temperature of maximum rate of decrease of $M_{2T}16$;

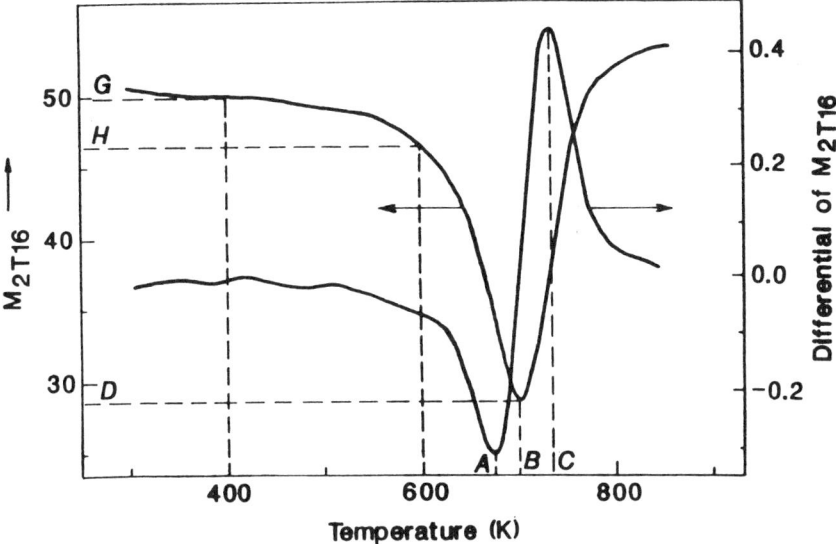

Figure 22. ^1H *NMR pyrogram of the second-moment parameter* $M_{2T}16$ *from analysis of a thermoplastic bituminous coal. The differential form of the pyrogram is plotted, and important secondary parameters are indicated. Definitions of useful secondary parameters are indicated by the letters A, B, and C (temperatures) and D, G, and H (*$M_{2T}16$ *values) that can be used to characterize the coal and predict property values as shown in Table I.*

- $T_{M2T}16(\text{max}+)$, the temperature of maximum rate of increase of $M_{2T}16$;
- $M_{2T}16$ values at selected temperatures.

These parameters have proven to be particularly useful for the characterization of coal materials (*60*).

A database of ^1H NMR thermal analysis and conventional analytical and other test data for a representative selection of Australian coals has been established (*60*). One- or two-parameter correlations between the ^1H NMR and conventional coal test data have been computed. Examples of useful correlations that have been established in this way are (1) between $T_{M2T}(\text{min})$ and the maximum vitrinite reflectance $R_{v,\text{max}}$ shown in Figure 23 and (2) between a linear combination of $T_{M2T}16(\text{max}+)$ and $M_{2T}40$ at 640 K and the Hardgrove grindability index (HGI) shown in Figure 24. Correlations established from this database constitute a powerful predictive tool to characterize Australian coals. Thus a wide range of characteristics used to specify coals can be estimated to a useful degree of accuracy by using only the data generated in a standard ^1H NMR thermal

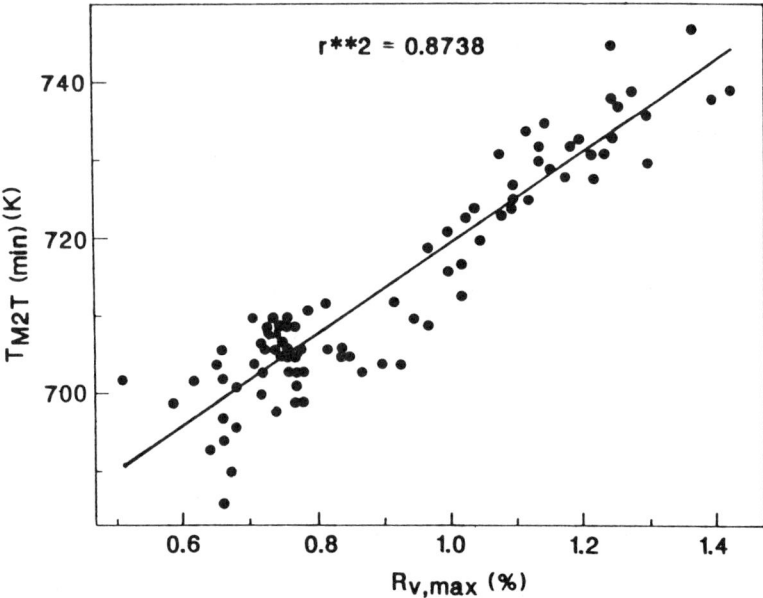

Figure 23. Correlation between the PMRTA parameter $T_{M2T}16$ (min) and the maximum vitrinite reflectance $R_{v,max}$ for a representative selection of Australian bituminous coals.

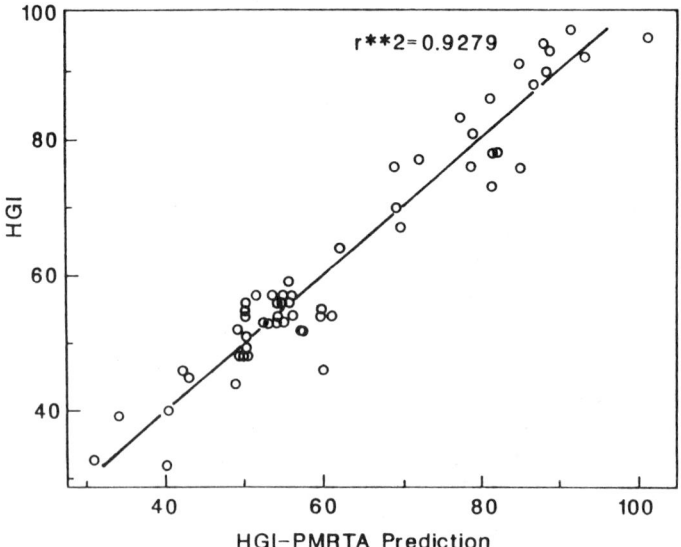

Figure 24. Correlation between the Hardgrove grindability index (HGI) and a linear combination of the PMRTA parameters $M_{2T}40$ at 640 K and $T_{M2T}(max+)$ for a representative selection of Australian coals.

Table I. Comparison of Coal Property Values of a High-Volatile Bituminous Coal Predicted from PMRTA and Conventional Methods of Analysis

Property[a]	PMRTA Prediction	Standard Error	Actual Value
Carbon content (% daf)	83.8	0.8%	83.4
Hydrogen content (% daf)	5.57	0.14%	5.59
Moisture content (% ad)	2.6	0.7%	2.9
Volatile matter (% daf)	38.9	1.5%	39.5
Specific energy (MJ/kg, daf)	34.77	0.3 MJ/kg	34.64
Crucible swelling number	6.0	1.5	7.0
Gray–King coke type	G4	2	G4
Hardgrove grindability index	57	6	54
$R_{v,\max}$ (all vitrinite) (%)	0.73	0.06%	0.73
Vitrinite content (% mmf)	71	10%	69
Gieseler plastometry			
Softening temperature (°C)	384	9°	380
log (maximum fluidity, ddpm)	2.67	0.6 units	2.78
Temperature of maximum fluidity (°C)	435	7°	430
Solidification temperature (°C)	466	7°	460
Audibert-Arnu dilatometry			
Softening temperature (°C)	382	12°	375
Temperature of maximum contraction (°C)	425	12°	425
Temperature of maximum dilation (°C)	459	7°	460

[a] ABBREVIATIONS: daf, dry, ash-free basis; mmf, mineral-matter free; and ddpm, dial divisions per minute.

analysis experiment. Table I compares the conventionally measured and PMRTA predicted values for such a range of characteristics of a high-volatile bituminous coal. Also listed are the standard errors for the individual predictions.

^{1}H NMR data measured during isothermal heat treatment of a material such as these shown in Figure 19 for a pitch specimen can be parameterized similarly to the nonisothermal data (61, 62). The time dependence of these parameters for a range of heat-treatment temperatures together with the corresponding nonisothermal pyrograms for a range of heating rates comprise a data set suitable for detailed kinetic modeling. The pyrolysis of an oil shale kerogen has been modeled by using this approach.

References

1. Poole, C. P., Jr. *Electron Spin Resonance: A Comprehensive Treatise on Experimental Techniques;* Interscience: New York, 1967.
2. Alger, R. S. *Electron Paramagnetic Resonance: Techniques and Applications;* Interscience: New York, 1968.
3. Retcofsky, H. L. In *Chemistry of Coal Utilization, Supplementary Volume;* Elliott, M. A., Ed.; John Wiley and Sons: New York, 1981; p 241.
4. Ingram, D. J. E.; Tapley, J. G.; Jackson, R.; Bond, R. L; Murnagham, A. R. *Nature (London)* **1954**, *174*, 797.
5. Brown, H. R.; Walters, P. L. *Fuel* **1966**, *45*, 17.
6. Thomann, H.; Silbernagel, B. G.; Jin, H.; Gebhart, L. A.; Tindall, P.; Dyrkacz, G. R. *Energy Fuels* **1988**, *2*, 333–339.
7. Ubersfeld, J.; Etienne, A.; Combrisson, J. *Nature (London)* **1954**, *174*, 615.
8. Petrakis, L.; Grandy, D. W. *Free Radicals in Coal and Synthetic Fuels;* Elsevier: Amsterdam, Netherlands, 1983.
9. Seehra, M. S.; Ghosh, B.; Mullins, S. E. *Fuel* **1986**, *65*, 1315.
10. Fowler, T. G.; Bartle, K. D.; Kandiyoti, R. *Fuel* **1987**, *66*, 1407.
11. Sprecher, R. F., Retcofsky, H. L. *Fuel* **1983**, *62*, 473.
12. Yokono, T.; Obara, T.; Iyama, S.; Sanada, Y. *Carbon* **1984**, *22*, 623.
13. Yokono, T.; Iyama, S.; Sanada, Y.; Makino, K. *Fuel* **1985**, *64*, 1014.
14. Yokono, T.; Iyama, S.; Sanada, Y.; Shimokawa, S.; Yamada, E. *Fuel* **1986**, *65*, 1701.
15. Murakami, K.; Yokono, T.; Sanada, Y. *Fuel* **1986**, *65*, 1709.
16. Bakr, M.; Yokono, T.; Sanada, Y. *Proc. Int. Conf. Coal Sci.* (Tokyo) **1989**, 217.
17. Seehra, M. S; Ghosh, B. *J. Anal. Appl. Pyrolysis* **1988**, *13*, 209.
18. Fowler, T. G; Bartle, K. D.; Kandiyoti, R; Snape, C. E. *Carbon* **1989**, *27*, 197.
19. Burnham, A. K.; Myongsook, S. O.; Richard, W. C. *Energy Fuels* **1989**, *3*, 42.
20. Maresch, G. G.; Kendrick, R. D.; Yannoni, C. S. *Rev. Sci. Instrum.* **1990**, *61*, 77–80.
21. Taulelle, F.; Coutures, J. P.; Massoit, D.; Rifflet, J. P. *Bull. Magn. Reson.* **1989**, *11*, 318–20.
22. Ploumbidis, D. *Exp. Tech. Phys.* **1982**, *30*, 323–34.
23. Carlos, W. E.; Taylor, P. C. *J. Phys. (Colloq.)* **1981**, *C4*, 725–7.
24. Pan, Linzhang; Xu, S.; Tian, F. *Bopuxue Zazhi* **1988**, *5*, 295–9.
25. Warren, W. W., Jr.; Dupree, R. *Phys. Rev. B: Condens. Matter* **1980**, *22*, 2257–75.
26. Ruenger, R.; Ploumbidis, D. *Phys. Chem. (Kosmophys)* **1984**, *39A*, 145–7.
27. Kruglov, V. F.; Verkhovskii, S. V.; Kleschev, G. V. *Fiz. Met. Metalloved.* **1983**, *55*, 617–9.

28. Jaeger, C.; Scheler, G. *Exp. Tech. Phys.* **1984**, *32*, 315–23.
29. Liu, S. B.; Stebbins, J. F.; Schneider, E.; Pines, A. *Geochim. Cosmochim. Acta* **1988**, *52*, 527–38.
30. Shimokawa, S.; Maekawa, H.; Yamada, E.; Maekawa, T.; Nakamura, Y.; Yokokawa, T. *Chem. Lett.* **1990**, *4*, 617–20.
31. Clark, W. G.; Wong, W. H.; Hines, W. A.; Lan, M. D.; MacLaughlin, D. E.; Fisk, Z.; Smith, J. L.; Ott, H. R. *J. Appl. Phys.* **1988**, *63*, 3890–92.
32. Adler, S. B.; Michaels, J. N.; Reimer, J. A. *Rev. Sci. Instrum.* **1990**, *61*, 3368–71.
33. Topic, B.; Haeberlen, U.; Blinc, R. *Z. Phys. Rev. B: Condens. Matter* **1990**, *79*, 275–8.
34. Shimokawa, S.; Yamada, E. *Kogakubu Kenkyu Hokoku (Hokkaido Daigaku)* **1980**, *102*, 155–9.
35. Shimokawa, S.; Yamada, E. *Kogakubu Kenkyu Hokoku (Hokkaido Daigaku)* **1983**, *116*, 93–8.
36. Shimokawa, S.; Yamada, E. *Kogakubu Kenkyu Hokoku (Hokkaido Daigaku)* **1983**, *116*, 87–92.
37. Shimokawa, S.; Yamada, E.; Makino, K. *Bull. Chem. Soc. Jpn.* **1983**, *56*, 412–15.
38. Miyazawa, K.; Yokono, T.; Sanada, Y. *Carbon* **1979**, *17*, 223–5.
39. Miyazawa, K.; Yokono, T.; Sanada, Y.; Yamada, E.; Shimokawa, S. *Carbon* **1981**, *19*, 143–4.
40. Miyazawa, K.; Yokono, T.; Sanada, Y. *Ext. Abstr. Program 15th Bienn. Conf. Carbon* **1981**, 126–7.
41. Shimokawa, S.; Yamada, E.; Yokono, T.; Iyama, S.; Sanada, Y. *Carbon* **1984**, *22*, 433–6.
42. Shimokawa, S.; Yamada, E.; Yokono, T.; Yamada, J.; Sanada, Y.; Inagaki, M. *Carbon* **1986**, *24*, 771–2.
43. Yokono, T.; Takahashi, N.; Kaneko, T.; Sanada, Y. *Fuel* **1990**, *69*, 796–8.
44. Azami, K.; Yokono, T.; Sanada, Y.; Uemura, S. *Ext. Abstr. Program 17th Bienn. Conf. Carbon.* **1985**, 157–8.
45. De Lopez, H.; Yokono, T.; Takahashi, N.; Sanada, Y. *Fuel* **1988**, *67*, 301–3.
46. Azami, K.; Yokono, T.; Sanada, Y.; Uemura, S. *Carbon* **1989**, *27*, 177–83.
47. Yokono, T.; Miyazawa, K.; Sanada, Y.; Marsh, H. *Fuel* **1981**, *60*, 603–6.
48. Yokono, T.; Obara, T.; Sanada, Y.; Miyazawa, K. *Carbon* **1984**, *22*, 169–71.
49. Yokono, T.; Iyama, S.; Sanada, Y.; Shimokawa, S.; Yamada, E. *Fuel* **1986**, *65*, 1701–4.
50. Shimokawa, S.; Yamada, E. *J. Magn. Reson.* **1983**, *51*, 103–9.
51. Webster, D. S.; Cross, L. F.; Lynch, L. J. *Rev. Sci. Instrum.* **1979**, *50*, 390.
52. Lynch, L. J.; Webster, D. S.; Barton W. A. In *Advances in Magnetic Resonance 12;* Waugh, J. S., Ed.; Academic: San Diego, CA, 1988; pp 385–421.
53. Webster, D. S.; Lynch, L. J. International Patent Application No. PCT/AU90/00519, 1990. NMR Thermal Analyser, Commonwealth Scientific and Industrial Research Organisation.
54. Lowe, I. J.; Tarr, C. E. *J. Phys. E* **1968**, *1*, 320.
55. Powles, J. D.; Mansfield, P. *Phys. Lett.* **1962**, *2*, 58.
56. Lynch, L. J.; Sakurovs, R.; Barton, W. A. *Fuel* **1986**, *65*, 1108.
57. Lynch, L. J.; Webster, D. S. *Fuel* **1979**, *58*, 235.

58. Barton, W. A.; Lynch, L. J. *J. Magn. Reson.* **1988**, *77*, 439–459.
59. Sakurovs, R.; Lynch, L. J.; Barton, W. A. In *Magnetic Resonance of Carbonaceous Solids;* Botto, R. E.; Sanada, Y., Eds.; Advances in Chemistry 229; American Chemical Society: Washington DC, 1993.
60. Lynch, L. J.; Sakurovs, R.; Maher, T. P.; Bannerjee, R. N. *Energy Fuels* **1987**, *1*, 167–172.
61. Lynch, L. J.; Webster, D. S.; Sakurovs, R.; Maher, T. P. *Proc. Int. Conf. Coal Sci.* (Tokyo) **1989**, 1107–1110.
62. Bacon, N. A.; Barton, W. A.; Lynch, L. J.; Webster, D. S. *Carbon* **1987**, *25*, 669.
63. Parks, T. J.; Lynch, L. J.; Webster, D. S. *Fuel* **1987**, *66*, 338.

RECEIVED for review March 1, 1991. ACCEPTED revised manuscript October 2, 1991.

NMR SPECTROSCOPY

8

Diffusion-Coupled Relaxation and the Submicroscopic Structure of Bituminous Coals

Wesley A. Barton, Leo J. Lynch, David S. Webster, and Greg Simms

Commonwealth Scientific and Industrial Research Organisation (CSIRO), Division of Coal and Energy Technology, P.O. Box 136, North Ryde, New South Wales 2113, Australia

The applicability of a novel approach to modeling the diffusion-coupled spin–lattice relaxation of magnetization in heterogeneous organic solids for resolving microdomain structure in solvent-swollen coals is examined. Three NMR selective-excitation experiments, one of them new, are considered. Methods based on existing theory for heat conduction in solids are outlined for analyzing the data obtained by these selective excitation techniques for particular two-phase systems in terms of the morphological and intrinsic relaxation properties of the separate phases. ^1H NMR results for pyridine-swollen bituminous vitrinites show that these materials contain very small domains of solvent-destabilized and rigid-lattice molecular structures and that magnetization transfer between protons in these two phases is rapid compared to the intrinsic spin–lattice relaxation processes.

THE MOLECULAR STRUCTURE AND PROPERTIES of bituminous coals have been investigated extensively with ^1H NMR techniques (1–3). In particular, the effects of swelling coals in specific solvents on their ^1H NMR spin–spin relaxation behavior have provided valuable insights into the molecular conformation and stability of coals (4–7).

^{1}H NMR transverse-magnetization signals measured at ambient temperatures for bituminous coals swollen by nucleophilic solvents, such as pyridine, which destabilize polar interactions contributing to the conformational stability of the molecular lattice (8), show that up to ~60% of the coal molecular structures can be sufficiently destabilized to become mobile (4, 5, 9). Thus, at least 40% of the coal hydrogen remains in rigid molecular structures and is characterized by an NMR signal component that is similar to the signal for the corresponding dry coal. This result requires the existence in the coals of regions (or domains) that are extensive on the molecular scale (i.e., at least an order of magnitude greater than chemical-bond lengths) and that are apparently impervious to these solvents. Optical studies of thin sections of bituminous vitrinite by Brenner (10, 11) have demonstrated that pyridine saturation swells the coal conformally and at the same time relaxes natural optical anisotropy. These results indicate that the domains that are impervious to the pyridine must be small on the scale of optical wavelengths (i.e., <1 μm).

We therefore have a two-phase concept of bituminous coal molecular structure: (1) regions in which molecular rigidity is stabilized by polar interactions and thus are susceptible to penetration and destabilization to a "rubbery" or mobile state by polar solvents and (2) regions that are impervious and therefore remain as a stable, rigid molecular lattice in the presence of such solvents. The ^{1}H NMR transverse-magnetization signal resolves these phases so that their extent can be estimated. However, the distribution of the material constituting these different phases is unclear.

In a heterogeneous or multiphase system such as coal, the observed ^{1}H NMR relaxation behavior will be a linear combination of the behavior of the separate phases only when a negligible exchange of magnetization occurs between the phases on the time scale of the measurement. Under this condition, the time constants of the observed relaxation process represent the intrinsic relaxation behavior of the separate phases. For both dry bituminous coals and those swollen with deuterated solvents, this situation generally applies to proton spin–spin relaxation, which occurs on a time scale of ~10^{-5}–10^{-3} s (4, 5, 9), but not to the spin–lattice relaxation process, which is characterized by times of ~10^{-1} s (2). For coals swollen by saturation with deuterated pyridine, two "mobile" hydrogen populations with significantly different mobilities and a "rigid" hydrogen population can be distinguished from the ^{1}H NMR transverse magnetization signals (5, 9). However, the observed proton spin–lattice relaxation behavior of the different molecular phases resolved by the ^{1}H NMR spin–spin relaxation of solvent-swollen coals is likely to differ greatly from the intrinsic relaxation behavior because of the averaging effects of spin and molecular diffusion and also because of magnetization exchange between the phases.

This chapter discusses theoretical models and associated NMR experimental techniques, including some novel approaches, whereby diffusion-coupled spin–lattice relaxation behavior of heterogeneous organic solids can be resolved, and the size and dispersion of the separate domains, in principle, can be determined. Because many parameters require definition in such models, extensive experimental data are necessary. NMR experiments that involve selective excitation of the spin systems of the different phases allow this data requirement to be achieved. These experiments are possible when the separate phases in a heterogeneous material give rise to distinct components in the transverse-magnetization signal, so that the Zeeman magnetization in these phases can be established at different levels of saturation. The subsequent relaxation behavior is modified by magnetization exchange between the domains of the separate phases via spin and molecular diffusion. The theories provide models by which the data obtained can be analyzed in terms of the intrinsic spin–lattice relaxation and morphological properties of the phases and the magnetic coupling between them.

The selective excitation techniques described here are derived from those developed by Edzes and Samulski (*12*) and by Goldman and Shen (*13*) to investigate two-phase systems in which the spin–spin relaxation times, T_{2R} and T_{2M}, for the two phases (rigid and mobile) are distinctly different, that is, $T_{2R} << T_{2M}$.

In the Edzes and Samulski experiment, a "soft" inversion pulse (i.e., with a length t_p such that $T_{2R} < t_p << T_{2M}$) is used to prepare the spin populations, characterized by these different spin–spin relaxation times, at different initial Zeeman magnetizations. The two spin populations subsequently relax via spin–lattice relaxation and concomitant cross-relaxation of the Zeeman magnetization between the two phases. Within each of the two phases of the material, spin or molecular diffusion is assumed to rapidly establish a uniform magnetization density at all times in the relaxation process. In this case, the relaxation process is theoretically described by a twin exponential behavior of the reduced magnetization of each spin population. The Edzes and Samulski experiment, in principle, enables the intrinsic spin–lattice relaxation rates for the two phases and the rates of cross-relaxation between the phases to be estimated.

In the Goldman–Shen experiment, two 90° pulses separated by a time t_1 such that $T_{2R} < t_1 << T_{2M}$ are used to establish different levels of Zeeman magnetization in the two spin populations, and the effects of magnetization diffusion between the two phases, preferably in the absence of significant spin–lattice relaxation, are monitored. The data are analyzed according to a model of magnetization diffusion between the discrete phases analogous to heat diffusion between phases of a material in a thermal conduction experiment. Free diffusion of the magnetization (i.e., not limited by domain boundaries) usually is assumed to occur in at least one of the two phases. The Goldman–Shen experiment, aided by independent estimations of certain intrinsic parameters, is intended to provide information about domain sizes.

Both the Edzes and Samulski and the Goldman–Shen experiments (and related experiments based on the same concept) have been applied extensively to the elucidation of the intrinsic proton-relaxation behavior and the domain morphology of the separate phases in hydrated macromolecular systems (e.g., collagen and keratin) (*12, 14–16*) and heterogeneous polymers (*17–23*). One investigation of domain structures in a coal vitrain by means of the Goldman–Shen technique has been reported (*18*). Efforts have also been made to directly analyze the effects of magnetization diffusion on the spin–lattice relaxation processes of the spin populations (*24*).

In the following sections, advances made in the theory to model the diffusion-coupled relaxation of magnetization after its selective saturation in the separate phase domains of a material are outlined and novel experimental techniques and methods for analyzing the data thereby obtained are considered. Some results for pyridine-swollen bituminous vitrinites will be presented to illustrate the applicability of such selective excitation techniques for defining coal structures at the molecular and submicroscopic levels.

Theory

Basic Concepts. A direct correspondence can be established between the magnetic parameters of a material in which magnetization diffusion occurs and the heat parameters of the analogous thermal-diffusion problem. This correspondence allows the considerable body of theoretical results available from analysis of heat conduction in solids to be transferred to the magnetic case.

The macroscopic Zeeman magnetization, M, can be considered to be analogous to the thermal parameter of heat content, Q. For a region A in which the spin density, ρ, and magnetic dipole moment, μ, are constant,

$$M = \rho\mu \int^{\text{region A}} v(x, y, z, t) \, dA \tag{1}$$

where the integration is over the entire volume of the region, and the Zeeman magnetization density v is the magnetic analogue of temperature. For a population of N spins in an applied magnetic field,

$$v = \frac{N^+ - N^-}{N} \tag{2}$$

is the excess spin fraction aligned parallel to the field. The Zeeman magnetization density, v, and hence M can have both positive and negative values. All relevant magnetic parameters and their thermal analogues are listed in Table I.

Thus mathematical expressions derived in modeling two-phase heat-conduction problems can be readily translated to the analysis of magnetic diffusion-coupled relaxation.

A phase is considered to behave as either a "spin fluid" or a "spin solid", depending on whether isomagnetic conditions can be achieved instantaneously. In a rigid organic solid with small domains, magnetization diffusion is usually sufficiently rapid for the spin fluid description to be an adequate approximation. Isomagnetic conditions in phase domains are also induced by rapid spin–lattice relaxation. The magnetization of a spin-fluid phase is given by

$$M_{\text{fluid}} = N\mu v \qquad (3)$$

In a phase that behaves as a spin solid, diffusion processes are either restricted or free, depending on whether they are affected by the boundaries of the medium. In a finite domain, free diffusion occurs for times insufficient to allow a quantity of magnetization introduced at the center of the domain to reach a boundary.

The boundary impedance H^* (Table I) is an inverse measure of the degree of magnetic contact between domains of the two phases. If the domains are in perfect contact ($H^* = 0$), no discontinuity in the magnetization density occurs at the boundary.

Table I. Correspondence Between Magnetic and Thermal Parameters

Magnetic Parameter	Symbol	Thermal Parameter	Symbol
Macroscopic Zeeman magnetization	M	Heat content	Q
Spin density	ρ	Mass density	ρ^*
Dipole moment	μ	Specific heat	c
Domain length	L	Dimension	L
Diffusion coefficient	K	Diffusivity	D
Boundary impedance	H^*	Contact resistance	H
Zeeman magnetization density	v	Temperature	T

Macroscopic Two-Phase Lamellar System. The microdomains of each phase are considered to exist in a repeating lamellar configuration (Figure 1) and to have either identical lengths L or a Gaussian distribution of lengths. In the Gaussian distribution, L is the average domain length.

When such a macroscopic system is at equilibrium in an applied magnetic field, its spin–spin relaxation behavior following a "hard" excitation pulse (i.e., $t_p \ll T_{2R}$) will delineate the phase populations if magnetization exchange is slow compared to the spin–spin relaxation rates. From eq 1, the observed macroscopic magnetization of each phase is given by

$$M = \mu S \sum_{i=1}^{n} \rho_i L_i v_i \qquad (4)$$

where the subscript i refers to the ith microdomain, the summation is over the n microdomains of the phase (assumed to be the same for both phases), and S is their cross-sectional area. Therefore, the initial amplitudes, A_1 and A_2, of the components corresponding to the two phases in the transverse-magnetization signal are in the ratio

$$\frac{A_1}{A_2} = \frac{\sum_{i=1}^{n} \rho_{1i} L_{1i} v_{1i}}{\sum_{i=1}^{n} \rho_{2i} L_{2i} v_{2i}} \qquad (5)$$

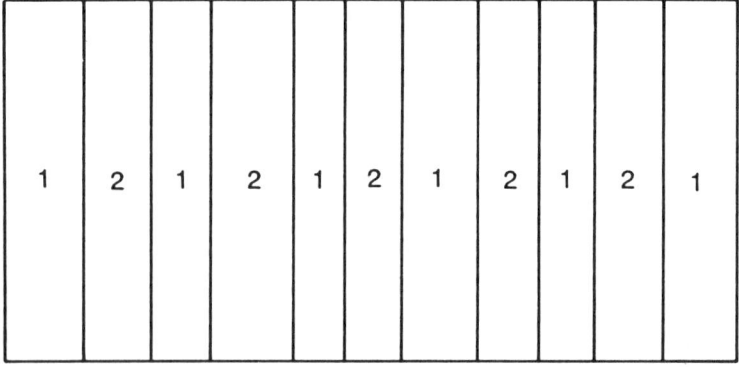

Figure 1. The macroscopic two-phase lamellar model with a distribution of domain lengths in each of the two phases (denoted by 1 and 2).

If the spin densities, ρ_{1i} and ρ_{2i}, and the magnetization densities, v_{1i} and v_{2i}, are independent of i and $v_1 = v_2$, this ratio becomes

$$\frac{A_1}{A_2} = \frac{\rho_1 L_1}{\rho_2 L_2} \tag{6}$$

where L_1 and L_2 are the average domain lengths in the two phases.

Conservative Diffusion Regimes. A conservative system is one in which no significant spin–lattice relaxation occurs during the diffusion process. In this case, the magnetization density, v, within a domain is described by the diffusion equation

$$\frac{\partial v(x, y, z, t)}{\partial t} = K \nabla^2 v(x, y, z, t) \tag{7}$$

where

$$\nabla^2 = \frac{\partial^2}{\partial x^2} + \frac{\partial^2}{\partial y^2} + \frac{\partial^2}{\partial z^2}$$

and K is the spin or molecular diffusion coefficient. Equation 7 states that the magnetization flux at a point in the domain is proportional to the magnetization density gradient at that point.

Nonconservative Diffusion Regimes. In a nonconservative system, spin–lattice relaxation can be regarded as a point source or sink phenomenon that provides the ability to control both the localized production and loss of magnetization within a domain and the magnetization density at the boundaries of an adjoining domain. For diffusion in a domain that contains localized point sources or sinks, the diffusion equation becomes

$$\frac{\partial v(x, y, z, t)}{\partial t} = K \nabla^2 v(x, y, z, t) - R \left[v(x, y, z, t) - v(x, y, z, \infty) \right] \tag{8}$$

where R is the intrinsic spin–lattice relaxation rate.

Reduced Two-Phase Model. For diffusion in a conservative two-phase system with a macroscopic configuration as shown in Figure 1, the magnetization-density distribution in each domain will always be symmetric about the center of the domain. Therefore, in the diffusion analysis a "unit cell" made up of two half domains, one from each phase, can be considered. The boundaries of the unit cell occur at the center lines of the domains at which the gradient of the magnetization density is zero. This boundary condition is the magnetic analogue of an insulated boundary. Extensive quantities in the two-phase system are related to quantities in an average unit cell by the factor $2n$ where n is the number of domains in each phase.

Experimental Techniques

In this section, three types of selective saturation experiments for investigation of diffusion-coupled relaxation in, and the morphology of, two-phase systems are considered. Soft excitation pulses (i.e., ones for which $T_{2R} < t_p << T_{2M}$) will be identified in the following discussions.

Edzes and Samulski Experiment. The selective saturation method devised by Edzes and Samulski (12) uses the pulse sequence:

$$180°_x \text{ (soft)} - t - 90°_x$$

where x denotes the phase of the pulses in the rotating frame and t is the evolution time prior to the final detection pulse, to establish different initial levels of Zeeman magnetization in two spin populations distinguished by different spin–spin relaxation rates. To determine the intrinsic spin–lattice relaxation rates for the two phases and the rates of cross-relaxation between them, at least two different initial magnetizations are required. These different initial magnetizations are achieved by varying the radiofrequency (rf) amplitude and hence the length, t_p, of the 180° soft pulse.

An alternative method of producing different initial levels of magnetization in each of the two phases is to use the Goldman–Shen pulse sequence (13):

$$90°_x - t_1 - 90°_{-x} - t - 90°_x$$

together with a modification of this sequence (hereafter referred to as the modified Goldman–Shen sequence):

$$90°_x - t_1 - 90°_x - t - 90°_{-x}$$

in which the phases of the second and third pulses have been reversed. In a given experiment, the selection time, t_1, is normally fixed at a value that exceeds the decay time of the signal component with the shorter spin–spin relaxation time.

For all three pulse sequences just described, the effects of spin or molecular diffusion and spin–lattice relaxation in the two-phase system are monitored by varying the evolution time t prior to the final detection pulse.

Dynamic Diffusion Experiments. These experiments involve the measurement of variations in the magnetization density over times for which the total Zeeman magnetization of the system also changes. One way of achieving a time-varying magnetization density is to induce spin–lattice relaxation by perturbing the magnetization from its value in equilibrium with the applied magnetic field. Selective saturation techniques allow this perturbation to be achieved in essentially only one of the two phases.

Spatial variations in the magnetization density of a phase domain due to, for instance, diffusion into or out of the domain, are reduced by the spin–lattice relaxation process so that for sufficiently rapid relaxation the magnetization density is approximately uniform throughout the domain. If good magnetic contact exists between the phase domains, this magnetization density is applied to the boundaries of adjoining domains of the second phase.

The particular technique required to selectively activate the spin–lattice relaxation depends on the relative rates of spin–spin relaxation and the intrinsic spin–lattice relaxation rates of the spin populations in the two phases. For the following discussion, the faster spin–spin relaxation is assumed to occur in phase 1. If the intrinsic spin–lattice relaxation rate in phase 1 is rapid compared to that for phase 2, the Goldman–Shen pulse sequence is used to establish approximately the initial (i.e., $t = 0$) Zeeman magnetization-density conditions

$$v_2(0) = v_2(\infty) \qquad v_1(0) = 0 \tag{9}$$

This situation produces spin–lattice relaxation of rate R_1 in phase 1 and thus a time-dependent magnetization density given by

$$v_1(t) = v_1(\infty)\left[1 - \exp(-R_1 t)\right] \tag{10}$$

If perfect magnetic contact occurs between the phase domains, the magnetization density at the boundaries of phase 2 is also given by eq 10.

For intrinsic spin–lattice relaxation that is slow in phase 1 compared to that in phase 2, the modified Goldman–Shen pulse sequence is used to align the magnetization vector of phase 2 antiparallel to the applied field. This alignment induces a maximum rate of change of magnetization density in phase 2 due to spin–lattice relaxation of rate R_2. For perfect contact between the phase domains, the time-dependent boundary condition

$$v_2(t) = v_2(0) \left[2 \exp(-R_2 t) - 1 \right] \tag{11}$$

is applied to the magnetization density in phase 1.

If there is a nonzero boundary impedance between the domains of the two phases, the boundary magnetization density for the phase with slower intrinsic spin–lattice relaxation also depends on the degree of contact between the phase domains.

Conservative Diffusion Experiment. The Goldman–Shen pulse sequence can be used to generate an approximately conservative diffusion regime in which negligible spin–lattice relaxation occurs during the diffusion process (18). This condition can be tested by monitoring the constancy of the total initial signal amplitude over the range of evolution times t used in the experiment. In the phase with the slower spin–spin relaxation, the second pulse in the Goldman–Shen sequence restores the magnetization density almost to the value corresponding to equilibrium with the applied field so that subsequent spin–lattice relaxation is minimal. The time during which the conservative diffusion approximation is valid is limited by the effects of spin–lattice relaxation in the second phase. Therefore, such diffusion experiments are most useful for materials in which this second phase has slow intrinsic spin–lattice relaxation.

At evolution times sufficiently short to avoid significant spin–lattice relaxation effects, the free diffusion approximation is more likely to be valid, particularly in the phase with the slower spin or molecular diffusion.

Methods of Data Analysis

The mathematical expressions describing the magnetization density produced in the phases by the selective saturation experiments outlined in the preceding sections are presented in the following sections together with methods by which parameters relating to the relaxation and structural properties of these phases can be extracted.

Edzes and Samulski Analysis. Combined data obtained from the Goldman–Shen and the modified Goldman–Shen pulse sequences for a number of evolution times t can be analyzed in the same way as data produced by the Edzes and Samulski experiment.

Edzes and Samulski (12) considered the reduced magnetization

$$m_i(t) = \frac{M_{i\infty} - M_i(t)}{2M_{i\infty}} \tag{12}$$

of each spin population i of a two-phase system where $M_i(t)$ is the Zeeman magnetization and $M_{i\infty}$ is its equilibrium value. If a uniform magnetization distribution is rapidly established within each phase at all times in the spin–lattice relaxation process, then this process is described by the coupled differential equations

$$\frac{dm_1(t)}{dt} = -R_1 m_1(t) - k_1 m_1(t) + k_1 m_2(t) \tag{13}$$

$$\frac{dm_2(t)}{dt} = -R_2 m_2(t) - k_2 m_2(t) + k_2 m_1(t) \tag{14}$$

where R_1 and R_2 are the intrinsic spin–lattice relaxation rates of the two spin populations (labeled 1 and 2), and k_1 and k_2 are cross-relaxation rates related by the dynamic equilibrium condition

$$N_1 k_1 = N_2 k_2 \tag{15}$$

where N_1 and N_2 are the spin populations of the two phases.

The general solution of these differential equations is a twin exponential mode of decay for the phase magnetizations, that is,

$$m_1(t) = C_1^+ \exp(-R^+ t) + C_1^- \exp(-R^- t) \tag{16}$$

$$m_2(t) = C_2^+ \exp(-R^+ t) + C_2^- \exp(-R^- t) \tag{17}$$

where

$$2R^{\pm} = R_1 + R_2 + k_1 + k_2 \\ \pm \left[(R_1 - R_2 + k_1 - k_2)^2 + 4k_1 k_2\right]^{1/2} \tag{18}$$

and

$$C_1^{\pm} = \frac{\pm m_1(0)\left[R_1 - R^{\mp}\right] \pm k_1\left[m_1(0) - m_2(0)\right]}{R^+ - R^-} \quad (19)$$

and similarly for $C_2^{\pm}$.

The preceding analysis in terms of intrinsic and cross-relaxation rates can also be carried out in terms of magnetization diffusion between two spin fluid phases with a contact impedance, H^*, between them and spin–lattice relaxation in each phase (Figure 2). For this diffusion analysis, the differential equation for phase 1 magnetization is

$$\frac{dM_1(t)}{dt} = -R_1\left[M_1(t) - M_{1\infty}\right] - \frac{1}{H^*}\left[v_1(t) - v_2(t)\right] \quad (20)$$

which can be expressed in the form

$$\frac{dM_1(t)}{dt} = -R_1\left[M_1(t) - M_{1\infty}\right]$$
$$- \frac{1}{N_1\mu H^*}\left[M_1(t) - \frac{N_1}{N_2}M_2(t)\right] \quad (21)$$

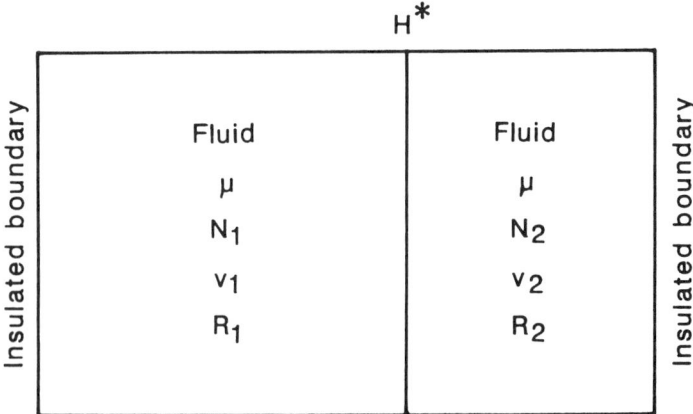

Figure 2. Reduced two-phase model of a spin-fluid system with contact impedance H^* between the phase domains and spin–lattice relaxation rates R_1 and R_2 in the two phases.

by using eq 3. In terms of the reduced magnetization $m_1(t)$, eq 21 becomes

$$\frac{dm_1(t)}{dt} = -R_1 m_1(t) - \frac{1}{N_1 \mu H^*} m_1(t) + \frac{1}{N_1 \mu H^*} m_2(t) \quad (22)$$

which is equivalent to eq 13 if

$$k_1 = \frac{1}{N_1 \mu H^*} \quad (23)$$

The corresponding analysis for phase 2 magnetization yields

$$k_2 = \frac{1}{N_2 \mu H^*} \quad (24)$$

The reduced-magnetization relaxation curves may be obtained from the measured transverse-magnetization signals by two methods. The first method, described by Edzes and Samulski (12), involves separation of the components corresponding to the two phases in each signal by curve-fitting of the signal. The second method uses a multiwindow technique to select data from two windows, one immediately after the final (detection) pulse dead time and the other after the signal component with the shorter spin–spin relaxation time has completely decayed.

Twin exponential fits to the reduced-magnetization relaxation data according to eqs 16 and 17 yield, in principle, the intrinsic spin–lattice relaxation rates and the cross-relaxation rates for the two phases.

Dynamic Diffusion Analysis. Duhamel (25) developed a theorem that can be used to determine the temperature distribution as a function of time in a solid slab, initially at a known temperature, whose surfaces are subjected to a time-varying temperature. This theorem can readily be adapted to describe the magnetization density in a domain of dimension L whose boundaries are subjected to a time-varying magnetization density $v(t)$ and thus to determine the quantity of magnetization that diffuses into the domain in the absence of significant spin–lattice relaxation in the corresponding phase of the material.

Consider a spin-solid phase (phase 1) with zero initial magnetization density in perfect contact with an external spin-fluid phase (phase 2) whose uniform magnetization density is assumed to have a time variation of the form

$$v_2(t) = v_2(0)\left[2\exp(-R_2't) - 1\right] \qquad (25)$$

where the rate R_2' is determined by the combined effects of spin–lattice relaxation and magnetization diffusion (Figure 3). This situation is analogous to the second dynamic diffusion case considered previously. The average magnetization density of phase 1 can be deduced from the analysis of the corresponding heat-conduction case (25) and is given by

$$v_1(t) = 8v_2(0)K_1/L_1^2 \sum_{m\ \text{odd}} \left[\frac{2\left[\exp(-R_2't) - \exp(-m^2\pi^2 K_1 t/L_1^2)\right]}{m^2\pi^2 K_1/L_1^2 - R_2'} + \frac{\exp(-m^2\pi^2 K_1 t/L_1^2) - 1}{m^2\pi^2 K_1/L_1^2}\right] \qquad (26)$$

The expression for $v_1(t)$ (eq 26) can be simplified by assuming that $\pi^2 K_1/L_1^2 \gg R_2'$, which is valid unless the domain dimension L_1 or the

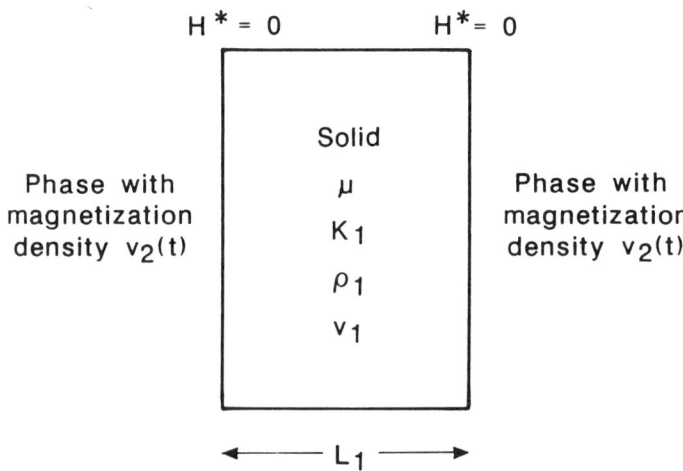

Figure 3. Model of a spin-solid phase in perfect contact with an external phase with magnetization density v(t).

observed magnetization relaxation rate R_2' is very large. Then, at times t for which only the $m = 1$ term in the sum

$$\sum_{m \text{ odd}} \left[\frac{\exp(-m^2\pi^2 K_1 t / L_1^2)}{m^2 \pi^2 K_1 / L_1^2} \right]$$

is significant,

$$v_1(t) = \frac{8v_2(0)}{\pi^2} \left[\frac{\pi^2(2\exp(-R_2't) - 1)}{8} - \exp(-\pi^2 K_1 t / L_1^2) \right] \quad (27)$$

By using eq 25 for the magnetization density of phase 2, the difference in magnetization density between the two phases is

$$v_1(t) - v_2(t) = -\frac{8}{\pi^2} v_2(0) \exp(-\pi^2 K_1 t / L_1^2) \quad (28)$$

A plot of $v_1(t) - v_2(t)$ versus time t can then be used to extract the ratio K_1/L_1^2. When phase 1 consists of a uniform rigid lattice, the spin-diffusion coefficient, K_1, can be estimated from (18)

$$K_1 \rho_1^{2/3} \sim \frac{0.2}{T_{2R}} \quad (29)$$

where T_{2R} is the spin–spin relaxation time. Then the domain dimension L_1 can be estimated by using the preceding analysis.

A result similar to eq 28 can also be derived for the dynamic-diffusion situation in which a spin-solid phase (phase 2) with initial magnetization density $v_2(0) = v_2(\infty)$ is in perfect contact with an external spin-fluid phase (phase 1) whose uniform magnetization density has a time variation of the form given by eq 10. In this case, the ratio K_2/L_2^2 can be evaluated.

Conservative Diffusion Analysis. Consider diffusion in a spin-solid phase (phase 2) that is in perfect contact with a spin-fluid phase (phase 1) (fluid by virtue of its very rapid spin diffusion). In a typical example, the spin-fluid phase would consist of rigid lattice domains, and the spin-solid phase would correspond to mobile molecular structures (*see* *also* the "Theory" section). The initial magnetization densities generated by the Goldman–Shen pulse sequence are $v_1(0) = 0$ and $v_2(0) \simeq v_2(\infty)$. In the conservative diffusion regime, the magnetization density in the spin-fluid phase is given by *(25)*

$$v_1(t) = v_1(\text{eq}) \left[1 - \exp\left(\frac{\rho_2^2 K_2 t}{\rho_1^2 L_1^2}\right) \text{erfc}\left(\frac{\rho_2}{\rho_1 L_1} K_2^{1/2} t^{1/2}\right) \right] \quad (30)$$

where erfc is the error function complement and $v_1(\text{eq})$ is the final equilibrium magnetization density of the phase in the absence of spin–lattice relaxation. This result has also been derived from the spin-diffusion equation (eq 7) by Cheung and Gerstein *(18)*.

For evolution times $t \ll (4\rho_1^2 L_1^2)/(\pi \rho_2^2 K_2)$, eq 30 may be written in the form

$$v_1(t) = v_1(\text{eq}) \left[C \frac{K_2^{1/2}}{L_2} t^{1/2} \right] \quad (31)$$

where

$$C = \left(\frac{2}{\pi^{1/2}}\right)\left(\frac{\rho_2 L_2}{\rho_1 L_1}\right) \quad (32)$$

can be estimated by using eq 6. Equation 31 corresponds to the free-diffusion approximation in the spin-solid phase (Figure 4). Hence the ratio $K_2^{1/2}/L_2$ can be calculated from the initial gradient of a plot of $v_1(t)$ against $t^{1/2}$. If the domain dimension, L_2, of the spin-solid phase has already been estimated, this conservative diffusion experiment allows the spin-diffusion coefficient of this phase to be calculated. In previous applications of this experiment to semicrystalline polymers and other materials *(17, 18, 20)*, indirect estimates of K_2 were used to provide values of L_2.

Application of Selective-Saturation Experiments to Bituminous Coals

The three types of selective saturation experiments described in the preceding sections were carried out at room temperature on protons in vit-

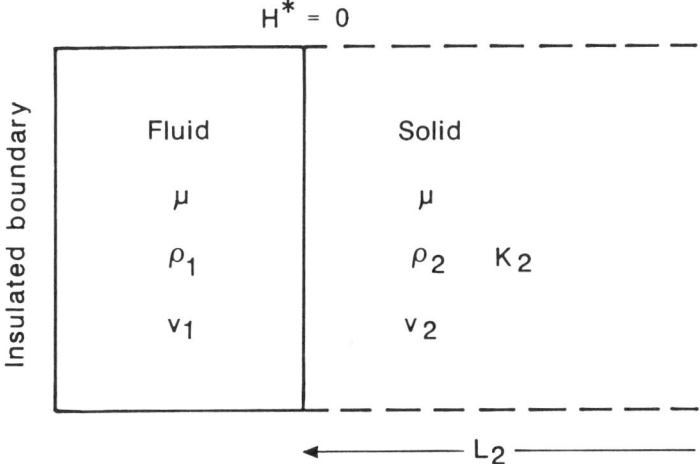

Figure 4. Model of a spin-fluid phase in perfect contact with a spin-solid phase in which free diffusion occurs.

Table II. Selected Properties of the Bituminous Vitrinites Investigated

Coal Seam	C	H	O	Maceral Analysis			Mobilized H
				Vitrinite	Liptinite	Inertinite	
Lithgow	82.4	5.76	9.0	79	8	13	32
Borehole	84.1	5.68	7.4	84	4	12	42
Bulli	85.2	5.46	7.0	79	3	18	29

NOTE: All values are given as percentages. Percent C, H, and O are given on a dry, ash-free basis; the values for the maceral analysis are given as volume percent, mineral-matter free; and the mobilized H is the proportion of hydrogen in coal structures that is mobilized by imbibed pyridine.

rinite-rich specimens of three bituminous coals that had been soaked in excess deuterated pyridine in sealed ampules for at least 18 months. Selected analytical and NMR data for these coals are shown in Table II.

Edzes and Samulski Approach. Data for this type of experiment were acquired by the alternative method, described in the "Experimental Techniques" section, that uses the Goldman–Shen and modified Goldman–Shen pulse sequences. A selection time (t_1) of 60 μs was chosen because it allows complete dephasing of transverse magnetization to occur in rigid-lattice domains but relatively little dephasing of the magnetization associated with mobile molecular structures. Because the signals thereby produced could not be easily separated into two well-defined components, reduced-magnetization relaxation curves were obtained by using the multiwindow technique referred to previously. By using windows immediately after the detection-pulse dead time and after the rigid-lattice signal component has completely decayed, reduced-magnetization values approximating those for the total proton population and for protons in mobile molecular structures, respectively, were deduced. Whereas the reduced-magnetization relaxation data for the mobile proton population can be satisfactorily fitted by a twin exponential curve, the relaxation of the magnetization of the total proton population during the 200-ms time interval considered is approximately described by single exponential behavior (Figure 5). Analysis of the Edzes and Samulski experiment outlined in a previous section of this chapter indicates that single exponential

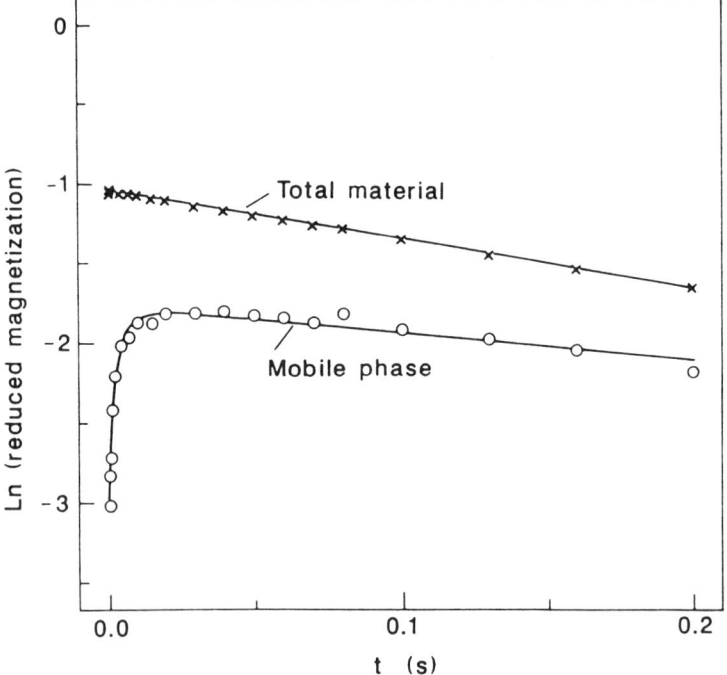

Figure 5a. Reduced proton-magnetization relaxation data for the mobile phase (○) and the total material (×) in pyridine-swollen Borehole vitrinite. The data were obtained from Goldman–Shen experiments, and the solid lines represent exponential fits to the data as described in the text.

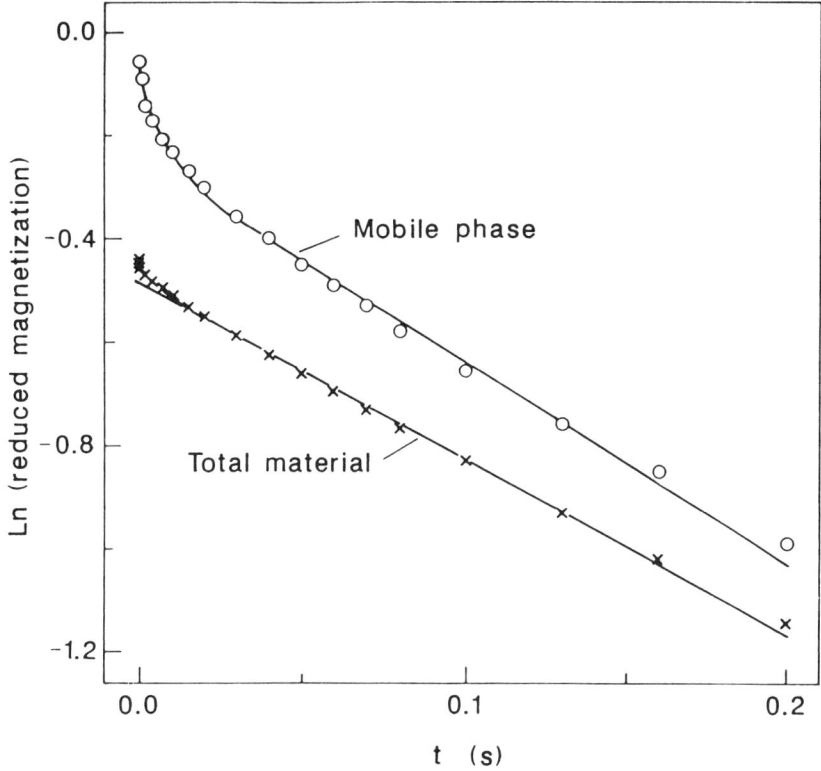

Figure 5b. Same as Figure 5a, except that data were obtained from modified Goldman–Shen experiments.

relaxation of the total reduced magnetization occurs when the intrinsic spin–lattice relaxation rates of the two proton populations are identical.

Values obtained from this analysis for the rates of cross-relaxation (k) between the two phases distinguished (i.e., molecular structures destabilized by pyridine and those that remain as a rigid molecular lattice) are an order of magnitude greater than those for the intrinsic spin–lattice relaxation rates (R), which are indeed similar for both phases and comparable to the average spin–lattice relaxation rates determined for the dry coals (*2*) (Table III). This similarity among the relaxation rates probably arises from the fact that proton relaxation in vitrinites is predominantly due to unpaired electrons rather than to molecular motions (*2, 26*). However, large uncertainties exist in the values calculated for the intrinsic spin–lattice relaxation rates, particularly for the rigid molecular phase for which the average relaxation rates measured for the dry coals may provide more accurate values.

Table III. Proton Relaxation Rates Estimated from Selective Saturation Experiments and for the Dry Coals

Coal Seam	R (rigid)	R (mobile)	R (dry coal)	k (rigid)	k (mobile)
Lithgow	2	8	4.0	21	66
Borehole	<1	10	5.7	27	58
Bulli	1	5	3.4	20	81

NOTE: All values are given in inverse seconds (s^{-1}).

Although the results shown, for example, in Figure 5 demonstrate the occurrence of rapid magnetization diffusion between the two phases in bituminous vitrinites, the reduced magnetizations of these two phases do not become equivalent as a result of this cross-relaxation process. This behavior is also illustrated in Figure 6, which shows the reduced-magnetization data for rigid and mobile proton populations in the Bulli vitrinite estimated from approximate two-component fits to the signals generated in a Goldman–Shen experiment. This nonequivalence of the reduced magnetizations of the two phases is a consequence of the similarity of the intrinsic spin–lattice relaxation rates. When the intrinsic spin–lattice relaxation rate for the mobile phase is appreciably greater than that for the rigid phase, the reduced magnetizations of the two phases will converge and become equivalent as observed in studies of hydrated biological systems (12, 15) and shown in Figure 7 for hydrated keratin.

Dynamic Diffusion Experiments. To deduce information about phase-domain sizes from application of Duhamel's theorem to the analysis of magnetization data obtained in dynamic-diffusion experiments, spin–lattice relaxation must be rapid in one phase relative to that in the other and the first phase must behave as a spin fluid. Such behavior occurs for protons in hydrated keratin in a modified Goldman–Shen experiment (Figure 7). In this material, the intrinsic spin–lattice relaxation of protons in the mobile phase is much faster than that in the rigid phase (largely due to the difference in average molecular mobility), and the rapid translational motions of water molecules in the mobile phase allow it to be treated as a spin fluid. The dimensions of the rigid phase domains in hydrated keratin have been estimated to be ~200 nm (15).

In the bituminous vitrinite–deuterated-pyridine systems studied here, plots of ln $[v_1(t)-v_2(t)]$ versus time t (see eq 28) are distinctly nonlinear for data obtained from both the Goldman–Shen and the modified Gold-

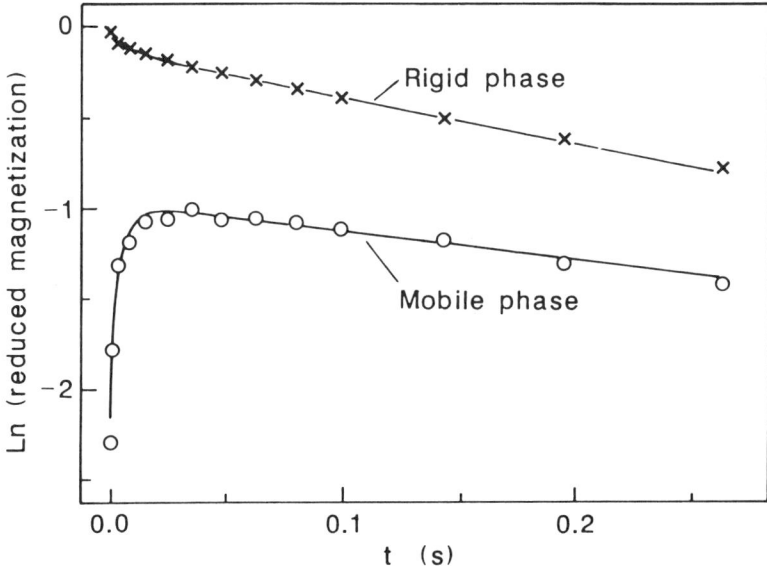

Figure 6. Reduced proton-magnetization relaxation data obtained from a Goldman–Shen experiment for the mobile (○) and rigid (×) phases in pyridine-swollen Bulli vitrinite. The solid lines represent twin exponential fits to the data.

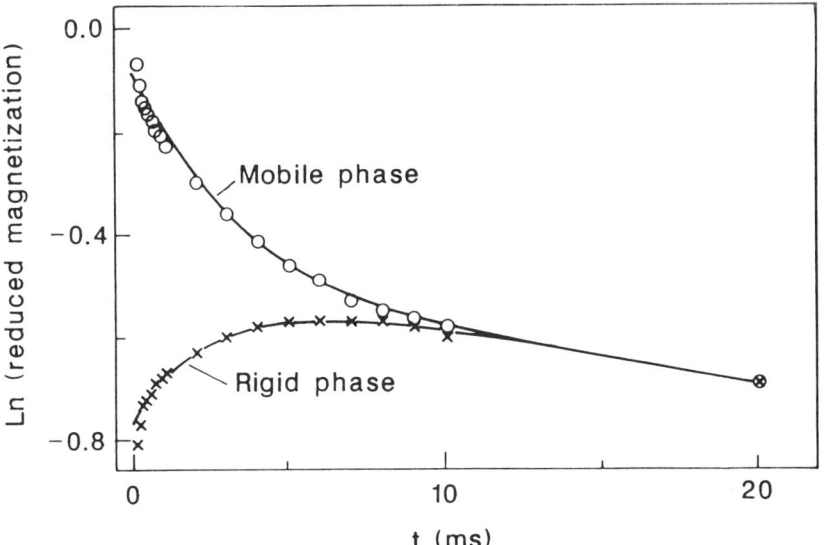

Figure 7. Reduced proton-magnetization relaxation data obtained from a modified Goldman–Shen experiment for the mobile (○) and rigid (×) phases in a hydrated keratin at 270 K. The solid lines represent twin exponential fits to the data.

man–Shen experiments. This behavior shows that at least one of the assumptions made in the Duhamel analysis of dynamic-diffusion experiments is not applicable to these coal–solvent systems and hence is consistent with the finding that the intrinsic spin–lattice relaxation rates of protons in the separate phases are not substantially different (Table III). Furthermore, because the mobile phase in these coal–solvent systems consists largely of coal structures undergoing reorientational rather than translational motions, it is not expected to behave as a spin-fluid phase with a uniform magnetization density.

An alternative approach is to consider pyridine-swollen coals to be comprised of more than two phases (perhaps three as indicated by the ^{1}H NMR transverse-magnetization signals). Although this assumption would increase the number of parameters to be defined, extensions of the selective saturation techniques described here can, in principle, provide information about the dispersion of domain sizes in a multiphase system (15).

Conservative Diffusion Experiment. Results from a Goldman–Shen experiment in which the evolution times t were sufficiently short that spin–lattice relaxation effects were negligible are shown in Figure 8. Significant recovery of the rigid-phase magnetization occurs before spin–lattice relaxation becomes evident, and this recovery is consistent with cross-relaxation rates between the two phases being much greater than the intrinsic spin–lattice relaxation rates (Table III). Although the smaller times t in the data set are also sufficiently short for the free-diffusion approximation (eq 31) to be valid for the pyridine-destabilized phase, the nonlinearity of the plot indicates that this behavior does not occur over the entire range of evolution times used. There is excellent agreement between the values of the ratio $K_2^{1/2}/L_2$ for the pyridine-destabilized phase in this vitrinite obtained from the initial gradient of the plot in Figure 8 and from a fit to all the data by using eq 30. These values ($\sim$24 s$^{-1/2}$) and those for the other two vitrinites investigated ($\sim$30–40 s$^{-1/2}$) are higher than the corresponding results ($\sim$14 s$^{-1/2}$) obtained by Cheung and Gerstein (18) for a Kentucky vitrain (whose ^{1}H NMR signal showed a significant, slowly decaying component without the addition of a solvent to the coal).

Information about average domain sizes in solvent-swollen bituminous coals can be deduced from these values of the ratio $K_2^{1/2}/L_2$. In the preceding analysis of data from a conservative diffusion experiment, magnetic diffusion was assumed to be more rapid in the rigid phase, that is, $K_1 > K_2$. The nature of the rigid-phase magnetization recovery (Figure 8) is

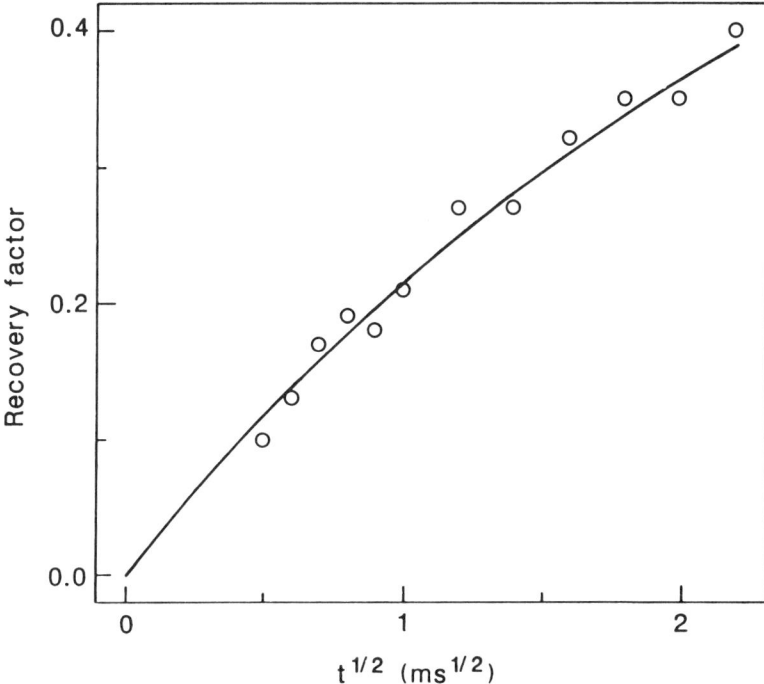

Figure 8. Recovery of the proton magnetization as a function of $t^{1/2}$ (evolution time) in the rigid phase of pyridine-swollen Borehole vitrinite. The data were obtained from a conservative diffusion experiment, and the solid line represents the best fit to the data by eq 30.

consistent with this assumption (18), which is also reasonable in the absence of rapid translational diffusion of coal molecules in the solvent-destabilized phase. Thus, with K_1 estimated from eq 29 to be 8×10^{-12} cm^{2}s^{-1}, the upper limit for L_2 is ~1 nm, and for L_1 (the average rigid domain size) the upper limit is ~4 nm (from eq 6).

These calculations indicate that phase domains in pyridine-swollen coals have dimensions that are close to the lower end of the molecular scale. Such small domain lengths are consistent with the rapid cross-relaxation between the two phases as illustrated in Figures 5 and 6. However, the domain lengths calculated from the one-dimensional lamellar-model analysis used here correspond to the shortest of the three dimensions of actual domains (18), and the other two length parameters may be much greater.

Conclusions

1. A direct correspondence exists between magnetic and thermal parameters, thereby providing the framework within which a wide variety of selective saturation experiments appropriate for the study of microdomain structure in organic solids such as coals can be analyzed by reference to existing theory for heat conduction in solids.
2. Application of this approach to the reduced-magnetization analysis of data produced by an Edzes and Samulski type experiment on a two-phase system establishes relationships between the cross-relaxation rates and the boundary impedance between domains of the two phases.
3. ^{1}H NMR experiments on pyridine-swollen bituminous vitrinites demonstrate the occurrence of rapid magnetization transfer between protons in solvent-destabilized and rigid-lattice molecular structures and indicate that these two types of structure exist in domains whose smallest dimensions are no larger than a few nanometers.
4. The intrinsic spin–lattice relaxation rates for the separate phases identified in these coals by solvent destabilization and ^{1}H NMR spin–spin relaxation are similar and are an order of magnitude less than the rate of magnetization transfer between these phases.

References

1. Davidson, R. M. "Nuclear Magnetic Resonance Studies of Coal," Report No. ICTIS/TR32; IEA Coal Research: London, 1986.
2. Barton, W. A.; Lynch, L. J. *Energy Fuels* **1989,** *3,* 402–11.
3. Sanada, Y.; Lynch, L. J. In *Magnetic Resonance of Carbonaceous Solids;* Botto, R. E.; Sanada, Y., Eds.; Advances in Chemistry 229; American Chemical Society: Washington, DC, 1993.
4. Jurkiewicz, A.; Marzec, A.; Pislewski, N. *Fuel* **1982,** *61,* 647–50.
5. Kamienski, B.; Pruski, M.; Gerstein, B. C.; Given, P. H. *Energy Fuels* **1987,** *1,* 45–50.
6. Sakurovs, R.; Lynch, L. J.; Barton, W. A. In *Coal Science II;* Schobert, H. H.; Bartle, K. D.; Lynch, L. J., Eds.; ACS Symposium Series 461; American Chemical Society: Washington, DC, 1991; pp 111–26.
7. Derbyshire, F.; Marzec, A.; Schulten, H.-R.; Wilson, M. A.; Davis, A.; Tekely, P.; Delpuech, J.-J.; Jurkiewicz, A.; Bronnimann, C. E.; Wind, R. A.; Maciel, G. E.; Narayan, R.; Bartle, K.; Snape, C. *Fuel* **1989,** *68,* 1091–1106.
8. Larsen, J. W.; Green, T. K.; Kovac, J. *J. Org. Chem.* **1985,** *50,* 4729–35.
9. Barton, W. A.; Lynch, L. J.; Webster, D. S. *Fuel* **1984,** *63,* 1262–68.

10. Brenner, D. *Nature (London)* **1983**, *306*, 772–73.
11. Brenner, D. *Fuel* **1984**, *63*, 1324–28.
12. Edzes, H. T.; Samulski, E. T. *J. Magn. Reson.* **1978**, *31*, 207–29.
13. Goldman, M.; Shen, L. *Phys. Rev.* **1966**, *144*, 321–31.
14. Wise, W. B.; Pfeffer, P. E. *Macromolecules* **1987**, *20*, 1550–54.
15. Simms, G. Ph.D. Thesis, University of New South Wales, Sydney, Australia, 1987.
16. Stanley, J. A.; Peemoeller, H. *J. Magn. Reson.* **1991**, *91*, 209–14.
17. Assink, R. A. *Macromolecules* **1978**, *11*, 1233–37.
18. Cheung, T. T. P.; Gerstein, B. C. *J. Appl. Phys.* **1981**, *52*, 5517–28.
19. Cheung, T. T. P. *J. Chem. Phys.* **1982**, *76*, 1248–54.
20. Packer, K. J.; Pope, J. M.; Yeung, R. R.; Cudby, M. E. A. *J. Polym. Sci. Polym. Phys.* **1984**, *22*, 589–616.
21. Havens, J. R.; VanderHart, D. L. *Macromolecules* **1985**, *18*, 1663–76.
22. Tanaka, H.; Nishi, T. *Phys. Rev. B* **1986**, *33*, 32–42.
23. Zhang, S.; Mehring, M. *Chem. Phys. Lett.* **1989**, *160*, 644–46.
24. Booth, A. D.; Packer, K. J. *Mol. Phys.* **1987**, *62*, 811–28.
25. *Conduction of Heat in Solids*, 2nd ed.; Carslaw, H. S.; Jaeger, J. C., Eds.; Clarendon: Oxford, England, 1959; pp 30, 104, 306.
26. Wind, R. A.; Jurkiewicz, A.; Maciel, G. E. *Fuel* **1989**, *68*, 1189–97.

RECEIVED for review June 8, 1990. ACCEPTED revised manuscript August 10, 1991.

9

^{13}C NMR Spectroscopy of Pyridine and Alkylpyridines Sorbed onto Coal

Anthony M. Vassallo

Division of Coal Technology, Commonwealth Scientific and Industrial Research Organization, P.O. Box 136, North Ryde, New South Wales, 2113, Australia

> *Heterocyclic compounds such as pyridine and alkylpyridines sorbed onto coals can be examined by cross-polarization (CP) nuclear magnetic resonance (NMR) techniques in which signals from both the coal and the pyridine are seen. The addition of one methyl group to pyridine has little effect on the intensity of the observed spectrum, but two methyl groups show a decreased signal intensity of the pyridine carbons, especially for the 2,5-substitution pattern. Larger alkyl groups result in a much-reduced signal intensity for the sorbed molecule. In the case of pyridine sorption of the coals studied, those coals with carbon contents greater than 85% showed the weakest pyridine signals in the CP spectrum. A proposed model for the pyridine bonding is one in which surface-immobilized molecules are bonded to a number of additional pyridine molecules. These bonded molecules constitute the bulk of the CP pyridine signal observed in the spectrum.*

COAL–SOLVENT INTERACTIONS ARE IMPORTANT in studies of the macromolecular structure of coal (1). In particular, these interactions are evident when certain solvents (e.g., pyridine) cause swelling of the coal structure. Those solvents that cause a large degree of swelling disrupt hydrogen bonding in the coal and allow the macromolecular network to expand to the limits allowed by the covalent cross-links (2). The degree of solvent swelling is not easily related to any one parameter of a solvent such as polarity because many highly polar solvents do not appreciably swell coal. For polymers, maximum swelling occurs when the solubility parameter, δ, of the solvent is equal to that of the polymer, and this fact has been applied to coal, although there are many limitations (3).

Pyridine bonds to the coal surface through the lone pair of electrons on the nitrogen atom, probably to acidic sites such as carboxylic and hydroxyl groups. Indeed, pyridine has been used to titrate hydroxyl groups in coals (low in carboxylic acid groups), and it was suggested (1) that one pyridine molecule bonds to each hydroxyl group. Alkylation of hydroxyl groups prevents pyridine bonding but does not affect the degree of solvent swelling (4) because the amount of swelling is a consequence of the extent of covalent cross-linking once hydrogen-bonded interactions are removed.

Pyridine bonded to the coal surface is greatly immobilized compared to free (liquid) pyridine. As a consequence of this immobilization, solid-state NMR techniques can be used to probe this bound pyridine. These techniques have been applied to pyridine bound to silica–alumina (5, 6), alumina (7–10), silica–gel (11), and zeolites (12) and have also recently been applied to coal (13–15). NMR techniques are useful for studying sorbed molecules because the hindered motion of the molecules influences many aspects of their NMR behavior, particularly relaxation. In addition, for molecules such as pyridine, enrichment of the magnetically active nuclei (^{13}C or ^{15}N) greatly increases sensitivity. This increased sensitivity enables the detection of very low surface coverages without excessive scan time.

Maciel and co-workers (5) have used magic-angle spinning (MAS) and cross-polarization (CP) solid-state NMR spectroscopy to study the type of bonding that occurs between pyridine and silica–alumina. They were able to show that at low surface coverages (~0.2 monolayers) Lewis acid–base complexes dominate, and at higher surface coverage hydrogen bonding is the dominant interaction. In coals, the surface sites are likely to be much less acidic than those of silica–alumina; however, Ripmeester and co-workers (15) showed that strongly acidic sites in oxidized coals are capable of protonating sorbed pyridine. These workers used ^{15}N-labeled pyridine and CP NMR spectroscopy with MAS. In the work reported in this chapter, CP NMR spectroscopy was used to study the sorption of pyridine and alkylated pyridines onto a selection of coals. In particular, the effects of coal rank and alkyl substitution of the pyridine on the sorption behavior were examined.

Experimental Details

The elemental analyses of the coals used in this work are shown in Table I. In a typical experiment, approximately 300 mg (accurately weighed) of dry, finely

Table I. Ultimate and Petrographic Analyses of Coals

Coal	C	H	N	S	O	Ash	R_v (max)[a]	vit[b]	lip[c]	pyd (wt%)
Morwell	69.0	4.8	0.6	0.3	25.3	2.3	nd[d]	nd	nd	62
Rosewood	79.3	6.1	1.6	0.6	12.4	13.6	0.66	92	8	47
Myuna	82.6	5.3	1.5	0.5	10.1	2.3	0.77	89	6	53
Liddell	84.0	5.3	1.7	0.7	8.3	1.0	nd	>90	nd	50
Wongawilli	87.3	5.2	1.9	0.6	5.0	1.4	1.09	98	nd	43
Tongarra	87.3	5.4	1.8	0.9	4.6	4.1	1.12	93	1	35

NOTE: All values are given as percents unless otherwise indicated. C, H, N, S, and O values were determined on a dry, ash-free basis, and percent O was calculated by difference.
[a] R_v (max) is vitrinite reflectance.
[b] vit is percent vitrinite.
[c] lip is percent liptinite.
[d] nd means not determined.

crushed (<100 μm) coal was placed in a 10-mm-diameter glass tube, and pyridine or another solvent was added to give a solvent-to-coal ratio of approximately 0.4, which should ensure an excess of pyridine over that which is bonded to the coal surface. For example, assuming a CO_2 surface area of 100 m^2/g of coal, the amount of pyridine that would theoretically cover the coal surface, if bound through the nitrogen and free to rotate, would be ~40 mg/g. Even allowing for surface-area increases due to solvent swelling, the maximum theoretical coverage would still be far less than the amount of pyridine added. In practice there would be fewer available bonding sites than the theoretical maximum. The actual ratios used are given in Table I. The tube was then capped, and the sample was allowed to equilibrate overnight, except for one experiment in which the NMR spectra were obtained as soon as possible after solvent addition.

^{13}C NMR spectra were obtained on a Bruker CXP300 spectrometer operating at 75.6 MHz. CP (16) and Bloch-decay techniques were used to obtain the spectra. MAS was not used. For the CP experiments, a 1-ms contact time was used, and 5 s was allowed between pulses. In the Bloch-decay experiments, a single pulse of 7 μs (90°) was used, and 60 s was allowed between pulses. In both cases the free induction decay (FID) was collected in the presence of high-power proton decoupling. An additional broad resonance to higher field of the pyridine was reported in earlier experiments (17), but this resonance was artifactual.

^{13}C spin–lattice relaxation times, T_1, were measured by using the pulse sequence of Torchia (18), and $T_{1\rho}^H$ values were measured by using standard pulse sequences (19).

Results and Discussion

Comparison of Bloch-Decay and CP Spectra.
The CP pulse sequence should not readily generate signals from mobile, liquidlike molecules because the transfer of spin magnetization depends on strong proton–carbon dipolar coupling. In solution, rapid tumbling of the molecules results in a weak average dipolar coupling, and cross-polarization is ineffective. In practice, if the CP pulse sequence is used on liquid pyridine, a signal that arises from residual magnetization left in the x–y plane after the carbon pulse may be observed. The use of phase alternation on the initial proton pulse and in data acquisition is usually sufficient to cancel this residual signal after a few scans. Simple Bloch-decay experiments should detect all carbons, whether mobile or not, if sufficient time is allowed between pulses for all carbons to relax. The CP and Bloch-decay spectra of Liddell coal mixed with pyridine are shown in Figure 1. The pyridine signals (three peaks between 125 and 150 ppm) are much more intense in the Bloch-decay spectrum than in the CP spectrum. Even allowing for a moderate degree of signal loss due to $T_{1\rho}^H$ in the CP spectrum, this result shows that some of the pyridine added to the coal is liq-

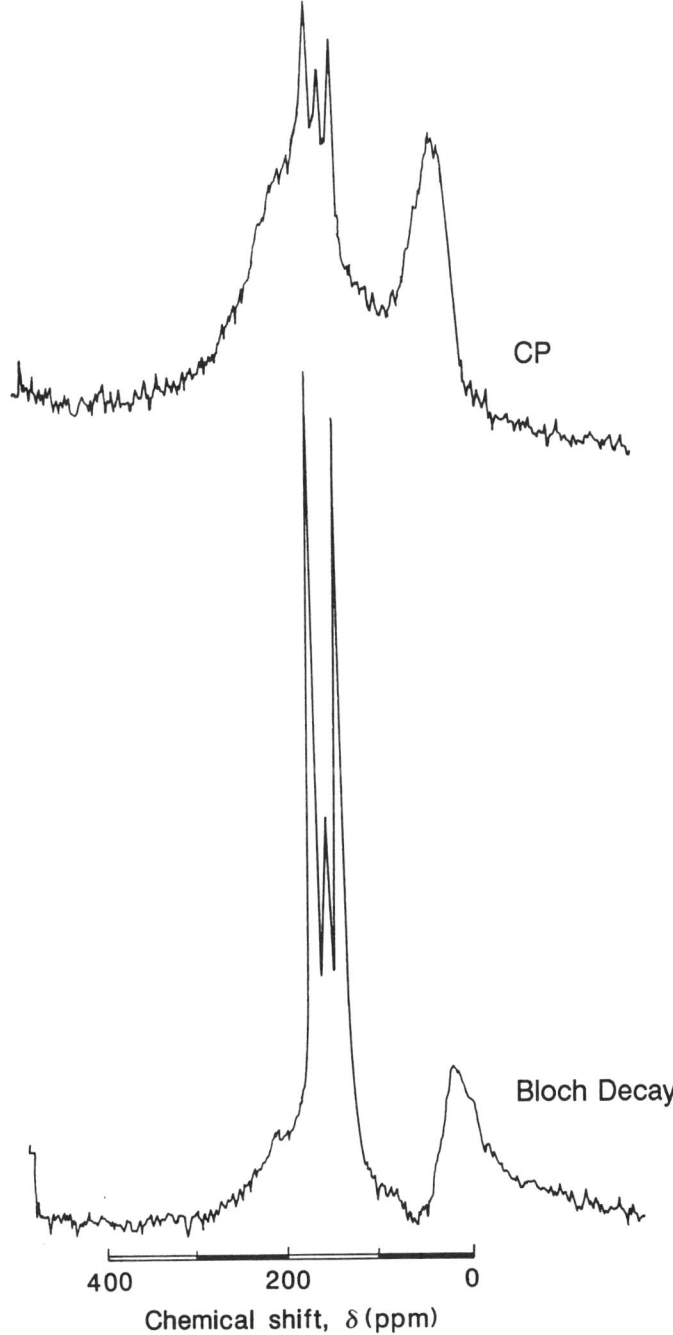

Figure 1. CP and Bloch-decay spectra of pyridine sorbed onto Liddell coal.

uidlike and is in excess of that which is bonded to the coal and observable with CP techniques. The chemical-shift values of the pyridine peaks in both the Bloch-decay spectrum and the CP spectrum are close to those reported for neat pyridine (124, 136, and 150 ppm, [20]). Protonated pyridine has resonances at 129, 142, and 148 ppm and so can be distinguished from the free base.

MAS and high-power proton decoupling remove line broadening that arises from chemical-shift anisotropy (CSA) and proton dipolar coupling, respectively. In the absence of MAS, CSA is likely to be a major cause of line broadening for rigid or immobilized molecules. Indeed, the aromatic resonances of the coal are broadened by such a mechanism, and the aliphatic resonances are not and thus are not greatly narrowed by MAS. The observation of relatively narrow lines for the pyridine resonances indicates that the CSA of the pyridine carbons is largely removed by motion. Other workers (21) have shown that aromatic molecules chemically bound to a surface still exhibit broad resonances that are similar to *powder patterns* in the absence of MAS, even though rapid reorientation around one or more axes is occurring. This observation, in combination with the chemical-shift data, indicates that the pyridine resonances observed under the conditions used here do not arise from those molecules directly bonded to the coal surface.

The effect of time on the CP signal intensity has been studied by measuring the total signal area at regular time intervals after pyridine addition. It is conceivable that the immobilization of pyridine is diffusion controlled and that the signal from cross-polarized pyridine would increase with time up to the natural limit. The results shown in Figure 2 demonstrate that after 800 s the signal intensity of the coal–pyridine mixture is reasonably constant, and no changes were observed in additional experi-

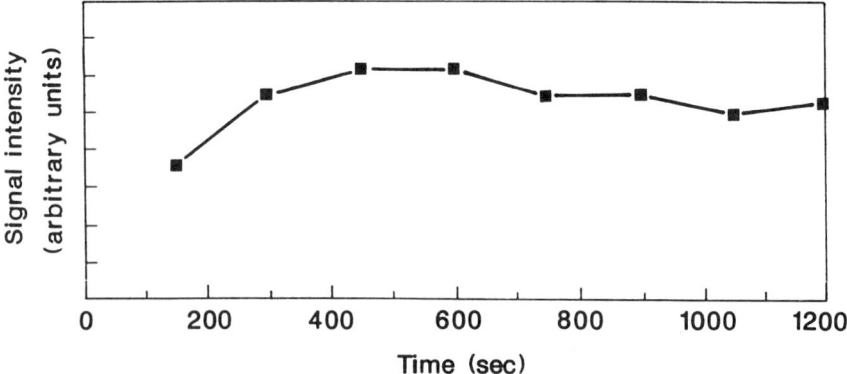

Figure 2. Effect of time on the CP signal intensity of pyridine sorbed onto Liddell coal.

ments after many days or weeks. Apparently the immobilization of pyridine occurs quite rapidly, and this result is consistent with the observation that swelling of thin sections of coal is complete within 60 s (22).

Effect of Temperature. NMR studies of pyridine sorbed onto silica–alumina (5) have shown that the CP signal of the pyridine changes in intensity and shape with decreasing temperature. Similar variable-temperature experiments (except without MAS) have been performed on pyridine–Rosewood coal mixtures, and the results are shown in Figure 3. As the temperature is lowered, the resonances from the cross-polarizable pyridine decrease in intensity and broaden somewhat, and the signal from the γ carbon shows the strongest effect. This result is similar to that observed for the pyridine–silica–alumina system (5). The broadening of the resonances in the coal–pyridine system can arise from a number of effects. First, in the absence of MAS, CSA is not spun out. As the temperature is lowered and the motional averaging of CSA becomes less effective, the resonances may broaden toward their static value. Another mechanism that may cause broadening is the reduction in effective proton decoupling caused by motion at the same frequency as the decoupling field (23). Furthermore, line broadening can result from a decrease in the chemical-exchange rates between different sites of the pyridine. Any of these mechanisms may contribute to the observed line broadening, and further work is needed to establish their relative contributions. The observation of a stronger effect for the γ carbon may indicate that a component of motion is the rotation of the molecule about its C_2 axis; this motion results in a type of "magic-angle spinning" for the α and β carbons but not for the γ carbon (24).

Effect of Coal Rank. The effect of coal rank on the CP signal of sorbed pyridine was investigated by measuring the coal–pyridine spectrum of a number of coals ranging in carbon content (dry, ash-free) from 69.0 to 87.3%. These spectra are shown in Figure 4, arranged in order of decreasing carbon content. On the basis of the intensity of the pyridine signals, the two highest rank coals (Tongarra and Wongawilli) appear to interact with pyridine in a very different manner than the other, lower rank coals. The amount of pyridine added to these two coals (Table I) is similar to that added to the other coals, but very little of this pyridine is seen by CP techniques under the conditions of this study. Bloch-decay experiments (not shown) reveal, as expected, large pyridine resonances as well as coal signals. Possible explanations of this behavior are given in the following sections.

Alkyl-Substituted Pyridines. The lone pair of electrons on the nitrogen in pyridine is an obvious means of bonding to acidic sites. In

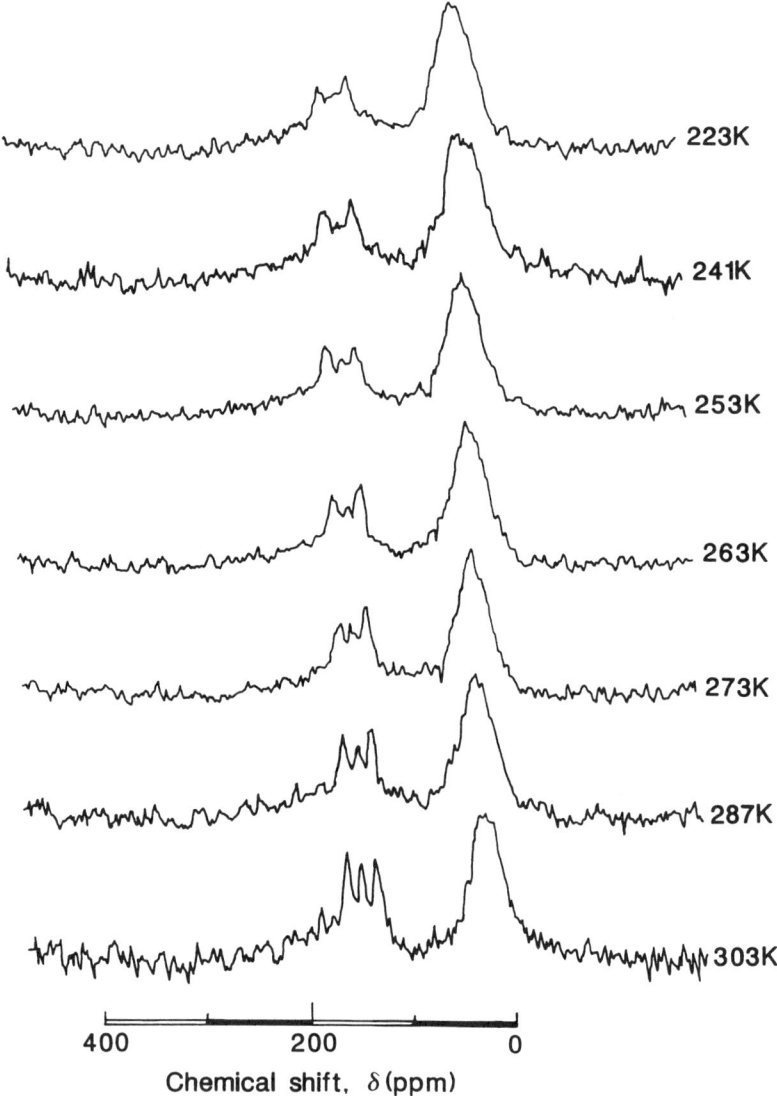

Figure 3. Effect of temperature on cross-polarizable pyridine sorbed onto Rosewood coal.

view of the expected role of nitrogen in the interaction of pyridine with coal, a series of alkylpyridines was used in similar experiments to determine whether steric hindrance of the nitrogen or of the molecule in general would affect the observed CP behavior of the solvent. The alkylpyridines included 2-, 3-, and 4-methylpyridine (picolines); 3,5- and 2,6-di-

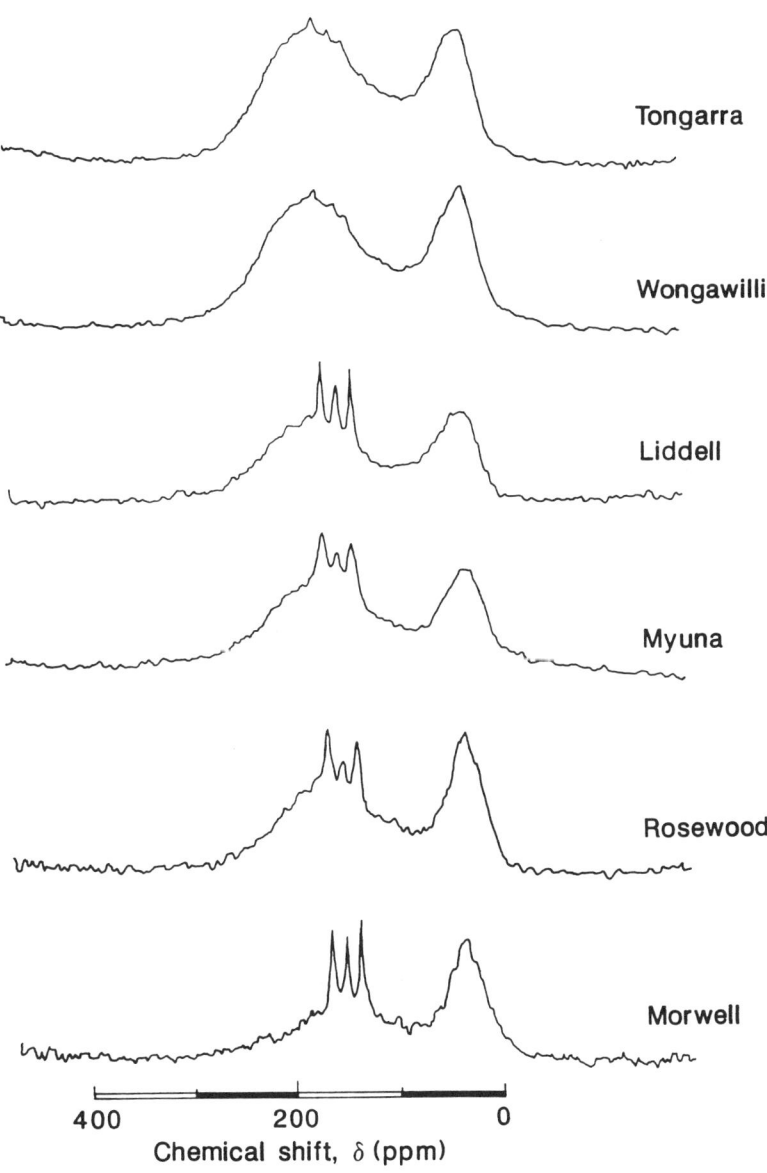

Figure 4. Cross-polarization of pyridine sorbed onto various coals.

methylpyridine (lutidines); 2-ethylpyridine; and 4-*tert*-butylpyridine. The CP spectra of these alkylpyridines sorbed onto Liddell coal are shown in Figures 5a and 5b. The substitution of a single methyl group onto the pyridine ring does not appear to greatly alter the intensity of the resonances. The addition of two methyl groups to the pyridine ring causes a decrease in signal intensity, especially for the 3,5-dimethylpyridine. If steric hindrance of the nitrogen were effective in decreasing the coal–solvent interaction, then the strongest effect should be observed with the 2,6-isomer, which is not the case. The spectrum of 2-ethylpyridine sorbed onto Liddell coal shows an even weaker signal, and the substitution of a bulky *tert*-butyl group onto the pyridine molecule results in an even weaker signal from both the ring and side-chain carbons. Unfortunately, simple correlations between CP signal intensity and sorbed molecule motion cannot be made because the observed signal will be influenced by a number of processes, such as the type of motion (e.g., vibrational or rotational) and its frequency and degree of anisotropy. Nevertheless, the observation that 3,5-dimethylpyridine behaves differently than 2,6-dimethylpyridine indicates that the position of the substituents relative to the nitrogen atom influences the type of bonding between the pyridine molecule and coal.

Relaxation Behavior of Sorbed Pyridine. Relaxation measurements are a means of interpreting motion on the NMR time scale. In the absence of other effects, motion with frequency components at the Larmor frequency will provide an efficient means of spin relaxation through the lattice, and this is termed *spin–lattice relaxation*, T_1. The T_1^Cs of the protonated ring carbons of pyridine and alkylpyridines sorbed onto Liddell coal were measured to determine whether carbon-relaxation measurements are useful in studying this type of interaction. Because the coal–pyridine system includes unbound "liquid" pyridine, the pulse sequence used for T_1 measurement must discriminate between free and cross-polarizable solvent. The pulse sequence of Torchia (*18*) achieves this discrimination by canceling unwanted signals from the free solvent. Because of the overlap of pyridine resonances with the broad aromatic band of the coal, estimation of the base line for measurement of the intensity of the pyridine signal is somewhat subjective. This base-line estimation will introduce some error, but the measurements are believed to be accurate to ±10%. The T_1^C values for pyridine and alkylpyridines sorbed onto Liddell coal are given in Table II. The T_1^C values for pyridine sorbed onto other coals are as follows:

- Morwell, 70 ms
- Rosewood, 151 ms
- Myuna, 232 ms

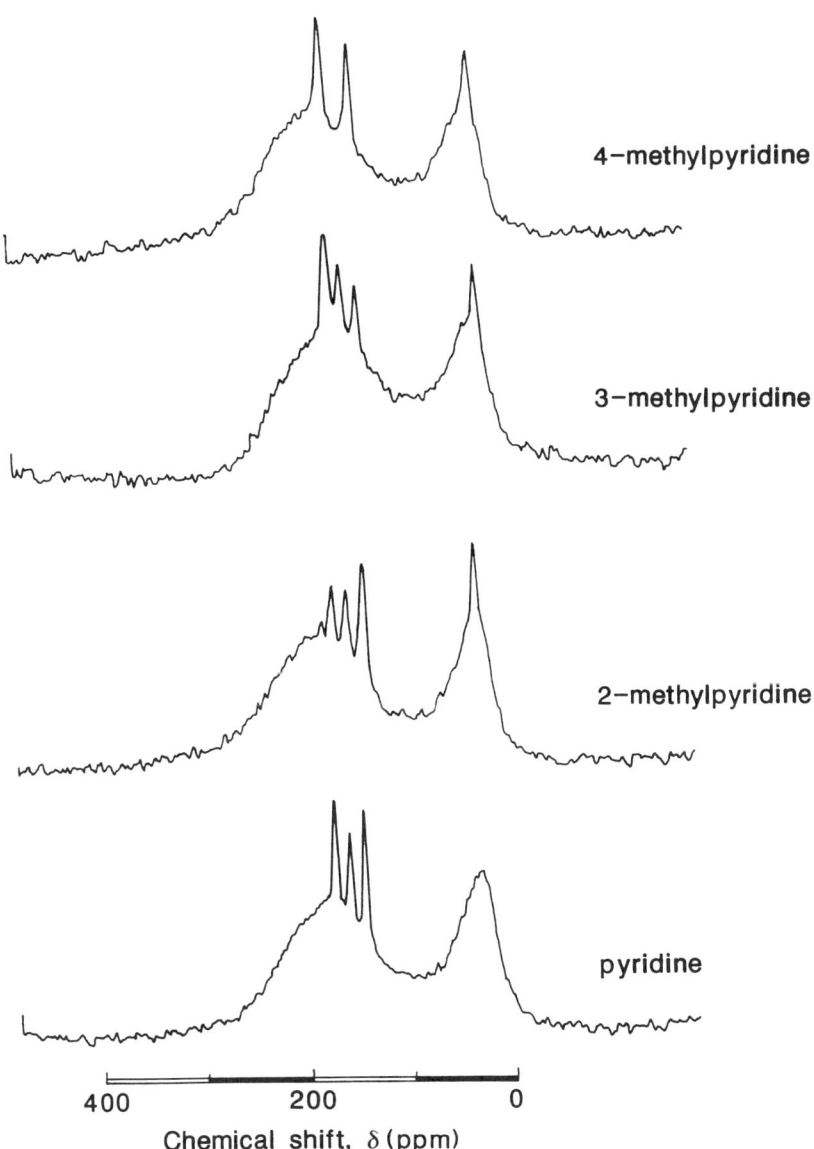

Figure 5a. Cross-polarization of pyridine and picolines sorbed onto Liddell coal.

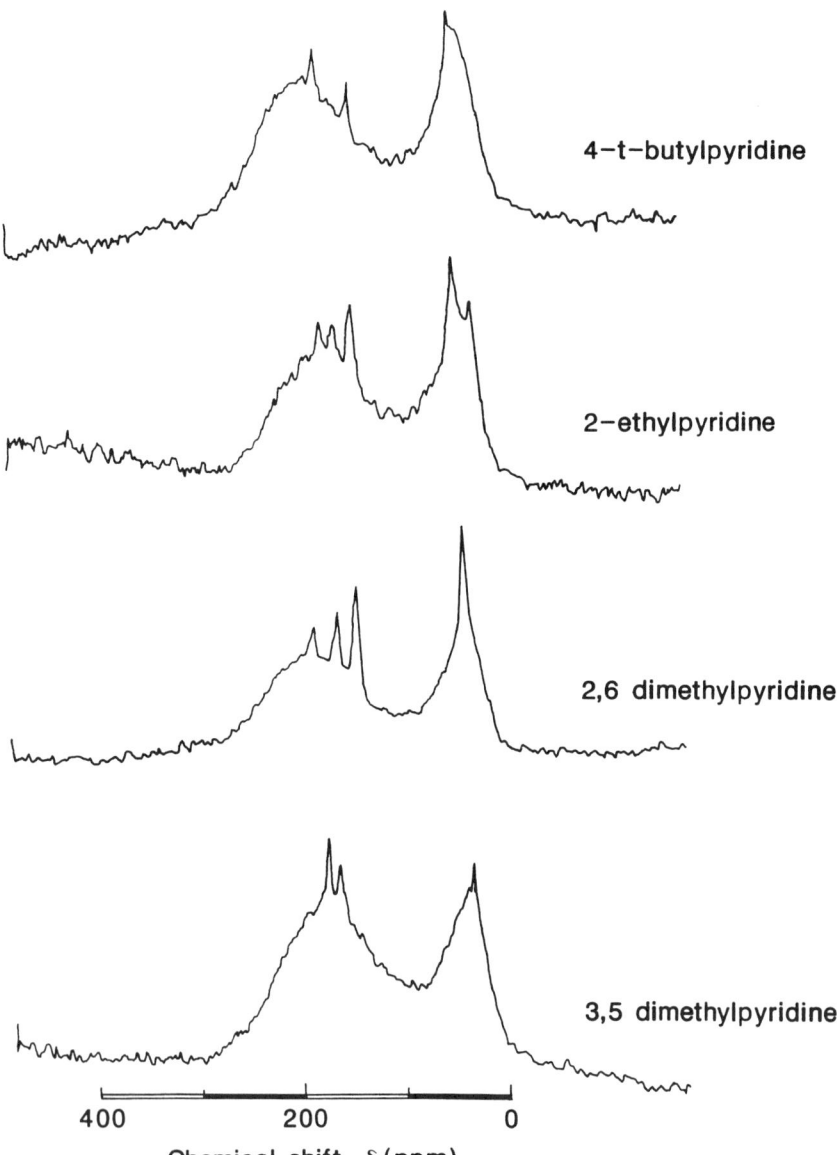

Figure 5b. Cross-polarization of dimethylpyridine, ethylpyridine, and tert-*butylpyridine sorbed onto Liddell coal.*

Table II. T_1^C Values of Cross-Polarized Pyridine and Alkylpyridines Sorbed onto Liddell Coal

Molecule	T_1^C (ms)
Pyridine	394
2-Methylpyridine	249
3-Methylpyridine	170
4-Methylpyridine	278
2,6-Dimethylpyridine	256
3,5-Dimethylpyridine	nd[a]
2-Ethylpyridine	224
4-*tert*-Butylpyridine	2380

NOTE: T_1^C values (in milliseconds) were measured for α carbons except where the position was substituted, in which case the nearest unsubstituted carbon was chosen. T_1 differences between different unsubstituted ring carbons were not large.

[a] nd means not determined.

The α-carbon T_1 of the sorbed pyridine increases with increasing coal carbon content within the range of the four coals studied here. The addition of alkyl substituents, except for *tert*-butyl, appears, not unexpectedly, to decrease the T_1. The very long T_1 of the *tert*-butylpyridine indicates a significantly different type of motion compared to the other molecules. A plot of $1/T_1$ versus the percent C in the coal (Figure 6) demonstrates the almost linear relationship between the observed relaxation rate of pyridine and the carbon content of these coals. The interpretation of this relationship is complicated by a number of factors. The observed relaxation rate will depend on the relative contribution of many different relaxation processes, which may include dipole–dipole relaxation (with coupling to protons or unpaired spins), spin-rotation relaxation, chemical-shift anisotropy, and scalar relaxation. Coal contains significant amounts of unpaired spins (on the order of 10^{19} spins per gram [25]), and relaxation due to coupling with these spins (or to paramagnetic adsorbed oxygen) may be the dominant mode. The T_1^H in coals with carbon contents less than 85% is predominantly determined by this mechanism (26). However, T_1^C values in coal are considerably longer (27), and coupling to carbon in coal is much less effective than coupling to protons. The extent of relaxation due to unpaired spins affecting the sorbed pyridine should be further decreased by the pyridine's remoteness to these spins. Moreover, the concentration of unpaired spins in coal tends to increase with rank (25), and the effect would be more rapid relaxation of sorbed pyridine with increas-

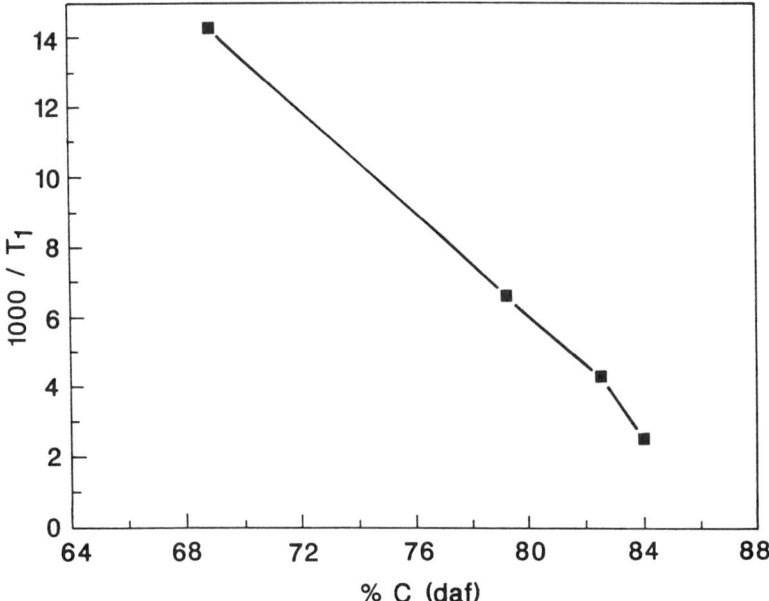

Figure 6. Spin–lattice relaxation of pyridine sorbed onto coal.

ing coal carbon content if this mechanism were dominant. As Figure 6 demonstrates, the observed relaxation rate decreases rather than increases with coal rank. Additional experiments at different magnetic fields will help in determining the sources of relaxation, and these experiments are currently under way.

Implications for Structure of Sorbed Pyridine. Previous work (1) has supported the concept that pyridine bonds to coal surfaces through the interaction of the lone pair of electrons on the nitrogen with hydroxyl or carboxyl groups on the coal. In the techniques used here, surface-bound pyridine would probably not be readily evident in the spectra because of the low relative concentration of pyridine (on the basis of surface-area calculations, the ratio of pyridine carbon to coal carbon would be <0.04) and the broadness of the resonance in the absence of MAS. The pyridine resonances that we observed with CP techniques must arise from immobilized pyridine in excess of surface- (monolayer)-bound molecules. This immobilization may arise from pyridine–pyridine hydrogen bonding, which results in additional layers of pyridine molecules, or pyridine trapped and immobilized in pores, or both. The latter process, immobilization in pores, has been studied in zeolites, and one case (28) demonstrated that as the pore size became larger, the T_1^C of the trapped tetramethylammonium ion decreased. Unfortunately, the pore-size distributions of the coals used in this work have not been measured; however, the average pore size should decrease with increasing carbon content. If this were the case, the coal–pyridine system shows the opposite behavior to the zeolite system. The observation that 3,5-dimethylpyridine has a weaker CP signal than 2,6-dimethylpyridine (Figure 5) supports the theory that much of the observable pyridine arises from a "piggyback" type of bonding to the surface-bound layer, because the 3,5-substitution would interfere with this type of bonding. The very weak signal from the *tert*-butylpyridine is also consistent with this manner of bonding, but the bulky *tert*-butyl group may also inhibit access of the pyridine molecule to surface sites, with similar results. As shown in Figure 4, the spectra of the two coals with the highest carbon content did not have much of a CP pyridine signal. This result is consistent with the proposed "piggyback" mechanism because of the expected lower hydroxyl content and pore size of these coals in comparison with the other coals.

References

1. Green, T. K.; Larsen, J. W. *Fuel* **1984**, *63*, 1538–1543.
2. Sanada, Y.; Honda, H. *Fuel* **1966**, *46*, 451–456.
3. Larsen, J. W.; Green, T. K.; Chiri, I. *Proc. Int. Conf. Coal Sci.* **1983**, 277–279.
4. Brenner, D. *Fuel* **1985**, *64*, 167–173.
5. Maciel, G. E.; Haw, J. F.; Chuang, I. S.; Hawkins, B. L., Early, T. A.; McKay, D. R.; Petrakis, L. *J. Am. Chem. Soc.* **1983**, *105*, 5529–5535.
6. Haw, J. F.; Chuang, I. S.; Hawkins, B. L.; Maciel, G. E. *J. Am. Chem. Soc.* **1983**, *105*, 7206–7207.
7. Petrakis, L.; Kiviat, F. E. *J. Phys. Chem.* **1976**, *80*, 606–611.
8. Pearson, R. M. *J. Catal.* **1977**, *46*, 279–288.
9. Dawson, W. H.; Kaiser, S. W.; Ellis, P. D.; Inners, R. R. *J. Phys. Chem.* **1982**, *86*, 867–868.
10. Ripmeester, J. A. *J. Am. Chem. Soc.* **1983**, *105*, 2925–2927.
11. Bernstein, T.; Kitaev, L.; Michel, D.; Pfeifer, H.; Fink, P. *J. Chem. Soc., Faraday Trans. 1* **1982**, *78*, 761–769.
12. Michel, D.; Germanus, A.; Pfeifer, H. *J. Chem. Soc., Faraday Trans. 1* **1982**, *78*, 237–254.
13. Silbernagel, B. G.; Ebert, L. B.; Schlosberg, R. H.; Long, R. B. In *Coal Structure;* Gorbaty, M. L.; Ouchi, K., Eds.; Advances in Chemistry 192; American Chemical Society: Washington, DC, 1981; pp 23–35.
14. Vassallo, A. M.; Wilson, M. A. *Fuel* **1984**, *63*, 571–573.
15. Ripmeester, J. A.; Hawkins, R. E.; MacPhee, J. A.; Nandi, B. N. *Fuel* **1986**, *65*, 740–742.
16. Pines, A.; Gibby, M. G.; Waugh, J. S. *J. Chem. Phys.* **1973**, *59*, 569–590.
17. Wilson, M. A.; Vassallo, A. M. *Org. Geochem.* **1985**, *8*, 299–312.

18. Torchia, D. A. *J. Magn. Reson.* **1978**, *30*, 613–616.
19. Wilson, M. A. *NMR Techniques and Applications in Geochemistry and Soil Chemistry;* Pergamon: Oxford, England, 1987.
20. Levy, G. C.; Nelson, G. L. *Carbon-13 Nuclear Magnetic Resonance for Organic Chemists;* Wiley Interscience: New York, 1972.
21. Slotfeldt-Ellingson, D.; Resing, H. A. *J. Phys. Chem.* **1980**, *84*, 2204–2209.
22. Brenner, D. *Fuel* **1984**, *63*, 1324–1328.
23. VanderHart, D. L.; Earl, W. L.; Garroway, A. N. *J. Magn. Reson.* **1981**, *44*, 361–401.
24. Gutowsky, H. S.; Pake, G. E. *J. Chem. Phys.* **1950**, *18*, 162–170.
25. Retcofsky, H. In *Coal Science,* Vol 1; Gorbaty, M. L.; Larsen, J. W.; Wender, I., Eds.; Academic: New York, 1982.
26. Wind, R. A.; Jurkiewicz, A.; Maciel, G. E. *Fuel* **1989**, *68*, 1189–1197.
27. Axelson, D. E; Parkash, S. *Fuel Sci. Tech. Int.* **1986**, *4*, 45–85.
28. Hayashi, S.; Suzuki, K; Hayamizu, K. *J. Chem. Soc., Faraday Trans. 1* **1989**, *85*, 2973–2982.

RECEIVED for review June 8, 1990. ACCEPTED revised manuscript December 26, 1990.

10

Distortion-Free ^{13}C NMR Spectroscopy in Coal

^{1}H Rotating-Frame Dynamic Nuclear Polarization and ^{1}H–^{13}C Cross-Polarization

Robert A. Wind[1]

Department of Chemistry, Colorado State University, Fort Collins, CO 80523

A ^{1}H–^{13}C cross-polarization (CP) experiment is described in which the ^{1}H magnetization, used in CP, is obtained via dynamic nuclear polarization (DNP) in the proton rotating frame (RF DNP). This experiment can be carried out in coal and other solids containing unpaired electrons. In this so-called RF DNP–CP experiment, interplay effects between the ^{1}H–^{13}C polarization-transfer times and the ^{1}H rotating-frame relaxation time are avoided; thus ^{13}C spectral distortions due to these effects are prevented. Moreover, multiple-contact RF DNP–CP experiments are possible, and these experiments reduce the measuring time of a ^{13}C spectrum. An application of the RF DNP–CP technique in a low-volatile bituminous coal is given.

SINCE THE INTRODUCTION of ^{1}H decoupling (*1*), cross polarization (CP) (*2*) and magic-angle spinning (MAS) (*3*), which made it possible to obtain high-resolution ^{13}C NMR spectra in solids in a relatively short

[1] Current address: Chemagnetics, Inc., 2555 Midpoint Drive, Fort Collins, CO 80525

measuring time, these techniques have been applied extensively in coal research (*4–7* and references cited therein). However, even in the very first article about ^{13}C CP NMR spectroscopy in coal (*8*) the authors reported that only 50% of the total amount of carbon atoms could be detected with this technique, and since then the issue of quantitative information that can be obtained from the ^{13}C spectra has been the topic of much debate (*9–16*). One of the major factors that can influence the quantitative aspects in CP is the interplay between T_{CH} and $T_{1\rho}$ during CP, where T_{CH} is the ^{1}H–^{13}C polarization-transfer time governing the buildup of the ^{13}C polarization, and $T_{1\rho}$ is the ^{1}H–^{13}C rotating-frame polarization time characterizing the decay of the ^{1}H polarization toward zero. When T_{CH} and $T_{1\rho}$ are of the same order of magnitude, the ^{13}C magnetization is decreased (*17*). Moreover, when there is a distribution of T_{CH} values resulting from different ^{13}C–^{1}H distances and molecular motions, distorted ^{13}C spectra are obtained (*18*). In coal, different T_{CH} values varying from tens of microseconds to more than a millisecond have been observed (*14*), and $T_{1\rho}$ values of a millisecond or less have been reported (*13, 14*), and thus distorted ^{13}C CP spectra can indeed be expected. Moreover, in many coals a distribution of $T_{1\rho}$ values is observed as well (*13, 14*). This distribution makes an analysis of the CP results very complicated as it requires, among other things, a detailed knowledge of the distribution of the $T_{1\rho}$ values over the aliphatic and aromatic protons. Therefore, at best the ^{13}C CP spectra can provide such qualitative information as the presence of different molecular fractions in coal.

The problem in CP associated with the interplay between T_{CH} and $T_{1\rho}$ arises from the fact that for the proton system during CP, a nonstationary situation exists, in which the ^{1}H magnetization, which is usually equal to the thermal equilibrium magnetization corresponding to the external field, B_o, is spin-locked along a radio frequency (rf) field, applied at the ^{1}H Larmor frequency, with an amplitude, B_R, much smaller than B_o. This nonstationary situation causes the ^{1}H magnetization to decay to zero. Hence, the problem in CP arising from this decay could be avoided if the ^{1}H magnetization along B_R could be made stationary. Then the matching time, t_{CP}, during which CP occurs could have an arbitrary length, allowing all ^{13}C nuclei to polarize fully and resulting in undistorted ^{13}C spectra. In a standard CP experiment, this result is impossible to achieve.

In this chapter, an alternative CP method is described in which a stationary ^{1}H magnetization along B_R is indeed obtained. This method makes use of the unpaired electrons that are present in coal by nature and generates this stationary magnetization via dynamic nuclear polarization (DNP). The principle of this technique and the application of this method in a low-volatile bituminous (LVB) coal are described in the following sections.

Dynamic Nuclear Polarization

In solids containing both magnetic nuclei and unpaired electrons, the nuclear magnetization can be enhanced by irradiating at or near the electron Larmor frequency, ω_e, yielding the DNP effect (19, 20). Traditionally, the nuclear magnetization aligned along B_0 in the laboratory frame is enhanced via DNP. This effect will be called laboratory-frame DNP (LF DNP). In LF DNP, after the onset of the irradiation the nuclear magnetization reaches a stationary value, M_{HL}, in a time governed by the nuclear Zeeman relaxation time. Figure 1a shows the ^{1}H LF DNP enhancement factor minus unity, $E_L - 1$, of an LVB coal (volatile matter = 19.8%, C = 90.2%, and H = 4.6% dry, mineral-matter free [dmmf]) as a function of the microwave offset frequency, $(\omega - \omega_e)$, for B_0 = 1.4 T. Here E_L is defined as

$$E_L = \frac{M_{HL}}{(M_{HL})_o} \quad (1)$$

In eq 1, $(M_{HL})_o$ is the laboratory-frame thermal-equilibrium magnetization in the absence of microwave irradiation. An elaborate description of LF DNP is given in the references (19, 20). Here we confine ourselves to remarking that the observed enhancement curve, $(E_L - 1)$ versus $(\omega - \omega_e)$ (Figure 1a, curve 1), can be decomposed into an antisymmetrical curve (curve 2) governed by static electron–proton interactions, and a symmetrical curve (curve 3) originating from time-dependent electron–proton interactions.

DNP can also be obtained in the nuclear rotating frame (21, 22). This rotating-frame DNP will be called RF DNP. In RF DNP, a simultaneous irradiation is applied of a microwave field with a frequency near ω_e and a strong rf field with an amplitude B_R and a frequency equal to the nuclear Larmor frequency. This irradiation results in a nuclear magnetization, M_{HR}, aligned along in the nuclear rotating frame, which becomes stationary in a time determined mainly by the nuclear rotating-frame relaxation time. Figure 1b shows the ^{1}H RF DNP enhancement factor, E_R, of the LVB coal as a function of $\omega - \omega_e$ for B_R = 1.4 mT. Here E_R is defined as

$$E_R = \frac{M_{HR}}{(M_{HL})_o} \quad (2)$$

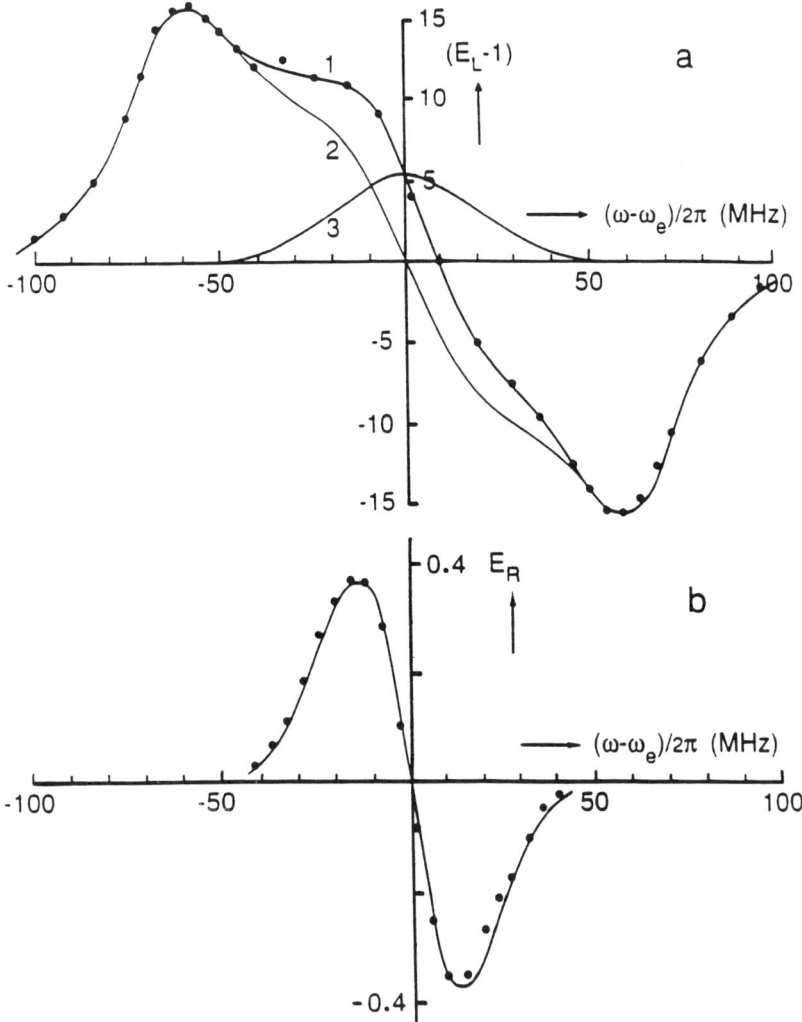

Figure 1. ^1H DNP signal enhancement in the laboratory frame (a) and in the ^1H rotating frame (b) as a function of the microwave offset frequency, $\omega - \omega_e$.

Compared to the LF DNP results, the following observations can be made:

1. In RF DNP, the observed enhancement curve is completely antisymmetrical, which means that in RF DNP the enhancement is due to static electron–proton interactions only.

2. In RF DNP, the maximal magnetization is a factor 2.7× smaller than the laboratory-frame thermal equilibrium magnetization and a factor 40× smaller than the maximal magnetization obtained via LF DNP.

However, despite this relatively small value of M_{HR}, RF DNP provides some major advantages. First, M_{HR} remains stationary along B_R as long as the simultaneous irradiation of the microwave field and the rf field is applied. Hence, if ^{1}H RF DNP is followed by ^{1}H–^{13}C CP, M_{HR} will remain constant during the polarization transfer, provided that the B_R field is not altered during CP. This condition means that in this so-called RF DNP–CP experiment the interplay between T_{CH} and $T_{1\rho}$ is avoided. Second, if during the ^{13}C signal acquisition the magnitude of the ^{1}H decoupling field is kept equal to B_R as well, M_{HR} will still be present after the signal acquisition. This result opens the possibility of applying multiple-contact CP strategies, where the next CP experiment is applied immediately after the ^{13}C signal acquisition. Finally, in RF DNP M_{HR} becomes stationary after a few times $T_{1\rho}$. Hence, even when multiple-contact RF DNP–CP experiments cannot be used (e.g., because of heating problems of the probe or the sample or because B_R had to be changed during the signal acquisition), the recycle delay can be made very short in an RF DNP–CP experiment (i.e., milliseconds), and the measuring time of such an experiment is reduced considerably.

The RF DNP–CP Experiment

The experimental setup has been described elsewhere (*20, 23*). The external magnetic field was 1.4 T, which corresponds to a ^{1}H Larmor frequency of 60 MHz, a ^{13}C Larmor frequency of 15 MHz, and an electron Larmor frequency of 40 GHz. A 10-W Varian fixed-frequency klystron, which provides a microwave field with an amplitude of about 0.03 mT, was used as a microwave source. The B_R field was 1.4 mT throughout the RF DNP–CP experiment. Figure 2 shows the pulse sequence employed in the RF DNP–CP experiment. The microwave irradiation was applied continuously at an offset frequency $(\omega - \omega_e)/2\pi = -15$ MHz, for which the ^{1}H RF DNP enhancement is maximal (*see* Figure 1b).

The main difference between RF DNP–CP and a standard CP experiment is that the ^{1}H 90° pulse, employed in the latter experiment to spin-lock the laboratory-frame ^{1}H magnetization along B_R, is replaced by an rf irradiation during a time, t_1. During t_1, the ^{1}H magnetization, M_{HR}, builds up along B_R. The value of t_1 must be a few times greater than $T_{1\rho}$ in order for M_{HR} to become stationary (for the LVB coal, $T_{1\rho} = 3$ ms, and a t_1 value of 6 ms was used). Between CP and the ^{13}C signal acquisition, the ^{13}C magnetization is flipped alternatively along the $+z$ and $-z$

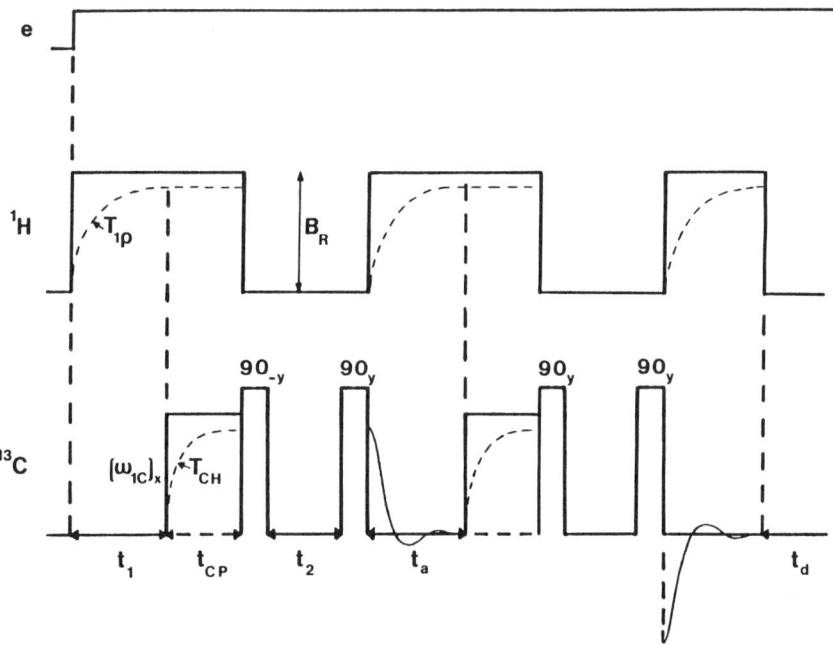

Figure 2. The pulse sequence for the RF DNP–CP experiment.

axes via a $(90°)_{-y}$ and a $(90°)_y$ pulse, respectively, followed by a short delay time t_2 (10 ms), followed by another $(90°)_y$ pulse, which brings the ^{13}C magnetization alternatively about the $+x$ and $-x$ axes. This sequence produces a spin–temperature alternation for the ^{13}C signal, and this alternation together with an add–subtract routine eliminates rf transient effects.

In a standard CP experiment this elimination is achieved by alternating the ^{1}H spin–temperature, that is by spin-locking the ^{1}H magnetization alternatively parallel or antiparallel to B_R. However, in RF DNP–CP spectroscopy this procedure is not possible because the ^{1}H magnetization will always be directed parallel to B_R.

A two-contact RF DNP–CP experiment was performed in which the ^{1}H decoupling field, applied during the first ^{13}C signal acquisition time, t_a, is used to generate the ^{1}H magnetization along B_R, which is employed in the second ^{1}H–^{13}C polarization transfer ($t_a = t_1 = 6$ ms). Between the two-contact RF DNP–CP sequences, a recycle delay, t_d, of 0.2 s was applied to prevent overheating of the probe and the sample.

10. WIND *Distortion-Free ^{13}C NMR Spectroscopy in Coal*   223

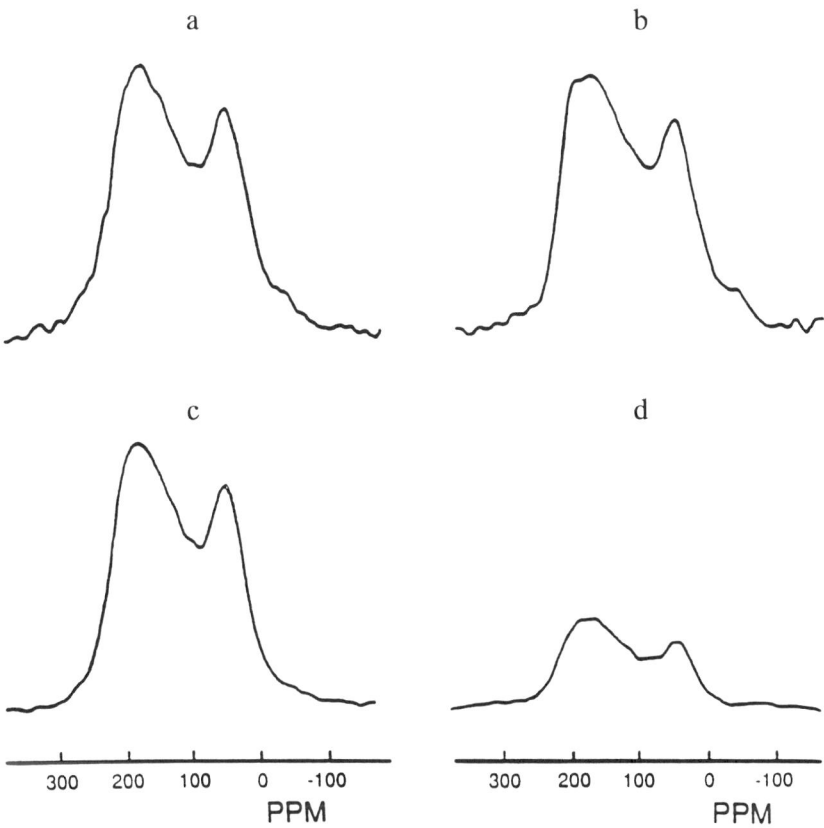

Figure 3. ^{13}C spectra of an LVB coal obtained via RF DNP–CP (a, b) and standard CP (c, d) experiments. The matching times, t_{CP}, were 1 ms (a, c) and 5 ms (b, d), respectively. (Reproduced with permission from reference 22. Copyright 1990.)

Results and Discussion

Figure 3 shows ^{13}C spectra of the LVB coal. These spectra were obtained via RF DNP–CP (Figures 3a and 3b) and standard CP (Figures 3c and 3d) NMR spectroscopy (in the latter experiment, the ^{1}H magnetization, prior to CP, was increased by using LF DNP in order to enhance the ^{13}C sensitivity as in ref. 14). The spectra were obtained for two values of the matching time t_{CP}: 1 ms (Figures 3a and 3c) and 5 ms (Figures 3b and 3d). In these experiments, MAS was avoided in order to prevent possible spectral distortions due to sample rotation (14, 16). Under this condition, the ^{13}C spectra consist of overlapping chemical-shift anisotropy (CSA) patterns, and the narrower CSA pattern, peaking at about 40 ppm, reflects the presence of aliphatic carbons, whereas the broad CSA pattern with a maximum at about 160 ppm is due to aromatic ^{13}C nuclei.

The following results can be deduced from Figure 3:

1. In the standard CP experiment, the ^{13}C signal intensity observed after t_{CP} = 5 ms is considerably less than that observed after t_{CP} = 1 ms, a result reflecting the short ^{1}H $T_{1\rho}$ value in this coal. In RF DNP–CP the ^{13}C signal intensities observed after the 1- and 5-ms matching times are about the same; this result means that ^{1}H $T_{1\rho}$ effects are eliminated in this experiment (the small decrease in the ^{13}C signal intensity after a matching time of 5 ms is due to ^{13}C $T_{1\rho}$ effects).

2. The ^{13}C aromaticity observed via RF DNP–CP spectroscopy for t_{CP} = 5 ms is larger than that observed after t_{CP} = 1 ms. This result means that a part of the aromatic ^{13}C nuclei are cross-polarizing with a polarization-transfer time, T_{CH}, of the order of a millisecond or more, a result indicating that interplay effects between ^{1}H $T_{1\rho}$ and T_{CH} play an important role in the standard CP experiment.

3. In the ^{13}C spectra observed via RF DNP–CP spectroscopy, shoulders are observed at about −50 ppm; these shoulders are not present in the spectra obtained via standard CP. These shoulders probably reflect the high-shielding limit of the CSA patterns corresponding to condensed aromatic ^{13}C nuclei such as bridge carbons in polycyclic aromatic compounds (10). These aromatic ^{13}C nuclei are nonprotonated, and the corresponding T_{CH} values are long. Hence, in a standard CP experiment the ^{13}C signal intensity due to these nuclei is seriously suppressed as a result of the interplay between T_{CH} and ^{1}H $T_{1\rho}$, which explains why these shoulders are not observed in the CP experiment.

In conclusion, the RF DNP–CP interference effects between T_{CH} and ^{1}H $T_{1\rho}$ are avoided; thus distortion-free ^{13}C spectra can be measured in coal. In this method, signal losses during CP due to ^{1}H $T_{1\rho}$ effects are avoided so that t_{CP} can be made as long as is needed for all the ^{13}C nuclei to polarize completely. The only restriction is that for long t_{CP} values signal losses occur as a result of the ^{13}C rotating-frame relaxation effects, but this is not a serious constraint because in coals the ^{13}C rotating-frame relaxation time is considerably longer than that of the protons (14).

Finally, the following remarks can be made:

1. It follows from Figures 3a and 3c that the apparent ^{13}C aromaticity obtained via a standard CP experiment for $t_{CP} = 1$ ms is approximately equal to that measured by means of RF DNP–CP spectroscopy. This result should be considered as a coincidence, and many different CP experiments (e.g., variable and fixed t_{CP} and a variable delay prior to CP) are needed in order to obtain a better estimate of the quantitative character of the ^{13}C CP spectra. (This issue will be discussed in detail in a separate paper.)

2. Although the ^{13}C spectra obtained by means of RF DNP–CP spectroscopy provide more quantitative information about, for example, the amount of different molecular groups present in coal than a standard CP experiment, still other factors, such as the broadening effects due to the presence of the unpaired electrons, can limit the quantitative integrity of the ^{13}C spectra *(14–16)*. This issue will also be addressed in a separate publication.

3. When the ^{1}H rotating-frame relaxation is fast, a distribution of relaxation times may arise as a result of insufficiently rapid spin diffusion among the protons *(24)*. Such a distribution has been observed in many coals *(13, 14)*. If this situation occurs, the RF DNP–CP technique will provide undistorted ^{13}C spectra only if the ^{1}H rotating-frame relaxation is determined completely by the electron–proton interactions. For example, if for a part of the protons the relaxation is determined by ^{1}H–^{1}H dipolar interactions, the RF DNP enhancement factor for these protons is suppressed, which results, after CP, in a reduction of the ^{13}C intensity of the ^{13}C nuclei cross-polarizing with these protons. However, for many coals of different ranks, the electron–proton interactions are indeed the dominant mechanism for ^{1}H $T_{1\rho}$ *(25)*, and this restriction probably does not impose serious consequences for the applicability of the RF DNP–CP approach. Moreover, in low-rank coals and large B_R fields, the ^{1}H rotating-frame relaxation times increase and can become so large that spin diffusion is capable of averaging out the relaxation rates. In this case, undistorted ^{13}C RF DNP–CP spectra will be obtained regardless of the origin of the ^{1}H rotating-frame relaxation.

4. Substantial ^{1}H RF DNP enhancement factors have also been observed in coals of considerably lower rank than the low-volatile bituminous coal discussed in this chapter. Hence, the RF DNP–CP technique should be applicable to coals of many different ranks.

5. The ^{1}H RF DNP enhancement factors can be enlarged by employing larger B_R fields or microwave fields than have been used in the ex-

periments discussed previously. Preliminary experiments indicate that it should be possible to obtain M_{HR} values comparable to the laboratory-frame thermal equilibrium magnetization, $(M_{HL})_0$. If this can be achieved, the number of scans necessary to obtain a ^{13}C spectrum with a good signal-to-noise ratio (100,000 in the experiment shown in Figures 3a and 3b) can be reduced substantially. A probe capable of handling a continuous irradiation of an rf field at the proton Larmor frequency with an amplitude B_R of at least 2 mT is under construction. If this probe can be achieved, it will enable researchers to perform multiple-contact RF DNP–CP experiments, which will reduce the measuring time of a ^{13}C RF DNP–CP spectrum from the 2.8 h it took to obtain each of the spectra shown in Figures 3a and 3b to a few minutes. Hence, RF DNP–CP opens the possibility of a fast characterization of coal by means of ^{13}C NMR spectroscopy.

Acknowledgments

I thank C. E. Bronnimann for suggesting the spin–temperature inversion sequence employed in the RF DNP–CP experiment.

References

1. Pines, A.; Gibby, M. G.; Waugh, J. S. *J. Chem. Phys.* **1972**, *56*, 1776.
2. Pines, A.; Gibby, M. G.; Waugh, J. S. *J. Chem. Phys.* **1973**, *59*, 569.
3. Stejskal, E. O.; Schaefer, J.; McKay, R. A. *J. Magn. Res.* **1977**, *25*, 569.
4. *Analytical Methods for Coal;* Karr, C., Jr., Ed.; Academic: New York, 1978.
5. *Coal Science;* Gorbaty, M. L.; Larsen, J. W.; Wender, I., Eds.; Academic: New York, 1982.
6. *Magnetic Resonance.* "Introduction: Advanced Topics and Applications to Fossil Energy"; Petrakis, L.; Fraissard, J. P., Eds.; D. Reidel: Dordrecht, Netherlands, 1984.
7. Axelson, D. E. *Solid State Nuclear Magnetic Resonance of Fossil Fuels: An Experimental Approach;* Multiscience: Montreal, Canada, 1985.
8. VanderHart, D. L.; Retcofsky, H. L. *Fuel* **1976**, *55*, 202.
9. Miknis, F. P.; Sullivan, M. J.; Bartuska, V. J.; Maciel, G. E. *Org. Geochem.* **1981**, *3*, 19.
10. Wemmer, D. E.; Pines, A.; Whitehurst, D. D. *Philos. Trans. R. Soc. London* **1981**, *A300*, 15.
11. Hagaman, E. W.; Woody, M. C. *Proc. Int. Conf. Coal Sci.;* Verlag Glueckauf GmbH: Essen, Germany, 1981; p 807.
12. Dudley, R. L.; Fyfe, C. A. *Fuel* **1982**, *61*, 651.

13. Packer, K. J.; Harris, R. K.; Kenwright, A. M.; Snape, C. E. *Fuel* **1983**, *62*, 999.
14. Wind, R. A.; Duijvestijn, M. J.; van der Lugt, C.; Smidt, J.; Vriend, J. *Fuel* **1987**, *66*, 876.
15. Botto, R. E.; Wilson, R.; Winans, R. E. *Energy Fuels* **1987**, *1*, 173.
16. Snape, C. E.; Axelson, D. E.; Botto, R. E.; Delpuech, J. J.; Tekely, P.; Gerstein, B. C.; Pruski, M.; Maciel, G. E.; Wilson, M. A. *Fuel* **1989**, *68*, 547.
17. Mehring, M. *High Resolution NMR in Solids;* Springer-Verlag: New York, 1983.
18. Alemany, L. B.; Grant, D. M.; Pugmire, R. J.; Alger, T. D.; Zilm, K. W. *J. Am. Chem. Soc.* **1983**, *105*, 2133.
19. Abragam, A. *The Principles of Nuclear Magnetism;* Clarendon: Oxford, England, 1961.
20. Wind, R. A.; Duijvestijn, M. J.; van der Lugt, C.; Manenschijn, A.; Vriend, J. *Prog. Nucl. Magn. Reson. Spectrosc.* **1985**, *17*, 33.
21. Wind, R. A.; Li, L.; Lock, H.; Maciel, G. E. *J. Magn. Reson.* **1988**, *79*, 577.
22. Wind, R. A.; Lock, H. *Adv. Magn. Opt. Reson.* **1990**, *15*, 51.
23. Wind, R. A.; Anthonio, F. E.; Duijvestijn, M. J.; Smidt, J.; Trommel, J.; de Vette, G. M. C. *J. Magn. Reson.* **1983**, *52*, 424.
24. Tse, D.; Hartmann, S. R. *Phys. Rev. Lett.* **1968**, *21*, 511.
25. Jurkiewicz, A.; Wind, R. A.; Maciel, G. E. *Fuel* **1990**, *69*, 830.

RECEIVED for review June 8, 1990. ACCEPTED revised manuscript December 18, 1990.

11

Proton Magnetic Resonance Thermal Analysis of Argonne Premium Coals

Richard Sakurovs, Leo J. Lynch, and Wesley A. Barton

Commonwealth Scientific and Industrial Research Organisation (CSIRO), Division of Coal and Energy Technology, P.O. Box 136, North Ryde, New South Wales 2113, Australia

> *The eight Argonne Premium coals were examined by using proton magnetic resonance thermal analysis (PMRTA). A new empirical parameter that is related to the second moment of the truncated frequency spectrum of the NMR signal is described and is used to quantify the fusion behavior of coal. The two lowest rank Argonne coals show the same fusion behavior as Australian brown coals of comparable atomic H–C ratios. The temperatures at which fusion of the Argonne bituminous coals occurs are not significantly different from those of Australian coals of the same carbon content. However, the clear relationship found between the fusibility of Australian bituminous coals and hydrogen content is not observed for the Argonne coals. The enhanced fusibility of some of the Argonne Premium coals as compared to Australian coals with the same hydrogen content may be due to the higher sulfur content of those Argonne coals.*

THE PHYSICAL AND CHEMICAL TRANSFORMATIONS that a coal undergoes when heated determine its utility as a fuel or a feedstock for the manufacture of cokes, carbons, and other products. The study of these transformations is complicated by the natural variability and heterogeneity of coals. Variability and heterogeneity ultimately depend on the type and density of covalent and noncovalent interactions that contribute to the stability of these solid materials. Thermal analysis methods with adequate time resolution allow in situ monitoring of transient intermediate states of a material. The information gathered by these methods can be used to specify the kinetics of, and identify the important metamorphic events in, the transformation of the material under investigation. The thermoplastic state attained by bituminous coals during heating to pyrolysis conditions is an important example of a transient intermediate state that determines the behavior of the pyrolysis residue in subsequent processing.

A method of proton magnetic resonance thermal analysis (PMRTA) has been developed in our laboratory to investigate thermal transformations in coal materials (*1*). The undecoupled proton magnetic resonance, which can simply be recorded as a time-domain transient or transverse-magnetization signal, is sensitive to the molecular structure and dynamics of organic solids. The influence of structure on the resonance signal occurs via the static magnetic interactions of the protons that are dominated by the short-range proton–proton dipolar fields. The molecular dynamics influence the proton magnetic resonance signal by decoupling this static dipolar interaction when molecular reorientations occur at a frequency equivalent to the strength of this interaction. Thus molecular reorientations on a time scale of ≤ 10 μs will affect the NMR signal; as the motional decoupling increases, so does the lifetime of the transverse-magnetization signal. The ability to perform the NMR measurements in situ during heating to pyrolysis temperatures and at a sufficient time resolution to monitor transient phenomena constitutes the technique of PMRTA. PMRTA has been applied to studies of the thermoplastic phenomenon of bituminous coals (*2*), to the pyrolysis of brown coals (*3*), and to surveys of Australian coals (*4–6*). These studies involved correlation analyses of the parameters derived from the PMRTA data with a wide range of chemical, proximate, petrographic, and carbonization test data on eastern Australian coals and both bituminous and subbituminous brown coals. A number of strong single- and two-factor correlations were obtained, and the results enable the Australian coal resource to be represented in a classification space such that an unknown coal's properties can be predicted in a comprehensive fashion and with a useful degree of accuracy from the PMRTA data (*6, 7*).

This study concerns PMRTA characterization of the eight Argonne Premium coals. The results are discussed in terms of the thermal-transformation properties revealed, and the data provided allow these North American coals to be compared with the well-characterized set of the mostly Permian Australian coals.

Experimental Details

The analytical data for the Argonne Premium samples were taken from Vorres (8). These eight coals are each largely composed of vitrinite, but there are considerable quantities of inertinite in all of the coals and significant amounts of liptinite in four of them (Table I). The reference set of 130 Australian coals have been characterized by proximate, ultimate, and maceral analyses according to the corresponding Australian standards (9–11).

For PMRTA experiments, coals were first crushed to pass a 0.25-mm sieve and washed in dilute hydrochloric acid at 60 °C for 1 h to remove any iron(II) carbonate. Iron(II) carbonate decomposes during pyrolysis to form magnetically ordered magnetite, which can interfere with NMR measurements (12). The resulting specimens were stored under nitrogen at −18 °C. Subsamples (∼0.2 g) were dried overnight at 105 °C under nitrogen immediately prior to being analyzed by PMRTA.

The experimental design and procedure of the PMRTA experiment are discussed in Chapter 7, and only a brief outline will be given here. The solid echo pulse sequence ($90°_x$–τ–$90°_y$) was used to generate a ^{1}H NMR transverse-magnetization signal that closely approximates the free induction decay (FID) produced by a single 90° pulse (13, 14). The pulse separation, τ, was set to 6 μs, the minimum value for which the echo peak was not obscured by the dead time of the spectrometer after the second pulse. The solid echo signal $I(t)$ was recorded at regular intervals (∼2 min) while the sample was being heated at a rate of 4 °C/min under nonoxidizing conditions from room temperature through the region of main pyrolysis to a maximum temperature of ∼600 °C.

Results and Data Analysis

The shape of the transverse-magnetization signal is determined by the magnetic interactions between the protons of the sample material and the extent to which these interactions are decoupled by molecular motions. In particular, the signal from heterogeneous specimens such as partially fused coals consists of components that include

- short-lived (the characteristic time constant, T_2, is ∼16–24 μs for coals) Gaussian-like components from hydrogen in structures in which molecular reorientations occur on a time scale of >10 μs (termed *immobile* or *rigid-lattice* structures)

- relatively long-lived (T_2 is ∼100–1000 μs for coals), exponential-like components from hydrogen in structures in which the molecular

Table I. Analytical Data for the Argonne Coals

Analysis	ND	WY	IL	UT	WV	PITT	UF	POC
Elemental[a]								
Carbon content (daf)	72.94	75.01	77.67	80.69	82.58	83.20	85.50	91.05
Hydrogen content (daf)	4.83	5.35	5.00	5.76	5.25	5.32	4.70	4.44
Total sulfur (d)	0.80	0.63	4.83	0.62	0.71	2.19	2.32	0.66
Organic sulfur (d)	0.63	0.43	2.01	0.35	0.52	0.81	0.54	0.48
Atomic H–C ratio	0.789	0.849	0.766	0.850	0.757	0.761	0.654	0.580
Carbon aromaticity, f_a		0.54	0.55	0.72	0.64	0.75	0.72	0.81
Petrographic[b]								
Vitrinite	—	89	85	87	73	85	91	89
Liptinite	—	<1	5	5	12	7	1	1
Inertinite	—	11	10	8	15	8	8	10
Gieseler plastometry[c]								
Softening	—	—	367	396	—	381	373	460
Maximum fluidity	—	—	417	409	—	430	450	474
Solidification	—	—	444	438	—	472	497	509
log (maximum fluidity)	—	—	1.9	0.5	—	4.5	4.5	1.6

PMRTA[d]								
Onset of fusion (3.4[d])	337	—	383	390	408	405	422	456
Maximum fusion (2.0[d])	368	379	410	416	434	431	452	483
Solidification (1.1[d])	423	431	441	451	464	462	485	513
Minimum $M_{2T}16$ (kHz^2) (maximum fusion) (0.8[d])	42.4	39.8	31.6	29.0	31.5	20.7	23.1	43.1

ABBREVIATIONS: ND, Beulah–Zap; WY, Wyodak–Anderson; IL, Illinois No. 6; UT, Blind Canyon; WV, Lewiston–Stockton; PITT, Pittsburgh No. 8; UF, Upper Freeport; POC, Pocahontas No. 3; daf, dry ash-free; and d, dried.

[a] Except for H–C ratio and aromaticity, values are given as percentages.
[b] Values are given as volume percentages.
[c] Values are given in degrees celsius.
[d] The values given are the mean difference between duplicates.

SOURCE: Data for elemental and petrographic analyses and Gieseler plastometry are taken from reference 8. Carbon aromaticity (f_a) data are taken from reference 18.

reorientations occur on a time scale of <10 µs (termed *mobile* structures)

The amplitude contributions of these components at zero time (i.e., at the peak of the solid echo) to the total signal $I(0)$ are directly proportional to the hydrogen populations of the corresponding structures, whether they are mobile or rigid. The raw PMRTA data for the eight Argonne Premium coals, in the form of stacked plots of the ^{1}H NMR transverse-magnetization signals $I(t)$, are adequately represented by the data sets for four of the coals: Wyodak–Anderson, Illinois No. 6, Blind Canyon, and Pocahontas No. 3 (Figures 1a–1d). At low temperatures, the signals decay relatively rapidly, as would be expected for solid organic materials. With increasing temperature, depending on the coal, an increasing fraction of the signals becomes relatively long-lived, providing direct evidence that molecular mobility is being thermally activated in parts of the structure. At even higher temperatures, both the contribution of this mobile fraction and the total signal amplitude $I(0)$ are reduced as hydrogen-containing volatile substances are evolved. At ~600 °C, the signals of the residual char or semicoke are reduced in amplitude and are relatively short-lived, a result that is characteristic of rigid solids with low H contents.

Generally, the ^{1}H NMR signals obtained for coals are not well fitted by the sum of one exponential component and one Gaussian component. No doubt this result is due to the fact that coal is heterogeneous at the molecular level, and thus the hydrogen in a given coal produces a number of signal components with a range of time constants. Therefore parameters that embody the shape of the whole signal are to be preferred.

One such approach to quantifying the ^{1}H NMR signal $I(t)$ is to obtain a frequency-domain spectrum $G(\nu)$ by Fourier transformation and then calculate the second moment, M_2 of that spectrum. The second moment captures the sensitivity of the NMR signal to the static magnetic interactions experienced by the resonant nuclei and can thus provide valuable information about the molecular dynamics and the concentration and distribution of spin species in the material (*15*). As properly defined in terms of an infinite integral over frequency, M_2 is a linear parameter in that its value for a spectrum composed of several components will be the population-weighted average of the values for the individual components. However, this parameter has the disadvantage that for the Lorentzian line shape arising from the transformation of a signal containing an exponentially decaying component, the infinite integral diverges (*15*). This difficulty has been avoided by truncating the integration at a convenient finite frequency ν_T, giving a truncated second moment $M_{2T}(\nu_T)$ defined as (*16*)

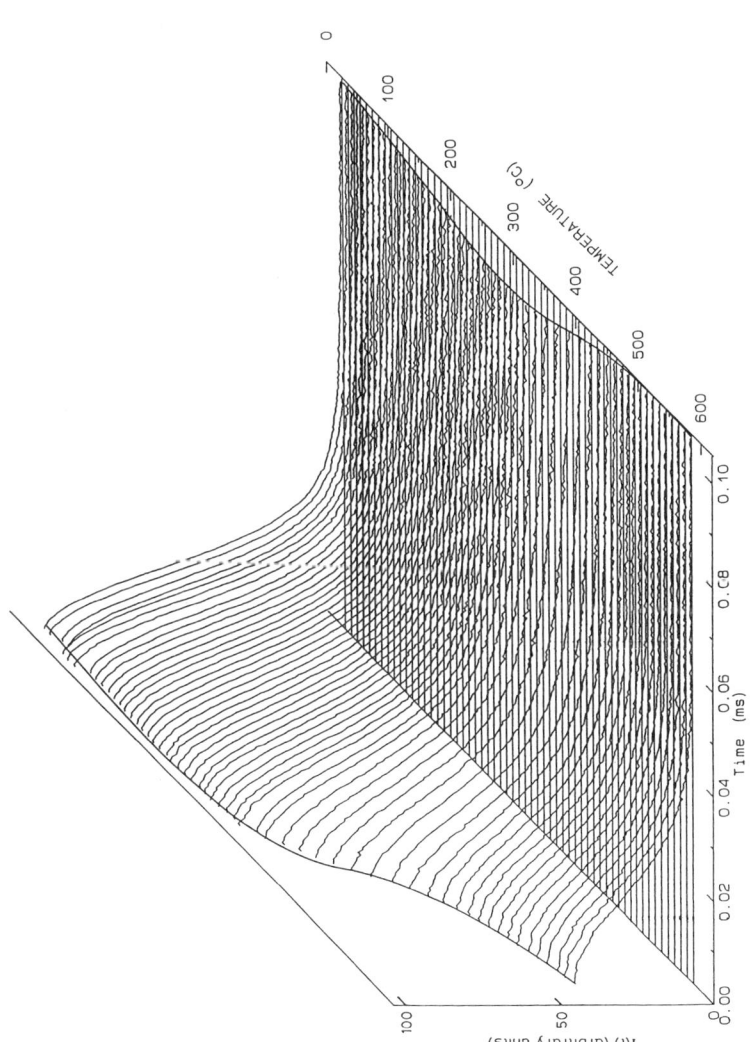

Figure 1a. Stacked plot of ^1H NMR time-domain signals obtained during pyrolysis of Wyodak–Anderson coal.

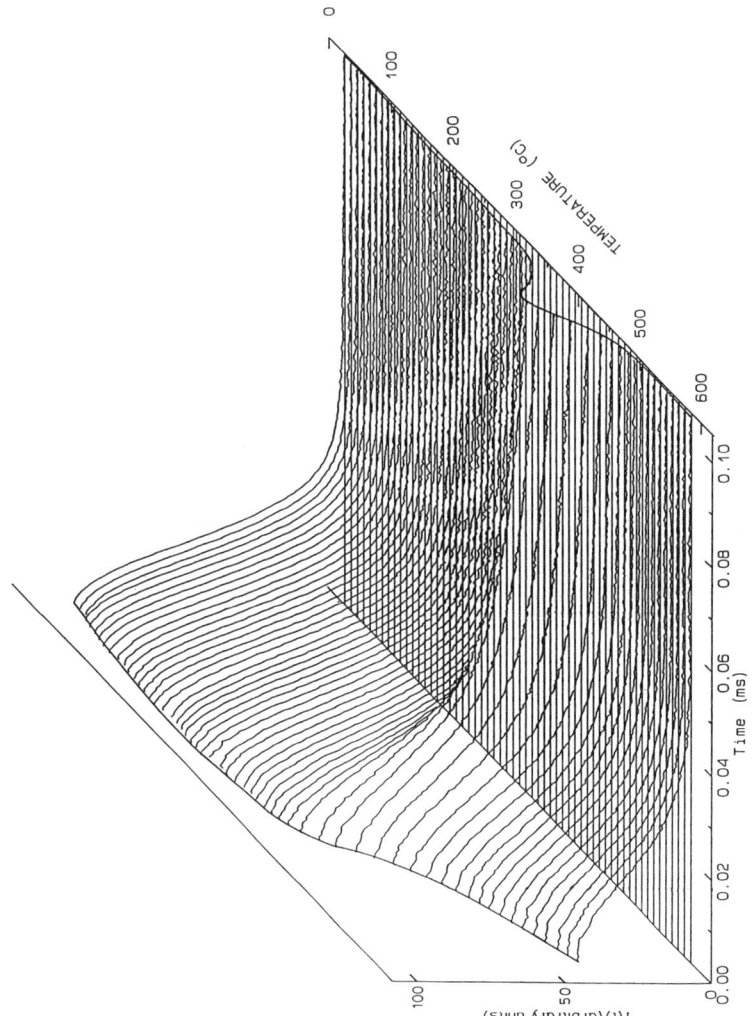

Figure 1b. Stacked plot of ^1H NMR time-domain signals obtained during pyrolysis of Illinois No. 6 coal.

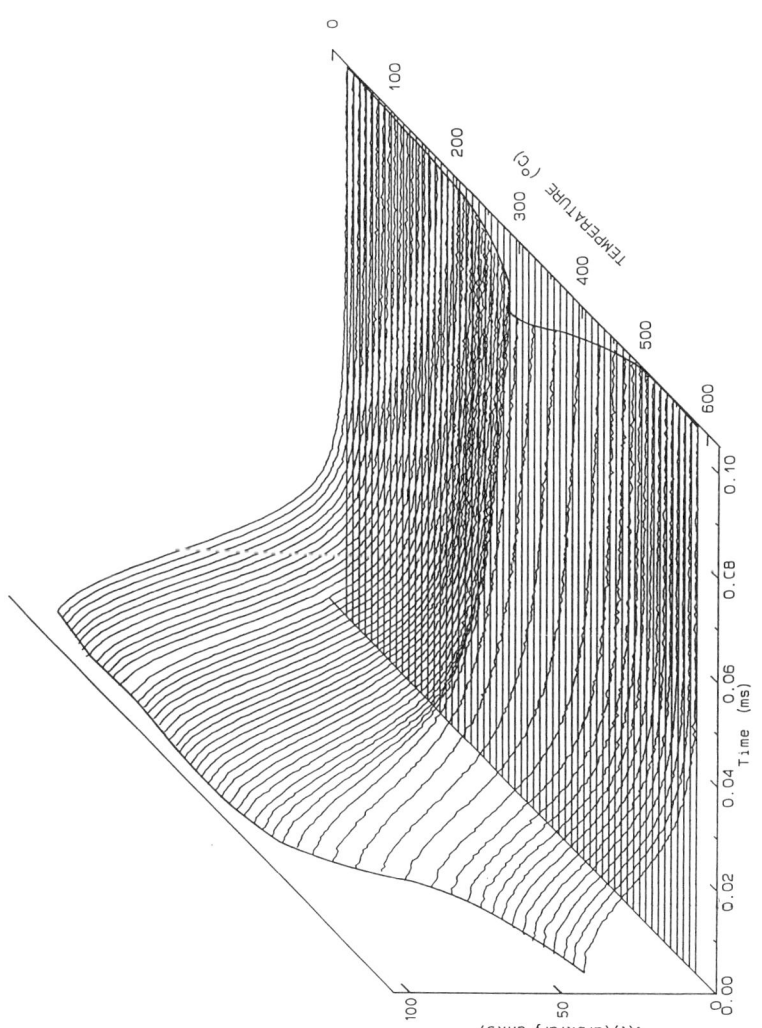

Figure 1c. Stacked plot of ^1H NMR time-domain signals obtained during pyrolysis of Blind Canyon coal.

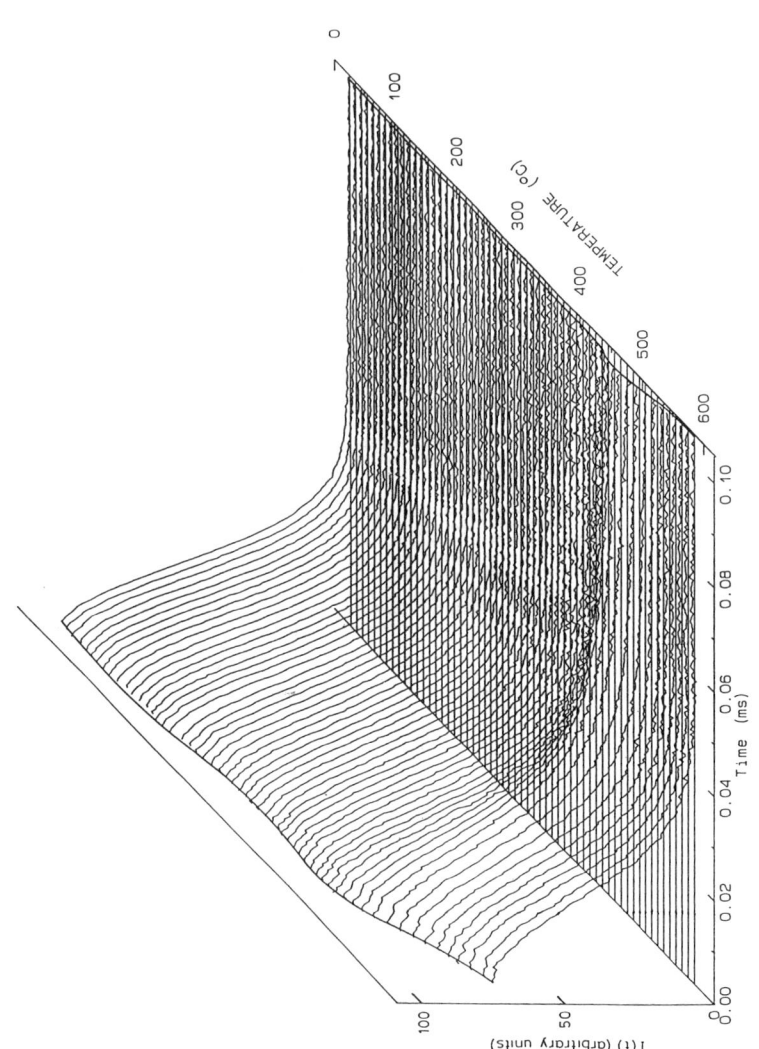

Figure 1d. Stacked plot of ^1H NMR time-domain signals obtained during pyrolysis of Pocahontas No. 3 coal.

$$M_{2T}(\nu_T) = \frac{\int_0^{\nu_T} \nu^2 G(\nu) \, d\nu}{\int_0^{\nu_T} G(\nu) \, d\nu} \quad (1)$$

where ν and ν_T are specified relative to the resonance frequency. A difficulty with the $M_{2T}(\nu_T)$ parameter so defined is that for a spectrum containing several different components, the overall $M_{2T}(\nu_T)$ is not necessarily a weighted average of the values for the individual components. This nonlinearity limits the usefulness of $M_{2T}(\nu_T)$ in describing interactions within heterogeneous systems such as coals.

If, however, an empirical approach is adopted and the parameter $M_{2T}(\nu_T)$ is defined as

$$M_{2T}(\nu_T) = \frac{\int_0^{\nu_T} \nu^2 G(\nu) \, d\nu}{\int_0^{\infty} G(\nu) \, d\nu} \quad (2)$$

where the denominator is equivalent to the initial amplitude $I(0)$ of the corresponding time-domain signal, then $M_{2T}(\nu_T)$ has the desirable properties of being both finite and linear. The $M_{2T}(\nu_T)$ parameter thus defined is not equivalent to the second moment of the frequency spectrum truncated at ν_T (which is given by eq 1). In the following discussion, $M_{2T}(\nu_T)$ refers to the empirical parameter defined according to eq 2.

Because $M_{2T}(\nu_T)$ is calculated by integration of the frequency spectrum, it is less affected by random and systematic errors than parameters obtained by fitting the time-domain signal. Typical random errors in the determination of $M_{2T}(\nu_T)$ for coals at truncation frequencies of 40 kHz or less are 0.3% of the "rigid" $M_{2T}(\nu_T)$ value. Larger errors occur at higher truncation frequencies because of the increasing contribution to $M_{2T}(\nu_T)$ from random noise in the time-domain signal. This is another reason why it is necessary, in practice, to truncate the integration at a suitable finite frequency.

$M_{2T}(\nu_T)$ values for a series of analytical functions with characteristic time constants similar to those of components in typical ^{1}H NMR signals obtained for coals are listed in Table II for three truncation frequency values. Factors influencing $M_{2T}(\nu_T)$ for a typical heated coal, whose ^{1}H NMR signal usually can be approximated by the sum of Gaussian compo-

Table II. $M_{2T}(\nu_T)$ Values Calculated According to Equation 2 for Various Analytical Functions with Characteristic Time Constants Typical of Components in ^{1}H NMR Signals Obtained for Coals

Decay Shape	Time Constant (T_2)	$M_{2T}5$	$M_{2T}16$	$M_{2T}40$
Gaussian ($\exp[-t/T_2]^2$)	14	2.03	50.84	231.81
	16	2.30	53.35	189.04
	18	2.56	54.60	153.69
	20	2.82	54.71	125.91
	22	3.06	53.87	104.48
	24	3.29	52.24	87.90
Exponential ($\exp[-t/T_2]$)	50	3.66	23.56	71.40
	100	3.03	13.84	38.06
	1000	0.48	1.60	4.03
Exponential plus Gaussian	100 (50%) + 20 (50%)	2.92	34.27	81.98
Typical coals (room temp.)		2.3–2.7	48–55	150–190
Typical coals (425 °C)		2.4–3.2	23–54	50–130

NOTE: The values of T_2 are given in microseconds, and the values of M_{2T} are given in kilohertz squared.

nents attributed to a rigid fraction of the coal structure and exponentially decaying components attributed to the mobilized fraction, are

- the extent to which the molecular structure is mobilized by fusion or other thermally activated processes and thus the proportion of the NMR signal with a slow exponential-type decay
- the degree of mobility of the mobile structures, which is reflected in the time constants of any exponential decays
- the hydrogen density of the rigid material and the local density of unpaired electrons (12), which can affect the time constants of the Gaussian-like decays.

Table II shows that $M_{2T}5$ (i.e., $\nu_T = 5$ kHz) increases appreciably with the time constant of the Gaussian component over the range of Gaussian time constants associated with coals (16–24 μs) but is not very sensitive to the proportion of the signal due to short-lived exponential components. Because for typical coals the time constant of the Gaussian-like component becomes longer with increasing temperature, due in part to thermal expansion of the coal (which increases interproton distances and hence reduces their magnetic interactions), $M_{2T}5$ is observed to increase steadily with temperature for most coals (Figure 2). In contrast, $M_{2T}40$ decreases with increasing time constant of the Gaussian component (Table II) and thus steadily decreases as a coal is heated (Figure 2).

$M_{2T}16$ is insensitive to the time constants of the Gaussian-like components corresponding to immobile structures in coals (Table II) but is more sensitive than either $M_{2T}5$ or $M_{2T}40$ to the fraction of hydrogen associated with mobile structures (Table II and Figure 2). Because of this sensitivity to the extent of molecular mobility, the value of $M_{2T}16$ provides a more appropriate measure of the extent to which the molecular lattice of a coal has been mobilized by thermal destabilization. For materials whose ^{1}H NMR signals have typical Gaussian time constants different from those for coals, the sensitivity of these $M_{2T}(\nu_T)$ parameters to molecular mobility and hydrogen density will be different.

Plots of $M_{2T}16$ values versus temperature are smoothed by a cubic spline fit to provide the characteristic $M_{2T}16$ pyrograms of the coal spec-

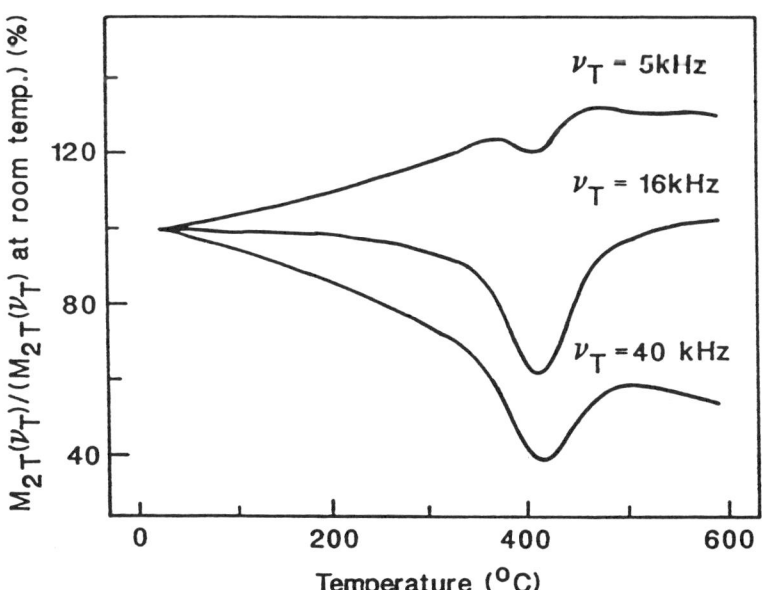

Figure 2. Plot of the variations of $M_{2T}(\nu_T)$ with temperature at truncation frequencies of 5, 16, and 40 kHz, expressed as a percentage of the values at room temperature, for Illinois No. 6 coal.

imens. The pyrograms for the eight Argonne Premium coals are plotted in Figures 3–5. Secondary parameters derived from these pyrograms are

1. the minimum value of $M_{2T}16$ reached during the pyrolysis and referred to as the maximum fusion or *fusibility* of the coal;
2. the temperature of maximum extent of molecular mobility (i.e., that at which the minimum $M_{2T}16$ value occurs);
3. the temperature at which $M_{2T}16$ decreases most rapidly (termed the *PMRTA fusion temperature*); and
4. the temperature at which $M_{2T}16$ increases most rapidly (termed the *PMRTA solidification temperature*).

Parameters 3 and 4 are obtained from the derivative of $M_{2T}16$ with respect to temperature. These four parameters describe the thermoplasticity of bituminous coal in a fashion analogous to the standard parameters obtained from Gieseler plastometry (*10*). The definitions and values of these parameters are illustrated for the Upper Freeport coal in Figure 5.

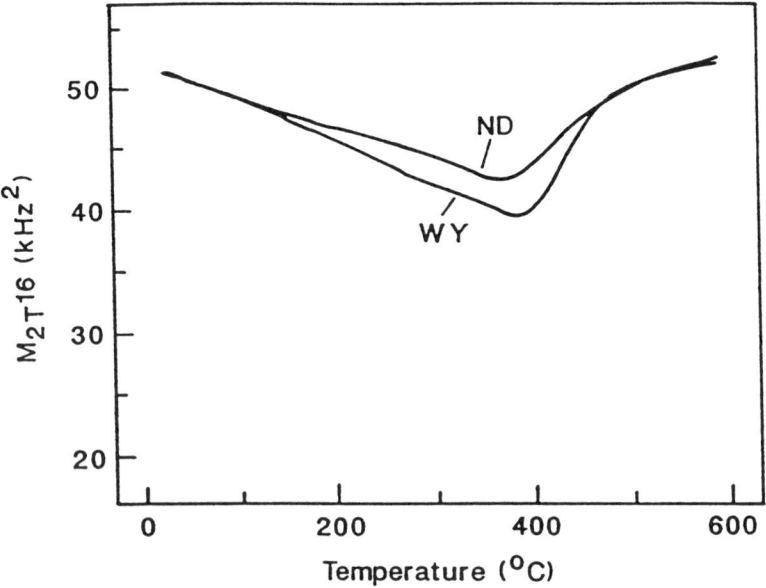

Figure 3. $M_{2T}16$ *pyrograms for Beulah–Zap (ND) and Wyodak–Anderson (WY) coals.*

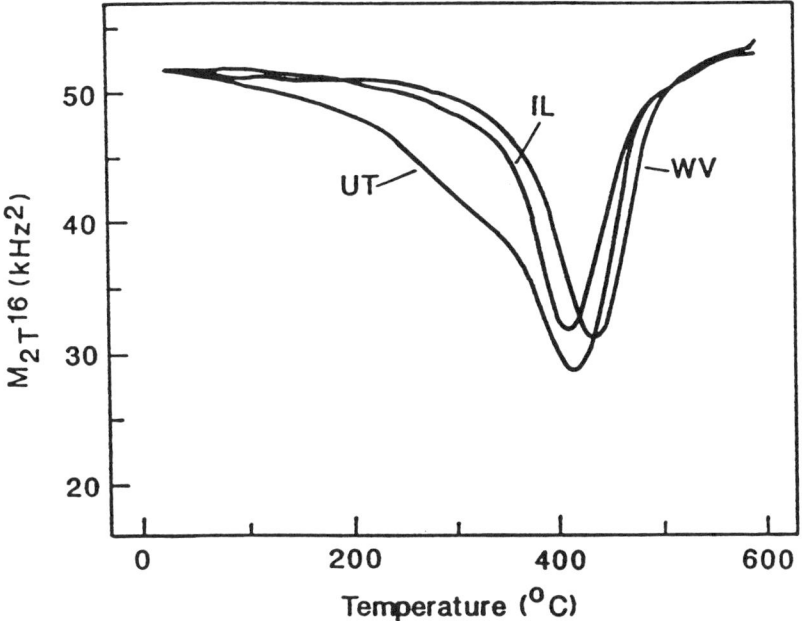

Figure 4. $M_{2T}16$ pyrograms for Illinois No. 6 (IL), Blind Canyon (UT), and Lewiston–Stockton (WV) coals.

Discussion

The PMRTA data identify two of the Argonne coals (Wyodak–Anderson and Beulah–Zap lignite) as nonthermoplastic coals of lower than bituminous rank and the other six coals as having the thermoplastic character typical of bituminous coals. As such, these results are consistent with the American Society for Testing and Materials (ASTM) classifications and carbon ranks of these coals except perhaps for the Illinois No. 6 coal, which has a high sulfur content (*see* the "Bituminous Coals" section).

Subbituminous Coals. The stacked plot of the ^{1}H NMR transverse-magnetization signals of the Wyodak–Anderson subbituminous coal in Figure 1a is qualitatively similar to that of the Beulah–Zap lignite, and the data for both coals fit the class of behaviors that includes the tertiary Victorian brown coals of Australia (*3*). The slow decrease in $M_{2T}16$ with increasing temperature shown by the pyrograms of these two lowest rank Argonne Premium coals (Figure 3) prior to significant loss of volatile material is characteristic of that found for the Australian brown coals examined by PMRTA (*3*).

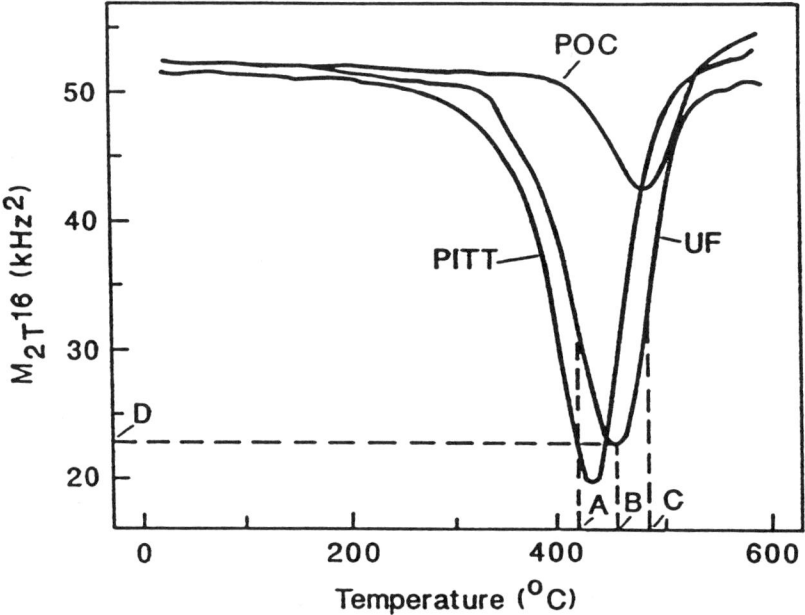

Figure 5. $M_{2T}16$ *pyrograms for Pittsburgh No. 8 (PITT), Upper Freeport (UF) and Pocahontas No. 3 (POC) coals. Definitions of secondary parameters derived from* $M_{2T}16$ *pyrograms are illustrated for Upper Freeport coal: A, PMRTA fusion temperature; B, PMRTA temperature of minimum* $M_{2T}16$ *(maximum fusion); C, PMRTA solidification temperature; and D, minimum* $M_{2T}16$ *(maximum fusion).*

This decrease in $M_{2T}16$ reflects the increase in the lifetime of the NMR signals, $I(t)$ that is apparent in the stacked plot of Figure 1a and that, in the case of the Victorian brown coals, has been demonstrated to be the consequence of the thermally activated mobilization of a "waxy" component material with a high H–C ratio, which can be extracted under relatively mild conditions (*17*). Indeed, a strong correlation exists between the extent to which this mobilization occurs and both the H–C ratio of the coals and the amount of extractable material. The correspondence between the extent of mobilization as revealed by the PMRTA data and the H–C value holds for the Beulah–Zap and Wyodak–Anderson coals. This correspondence is shown in the plot of the minimum $M_{2T}16$ value versus H–C ratio in Figure 6, which includes these data with those of the set of Australian subbituminous and brown coals. Correspondingly, these PMRTA data predict a greater extractable component (*17*) for the Wyodak–Anderson coal (~11%) than for the Beulah–Zap lignite (~7%).

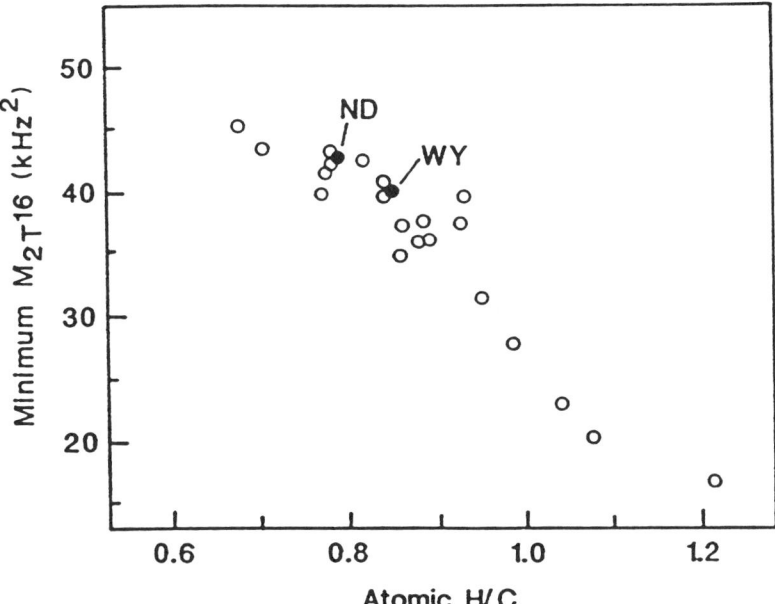

Figure 6. Plot of minimum $M_{2T}16$ vs. atomic H–C ratio for Australian (○) and Argonne (●) subbituminous and brown coals. The abbreviations are the same as those used in Table I.

The profile of the initial signal amplitude, $I(0)$, with increasing temperature, which is delineated in the stacked plot diagrams, indicates the onset and progress of loss of volatile material from the coal specimen. This profile is exemplified by the Wyodak–Anderson coal data shown in Figure 1a, and onset of significant volatile material loss coincides with the onset of loss of molecular mobility of the heated material. Thus this loss of material, at least initially, can be attributed to volatilization of the mobilized component of the coal.

The residues of the extracted Australian brown coals are considered to be lignin-derived materials (*17*) and were found by PMRTA to be non-fusible on heating to pyrolysis conditions (*3*). This infusibility is indicative of a highly cross-linked macromolecular nature, and therefore this material is, by inference, the precursor of the char formed in the pyrolysis of these low-rank coals.

The considerable degree of molecular mobility attained by these coals when heated as revealed by PMRTA must be reconciled with their known lack of thermoplastic behavior. The fact that these coals do not flow when subjected to shear at any stage during heating to pyrolysis conditions as, for example, in the Gieseler plastometer test, can be explained if the fusible "waxy" material is dispersed in an infusible macromolecular matrix.

Bituminous Coals. The thermoplastic character of the six bituminous Argonne coals is apparent from the increase in lifetime of their NMR signals (e.g., Figures 1b–1d) and hence in their molecular mobility as the temperature approaches 400 °C. This effect is clearly delineated in each of the six $M_{2T}16$ pyrograms derived from these data (Figures 4 and 5). PMRTA studies (2) have demonstrated that the onset of this mobilization of bituminous coal structure is heating-rate independent and that the process involves thermal destabilization of relatively weak physical interactions rather than scission of covalent bonds between the molecular units. The mobilization process is therefore considered to be a fusion or physical melting.

The Cretaceous Blind Canyon coal is distinguished from the other bituminous Argonne coals in that its PMRTA data reveal two stages of molecular mobilization. The gradual reduction in $M_{2T}16$ at temperatures below that of the thermoplastic fusion is somewhat akin to the mobilization of the subbituminous coals but occurs at higher temperatures. Experience in the application of PMRTA points to the probability that this first-stage mobilization is of the aliphatic-rich regions of the coal structure, with contributions from both liptinite and vitrinite macerals. Indeed, statistical analyses of PMRTA data of Australian coals (6) have demonstrated that the "average" liptinites of the lower rank bituminous coals (80–85% carbon) exhibit significant molecular mobility above ~200 °C. The Blind Canyon coal has a relatively high atomic H–C ratio of 0.85 and a carbon aromaticity (estimated from ^{13}C NMR spectroscopy) of 0.64 (18) but contains only 5% liptinite. These properties imply that the vitrinite in this coal contains a relatively large proportion of aliphatic or hydroaromatic structures for a coal of its carbon rank. However, the Carboniferous Lewiston–Stockton coal, which contains 12% liptinite, shows little evidence of fusion below ~300 °C. Therefore, the liptinite in this coal has greater thermal stability than that found for the average liptinite in the Australian coals of similar carbon content (6) and thus may fuse in a temperature interval similar to that of its vitrinite counterpart.

The relationships between the values of the PMRTA and Gieseler parameters defining thermoplasticity that were established for the extensive suite of Australian bituminous coals (7) are followed by the Argonne coals for which plastometry data are available (Table I). There are good correlations between the PMRTA and Gieseler parameters defining the temperatures of maximum extent of fusion and maximum Gieseler fluidity and the temperatures of PMRTA and Gieseler solidification, respectively; a poor correlation of the PMRTA maximum fusion and Gieseler maximum fluidity parameters; and no apparent correlation between the PMRTA fusion and Gieseler softening temperatures.

The temperature at which the fusion of a bituminous coal reaches a maximum increases with increasing carbon content of the coal (19). The relationship for the Australian bituminous coals is followed and extended by the Argonne bituminous coals (Figure 7). PMRTA fusibility is found to correlate reasonably well with hydrogen content for Australian bituminous coals (4). The plot of minimum $M_{2T}16$ against hydrogen content for the Australian and Argonne bituminous coals (Figure 8) shows that the linear relationship found for the Australian coals is not obeyed by the Argonne Premium coals. The Argonne coals in general are significantly more fusible than Australian coals of comparable hydrogen content, and the correlation between minimum $M_{2T}16$ and hydrogen content is also far weaker for the Argonne coals than for the Australian coals. In particular, the two highly thermoplastic coals, Upper Freeport and Pittsburgh No. 8, show major deviations from the behavior of Australian coals. These observations suggest a major systematic difference between the fusibility, and hence the underlying molecular properties, of these two groups of coals.

One other difference between the two groups of coals is the relatively high sulfur contents of the Argonne coals. Three of the eight Argonne coals have total sulfur contents of greater than 2% (Table I), and all three have anomalously high fusibility for their hydrogen content. Only two of

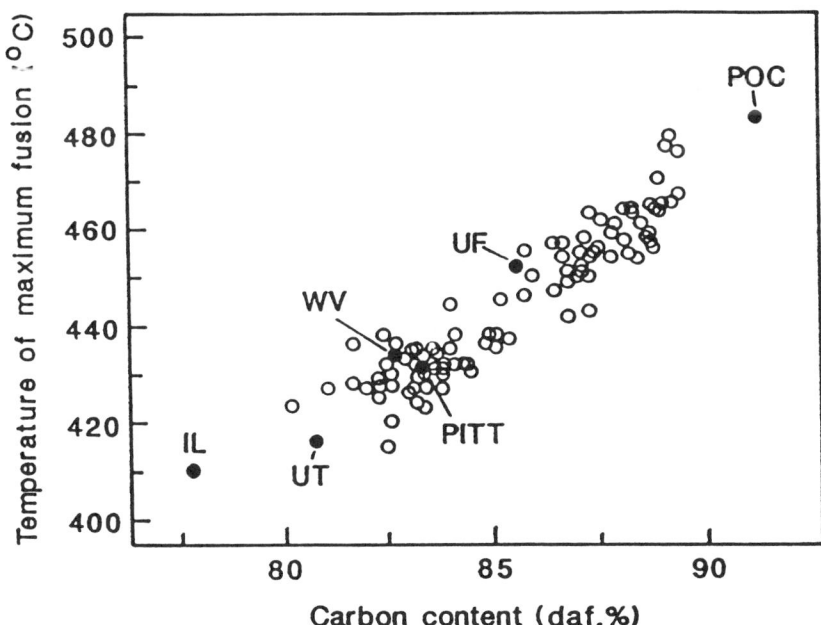

Figure 7. Plot of the temperature of minimum $M_{2T}16$ vs. carbon content for Australian (○) and Argonne (●) bituminous coals. The abbreviations are the same as those used in Table I.

the 110 Australian bituminous coals examined have total sulfur contents greater than 2% (Figure 8) and, of these two, the one with the higher total sulfur content (4.5%, dried basis) has anomalously high fusibility for its hydrogen content. These results suggest that some interaction between sulfur-containing materials and the rest of the coal has occurred, either during the time that the coal was thermoplastic or during the coalification process. However, the magnitude of the increase in fusibility is not simply proportional to sulfur content because, of the Argonne coals with >2% sulfur, the one richest in both organic and pyritic sulfur (Illinois No. 6) shows the smallest deviation from the trend established by the low-sulfur Australian coals. If the high sulfur content does increase fusibility, then pyritic sulfur is responsible because the Upper Freeport coal, which shows marked deviation from the fusibility trend of the low-sulfur Australian coals, does not have an unusually high organic sulfur content (*see* Table I and Figure 8). Any chemical interaction between sulfur-containing materials and the coal that enhances the fusibility is unlikely to involve elemental sulfur as an intermediate because addition of elemental sulfur to bituminous coals reduces their fusibility (*20*) and impairs their coking performance (*21*).

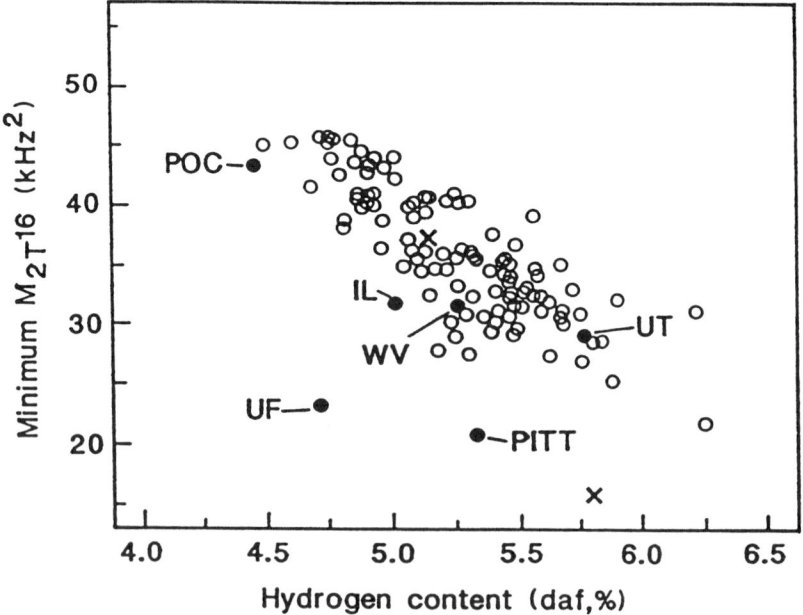

Figure 8. Plot of minimum $M_{2T}16$ vs. hydrogen content for Australian (○) and Argonne (●) bituminous coals. The abbreviations are the same as those used in Table I. The results for the two high-sulfur Australian coals are indicated by ×.

Enhancement of fusibility of coal by sulfur-containing materials does not explain the relatively high fusibility of Pocahontas No. 3 coal for its hydrogen content as compared to the Australian coals (Figure 8), and thus another reason must be sought in this case. Differences between the Argonne Carboniferous and Cretaceous and Australian Permian bituminous coals may stem from the very different origins of these two groups of coals (22, 23). The Carboniferous Laurasian coals were deposited in moist, warm, swampy conditions, whereas Permian Gondwana coals were deposited in relatively cooler and far more seasonal conditions—possibly freezing winters in some cases.

These differences in source material and depositional environment are known to have affected the maceral composition of the coals formed from them. Permian Gondwana coals are generally richer in macerals from the relatively infusible inertinite group than the Carboniferous Laurasian coals (23). Some properties of the inertinites contained in these two groups of coals are also different (23, 24). Substantial portions of the inertinites of Gondwana coals soften and become integrated with the rest of the coke matrix on heating, whereas inertinites from Carboniferous coals are less fusible on heating (24). The inertinites from Gondwana coals are generally more fluorescent than their Laurasian counterparts for a given reflectance (23). There are also systematic differences in the free radical concentrations of Gondwana and Laurasian coals (25). Therefore, the relationships of fusibility properties with other properties of the coals might be expected to show systematic differences. Perhaps the first question asked in comparisons between coals obtained from different continents should not be "Is there a systematic difference in the properties of these coals?", but rather "In view of the fact that these coals were formed under very different conditions in different geological ages from different precursor plant materials, why are their properties so similar?"

Summary

The thermal-transformation behavior during heating to pyrolysis temperatures of each of the Argonne Premium coals, as determined by PMRTA, was detailed. The two subbituminous Argonne coals, Beulah–Zap lignite and Wyodak–Anderson, have thermal-transformation properties that are similar to Australian coals of the same atomic H–C ratio.

The Argonne coals of bituminous rank have characteristic temperatures describing their fusibility that are similar to Australian bituminous coals of the same carbon content. The relationships between the results from Gieseler plastometry and PMRTA for the Argonne coals are similar to those obtained for Australian coals.

The strong relationship between hydrogen content and fusibility that is observed for Australian bituminous coals is not followed by Argonne coals. Three of the four Argonne coals with greater fusibility for their hydrogen content than expected from the relationship found for Australian coals have high sulfur contents, although other differences between the coals that could explain their different fusibilities may occur.

An empirical NMR parameter similar to the second moment of the truncated frequency spectrum has been defined, and it, unlike previous definitions, is both finite and linear. The truncation frequency can be chosen so that this parameter is sensitive primarily to the extent of molecular mobility of a coallike material.

Acknowledgment

The Argonne Premium coals were provided by Karl S. Vorres (Argonne National Laboratory). Neil Thomas, Zenta Lauks, and Duc Phan performed the PMRTA measurements.

References

1. Lynch, L. J.; Webster, D. S.; Barton, W. A. In *Advances in Magnetic Resonance;* Waugh, J. S., Ed.; Academic: New York, 1988; Vol. 12, pp 385–421.
2. Lynch, L. J.; Webster, D. S.; Sakurovs, R.; Barton, W. A.; Maher, T. P. *Fuel* **1988**, *67*, 579–83.
3. Lynch, L. J.; Sakurovs, R.; Webster, D. S.; Redlich, P. J. *Fuel* **1988**, *67*, 1036–41.
4. Sakurovs, R.; Lynch, L. J.; Maher, T. P.; Banerjee, R. N. *Energy Fuels* **1987**, *1*, 167–72.
5. Webster, D. S.; Sakurovs, R.; Lynch, L. J.; Maher, T. P. *Proc. Int. Conf. Coal Sci.* **1989**, *2*, 1107.
6. Sakurovs, R.; Lynch, L. J.; Barton, W. A. In *Coal Science II;* Schobert, H. H.; Bartle, K. D.; Lynch, L. J., Eds.; ACS Symposium Series 461; American Chemical Society: Washington, DC, 1991; pp 111–26.
7. Sakurovs, R.; Lynch, L. J.; Webster, D. S. *J. Coal Qual.* **1991**, *10*, 37–41.
8. Vorres, K. S. *Users Handbook for the Argonne Premium Coal Sample Program;* ANL/PCSP–89/1; Argonne National Laboratory: Argonne, IL, 1989.
9. Australian Standard AS 1038, *Methods for the Analysis and Testing of Coal and Coke,* 1986.
10. Australian Standard AS 2137, *Hard Coal: Determination of Plastic Properties Using the Gieseler Plastometer,* 1981.
11. Australian Standard AS 2856, *Coal: Maceral Analysis,* 1986.
12. Lynch, L. J.; Sakurovs, R.; Barton, W. A. *Fuel* **1986**, *65*, 1108–11.
13. Powles, J. G.; Strange, J. H. *Proc. Phys. Soc. (London)* **1963**, *82*, 6–15.
14. Barton, W. A.; Lynch, L. J.; Webster, D. S. *Fuel* **1984**, *63*, 1262–68.

15. Abragam, A. *The Principles of Nuclear Magnetism;* Clarendon: Oxford, England, 1961; p 106.
16. Srinivasan, R.; Jagannathan, N. R. *Indian J. Pure Appl. Phys.* **1982**, *20*, 879–82.
17. Redlich, P. J.; Jackson, W. R.; Larkins, F. P. *Fuel* **1985**, *64*, 1383–90.
18. Brown, H. R.; Hesp, W. R.; Waters, P. L. *J. Inst. Fuel* **1964**, *37*, 130.
19. Solum, M. S.; Pugmire, R. J.; Grant, D. M. *Energy Fuels* **1989**, *3*, 187–93.
20. Sakurovs, R.; Lynch, L. J.; Ito, O. *Proc. 4th Aus. Coal Sci. Conf.* **1990**, 1.
21. Berkovitch, I.; McCulloch, A. *Fuel* **1946**, *25*, 36–41, 69–78.
22. Martini, I. P.; Johnson, D. P. *Int. J. Coal Geol.* **1987**, *7*, 365–88.
23. Taylor, G. H.; Liu, S. Y.; Diessel, C. F. K. *Int. J. Coal Geol.* **1989**, *11*, 1–22.
24. Diessel, C. F. K.; Wolff-Fischer, E. *Int. J. Coal Geol.* **1987**, *9*, 87–108.
25. Wind, R. A.; Duijvestijn, M. J.; van der Lugt, C.; Smidt, J.; Vriend, H. *Fuel* **1987**, *66*, 876–85.

RECEIVED for review June 7, 1990. ACCEPTED revised manuscript July 11, 1991.

12

Multivariate Statistical Analysis and Pattern Recognition

Principal-Component Analysis of Coal Solid-State NMR Relaxation Times

David E. Axelson[1]

NMR Technologies, Inc., Advanced Technology Center, 9650–20 Avenue, Edmonton, Alberta T6N 1G1, Canada

> *Applications of multivariate statistical analysis are introduced by using a suite of 18 Canadian coals, ranging in rank from lignite to semi-anthracite. These samples were characterized by conventional proximate and ultimate analyses, as well as by numerous solid-state NMR relaxation times (proton rotating-frame spin–lattice relaxation, cross-polarization relaxation, and dipolar-dephasing relaxation). The purpose is to introduce a different approach to the analysis of such data and to determine whether the concept of rank as defined by more classical measurements (reflectance, percent carbon) can be compared to NMR-derived parameters in some systematic way.*

COALS AND THEIR DERIVATIVES are widely studied and used in commercial endeavors, but much of the work involved in optimizing coal-related processes involves costly, labor-intensive, trial-and-error experimentation. Many conventional testing protocols may not be sufficiently

---

[1]Current address: Petroleum Recovery Institute, 3512 33rd Street, NW, Calgary, Alberta T2L 2A6, Canada

accurate to allow proper identification of the most appropriate processing conditions because of complex synergistic and antagonistic interactions among the various components. This problem arises, in part, from the lack of explicit and unambiguous structure–property relationships that relate chemical structure to processability.

The development of sophisticated instrumental tools to define complex chemical characterization problems has not necessarily been accompanied by a corresponding general appreciation of the need for more sophisticated methods of data analysis. Problems arise from the acquisition of massive amounts of data that are not effectively interpreted or incorporated into proper mechanistic descriptions and predictive models of the system(s) under investigation. In complex systems, significant patterns or correlations are not always evident when the data are measured one variable at a time. Interactions among measured chemical variables tend to obscure the data. The need for multivariate techniques is apparent.

The newly emerging discipline of *chemometrics*, that is, the application of statistical analytical and pattern-recognition techniques to chemical problems, has begun to address this "information overload" problem (*1–23*).

The first step in the application of these techniques, which can be denoted as exploratory data analysis, is designed to uncover three main aspects of the data:

1. anomalous samples or measurements
2. significant relationships among the measured variables
3. significant relationships or groupings among the samples

The second step, applied pattern recognition, tests the strength of these basic relationships and other presumed relationships by developing classification–prediction models and determining the accuracy of these models. Principal-component-analysis (PCA) and factor-analysis (FA) techniques will be briefly discussed in this regard.

The purpose of this chapter is not to dwell on the intricacies of the mathematical methods themselves, but rather to illustrate, in a somewhat simplistic manner, the possible merits of this method of analysis. The philosophical significance of such calculations must also be addressed. PCA is a mathematical manipulation rather than a strict statistical procedure; it belongs to that category of techniques, including cluster analysis, in which utility is judged by performance and not by theoretical considerations (*4*). Other considerations are

1. There may be many meaningful groupings, not necessarily only one "right" answer.

2. A single classification may give a distorted view of a multifaceted data set.
3. The context of the problem is important.
4. The same classification may be achieved in several different ways.
5. A variety of analytical techniques may be needed to show meaningful groupings.

These analyses are primarily exploratory devices wherein the classifications (or clusters) are not interesting in themselves, but rather are interesting for what inferences and new insights can be deduced about the structure of the data. For instance, the calculation may indicate relationships or principles that were previously unnoticed.

One recent application involved ^{13}C cross-polarization magic-angle spinning NMR (CP–MAS) chemical shifts obtained from an extensive series of peat size fractions (5). When combined with the PCA and FA, the NMR data could be used to complement or replace classical analytical methods for the determination of the degree of decomposition in these materials. Additional qualitative and quantitative insights into the origins of the mechanism(s) of degradation were gained by this approach.

Although these techniques are of general utility, this discussion will focus on the applications of PCA and FA to coal in general, and to NMR-derived measurements in particular. When combined with extensive spectroscopic characterization, and with physical, mechanical, or process-related test data, other general benefits of applying this approach include

- establishing a comprehensive database for ongoing fundamental studies
- establishing the utility of NMR (and non-NMR) analyses for these fundamental studies and quantifying the relationships among various test methods
- minimizing laboratory testing and material requirements, and enabling significant reduction in the time required to optimize conditions
- monitoring chemical changes (e.g., degradation) due to long-term storage, exposure to light or heat, mechanical degradation, or processing conditions, and being more readily able to assess the nature and extent of synergistic and antagonistic interactions
- determining mechanisms, because a testing program would be based on a fundamental knowledge of chemical structure
- being able to correlate NMR data with other analytical techniques that can be brought into use to further minimize testing procedures

- extending correlations to untested formulations or situations through the development of predictive models and classification procedures

Another benefit of these calculation methods is that PCA–FA analyses allow the determination of the number of significant components in a complex system. With NMR chemical-shift data as the input, the NMR spectra of these components may be obtained without having to resort to physical separations (which may not be feasible anyway).

Bivariate Correlations

Although the primary emphasis of this chapter involves multivariate statistical analysis, selected bivariate plots for NMR-derived data versus a variety of conventional, commonly measured parameters are presented for comparison. The correlations discussed in this chapter may not necessarily be extended to coals derived from other sources and geological conditions.

A plot of reflectance versus percent carbon is shown in Figure 1 (the correlation coefficient was 0.916). Proton rotating-frame spin–lattice relaxation times for the aromatic carbons exhibit a much lower correlation with aromaticity (not shown, correlation coefficient (r): −0.541). The aromatic carbon dipolar-dephasing relaxation times (Lorentzian component only) are shown plotted against percent carbon in Figure 2 ($r = 0.923$). Correlation coefficients between cross-polarization relaxation times and reflectance or percent carbon are much lower, with $|r| < 0.3$. In all cases, both the aromatic and aliphatic carbon relaxation-time correlations with various parameters were very similar, a result that indicates comparable basic structural features. Trends are not expected to be linear in nature, and both monotonic and nonmonotonic trends have been observed when dealing with NMR data and coal structure.

Principal-Component Analysis (PCA)

PCA consists of a linear transformation of m original variables to m new variables, where each new variable is a linear combination of the old. The process is done in a fashion that requires that each new variable accounts for, successively, as much of the total variance as possible. When m new variables have been computed, all of the original variance will be accounted for.

Two major steps must be considered in these analyses. First, the PCA itself, which can be generally described as eigenanalysis, is a pro-

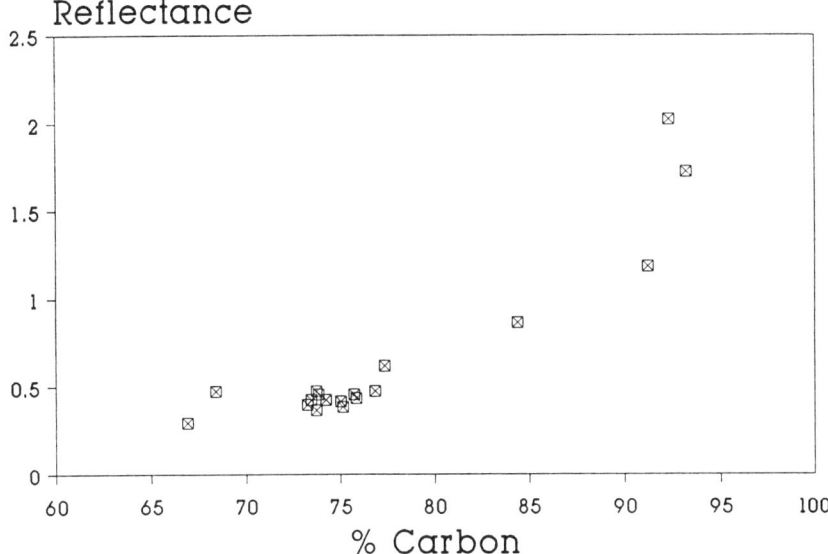

Figure 1. Reflectance vs. percent carbon.

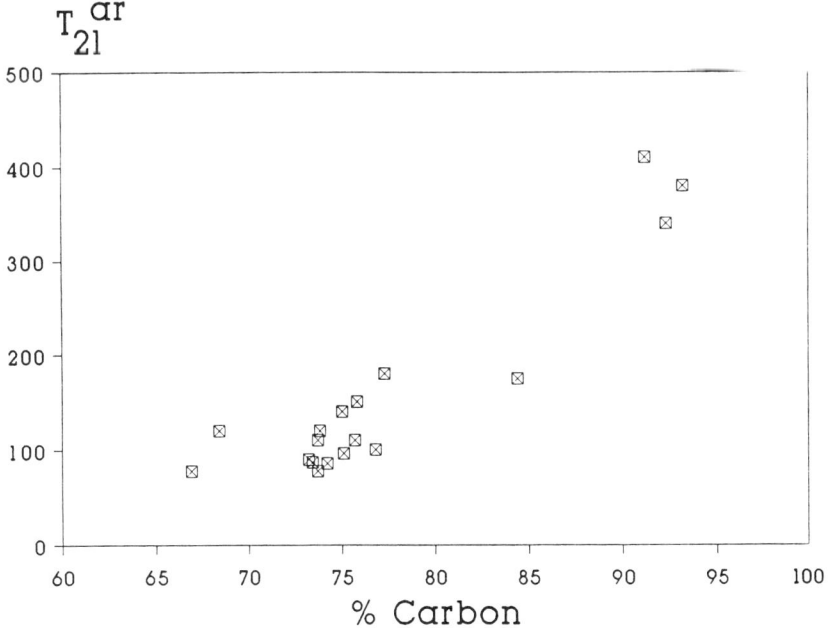

Figure 2. Aromatic carbon dipolar-dephasing (Lorentzian decay) times (T_{2l}^{ar}) vs. percent carbon.

cedure by which the best, mutually independent axes (dimensions) that describe the data set are selected. These axes are the so-called *factors* or *principal components* (*PCs*). The PC's are nothing more than the eigenvectors of a variance–covariance matrix (a standardized variance–covariance matrix is just the correlation-coefficient matrix).

The utility of constructing a new set of axes to describe the data is that most of the total variance (information) in the data set may be concentrated into a few derived variables. The process can be viewed as projecting the original data from a multidimensional representation into two dimensions. As with any projection, information is lost; but this technique maximizes the retention of information and quantifies the amount of information contained within each projection. For most chemical systems 80–90% of the total information can be depicted in fewer than half a dozen plots (*4*).

The second step is interpreting the principal components or factors. This step is done by examining the contribution that each of the original variables makes to the linear combination describing the factor axis. These contributions are called the *factor loadings*. When several variables have large loadings on a factor, the variables may be considered to be associated. From this association, chemical or physical interactions may be inferred, and these interactions may then be interpreted in a mechanistic sense.

As noted, the major analytical tool to be considered in this study is ^{13}C solid-state NMR spectroscopy. The variables discussed include cross-polarization relaxation times, dipolar-dephasing relaxation, proton rotating-frame spin–lattice relaxation, and aromaticity measurements (*24, 25*). Ultimate- and proximate-analysis data are also considered (*24, 25*).

Before performing the analysis, the form of the data to be analyzed must be considered. A collection of data can be converted to standardized, or unitless, form by subtracting from each observation the mean of the data set and dividing by the standard deviation. The new, transformed variables will then have a mean of 0 and a variance of 1. This procedure is useful if the distribution of one variable is to be compared to that of another variable when the two variables are expressed in different units of measurement.

Standardization may have a tremendous influence on the structure of the variance–covariance matrix and consequently on the results of the analysis. In many instances, there is no alternative but to standardize the data, because the raw (or unstandardized) matrix of variances and covariances contains such a mixture of measurement units that logical interpretation would be impossible. In other problems, measurement units are consistent across all variables, and there seems to be no good reason to standardize. In the present case, the correlation-coefficient matrix was used for the calculations because many different units and magnitudes of numbers were in the data set. Table I summarizes the data used for the present analysis.

Table I. Characterization Data

No.	T_{CH}^{al}	$T_{1\rho}^{Har}$	T_{CH}^{ar}	$T_{1\rho}^{Hal}$	R_0	C	H	O	f_a	T_{2l}^{ar}	T_{2g}^{al}	T_{2l}^{al}	T_{2g}^{ar}	Rank
1	22.9	5.48	262	4.55	0.29	66.9	4.6	27.2	0.61	78	14	180	15	lig
2	32.4	5.86	326	5.31	0.39	73.2	4.6	19.9	0.70	90	15	43	13	lig
3	31.1	4.11	254	4.22	0.36	73.7	4.6	19.2	0.67	78	12	64	12	subC
4	19.8	4.37	446	3.88	0.47	73.7	4.6	19.2	0.75	110	17	80	14	subC
5	n.m.	n.m.	n.m.	n.m.	0.42	73.4	4.4	19.4	0.73	87	10	270	16	subC
6	n.m.	n.m.	n.m.	n.m.	0.38	75.1	4.1	19.5	0.73	96	21	160	16	subC
7	30.4	4.03	195	4.52	0.42	74.2	4.5	19.3	0.67	86	17	170	16	subC
8	27.8	6.64	361	4.82	0.45	75.7	4.1	18.5	0.75	110	8	80	21	subB
9	26.6	4.92	247	4.26	0.47	68.4	5.1	24.4	0.72	120	17	89	16	subB
10	42.2	3.11	310	3.73	0.41	75.0	4.4	19.3	0.79	140	17	63	12	subB
11	62.7	4.66	341	5.06	0.47	76.8	5.1	15.6	0.65	100	8	55	11	subB
12	23.5	4.66	208	3.71	0.45	73.8	4.5	18.7	0.64	120	9	300	13	subB
13	20.4	4.35	294	4.11	0.43	75.8	4.2	18.4	0.74	150	16	35	12	subB
14	31.1	5.56	386	5.63	0.61	77.3	4.9	16.5	0.74	180	16	65	16	HVBC
15	16.3	4.77	301	5.54	0.86	84.4	5.6	5.4	0.71	175	13	19	19	HVBA
16	34.5	4.02	424	3.76	1.18	91.2	5.0	2.2	0.80	410	16	100	17	MVB
17	19.0	1.47	430	2.65	1.72	93.2	4.6	0.2	0.82	380	17	180	16	LVB
18	24.0	3.05	191	3.11	2.02	92.3	4.0	1.5	0.90	340	15	390	11	sa

NOTE: C, H, and O are given as percents. Values that were not measured are denoted by n.m. Error estimates are as follows: T_{CH}, ±10%; $T_{1\rho}^{H}$, ±10%; and f_a, ±5%.

ABBREVIATIONS: al, aliphatic component; ar, aromatic component; T_{CH}, cross-polarization relaxation time (microseconds); $T_{1\rho}^{H}$, proton rotating-frame spin–lattice relaxation time (milliseconds); T_{2g}, T_{2l}, Gaussian and Lorentzian dipolar-dephasing relaxation times, respectively (microseconds); R_0, reflectance; f_a, aromaticity; lig, lignite; sub, subbituminous; HVB, high-volatile bituminous; MVB, medium-volatile bituminous; LVB, low-volatile bituminous; and sa, semianthracite.

SOURCE: Data were taken from references 25 and 26.

Eigenvalues and Eigenvectors. Consider an example involving the analysis of several NMR-derived parameters on a series of coals ranging in rank from lignite to semi-anthracite. Here, m variables (NMR parameters) are measured on a collection of n objects (coals), where m is 5 (number of columns) and n is 16 (number of rows). The variables studied were as follows:

1. $T_{1\rho}^{H\ al}$ (in milliseconds, aliphatic component, longest decay, proton rotating-frame spin–lattice relaxation time)
2. $T_{1\rho}^{H\ ar}$ (in milliseconds, aromatic component, longest decay)
3. f_a (aromaticity)
4. T_{2l}^{ar} (in microseconds, aromatic component, dipolar-dephasing relaxation time, Lorentzian component)
5. T_{2l}^{al} (in microseconds, aliphatic component, Lorentzian component)

Because data representing different units and relative magnitudes are being compared, an $m \times m$ matrix of standardized variances and covariances (i.e., the correlation-coefficient matrix, Table II) is computed. By no means do these variables constitute the only ones that could be used for this analysis. They were chosen as much to illustrate the procedures involved in the calculations as for their relevance to the problem at hand. In those cases in which the relative importance of the variables or the nature of the interactions among the variables is not known, then the analysis would begin with a calculation using all available data. Guided by the results of such a preliminary analysis, the problem would then be reduced to the most relevant variables.

The elements in an $m \times m$ matrix can be regarded as defining points lying on an m-dimensional ellipsoid. The eigenvectors of the matrix yield the principal axes of the ellipsoid, and the eigenvalues represent the lengths of the principal axes. The PCA method is used to find these axes and measure their magnitudes. Tables III and IV show the m eigenvectors and m eigenvalues, respectively. Because a variance–covariance matrix is always symmetrical, these m eigenvectors will be *orthogonal*, or oriented at right angles, to each other.

The total variance in our data set can be defined as the sum of the individual variances. Because these variances are located along the diagonal of the variance–covariance matrix, this sum is equivalent to finding the trace of the matrix. In this example, the total variance is 5.0. The sum of the eigenvalues of a matrix is equal to the trace of the matrix, so the total of the five eigenvalues is equal to 5.0. These eigenvalues represent the lengths of the principal axes. Therefore, the principal axes also represent the total variance of the data set, and each accounts for an amount of the total variance equal to the eigenvalue divided by the trace. The first prin-

Table II. Correlation-Coefficient Matrix from PC Analysis

Variable	m	1	2	3	4	5
$T_{1\rho}^{H\ al}$	1	1.0	0.79	−0.53	−0.61	−0.36
$T_{1\rho}^{H\ ar}$	2	0.79	1.0	−0.54	−0.56	−0.59
f_a	3	−0.53	−0.54	1.0	0.77	0.26
T_{21}^{ar}	4	−0.61	−0.56	0.77	1.0	0.35
T_{21}^{al}	5	−0.36	−0.59	0.26	0.35	1.0

NOTE: 1 through 5 are the eigenvectors.

Table III. Eigenvectors from PC Analysis

Variable	m	1	2	3	4
$T_{1\rho}^{H\ al}$	1	0.47	−0.01	0.64	−0.22
$T_{1\rho}^{H\ ar}$	2	0.49	−0.28	0.32	0.35
f_a	3	−0.44	−0.49	0.33	−0.60
T_{21}^{ar}	4	−0.47	−0.37	0.30	0.66
T_{21}^{al}	5	−0.34	0.73	0.51	0.06

NOTE: 1 through 4 are the eigenvectors.

Table IV. Eigenvalues from PC Analysis

Principal Component	Eigenvalue	Percent of Trace	Cumulative Percent of Trace
1	3.19	63.9	63.9
2	0.88	17.5	81.4
3	0.55	11.0	92.4
4	0.24	4.8	97.2
5	0.14	2.8	100.0

cipal axis contains 3.19/5.00 or approximately 64% of the total variance, and the second principal axis represents 0.87/5.00 or approximately 18% of the total variance. In other words, if the variation in this data set is measured along the first principal axis, 2/3 of the total variation is represented in the observations. At least one principal axis will usually be more efficient (in terms of accounting for total variance) than any of the original variables. On the other hand, at least one of the axes must be less efficient than any of the original variables.

From these data we can focus on the first and second principal components, because together they account for approximately 81% of the total variance in the data set. Again, in more complex problems involving dozens of variables half a dozen PCs may have to be retained for consideration. Given that the problem lends itself to PCA or FA through the fact that numerous variables are (highly) correlated, even in this situation the complexity of the problem and the subsequent interpretation(s) will have been considerably reduced.

Scores and Loadings. In Table V, each original observation has been converted into what is called a *score* by projecting the observation onto the principal axes. This projection also provides a convenient method for visualizing the data for the purposes of interpretation.

The elements of the eigenvectors that are used to compute the scores of observations are called *loadings*. The loadings are nothing more than the coefficients of the linear equation that the eigenvector defines. Therefore eigenvector 1 (PC1) is comprised of the elements 0.47, 0.49, −0.44, −0.47, and −0.34, corresponding to variables 1 through 5, respectively. A loading plot yields information about the importance of each variable to the interpretation of the PC and about the relationship among variables in that PC. A variable's contribution to a PC is directly proportional to the square of the loading. The distance of a variable to the origin along a PC is a quantitative measure of the importance of that variable in the PC. Variables grouped together contain the same information in the PC.

Figure 3 is a loading plot of the data for the first (PC1) and second (PC2) principal components. The $T_{1\rho}^H$-related variables yield positive loadings, and the remaining variables have negative correlations with the loadings in PC1. The major contributor to PC2 is the aliphatic dipolar-dephasing relaxation time.

Figure 4 is a score plot derived from the NMR data in Table V. The samples are numbered from 1 through 18, the higher numbers being associated with the coals of higher rank. The relatively large number of lower rank coals leads to greater scatter in these plots; this scatter also reflects the diversity of composition and structure characterizing these materials. Based on the NMR results, such plots graphically illustrate the relationships among the samples. For instance, lignite 1 bears a stronger correlation to subC 3 than to lignite 2. The highest rank samples form a cluster to the far left side of the plot, and the lower rank samples are found on the right side. As can be seen in the loading plots, this basic separation is due to some factor(s) in PC1. The bivariate plots shown previously indicated that some relaxation parameters were highly correlated with conventional measures of rank. This same information is reflected in the score plots, but in a manner that invites further deliberation on the nature of the relationships among the samples. The score-plot technique may be particularly useful when considering both NMR data and process-related information.

Table V. Score Plot		
Sample Number	PC1	PC2
1	1.43	1.43
2	1.89	−0.45
3	0.74	0.32
4	−0.036	−0.09
7	0.49	0.96
8	1.409	−0.42
9	0.51	0.01
10	−0.917	−0.51
11	1.50	0.03
12	−0.14	2.24
13	0.15	−0.56
14	1.27	−0.97
15	1.26	−1.05
16	−1.88	−1.26
17	−3.77	−0.32
18	−3.90	0.63

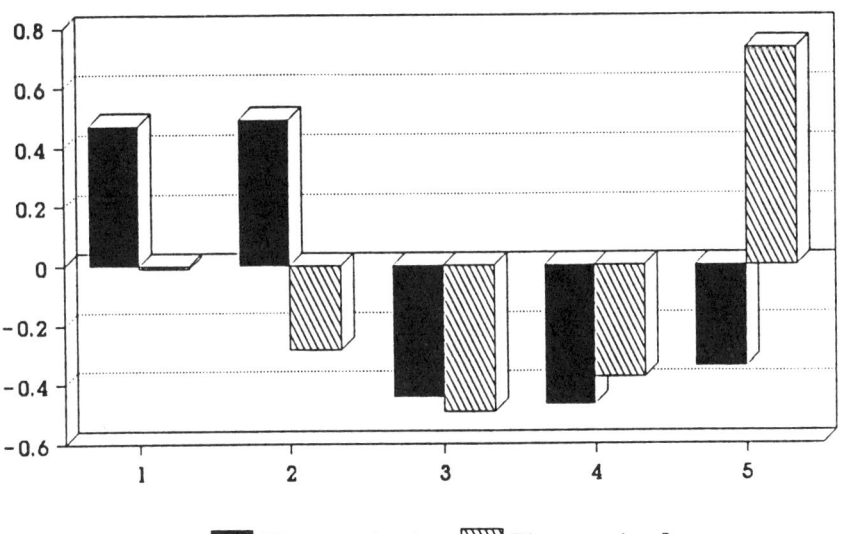

Figure 3. Loading plot for NMR PCA. The coefficients of both eigenvectors 1 and 2 are plotted from data in Table III.

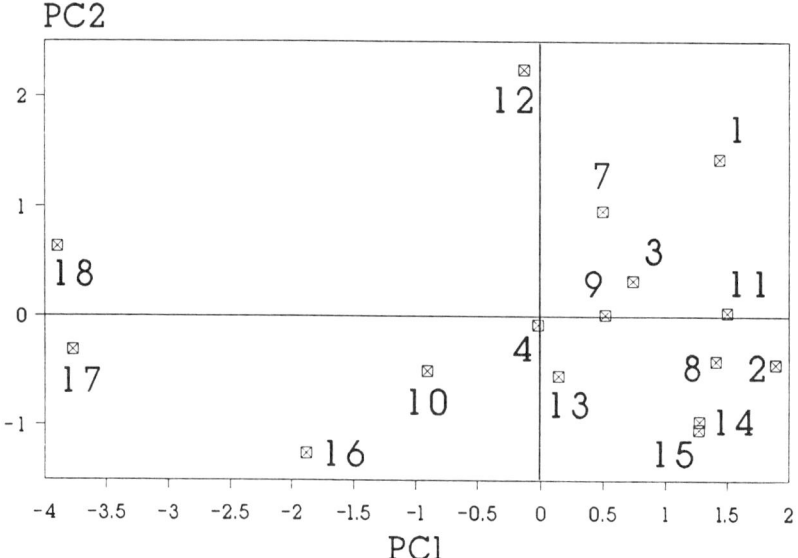

Figure 4. Score plot for PCA involving only NMR parameters (column 1 vs. column 2 in Table V).

Q-Mode Factor Analysis. A form of factor analysis in which the roles of the samples and variables are reversed is called *Q-mode analysis*. Rather than investigating interrelationships between variables, Q-mode analysis is concerned with interrelationships between samples. The objectives of Q-mode analysis are much the same as the objectives of cluster analysis: to arrange a suite of samples into a meaningful order so the relationship between one sample and another may be deduced (4). The first step in Q-mode analysis is to create an $n \times n$ matrix of similarities between samples. The measure of similarity may be the correlation coefficient or some other measure, provided the similarity coefficient does not exceed the range ±1.

Q-mode factor analysis consists of finding the principal axes of an n-dimensional hyperellipsoid that is defined by the interrelationships between n sample vectors. After the $n \times n$ matrix of similarities has been computed, the principal axes are determined. This determination is done by extracting the eigenvectors and eigenvalues. Because only a few factors will be retained, it may be unnecessary to extract all possible eigenvalues. The factors may be considered to represent "idealized" end members of the sample suite, and the samples themselves may be regarded as "mixtures" of these idealized extreme samples.

The results of these calculations are shown in Figures 5 and 6 for the NMR-derived data and the conventional data, respectively. The factor

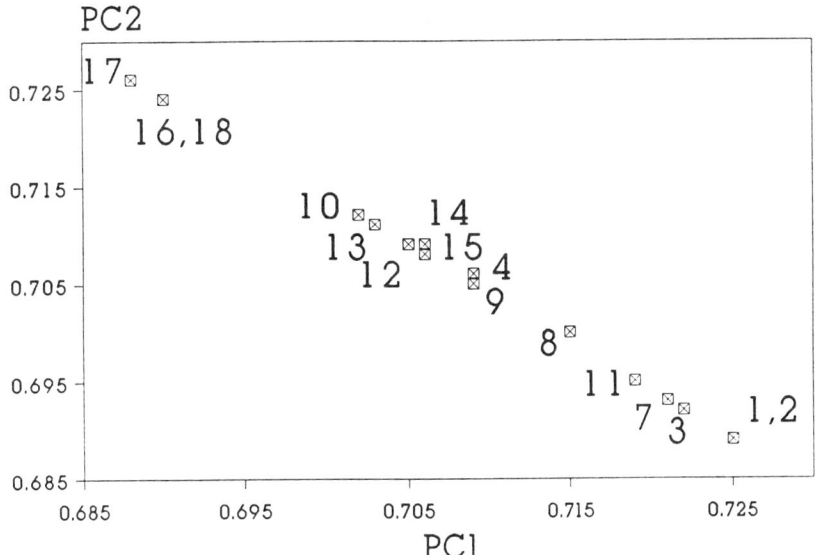

Figure 5. Q-mode factor-analysis results for the NMR-derived data set. The loadings for the first two factors are plotted with respect to one another.

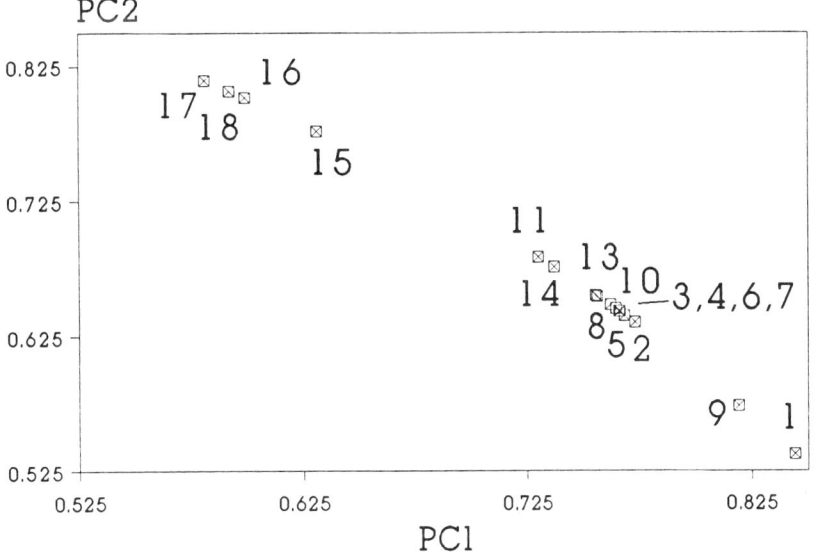

Figure 6. Q-mode factor-analysis results for conventional coal parameters. The loadings for the first two factors are plotted with respect to one another.

loadings for the first two factors have been plotted against one another. For both the NMR- and conventional-parameter-derived plots, the end members are the low-rank (lignite) sample and the high-rank (low-volatile bituminous, semi-anthracite) samples, with the remaining data spread across the whole range between these samples.

The NMR data provide a greater degree of discrimination among the samples than the normal variables usually chosen to characterize coal rank. For example, samples 3 and 4 are indistinguishable in the coal-parameter-derived plot, but they are quite well separated in the NMR-derived plot because their NMR parameters vary more than the values obtained from ultimate and proximate analyses.

The relevance of these observations lies primarily in the relationship of this plot to real-life, practical considerations. These considerations might include the relative ability of these coals to be modified by various low-rank-coal-upgrading processes or the response of the coals to carbonization, liquefaction, or gasification processes. Thus, if the measured NMR relaxation times accurately reflect systematic differences in structure or composition that are not reflected as accurately by such traditional variables as reflectance and elemental composition, more valuable (and predictive) correlations between processability and NMR-derived parameters may be generated. The present analysis has established that a good correlation exists between coal rank and NMR relaxation times in this suite of samples; however, the results shown in Figure 5 could not be derived from the more traditional bivariate correlations with the NMR values.

References

1. Gorsuch, R. L. *Factor Analysis;* W. B. Saunders: Philadelphia, 1974. Harper, A. M.; Duewer, D. L.; Kowalski, B. R.; Fasching, J. L. In *Chemometrics: Theory and Applications;* Kowalski, B. R., Ed.; ACS Symposium Series 52, American Chemical Society: Washington, DC, 1977.
2. Sharaf, M.; Ellman, E. L.; Kowalski, B. R. *Chemometrics;* Wiley: New York, 1986.
3. Malinowski, E. R.; Howery, D. G. *Factor Analysis in Chemistry;* Wiley Interscience: New York, 1980.
4. Davis, J. C. *Statistics and Data Analysis in Geology;* Wiley: New York, 1973.
5. Preston, C. M.; Axelson, D. E.; Levesque, M.; Mathur, S. P.; Dinel, H.; Dudley, R. L. *Org. Geochem.* **1989**, *14*, 393.
6. Bracewell, J. M.; Robertson, G. W. *Geoderma* **1987**, *40*, 1.
7. Windig, W.; Kistemaker, P. G.; Haverkamp, J. *J. Anal. Appl. Pyrolysis* **1981**, *3*, 199.
8. Devaux, M. F.; Bertrand, D.; Robert, P.; Morat, J. L. *J. Chemom.* **1987**, *1*, 103.

9. Herron, M. M. *Clays Clay Miner.* **1986,** *34,* 204.
10. Lowe, L. E.; Scagel, A. M.; Klinka, K. *Can. J. Soil Sci.* **1987,** *67,* 383.
11. Kvalheim, O. M.; Aksnes, D. W.; Brekke, T.; Eide, M. O.; Sletten, E.; Telnaes, N. *Anal. Chem.* **1985,** *57,* 2858.
12. Metcalf, G. S.; Windig, W.; Hill, G. R.; Meuzelaaar, H. L. *Int. J. Coal Geol.* **1987,** *7,* 245.
13. Kvalheim, O. M.; Telnaes, N. *Anal. Chem.* **1986,** *191,* 87, 97.
14. Telnaes, N.; Dahl, B. *Org. Geochem.* **1986,** *10,* 425.
15. Norden, B.; Albano, C. *Fuel* **1989,** *68,* 771.
16. Wallbacks, L.; Edlund, U.; Norden, B. *J. Wood Chem. Technol.* **1989,** *9,* 235.
17. King, A. G.; Deane, J. M. *Petroanal.* **1988,** 261.
18. Hearmon, R. A.; Scrivens, J. H.; Jennings, K. R.; Farncombe, M. J. *Chemom. Intell. Lab. Syst.* **1987,** *1,* 167.
19. Gillette, P. C.; Lando, J. B.; Koenig, J. L. *Anal. Chem.* **1983,** *55,* 630.
20. Malinowski, E. R. *Anal. Chim. Acta* **1982,** *129,* 134.
21. Kormos, D. W.; Waugh, J. S. *Anal Chem.* **1983,** *55,* 633.
22. Arunachalam, J., Gangadharan, A. *Anal Chim. Acta* **1984,** *157,* 245.
23. Deane, J. M.; MacFie, H. J. M.; King, A. G. *J. Chemom.* **1989,** *3,* 359.
24. Theriault, Y.; Axelson, D. E. *Fuel* **1988,** *67,* 62.
25. Axelson, D. E. *Fuel Process. Technol.* **1987,** *16,* 257.

RECEIVED for review June 7, 1990. ACCEPTED revised manuscript December 13, 1990.

13

Structural Parameters of Argonne Coal Samples

Solid-State ^{13}C NMR Spectroscopy

Yoshio Adachi and Minoru Nakamizo

Government Industrial Research Institute, Kyushu, Shuku-machi Tosu-shi Saga-ken 841, Japan

Solid-state ^{13}C NMR spectroscopy was applied to the structural characterization of the Argonne standard coal samples. The apparent ratio of quaternary to tertiary aromatic carbons was determined by using combined dipolar-dephasing and cross-polarization–magic-angle spinning (CP–MAS) NMR experiments. By using the apparent ratio and the value of the fraction of aromatic carbon, several parameters can be obtained on the average chemical structure of coals. Among the Argonne samples, a low-volatile bituminous coal consists of approximately four aromatic rings with a few ethyl groups as alkyl substituents. High-volatile bituminous coals are composed of two aromatic rings with approximately two propyl groups and one hydroxyl group as substituents. Both the subbituminous and lignite coals have approximately four aromatic rings with approximately three pentyl groups and two oxygen-containing substituent groups, such as hydroxyl or carboxyl groups.

SOLID-STATE HIGH-RESOLUTION NMR SPECTROSCOPY can be performed by using cross-polarization (CP) and magic-angle spinning (MAS) (*1*). On the CP–MAS NMR spectra of solid carbonaceous fuels, aromatic and aliphatic absorption bands can be observed separately (*2–4*). Two problems are associated with the reliability of the quantitative measurements of aromaticity of coal by CP–MAS spectroscopy. One problem is the difference in the rate of polarization transfer from ^{1}H to ^{13}C spins between aromatic and aliphatic carbons during CP (*5–8*). Therefore, the selection of contact time for CP is very important for a precise determination of aromatic- and aliphatic-carbon contents. Another problem arises from spinning side bands (SSB) of aromatic bands with high-field measurement of coals because SSB overlap with aliphatic bands. The TOSS (total suppression of side bands) pulse sequence is a powerful technique for reducing the intensity of SSB, but careful treatments of instrumental factors are needed when the signal intensity of a TOSS spectrum is discussed (*9–13*). In spite of these problems, detailed chemical structures of coals have been investigated by using the analysis of carbon functional groups (*14–18*) and the dipolar-dephasing (DD) technique (*19–25*).

In this chapter, the structural parameters of eight coals from the Argonne National Laboratory Premium Coal Sample Program were determined by using conventional ^{13}C NMR spectroscopy and DD.

Experimental Details

Eight of the coals from the Premium Coal Sample Program at Argonne National Laboratory were used in this study. The elemental analyses of the Argonne coals were published previously (*26*).

Solid-state ^{13}C NMR spectra were obtained on a Brucker AC-200 NMR spectrometer with a double air-bearing CP–MAS probe head and ^{1}H and ^{13}C high-power amplifiers. A sample was packed in a ceramic capsule, and the capsule was spun at rates of 3 and 4 kHz. The acquisition parameters for all spectra were as follows: 2-ms CP contact time, 1024–2048 accumulations with a 4-s repetition time, 3.5-µs 90° ^{1}H pulse width, 31.25-kHz spectral width, 2048 data points, and 50-Hz line broadening. The contact time of 2 ms was chosen for an accurate determination of the fraction of aromatic carbon, f_a, in the solid-state measurements. The f_a value determined from the solution spectrum of coal-tar pitch in chloroform-*d* was nearly equal to that of the same sample measured in the solid state with the 2-ms contact time. Chemical shifts were calibrated with respect to tetramethylsilane using glycine as a secondary standard.

The f_a values of the coal samples were calculated from the integrated intensities of aromatic bands and associated SSB in conventional CP–MAS spectra that were obtained with a 3-kHz spinning rate. The contents of aliphatic carbons in C, CH, CH_2, and CH_3 groups were also calculated with a 3-kHz spinning rate. The contents of carboxyl, oxygenated, alkyl-substituted, and protonated and bridgehead aromatic carbons were calculated from the integrated intensities of the aromatic bands of conventional CP-MAS spectra obtained with a 4-kHz spinning rate.

Fourier transform infrared (FTIR) spectra of coal samples were measured on a Digilab FTS-60 system.

Results and Discussion

^{13}C NMR Spectra of Coals.

The ^{13}C NMR spectra of the Illinois No. 6 coal used in this study are shown in Figure 1 with two different spinning rates, 3 and 4 kHz. Each spectrum is composed of aromatic (90–165 ppm) and aliphatic (0–50 ppm) bands. In the conventional CP-MAS spectra, several SSB appear, and one of the side bands overlaps completely with the aliphatic bands in the spectrum obtained with a 4-kHz spinning rate. The intensity ratios of aliphatic to aromatic bands obtained

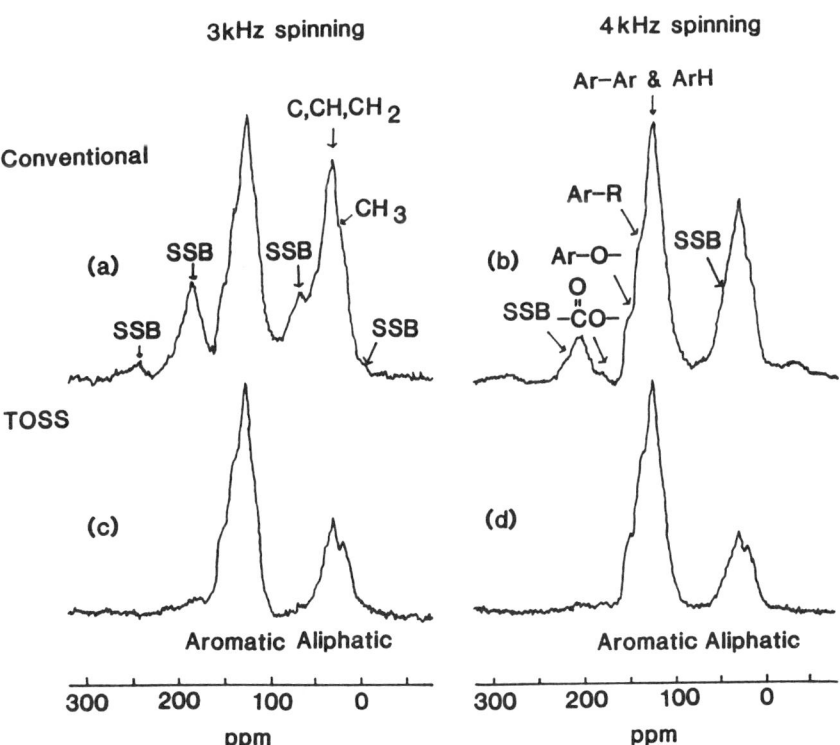

Figure 1. 50-MHz ^{13}C NMR spectra obtained with different spinning rates of Illinois No. 6 coal.

with the TOSS sequence are always smaller than those obtained with the conventional CP–MAS measurements irrespective of the spinning rates used. Because the SSB cannot be completely eliminated even if the TOSS sequence is employed, the unextinguished SSB overlap partly with the absorption bands of the carboxyl (165–190 ppm) and ether (50–90 ppm) carbons at a spinning rate of 3 kHz and with those of the carbonyl (190–210 ppm) carbons at 4 kHz (*16, 18*).

In the conventional CP–MAS measurements with a 3-kHz spinning rate on Illinois No. 6 coal, the aliphatic bands around 30 ppm can be observed separately from the SSB. One shoulder band observed in the 3–25-ppm region of aliphatic band in Figure 1a is assigned to methyl carbons. In the spectrum obtained with a 4-kHz spinning rate, the carboxyl-carbon (–COO) signal is observed at 180 ppm as shown in Figure 1b. In the aromatic band, there are two shoulders; the shoulder band at approximately 155 ppm is assigned to the oxygenated aromatic carbons, Ar–O, such as phenol (*17*). The second shoulder band in the region of 135–148 ppm is assigned to the alkyl-substituted aromatic carbons, Ar–R. Aromatic bands around 127 ppm are assigned to the aromatic protonated (Ar–H) and bridgehead (Ar–Ar) carbons (*18*).

In Table I, the carbon distributions of the Argonne coal samples are shown. These distributions were obtained from integrated intensities of the conventional CP–MAS spectra without curve-fitting techniques. Carbonyl and ether carbons, which may exist to a slight extent, were ignored because their absorption bands overlapped with the SSB in the CP–MAS spectra. Discounting the carbonyl and ether carbons does not seem to cause serious errors in the estimation of the aromatic-carbon contents.

Table I. Carbon Distributions of Argonne Coal Samples by Conventional CP–MAS Spectroscopy

Coal	f_a	–COO	Ar–O	Ar–Al	Ar–H and Ar–Ar	C, CH, and CH_2	CH_3
Pocahontas	0.834	0.8	2.3	14.0	66.2	7.4	9.2
Upper Freeport	0.767	0.1	2.4	12.8	61.4	13.0	10.2
Stockton	0.723	1.4	4.7	14.0	52.2	19.3	8.4
Pittsburgh	0.703	0.9	5.3	14.9	49.2	19.2	10.5
Illinois No. 6	0.673	1.1	4.4	13.6	48.3	23.7	9.0
Blind Canyon	0.610	1.6	5.9	10.9	42.6	28.2	10.8
Wyodak	0.613	3.8	7.0	10.1	40.4	29.9	8.8
Beulah–Zap	0.692	4.7	6.2	11.3	47.0	22.9	7.9

NOTE: All values are given as percents.

The contents of carboxyl and oxygen-substituted aromatic carbons decrease with increasing f_a (Table I). However, the content of methyl carbons does not vary greatly among the coal samples studied. The values listed in Table I were used in the calculation of structural parameters of coals.

Dipolar Dephasing. In Figure 2, DD spectra are shown for Illinois No. 6 coal. The spinning rate was 3 kHz, and the delay time, t_1, was ≤ 120 μs. The intensity of the aliphatic-carbon signal decreases more rapidly than the intensity of the aromatic-carbon signal with increasing t_1. In the aliphatic region, the intensity of methyl-carbon signal in the 3–25-ppm range decreases more slowly than that of other aliphatic carbons. In the aromatic region, the signal intensity of substituted aromatic carbons (135–165 ppm) decreases more slowly than that of the protonated and bridgehead aromatic carbons around 127 ppm as is evident in Figure 2 from DD spectra taken at delay times longer than 70 μs. The signal intensities of the SSB decrease more slowly than the intensities of the main aromatic bands. The results indicate that the chemical-shift anisotropy of the quaternary aromatic carbons is larger than that of the tertiary aromatic carbons. In the spectra that were obtained with long t_1's, a part of the secondary SSB overlaps with the aliphatic bands, and the contribution of the SSB to the intensity of the aliphatic-carbon signal increases with t_1.

In Figure 3, the signal intensities of aromatic carbons in the model compound 1-methylfluorene are plotted against the square of t_1. The intensities of the tertiary-carbon signals and the quaternary-carbon signals in the aromatic region decay as a second-order exponential of t_1. Similarly, in the aliphatic region the signal intensities of methyl and methylene carbons also decay as a second-order exponential of t_1.

The ratio of quaternary to tertiary aromatic carbons was obtained from a plot of the second-order exponential decay as shown in Figure 4. In Table II, the ratio of quaternary to tertiary aromatic carbons, $f_{Q/T}^{ar*}$; the ratio of aliphatic primary and quaternary carbons to secondary and tertiary carbons, $f_{PQ/ST}^{al}$; and the decay constants, T_{DD} are shown. The aromatic $f_{Q/T}^{ar*}$ values of Wyodak and Beulah–Zap coals are markedly larger than those of other coals. The decay constant of aromatic quaternary carbons, $T_{DD}(Q)$, has a tendency to increase in high-rank coals. There is no remarkable difference in the decay constants of aromatic tertiary carbons, $T_{DD}(T)$, among the coal samples studied. The values of $f_{PQ/ST}^{al}$ and the decay constants of aliphatic carbons, $T_{DD}(PQ)$ and $T_{DD}(ST)$, may include errors to some extent because a part of the secondary SSB overlaps with the aliphatic bands, and the contribution of the SSB to the integrated intensity of aliphatic carbons increases with increasing t_1 as shown in Figure 2.

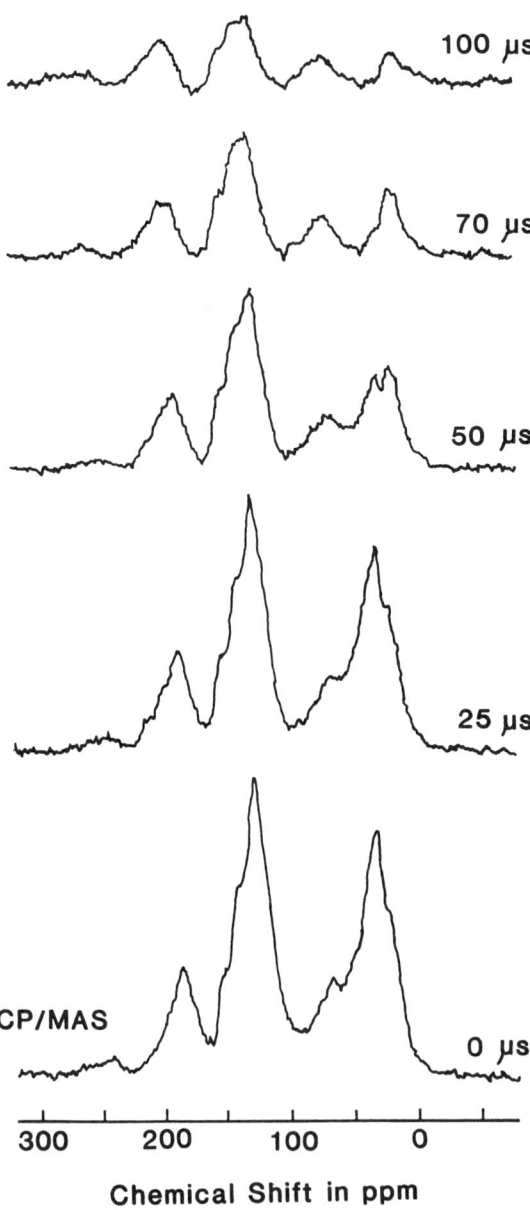

Figure 2. Dipolar-dephasing spectra of Illinois No. 6 coal. The different delay times are noted on the figure.

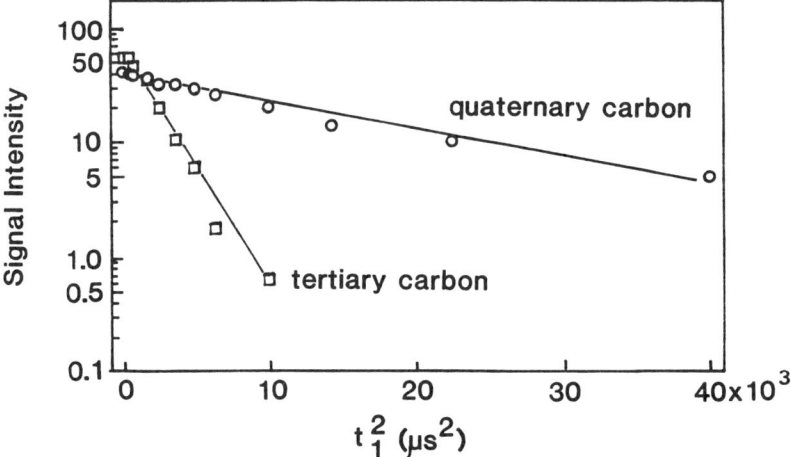

Figure 3. Plot of the signal intensities of the aromatic carbons in 1-methylfluorene as a function of the square of the delay time.

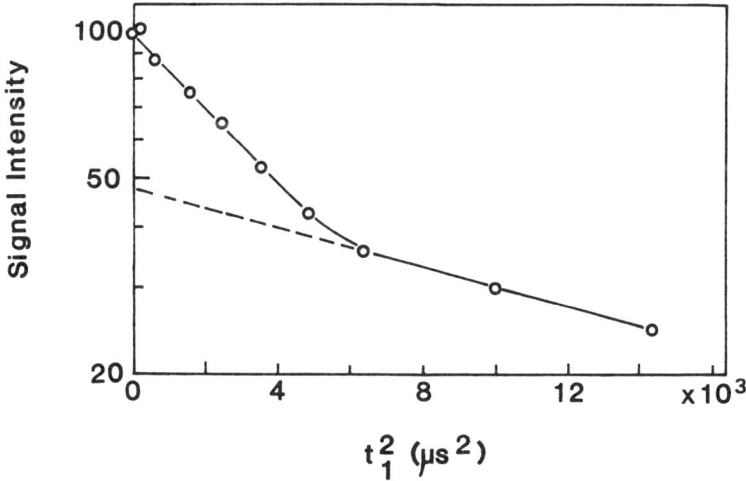

Figure 4. Plot of the signal intensity of aromatic carbons in Illinois No. 6 coal as a function of the square of the delay time.

Structural Parameters of Argonne Coal Samples. The structural parameters of the coals were obtained according to the method proposed by Ollivier and Gerstein (27). As indicated in the previous section, the ratios of quaternary to tertiary aromatic carbons, $f_{Q/T}^{ar*}$, for Wyodak and Beulah–Zap coals were large compared to the other coals

Table II. Results of Dipolar-Dephasing Experiments on Argonne Coal Samples

Coal	$f_{Q/T}^{ar*}$	$T_{DD}(Q)$	$T_{DD}(T)$	$f_{PQ/ST}^{al}$	$T_{DD}(PQ)$	$T_{DD}(ST)$
Pocahontas	1.30	134.8	29.4	0.51	191.0	23.1
Upper Freeport	1.05	119.0	33.0	0.70	81.2	27.7
Stockton	0.82	135.0	35.2	0.38	70.0	32.1
Pittsburgh	1.33	105.7	29.6	0.34	91.3	33.0
Illinois No. 6	0.99	101.5	35.8	0.23	86.6	34.7
Blind Canyon	1.71	90.8	39.0	0.79	73.7	39.8
Wyodak	3.69	77.6	33.5	0.46	65.3	32.1
Beulah–Zap	2.65	77.3	32.1	0.34	71.1	33.3

NOTE: The decay constants, T_{DD}, are given in microseconds.

examined. This result can be partly ascribed to an overlapping of the carboxyl-carbon band with that of the aromatic quaternary carbons because of the high content of carboxyl groups in these two coals. The aromatic carbons in the coal samples consist of protonated carbons, substituted carbons, and bridgehead carbons. Therefore, the true ratio, $f_{Q/T}^{ar}$, of the quaternary aromatic carbons to tertiary aromatic carbons must be corrected for the overlapping carboxyl-carbon band as follows:

$$f_{Q/T}^{ar} = f_{Q/T}^{ar*} - \frac{C_C}{C_P^{ar}} \tag{1}$$

where C_C and C_P^{ar} express the numbers of carboxyl carbons and protonated aromatic carbons, respectively. The C_C/C_P^{ar} value can be obtained from the following equation.

$$\frac{C_C}{C_P^{ar}} = \frac{1}{2}\left(\frac{C_C}{C^{ar}}\right)\left(\frac{f_a}{1-f_a}\right)\left(\frac{H^{al}}{H^{ar}}\right) \tag{2}$$

where C^{ar} is the number of aromatic carbons. The ratio of aliphatic to aromatic hydrogens, H^{al}/H^{ar}, was obtained from the absorption-intensity ratio of aliphatic to aromatic C–H stretching vibrations, D_{al}/D_{ar}, in the range of 2750–3100 cm^{-1} in the FTIR spectra of the coal samples as expressed in eq 3 (28).

$$\frac{H^{al}}{H^{ar}} = \left(\frac{e_{ar}}{e_{al}}\right)\left(\frac{D_{al}}{D_{ar}}\right) \tag{3}$$

The ratio of extinction coefficients of aromatic to aliphatic C–H absorption bands, e_{ar}/e_{al}, is assumed to be 0.5 in most cases (28, 29).

The substitution coefficient, R_S, was obtained from the aromatic bands of the CP–MAS spectra that were obtained with a 4-kHz spinning rate. If the ratio of the substituted aromatic carbons to the protonated and bridgehead aromatic carbons is defined as $f_{S/PB}^{ar}$, R_S can be determined by using the following equation.

$$R_S = \frac{C_S^{ar}}{C_P^{ar}} + C_S^{ar} = \frac{1 + f_{Q/T}^{ar}}{2 + f_{Q/T}^{ar} + \dfrac{1}{f_{S/PB}^{ar}}} \tag{4}$$

where

$$f_{S/PB}^{ar} = \frac{C_S^{ar}}{C_P^{ar} + C_B^{ar}} \tag{5}$$

where C_B^{ar} and C_S^{ar} are the number of bridgehead and substituted aromatic carbons, respectively.

The ratio $f_{S/PB}^{ar}$, can be calculated from the intensity ratio of the aromatic shoulder band (135–165 ppm) to the aromatic band (95–135 ppm). Because the band of the carboxyl-substituted aromatic carbons overlaps with those of the protonated and bridgehead aromatic carbons, the intensity of the aromatic carbons must be corrected for the overlapping of both bands. In this case, the contribution of the aliphatic carboxyl group to the intensity of the aromatic carbons was assumed to be negligible.

In Table III, the structural parameters that were obtained from the lozenge model of Ollivier and Gerstein (27) are shown for the Argonne coal samples. The number of rings, N_R, for the Argonne coal samples is not large, ranging from 1.6 to 4.4. The number of substitution, n, is 2.3–6.3. The average chain length of aliphatic carbons, n_{al}, was obtained from the intensity ratio of all the aliphatic carbons to the methyl carbons and is 1.8–4.4.

A low-volatile bituminous coal (Pocahontas) consists of approximately four aromatic rings with a few ethyl groups as alkyl substituents. A medium-volatile bituminous coal (Upper Freeport) consists of approximately three rings with a few substituent ethyl or propyl groups. Three high-volatile bituminous coals (Illinois No. 6, Pittsburgh, and Stockton) are composed of two aromatic rings with two propyl groups and one hydroxyl group. Wyodak subbituminous coal consists of approximately four aromatic rings and three pentyl, two hydroxyl, and one carboxyl groups. Beulah–Zap lignite has approximately three aromatic rings with two or three pentyl, hydroxyl, and carboxyl groups.

Table III. Structural Parameters of Argonne Coal Samples

Coal	$f_{Q/T}^{ar}$	$f_{S/PB}^{ar}$	R_S	N_R	n	n_{al}
Pocahontas	1.27	0.26	0.32	3.6	3.1	1.8
Upper Freeport	1.04	0.25	0.29	3.0	2.6	2.3
Stockton	0.73	0.40	0.33	1.6	2.3	3.3
Pittsburgh	1.24	0.44	0.41	2.4	3.3	2.8
Illinois No. 6	0.88	0.40	0.35	1.8	2.6	3.6
Blind Canyon	1.46	0.45	0.43	2.7	3.7	3.6
Wyodak	3.17	0.57	0.60	4.4	6.3	4.4
Beulah–Zap	2.10	0.53	0.52	3.3	4.8	3.9

Conclusion

Structural characterization of the Argonne standard coal samples was made by using a combined technique of conventional CP–MAS NMR spectroscopy and DD. Most of the Argonne coal samples consist of condensed aromatic compounds with two to four benzene rings and a few alkyl substituents ranging from ethyl to pentyl groups. Some bituminous coals have oxygen-containing substituents such as hydroxyl and carboxyl groups. These results are approximately in good agreement with those from Solum et al. (25).

References

1. Schaefer, J.; Stejskal, E. O. *J. Am. Chem. Soc.* **1976,** *98,* 1031.
2. Maciel, G. E.; Bartuska, V. J.; Miknis, F. P. *Fuel* **1979,** *58,* 391.
3. Zilm, K. W.; Pugmire, R. J.; Larter, S. R.; Allan, J.; Grant, D. M. *Fuel* **1981,** *60,* 717.
4. Russell, N. J.; Wilson, M. A.; Pugmire, R. J.; Grant, D. A. *Fuel* **1983,** *62,* 601.
5. Miknis, F. P.; Maciel, G. E.; Bartuska, V. J. *Prepr. Am. Cem. Soc. Div. Fuel Chem.* **1979,** *24*(2), 327.
6. Sullivan, M. J.; Maciel, G. E. *Anal. Chem.* **1982,** *54,* 1606.
7. Yoshida, T.; Maekawa, Y.; Fujito, T. *Anal. Chem.* **1983,** *55,* 388.
8. Botto, R. E.; Wilson, R.; Hayatsu, R.; McBeth, R. L.; Scott, R. G.; Wonans, R. E. *Prepr. Am. Chem. Soc. Div. Fuel Chem.* **1985,** *62*(3), 187.
9. Dixon, W. T.; Schaefer, J.; Sefcik, M. D.; Stejskal, E. O.; Mckay, R. A. *J. Magn. Reson.* **1982,** *49,* 341.
10. Olejniczak, E. T.; Vega, S.; Griffin, R. G. *J. Chem. Phys.* **1984,** *81,* 4804.
11. Axelson, D. E. *Fuel* **1987,** *66,* 195.
12. Axelson, D. E. *Fuel Process. Technol.* **1987,** *16,* 257.

13. Tekely, P.; Nicole, D.; Delpuech, J. J.; Totino, E.; Muller, J. F. *Fuel Process. Technol.* **1987**, *15*, 225.
14. Dereppe, J. M.; Boudou, J. P.; Moreaux, C.; Durand, B. *Fuel* **1983**, *62*, 575.
15. Newman, R. H.; Davenport, S. J. *Fuel* **1986**, *65*, 533.
16. Yoshida, T.; Maekawa, Y. *Fuel Process. Technol.* **1987**, *15*, 385.
17. Newman, R. H.; Sim, M. N.; Johnston, J. H.; Collen, J. D. *Fuel* **1988**, *67*, 420.
18. Supaluknari, S.; Larkins, F. P.; Redlich, P.; Jackson, W. R. *Fuel Process. Technol.* **1989**, *23*, 47.
19. Opella, S. J.; Frey, M. H. *J. Am. Chem. Soc.* **1979**, *101*, 5854.
20. Murphy, P. D.; Gerstein, B. C.; Weinberg, V. L.; Yen, T. F. *Anal. Chem.* **1982**, *54*, 522.
21. Murphy, P. D.; Cassady, T. J.; Gerstein, B. C. *Fuel* **1982**, *61*, 1233.
22. Alemany, L. B.; Grant, D. M.; Alger, T. D.; Pugmire. R. J. *J. Am. Chem. Soc.* **1983**, *105*, 6697.
23. Alemany, L. B.; Grant, D. M.; Pugmire, R. J.; Stock, L. M. *Fuel* **1984**, *63*, 513.
24. Theriault, Y.; Axelson, D. E. *Fuel* **1988**, *67*, 62.
25. Solum, M. S.; Pugmire, R. J.; Grant, D. M. *Energy Fuels* **1989**, *3*, 187.
26. Vorres, K. S. *Prepr. Am. Chem. Soc. Div. Fuel Chem.* **1987**, *32*(4), 221.
27. Ollivier, P. J.; Gerstein, B. C. *Carbon* **1986**, *24*, 151.
28. Brown, J. K. *J. Chem. Soc.* **1955**, 744.
29. Yoshida, R.; Maekawa, Y.; Yokoyama, S.; Takeya, G. *Nenkau* **1975**, *54*, 332.

RECEIVED for review June 7, 1990. ACCEPTED revised manuscript December 11, 1990.

14

Characterization of Coal Liquefaction Residues by Solid-State ^{13}C NMR Spectroscopy

Natsuko Cyr[1,3] and Nosa O. Egiebor[2]

[1]Alberta Research Council, P.O. Box 8330, Postal Station F, Edmonton, Alberta T6H 5X2, Canada
[2]Department of Mining Metallurgical and Petroleum Engineering, University of Alberta, Edmonton, Alberta T6G 2G6, Canada

> *Feeds and residues from coal liquefaction processes under a variety of conditions were studied by using solid-state ^{13}C NMR spectroscopy and other analytical methods. The conversion yields were found to be dependent not only on the aliphatic content of the feed coal but also on the aromatic hydrogen content and possibly on the presence of other heteroatoms such as sulfur. The high aromatic hydrogen content observed in Illinois No. 6 coal suggested a relatively low degree of aromatic condensation in this coal. Residues from high-conversion runs were found to contain more aliphatic (fewer aromatic) carbons than those from low-conversion runs. This result was attributed to methyl groups and hydroaromatic structures produced by pyrolysis and subsequent hydrogenation at peripheral aromatic carbons.*

CONVENTIONAL COAL LIQUEFACTION PROCESSES involve the partial breakdown of large complex coal structures into smaller molecules.

[1]Current address: Power Scientific, 3208 91 Street S.W., Edmonton, Alberta T6X 1A2, Canada

The degradative process, which is primarily thermal in nature, is normally achieved in the presence of high-pressure molecular hydrogen and a donor solvent. The reaction of coal in hydrogen donor solvents such as 1,2,3,4-tetrahydronaphthalene (Tetralin) at elevated temperature and pressure produces gas and liquid products and leaves residual solid matter.

The generally accepted mechanism for coal liquefactions was first described by Curran et al. (*1, 2*) in 1966. The mechanism involves a free radical process. The reaction is initiated by the thermal cleavage of C–C bonds to form small coal radicals. Some of these radicals are stabilized by abstraction of hydrogen from Tetralin to form stable small molecules that are liquid or that are soluble in common organic solvents such as toluene. The donor molecule is subsequently hydrogenated by molecular hydrogen, and the liquefaction process continues in a cyclic manner. This mechanism has since been expanded upon by several other investigators (*3–5*).

Most of the current knowledge of coal-liquefaction chemistry has been derived from studies of model compounds as well as the analysis of fluid conversion products under various process conditions. For example, the mechanisms of hydrogen transfer from Tetralin to coal and coal products were studied in 1982 by using deuterated Tetralin and NMR spectroscopy (*6*). However, little attention has been given to the study of residues from coal liquefaction to elucidate a possible mechanistic pathway or process chemistry. One of the reasons for this lack of attention has been the absence of appropriate analytical methods for the structural characterization of carbonaceous solids. Since the late 1970s, advances in high-resolution solid-state ^{13}C NMR spectroscopy have made it possible to obtain structural information from insoluble carbonaceous solids. Residues from retorting oil shales have been studied with solid-state ^{13}C NMR spectroscopy, and the oil yields have been correlated to the original kerogen organic structure (*7–9*). The previous study on residue samples from the coal-liquefaction process was performed by Wilson et al. on Liddell coal (*10, 11*).

The objectives of this study were, first, to identify some of the chemical structural factors that resulted in high conversions and, second, to learn the liquefaction mechanisms by studying residues with high-resolution solid-state NMR spectroscopy.

Experimental Details

Coal Samples. Three coal samples were used in this study. Wyodak–Anderson and Illinois No. 6 coals were obtained from the Argonne Premium Coal Sample Program. Highvale coal was obtained from the Alberta Research Council Coal Sample Bank. All three samples (~100 mesh size) were dried at 105 °C before use.

Hydrogenation Process. Liquefaction experiments were performed under nitrogen, hydrogen, and methane gas pressures in a small stainless steel batch tubing bomb reactor (10-mm i.d., 10 cm long). Approximately 2 g of coal and 3 mL of Tetralin were used. An Fe_2O_3 catalyst (10 wt% of coal) was used for some experiments. The initial gas pressure at room temperature was 6.95 MPa (~1000 psi). The reactor was quickly heated to 450 °C and was kept at this temperature for 1 h with vertical agitation. After the reaction, the gases were collected for analysis, and the condensed products were extracted in 200 mL of toluene for 48 h in a Soxhlet apparatus. Toluene was evaporated, and the weight of the residue was used to calculate the conversion yield.

^{13}C NMR Spectroscopy. Cross-polarization–magic-angle spinning (CP–MAS) ^{13}C NMR spectra of samples were measured with a Bruker CXP-200 spectrometer operating at 50.3 MHz. The contact time was 1 ms, the proton 90° pulse was 6 μs, and the repetition time was 1 s. For dipolar-dephasing (DD) spectra, a 40-μs delay was used as before (12). To collect free induction decay (FID) signals, 1024 data points were used, and the data were zero-filled to 8192 data points before Fourier transformation. The sample was spun at approximately 4.5 kHz in a cylindrical (7-mm i.d.) sapphire rotor with a cap and a turbine made of a homopolymer of chlorotrifluoroethene (Kel–F). A Doty Scientific probe was used. For most of samples, 4000 transients were sufficient to obtain a good signal-to-noise ratio. Apparent aromaticities, f_a, were obtained from the integral of the aromatic peak between 90 and 165 ppm and associated spinning side bands (SSB) assuming that the SSB on both sides of the peak were equal in size (13). The reproducibility of the f_a values was confirmed by measuring one sample (treated Illinois No. 6 coal under N_2 gas) at various spinning rates (3.7–4.5 kHz). The f_a values ranged from 81.8 to 83.1%.

Thermal Analysis. The moisture content, volatile matter (VM), fixed carbon (FC), and ash in the sample were determined by using a Du Pont thermal gravimetric analyzer model 951 equipped with a data station. The heating rate was 100 °C/min. Samples were heated to approximately 950 °C in nitrogen gas and held at this temperature for 5 min. The ash content was obtained by subsequently introducing oxygen gas to burn the remaining organic materials.

Elemental Analysis. Carbon, hydrogen, and nitrogen were analyzed in the Microanalytical Laboratory, Chemistry Department, University of Alberta, using a Perkin-Elmer 240B CHN analyzer. Total sulfur contents were obtained with a LECO model SC-132 sulfur determinator. The pyritic sulfur data were obtained from the Users Handbook for the Argonne Premium Coal Sample Program (14).

Results and Discussion

Feed Coals. The rank and other chemical properties of the feed coals are shown in Table I. Illinois No. 6 coal had the highest ash content, and Wyodak–Anderson coal had the lowest. The fixed-carbon content for the feed coals varied from 53.8% (Wyodak–Anderson) to 60.0% (Highvale). The H–C ratio for Highvale coal was the lowest of the three coals studied; Wyodak–Anderson and Illinois No. 6 coals showed significantly higher H–C ratios than Highvale coal. Wyodak–Anderson and Highvale coal samples had significantly lower total and pyritic sulfur contents when compared to Illinois No. 6 coal. This result is in agreement with the high ash content observed for Illinois No. 6 coal. The apparent carbon aromaticity data indicated that Highvale coal contained the greatest amount of aromatic carbon (0.69), and Wyodak–Anderson coal had the least (0.59).

Stack plots of ^{13}C CP–MAS and DD NMR spectra of feed coals are shown in Figure 1. Only nonprotonated carbons and carbons in very mobile side groups were detected in the DD spectra. The aliphatic carbon content varied among the samples. Wyodak–Anderson coal had the greatest amount and Highvale coal had the least amount of aliphatic carbons. The spectra of all three feed coals show the presence of some carbons directly associated with oxygen. The peaks at approximately 180 ppm are due to carboxyl carbons. Absorption signals centered around 145 and 155 ppm are due to aromatic carbons directly bonded to oxygens. If

Table I. Data on Feed Coals

Property	Wyodak–Anderson	Highvale	Illinois No. 6
ASTM[a] rank	subbituminous	subbituminous	bituminous
Ash	5.2	11.6	13.1
FC[b]	53.8	60.0	57.2
VM[c]	46.2	40.0	42.6
H–C ratio	0.85	0.71	0.85
O	20.8	20.2	10.9
Total S	0.8	0.4	6.4
FeS$_2$	0.1	trace	5.5
f_a	0.59	0.69	0.67
Aliphatic C	39	28	32

NOTE: All values, except for H–C ratio and f_a, are given as percents.
[a] American Society for Testing and Materials.
[b] Fixed carbon.
[c] Volatile material.

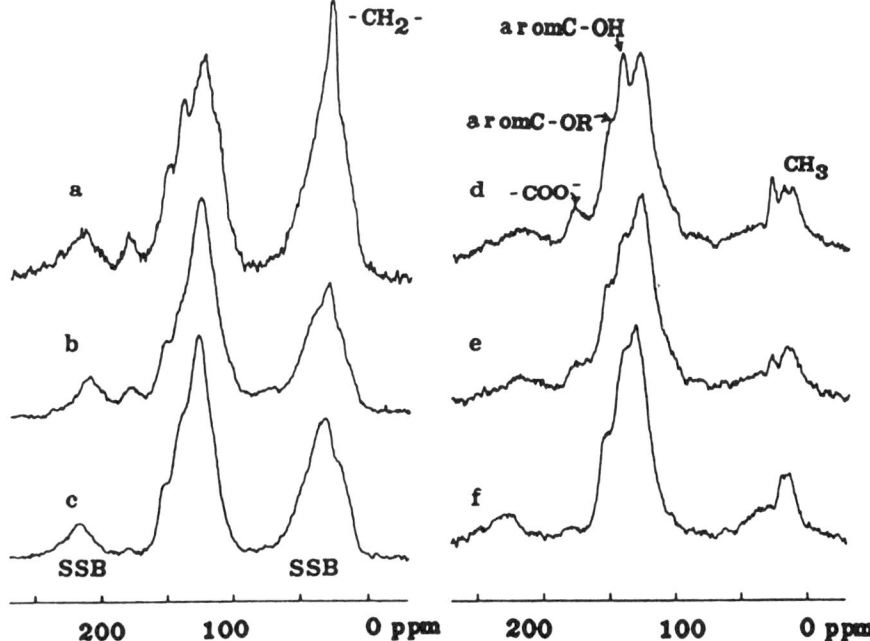

Figure 1. ^{13}C CP–MAS and DD NMR spectra of feed coals: (a) Wyodak–Anderson CP–MAS, (b) Highvale CP–MAS, (c) Illinois No. 6 CP–MAS, (d) Wyodak–Anderson DD, (e) Highvale DD, and (f) Illinois No. 6 DD.

the oxygen is in the form of a free phenolic group, the aromatic carbons absorb at approximately 148 ppm, but if the oxygen is ether-linked, the aromatic carbon resonance shifts slightly downfield to approximately 152 ppm (15). Apparently, the relative quantities of aromatic carbons associated with oxygens (obtained from the signals between 140 and 160 ppm) do not vary among the three feed coals. The signals from α carbons in aliphatic alcohols are normally observed between 60 and 80 ppm, but the presence of these peaks was not obvious except in the spectrum of the Highvale coal, which showed some indication of the presence of this group. Carbonyl carbons normally appear at approximately 200 ppm (16), but their presence could not be confirmed because of overlapping SSB.

Residues from Liquefaction under N_2 Gas. In the present study, the three different coals were first processed under inert nitrogen gas. Without hydrogen gas to regenerate the donor solvent Tetralin, the extent of conversion of coal to liquid and gaseous products is limited by the available hydrogen from Tetralin only. Figure 2 shows the CP–MAS

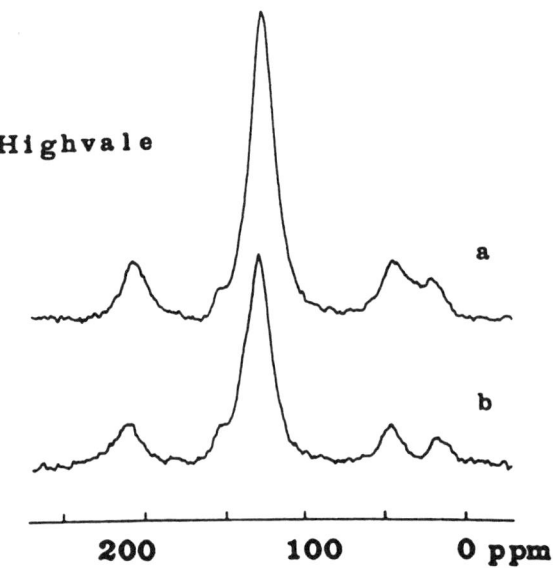

Figure 2. ^{13}C *CP–MAS (a) and DD (b) NMR spectra of the residue from Highvale coal.*

and DD spectra of the residue sample from Highvale coal after liquefaction under nitrogen. A high aromatic carbon content characterizes the residue spectrum. A comparison of the residue spectra with the spectra of the feed coal shows that the following types of carbons have disappeared during the liquefaction process under nitrogen:

- most of the aliphatic carbons (0–40 ppm)
- some protonated aromatic carbons (100–128 ppm)
- most of aromatic carbons directly bonded to oxygen (140–160 ppm)
- most of carboxyl carbons (~180 ppm)

The remaining aliphatic-carbon peaks were very small and were observed in the chemical-shift range of 5–45 ppm. The DD spectra of the residues indicated that these small peaks were due to the methyl carbon directly bonded to aromatic carbons and possibly due to hydroaromatic carbons (*17*). The same types of methyl groups have also been detected in the organic residue deposited on the surface of spent catalysts from hydroprocessing of heavy oils at 360–400 °C (*18, 19*).

The half-height line widths of the aromatic carbon peaks centered at approximately 128 ppm varied among coal samples. The broadening of the peaks may be due to paramagnetic iron in the sample. Illinois No. 6 coal showed the greatest line broadening, and this broadening was in agreement with the highest content of pyritic sulfur.

The percent yield, aromaticity, and fixed carbon data for residues from liquefaction experiments under nitrogen, hydrogen, and methane are listed in Table II. The solid yields were calculated from the weights of the residue on a dry, ash-free (daf) basis, and the total conversion to gases and toluene-soluble compounds is 100 minus the solid yield. Under nitrogen, Illinois No. 6 coal gave the highest conversion and Highvale coal gave the lowest.

Because the aliphatic portions of coal and of other carbonaceous solids are believed to be more easily pyrolyzed under liquefaction conditions, a positive correlation between the aliphatic content and the total conversion under similar reaction conditions would be expected. For example, the oil yield from oil shale has been found to be linearly dependent on the aliphatic carbon content (7, 8). In the present study, although the aromaticity values show a substantial increase in the residue samples when compared to the feed coal, the total conversion does not seem to depend on only the aliphatic carbon content but also on the amount of fixed carbon in the feed coal. This observation indicates that other structural properties of coal are also important factors in liquefaction conversion.

Illinois No. 6 coal, which gave the highest conversion, contained the greatest amount of organic and inorganic sulfur. The presence of sulfur is difficult to detect with ^{13}C NMR spectroscopy, but if sulfur or thio-ether

Table II. Liquefaction Data

Feed Coal	Gas	Catalyst	Solid Yield (%)[a]	f_a	Fixed Carbon (%)[a]
Wyodak–Anderson	N_2	no	36.3	0.87	74.4
	H_2	no	35.7	0.88	79.9
	H_2	yes	26.2	0.84	76.3
	CH_4	no	25.3	0.83	80.3
	CH_4	yes	42.7	0.87	76.8
Highvale	N_2	no	46.7	0.91	77.8
	H_2	no	42.1	0.88	77.2
	H_2	yes	47.3	0.92	80.4
	CH_4	no	46.4	0.89	74.2
	CH_4	yes	54.8	0.91	80.0
Illinois No. 6	N_2	no	25.6	0.84	72.2
	H_2	no	24.0	0.82	72.0
	H_2	yes	31.5	0.83	71.5
	CH_4	no	30.1	0.81	71.2
	CH_4	yes	27.5	0.77	70.2

[a] Dry ash-free basis.

linkages are present in coal, these may be more easily broken during pyrolysis than C–C linkages. Furthermore, the high pyrite and ash contents of Illinois No. 6 coal may have contributed to the high conversion as a result of catalytic activity.

Although the apparent carbon aromaticity may not represent the true carbon aromaticity of coal (20, 21), the variations of f_a values among similar samples should be comparable as long as the spectral conditions are kept constant.

If the pyrolysis did not cause aromatization of aliphatic carbons originally present in feed coals, approximate amounts of lost aromatic and aliphatic carbons can be calculated. The remaining aromatic carbons can be obtained by multiplying the solid yield (as a fraction of the original feed coal) by the apparent aromaticity value, f_a, of the residue. Such data are shown in Table III for liquefaction experiments under nitrogen. The results clearly show that Illinois No. 6 coal lost many more aromatic carbons than the other two feed coals. The aliphatic content of feed coal was not a single factor affecting the conversion yield. The presence of aromatic structures that are easily pyrolyzed to small molecules seems to contribute significantly to the high conversion yield of coal.

From the preceding discussion, this question arises: "What is the difference in the aromatic structure of Illinois No. 6 coal as compared to the other two coals?" In the following discussion, we will attempt to answer this question.

The H–C ratio of the aromatic portion can be calculated if the aliphatic hydrogen content in coal can be estimated. We assumed that the average aliphatic carbons were methylene carbons (22), and therefore the average H–C ratio in aliphatic portions is 2. The H–C ratios from the elemental analysis and the carbon aromaticity data provide the original distribution of hydrogens between aliphatic and aromatic structures. In this calculation, contributions from phenolic and carboxylic hydrogens were assumed to be negligible. Results for the aromatic H–C ratios are shown in Table IV. Even if the assumptions that $(H–C)_{al} = 2$ and $f_a = C_{ar}/100$ are not absolutely correct, a clear trend can be observed. In Illi-

Table III. Aromatic Carbon Distribution

Coal	Solid Yield	Aromatic Carbon[a]			
		Feed	Residue	Remaining	Lost
Wyodak–Anderson	0.363	59	87	32	27
Highvale	0.467	69	91	42	27
Illinois No. 6	0.256	67	84	22	46

[a]The number given is the number of aromatic carbons per every hundred carbons.

Table IV. Hydrogen Distribution in Coals

Coal	H_{tot}^{a}	C_{al}^{b}	H_{al}^{c}	H_{ar}^{d}	H_{ar}/C_{ar}
Wyodak–Anderson	85	39	78	7	0.12
Residue	56	11	22	34	0.39
Highvale	71	28	56	15	0.22
Residue	54	7	14	40	0.44
Illinois No. 6	85	32	64	21	0.31
Residue	59	13	26	33	0.39

NOTE: All values are given per 100 carbon atoms. The subscript abbreviations are as follows: tot, total; al, aliphatic; and ar, aromatic.
[a] Values obtained from H–C ratios in Table I.
[b] Values obtained by using the following equation: $100 - f_a - \%$ carboxyl carbon.
[c] Values obtained by multiplying the aliphatic carbon value by 2.
[d] Values obtained by subtracting H_{al} from H_{tot}.

nois No. 6 coal, more aromatic carbons are protonated than in the other two feed coals. This observation suggests that a greater availability of protonated aromatic carbons is a factor contributing to the high conversion of aromatic carbons. Many condensed aromatic moieties are probably linked through simple C–C bonds, methylene bridges, ether bonds, and thioether bonds in the coal chemical structure (23). If each of the aromatic moieties is less condensed and smaller, a higher H–C ratio would be expected when more peripheral aromatic carbons are available. The data in Table IV suggest that the degree of aromatic condensation in Illinois No. 6 coal is lower than in the other two coals. The abundance of small aromatic units leads to greater production of small fragments by pyrolysis and to high conversion.

The H–C ratios in the aromatic structures in the process residues were also calculated, and the results are shown in Table IV. Interestingly, the ratios of the residues are greater than those of the feed coal for all three samples. This observation suggests that some hydrogenation of peripheral carbons of condensed aromatic structures, which constitute the bulk of the residue, occurs during liquefaction.

Residues from Process under H_2 and CH_4 Gases.

All three feed coals were also processed under hydrogen or methane gas with and without a catalyst. These liquefaction experiments were performed to study the relationship between the chemical characteristics of the unconverted residue and the conversion yield under different conditions. The yield, the aromaticity, and the fixed carbon content of the residue derived from liquefaction experiments in hydrogen and methane are listed in Table II.

The residues obtained from the liquefaction experiment with a catalyst were ferromagnetic, and more than six sets of SSB were observed after overnight scanning. Hence, the ferromagnetic iron had to be removed before analysis. These residues were extracted overnight in a 10% HCl solution at room temperature, filtered, washed, and dried at 85 °C. The extraction procedure had to be repeated three times before the effect of the ferromagnetic iron became negligible. Feed coal and some residue samples obtained from the processes without catalysts were also extracted in a 10% HCl solution to study the effect of the acid on organic materials. NMR spectra obtained before and after the extraction were compared, and no detectable differences were observed for any of the residue samples. Therefore, the organic matter was not altered by washing with a 10% HCl solution at room temperature. The measured f_a value for Illinois No. 6 coal after extraction was larger by 0.03 than that measured before extraction. Line broadening due to the relatively high iron content of this coal before extraction probably caused the slightly smaller apparent carbon aromaticity (24). The f_a value after the extraction was used for Illinois No. 6 feed coal. No changes were observed in the spectra of the other two feed coals.

Figure 3 presents a plot of apparent carbon aromaticities versus fixed carbon contents for all samples, including residues and feed coals. A good correlation is clearly indicated irrespective of the sample source. This

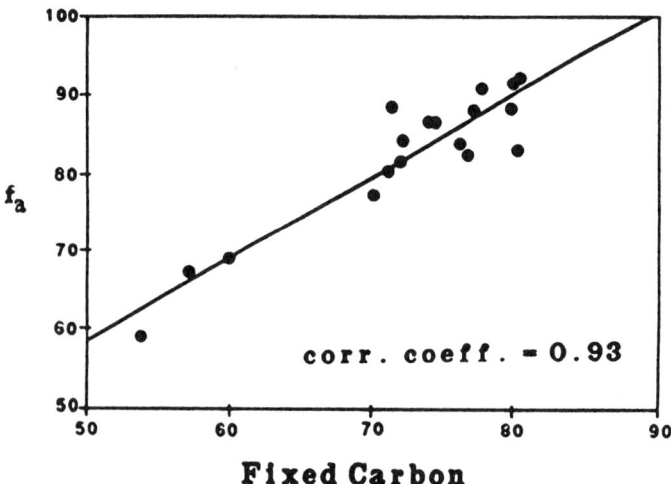

Figure 3. Plot of apparent carbon aromaticity vs. fixed carbon content.

result is in agreement with many previous studies (13, 25–29), and it confirms that fixed carbons are generally highly aromatic.

Figure 4 shows a plot of aromaticities versus solid yields for all the residue samples. A good correlation is also indicated here, regardless of the nature of the feed coal or the process condition. A lower aromaticity value was observed for the residues with higher conversion (i.e., low solid yield). This result is contrary to earlier observations with coal hydrogenation residues (10, 11). The process that resulted in high conversions left residues with low aromaticity because nonvolatile fixed carbons are highly aromatic. However, this low aromaticity can be attributed to the hydrogenation of more condensed aromatic structures and to the creation of more directly bonded methyl groups during coal liquefaction for higher-yield experiments. In addition to the pyrolysis of aliphatic chains linking large aromatic moieties, direct hydrogenation of these large aromatic structures also occurs during coal liquefaction. Such hydrogenation reactions should lead to the formation of hydroaromatic and naphthenic groups that are subsequently pyrolyzed to increase the liquefaction conversion. The explanation for the relative increase of aliphatic carbons with increasing conversion may be given by the proposed simplified mechanism of pyrolysis and hydrogenation of peripheral aromatic carbons shown in Scheme I. As the peripheral carbons in condensed aromatic clusters become hydrogenated, hydroaromatic structures that are susceptible to pyrolysis are formed. Subsequent breakage of the aliphatic C–C bonds within the hydroaromatic structures leads to the formation of methyl groups that are directly bonded

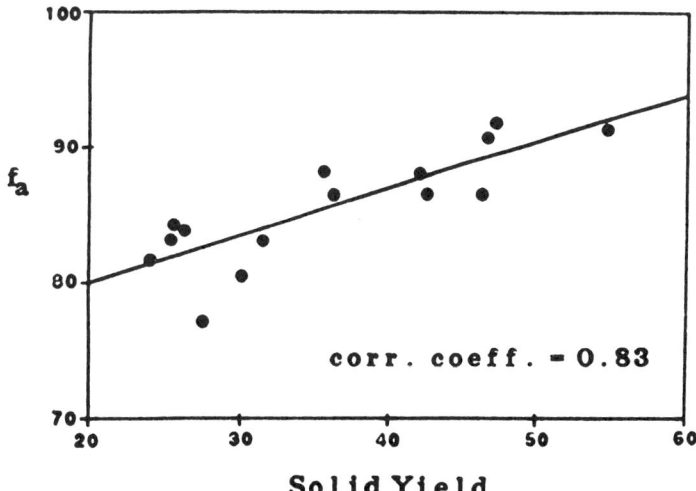

Figure 4. Plot of aromaticities vs. solid yields.

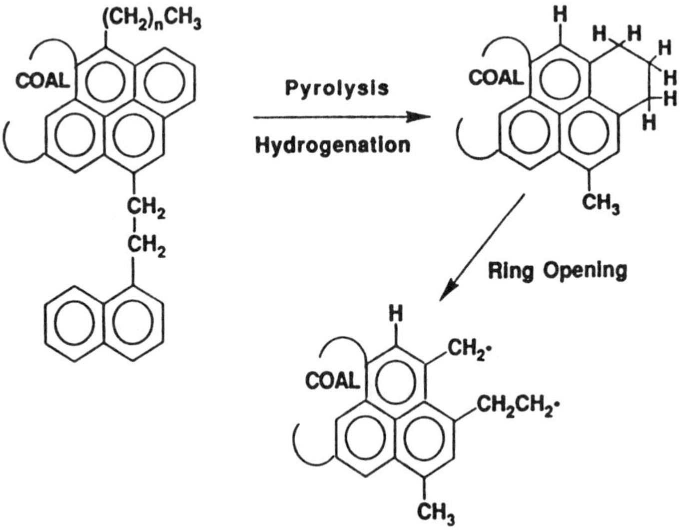

Scheme I. Proposed coal liquefaction mechanism.

to aromatic compounds. Other aliphatic chains that are formed in this process can then be removed by dealkylation.

Conclusions

The results of this study show that solid-state ^{13}C NMR spectroscopy can provide a better understanding of the chemical structure of feed coals and their liquefaction residues as well as of the liquefaction process chemistry and mechanism. On the basis of the results presented in this chapter, the following conclusions can be made:

1. The conversion yield of a given coal under a given liquefaction condition is a function of several chemical characteristics of the coal: the aliphatic carbon content, the aromatic H–C ratio (or the average size of the condensed aromatic structures), and possibly the sulfur content.

2. A majority of the aliphatic carbons and most of the functional groups involving oxygen were lost during the liquefaction process, a result that left residues with very high carbon aromaticities.

3. A high conversion was associated with low carbon aromaticity in the residue. This association indicated that in addition to thermal pyrol-

ysis of C–C bonds, direct hydrogenation of condensed aromatic structures was the important process in coal liquefaction.

Acknowledgments

Financial support from the Natural Sciences and Engineering Research Council of Canada for N. O. Egiebor is gratefully acknowledged. We also acknowledge S. Chakrabartty and M. Selucky for valuable discussions, K. F. Schulz for suggesting the ferromagnetism of some samples, and Jilu Zou and Wendy Wade for technical assistance.

References

1. Curran, G. P.; Struck, R. T.; Gorin, E. *Proc. Div. Petrol Chem. ACS.* **1966**, C130–C148.
2. Curran, G. P.; Struck, R. T.; Gorin, E. *Ind. Eng. Chem. Process Des. Dev.* **1967**, 6, 166.
3. Wiser, W. H. *Fuel* **1968**, 47, 475–486.
4. Neavel, R. C. *Fuel* **1976**, 55, 237.
5. Vernon, L. W. *Fuel* **1980**, 59, 102–106.
6. Franz, J. A.; Camaioni, D. M. *Fuel* **1984**, 63, 990–1001.
7. Maciel, G. E.; Bartuska, V. J.; Miknis, F. P. *Fuel* **1978**, 57, 505.
8. Maciel, G. E.; Bartuska, V. J.; Miknis, F. P. *Fuel* **1979**, 58, 155–156.
9. Miknis, F. P.; Szevereny, N. M.; Maciel, G. E. *Fuel* **1982**, 61, 341–345.
10. Barron, P. F.; Stephens, J. F.; Wilson, M. A. *Fuel* **1981**, 60, 547–549.
11. Wilson, M. A.; Pugmire, R. J.; Vassallo, A. M.; Grant, D. M.; Collin, P. J.; Zilm, K. W. *Ind. Eng. Chem. Proc. Des. Dev.* **1982**, 21, 477–482.
12. Cyr, N.; Selucky, M. L. *Liq. Fuels Technol.* **1985**, 3, 377–396.
13. Frimsky, E.; Ripmeester, J. A. *Fuel Process. Technol.* **1983**, 7, 191–202.
14. Vorres, K. S. In *Users' Handbook for the Argonne Premium Coal Sample Program;* Argonne National Laboratory: Chicago, IL, 1989.
15. Cyr, N.; Elofson, R. M.; Ripmeester, J. A.; Mathison, G. W. *J. Agri. Food Chem.* **1988**, 36, 1197–1201.
16. Yoshida, T.; Tokuhashi, K.; Narita, H.; Hasegawa, Y.; Maekawa, Y. *Fuel* **1984**, 63, 282–284.
17. Cyr, N.; Gawlak, M.; Carson, D. W.; Ignasiak, B. S. *Fuel* **1983**, 62, 413–416.
18. Egiebor, N. O.; Gray, M. R.; Cyr, N. *Chem. Eng. Commun.* **1989**, 77, 125–133.
19. Egiebor, N. O.; Gray, M. R.; Cyr, N. *Appl. Catal.* **1989**, 55, 81–91.
20. Dudley, R. L.; Fyfe, C. A. *Fuel* **1982**, 61, 651–657.
21. Packer, K. J.; Harris, R. K.; Kenwright, A. M.; Snape, C. E. *Fuel* **1983**, 62, 999–1002.
22. Berkowitz, N. In *The Chemistry of Coal;* Elsevier: Amsterdam, Netherlands, 1985.
23. Berkowitz, N. In *An Introduction to Coal Technology;* Academic: New York, 1979.

24. Pfeffer, P. E.; Gerasimowics, W. V.; Piotrawski, E. G. *Anal. Chem.* **1984,** *56,* 734–741.
25. Maciel, G. E.; Bartuska, V. J.; Miknis, F. P. *Fuel* **1979,** *58,* 391–394.
26. Pugmire, R. J.; Zilm, K. W.; Grant, D. M.; Larter, S. R.; Allen, J.; Senfite, J. T.; Davis A.; Sparckman, W. In *New Approaches in Coal Chemistry;* Blaustein, B. D.; Bockrath, B. C.; Friedman, S., Eds.; ACS Symposium Series No. 169; American Chemical Society: Washington, DC, 1981; Chapter 2, p 23.
27. Miknis, F. P.; Sullivan, M.; Bartuska V. J.; Maciel, G. E. *Org. Geochem.* **1981,** *3,* 19–28.
28. Painter, P. C.; Kuehn, D. W.; Starsinic, M.; Davis, A.; Havens, J. R.; Koenig, J. L. *Fuel* **1983,** *62,* 103–111.
29. Sfihi, H.; Quinton, M. F.; Legrand, A.; Pregermain, S.; Carson, D.; Chiche, P. *Fuel* **1986,** *65,* 1006–1011.

RECEIVED for review June 7, 1990. ACCEPTED revised manuscript May 28, 1991.

15

^{1}H NMR Spectroscopy and Spin−Lattice Relaxation of Argonne Premium Coals

Kikuko Hayamizu, Shigenobu Hayashi, Kunio Kamiya, and Mitsutaka Kawamura

National Chemical Laboratory, Tsukuba, Ibaraki 305, Japan

Solid-state ^{1}H NMR spectra and spin−lattice relaxation times (T_1) are reported for eight Argonne Premium coals that were prepared under several different conditions. Measurements were performed on pristine coals, degassed and dried samples, dried coal samples stored in capped vials for about 1 year, and coals treated with two different acids, HCl and HF−HCl. In general, nonexponential relaxation curves were analyzed by fitting the data to two exponential decay constants. With acid treatment, an increase in T_1 was observed for the coals with carbon contents of less than 77%, and a decrease was observed for the higher rank coals. Electron spin resonance (ESR) measurements were performed on coals with and without acid treatment to assess the ^{1}H spin−lattice relaxation mechanisms. The important ^{1}H NMR relaxation mechanism in lignite and subbituminous coals is due to the presence of paramagnetic metal ions, whereas that in bituminous coals is due to organic free radicals. The effects of various treatments on the ^{1}H NMR spectra and the T_1 values of coals are discussed.

ONE OF THE MOST POWERFUL METHODS for characterizing the gross molecular structures and molecular motions in organic solids is ^{1}H NMR spectroscopy. A great number of NMR studies of the structure of coal have been published during the past two decades (*1−12*). The most important mechanism for proton spin–lattice relaxation in coals is assumed to be due to the effects induced by inherent organic free radicals and paramagnetic metal ions or by paramagnetic oxygen gas that is introduced into fresh coals. The correlation between the ^{1}H relaxation times and the organic free radicals has been widely acknowledged, and relationships have been found between the unpaired-electron spin density and the proton T_1 for bituminous and higher rank coals. However, a clear interpretation of proton relaxation has not been forthcoming, especially in the lower rank coals.

To assess the effects of paramagnetic ions on the relaxation times, we measured the ^{1}H NMR parameters before and after metal ions were removed from the coals by acid treatment. Two different acid-treated samples were prepared for each of the eight Argonne Premium coals. Four additional samples were prepared from these coals for analysis (i.e., fresh, degassed, dried, and aged and dried samples for each coal). Insights into the relaxation mechanisms were obtained by comparing the ^{1}H NMR spectra and spin–lattice relaxation times (T_1) for these six differently treated samples of each coal.

For the measurement of ^{1}H NMR spectra, removal of moisture is another important problem. In this chapter, using samples prepared under various conditions, the effects of H_2O on ^{1}H spectra are discussed. To complement the ^{1}H NMR results, electron spin resonance (ESR) spectra were recorded for the aged and dried samples and for the acid-treated and dried samples.

Experimental Details

The eight coals used in this study were obtained from the Premium Coal Sample Program at Argonne National Laboratory. Six samples were prepared for each Argonne Premium coal. Sample A for each coal was prepared by opening the ampules and rapidly transferring a sample to a 10-mm NMR sample tube, which was immediately capped. The T_1 measurements of these samples were carried out within 1 h. B samples were prepared by pumping out air while cooling the sample in liquid nitrogen before sealing. C samples were dried at 110 °C under vacuum for more than 8 h before sealing. D samples were prepared from coals stored in capped bottles for about 1 year before drying at 110 °C. Samples E and F were prepared by treating sample D coals with HCl and HF–HCl, respectively, prior to drying at 110 °C.

The acid-treated coals were prepared according to the following procedures: About 2 g of coal (sample D) was refluxed in 50 mL of 10% HCl at 100 °C for 4 h. The sample was washed well with water and dried at 60 °C under vacuum for about 24 h, producing sample E. The amount of iron removed was determined by calorimetric analysis of the combined solutions. Sample F was prepared from about 1.2 g of sample E. The coal was suspended in 10 mL of an aqueous solution consisting of 47.5% HF, 2.5% HCl, and 50% H_2O by volume and was then heated at 50 °C for 4 h. The HF–HCl-treated coal was filtered, washed with H_2O, and dried.

The ^1H NMR spectra were observed at the resonance frequency of 90 MHz on a Bruker CXP-100 spectrometer. The spin–lattice relaxation times were obtained by the usual 180°–τ–90° pulse sequence. The width of the 90° pulse was 2.6 μs. The free induction decays (FIDs) were accumulated 32 times for each successive interval time, τ, between 180° and 90° pulses. When the relaxation curves demonstrated single exponential behavior, more than six τ values were employed in the analysis. More than 20 different τ values were utilized to analyze curves with two-component relaxation behavior. Fourier transformed spectra were obtained by using a spectral width of 125 kHz.

Proton relaxation times were calculated by using a nonlinear least-squares analysis according to eq 1:

$$\frac{M_o - M(\tau)}{M_o} = N \sum C_j \exp \left| \frac{-\tau}{T_{1j}} \right| \quad (1)$$

where M_o is the magnetization at equilibrium, $M(\tau)$ is the magnetization at time τ, C_j is the magnetization fraction characterized by T_{1j} for the component j, and N is the normalization factor. The average relaxation time, T_{1a}, was obtained according to eq 2:

$$\frac{1}{T_{1a}} = \sum \frac{C_j}{T_{1j}} \quad (2)$$

ESR spectra were observed with a JEOL JER–RE spectrometer.

Results and Discussion

Proton NMR spectra of A (fresh), B (degassed), and C (dried) samples are shown for North Dakota Beulah–Zap (ND), Utah Blind Canyon (UT), and Pocahontas No. 3 (POC) coals in Figures 1, 2, and 3, respectively. The H_2O contents of fresh ND, UT, and POC coals are 32, 4.6, and 0.65%, respectively (13).

Spectra of the A and B samples of the ND coal in Figure 1 exhibit very sharp resonances that are assumed to originate from H_2O. Increasing the vertical scale in these spectra reveals a broader component that can be attributed to protons of the coal. The line width of the water resonance

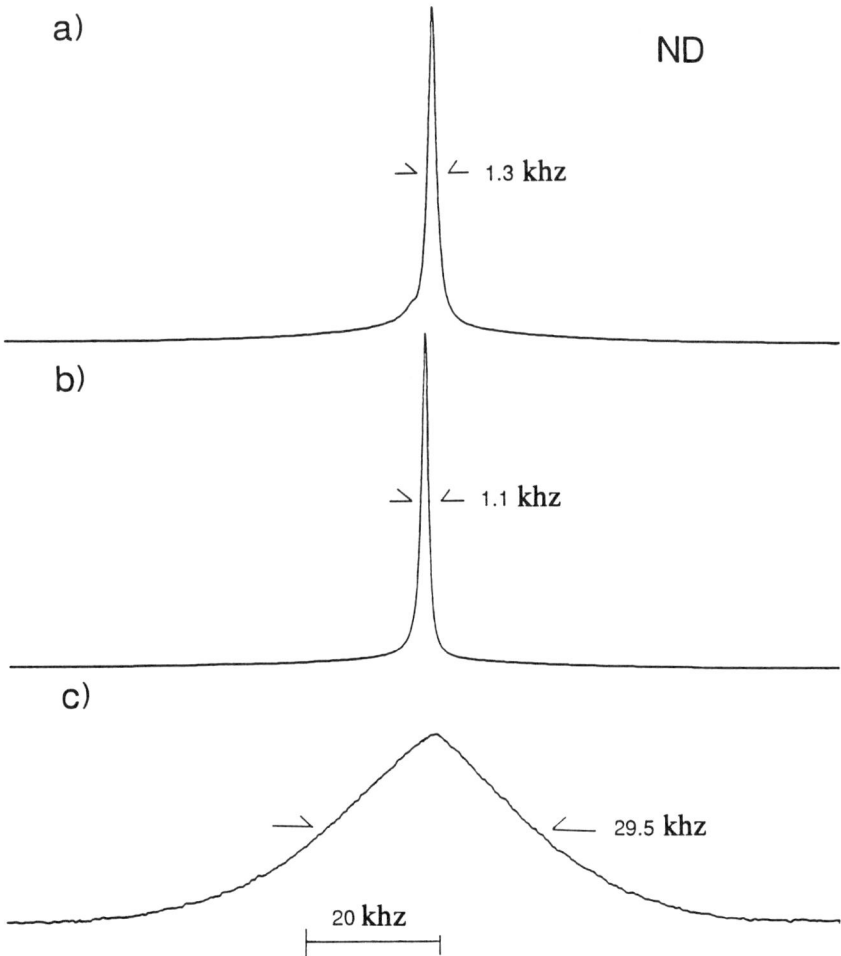

Figure 1. ^1H NMR spectra of (a) A (fresh), (b) B (degassed), and (c) C (dried) samples of Beulah–Zap North Dakota (ND) coal.

in sample A is 1.3 kHz, and that in sample B is 1.1 kHz. The water resonances in these coals appear at higher field than the resonance for an aqueous solution of 5% $CuSO_4$. The ^1H resonance of sample A is upfield of that of sample B by about 12 ppm, and the latter resonance is upfield of a 5% $CuSO_4$ solution by about 15 ppm. The degassing procedure used in the preparation of sample B may have lead to a significant loss of H_2O, together with the loss of various trapped gases. However, the intense H_2O signal observed in sample B suggests that a significant amount of H_2O remains in the sample after degassing.

15. HAYAMIZU ET AL. *¹H NMR Spectroscopy & Spin–Lattice Relaxation* 299

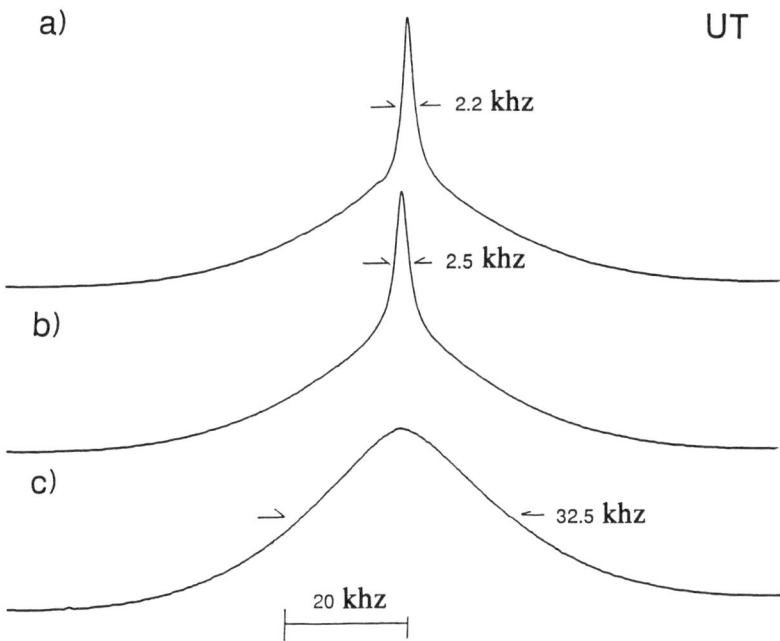

Figure 2. ¹H NMR spectra of (a) A (fresh), (b) B (degassed), and (c) C (dried) samples of Blind Canyon Utah (UT) coal.

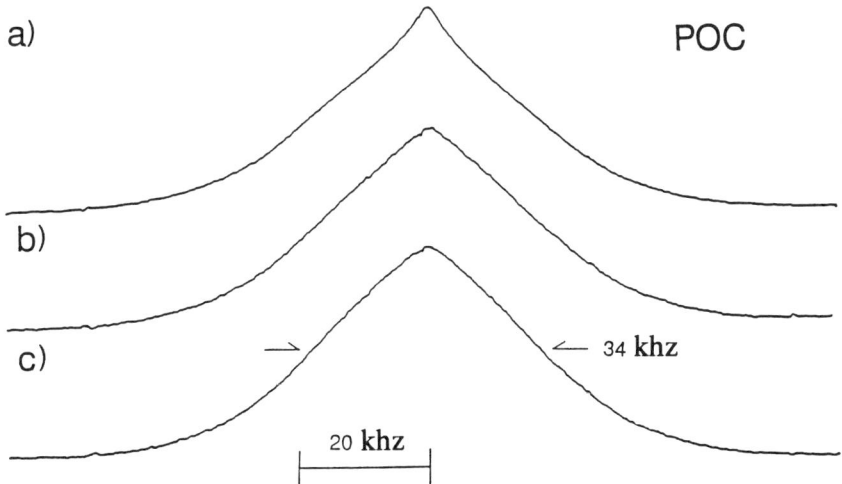

Figure 3. ¹H NMR spectra of (a) A (fresh), (b) B (degassed), and (c) C (dried) samples of Pocahontas No. 3 (POC) coal.

The significantly broad resonance in the spectrum of sample C shown in Figure 1c can be attributed to protons in the coal. Moreover, this pattern does not change for samples D, E, and F. In fact, ^{1}H spectra are remarkably similar for all Argonne Premium coal samples that have been thoroughly dried.

Similar behavior was observed in samples A, B, and C of the Wyodak–Anderson (WY) and Illinois No. 6 (IL) coals, whose H_2O contents are 28 and 8.0%, respectively. The spectra of WY and ND coal samples are nearly the same. Although the IL coal contains less H_2O than the WY and ND coals, the spectra of samples A and B exhibit sharp resonances with line widths of 1.8 and 1.3 kHz, respectively. Also, the H_2O signal in sample A's spectrum appears at higher field than that of B. Differences observed in proton line widths and chemical shifts of samples A and B may originate from exchange effects between "free" and "bound" water, which may be altered upon evacuation of the coals.

Proton spectra for the UT (Blind Canyon) coal in Figures 2a and 2b show the typical pattern for mobile H_2O; the line widths of the narrow components in samples A and B of this coal are 2.2 and 2.5 kHz, respectively. The narrow resonances in the spectra of samples A and B of the Lewiston–Stockton (WV) coal, which has less H_2O content (2.4%) than UT (4.6%), are less intense and are characterized by considerably broader line widths of 3.9 and 5.1 kHz, respectively. For Pittsburgh No. 8 (PITT), which has an H_2O content of 1.65%, the narrow resonance in samples A and B is proportionately smaller, and the line widths are too broad to be determined by simple first-order analysis.

In summary, the line widths of the narrow H_2O resonances increase with increasing carbon content of the coal. This trend parallels a decrease in H_2O content. Moreover, chemical-shift differences observed between samples A and B for all coals are about 12 ppm, and the narrow H_2O signal of sample A always appears at higher field.

The vapor pressure of water is reduced by the degassing procedure. Thus, exchange effects between free and bound water cannot be neglected, and the corresponding line widths of the water resonances depend on the H_2O content of the coal. Also, the shifts of water to higher field in A samples relative to B samples for UT, WV, and PITT coals are probably due to exchange effects. However, more detailed experiments are necessary to elucidate these phenomena in greater detail, including analyses of the ratios of free and bound water, mobility of the water, macro- and micropore structures of coal, etc.

The three spectra of POC (Pocahontas No. 3) in Figure 3 are remarkably similar; however, the narrow component is observed in sample A only. The spectra of Upper Freeport coal (UF) show similar behavior. Simple comparison of the spectra of samples A and B suggests that the moisture in UF and POC, which presumably is present as free water, was lost upon evacuating the samples.

The line widths at half-height for the dried coals (C samples) are plotted against carbon contents in Figure 4. An increase in the line width of the coal-proton signal is observed with increasing carbon content. Although it is not certain whether trace amounts of tightly bound H_2O remain in the dried samples, ^1H NMR spectra appear to be devoid of signals due to mobile protons. The lower rank coals, ND and WY, exhibit significantly narrower line widths. Possible reasons for the reduction in line width in these coals include a greater abundance of mobile side chains, weaker network structures due to smaller aromatic rings, and the presence of chemically bound water.

The ^1H spin–lattice relaxation times (T_1) and the relative fractions of each component were calculated from eq 1. Most of the relaxation data were analyzed by the sum of two exponential decay constants, but a few could be analyzed as single exponential relaxation decay. Results for WY (Wyodak–Anderson), WV (Lewiston–Stockton), and UF (Upper Freeport) are shown in Figures 5, 6, and 7, respectively. In these figures, the areas of the open circles express the relative fractions of the components normalized to unity.

All WY samples exhibit two-component spin–lattice relaxation as shown in Figure 5. The fractional amplitudes of longer T_1 components are generally greater than those with shorter T_1s, except for sample A in which the component fractions are nearly equal. Because the water content of the WY coal is 28%, proton relaxation in samples A and B originates from H_2O. Because the ratios of the two relaxation components are similar for samples A, B, and C (dried), the relaxation mechanisms of

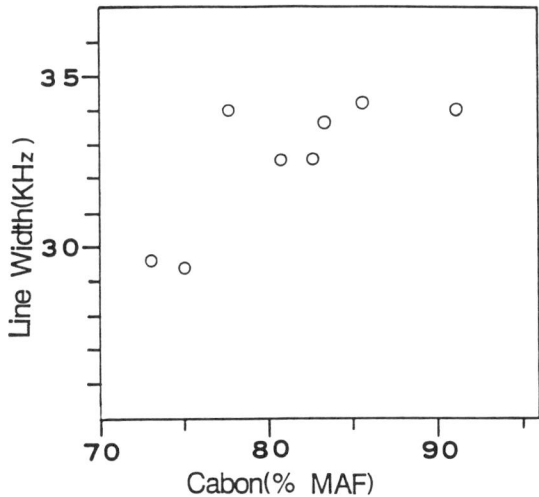

Figure 4. Plot of line widths of dried Argonne Premium coals (C samples) vs. percent carbon.

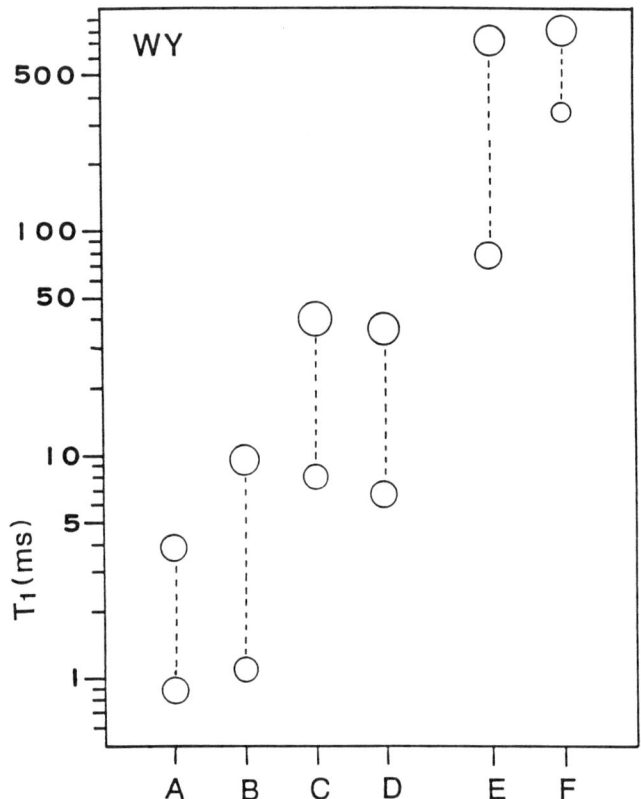

Figure 5. ^1H *relaxation times of the six differently prepared samples of the Wyodak–Anderson (WY) coal. The areas of the open circles correspond to the relative amplitudes of the components.*

these samples must be similar. The somewhat shorter T_1 for sample D (aged, stored in a bottle) may be due to mild oxidation of this sample, which may lead to an increase in free radical content (6). With acid treatment, the T_1 values become much longer for both components. The T_1 values of the longer component are similar (ca. 800 ms) for the two acid-treated samples, and an increase in T_1 is apparent for the shorter component with additional HF–HCl treatment. Similar behavior was obtained for the ND (Beulah–Zap) lignite.

In Figure 6, the ^1H relaxation behavior of the WV (Lewiston–Stockton) coal is shown. A single exponential decay in proton magnetization is observed for sample B. Samples A and B contain considerable H_2O, which is clearly shown in the proton spectra. Dried samples exhibit a two-component decay. The weighted-average value of T_{1a}, calculated by

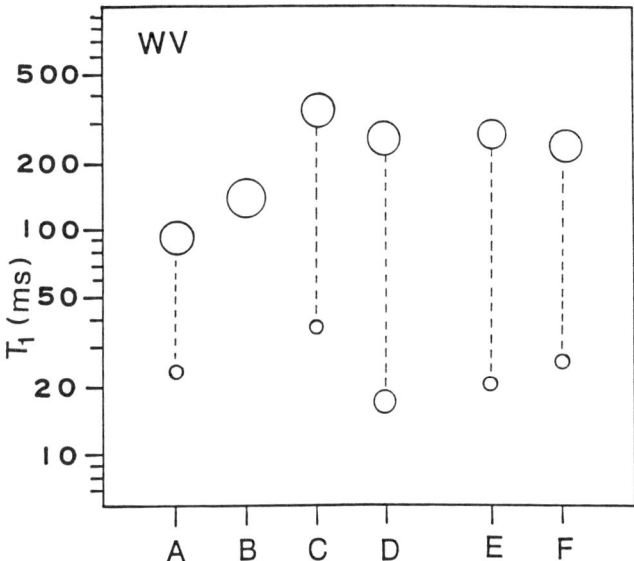

Figure 6. ^1H relaxation times of the six differently prepared samples of the Lewiston–Stockton (WV) coal. The areas of the open circles correspond to the relative amplitudes of the components.

using eq 2, of sample C equals the T_1 value of sample B within experimental error. Thus, translational diffusion of water molecules inside the coal may play a role in averaging T_1 of sample B to a single time constant. The T_1 value for sample D, which had been stored in a capped vial, was decreased. Acid treatment changes the relaxation behavior little, a result indicating that the relaxation is not governed by the presence of paramagnetic metal cations because the iron contents of samples E and F are significantly reduced compared to that of sample C. The PITT (Pittsburgh No. 8) samples show relaxation behavior that is similar to that for the WV coal.

The T_1 relaxation results for UF coal samples are shown in Figure 7. Samples A, C, and D exhibit single exponential decay of proton magnetization, and Samples B, E, and F can be analyzed for two components. After degassing, the long T_1 components increase significantly. On the other hand, acid treatment induces a shortening of the T_1 relaxation of both components. Clearly, ^1H relaxation in the dried coal is not governed by the paramagnetic metal ions. However, treating WV coal with acids may have caused some degradation in coal structure, resulting in an increased free radical content.

In Figure 8, the proton T_1 relaxation times of dried Argonne Premium coals (C samples) are plotted against carbon content. In this figure,

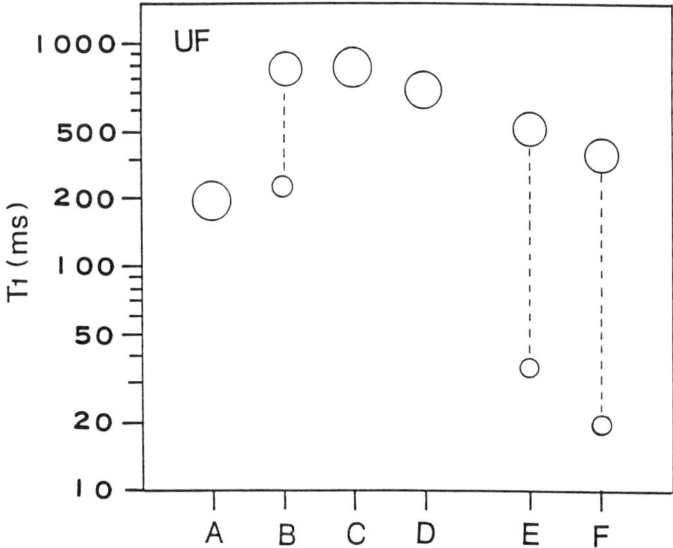

Figure 7. ^{1}H relaxation times of the six differently prepared samples of the Upper Freeport (UF) coal. The areas of the open circles correspond to the relative amplitudes of the components.

the areas of the circles (total area normalized to unity) are proportional to the relative amplitudes of the T_1 components. In general, the longer T_1 components always have the greater amplitudes for all coals, and the ratio of long-to-short T_1 components increases with carbon content of the coal. The UT and UF coals exhibit single exponential T_1 decay. Wind et al. (11) reported similar results on coals dried at ambient temperature under high vacuum. They observed the ^{1}H relaxation curves for eight Argonne Premium coals and five other coals at 60 and 187 MHz and analyzed their data by assuming two relaxation components in the same way as the present study. Although the drying procedures are quite different, our results obtained at 90 MHz and their results at 60 MHz agree well within experimental error, including the T_1 values and the relative amplitudes of the T_1 components. A difference is seen for UT coal only, for which they observed a two-component T_1 decay.

The average T_{1a} values of samples A (fresh), B (degassed), and C (dried) are plotted against the carbon content of the coals in Figure 9. When the T_1 values are compared for the various samples of the same coal, generally sample A has the shortest T_1, and sample C has the longest T_1. Because the degassing procedure (sample B) was carried out at liquid-nitrogen temperature, the loss of moisture is not expected to be great. The differences in the T_{1a} values of samples A and B are mainly

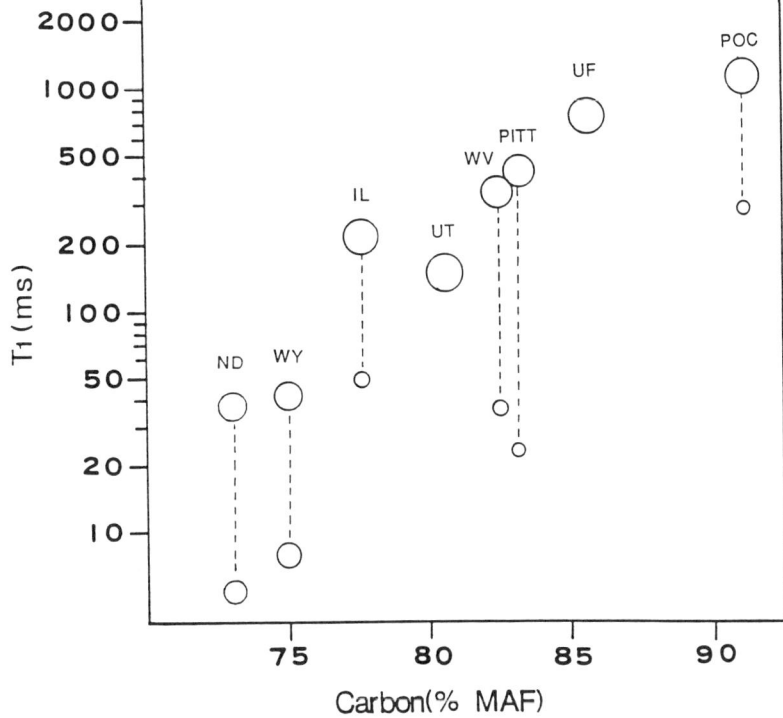

Figure 8. Plot of relaxation times for dried Argonne Premium coals (C samples) vs. percent carbon. The areas of open circles correspond to the relative amplitudes of the components.

due to the effect of some water removal as a function of overall water content of the coals, which is consistent with observations in the ^{1}H spectra discussed previously. After drying, however, the ^{1}H spectra of the C samples do not exhibit a narrow resonance that can be attributed to water. Although we are unsure whether tightly bound water remains in these dried coals, it is certain that the observed increase of T_{1a} from sample A to sample C is induced by the removal of mobile H_2O. The T_{1a} values for dried Argonne Premium coals can be classified in three groups:

1. lignite (ND) and subbituminous (WY) coals with the shortest T_{1a} values (10–20 ms)

2. high-volatile bituminous coals (IL, UT, WV, and PITT) with intermediate T_{1a} values (140–160 ms)

3. medium- and low-volatile bituminous coals (UF and POC) with the longest T_{1a} values (730–780 ms).

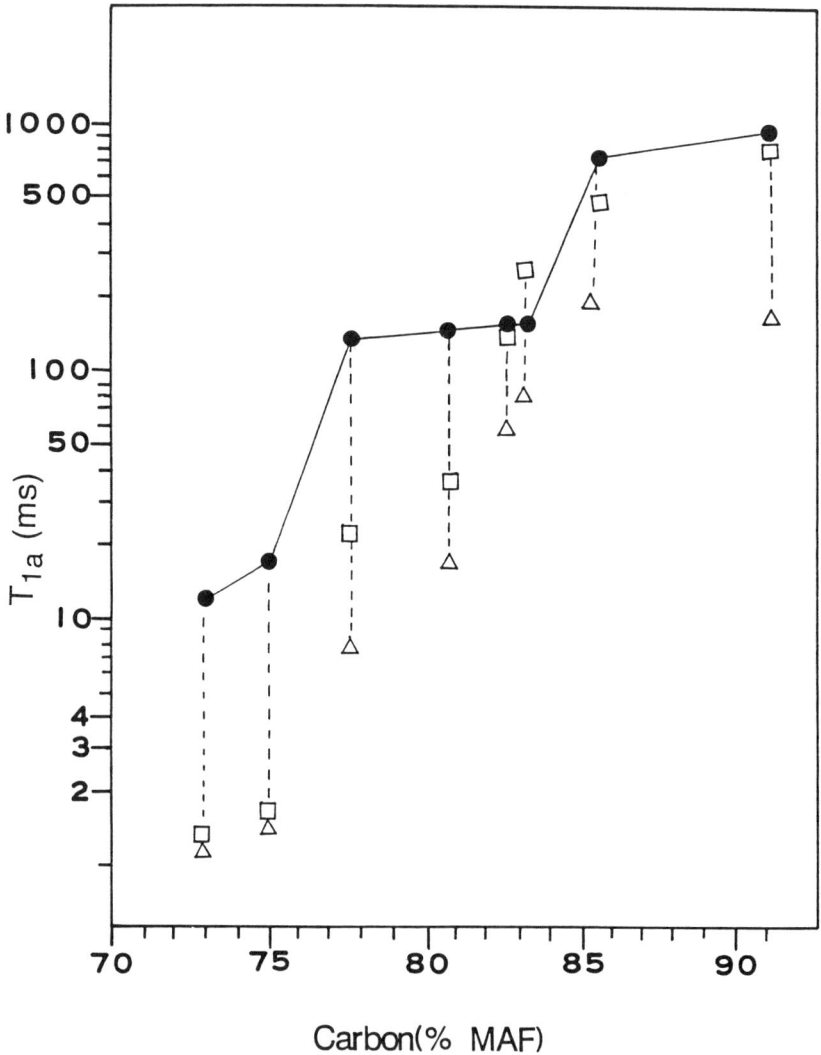

Figure 9. Plots of the average T_{1a} *values vs. percent carbon for samples A (fresh, △), B (degassed, □), and C (dried, ●).*

Similarly, T_{1a} values of the dried and acid-treated coal samples can be plotted against the coal carbon content (Figure 10). As expected, large increases in T_1 values for ND and WY coals occur after the acid treatment. Clearly the dominant relaxation mechanism for these coals is electron–nuclear interactions involving paramagnetic metal ions, which are subsequently removed by acid treatment. Iron was always detected in

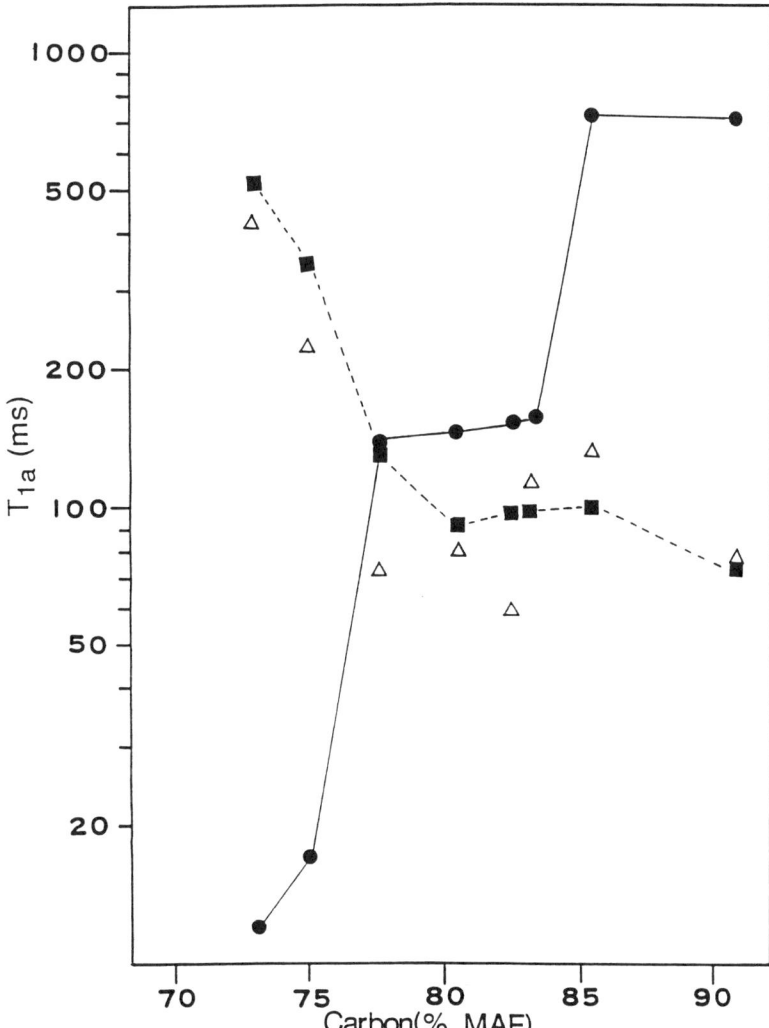

Figure 10. Plots of the average T_{1a} values vs. percent carbon for samples C (dried, ●), E (HCl treated, △), and F (HCl–HF treated, ■).

the filtrates from the first acid-treatment step for all coals, but none was detected after subsequent treatments with HCl–HF. Because the Fe contents of samples E and F are similar, the increase of T_{1a} for sample F of these two coals may be caused by other paramagnetic metals such as Mn, Ni, Cu, or V *(13)*.

For the six higher rank coals, T_{1a} values become shorter after acid treatment. Because the content of Fe and other paramagnetic metal ions decreases upon acid washing, the contribution of paramagnetic ions to the relaxation mechanism is unimportant in higher rank coals.

To confirm these observations concerning the ^{1}H spin–lattice relaxation mechanisms, ESR spectra were measured for aged and dried (same as sample D, but not sealed) and acid-treated and dried (same as sample F, but not sealed) coal samples. Area intensities of the sharp ESR peaks were estimated as shown in Figure 11. For ND and WY coals, a small decrease in spin density was observed after the acid treatment, and the T_{1a} values increased greatly, as shown in Figure 10. Therefore, the ^{1}H relaxation times of ND and WY coals are determined by magnetic interactions with paramagnetic metal ions. The other Argonne Premium coals show an increase in spin density and a decrease in the T_1 values after acid treatment. This result indicates that the main relaxation mechanism involves

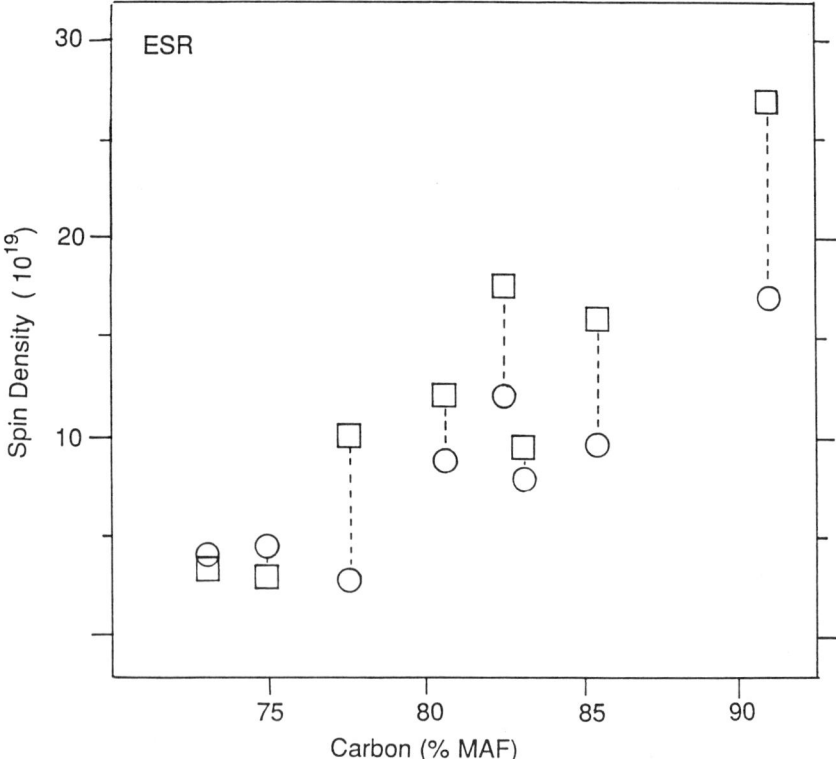

Figure 11. Variation in the ESR free radical spin densities before (○) and after (□) acid treatment.

the organic free radicals in these higher rank coals. For the IL coal, however, a large increase in spin density was observed after acid treatment, whereas T_{1a} values remained nearly the same. This result suggests both paramagnetic metal ions and organic free radicals are effective for ^{1}H relaxation in the IL coal. The increases in radical population with acid treatment in the high-rank coals could be due to changes in the coal structure.

Two factors determine ^{1}H relaxation: (1) the type of magnetic interaction and (2) the degree of molecular mobility. The higher rank coals tend to be more homogeneous and are highly aromatic. Although the spin densities of the organic radicals in these coals are slightly larger than those of the lower rank coals, the molecular structure is more rigid. This rigidity results in the longer relaxation times of the dried UF and POC coals. However, when the two high-rank coals are demineralized, their free radical contents increase considerably, and the ^{1}H relaxation is dominated by magnetic interactions with the free-electron spins.

Because ^{1}H spin–lattice relaxation can be analyzed into two components for most of the samples under study, a multiple-phase coal structure model can be considered. Because similar changes in the T_1 values of these components occur with sample preparation, the major relaxation mechanism may be the result of their different molecular mobilities. It has been proposed (14) that coal is a three-dimensional macromolecular network that forms a "rigid" structure containing macro- and micropores, where many smaller "mobile" molecules exist. This model is consistent with our results for higher rank coals. Because ^{1}H relaxation times for H_2O are similar to those for coal protons (for the lower rank coals), coal structure, as well as the roles of water and paramagnetic ions, must be considered to explain the results obtained for the lower rank coals.

References

1. Gerstein, B. C.; Chow, C.; Pembleton, R. G.; Wilson, R. C. *J. Phys. Chem.* **1977**, *81*, 565.
2. Retcofsky, H. L.; Friedel, R. A. *Fuel* **1968**, *47*, 391.
3. Yokono, T.; Mayazawa, K.; Sanada, Y.; Marshall, H. *Fuel* **1979**, *58*, 896.
4. Lynch, L. J.; Webster, D. S. *J. Magn Reson.* **1980**, *49*, 259.
5. Webster, D. S.; Lynch, L. J. *Fuel* **1981**, *60*, 549.
6. Ripmeester, J. A.; Couture, C.; MacPhee, J. A.; Nandi, B. N. *Fuel* **1984**, *63*, 522.
7. Wind, R. A.; Duijvestijn, M. J.; van der Lugt, C.; Smidt, J.; Vriend, H. *Fuel* **1987**, *66*, 876.
8. Kuriki, Y.; Hayamizu, K.; Yumura, M.; Ohshima, S.; Kawamura, M.; In *1987 International Conference on Coal Science;* Moulijn, J. A.; Nater, K. A.;

Chermin, H. A. G., Eds.; "Coal Science and Technology 11"; Elsevier: Amsterdam, Netherlands, 1987; p 390.
9. Jurkiewicz, A.; Idziak, S.; Pislewski, N. *Fuel* **1987**, *68*, 1066.
10. Barton, W. A.; Lynch, L. J. *Energy Fuels* **1989**, *3*, 402.
11. Wind, R. A.; Jurkiewicz, A.; Maciel, G. E. *Fuel* **1989**, *68*, 1189.
12. Hayamizu, K.; Yanagisawa, M.; Yumura, Y.; Kamiya, K.; Kuriki, Y.; Ohshima, S.; Kawamura, M. In *1989 International Conference on Coal Science;* New Energy and Industrial Technology Development Organization: Tokyo, Japan, 1989; Vol. 1, p 29.
13. Vorres, K. S. *Energy Fuels* **1990**, *4*, 420.
14. Given, P. H.; Marzec, A.; Barton, W. A.; Lynch, L. J.; Gerstein, B. C. *Fuel* **1986**, *65*, 155.

RECEIVED for review June 7, 1990. ACCEPTED revised manuscript September 25, 1991.

16

High-Field NMR Studies of Argonne Premium Coals

Jiangzhi Hu, Liyun Li, and Chaohui Ye[1]

Laboratory of Magnetic Resonance and Atomic and Molecular Physics, Wuhan Institute of Physics, Academia Sinica, Wuhan 430071, Peoples Republic of China

> *Seven Argonne Premium coals were studied by cross-polarization (CP)–combined rotation and multiple-pulse spectroscopy (CRAMPS) in a 9.4-T magnetic field. The apparent carbon aromaticities of the coals obtained at high field via static CP spectra agreed well with those measured at low field, that is, 2.3 T. A detailed discussion of the spectral distortion in the total suppression of sidebands (TOSS) experiment at low spinning rates is given. The experimental evidence presented in this chapter shows that static CP measurements are promising for coal studies.*

SOLID-STATE NMR TECHNIQUES such as cross polarization (CP) and magic-angle spinning (MAS) (*1*), combined rotation and multiple-pulse spectroscopy (CRAMPS) (*2*), and dynamic nuclear polarization (DNP) (*3*) have been useful tools for the study of solid fossil fuels. These studies have been continually carried out in a relatively low, generally below 4.7 T, magnetic field because NMR measurements of solid fossil-fuel samples in a high field do not gain much in resolution, as would be expected for liquids. Moreover, in a higher magnetic field, a faster sample-spinning rate would be required to meet the so-called rapid-rotation condition (*4*) in order to prevent spinning sidebands (SSB). Therefore, a minimum rate

---

[1] Corresponding author

of 8 kHz at 4.7 T and 16 kHz at 9.4 T is required for MAS studies of coals in order to prevent the first SSB of the aromatic carbons from overlapping the resonances of the aliphatic carbons. These higher spinning rates have not been routinely accessible. Extensive efforts to obtain higher spinning rates have been made, and the fastest rate, about 23 kHz, was achieved recently (5). In addition, MAS probe suppliers (e.g., Doty Scientific, Columbia, SC and Chemagnetics, Fort Collins, CO) currently provide MAS probes that can spin at 10 kHz.

An attractive feature of the high-field experiment is its high sensitivity. The NMR signal-to-noise ratio increases with the 7/4 power of the field strength (6). For example, the sensitivity at 9.4 T is nearly 17 times larger than that at 1.9 T; the latter case involves an 80-MHz proton resonant frequency. In other words, a time-saving factor of 280 would be gained by obtaining the measurements at 9.4 T rather than that at 1.9 T, because the signal-to-noise ratio is proportional to the square of the signal-accumulation numbers.

The apparent ^{13}C aromaticities obtained from cross polarization with magic-angle spinning (CP–MAS) spectroscopy are apparently smaller than those obtained from CP experiments (7, 8). This result is partially because MAS eliminates some weakly dipolar couplings so that the CP mechanism is partially broken in the process. The aromatic portions of coals contain more carbons that are remote from protons than do the aliphatic portions; therefore, MAS reduces the apparent aromaticity in comparison with static CP measurements. Conducting NMR studies of coal at high field with a static-CP approach would be expected to result in an increased detection sensitivity without spinning problems.

We report here the measurements of seven Argonne Premium coals in a 9.4-T magnetic field via CP, CP–MAS, and CRAMPS techniques. Dipolar dephasing (DD) was also used to extract structural parameters. The ^{13}C aromaticities that we obtained in the CP experiment agreed well with those obtained at 2.3 T by the Utah group (9) using CP–MAS. However, the aromaticities obtained with the CP–MAS experiment are significantly smaller than those obtained with CP measurements at high field, and this difference is due to spectral distortion in the total suppression of sidebands (TOSS) technique (10). The distortion will be discussed in this chapter.

Experimental Details

The seven coal samples were obtained from the Argonne Premium Coal Sample Program. All of the coal samples were packed into rotors in a nitrogen environment, and the experimental measurements were performed immediately. The CP, CP–MAS, and CRAMPS spectra were obtained with a Bruker MSL-400 spectrometer with a proton frequency of 400.13 MHz and a ^{13}C frequency of 100.63

MHz. The MAS rate was approximately 4 kHz, and the TOSS sequence was used to eliminate SSB. The 90° pulse width was 4 μs, and the proton-decoupling strength was 64 kHz in both the CP and CP–MAS experiments. In the CRAMPS experiments, the 90° pulse width was 1.95 μs, and the MREV-8 homonuclear-decoupling sequence described in ref. 11 was used. (MREV is an acronym created from the surnames of the originators of the method.) The contact time for the CP and CP–MAS experiments was 1 ms.

We carried out the ^{13}C CP–MAS experiments and extracted the structural parameters of the coals in a manner similar to that described by Solum et al. (9), and the same symbols that they used are also presented in this chapter for convenient comparison.

Results and Discussion

The static CP ^{13}C spectra were obtained within 10 min with a reasonable signal–noise ratio at 9.4-T high-field measurements. The static CP aromaticities were obtained by digital subtraction of the spectrum from a standard sample. The standard sample had been carefully studied at a lower field, and its aromaticity was 0.86. The digital-subtraction procedure was first described by Wind et al. (7). In Figure 1, such a measurement is shown. The ^{13}C spectrum of an Argonne Premium coal (top) was digitally added to the standard spectrum in an inverse-phase mode (middle), and the result is a difference spectrum (bottom). The apparent aromaticity (the fraction of aromatic carbons, f_a) can then be easily calculated from the following equation:

$$f_a = [f_a(0)(S - B - A) + B]/S \quad (1)$$

where $f_a(0)$ is the aromaticity of the standard sample, S is the spectral integration of the sample to be measured, B is the spectral integration of the difference of the aromatic portions, and A is the integration of the difference of the aliphatic portions.

Figure 2 shows the stack plot of the ^{13}C CP spectra of the seven coals. The aromaticities of these coals as determined by CP–MAS–TOSS and CP measurements only are listed in Table I. The relevant values obtained by the Utah group (9) are also included in the table for comparison. The carbon structural parameters that were obtained with the heteronuclear-dephasing and variable-contact-time experiments are shown in Table II.

The proton aromaticities of coals are generally obtained from their CRAMPS spectra, which are digitally decomposed into aromatic and aliphatic portions. Figure 3 shows a stack plot of the CRAMPS spectra of

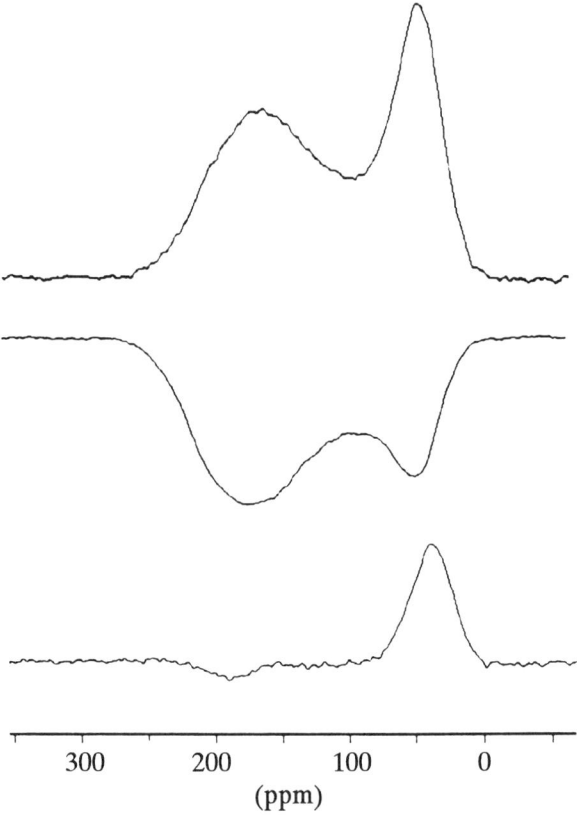

Figure 1. The aromaticity-extraction process using the CP spectrum. The ^{13}C static CP spectrum of a coal (top) was added to the inverse-phase spectrum of the standard sample (middle) to obtain a difference spectrum (bottom). Spectral integration was then used to determine aromaticity.

the seven coals. Their resulting proton aromaticities H_a (CRAMPS) are listed in Table III. Determination of proton aromaticities of coals by digital decomposition of the CRAMPS spectrum is very tricky and can be fraught with errors because the proton spectra of coals are usually poorly resolved. However, the proton aromaticities, can be roughly estimated, as H_a^*, by using the carbon structural parameters in Table II and the following equation:

$$H_a^* = f_a^H / [f_a^H + 3f_{al}^* + 2f_{al}^H + Kf_{al}^O] \tag{2}$$

where K is a parameter between 1 and 3 depending on the structure of

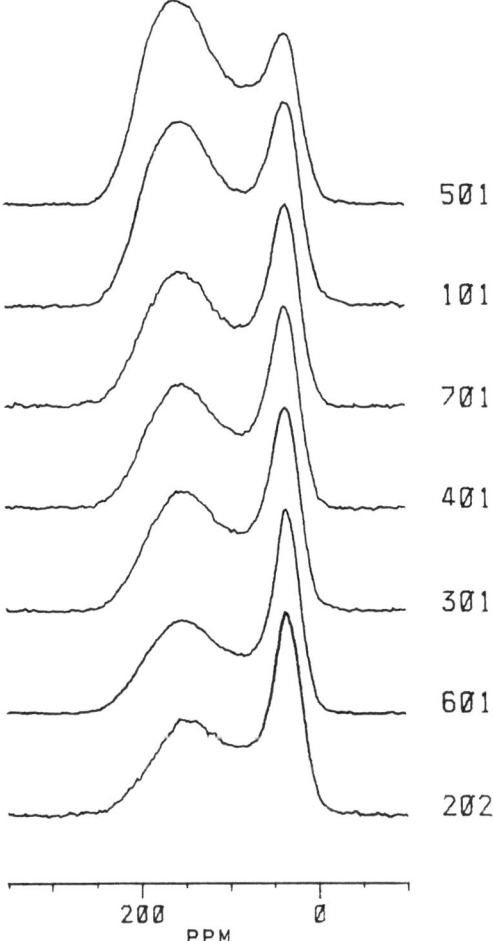

Figure 2. ^{13}C *CP NMR spectra of the seven Argonne Premium coals. The spectra are stacked from high aromaticity (top) to low aromaticity (bottom). The numbers are sample numbers; they are explained in Table I.*

coal, H_a^* is the hydrogen aromaticity from carbon NMR spectra, f_a^H is the fraction of protonated aromatic carbon, f_{al}^H is the fraction of methine or methylene carbon, f_{al}^* is the fraction of methyl or nonprotonated carbon, and f_{al}^O is the fraction of carbon bonded to oxygen. We set the value of K at 1.5. The H_a^* values of the seven coals are also presented in Table III.

^{13}C aromaticity is an important parameter for the characterization of coal structure. Many investigations have been undertaken to determine f_a via ^{13}C NMR spectroscopy and to relate this parameter to the coal rank. The quantitative accuracy of the measured aromaticity value has long been

Table I. Apparent Aromaticities of the Argonne Premium Coals

Sample No.	Coal	Rank	CP–MAS–TOSS	CP	CP–MAS, low field[a]
501	Pocahontas No. 3	LVB	0.77	0.86	0.86
101	Upper Freeport	MVB	0.66	0.80	0.81
701	Lewiston–Stockton	HVB	0.63	0.74	0.75
401	Pittsburgh No. 8	HVB	0.61	0.72	0.72
301	Illinois No. 6	HVB	0.60	0.70	0.72
601	Blind Canyon	HVB	0.53	0.67	0.65
202	Wyodak–Anderson	SB	0.58	0.65	0.63
801	North Dakota	L	0.60	0.65	0.61

NOTE: The error estimate for f_a determination via CP was ±0.03.
ABBREVIATIONS: LVB, low-volatile bituminous; MVB, mid-volatile bituminous; HVB, high-volatile bituminous; SB, subbituminous; and L, lignite.
[a] These f_a values were obtained from reference 9.

Table II. Carbon-Structure Distribution Parameters of the Argonne Premium Coals

Sample	f_a^C	f_a^H	f_a^N	f_a^P	f_a^S	f_a^B	f_{al}	f_{al}^H	f_{al}^*	f_{al}^O
501	0	0.35	0.51	0.03	0.12	0.36	0.14	0.08	0.10	0.0
101	0	0.33	0.47	0.05	0.15	0.27	0.20	0.08	0.10	0.02
701	0	0.28	0.46	0.07	0.13	0.26	0.26	0.14	0.11	0.01
401	0	0.27	0.45	0.07	0.14	0.24	0.28	0.14	0.11	0.03
301	0.01	0.21	0.48	0.08	0.12	0.28	0.30	0.15	0.12	0.03
601	0.01	0.23	0.43	0.09	0.13	0.21	0.33	0.18	0.13	0.02
202	0.04	0.18	0.43	0.10	0.18	0.15	0.35	0.22	0.10	0.02

NOTE: The error estimates are as follows: f_a^P and f_a^S, ±0.01; f_{al}^H and f_{al}^*, 0.02; $f_a^C, f_a^H, f_a^N, f_a^B, f_{al}$, and f_{al}^O, ±0.03.
ABBREVIATIONS: f_a^C, fraction of aromatic carbon; f_a^N, fraction of nonprotonated, aromatic carbon; f_a^P, fraction of phenolic or phenolic ether carbon; f_a^S, fraction of alkylated aromatic carbon; f_a^B, fraction of aromatic bridgehead carbon, and f_a^H, fraction of aliphatic carbon.

a subject of debate (13). Significant errors can arise in the measurements of aromaticity by CP–MAS ^{13}C NMR spectroscopy due to spin dynamics in coals in the rotating frame where the CP processes that occur at the aromatic and aliphatic portions are inhomogeneous. In addition, the paramagnetic centers in coals create unequal relaxation effects on the spins of

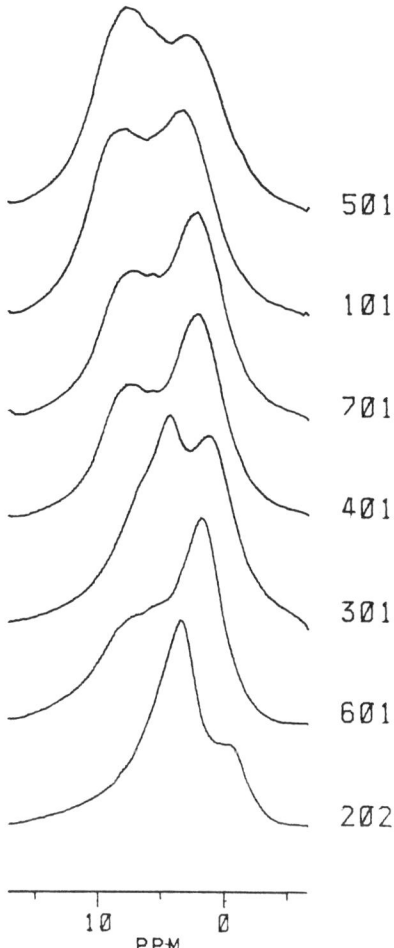

Figure 3. ^{1}H CRAMPS NMR spectra of the seven Argonne Premium coals. The series sequence of the stack plot is the same as that in Figure 2.

the aromatic and aliphatic portions, and the concentration of the paramagnetic centers varies with coal rank.

As shown in Table I, the aromaticities obtained by CP only at 9.4 T are comparable to those obtained at 2.3 T (9). However, the CP–MAS–TOSS values are much lower. As mentioned by Snape et al. (13), and Botto and Axelson (14), in some cases the values of aromaticity obtained with the TOSS sequence are almost identical to those obtained at a low magnetic field without TOSS. In our case, the significant differences be-

Table III. Proton Aromaticity
of the Argonne Premium Coals

Sample	H_a (CRAMPS)	H_a^* (CP–MAS)
501	0.55 ± 0.04	0.48 ± 0.08
101	0.45 ± 0.03	0.39 ± 0.05
701	0.42 ± 0.04	0.30 ± 0.05
401	0.40 ± 0.04	0.28 ± 0.04
301	0.33 ± 0.06	0.22 ± 0.03
601	0.27 ± 0.03	0.22 ± 0.03
202	0.26 ± 0.03	0.23 ± 0.03

tween CP and CP–MAS–TOSS aromaticities are mainly due to the spectral distortion that is inherent in the TOSS technique when a low spinning rate is used. We will discuss some of our analytical results in the following paragraphs.

The TOSS sequence (10) can, in principal, partially restore the spectral intensity of centerbands. Olejniczak et al. (15) determined that the loss of the centerband intensity occurs when the spinning speed is slow with respect to the chemical-shift anisotropy. Recently, Raleigh et al. (16) discussed the sideband suppression experiment. We (17) discussed the intensities of spinning sidebands for inhomogeneous interactions in rotating solids via rotating echoes. Similarly, the centerband intensity can be calculated as a function of the ratio of chemical anisotropy $\omega_0 \delta$ to spinning rate ω_r. The calculations will not be discussed in detail here; however, the equation that describes the restoration of centerband intensity with the TOSS sequence by the rotating-echo intensity ratio $I_T(0)/I_0(0)$ is as follows:

$$\frac{I_T(0)}{I_0(0)} = \frac{\sum_l \int_0^{2\pi} d\gamma \int_0^{\pi} J_{-2l}(A) J_l(B) \exp[i2l(\phi_2 - \phi_1)] \sin\beta \, d\beta}{\int_0^{2\pi} d\gamma \int_0^{\pi} |\sum_k J_k(B) J_{-2k}(A) \exp[i2k(\phi_2 - \phi_1)]|^2 \sin\beta \, d\beta} \quad (3)$$

In eq 3, $I_T(0)$ is the intensity of the centerband under the TOSS experiment, and $I_0(0)$ is the centerband intensity under MAS only. The variable J is the Bessel function of the first kind, and all the parameters in equation 3 were defined in reference 17. The integrations represent powder averaging over the whole spin system, the powder molecules having random orientation in space with an equal probability in each unit solid angle.

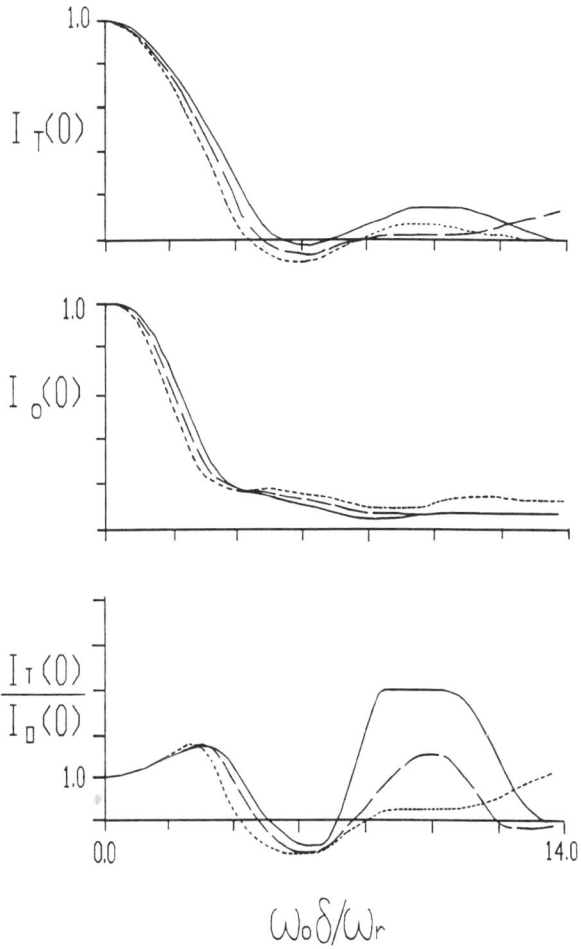

Figure 4. The intensity of the centerband vs. the ratio between chemical anisotropy and spinning rate. The values of η are as follows: 0, solid line; 0.5, dashed line; 1, dotted line.

In Figure 4, plots of $I_T(0)$, $I_0(0)$, and $I_T(0)/I_0(0)$ versus $\omega_0\delta/\omega_r$, with respect to various asymmetric parameters, η, are shown. The relative intensity of the centerband $I_0(0)$ drops rapidly when the relative inverse spinning rate, $\omega_0\delta/\omega_r$, increases, that is, when the spinning rate decreases. However, $I_0(0)$ is always positive when $\omega_0\delta/\omega_r$ changes. On the other hand, $I_T(0)$ (the centerband intensity with TOSS) decreases more slowly than $I_0(0)$ does when $\omega_0\delta/\omega_r$ is less than 4. This means that the TOSS sequence restores to some extent the centerband intensity in comparison with the MAS spectrum. The effect has been predicted in the original article (*10*). The value of $I_T(0)$ becomes negative when $\omega_0\delta/\omega_r$ reaches a point between 4 and 6 with respect to the value of η. This condition makes the use of the TOSS approach very critical, especially for coal studies when the static magnetic field is high and the spinning rate is relatively low.

In order to visualize the complicated TOSS restoration function in regard to MAS, the curves of $I_T(0)/I_0(0)$ versus $\omega_0\delta/\omega_r$ are shown in Figure 4. The ratio increases, a result indicating that the TOSS spectral restoration of the centerband occurs only at a relatively high spinning rate (i.e., the safest region for spectral restoration by TOSS is that in which $\omega_0\delta/\omega_r$ is less than 3). Beyond this region when the spinning rate is lower, the ratio exhibits a funny behavior. Therefore, severe spectral distortion will appear for those samples that cover a wide range of chemical shifts. The matter is complicated even more by the dependence of the chemical shift on the asymmetric parameter.

In our study, $\omega_0\delta/\omega_r$ is less than 4 for the aromatic portions of coal when $\omega_r = 4$ kHz at 9.4-T high field, and the aromatic portions of the coal do not meet the spectral-restoration condition, but the aliphatic ones do. As a result, the apparent aromaticities obtained from the TOSS measurements appear substantially lower than those obtained from static CP.

We conclude that the ratio of $\omega_0\delta/\omega_r$ equal to 3 represents a criterion of magnetic-field strength and spinning rate for the quantitative use of TOSS. The requirements of a high spinning rate to alleviate the TOSS spectral distortion may create a problem because a faster spinning rate eliminates the CP mechanism to a greater extent and hence also affects the measured aromaticity. In summary, from the evidence presented in this chapter, we suggest that a combination of a static CP experiment with a high magnetic-field strength be used for aromaticity measurements of coals.

The carbon-structure distribution parameters listed in Table II were also determined in a manner similar to that described by Solum et al. (*9*), and the symbols that they used are also presented for convenient comparison. In our CP–MAS–TOSS data extraction, the intensity ratio of the aromatic portion to the aliphatic portion was corrected in accordance with the static CP experiments. Therefore, the intense distortion from the TOSS approach in a high magnetic field was eliminated. The results listed in Table II are comparable to those of Solum et al. (*9*), and this fact again supports our static CP experiments in a high magnetic field.

The proton aromaticities, H_a, obtained from the CRAMPS spectra are comparable to the H_a^* values (estimated by the carbon parameters with eq 2, as shown in Table III). Hence, eq 2 may provide a means to indirectly measure proton aromaticity. However, more relevant comparisons should be done with the estimation approach because of the usually tricky decomposition of the CRAMPS spectrum, as mentioned previously.

DNP (*18*) has been proven to be a powerful approach for the study of coal. DNP enhances the NMR signal by irradiating at or near the electron Larmor frequency, and this aspect of DNP makes the measurements in a low magnetic field much easier. DNP measurements also provide useful information on electronic structures in solid coals. A homemade 82 MHz–54 GHz DNP spectrometer was recently developed in our laboratory. In Figure 5, the ^{13}C spectra of an anthracite with an aromaticity of 0.95 are shown. With a sample volume of 0.1 mL, the ^{13}C CP spectrum (bottom trace) was obtained with 26,000 scans in approximately 7.3 h with a 1-s recycling time. Because the proton signal had been enhanced by DNP, the ^{13}C DNP–CP spectrum (middle trace) required only 1000 scans and approximately 14 min, and a slightly better signal-to-noise ratio was obtained. The ^{13}C signal can also be directly enhanced by DNP. The ^{13}C

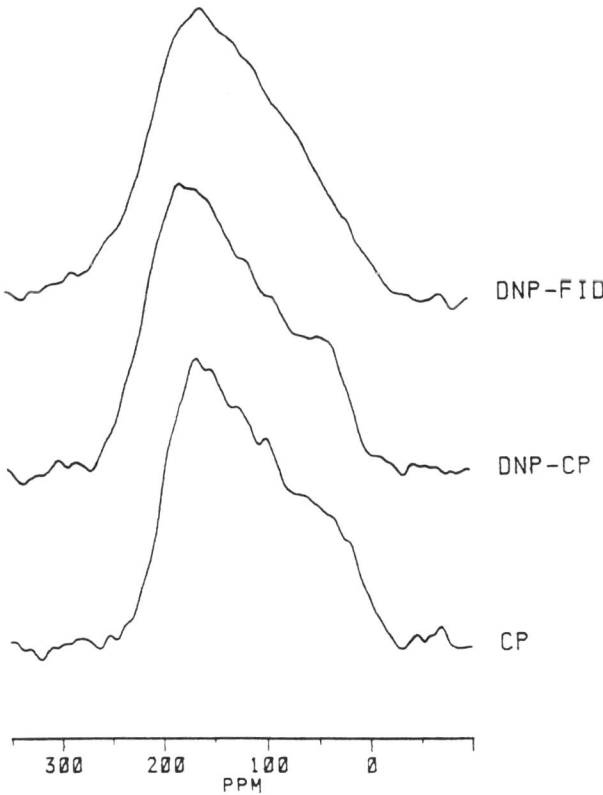

Figure 5. ^{13}C *spectra obtained with a 1.9-T magnetic field of an anthracite. The aliphatic portion in the spectrum was largely missed because the paramagnetic centers were mainly situated near the aromatic portions in the solid coal.*

DNP spectrum (top trace) was obtained with 12 scans in 6 min with a recycling time of 30 s. The substantial difference between the DNP and DNP–CP spectra reveals that there are more unpaired electrons located near the aromatic portions of the coal because the enhancement of the aromatic portion was much greater than that of the aliphatic portion. We are currently studying the Argonne Premium Coals with DNP.

Acknowledgment

This study was supported by the National Natural Science Foundation of China. Karl S. Vorres provided the coal samples. X. Zhang helped with the preparation of this manuscript.

References

1. Schaefer, J.; Stejskal, E. O. *J. Am. Chem. Soc.* **1976**, *98*, 1031.
2. Gerstein, B. C.; Chou, C.; Pembleton, R. G.; Wilson, R. C. *J. Phys. Chem.* **1977**, *81*, 565.
3. Wind, R. A.; Trommel, J.; Smidt, J. *Fuel* **1979**, *58*, 900.
4. Maricq, M.; Waugh J. S. *J. Chem. Phys.* **1979**, *70*, 3300.
5. Dec, S. F.; Wind, R. A.; Maciel, G. E.; Anthonio, F. E. *J. Magn. Reson.* **1986**, *70*, 355.
6. Hoult, D. I.; Richards, R. E. *J. Magn. Reson.* **1976**, *24*, 71.
7. Wind, R. A.; Duijvestijn, M. J.; van der Lugt, C.; Smidt, J.; Vriend, J. *Fuel*, **1987**, *66*, 876.
8. Ye, C.; Wind, R. A.; Maciel, G. E. *Sci. China, Ser. A* **1988**, *29*, 968.
9. Solum, M. S.; Pugmire, R. J.; Grant, D. M. *Energy Fuels* **1989**, *3*, 187.
10. Dixon, W. T. *J. Chem. Phys.* **1982**, *77*, 1800.
11. Gerstein, B. C.; Dybowski, C. R. *Transient Techniques in NMR in Solid;* Academic: New York, 1985.
12. Bronnimann, C. E.; Zeigler, R. C.; Maciel, G. E. *J. Am. Chem. Soc.* **1988**, *110*, 2023.
13. Snape, C. E.; Axelson, D. E; Botto, R. E; Delpuech, J. J.; Tekeley, F.; Gerstein, B. C.; Fruski, M.; Maciel, G. E.; Wilson, M. A. *Fuel* **1989**, *68*, 547.
14. Botto, R. E.; Axelson, D. E. *Prepr. Fuel Chem. Div. Am. Chem. Soc.* **1988**, *33*(3), 50.
15. Olejniczak, E. T.; Vega, S.; Griffin, R. G. *J. Chem. Phys.* **1984**, *81*, 4804.
16. Raleigh, D. P.; Olejniczak, E. T.; Griffin, R. G. *J. Chem. Phys.* **1988**, *89*, 1333.
17. Ye, C.; Sun, B.; Maciel, G. E. *J. Magn. Reson.* **1986**, *70*, 241.
18. Wind, R. A.; Duijvestijn, M. J.; van der Lugt, C.; Manenschijn, A.; Vriend, J. *Prog. Nucl. Magn. Reson. Spectrosc.* **1985**, *17*, 33.

RECEIVED for review June 8, 1990. ACCEPTED revised manuscript December 17, 1990.

17

Solid-State ^{13}C NMR Studies on Coal and Coal Oxidation

J. Anthony MacPhee[1], Hiroyuki Kawashima[2], Yasumasa Yamashita[2], and Yoshio Yamada[2]

[1]CANMET Energy Research Laboratories, c/o 555 Booth Street, Ottawa, Ontario, K1A 0G1, Canada
[2]National Institute for Resources and Environment, Onogawa 16-3, Tsukuba, Ibaraki 305, Japan

> *Conventional and dipolar-dephasing ^{13}C NMR spectra with cross-polarization and magic-angle spinning (CP–MAS) were recorded for the eight Argonne Premium coal samples at 300 MHz using a Bruker AC 300 spectrometer with 4-kHz sample spinning. Measured apparent aromaticities correlate well with the atomic H/C ratio (r = 0.96). Certain of these coals were subjected to oxidation in air at 100 °C for 168 h. For the fresh coals, the aromatic carbons exhibit a two-component (Gaussian–Lorentzian) decay in the dipolar dephasing experiment. The oxidized coals, in the range of dipolar dephasing delays investigated, exhibit only a Gaussian decay. Aliphatic carbons of both fresh and oxidized coals exhibit only a single Gaussian component. This behavior is explained in terms of the increased radical concentration in the oxidized coals. Studies of the dipolar dephasing behavior of model aromatic compounds (e.g., anthracene and naphthalene) were carried out, and the results of these help to interpret the results obtained for oxidized coals.*

THE SYSTEMATIC STUDY OF COAL STRUCTURE and chemistry has, in the past, been thwarted by the difficulty of obtaining representative and pristine samples upon which to perform experiments. Results obtained in one laboratory often prove difficult to verify in another because of the unavailability of identical samples. The laudable effort of the Argonne Premium Sample Bank has changed this situation and will undoubtedly allow a qualitative step forward in coal science in the coming decade.

We studied the eight Argonne Premium coal samples by means of ^{13}C cross-polarization–magic-angle spinning (CP–MAS) solid-state NMR spectroscopy with several objectives in mind. The first was to assess the aromaticity data obtained at high field (300 MHz) and moderate sample spinning rate (4 kHz) to allow comparisons with data from other laboratories both under rigorously identical conditions and at higher spinning rates. Our second objective was to examine, within the limits imposed by these conditions, the dipolar-dephasing behavior of these coals; and the third objective was to examine oxidized samples of some of these coals in order to gain some information on the mechanism of coal oxidation. There remain many caveats concerning the interpretation of ^{13}C CP–MAS solid-state NMR spectra of coals (*1*). A consensus will undoubtedly be long in coming.

Experimental Details

Coal Samples. The coal samples used were the eight samples provided by the Argonne Premium Coal Sample Program. These samples were provided as 100-mesh powders and were examined without further crushing. Care was taken with the fresh samples to avoid undue exposure to air by running the spectra as quickly as possible after opening the sample vials. Four samples were subjected to oxidation at 100 °C for 168 h. The analytical data for the oxidized samples as well as data relevant to the fresh coals are presented in Table I. Complete analytical data are available for the Argonne Premium coal samples elsewhere (*2*) and will not be included here.

Model Compounds. In addition to the coal samples, the dipolar-dephasing behavior of two model compounds, anthracene and naphthalene, was investigated in order to aid in the interpretation of the results for the coals. Because impurities seem to play a role in the relaxation behavior of such compounds, ultrapure samples of zone-refined compounds, obtained from the Tokyo Kasei Kogyo Co. Ltd., were used. Oxidized samples of anthracene and naphthalene were prepared by heating the samples in a sealed tube with air at 100 °C for 168 h. Spectra were obtained with a contact time of 2 ms, a repeat time of 120 s, and an accumulation of 100 scans.

Table I. Analytical Data for Coal Samples

Coal No.	Coal Name	C^a	H^a	O^b	Spins/g ($\times 10^{19}$)
101 fr	Upper Freeport	85.50	4.70	5.70	2.6
101 ox		82.33	4.43	8.57	4.1
202 fr	Wyodak–Anderson	75.01	5.35	19.28	3.8
202 ox		66.22	3.67	28.09	1.0
301 fr	Illinois No. 6	77.67	5.00	11.27	2.9
301 ox		69.26	4.03	18.92	1.3
501 fr	Pocahontas No. 3	91.05	4.44	3.44	3.9
501 ox		87.18	3.98	7.04	8.5

NOTE: All values are given as moisture- and ash-free (MAF) weight percents. ABBREVIATIONS: fr, fresh; and ox, oxidized.
[a] C and H analyses for the fresh coals were obtained from the Argonne Premium Coal Sample Program User's Handbook.
[b] All O analyses were determined directly.

Electron Spin Resonance Experiments. Electron spin resonance (ESR) experiments were carried out at room temperature with a JEOL JES-FE1X spectrometer. Spin concentrations were measured in vacuo, and 1,1-diphenyl-2-picrylhydrazyl (DPPH) was used as a calibrant.

NMR Experiments. The ^{13}C CP–MAS solid-state NMR spectra were measured on a Bruker AC 300 spectrometer at 75.46 MHz. A Bruker double air-bearing–magic-angle-spinning solid-state probe was used. Ceramic spinners with an internal volume of ~250 mL were used at a spinning rate of ~4 kHz. For the CP experiments, the following operating parameters were used: a spectral width of 30 kHz, a 90° proton pulse of 5 μs, an acquisition time of 30 ms, a pulse repetition time of 4 s, and an accumulation of 1000 scans. The interferogram was multiplied by a sinebell window function, supplied by Bruker software, before Fourier transformation of the data. (The sinebell window function is an exponential function to remove the noise of NMR signals.) Apparent aromaticities, f_a', were determined by using integrated signal intensities for aromatic and aliphatic carbons. For purposes of integration, the base line was corrected by using standard software provided by Bruker. The treatment of spinning side bands (SSB) is a bit more involved and is discussed in detail in the following section. Variable-contact-time experiments were carried out with contact times between 0.5 and 5 ms. Hartmann–Hahn conditions and the magic angle were adjusted with a glycine sample.

Dipolar-dephasing experiments were carried out by using the pulse sequence of Alla and Lipmaa, with a 180° pulse on each nucleus in the middle of the dipolar-dephasing period in order to remove linear-phase distortion (*3*). This dipolar-dephasing period is of particular importance at the low spinning rates used in this study. Dipolar-dephasing times up to 80 µs were used with a contact time of 2 ms and a repeat time of 3 s. We considered it unreliable, under the present circumstances, to use dipolar-dephasing times longer than ~80 µs because of the difficulty of establishing a base line for integration.

Results and Discussion

Aromaticity Determination. Calculating the aromaticity of a coal from a ^{13}C CP–MAS spectrum in the presence of significant side bands is not a trivial matter, as is shown in Figure 1a, where the spectrum of coal 501 is given. Slow spinning (4 kHz) and high field (300 MHz) produce large SSBs, and the result is a loss of intensity of the peak corresponding to sp^2 carbons. The integrated intensity of these side bands must be determined and added to the intensity of the sp^2 carbon band. To obtain the intensity of the sp^3 band, any intensity arising from the overlapping sp^2 SSB must be subtracted. These values are then used to obtain the apparent aromaticity, f_a'. The manner in which this is done depends on the how the side bands are distributed with respect to the main peak from which they are derived, but a priori, this is not known; therefore model compounds must be observed. The ^{13}C CP–MAS NMR spectrum of anthracene is shown in Figure 1b. The anthracene bands are labeled a, b, c, d, and e. In the coal spectrum (Figure 1a) all of these peaks can be integrated except that corresponding to sideband e. Because the intensities of peaks a and e are approximately equal in the anthracene spectrum, we have chosen to calculate the apparent aromaticity, f_a', assuming this to be the case for coal. In another study (*4*), the authors opted to correct for SSB in a slightly different manner based on the side-band distribution of coronene. Their method gives slightly higher aromaticities because they assume that SSBs d and e are equivalent in the coal spectrum. To check the validity of our approach we have obtained a single spectrum at 100 MHz with a 4-kHz spinning rate in which the SSBs were not a problem. For coal 301 at a cross-polarization time of 5 ms, we obtained an f_a' value of 0.69 (this value is consistent with the data shown in Figure 3a). The f_a' values obtained in this way are reproducible to <0.02.

Cross-Polarization Times. Spectra corresponding to cross-polarization times from 0.5 to 5 ms were recorded for all eight fresh coals and for the four oxidized samples prepared for this study. The variations

17. MACPHEE ET AL. *Solid-State ^{13}C NMR Studies* 327

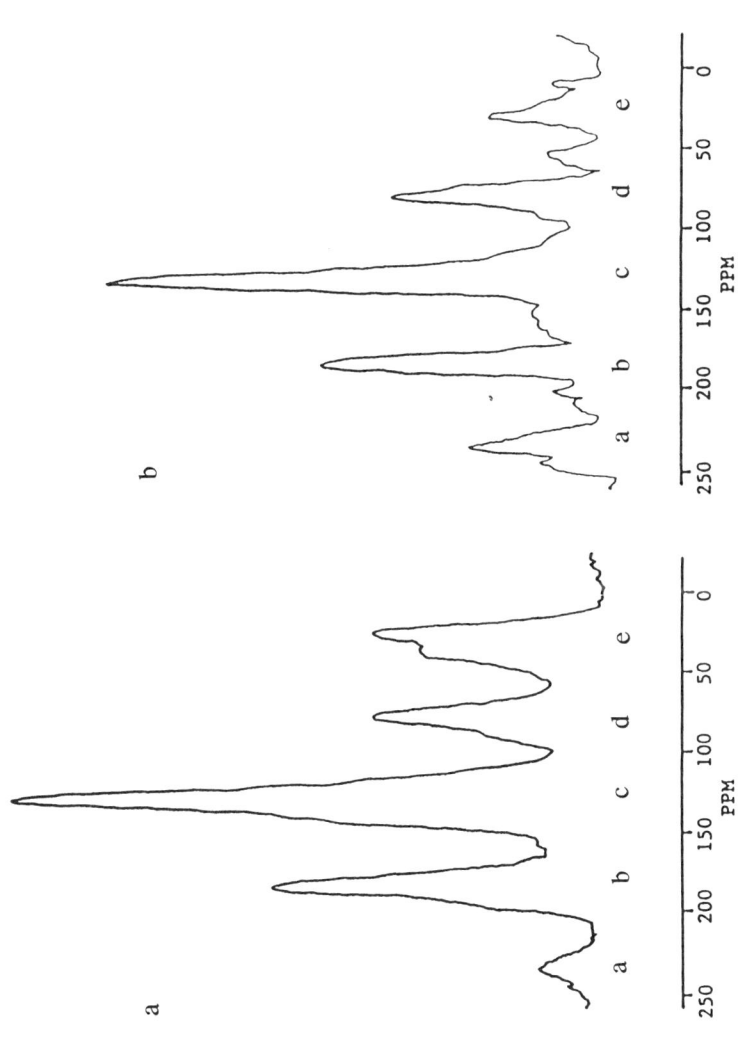

Figure 1. ^{13}C CP–MAS NMR spectra of coal 501 (Pocahontas No. 3) (a) and of zone-refined anthracene (b).

of integrated intensities for the aromatic and aliphatic peaks of two fresh and the corresponding oxidized coals are shown in Figure 2. The maximum in aliphatic intensity does not, in general, correspond to the maximum in aromatic intensity. Because the aromaticity is determined at a particular cross-polarization value, this difference in maxima creates the problem of choosing a meaningful cross-polarization time for a given set of spectral parameters. These difficulties have recently been discussed in some detail without arriving at a consensus (5).

The cross-polarization dynamics are a function of polarization-transfer times (T_{CH}) and the spin relaxation in the rotating frame of the

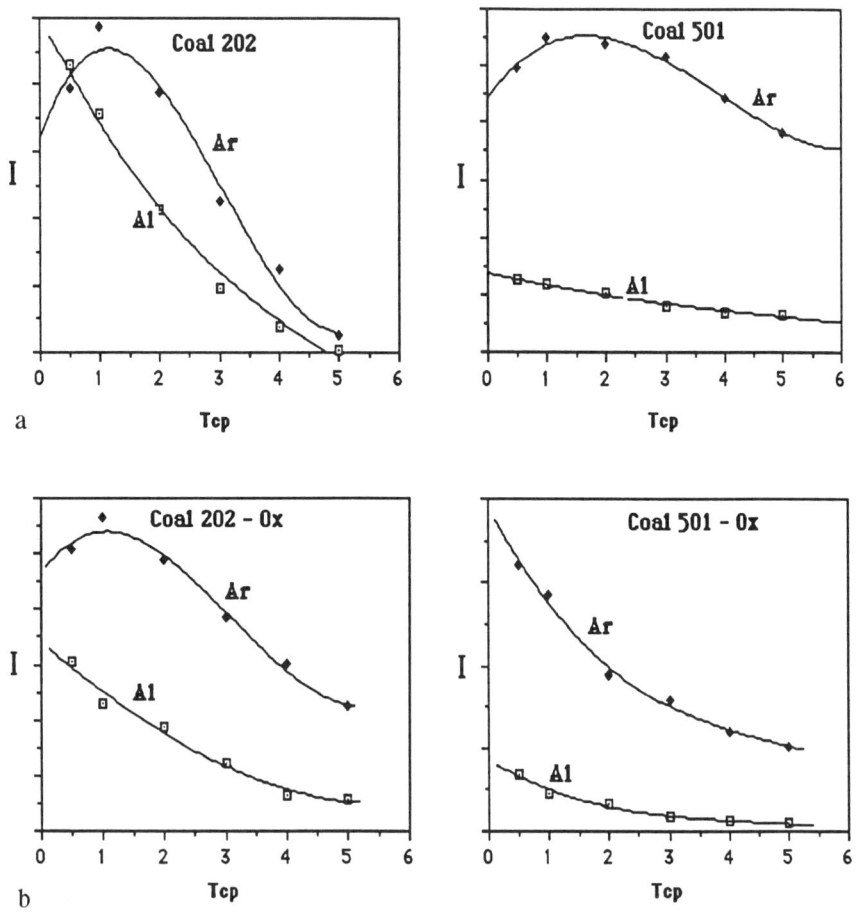

Figure 2. Typical plots of integrated intensity (arbitrary units) vs. cross-polarization time (milliseconds) for fresh (a) and oxidized (100 °C, 168 h in air) (b) coals.

proton reservoir ($T_{1\rho}^H$). The aliphatic and aromatic intensity data may be approximated (6) by the product of two exponential functions:

$$M_t = M_0 \exp(-t_{cp}/T_{1\rho}^H)[1 - \exp(bt_{cp}/T_{CH})] \quad (1)$$

where M_t is an observed intensity, M_0 is the initial amplitude for the carbon polarization, and $b = 1 - T_{CH}/T_{1\rho}^H$.

For all samples the value of $T_{1\rho}^H$ can be determined by fitting an exponential decay to the tail of the magnetization curve, where only $T_{1\rho}^H$ is operating. This value is then substituted into eq 1 as a constant, and the equation is then solved for the remaining parameters. The cross-polarization data reported here do not allow an accurate determination of T_{CH} for all the coals examined because in some cases the maximum value of the magnetization occurs at t_{cp} values shorter than 0.5 ms. Table II lists the values for T_{CH} and $T_{1\rho}^H$ that were calculated from the data shown in Figure 2.

Table II. Values for Polarization-Transfer (T_{CH}) and Proton Rotating-Frame Spin–Lattice Relaxation ($T_{1\rho}^H$) Times Calculated from Equation 1

Coal No.	Aromatic		Aliphatic	
	$T_{1\rho}^H$	T_{CH}	$T_{1\rho}^H$	T_{CH}
101 fr	6.62	0.377	6.74	0.236
101 ox	3.79	0.251	3.73	—
202 fr	3.18	0.437	3.92	—
202 ox	5.06	0.341	4.61	—
301 fr	5.20	0.417	4.78	0.185
301 ox	13.01	0.280	7.65	0.135
401 fr	7.55	0.371	6.40	0.200
501 fr	10.80	0.286	6.09	0.740
501 ox	3.85	—	2.21	—
601 fr	5.12	0.417	4.12	0.282
701 fr	4.85	0.372	4.75	1.09
801 fr	4.12	0.357	3.49	0.077

NOTE: All values are given in milliseconds.
ABBREVIATIONS: fr, fresh; and ox, oxidized.

Aromaticities. In Figure 3a, the f_a' values for the eight Argonne samples are given as a function of cross-polarization time; similar data for the oxidized coal samples are shown in Figure 3b. Some scatter in the data is evident. This scatter is probably due to the combination of high field (300 MHz) and relatively slow spinning rate (4 kHz) used in this study. Arbitrarily, we chose to consider the aromaticities of the coals at a cross-polarization time of 2 ms. This time is near the point of maximum aromatic signal intensity for virtually all of the samples. These aromaticity values are given in Table III along with analytical parameters that are useful for the correlation of these data.

Coals consist of different petrographic constituents (vitrinite, exinite, inertinite) known as *macerals*. Maceral aromaticity decreases in the order inertinite > vitrinite > exinite (7), and the overall aromaticity as meas-

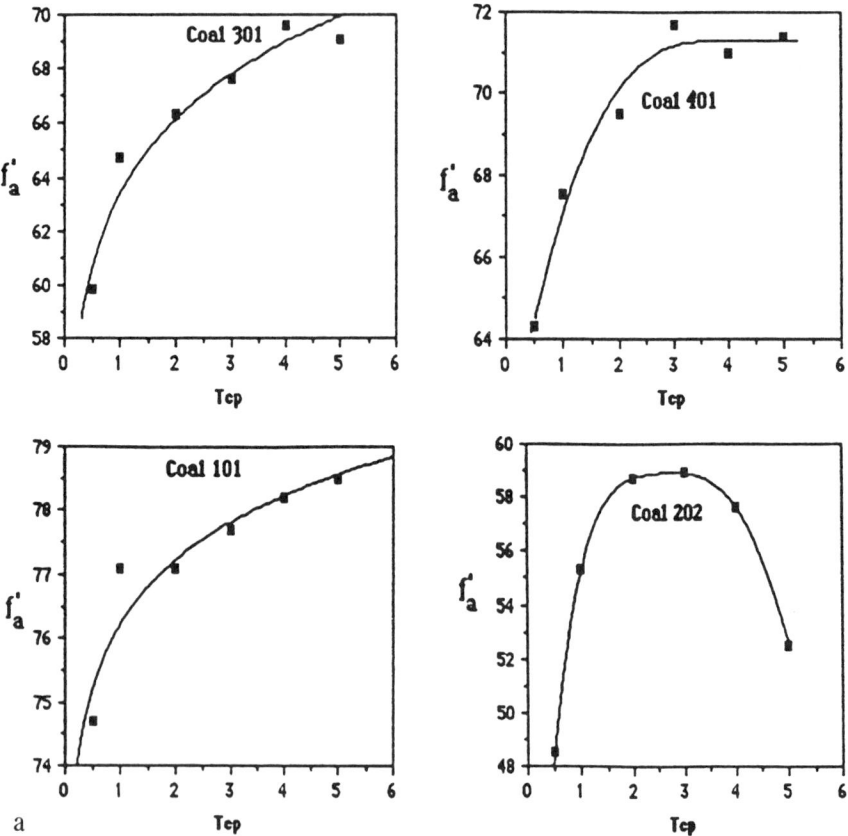

Figure 3a. Relationship between aromaticity and cross-polarization times (milliseconds) for the fresh coals.

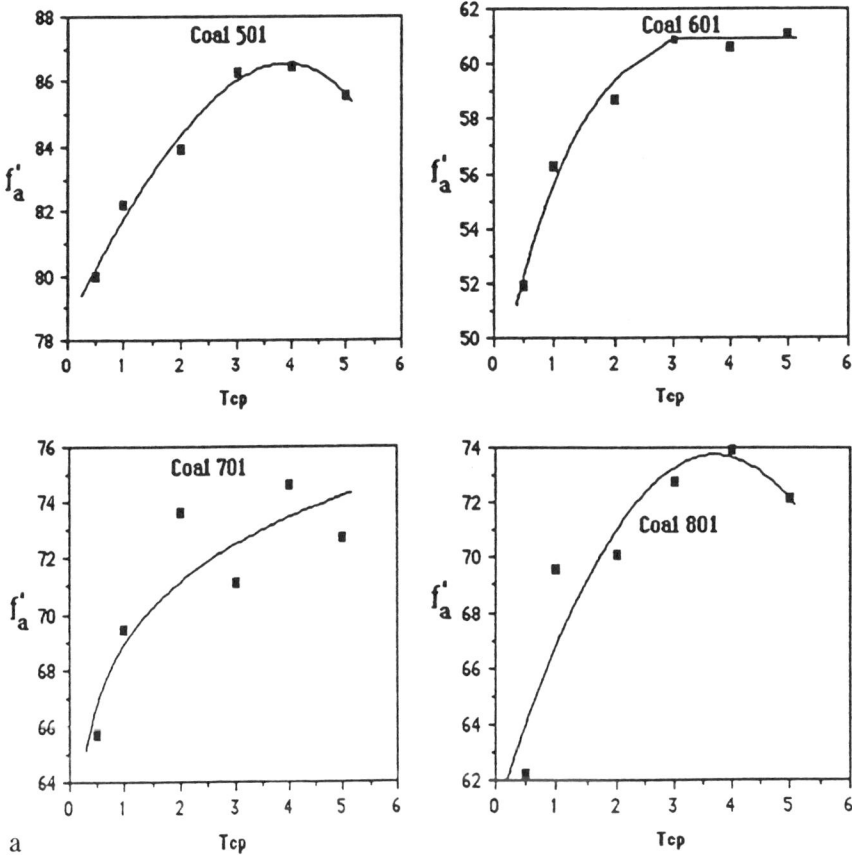

Figure 3a. Continued. Relationship between aromaticity and cross-polarization times (milliseconds) for the fresh coals.

ured here need not necessarily reflect coal rank. The overall aromaticity is the weighted average of the individual maceral aromaticities, which are unknown in this case. Nevertheless, a plot of aromaticity versus the H/C ratio yields a good straight line (Figure 4), a result indicating that one maceral (e.g., vitrinite) predominates in these coals. The straight line shown in Figure 4 was calculated for the fresh coals only. For comparison purposes the four oxidized coals are plotted as well. The points corresponding to three of the oxidized coals follow the correlation established for the fresh coals; that is, the increase in f_a' with oxidation is accompanied by a corresponding decrease in atomic H/C ratio. For one of the coals, Wyodak–Anderson (202), the data point is significantly below the line because the Wyodak–Anderson coal has a higher level of oxidation than the other three higher rank coals. Apparently changes in aro-

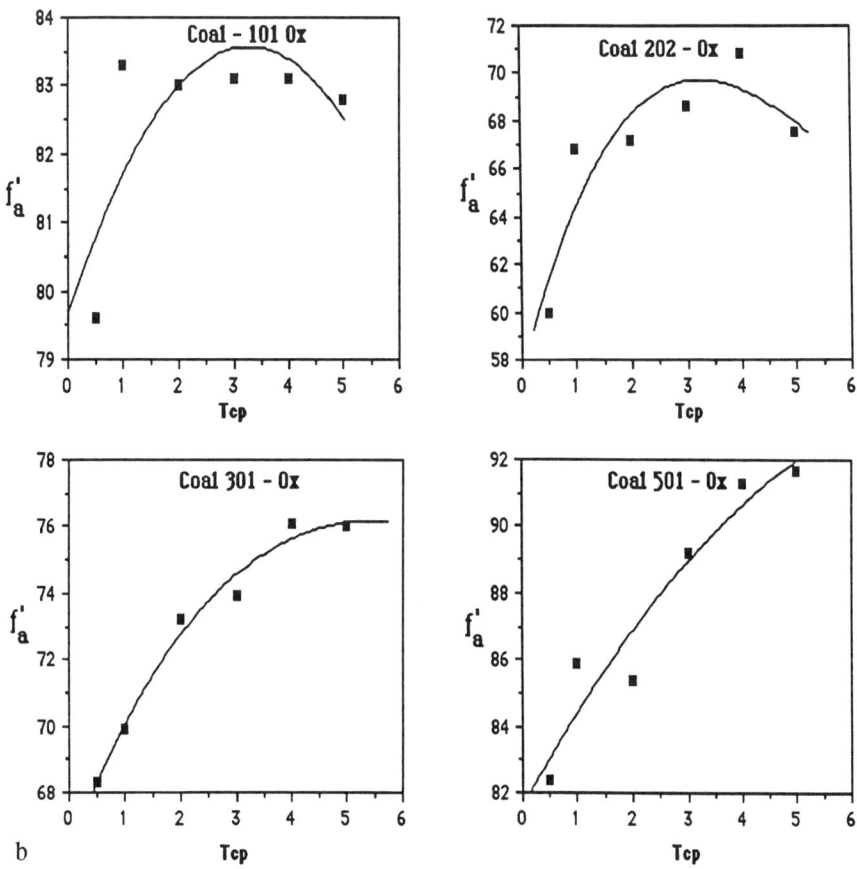

Figure 3b. Relationship between aromaticity and cross-polarization times (milliseconds) for coals oxidized at 100 °C for 168 h in air.

maticity cannot generally be used to detect small changes in the degree of oxidation of coals. A classical plot (8) of H/C versus O/C ratios (Figure 5) seems to be more useful in detecting the oxidized coals in a suite of fresh and oxidized coals. Correlation of f_a' with volatile matter is less successful than the correlation with the H/C ratio (correlation coefficient $r = 0.88$) mainly because of the low-rank coals in the series, and correlation with the O/C ratio exhibits considerable scatter ($R = 0.64$). Multiple correlations with combinations of the parameters shown in Table III do not lead to statistically significant improvements over the correlation shown in Figure 4.

Table III. Apparent Aromaticity and Compositional Data Used in Correlations

Coal No.	f_a'	H/C	O/C (× 100)	VM (daf)[a]
101 fr	0.775	0.655	6.6	31.62
101 ox	0.830	0.631	7.7	14.37
202 fr	0.588	0.850	18.0	49.03
202 ox	0.681	0.621	29.9	24.09
301 fr	0.662	0.767	13.1	47.39
301 ox	0.729	0.663	19.6	21.46
401 fr	0.702	0.762	8.0	41.67
501 fr	0.844	0.581	2.3	19.53
501 ox	0.868	0.546	6.0	12.71
601 fr	0.592	0.850	10.8	48.11
701 fr	0.709	0.758	8.9	37.64
801 fr	0.709	0.789	20.9	49.78

ABBREVIATIONS: fr, fresh; and ox, oxidized.
[a] Volatile matter (VM) is given on a dry, ash-free basis.

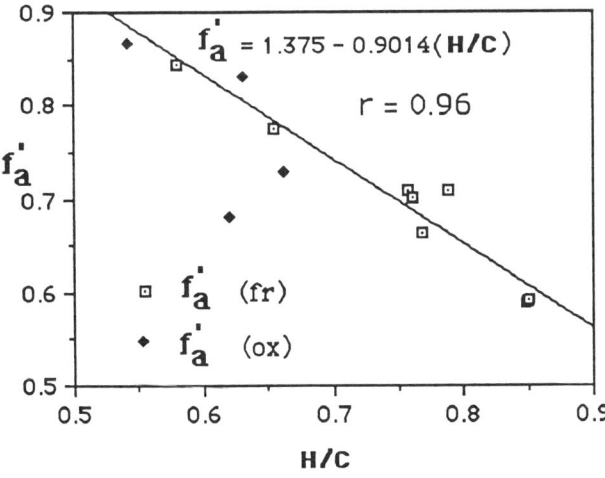

Figure 4. Relationship between aromaticity and H/C ratio for fresh and oxidized coals. The straight line corresponds to fresh coals only.

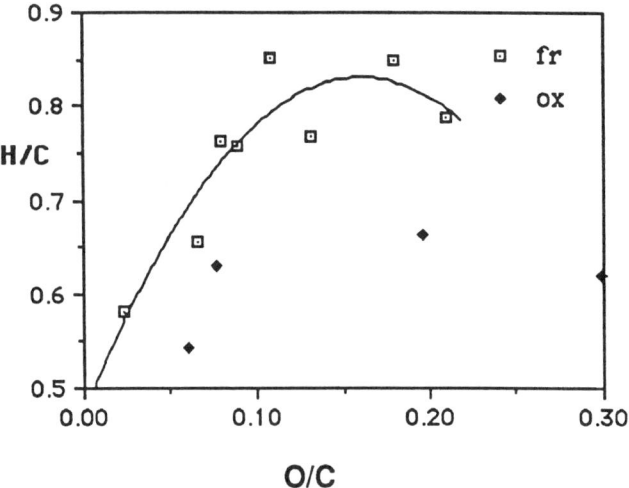

Figure 5. H/C ratio vs. O/C ratio for fresh and oxidized coals. The curve is traced for only the fresh coals.

Dipolar Dephasing. Dipolar-dephasing techniques permit, in principle, a discrimination between different types of carbons. For aromatic carbons, protonated and nonprotonated centers can be distinguished, and for aliphatic carbons, a distinction can be made between quaternary carbons, rotating methyls, and other highly mobile carbons and rigid methyls, methines, and methylenes (9). For coals the situation is not quite as clear as for model compounds. The aromatic component of coals exhibits a rapid (Gaussian) decay followed by a slower (Lorentzian) process (10, 11). The aliphatic component apparently exhibits only Gaussian decay for the delay times used in this study and elsewhere, and as a result scant structural information can be obtained. Dipolar-dephasing experiments were performed on all of the fresh coals and on the four oxidized coals (101, 202, 301, and 501). The results for two of the fresh coals (202 and 501) and the corresponding oxidized coals are shown graphically in Figures 6a and 6b, respectively.

The estimation of the fraction of aromatic carbon that is protonated ($f_a^{a,H}$) is accomplished in an approximate way by using a single dipolar-dephasing delay during which the signal corresponding to protonated carbon is assumed to have disappeared while that corresponding to nonprotonated carbon remains. A more rigorous, and much more time-consuming, way of calculating $f_a^{a,H}$ is by extrapolating the lines corresponding to the slower process (Lorentzian) in Figure 6a to zero time (12). These values are given in Table IV except for cases in which the extrapolated value is equal to or higher than the total aromatic intensity.

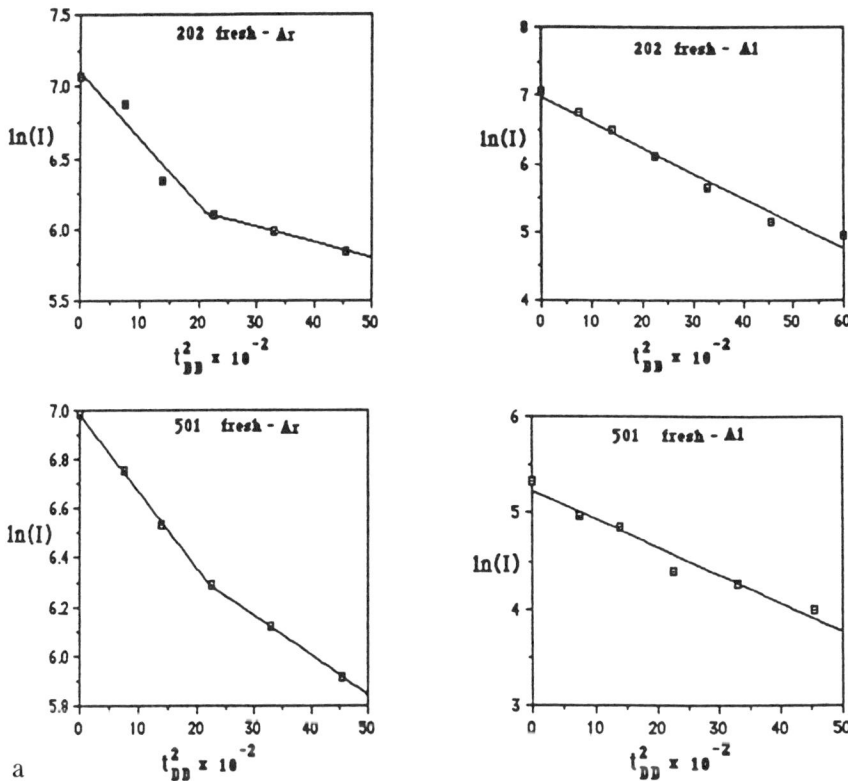

Figure 6a. Plot of ln (I) vs. t_{DD}^2 (microseconds squared) for the aliphatic and aromatic portions of coals 202 and 501 in fresh states.

This approach is valid for model compounds (9) and is assumed to be valid for coals. We were interested in a comparison of the relative amounts of protonated to nonprotonated aromatic carbons as a function of oxidation. Such structural information would have a bearing on the reaction mechanism. Surprisingly, for three of the oxidized coals studied, the slow relaxation process appears to be absent for aromatic carbons up to the 80-μs limit on dipolar-dephasing delays. This result means that a new rapid-relaxation mechanism becomes available to nonprotonated carbons after oxidation. It may also mean that so-called fresh coals may not behave like the model compounds in dipolar dephasing.

Tougne et al. (13) cast doubt on the validity of the dipolar-dephasing approach to the determination of the ratio of protonated to nonprotonated carbons. Even for model compounds (e.g., naphthalene and anthracene, among others) the fraction of nonprotonated carbons could not be estimated. Our work on the oxidized coals has prompted us to consider the effect of mild oxidation on the dipolar-dephasing characteristics

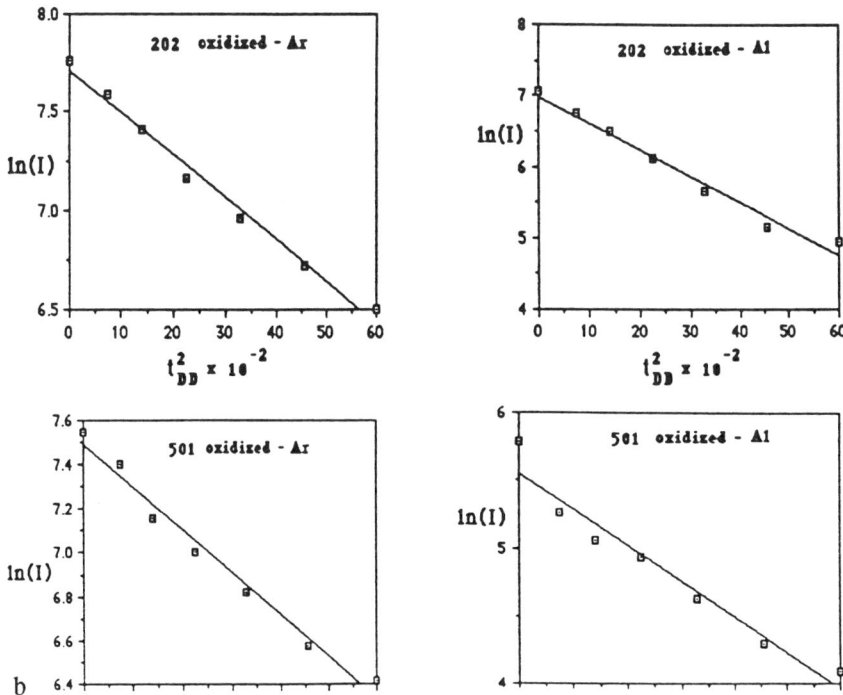

Figure 6b. Plot of ln (I) vs. t_{DD}^2 (microseconds squared) for the aliphatic and aromatic portions of coals 202 and 501 in oxidized states.

of naphthalene and anthracene. We obtained high purity samples of zone-refined compounds, and the samples were oxidized as indicated in the experimental section. The dipolar-dephasing results are shown in Figure 7 for the fresh and oxidized samples. The results parallel those for fresh and oxidized coals, that is, a two-process relaxation for the fresh samples and a single-process relaxation for the oxidized ones. The two-process relaxation allows an estimation of the nonprotonated carbon atom fraction for these compounds. For anthracene a value of 0.26 (theoretical 0.29) was obtained, and for naphthalene a value of 0.24 (theoretical 0.20) was obtained. Presumably the high purity of our samples made this estimation possible because after only slight oxidation the rapid-relaxation process dominates. The fresh and oxidized samples were examined by using Fourier transform infrared (FTIR) spectroscopy, which failed to show any detectable differences, a result that indicates that the oxidation was very slight.

Apparently, the examination of coal samples with dipolar-dephasing techniques may not be generally valid. Free radical concentration can have an important effect on the measurement of aromaticity (14). Therefore

Table IV. Gaussian and Lorentzian Decays for Aromatic and Aliphatic Carbons and the Fraction of Protonated Aromatic Carbons

Coal No.	T_2' (ar, Gaussian)	T_2' (ar, Lorentzian)	T_2' (al, Gaussian)	$f_a^{a,H}$
101 fr	38.9	94.9	34.5	0.19
101 ox	32.0	63.6	27.2	0.19
202 fr	32.8	80.1	33.9	0.34
202 ox	43.2	—	34.3	—
301 fr	38.6	85.8	33.2	0.13
301 ox	44.3	—	34.6	—
401 fr	42.0	205.2	34.4	0.46
501 fr	40.0	54.4	41.4	—
501 ox	46.9	—	42.9	—
601 fr	40.9	50.4	33.3	—
701 fr	40.1	147.5	35.0	0.46
801 fr	42.3	110.2	32.3	0.33

NOTE: Values of T_2 are given in microseconds and refer to Gaussian and Lorentzian decays for aromatic (ar) carbons and to Gaussian decays for aliphatic (al) carbons. The $f_a^{a,H}$ values are the fraction of protonated aromatic carbons. ABBREVIATIONS: fr, fresh; and ox, oxidized.

we measured the free radical concentrations of the four pairs of fresh and oxidized coals considered in this study. The results of this analysis are given in Table I. The magnitude of the spin concentrations agrees with previously reported values (15). For two of the coals, 101 (Upper Freeport) and 501 (Pocahontas No. 3), an approximately two-fold increase in radical concentration occurs with oxidation. For the other two coals, 202 (Wyodak–Anderson) and 301 (Illinois No. 6), which are lower in rank, radical concentration decreases with oxidation. Our previous work (15) indicated that early stages of coal oxidation are associated with increased concentrations of free radicals and that the free radical concentration decreases at more advanced stages of oxidation; the same phenomenon probably occurred in this study. The higher rank coals (101 and 501), which are less reactive to oxidation, are oxidized to a lesser extent than the lower rank coals (202 and 301) under similar reaction conditions. The oxygen data reported in Table I bear this out. For all of the coal samples, however, the free radical concentration is relatively high. The influence of such high free radical concentrations on dipolar relaxation in both pristine and oxidized coals is an important point that needs further clarification.

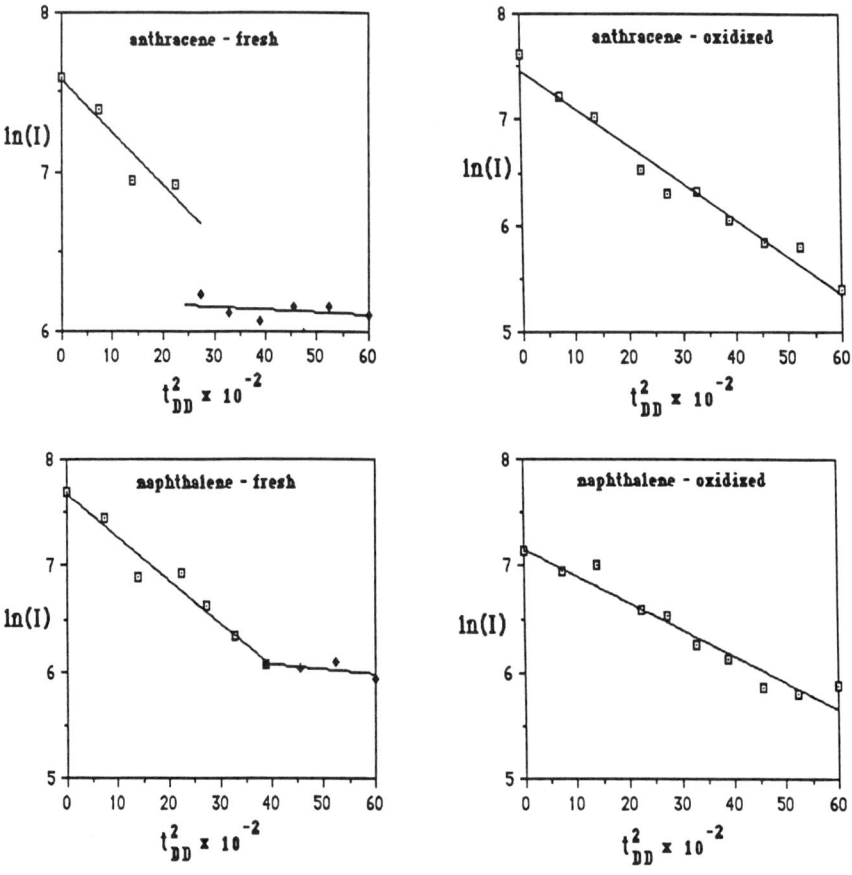

Figure 7. Dipolar-dephasing behavior of zone-refined anthracene and naphthalene before and after mild oxidation.

Summary

The aromaticities of the Argonne Premium coal samples were investigated by using ^{13}C CP–MAS NMR spectroscopy at high field (300 MHz) and moderate spinning rate (4 kHz). The aromaticity results are dependent on the set of instrumental parameters as well as on the manner in which the SSB are taken into account. The aromaticities are rank dependent; they correlated with atomic H/C values for the relatively wide range of coal ranks in this suite of samples. Results of dipolar-dephasing experiments on these coals were as expected, but for the oxidized coals the aromatic carbons exhibit a single relaxation process. Studies of model compounds indicate that this behavior may be due to changes in free radical concentration and could invalidate quantitative results because all coals contain high concentrations of free radicals.

Acknowledgments

J. A. MacPhee would like to thank the Japanese Agency of Industrial Science and Technology for the award of a fellowship tenable at the National Research Institute for Pollution and Resources (now the National Institute for Resources and Environment), Tsukuba, Japan, in November and December 1988. He also thanks the staff of the National Research Institute for Pollution and Resources for their warm welcome and the climate of scientific excellence that they provided.

References

1. Davidson, R. M. *Nuclear Magnetic Resonance Studies of Coal;* IEA Coal Research, Report No. ICTIS/TR32, January 1986.
2. Vorres, K. S. *Energy Fuels* **1990,** *4,* 420–426.
3. Alla, M.; Lipmaa, E. *Chem. Phys. Lett.* **1979,** *37,* 260.
4. Ito, O.; Akiho, S.; Nozawa, T.; Hatano, M.; Iino, M. *Fuel* **1989,** *68,* 335–340.
5. Snape, C. E.; Axelson, D. E.; Botto, R. E.; Delpuesch, J. J.; Tekely, P.; Gerstein, B. C.; Pruski, M.; Maciel, G. E.; Wilson, M. A. *Fuel* **1989,** *68,* 547–560.
6. Botto, R. E.; Wilson, R.; Winans, R. E. *Energy Fuels* **1987,** *1,* 173–181.
7. Pugmire, R. J.; Woolfenden, W. R., Mayne, C. L.; Karas, J.; Grant, D. M. In *Chemistry and Characterization of Coal Macerals;* Winans, R. E.; Crelling, J. C., Eds.; ACS Symposium Series 252; American Chemical Society: Washington, DC, 1984; pp 79–97.
8. Van Krevelen, D. W. *Coal;* Elsevier: Amsterdam, Netherlands, 1961.
9. Alemany, L. B.; Grant, D. M.; Alger, T. D.; Pugmire, R. J. *J. Am. Chem. Soc.* **1983,** *105,* 6697–6704.
10. Theriault, Y.; Axelson, D. E. *Fuel* **1988,** *67,* 62–66.
11. Axelson, D. E. *Solid State Nuclear Magnetic Resonance of Fossil Fuels: An Experimental Approach;* Multiscience: Montreal, Canada, 1985.
12. Wilson, M. A.; Pugmire, R. J.; Karas, J.; Alemany, L. B.; Woolfenden, W. R.; Grant, D. M. *Anal. Chem.* **1984,** *2,* 933–943.
13. Tougne, P.; Sfihi, H.; Legrand, A. P. *Proc. Int. Conf. Carbon* (University of Newcastle upon Tyne, Sept. 18–23, 1988); IOP: Bristol, United Kingdom, 1988; pp 352–354, 525–527.
14. Muntean, J. V.; Stock, L. M.; Botto, R. E. *Energy Fuels* **1988,** *2,* 108–110.
15. Ripmeester, J. A.; Couture, C.; MacPhee, J. A.; Nandi, B. N. *Fuel* **1984,** *63,* 522–524.

RECEIVED for review June 8, 1990. ACCEPTED revised manuscript July 15, 1991.

18

Measurement of Spin−Lattice Relaxation in Argonne Premium Coal Samples

Chihji Tsiao and Robert E. Botto[1]

Chemistry Division, Argonne National Laboratory, Argonne, IL 60439

Eight Argonne Premium coals and three weathered Argonne coal samples were investigated by using ^{13}C cross-polarization−magic-angle spinning (CP−MAS) NMR spectroscopy. Proton and carbon spin−lattice relaxation measurements were performed on each coal. The carbon spin−lattice relaxation time, T_1^C; the carbon spin−lattice relaxation time in the rotating frame, $T_{1\rho}^C$; and the proton spin−lattice relaxation time in the rotating frame, $T_{1\rho}^H$, are reported for aromatic and aliphatic carbons of the 11 coal samples. In general, the proton and carbon spin−lattice relaxation data can be evaluated as the sum of two exponential decays. The longer components of carbon relaxation times in the laboratory and rotating reference frames vary in a systematic way with coal rank as expressed by percent carbon. The trends can be explained in terms of motional properties of the coals and the presence of paramagnetic species. Marked changes in the relaxation parameters have been observed between pristine and weathered coals. Reduction in proton $T_{1\rho}$ values upon weathering is shown to have an adverse effect on quantitation with CP.

[1] Corresponding author

THE COMPLEX MOLECULAR DYNAMICS in synthetic polymers has been studied extensively with NMR spectroscopy during the past decade (*1–5*). However, such investigations of coals have been limited. Work in this area has been hampered by the complex nature of coal. Interpretation of nuclear relaxation, which underlies the study of molecular motion, is complicated by the existence of discrete heterogeneous domains in coals and by relatively large quantities of paramagnetic centers that are present either in the form of paramagnetic inorganic ions or as organic free radicals.

To date, an overwhelming majority of papers dealing with relaxation measurements on coals has focused on the study of ^{1}H NMR spin–lattice relaxation times both in laboratory (T_1) and rotating ($T_{1\rho}$) reference frames. Early ^{1}H NMR measurements (*6–9*) indicated that proton T_1's (T_1^H) in coals can display either single exponential or nonexponential relaxation behavior. The observation by Yokono et al. (*8*) that T_1^H varied linearly with the square root of resonance frequency for several bituminous coals was consistent with diffusion-limited relaxation to paramagnetic centers. Webster and Lynch (*10*), Ripmeester et al. (*11*), and Wind et al. (*12*) later demonstrated similar behavior for a wide variety of evacuated coal samples. The work indicated that spin–lattice relaxation may be a fundamental property of coals themselves and independent of either their oxygen or moisture content. In two specific papers (*12, 13*) the authors independently proposed that proton relaxation in bituminous coals was influenced by differences in molecular mobilities rather than by the relative concentrations of paramagnetic species.

A difficulty with the interpretation of spin–lattice relaxation of abundant proton spins in coals rests with the ability to separate molecular-mobility contributions from magnetization spin diffusion to paramagnetic centers within different phase (maceral) boundaries. Two thorough investigations of ^{1}H NMR spin–lattice relaxation in coals by Barton and Lynch (*14*) and Wind et al. (*15*) attempted to address this issue. The authors independently concluded that simple correlations of proton relaxation with other coal parameters are not easily realized. The presence of residual amounts of molecular oxygen in evacuated samples and the effects of different concentrations of unpaired electrons within different domains in coals were thought to have a profound, yet undeterminable, influence on the relaxation times. Solum et al. (*16*) reported proton relaxation data for eight Argonne Premium coals and three oxidized coals and showed that oxidized samples have shorter T_1^H values.

The problems are not confined to T_1^H measurements. Dudley and Fyfe (*17*) discussed proton $T_{1\rho}$ values for a pitch and three Canadian coals, and they emphasized the effects of paramagnetics, including oxygen, on their experimental results. Earlier studies performed in our laboratory indicated a general trend, but a lack of any definitive correlation, between

$T_{1\rho}^{H}$ values and radical concentrations for a series of "more homogeneous" maceral concentrates (18). We concluded that $T_{1\rho}^{H}$ values in macerals are clearly complicated by having two competing mechanisms for relaxation, molecular motion and heterogeneous spin diffusion to paramagnetic centers, and without any knowledge of the relative contributions of each to the overall relaxation times in the individual samples, any meaningful interpretation of the data would be impossible. For a homogeneous resinite sample doped with increasing amounts of a stable organic radical, however, an excellent correlation between $T_{1\rho}^{H}$ and radical concentration could be established (19).

In contrast with proton relaxation, relatively little work has been done on the measurement of ^{13}C relaxation times in coals. Sullivan and Maciel (20, 21) studied ^{13}C spin–lattice relaxation in a bituminous coal from the Powhatan No. 5 mine. They found that the aromatic resonances in this sample decayed with a single time constant, and the aliphatic resonances could be separated into two distinct regions that had markedly different time constants for decay. Differences observed in the T_1^{C} values for different spectral regions were interpreted in terms of molecular motion, although the contribution of free radicals to spin relaxation is important as well. Botto and Axelson (22) investigated the effect of static-field strength on the T_1^{C} relaxation parameters of five Argonne Premium coals and five Canadian coals and correlated the relaxation times with various coal properties.

In this chapter, we present the first comprehensive study of laboratory- and rotating-frame ^{13}C and ^{1}H spin–lattice relaxation times (T_1^{C}, $T_{1\rho}^{C}$, and $T_{1\rho}^{H}$) in coals. This study emphasized the suite of Argonne Premium coals in their pristine state and three intentionally weathered samples. Details of the investigation were directed to the issue of establishing a relationship between the relaxation data and structural and motional properties of the various coals.

Experimental Details

All of the coal samples were obtained from the Argonne Premium Coal Sample (APCS) Program. Table I presents the origin, rank, and elemental composition (weight percent) of the coals. In order to avoid exposure to oxygen, the coals were dried and transferred into sealed NMR rotors in a nitrogen-filled glove box. Weathered samples of APCS Nos. 2, 3, and 8 were prepared by exposing the coals to the atmosphere at ambient temperature for several months.

Solid-state ^{13}C cross-polarization–magic-angle spinning (CP–MAS) spectra were recorded at 2.3 T (25.18 MHz for ^{13}C) with a Bruker CXP-100 spectrometer in the pulse Fourier transform mode with quadrature phase detection. The ceramic sample spinners had an internal volume of 250 μL and were spun at a

Table I. Argonne Premium Coal Samples and Their Compositions

APCS No.	Coal	Rank	C	H	O	S_{org}
1	Upper Freeport	MVB	85.50	4.70	7.51	0.74
2	Wyodak–Anderson	SB	75.01	5.35	18.02	0.47
3	Illinois No. 6	HVB	77.67	5.00	13.51	2.38
4	Pittsburgh No. 8	HVB	83.20	5.32	8.83	0.89
5	Pocahontas No. 3	LVB	91.05	4.44	2.47	0.50
6	Blind Canyon	HVB	80.69	5.76	11.58	0.37
7	Lewiston–Stockton	HVB	82.58	5.25	9.83	0.65
8	Beulah–Zap	L	72.94	4.83	20.34	0.70

NOTE: The values for C, H, O, and organic S are given as weight percents.
ABBREVIATIONS: LVB, low-volatile bituminous; MVB, medium-volatile bituminous; HVB, high-volatile bituminous; SB, subbituminous; and L, lignite.
SOURCE: Reproduced from reference 23. Copyright 1990.

rate of 4 kHz. Each spectrum used for T_1 measurement was a total accumulation of 400–1000 transients with a recycle delay of 2 s. The ^{13}C chemical shifts were referenced to tetramethylsilane (TMS) by using tetrakis(trimethylsilyl)silane (TKS) as the secondary reference (24).

The ^{13}C spin–lattice relaxation times were measured with CP–MAS by using the T_1 pulse sequence described previously by Torchia (25). The pulse sequences used for $T_{1\rho}^C$ and $T_{1\rho}^H$ measurements are illustrated in Figure 1 (26–28). In both cases, the protons were allowed to come to thermal equilibrium with the lattice prior to spin-locking along the y axis in the rotating frame by using a 90° pulse from the radio frequency (rf) field (56 kHz). In the $T_{1\rho}^C$ experiment shown in Figure 1a, the proton rf field was turned off immediately following a matched Hartmann–Hahn generation of a carbon polarization. During the variable delay period, free induction decays (FIDs) representing carbon magnetization held in the rotating frame were acquired and Fourier transformed. In the $T_{1\rho}^H$ experiment shown in Figure 1b, the proton magnetization was spin-locked during a variable delay period prior to the contact period during which carbon polarization was established. In this case, variation in carbon magnetization was used to monitor the decay of proton magnetization held in the rotating frame as a function of delay time.

Results and Discussion

Carbon T_1 Relaxation. For carbonaceous solids placed in a static magnetic field, the alignment of carbon magnetic moments gives rise to a net macroscopic magnetization M_t whose value is M_o when the nu-

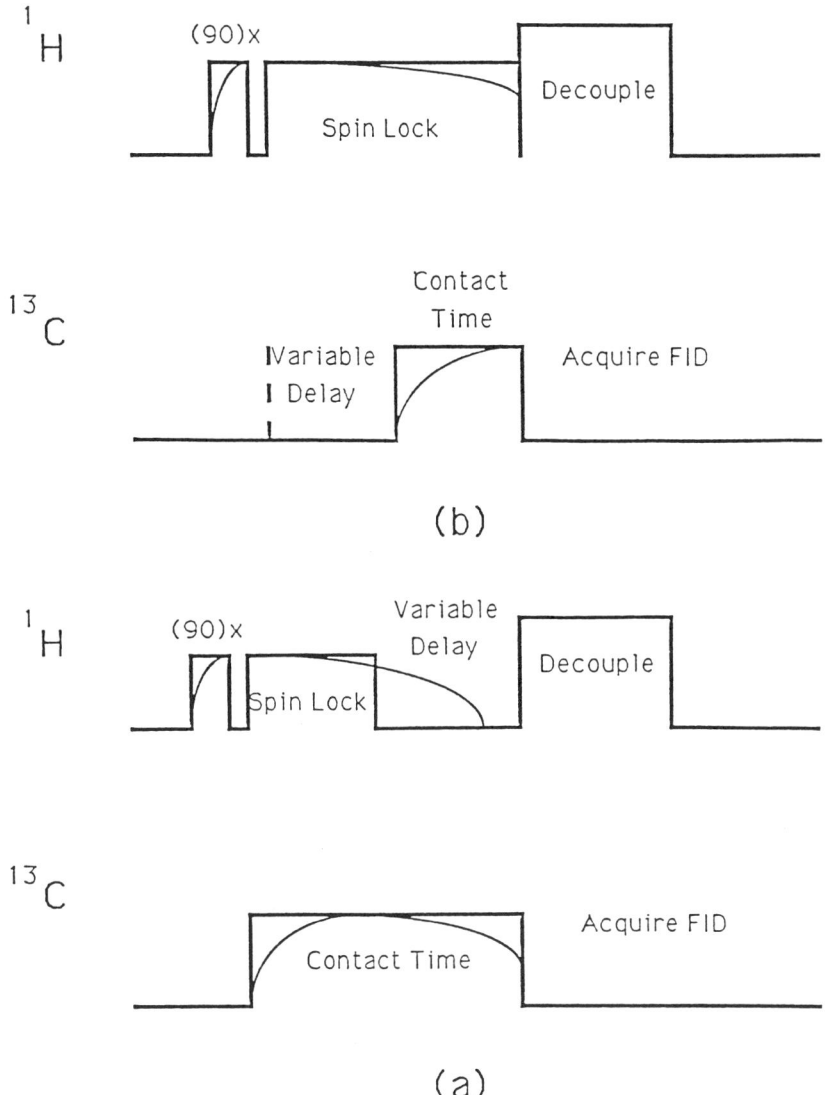

Figure 1. Pulse sequences for (a) $T_{1\rho}^C$ and (b) $T_{1\rho}^H$ measurements.

clear moments and the molecular lattice are in thermal equilibrium. A characteristic time constant, or carbon spin–lattice relaxation time (T_1^C), governs the restoration of magnetization M_o for individual carbon isochromats within the sample when the equilibrium condition is perturbed by an external rf field.

Recovery of magnetization M_t with time τ for the carbons in coal samples appears to be more complex. Integrated signal intensities for aromatic and aliphatic resonances were fitted with a Simplix algorithm (29) to the sum of two exponential decays by the following equation:

$$M_t/M_o = A_L \exp(-\tau/T_{1,L}) + A_S \exp(-\tau/T_{1,S}) + A_C \quad (1)$$

where A_L and A_S are the fractional amplitudes for long ($T_{1,L}$) and short ($T_{1,S}$) relaxation time constants, respectively, and A_C is the amplitude-correction factor. These parameters are extracted from the data by means of a nonlinear least-squares fitting procedure, which is applied in the following manner. Initially, an exponential fit is made to the more slowly decaying part of the data. With $T_{1,L}$ held constant in eq 1, a second exponential decay is fit by using the entire data set. Parameters obtained from the individual fits are then optimized by allowing the initial values obtained for $T_{1,L}$, $T_{1,S}$, A_L, and A_S to vary such that the sum of the squared deviations between data points and the fitted curve is minimized. Relaxation times determined in this manner are estimated to be accurate to within ±15%.

The ^{13}C spin–lattice relaxation times and fractional amplitudes for the aromatic and aliphatic resonances in Argonne coals and weathered Argonne coal samples are summarized in Table II. The weighted-average spin–lattice relaxation time, T_1^W, where

$$(T_1)^{W-1} = A_L(T_{1,L})^{-1} + A_S(T_{1,S})^{-1} \quad (2)$$

is defined to characterize the average relaxation behavior. The calculated T_1^W values are also presented in Table II. The relaxation times of the aromatic carbons are consistently longer than the relaxation times of the corresponding aliphatic carbons in a given sample, with the exception of Beulah–Zap lignite (APCS No. 8). Values for the aromatic carbons fall within two distinct ranges: 1.8–20.9 s for the long T_1^C components and 0.3–8.9 s for the short T_1^C components.

Ranges for the corresponding two components of aliphatic T_1^C values are 1.0–4.6 s and 0.1–0.6 s, respectively. The shorter relaxation times of the aliphatic carbons compared with aromatic carbons may reflect their greater molecular mobility. The T_1^C values of the three weathered coals are shorter as a result of paramagnetic molecular oxygen that is introduced. The values reported here are similar to those reported previously (22). The higher T_1^C values observed for some coals, in particular the Upper Freeport sample, may be the result of careful handling of the fresh coal samples.

Table II. Carbon T_1 and T_1^W Relaxation Times
of Aromatic and Aliphatic Resonances in Coals

APCS No.	Aromatic Resonances			Aliphatic Resonances		
	Long T_1 Component	Short T_1 Component	T_1^W	Long T_1 Component	Short T_1 Component	T_1^W
1	20.9 (49.1%)	2.8 (50.9%)	4.9	2.7 (52.1%)	0.5 (47.9%)	0.9
2	2.9 (44.7%)	0.6 (53.3%)	1.0	1.3 (34.6%)	0.3 (65.4%)	0.4
	2.0 (62.5%)	0.5 (37.5%)		0.8 (45.9%)	0.3 (54.1%)	
3	8.9 (66.0%)	3.0 (34.0%)	5.3	2.6 (100%)	—	2.6
	4.6 (65.9%)	1.0 (34.1%)		1.4	—	
4	20.0 (17.2%)	8.9 (82.8%)	9.9	4.2 (100%)	—	4.2
5	8.9 (45.3%)	1.7 (54.7%)	2.7	1.0 (48.6%)	0.6 (51.4%)	0.7
6	20.5 (38.3%)	1.8 (61.7%)	2.7	4.6 (35.8%)	0.3 (64.2%)	0.4
7	11.4 (80.3%)	1.6 (19.7%)	5.2	1.3 (68.9%)	0.2 (31.1%)	0.5
8	1.8 (41.6%)	0.3 (58.4%)	0.4	1.8 (51.4%)	0.1 (48.6%)	0.2
	1.1 (35.9%)	0.7 (64.1%)		1.8 (41.0%)	0.1 (59.0%)	

NOTE: Relaxation times are given in seconds, and the fractional amplitudes are in parentheses. Where two values are listed, the second value is that of the weathered coal sample.

Plots of long T_1^C components of aromatic and aliphatic carbons versus coal rank as indicated by percent carbon (see Figures 2 and 3) show an initial increase in T_1^C up to a maximum value of approximately 81–85% carbon, followed by a decrease. A regular trend is observed for the suite of coals, with the exception of the Lewiston–Stockton sample, whose T_1^C values typically lie off the correlation lines by a wide margin. The deviation may be due, at least in part, to the high contents of inertinite and liptinite macerals present in the Lewiston–Stockton coal compared with other Argonne coals, which have high vitrinite contents (>85%). Figure 4 shows the plots of the weighted-average carbon spin–lattice relaxation times (T_1^W) as a function of coal rank. Although significantly greater scatter is observed in these plots, the trends with rank are similar to those found for long T_1^C components.

The observation of multicomponent magnetization decays in macromolecular systems has generally been attributed to a distribution of relaxation times. Such T_1^C distributions are common in the ^{13}C spectra of solids (27, 30). The distributions arise because of different spatial orientations of carbon–hydrogen dipoles in a solid relative to the static magnetic field, and they are observable as a result of the inherent isolation of the rare carbon spins. In the case of coals, two additional factors may influ-

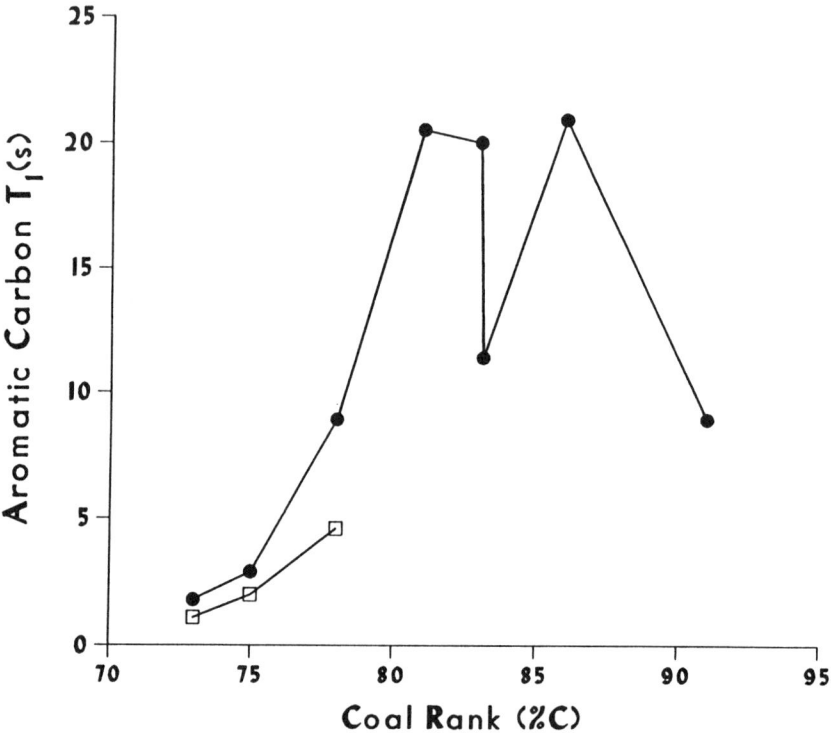

Figure 2. Plots of the long T_1^C components vs. coal rank for aromatic resonances of coals: ●, *pristine samples;* □, *weathered samples.*

ence this distribution as well. Variations in the levels of paramagnetic species and maceral compositions may affect the T_1^C distributions in an unpredictable manner. No obvious trends are observed for the variation in relative amplitudes of the long and short T_1^C components with rank, contrary to earlier observations (22). Furthermore, the shorter T_1^C components of both aromatic and aliphatic carbons generally show a poorer correlation with rank, leading to the large scatter found when the values of T_1^W are plotted against carbon content for the coals. This poor correlation would suggest that factors such as unpaired electrons are manifest in shorter T_1^C components to a greater degree and that these factors could mask a real influence of molecular mobility. Conversely, the long T_1 components are less affected by paramagnetic species, and hence provide better estimates of molecular structure.

Previous variable-field carbon relaxation experiments (22) on coals spanning a wider range of carbon contents than those investigated here support the foregoing conclusions. The ratios of the long T_1^C values at

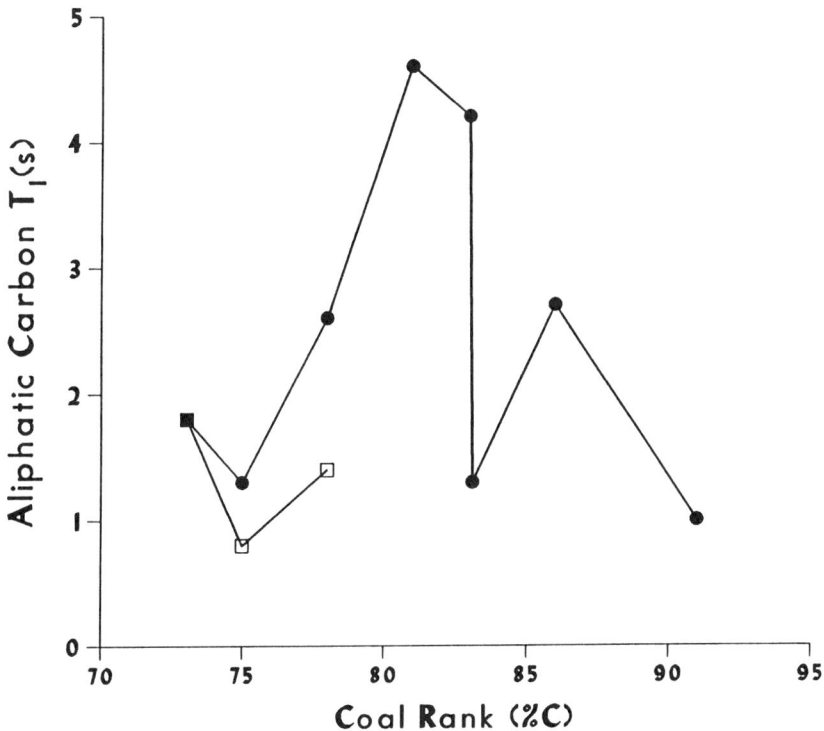

Figure 3. Plots of the long T_1^C components vs. coal rank for aliphatic resonances of coals: ●, *pristine samples;* □, *weathered samples.*

4.7 and 2.3 T for subbituminous through low-volatile bituminous coals are on the order expected (1.7–5.1) for a square dependence on field, as in the case of motional effects contributing to the relaxation processes. No such dependence, however, could be discerned for the short components of the T_1^C decays or for T_1^C values of two lignite coal samples whose relaxation times were clearly dominated by the presence of high concentrations of paramagnetic metal ions chelated to oxygen functionalities.

Distinct maxima seen in the plots of the more slowly relaxing T_1^C components and T_1^W values against rank show a parallel with several other methods of analysis. Proton spin–lattice relaxation times measured independently by Barton and Lynch (14), Wind et al. (15), and Yokono and Sanada (31) for coals varying widely in origin show evidence of a broad maximum in the 85–90% carbon range. The carbon and proton relaxation data are consistent with established relationships between coal structural properties and the degree of maturation and reflect the major changes that occur during coal evolution.

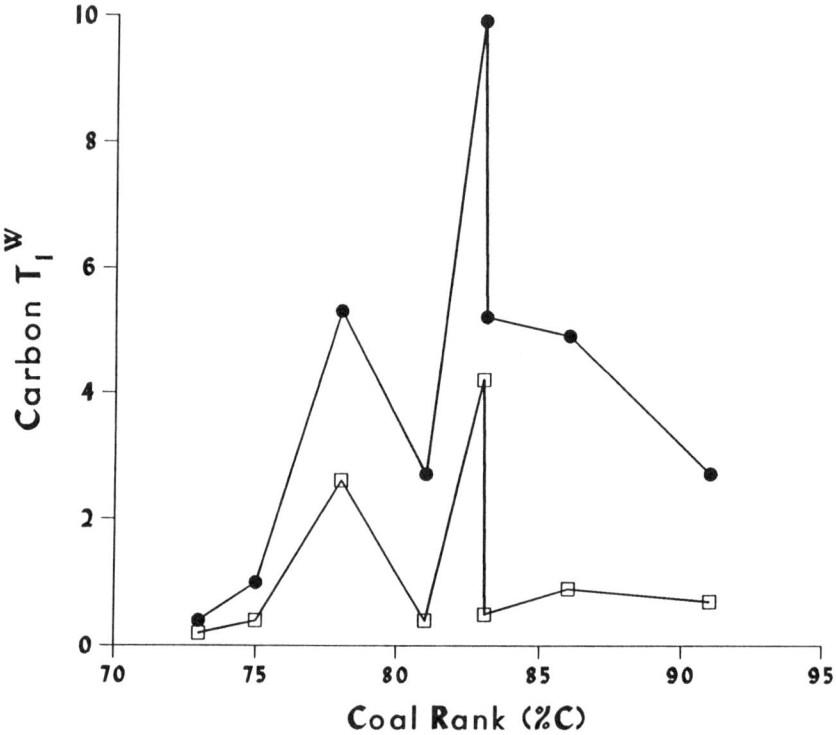

Figure 4. Plots of carbon T_1^W (in seconds) vs. coal rank for the Argonne Premium coals: ●, *aromatic carbons;* □, *aliphatic carbons.*

Coal is a cross-linked, three-dimensional macromolecular network. Data from solvent swelling of coal (*32, 33*) and stress–strain measurements (*34, 35*) have been used to obtain information about the elastic modulus of the macromolecular network. This modulus can then be used to calculate the average molecular weight between cross-links, M_C, a parameter that can be directly correlated with molecular mobility of the network. The shape of the curves of elastic modulus (*34*) or solvent swelling (*32*) plotted against coal rank resembles those plots obtained from T_1^C data. The elastic constants show similar behavior to T_1^C as a function of rank with a relative maximum at 85% carbon and a relative minimum around 90% carbon. At greater values than 95% carbon the modulus rises sharply, a result that is consistent with the development of more graphitelike structures. A higher value of the modulus thus corresponds to a maximum in the cross-link density of the coals, and the results are in good agreement with the T_1^C results, which indicate that the lowest molecular mobility occurs at the highest cross-link density. Apparently, the cross-link density increases with increasing coal rank up to the medium-volatile bituminous (MVB) stage and subsequently decreases through the low-volatile bituminous (LVB) stage until rapid graphitization occurs at higher rank.

Carbon $T_{1\rho}$ Relaxation. In their pioneering work on glassy polymers, Schaefer et al. (27) pointed out the utility of $T_{1\rho}$ measurements for probing molecular dynamics as a potential source of information on the mechanical and physical properties of polymers on the molecular level. Later, VanderHart and Garroway (36) outlined the complications that are involved in extracting molecular-motion information from $T_{1\rho}$ experiments. Both spin–spin and spin–lattice relaxation processes are known to contribute to carbon $T_{1\rho}$. For highly crystalline materials, spin–spin processes (mutual spin flips between spin-locked protons and carbons) are usually dominant, and the $T_{1\rho}^C$ values do not adequately describe the molecular dynamics of the system. For most glassy and amorphous materials, however, in which strong ^1H–^1H static dipolar interactions are negligible because of molecular motion, large interproton distances, or some combination of these, $T_{1\rho}$ is largely spin–lattice in character and can be interpreted in terms of localized segmental motions in the tens of kilohertzes frequency range (37).

The $T_{1\rho}$ values of aromatic and aliphatic carbons in Argonne Premium coals and three weathered Argonne coals are presented in Table III. The data were evaluated by using the two-component fitting procedure that was described previously. For aromatic carbons, $T_{1\rho}^C$ values range from 75 to 155 ms for the longer components and from 15 to 105 ms for the shorter components. The corresponding values for the aliphatic carbons are considerably shorter: 22–52 ms for the long components and 9–34 ms for the short components. $T_{1\rho}^C$ values of weathered versus pristine samples show some interesting trends: relaxation times of the aromatic carbons in weathered coals are longer by a factor generally greater than 2, and those of the aliphatic carbons are approximately the same. Clear deviations from ideal exponential behavior of the magnetization decay are observed in most cases. Nonexponential behavior may be interpreted as distributions of relaxation times due to the heterogeneity of the local environment.

Plots of the aromatic and aliphatic $T_{1\rho}^C$ values (long components) against coal rank are shown in Figures 5 and 6. The general shapes of the curves resemble those obtained previously for the T_1^C values, with one notable difference. The $T_{1\rho}^C$ values increase with increasing rank until 80% carbon content and then remain constant thereafter; no decrease is seen for the higher rank coals and hence no definitive maxima are observed.

Table III. Carbon $T_{1\rho}$ Relaxation Times
of Aromatic and Aliphatic Resonances in Coals

APCS No.	Aromatic Resonances		Aliphatic Resonances	
	Long $T_{1\rho}$ Component	Short $T_{1\rho}$ Component	Long $T_{1\rho}$ Component	Short $T_{1\rho}$ Component
1	151 (55.3%)	69 (44.7%)	48	34
2	76 (75.2%)	19 (24.8%)	25	17
	172 (69.9%)	24 (30.1%)	24	—
3	100 (64.3%)	88 (35.7%)	33	28
	187 (84.2%)	99 (15.8%)	30	—
4	150 (60.8%)	106 (39.2%)	44	28
5	141 (50.7%)	66 (49.3%)	52	20
6	156 (25.0%)	61 (75.0%)	50 (19.7%)	20 (80.3%)
7	125 (61.7%)	46 (38.3%)	50 (52.7%)	16 (47.3%)
8	80 (64.1%)	15 (35.9%)	22 (31.2%)	9 (68.8%)
	150 (61.9%)	51 (38.1%)	23	—

NOTE: Relaxation times are given in milliseconds, and fractional amplitudes are in parentheses. Where two values are listed, the second value is that of the weathered coal sample. Values of aliphatic resonances reported for APCS Nos. 1–5 correspond to single exponential decay constants for methylene (~30 ppm) and methyl (~20 ppm) resonances, respectively.

As discussed previously, the interpretation of $T_{1\rho}{}^C$ data can be complicated by the fact that spin–spin cross-relaxation processes as well as rotating-frame spin–lattice processes may contribute to the relaxation. Evidence to support the dominance of spin–lattice contributions to $T_{1\rho}{}^C$ is given in Table IV, which shows the dependence of $T_{1\rho}{}^C$ as a function of the rotating-frame field, B_1, for a weathered Illinois No. 6 coal (APCS No. 3). A square dependence of $T_{1\rho}{}^C$ on the B_1 field occurs for the long components of aromatic carbons and the aliphatic carbons in the coal, as would be expected in the case of the domination of $T_{1\rho}$ processes by molecular motion. If in fact spin–spin interactions were to dominate the relaxation, then an exponential dependence on the B_1 field would be expected (38).

The foregoing arguments reinforce the notion that changes observed in $T_{1\rho}{}^C$ for the coals provide insight into differences in macromolecular chain dynamics within the local environment of individual carbons. In-

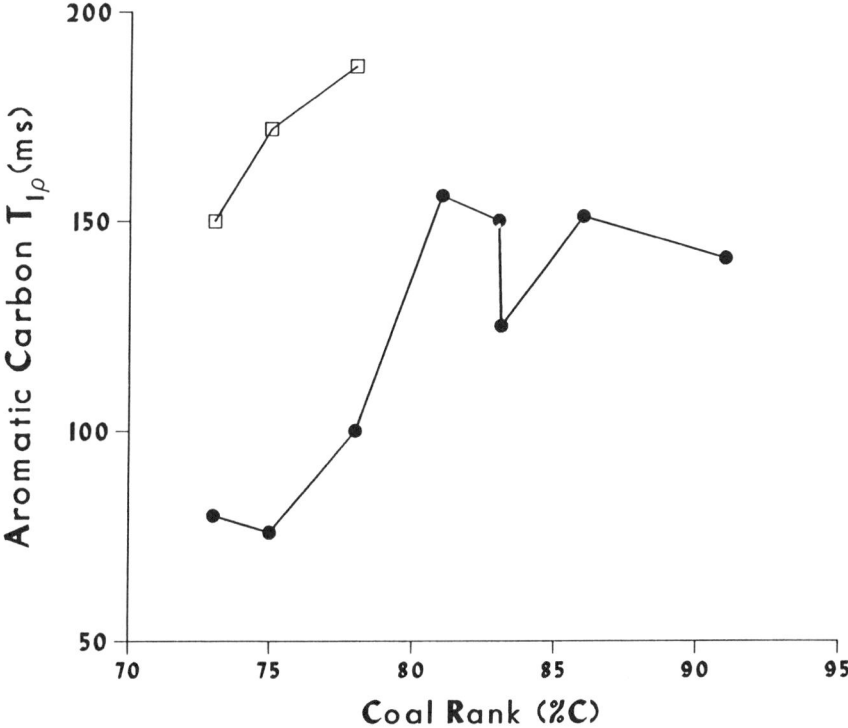

Figure 5. Plots of the long $T_{1\rho}^C$ components vs. coal rank for aromatic resonances of coals: ●*, pristine samples;* □*, weathered samples.*

creases in the relaxation times as a function of rank reflect decreases in molecular mobilities of the coal macromolecular networks. Changes observed upon coal weathering suggest decreases in molecular mobility as well. Moreover, weathering effects seem to be confined to the aromatic structures in coal, where large decreases in molecular mobility are observed. Oxidation of aromatic structures enhances their electron donor–acceptor properties. The increase in strength of the noncovalent interactions within the macromolecular network might, in fact, explain loss of mobility of aromatic structures with weathering. Negligible changes in $T_{1\rho}^C$ of the aliphatic carbons suggest that chain motions of the aliphatic structures are largely unaffected by weathering.

Differences in T_1^C and $T_{1\rho}^C$ values and their trends with rank and weathering are informative. The magnitudes of these two characterizations of mobility reflect, among other things, differences in the time scales for segmental motion within the macromolecular network. We propose that the decreases in T_1^C values observed with weathering and in the higher

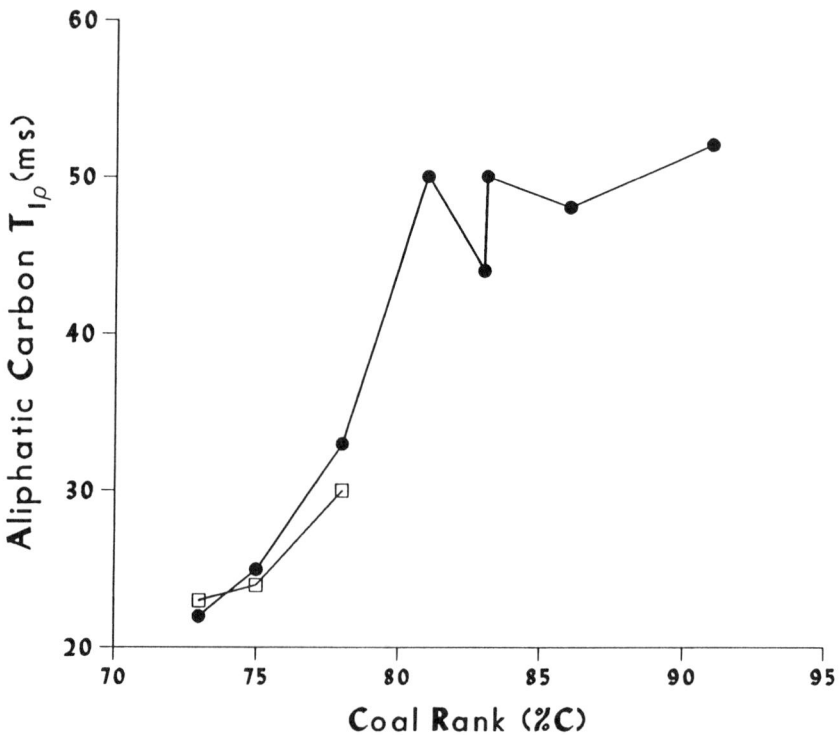

Figure 6. Plots of the long $T_{1\rho}{}^C$ components vs. coal rank for aliphatic resonances of coals: ●, pristine samples; □, weathered samples.

Table IV. Dependence of $T_{1\rho}{}^C$ on B_1 Field for Weathered Illinois No. 6 Coal (APCS No. 3)

B_1 (kHz)	Aromatic Carbon Component		Aliphatic Carbon
	Long	Short	
67	264	101	41
56	187	99	29
45	133	86	22

rank coals can be related to the presence of paramagnetic species. In weathered coals, introduction of molecular oxygen and its rapid translational diffusion throughout the solid matrix provides an effective pathway for relaxation. In coals of higher rank, the development of free radicals in large polycyclic aromatic arrays would extend the effects of free radicals over larger distances and thus facilitate rapid relaxation of a greater fraction of carbons. Dissimilar behavior of the two relaxation time constants emphasizes that T_1^C is complicated by having two competing mechanisms for relaxation, and in the absence of a priori knowledge of the relative contributions from motional and paramagnetic processes, T_1^C is apparently less reliable as a parameter for molecular structure.

Proton $T_{1\rho}$ Relaxation. Because proton $T_{1\rho}$ relaxation times were measured indirectly via the decay of ^{13}C magnetization, only those protons affecting the observable carbon magnetization are actually measured. The $T_{1\rho}^H$ values of aromatic and aliphatic carbons in the Argonne Premium coals and three weathered Argonne coal samples are presented in Table V. The decay of magnetization of the aromatic carbons is nonexponential,

Table V. Proton $T_{1\rho}$ Relaxation Times of Aromatic and Aliphatic Resonances in Coals

APCS No.	Long $T_{1\rho}$ Component	Short $T_{1\rho}$ Component	Methylene Resonances (~30 ppm)	Methyl Resonances (~20 ppm)
1	15 (37.3%)	4.4 (62.3%)	5.8	7.2
2	19 (27.9%) 5.0	2.3 (72.1%)	6.1 6.2	— —
3	19 (44.7%) 4.1	2.7 (55.3%)	5.5 5.1	4.9 —
4	15 (34.4%)	4.3 (65.6%)	6.4	5.7
5	16 (29.0%)	3.4 (71.0%)	3.3	4.5
6	22 (19.0%)	2.9 (81.0%)	5.6	2.9
7	20 (38.3%)	1.4 (61.7%)	2.0	—
8	3.7 4.0		3.0 2.8	— —

NOTE: Relaxation times are given in milliseconds, and fractional amplitudes are in parentheses. Where two values are listed, the second value is that of the weathered coal sample.

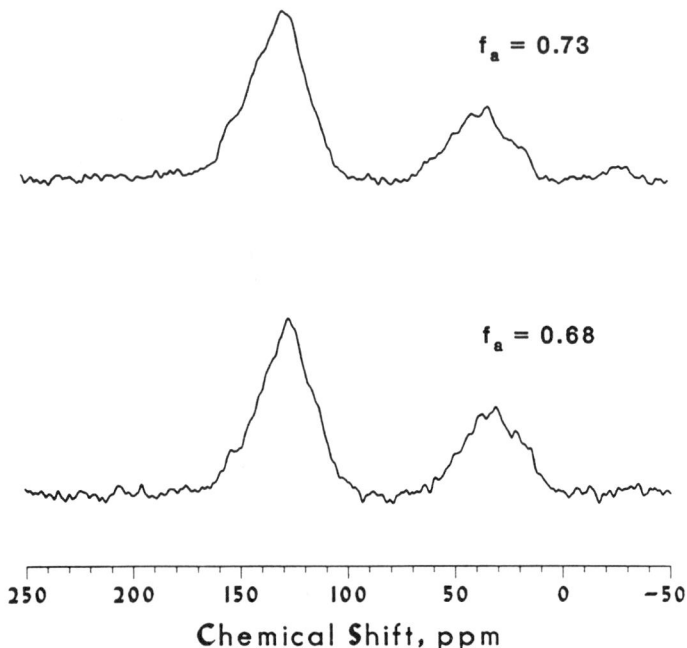

Figure 7. ^{13}C *CP–MAS spectra of pristine (top) and weathered (bottom) samples of Illinois No. 6 coal (APCS No. 3).*

and the proton $T_{1\rho}$ values fall into two distinct ranges: 15–22 ms for the long components and 1.4–4.4 ms for the short components, with the exception of the Beulah–Zap lignite, which has an extremely short $T_{1\rho}^H$ value of 3.7 ms. The fact that the lignite coal exhibits a significantly shorter $T_{1\rho}^H$ than the other coals implies that paramagnetic species play a major role. Previous studies (22) have indicated that significant amounts ($>10^{19}$ spins/g) of Fe^{3+} species are chelated to oxygen-rich aromatic structures in this coal.

Relaxation of the aliphatic resonances appears to follow exponential behavior. Aliphatic $T_{1\rho}^H$ values are markedly shorter than those of the aromatic carbons: 2.0–6.1 ms for the methylene region and 2.9–7.2 for the methyl region. Apparently there is no correlation between either the aromatic or aliphatic $T_{1\rho}^H$ values and rank.

Weathering of the Wyodak–Anderson subbituminous and Illinois No. 6 bituminous coals causes substantial reduction of aromatic $T_{1\rho}^H$ values. Presumably the effect is due to the introduction of paramagnetic oxygen or to the creation of paramagnetic sites in the sample during the weathering process. However, the aliphatic $T_{1\rho}^H$ values are largely unaffected by weathering.

More importantly, weathering of the samples appears to have an adverse effect on the analytical outcome of the CP experiment, as indicated by the spectra shown in Figure 7. A reduction in carbon aromaticity (f_a) of 0.05 units is observed in the spectrum of the weathered coal sample, corresponding to a decrease in aromaticity of about 7%. Closer inspection of the two spectra reveals a narrowing of the aromatic resonance band for the weathered coal with a concomitant loss of signal intensity in the shoulders appearing at 145 and 155 ppm. This change corresponds to the specific loss of carbon resonances for the more slowly cross-polarizing C-substituted and O-substituted aromatic carbons in the weathered sample brought about as a result of the large decrease in aromatic $T_{1\rho}^H$ values.

The foregoing results point to the need for careful handling of coal samples prior to NMR analysis. Reduction in $T_{1\rho}^H$ values as a result of weathering may cause intensity distortions in CP–MAS spectra, a result leading to an underestimation of carbon aromaticity values. Indeed, weathering effects as a result of poor sample handling may be responsible for the discrepancies in the literature regarding carbon aromaticities of similar coals.

Acknowledgments

This work was performed under the auspices of the Office of Basic Energy Sciences, Division of Chemical Sciences, U.S. Department of Energy, under Contract No. W–31–109–ENG–38. The authors express appreciation to Karl S. Vorres for providing the Argonne Premium Coal Samples.

References

1. Slichter, W. P. *NMR: Basic Princ. Prog. 1971,* 4, 209.
2. McCall, D. W. *Acc. Chem. Res.* **1971,** 4, 223.
3. Schaefer, J. *Topics in Carbon-13 NMR Spectroscopy;* Levy, G. C., Ed.; Wiley: New York, 1974; Vol. 1, p 149.
4. Schaefer, J.; Stejskal, E. O. *Topics in Carbon-13 NMR Spectroscopy;* Levey, G. C., Ed.; Wiley: New York, 1979; Vol. 3, p 284.
5. Bovey, F. A.; Jelinski, L. W. *J. Phys. Chem.* **1985,** 89, 571.
6. Retcofsky, H. L.; Friedel, R. A. *Fuel* **1968,** 47, 391.
7. Gerstein, B. C.; Chow, C.; Pembleton, R. G.; Wilson, R. C. *J. Phys. Chem.* **1977,** 81, 565.
8. Yokono, T.; Miyazawa, K.; Sanada, Y.; Marsh, H. *Fuel* **1979,** 58, 896.
9. Lynch, L. J.; Webster, D. S. *J. Magn. Reson.* **1980,** 40, 259.
10. Webster, D. S.; Lynch, L. J. *Fuel* **1981,** 60, 549.

11. Ripmeester, J. A.; Couture, C.; MacPhee, J. A., Nandi, B. N. *Fuel* **1984**, *63*, 522.
12. Wind, R. A.; Duijvestijn, M. J.; van der Lugt, C.; Smidt, J.; Vriend, H. *Fuel* **1987**, *66*, 876.
13. Sullivan, M. J.; Szeverenyi, N. M.; Maciel, G. E.; Petrakis, L.; Grandy, D. W. In *Magnetic Resonance. Introduction, Advanced Topics and Applications to Fossil Energy;* Petrakis, L., Fraissard, J. P., Eds.; NATO ASI Series C124; Reidel: Dordrecht, Netherlands, 1984; p 607.
14. Barton, W. A.; Lynch, L. J. *Energy Fuels* **1989**, *3*, 402.
15. Wind, R. A.; Jurkiewicz, A.; Maciel, G. E. *Fuel* **1989**, *68*, 1189.
16. Solum, M. S.; Pugmire, R. J.; Grant, D. M. *Energy Fuels* **1983**, *3*, 187.
17. Dudley, R. L.; Fyfe, C. A. *Fuel*, **1982**, *61*, 651.
18. Botto, R. E.; Wilson, R.; Winans, R. E. *Energy Fuels* **1987**, *7*, 173.
19. Snape, C. E.; Axelson, D. E.; Botto, R. E.; Delpuech, J. J.; Tekely, P.; Gerstein, B. C.; Pruski, M.; Maciel, G. E.; Wilson, M. E. *Fuel* **1989**, *68*, 547.
20. Sullivan, M. J.; Maciel, G. E. *Anal. Chem.* **1982**, *54*, 1606.
21. Sullivan, M. J.; Maciel, G. E. *Anal. Chem.* **1982**, *54*, 1615.
22. Botto, R. E.; Axelson, D. E. *Prepr. Am. Chem. Soc. Div. Fuel Chem.* **1988**, *33*(3), pp 50–57.
23. Vorres, K. S. *Energy Fuels* **1990**, *4*, 420.
24. Muntean, J. V.; Stock, L. M.; Botto, R. E. *J. Magn. Reson.* **1988**, *76*, 540.
25. Torchia, D. A. *J. Magn. Reson.* **1978**, *30*, 613.
26. Alla, M.; Lippmaa, E. *Chem. Phys. Lett.* **1976**, *37*(2), 260.
27. Schaefer, J.; Stejskal, E. O.; Steger, T. R.; Sefcik, M. D.; McKay, R. A. *Macromolecules* **1980**, *13*, 1121.
28. Jelinski, L. M.; Melchior, M. T. In *NMR Spectroscopy Techniques;* Dybowski, C. R.; Licher, R. L., Eds.; Marcel Dekker: New York and Basel, 1987; pp 311–334.
29. Noggle, J. H. *Physical Chemistry on a Microcomputer;* Little Brown: Boston, MA, 1985.
30. Gibby, M. G.; Pines, A.; Waugh, J. S. *Chem. Phys. Lett.* **1972**, *16*, 296.
31. Yokono, T.; Sanada, Y. *Fuel* **1978**, *57*, 334.
32. Green, T.; Kovac, J.; Brenner, D.; Larsen, J. W. In *Coal Structure;* Meyers, R. A., Ed.; Academic: New York, 1982; pp 199–282.
33. Sanada, Y.; Honda, H. *Fuel* **1966**, *45*, 295.
34. Schuyer, J.; Djkstra, H.; van Krevelen, D. W. *Fuel* **1954**, *33*, 409.
35. van Krevelen, D. W.; Chermin, H. A. G.; Schuyer, J. *Fuel* **1959**, *38*, 438.
36. VanderHart, D. L.; Garroway, A. N. *J. Chem. Phys.* **1979**, *71*, 2773.
37. Hester, R. K.; Ackerman, J. L.; Neff, B. L.; Waugh, J. S. *Phys. Rev. Lett.* **1976**, *36*, 1081.
38. Fleming, W. W.; Lyerla, J. R.; Yannoni, C. S. In *NMR and Macromolecules: Sequence, Dynamic, and Domain Structure;* Randall, J. C., Jr., Ed.; ACS Symposium Series 247; American Chemical Society: Washington, DC, 1984; pp 83–94.

RECEIVED for review June 8, 1990. ACCEPTED revised manuscript January 30, 1991.

19

Quantitation of Protons in the Argonne Premium Coals by Solid-State ^{1}H NMR Spectroscopy

Luisita dela Rosa, Marek Pruski, Bernard Gerstein[1]

Institute for Physical Research and Technology and Department of Chemistry, Iowa State University, Ames, IA 50011

> Quantitation of protons and moisture in the eight coals from the Argonne Premium Coal Sample Program was performed by using ^{1}H NMR spectroscopy. The solid echo was used to determine the true line shape of the on-resonance ^{1}H free induction decay (FID). A superposition of Gaussian and Lorentzian decay functions adequately described the FID of dry coals. The Gaussian fraction corresponds to rigid protons in the macrostructure of coal. The Lorentzian fraction is attributed to fragments in the coal exhibiting hindered molecular mobility. For wet coals, an additional slowly decaying Lorentzian fraction must be added to the description of the FID. Identification of the species responsible for the longest decay was made by liquid-state NMR spectroscopy of the condensate obtained by heating the sample at 100 °C under static vacuum, and high-resolution solid-state ^{1}H NMR spectra of the coals were obtained before and after removing the volatile matter at 100 °C. The NMR quantitation results compared favorably with the results that were obtained by chemical and thermogravimetric analyses.

[1] Corresponding author

THE ASSAY OF HYDROGEN by the American Society for Testing and Materials (ASTM) procedure D 3178 and of moisture by ASTM procedure D 3173 in samples of coal is carried out under carefully controlled conditions (*1*).

Because setting up and complying with the requirements of these procedures are tedious and time-consuming tasks, alternative procedures, which most laboratories could perform, are used instead. A fast, analytical procedure for determining the concentration of organic hydrogen in the coal matrix uses an elemental analyzer. However, this analytical method requires very small (<10 mg) samples and may not reflect the inherent heterogeneity of coal. The usual routine moisture analysis used by most laboratories is an indirect determination by mass loss upon heating the sample for 1 h at 107 ± 3 °C in a drying oven. Unfortunately, a number of possible sources of errors exist with this method. The loss of mass due to the desorption of gases such as CH_4 and CO is negligible (*2*), but the loss of CO_2 due to decarboxylation at temperatures as low as 60 °C, especially for low-rank coals (*3*), may be significant and thus may result in an overestimation of the moisture concentration. On the other hand, when the heating is not carried out in an inert atmosphere, oxidation (*4*) taking place during the drying period will result in underestimating the value of the moisture reported. A serious drawback of this procedure is the readsorption of moisture in the desiccator during the cooling process (*5*).

An isotope-dilution technique was reported (*6*) in which ^{18}O-enriched water added to the coal sample was allowed to exchange with the natural abundance water, and the isotope ratio was then analyzed by mass spectrometry. This technique is time-consuming because equilibration requires 16 h before the exchange is deemed to be complete. The main source of error in this technique results from the assumption that the only exchangeable oxygen present in the coal is that bound in water.

Pulsed NMR spectroscopy offers an alternative analytical technique for the analysis of hydrogen and moisture in coals. This method is fast and is not subject to the errors of the aforementioned methods. Quantitative elemental analysis by NMR spectroscopy relies on the simple proportionality between the observed signal intensity and the number of spins present in the sample. The initial value of the free induction decay (FID) of the resonant protons observed after a transient radio frequency (rf) pulse excitation can be used as a quantitative measure of the total hydrogen concentration in coals (*7*). Because the decay characterized by constants of the order of 10 μs contributes significantly to the ^1H NMR in coals, the pulse width used should be $<T_2/4$ (T_2 is the spin–spin relaxation time), and the recovery time of the NMR receiver should be of the order of the pulse width. The relative concentration of protons in the sample could be determined by a sample-transfer method if the coal sam-

ple and the calibration standard were exposed to the same homogeneous rf field and the measurements were made under the same experimental conditions. A plot of the logarithm of the signal amplitude versus the square of time, t, was used to extrapolate the FID to $t = 0$.

Using pulsed NMR spectroscopy, Lynch and Webster (8) studied water associated with brown coal and showed that the observed FID of the protons of coal samples not containing free water could always be resolved into two components: (1) a rapidly relaxing component that was attributed to organic hydrogen atoms and to chemisorbed water and (2) a slowly relaxing component attributed to physisorbed water.

Riley (9) investigated the suitability of a commercial 20-MHz pulsed-NMR spectrometer for routine total hydrogen and moisture determination in coals. Instead of measuring the initial amplitude of the FID, the amplitudes at 14 and 70 μs after the pulse were measured to determine the total hydrogen and moisture concentrations, respectively. For quantitation, a calibration curve containing corresponding amplitudes measured for reference coals was used. The reproducibility of the results obtained was good.

Cutmore et al. (10) investigated the possibility of using NMR spectroscopy for routine on-line analysis of water in coals by studying the effect of rank, particle size, and sample dimensions using ^1H NMR at proton resonance frequencies ranging from 6.5 to 60 MHz and found that the NMR results were independent of these factors.

In a study by Graebert and Michel (11) ^1H NMR spectroscopy was used for quantitative measurements of protons in a series of German brown coals. The water contents determined in this study compared favorably with the results obtained using thermogravimetry, near-infrared spectroscopy, and titration.

Using ^1H NMR spectroscopy as a technique for quantitation of hydrogen and moisture in coals implies that (1) all of the protons in the sample are observed by NMR, (2) the time dependence of the FID near $t = 0$ is known, and (3) the physical meaning of the different components of the FID is well understood. The validity of these assumptions will be examined in this chapter, using the Argonne Premium coals, by determining the "true" line shape of the whole FID using the solid echo, mathematical analysis of the full FID, and spin counting.

Experimental Details

The coals used in this study are from the Argonne Premium Coal Sample Program and are listed in Table I, where the coal identification number, seam, and

Table I. Proximate and Ultimate Analyses of the Argonne Premium Coals

Identification Number	Seam	Rank	H_2O^a	Ash^a	C^b	H^b	C^c	H^c
501	Pocahontas No. 3 (VA)	LVB	0.65	4.74	86.71	4.23	91.05	4.44
101	Upper Freeport (PA)	MVB	1.13	13.03	74.23	4.08	85.50	4.70
401	Pittsburgh No. 8 (PA)	HVB	1.65	9.10	75.50	4.83	83.20	5.32
701	Lewiston–Stockton (WV)	HVB	2.42	19.36	66.20	4.21	82.58	5.25
601	Blind Canyon (UT)	HVB	4.63	4.49	76.89	5.49	80.69	5.76
301	Illinois No. 6 (IL)	HVB	7.97	14.25	65.65	4.23	77.67	5.00
202	Wyodak–Anderson (WY)	SB	28.09	6.31	68.43	4.88	75.01	5.35
801	Beulah–Zap (ND)	L	32.24	6.59	65.85	4.36	72.94	4.83

ABBREVIATIONS: LVB, low-volatile bituminous; MVB, medium-volatile bituminous; HVB, high-volatile bituminous; SB, subbituminous; and L, lignite.
[a] Values are for as-received coals.
[b] Values are for dry coals.
[c] Values are for moisture- and ash-free coals.

rank are given. Only a partial list of the results of proximate and ultimate analyses performed in the laboratory of Commercial Testing and Engineering, Lombard, Illinois, are included in the table (12). Moisture concentration was determined by mass loss upon heating the sample in a convection oven at 105 °C and 40% relative humidity. Hydrogen concentration was estimated by using an elemental analyzer (Engelke, G., Commercial Testing and Engineering, Inc., personal communication, 1989).

The samples, prepared under humid nitrogen atmosphere, were obtained in brown glass ampules. Each ampule was opened in a glove box with a dry helium atmosphere; subsequently 50–100 mg of each coal were transferred from the ampules to preweighed 5-mm (o.d.) NMR tubes in less than 2 min. The tubes were sealed to preserve the samples and weighed. A reference water sample of the same geometry was prepared for ^{1}H spin counting. The sample lengths were approximately 50% of the length of the NMR coil.

All measurements were carried out at room temperature. A home-built solid-state NMR spectrometer operating at the ^{1}H resonance frequency of 100.06 MHz was used for line-shape and solid-echo determination. A proton-free probe was used to avoid background absorption. A Nicolet 2090 IIIA digital oscilloscope recorded 2048 data points using a 0.5-μs dwell time. The combined probe ringdown and receiver deadtime were 6.5 μs.

Before line-shape determination, the longitudinal relaxation times, T_1, were measured by using the inversion-recovery pulse sequence. The 100-point data set for each coal was analyzed by nonlinear least-squares fitting. The longer of the two T_1 constants extracted was used for the determination of the repetition rate for data averaging. A 90° pulse width of 1.0 μs was chosen for line-shape measurements. The method of spin temperature inversion on alternate pulses, with the subtraction of alternate scans in the averaging process (13), was necessary to eliminate base-line distortions from the NMR signal. The solid echo was produced by the pulse sequence $90_x°–\tau–90_y°$, with $\tau = 30$ μs. Twenty scans were collected in each of the NMR experiments.

The ^{1}H NMR combined rotation and multiple pulse decoupled (CRAMPS) spectra for the as-received and vacuum-dried coal samples were measured by using a Bruker MSL 300 spectrometer. The samples were spun in an alumina rotor at ~3.5 kHz, and the MREV−8 multipulse sequence (as described in ref. 14) used 90° pulses with widths of 1.5 μs. Free radical concentrations were determined by using a Bruker ER2000–SRC electron spin spectrometer at a frequency of 9.77 GHz. The coals (~20 mg) were air-dried and placed in 2-mm quartz tubes.

Thermogravimetric analyses (TGA) of the Argonne Premium coals were carried out using a Perkin-Elmer 7 series thermal analyzer system with argon as the purge gas. Approximately 50 mg of each coal was transferred to the platinum sample cup. The temperature was ramped from 40 to 110 °C at a rate of 1 °C/min and then held at 110 °C for 30 min.

Results

The Solid Echo. To measure the number of spins (protons) in a sample using NMR spectroscopy, the initial amplitude of the FID signal that occurs at the center of the 90° pulse must be known. However, in a single-pulse experiment, the first few microseconds of the protons' response to a transient rf pulse is always obscured by the probe ringdown and receiver deadtime. This loss of information cannot be completely eliminated, but it can be minimized by using a low-Q probe and a fast-recovery receiver. The solid echo (15), produced by the pulse sequence $90°_x-\tau-90°_y$ with the shortest possible pulse separation, makes it possible to circumvent the problem (11, 16). As was shown by Boden et al. (15), in solids in which dipolar interactions dominate the spectra an echo is formed following the $90°_x-\tau-90°_y$ sequence at $t = \tau$ after the second pulse. The τ dependence of the solid-echo amplitude, which results from the heteronuclear dipolar interactions (in systems containing two dipolar-coupled spin 1/2 species) or the interpair homonuclear dipolar interactions (in systems of loosely coupled spin 1/2 pairs), was not the subject of the present study. However, the decay of the solid echo for a fixed and sufficiently short τ is primarily due to the homonuclear dipolar interactions. Therefore, in coals, in which the proton–proton dipolar interactions dominate the ^{1}H NMR line shape, analysis of the signal following the top of the solid echo can be used to estimate the short-time behavior of the FID.

The solid-echo measurements were made by using the pulse separation $\tau = 30$ μs. Nonlinear least-squares fitting of the echo decay was performed by using either the Gaussian or the Lorentzian term alone or a superposition of Gaussian and Lorentzian terms:

$$A_G(t) = A_G(0) \exp\left[-0.5\left[\frac{t}{T_{2G}}\right]^2\right] \quad (1)$$

and

$$A_L(t) = A_L(0) \exp\left[\frac{-t}{T_{2L}}\right] \quad (2)$$

where $A_G(t)$ and $A_L(t)$ are the amplitudes at time t, with $t = 0$ at the top of an echo, and T_{2G} and T_{2L} are the transverse relaxation times of the Gaussian and Lorentzian components, respectively.

The analysis proved that the echoes of most of the coals in the set could be described by the superposition of a Gaussian and two Lorentzian forms. For coal 501, a low-volatile bituminous (LVB) coal, the superposition of two Gaussian forms described the echo best; for coal 101, a medium-volatile bituminous (MVB) coal, however, the superposition of Gaussian and Lorentzian forms was a better fit.

Only the types of line shapes found from this experiment were used in the analysis of the FID of the coals. Although the decay constants describing the solid echo agreed well with those obtained later from the analysis of the FID, the relative fractions of Gaussian and Lorentzian decays, as calculated from the solid echo, changed as a function of τ, perhaps because of the differences in the mobilities of these fractions.

Components of the FID. The same fitting routine was used in the computer analysis of the on-resonance FID of the studied coals. In Figure 1, the experimentally observed ^{1}H NMR signal and the nonlinear

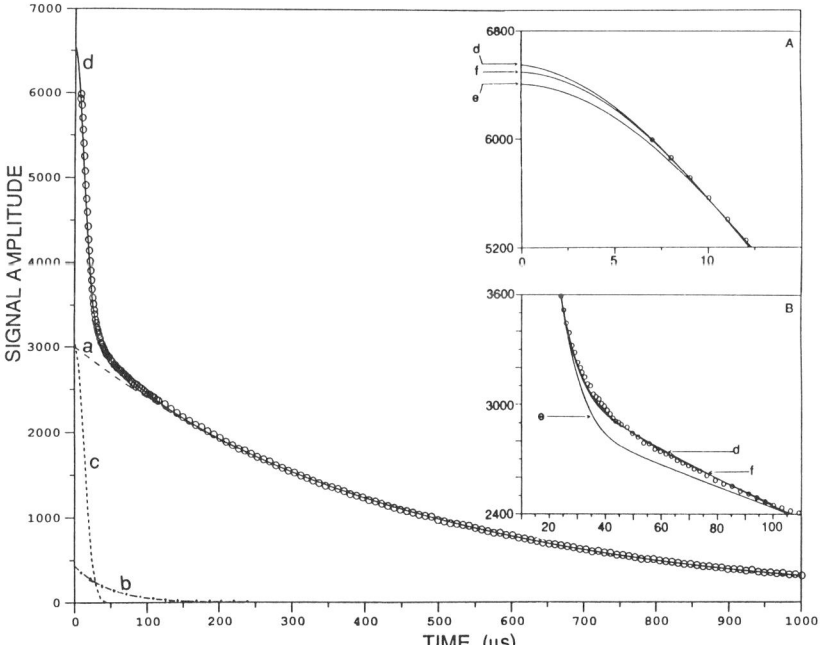

Figure 1. The numerical nonlinear least-squares fit of the ^{1}H FID of Argonne Premium coal No. 202. The experimental data points are denoted by open circles. (a) Lorentzian decay with T_{2L} = 435 µs; (b) Lorentzian decay with T_{2L} = 29 µs; (c) Gaussian decay with T_{2G} = 12.6 µs; (d) the superposition of curves a, b, and c; (e) numerical fit with one Gaussian and one Lorentzian decay function; and (f) numerical fit with two Gaussian and one Lorentzian decay function. Insets A and B are expansions of the initial portion of the FID.

least-squares fit are shown for coal 202. A slowly decaying component can be well fitted with a Lorentzian form (*see* eq 2) by analyzing the tail of the FID starting 200 μs after the excitation. Here the contribution from the shorter decays is negligible (*see* Figure 1, curve a). Subsequently, the contribution from the long decay was subtracted from the total FID, and the data were further analyzed for the shorter decays by using superpositions of Gaussian and Lorentzian forms as found from the analysis of the solid echo (Figure 1, curves b and c). The resulting decay constants and the corresponding fractions of the FID are compiled in Table II.

The superposition of the three functions (curves a, b, and c) is shown in Figure 1 as curve d. Curve d was compared with the fits for which Gaussian and Lorentzian decay functions (curve e) or two Gaussian and one Lorentzian decay functions (curve f) were used. Curve e does not match the experimental points in the middle of the decay. However, both curves d and f describe well the known portion of the FID, and only by the use of the solid echo could an unambiguous choice of the fitting functions be made. Also, because of the negligible intensity of the slowly decaying signal, the three-component decomposition of the FID could not be successfully performed for coals 501 and 101. For these two coals only two-component fits could be used.

Table II. ^{1}H FID Parameters for the Argonne Premium Coals

Coal No.	T_2			Fraction of Component		
	Fast[a]	Medium	Slow	Fast	Medium	Slow
501	12.7	52(4)[b]	—	0.969	0.031[b]	—
101	12.6	52(4)[b]	—	0.939	0.061[b]	—
401	12.5	28(8)	63(1)	0.912	0.016	0.072
701	12.4	22(1)	76(1)	0.850	0.089	0.061
601	12.2	24(2)	188(2)	0.777	0.138	0.084
301	12.3	37(3)	298(2)	0.778	0.065	0.157
202	12.6	29(1)	435(1)	0.447	0.085	0.468
801	12.7	29(1)	407(1)	0.364	0.092	0.544

[a] Average standard deviation was 0.04.
[b] Parameters were obtained by using the superposition of only two functions.

Spin Counting. The absolute number of proton spins in coals was found by comparing the extrapolated initial intensity of the FID to that of the reference sample that was measured several times during the course of the experiment. The results of the spin counting are presented in Table III where the total spin concentration per 1 g of sample, N_T, the concentration of spins exhibiting short and medium T_2 values, N_c, and the concentration of spins with the longest T_2 values, N_w, are given. From the proximate and ultimate analyses results of Table I, the total concentration of hydrogen atoms, the concentration of organic hydrogen atoms, and the concentration of hydrogen atoms in water present in the coal samples were calculated and are listed in Table III. Table III also gives the ratio of the proton spin counting results obtained by NMR spectroscopy to the corresponding numbers of hydrogen atoms calculated from the proximate and ultimate analyses of the Argonne Premium coals. In addition, the free radical concentrations as measured by electron spin resonance (ESR) spectroscopy are listed in Table III. The results of NMR measurements from Table III were used to calculate the weight-percent concentrations of the corresponding hydrogens in coals, which are listed in Table IV. The total weight loss (in percent) of the samples during TGA is also listed in Table IV.

Table III. Hydrogen and Free Radical Concentrations of Argonne Premium Coals

Coal No.	NMR Analysis[a]			Chemical Analysis[a]			Ratio[b]			ESR Analysis[c]
	N_T	N_c	N_w	N_T	N_c	N_w	N_T	N_c	N_w	N_e
501	2.47	—	—	2.55	2.51	0.043	97	—	—	1.5
101	2.52	—	—	2.48	2.41	0.074	102	—	—	1.3
401	2.91	2.70	0.21	2.95	2.84	0.11	99	95	(191)	1.4
701	2.57	2.41	0.16	2.63	2.47	0.16	98	98	100	2.2
601	3.55	3.25	0.30	3.44	3.13	0.31	103	104	97	1.8
301	2.75	2.32	0.43	2.85	2.40	0.45	97	97	96	1.1
202	3.58	1.92	1.66	3.97	2.09	1.88	90	92	88	1.7
801	3.44	1.57	1.87	3.92	1.76	2.13	88	89	87	1.5

[a] $\times 10^{-22}$/g of coal.
[b] Values are given as the ratio of $N \times 100$ from NMR analysis divided by N from chemical analysis.
[c] ESR analysis was used to measure free radical concentration, $\times 10^{-19}$/cm^{-3} (dry, mineral-matter free).

Table IV. Weight-Percent Concentration of Water and Organic Hydrogen in the Argonne Premium Coals

Coal No.	H_2O			Organic H	
	NMR Analysis	Proximate Analysis	TGA	NMR Analysis	Ultimate Analysis[a]
501	—	0.65	0.84	4.00	4.20
101	—	1.13	1.18	3.97	4.03
401	3.13	1.65	1.95	4.55	4.75
701	2.35	2.42	2.87	4.03	4.13
601	4.46	4.63	5.05	5.44	5.24
301	6.46	7.97	7.80	3.88	3.88
202	25.05	28.09	27.11	3.21	3.50
801	27.99	32.24	33.83	2.63	2.95

[a] Weight percent of organic hydrogen is determined with ultimate analysis as follows: (H [percent, dry] × 100)/(100 − H_2O [percent]).

Discussion

The Components of the FID. The analysis of the FID showed that a minimum of three components were present in most of the coals studied. The term minimum is used to indicate that two-component fits are usually not sufficient to fit the data. Assuming an arbitrary number of components to the decay, however, is not productive for understanding their nature.

The slowly decaying component of the FID corresponding to protons with high isotropic rotational mobilities suggests their complete detachment from the rigid coal framework. Any mobile water (moisture) present in coals would contribute to this component. Additional experiments indicated no significant contributions to the slowly relaxing component from other proton-containing species in the sample. First, the Fourier transform spectra of all the coals studied show only one narrow peak that has the same chemical shift as water (Figure 2). The height of the peak increased and its width decreased for the lower rank, wet coals. Second, the identity of the slowly decaying component was established for selected coals (202 and 801) by dry distillation of ~1.5 g of the coal under vacuum at 100 °C. Under these conditions, 3 h of evacuation was sufficient to completely remove the mobile species from the coal. A distillate, col-

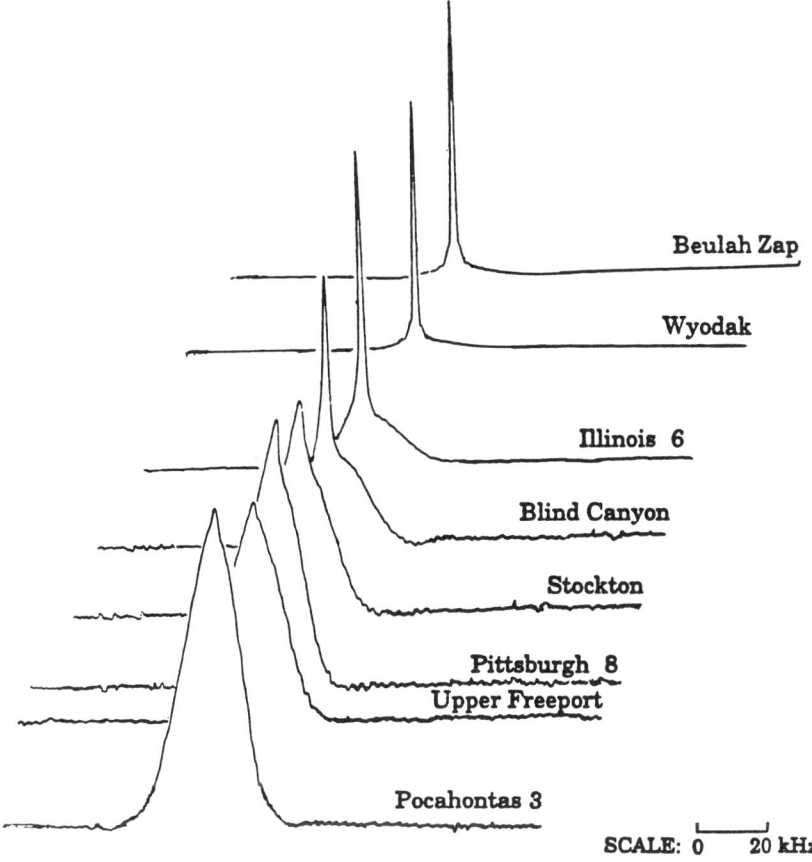

Figure 2. ^1H NMR spectra of the eight Argonne Premium coals.

lected in a liquid nitrogen trap, exhibited at room temperature an NMR spectrum identical to that of liquid water. No other peaks were present in the distillate spectrum. A small fraction of distillate, however, instantaneously evaporated upon the removal of the liquid nitrogen bath prior to the NMR experiment. Earlier studies by mass spectrometric analysis showed that the only gases evolved from a fresh sample of Illinois No. 6 coal upon outgassing at a pressure of 10^{-6} torr (1.33×10^{-4} Pa) at room temperature were H_2O and CO_2 (*17*). The intensities of the rapidly decaying components of the FID remained unchanged after the aforementioned treatment. Also, no detectable hydrogen-containing species besides water were found in our laboratory by using the TGA of the samples.

A further confirming identification of water as the material present before distillation is found in the ^1H NMR CRAMPS spectra of the coals. The CRAMPS spectrum of protons in coal 202 before distillation is shown in Figure 3a. A major peak is present at ~5 ppm, (the position characteristic of protons in water) with a shoulder near 7 ppm. Upfield at 2 ppm is a peak identified with the aliphatic portion of hydrogen in the coal. After vacuum-drying, the peak centered at 5 ppm disappeared, leaving peaks at 2 and 7.5 ppm (*see* Figure 3b), identified with aliphatic and aromatic protons, respectively. Thus, the slowly decaying component of the FID represents water, which in coals is mostly physisorbed and trapped in the microcapillary pores (*18*).

The CRAMPS technique has been used previously to probe chemical functionalities of hydrogen in coals (*14, 19, 20*). We will report the results of high-resolution solid-state NMR spectroscopy of ^{13}C and ^{1}H in the Argonne Premium coals at a later date.

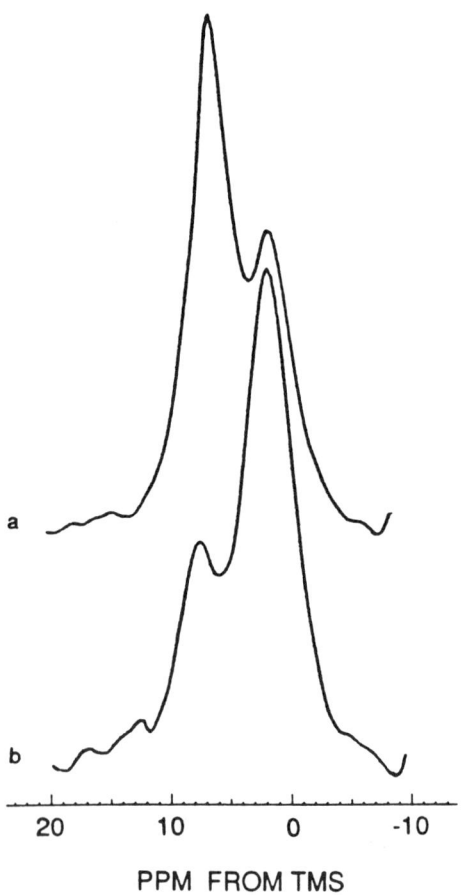

Figure 3. ^{1}H *CRAMPS spectra of Argonne Premium coal No. 202: (a) as received and (b) vacuum-dried at 100 °C.*

As indicated earlier, the rest of the FID, representing the "dry" coal, was fitted with the superposition of Gaussian ($T_{2G} \sim 12$ μs) and Lorentzian ($T_{2L} \sim 20\text{–}40$ μs) decay functions (Table II). An analysis of these two components was described for a series of coals of different ranks in studies by Jurkiewicz et al. (21) and Kamienski et al. (22). Theory (23) predicts that for spin systems with strongly coupled protons in the rigid-lattice limit and for a correlation time of motion $\tau_c >> T_2$, the transverse relaxation of protons immediately following a transient excitation can be approximated by a Gaussian function. The rapidly decaying component of the FID, therefore, describes the response of the protons in the rigid macrostructure of coal (e.g., protons attached to polynuclear aromatic rings). Because only 35–60% of protons in Argonne Premium coals are aromatic (as calculated from the CRAMPS spectra of the dry coals), the macrostructure contains nonaromatic protons that are also "rigid", such as protons in a polyadamantane type of structure (24). Thus, the high fraction of the Gaussian component reflects the rigidity of the macromolecular structure of coal. The average interproton distance corresponding to the values of T_{2G} in these samples is ~0.18 nm.

The Lorentzian portion of the decay with the medium T_2 represents protons for which anisotropic motion with $\tau_c < 10^{-5}$ s exists. Although some intramolecular mobility, such as the rotation of methyl groups, is expected in coals, it was proposed (21, 22) that this component represents molecules detached from the rigid macromolecular network. Strongly bound water may contribute partially to this signal, so the results contained in the present work may be viewed as a lower limit of the total water content.

Quantitative Measurements. As is evident from Figures 4 and 5, which illustrate the results of Tables III and IV, there is a very good agreement between the results of hydrogen spin counting by NMR spectroscopy and the hydrogen concentrations obtained via chemical analyses. Only for coals 202 and 801 did the results of the two methods differ by more than 5%. The repeatability of the results obtained by NMR spectroscopy was excellent in our experiments. Five determinations of the initial amplitude of the same sample had a deviation of only 0.5%, and the determination of the initial amplitude of the FID for five different 80-mg samples of the same coal had a deviation of ~4%. The error estimates for the results of the proximate analysis of moisture and the ultimate analysis for hydrogen were not available.

A comparison of the fraction N_w/N_T as calculated from NMR data and from the proximate analyses for six Argonne coals is shown in Figure 4a. Good agreement exists between the two quantities for most of the coals, and the agreement improved as the moisture content increased (especially for coals 202 and 801) despite the discrepancies in the concen-

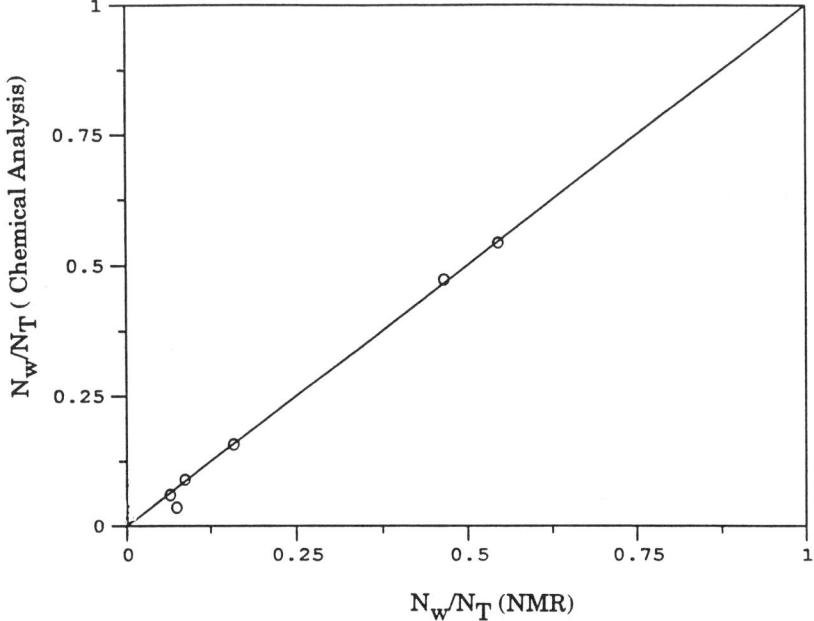

Figure 4a. The correlation of the results for moisture analysis by NMR and by mass loss upon heating of the sample: The plot of N_w/N_T from NMR spectroscopy vs. N_w/N_T from proximate analysis.

trations of hydrogen, N_w, and N_T determined by the two methods for these coals. A similar correlation is presented in Figure 4b, where the weight-percent concentration of the moisture determined from NMR spectroscopy is compared with the results of the proximate and TGA analyses (*see also* Table IV). Some discrepancies in the results are evident, especially for coals 202 and 801. Because the concentrations of carbonyl groups (25) and oxygen are highest in coals 202 and 801, the results of the proximate and TGA analyses for moisture reported for these two coals may be overestimated because of decarboxylation at temperatures above 60 °C. As was mentioned previously for coals 501 and 101, which have the lowest moisture contents of 0.65 and 1.13%, respectively, the three-component deconvolution of the FID could not be performed. For these two coals, the signals from water are included in the medium components of the decays, which consequently have extended T_2 values (~52 μs).

A similar comparison for the organic hydrogen content of the coal is illustrated in Figures 5a and 5b. Again, the fraction N_c/N_T calculated from NMR data showed very good agreement with the corresponding ultimate analysis data. However, the weight-percent concentrations of hydro-

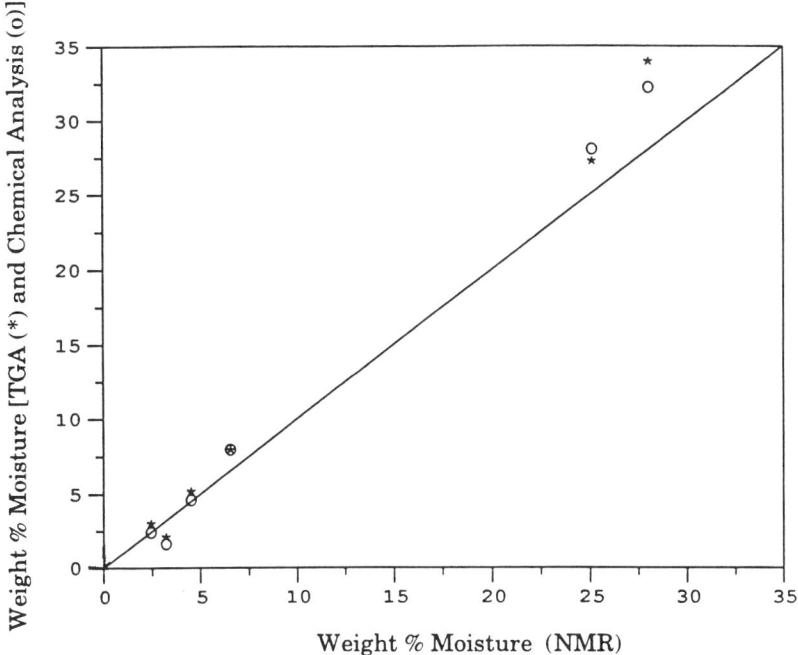

Figure 4b. The plot of the weight percent of moisture determined by NMR spectroscopy vs. that obtained from proximate (○) and ultimate (*) analyses.

gen as inferred from NMR data are lower (<10%) than the results of the elemental analysis (*see also* Table IV). A possible source of systematic loss of the NMR signal is the broadening of the resonance line by the dipolar interaction between protons and unpaired electrons present in organic free radicals in coals. The maximum number of protons made "invisible" by the presence of unpaired electrons can be estimated by using the concentration of the free radicals determined by ESR spectroscopy (*see* Table III). Assuming the diffusion barrier describing the diffusion of spin polarization to be ~0.8 nm (*26*) and assuming a uniform distribution of protons and isolated unpaired electrons in the coal, 3–6% of protons in the coals were subject to a proton–electron dipolar interaction that would move their resonances beyond observation. Because of the heterogeneity of coal, a nonuniform distribution of free radicals is likely, and this nonuniform distribution would thus decrease the estimate of proton content. However, no correlation exists between the concentration of the free radicals in the coals as determined by ESR spectroscopy and the difference observed between the two methods of analysis.

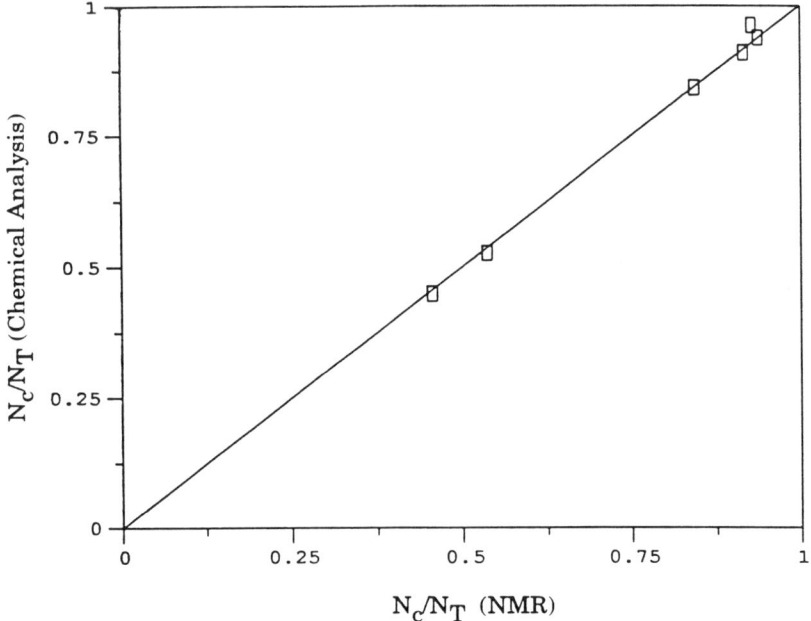

Figure 5a. The plot of N_c/N_T obtained by NMR spectroscopy vs. N_c/N_T obtained by ultimate analysis.

Conclusions

Solid-echo experiments allow the determination of line shapes and initial intensities of ^{1}H NMR in Argonne Premium coals. A superposition of Gaussian ($T_2 \sim 12$ μs) and Lorentzian ($T_2 \sim 20$–40 μs) terms adequately described the on-resonance FID of the organic protons in coals. The Gaussian fraction corresponds to rigid protons in the macrostructure of coal. The Lorentzian fraction is ascribed to protons that belong to fragments of the coal framework exhibiting some molecular mobility. For wet coals (>2% moisture content), an additional slowly decaying Lorentzian function must be added to the description of the FID.

The intensities obtained by NMR spectroscopy were compared with concentrations of hydrogen and water obtained via chemical and TGA analyses. Despite the shortcomings of all three methods, the results were in very good agreement. The advantages of NMR spectroscopy are (1) the simplicity of the analysis in which not only the concentrations but also the mobilities of all types of protons in the sample can be analyzed simultaneously and (2) the possibility of using NMR spectroscopy for on-line measurements where large numbers of samples with considerable volumes are analyzed. The development of a portable ^{1}H NMR spectrometer could combine these advantages with capability of on-site measurement.

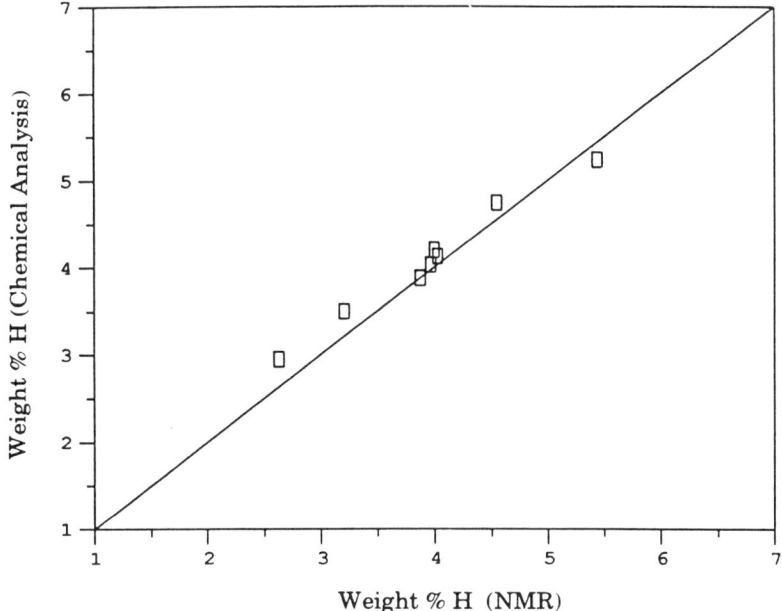

Figure 5b. The plot of the weight percent of organic hydrogen in coals obtained from NMR spectroscopy vs. that obtained from ultimate analysis.

Acknowledgments

We thank Dieter Michel for helpful discussions and Terry King for the use of the Perkin-Elmer thermogravimetric analyzer. The Institute for Physical Research and Technology is supported by the U.S. Department of Energy (Basic Energy Sciences Program, Chemical Science Division) under Contract No. W–7405–Eng–82.

References

1. Gould, G.; Visman, J. In *Coal Handbook;* Meyers, R. A., Ed.; Marcel Dekker: New York, 1981; Chapter 2.
2. Melton, C. E.; Giardini, A. A. *Fuel* **1976**, *55*, 155.
3. Allardice, D. J.; Evans, D. G. *Fuel* **1971**, *50*, 201.
4. Gethner, J. S. *Appl. Spectrosc.* **1987**, *41*, 60.
5. Belcher, R.; Spooner, C. E. *Fuel* **1947**, *26*, 55.
6. Finseth, D. *ACS Div. Fuel Chem. Prepr.* **1987**, *32*(4), 260.
7. Gerstein, B. C.; Pembleton, R. G. *Anal. Chem.* **1977**, *49*, 75.
8. Lynch, L. J.; Webster, D. S. *Fuel* **1979**, *58*, 429.

9. Riley, J. T. *Am. Lab. (Fairfield, Conn.)* **1983**, *15*(8), 17.
10. Cutmore, N. G.; Sowerby, B. D.; Lynch, L. J.; Webster, D. S. *Fuel* **1986**, *65*, 34.
11. Graebert, R.; Michel, D. *Fuel* **1990**, *69*, 826.
12. Vorres, K. S. *Energy Fuels* **1990**, *4*, 420.
13. Stejskal, E. O.; Schaefer, J. J. *Magn. Reson.* **1975**, *18*, 560.
14. Gerstein, B. C. *Philos. Trans. R. Soc. London* **1981**, *A299*, 521.
15. Boden, N.; Gibb, M.; Levine, Y. K.; Mortimer, M. *J. Magn. Reson.* **1974**, *16*, 471.
16. Lynch, L. J.; Sakurovs, R.; Barton, W. A. *Fuel* **1986**, *65*, 1108.
17. Fuller, E. L. In *Coal Structure;* Gorbaty, M. L.; Ouchi, K., Eds.; American Chemical Society: Washington, DC, 1981; Chapter 19.
18. Allardice, D. J.; Evans, D. G. *Fuel* **1971**, *50*, 236.
19. Jurkiewicz, A.; Bronnimann, C. E.; Maciel, G. E. *Fuel* **1989**, *68*, 871.
20. Derbyshire, F.; Marzec, A.; Schulten, H.; Wilson, M.; Davis, A.; Tekely, P.; Delpuech, J.; Jurkiewicz, A.; Bronnimann, C.; Wind, R.; Maciel, G.; Narayan, R.; Bartle, K.; Snape, C. *Fuel* **1989**, *68*, 1091.
21. Jurkiewicz, A.; Marzec, A.; Pislewski, N. *Fuel* **1982**, *61*, 647.
22. Kamienski, B.; Pruski, M.; Gerstein, B. C.; Given, P. H. *Energy Fuels* **1987**, *1*, 45.
23. Abragam, A. *The Principles of Nuclear Magnetism;* Clarendon: Oxford, England, 1961.
24. Davidson, R. M. "Molecular Structure of Coal"; Report No. ICTIS/TR 08; IEA Coal Research: London, 1980.
25. Solum, M. S.; Pugmire, R. J.; Grant, D. M. *Energy Fuels* **1989**, *3*, 187.
26. Wind, R. A.; Jurkiewicz, A.; Maciel, G. A. *Fuel* **1989**, *68*, 1189.

RECEIVED for review June 8, 1990. ACCEPTED revised manuscript December 21, 1990.

20

Bloch-Decay and Cross-Polarization–Magic-Angle Spinning ^{13}C NMR Study of the Argonne Premium Coals

Effects of High-Speed Spinning

James A. Franz and John C. Linehan

Battelle Northwest, P.O. Box 999, Richland, WA 99352

Bloch-decay and variable-contact-time ^{13}C cross-polarization–magic-angle spinning (CP–MAS) NMR data at a spinning rate of 13 kHz are presented for the Argonne Premium coals. High-field (7.05 T, 75 MHz, 13-kHz MAS) aromaticities are in satisfactory agreement with low-field CP–MAS (2.3 T, 25 MHz, 4-kHz MAS) results, with the exception of the high oxygen containing Wyodak–Anderson coal. Bloch-decay aromaticities obtained at 13-kHz MAS agree with CP–MAS results at 25 MHz (4-kHz MAS) for all of the bituminous coals, but CP–MAS gives significantly lower aromaticities than Bloch decay for the lignite (Beulah–Zap) and subbituminous (Wyodak–Anderson) coals. A Bloch-decay and variable-contact-time CP–MAS study of the ratios of protonated, nonprotonated, and isolated aromatic carbons of acenaphthene revealed small systematic errors for individual carbon intensities, but satisfactory average relative intensities for different structural groups. High-speed MAS leads to measured aromaticities within 3–5% of those measured by low-speed MAS for medium- and high-rank coals. However, significant errors in CP–MAS aroma-

ticities of low-rank coals, together with the failure to detect an appreciable fraction of oxygenated and quaternary carbons, suggest that Bloch decay is the method of choice for determination of structural distributions. Sources of disagreement between Bloch decay and CP–MAS aromaticities for Wyodak–Anderson and Beulah–Zap coals are discussed.

THE COMBINED USE of magic-angle spinning (MAS), high-power dipolar decoupling, and polarization transfer in ^{13}C cross-polarization–magic-angle-spinning (CP–MAS) solid-state NMR spectroscopy provided the first effective tool for statistical determination of the structure of coal (1–9). CP–MAS ^{13}C NMR spectroscopy has since been applied to the characterization of a wide range of coals, macerals, and organic geochemical materials (10–22). The large sensitivity gain provided by the CP–MAS method compared to the Bloch-decay method has led to the examination of the quantitative limitations of CP–MAS for determination of coal structure, in view of the special problems of structural heterogeneity and mobility and the free radical content of coal.

The quantitative reliability of ^{13}C CP–MAS NMR spectroscopy for measurement of ratios of categories of organic structure has been defined in model-compound studies (23, 24) and in careful comparisons of Bloch-decay and CP–MAS results for whole coals and macerals (25, 26). The mechanisms by which molecular motion affects polarization transfer and dipolar decoupling were discussed in a useful essay by Earl (22). Finally, spin-counting studies (12, 25–27) revealed the amount of carbon in coals rendered undetectable by paramagnetic centers. The quantitative reliability of ^{13}C CP–MAS NMR spectroscopy for the determination of coal structure has been reviewed (25) and debated (28).

In this study, we determined carbon aromaticities and structural distributions of the Argonne Premium coals by Bloch-decay and by ultrahigh-speed CP–MAS NMR spectroscopy. The ultrahigh-speed CP–MAS and Bloch-decay results are compared with conventional-spinning-speed CP–MAS data to evaluate the level of additional error introduced by ultrahigh-speed MAS.

The conventional model for growth of magnetization during a CP contact period is given by eqs 1 and 2 (29, 30)

$$M(t) = M(0)\lambda^{-1}\left[1 - \exp\left(\frac{-\lambda t}{T_{\text{CH}}}\right)\right]\exp\left(\frac{-t}{T_{1\rho}^{\text{H}}}\right) \quad (1)$$

where

$$\lambda = 1 + \left(\frac{T_{\mathrm{CH}}}{T_{1\rho}{}^{C}}\right) - \left(\frac{T_{\mathrm{CH}}}{T_{1\rho}{}^{H}}\right) \approx 1 - \frac{T_{\mathrm{CH}}}{T_{1\rho}{}^{H}} \tag{2}$$

$M(0)$ is the maximum equilibrium magnetization of the ^{13}C system in contact with the ^{1}H system in the absence of relaxation processes, t is the cross-polarization contact time (seconds), $1/T_{\mathrm{CH}}$ is the cross-polarization rate during the contact period, $T_{1\rho}{}^{H}$ is the proton spin–lattice relaxation time in the rotating frame, $T_{1\rho}{}^{C}$ is the carbon rotating-frame relaxation time, and $M(t)$ is the resulting time-dependent carbon magnetization.

Successful quantitation of structural bands in amorphous mixtures such as coal requires that all individual carbons of a mixture be polarized before significant carbon or proton rotating-frame relaxation occurs and that $T_{1\rho}{}^{H}$ values be sufficiently uniform such that the decay portion of the contact-time curve reflects the composite magnetization, $M(0)$, of each structural category in the sample. Although the requirement for successful quantitative measurements, $T_{1\rho}{}^{C} > T_{1\rho}{}^{H} \gg T_{\mathrm{CH}}$, holds for ideal nonmobile model compounds lacking carbon atoms isolated from protons, a number of simple organic compounds provide notorious examples where $T_{1\rho}{}^{H} < T_{\mathrm{CH}}$ (23, 24). The inequality $T_{1\rho}{}^{C} > T_{1\rho}{}^{H} \gg T_{\mathrm{CH}}$ and the requirement that values of $T_{1\rho}{}^{H}$ fall in a narrow range are likely to be only partially met with coals. Nevertheless, assuming $T_{1\rho}{}^{C} > T_{1\rho}{}^{H} \gg T_{\mathrm{CH}}$, eq 1 reduces to eq 3, and a composite $T_{1\rho}{}^{H}$ is determined from the decay portion of a variable-contact-time experiment (eqs 4 and 5).

$$M(t) = M(0)\left[\exp\left(\frac{-t}{T_{1\rho}{}^{H}}\right) - \exp\left(\frac{-t}{T_{\mathrm{CH}}}\right)\right] \tag{3}$$

$$M(t) = M(0)\left[\exp\left(\frac{-t}{T_{1\rho}{}^{H}}\right)\right] \tag{4}$$

$$\ln M(t) = \ln M(0) - \frac{t}{T_{1\rho}{}^{H}} \tag{5}$$

In current practice (25), a portion of the variable-contact-time curve is fit to eq 3 by using the value of $T_{1\rho}{}^{H}$ determined according to eqs 4 and 5 to obtain the composite magnetization, $M(0)$ and a composite CP (T_{CH}) time for a spectral region. This approach yields aromaticities that occasionally agree with Bloch-decay results for well-behaved mid- to high-rank coals, but more typically are a few percent lower than Bloch-decay aromaticities for coals and macerals (25, 26). The weakness of eqs 1 and 2 as a statistical model for coal is revealed by their inability to model the entire contact curve of a coal. These equations also neglect treatment of dipolar oscillation, which is observed in most coal contact curves.

High-Speed Spinning and CP–MAS Quantitation

Observation of ^{13}C at high fields requires spinning at sufficiently high speeds (e.g., 13-kHz MAS for detection of carbon at 75-MHz) to eliminate overlap of residual spinning sidebands of the aromatic carbons over the aliphatic band in the spectra of complex mixtures. Commercially available probes spin 5-mm sample rotors at up to 15 kHz with air, and spinning speeds up to 23 kHz have been reported for small rotors (*31*). The achievement of true high-speed spinning would seem to have solved the problem of obtaining high-field CP–MAS ^{13}C NMR spectra. However, Wind et al. (*32*) concluded that the modulation of the Hartmann–Hahn mismatch curve caused by high-speed MAS leads to distortions in peak intensities for protonated versus nonprotonated carbons. Figure 1A shows a 13-kHz MAS Hartmann–Hahn mismatch curve for the Lewiston–Stockton coal, which illustrates the effects of spinning at speeds approaching the proton–proton dipolar interaction. This phenomenon was first demonstrated for adamantane by Stejskal et al. (*33*). Amplitude-modulation of H_{IS}, the time-dependent Hamiltonian describing polarization trans-

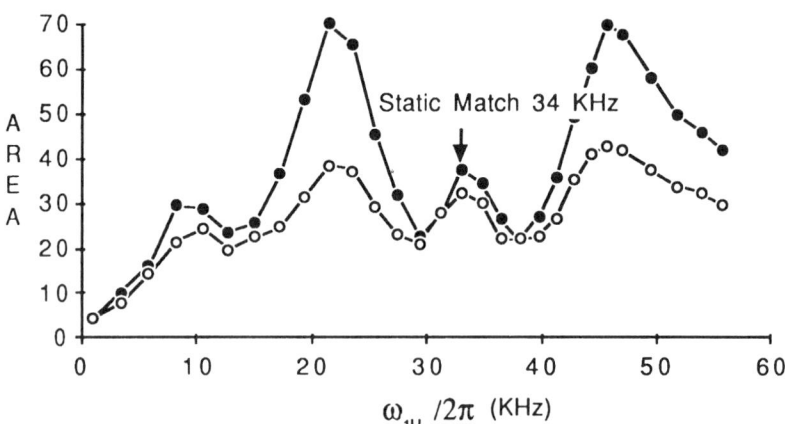

Figure 1A. Hartmann–Hahn matching curve of Lewiston–Stockton coal determined at 13-kHz MAS with a contact time 200 μs. The decoupler field (ω_{1H}) was varied at a constant carbon field ($\omega_{13C}/2\pi$ = 34 kHz). Variable-contact-time data were collected at $\omega_{1H}/2\pi$ = 21 kHz in this study unless otherwise indicated. Closed circles are aromatic carbon intensities, and open circles are aliphatic intensities.

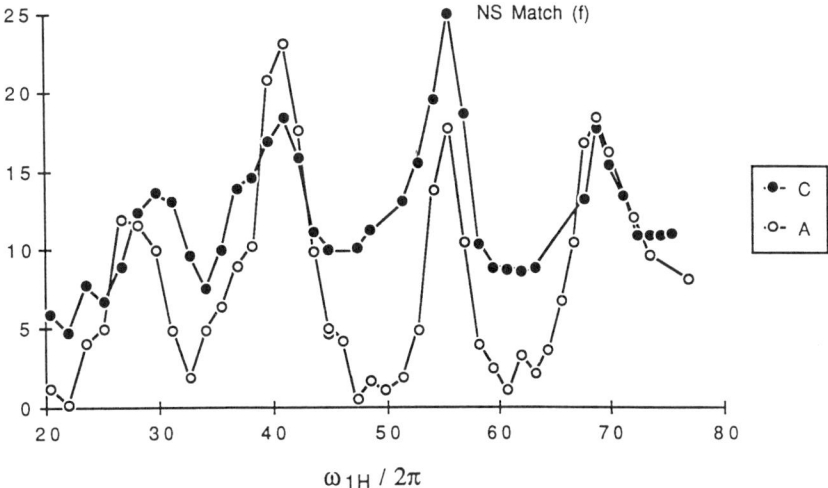

Figure 1B. Hartmann–Hahn curve at 13-kHz MAS for 4,4'-dimethoxybibenzyl with $\omega_{13C}/2\pi = 54$ kHz while varying $\omega_{1H}/2\pi$ from 20 to 74 kHz. Solid circles are carbon C (see structure 1 on page 382), and open circles are carbon A.

fer from the abundant to the rare spin reservoir, results in polarization-transfer peaks at $f(\pm\omega_1)$ and $f(+2\omega_r)$ and suppression of polarization transfer at the static (nonspinning) match, f (33, 34). The carbon contact power at the nonspinning Hartmann–Hahn match is given in frequency units by f; $f(\pm\omega_r)$ and $f(\pm2\omega_r)$ are the first and second pairs of modulation maxima in the Hartmann–Hahn matching curve, with a spinning rate of $\omega_r/2\pi$. The peaks at $\pm\omega_r$ and $\pm2\omega_r$ result from amplitude modulation by rotation of C–H internuclear vectors at ω_r.

At spinning speeds that are comparable to or larger than the proton line width in the rotating frame, the frequency modulation of H_{II}, the Hamiltonian for interaction between pairs of protons, results in frequency modulation of the rate of H–H mutual spin flips. This modulation is imposed on the amplitude-modulation (AM) bands at $f(\pm\omega_r)$ and $f(\pm2\omega_r)$ and restores intensity at the static match, f. This modulation leads to a smaller number of modulation sidebands than with low-speed spinning, and these modulation sidebands lie outside of the proton line width (7). This technique requires an offset of the carbon or proton channel power from the static match (f) to achieve optimal polarization-transfer rates. From available theoretical treatments of the polarization rate in the presence of MAS (34), it is not readily apparent that high-speed spinning will affect the ultimate equilibrium magnetization transferred to carbon in the absence of relaxation.

In the study of Wind et al. (*32*), a sample of the polymer Torlon (a poly(amide–imide) manufactured by Amoco) was examined at 1.4 T and at MAS speeds of 2.5, 7.3, 11.3, and 15.6 kHz at a single contact time of 2 ms as a function of the Hartmann–Hahn mismatch. The ^{13}C spectrum of Torlon shows a carbonyl band (159 ppm) and partially resolved nonprotonated (134 ppm) and protonated (125 ppm) aromatic carbon bands. The ratios of protonated to nonprotonated aromatic ^{13}C nuclei, which were equal in intensity at 2.5-kHz MAS, appeared with different intensities at the higher MAS speeds. At the $f(\pm\omega_r)$ and $f(\pm 2\omega_r)$ mismatch maxima, nonprotonated intensities were emphasized, whereas at the static match, protonated carbon intensities were emphasized. The widths of the nonprotonated modulation peaks were narrower than the protonated peaks. From these observations, Wind et al. (*32*) concluded that quantitative results could not be straightforwardly obtained with high-speed MAS. The effects described by Wind can be seen in Figure 1B for 4,4′-dimethoxybibenzyl (structure **1**).

With a carbon matching power of 54 kHz, the proton contact power is varied from 20 to 74 kHz with a 1.5-ms contact time at 13-kHz MAS. The intensities of one of the carbons ortho to the methoxy group (filled circles, 110 ppm, carbon C) are compared to the nonprotonated oxygen-substituted aromatic carbon at 158 ppm (carbon A). The protonated aromatic carbon has greater intensity at f (nonspinning match), but the intensities invert at the $f - \omega_r = 41$ kHz peak and are about equal in intensity at the $f - 2\omega_r$ and $f + \omega_r$ peaks.

The results of Wind et al. (*32*) show that polarization-transfer rates vary for protonated and nonprotonated carbons at different modulation maxima in a single-contact-time experiment. The key question that remains to be addressed for coals and model compounds is whether high-speed spinning will affect the equilibrium magnetization achieved in a variable-contact-time experiment for protonated versus nonprotonated carbons within a modulation maximum. For coals, polarization transfer will also be affected by molecular motion, dielectric effects of ionic domains, and inorganic and organic paramagnetic species.

Thus, in this study Bloch-decay and high-speed CP–MAS data were collected for the Argonne Premium coals for comparison with a complete recent study of the Argonne coals by Solum et al. (*35*). Bloch-decay and variable-contact-time CP–MAS results for acenaphthene, which possesses a selection of protonated aliphatic, protonated aromatic, nonprotonated aromatic, and isolated nonprotonated aromatic carbons, were examined to reveal the magnitude of distortions in carbon intensities induced by high-speed MAS. Acenaphthene provides examples of nonprotonated aromatic carbons (D and F in structure **2**) and isolated nonprotonated aromatic carbon (carbon E in structure **2**). 4,4′-Dimethoxybibenzyl was examined to determine the effects of contact power on carbon intensities and rotating-frame relaxation.

2

Experimental Details

Coals. The eight Argonne Premium coals used in this study were obtained from the Argonne Premium Coal Sample Program. The vials were opened under nitrogen, dried under vacuum (at 100 °C and 0.05 mm Hg for 4–8 h), and loaded into sample rotors in a nitrogen glove bag. Beulah–Zap and Wyodak–Anderson coals were also dried for 8- and 24-h periods, with no resulting difference in spectroscopic behavior. The samples were spun with nitrogen gas.

CP–MAS Experiments. ^{13}C NMR experiments were carried out with a Varian VXR-300 instrument using Doty Scientific, Inc., high-speed CP–MAS probes. The rotors were 5-mm silicon nitride cylinders with Vespel polymer end caps, packed with 75–80 mg of coal. Most CP–MAS experiments were run with a 21-kHz proton field, corresponding to the spinning modulation maximum at 13 kHz below the nonspinning match at 34 kHz. CP–MAS spectra were obtained by using a 4.5-μs, 90° proton pulse with 74-kHz dipolar decoupling gated on during free induction decay (FID) acquisition. The coals were initially examined with relaxation delays of 2, 5, 10, and 20 s at fixed contact times of 400 μs. An increase in aromaticity of 5–7% was observed from 2 to 5 s, with no change between 5 and 10 s for all coals except Pocahontas No. 3, which showed a slight increase between 10 and 15 s. A 5-s recycle delay was then used in variable-contact-time CP–MAS experiments for all coals except Pocahontas No. 3, which was examined with a 20-s relaxation delay. Contact-time curves were determined for 15–23 values between 10 μs and 35 ms. The areas of the aromatic sidebands, amounting to typically 4% of the center band, were added to the aromatic integral. Hartmann–Hahn matching curves were checked for each of the coals to allow careful match of the carbon and proton fields. After most of the data of this study were collected, the original ~100-W broadband pulse amplifier was augmented with a 300-W American Microwave Technology linear broadband pulse amplifier. This amplifier provided variable power and a significantly improved pulse shape, an important requirement for ultrahigh-speed CP–MAS experiments. Acenaphthene, 4,4′-dimethoxybibenzyl, and several coals were examined at higher proton and carbon matching fields to examine the effects of higher power levels and improved carbon pulse shape on carbon ratios, coal aromaticities, and $T_{1\rho}^H$ values.

Bloch-Decay Experiments. In Bloch-decay experiments, 450–850 transients were collected by using a 4.5-μs (90°) carbon pulse, 74-kHz dipolar decoupling field gated on during FID acquisition, and a 200-s relaxation delay. For all experiments, 64-ms acquisition times with a 100-kHz spectral window were employed. Under Bloch-decay conditions, the rotor caps and/or probe components contributed about 20% of the total signal, and this contribution was corrected by careful FID subtractions.

Bloch-Decay, CP–MAS, and T_1 Characterization of Acenaphthene. Carbon spin–lattice relaxation times (T_1) were determined by the inversion-recovery method, using a 400-s recycle time with 74-kHz dipolar decoupling during acquisition. Bloch-decay spectra were acquired by using a 400-s delay at 13-kHz MAS. Variable-contact-time CP–MAS spectra were acquired by using a 60-s recycle delay because of long proton spin–lattice relaxation times (T_1^H). The unprotonated aromatic carbons (D, E, and F) (*see* structure **2**) exhibited only modest dipolar oscillation and gave values of T_{CH}, $M(0)$, and $T_{1\rho}^H$, using method B (*see* the "Results" section), in good agreement with values calculated by the simplex-fit of all data between 10 μs and 20 ms in which $M(0)$, $T_{1\rho}^H$, and T_{CH} were varied with $T_{1\rho}^C$ held constant at 10 s. The protonated aromatic (A, B, and C) and aliphatic (G) carbons exhibited pronounced dipolar oscillation. The contact data were therefore edited to remove points between 150 μs and 3 ms before simplex fit to eq 1.

Results

The carbon magnetization, $M(0)$ and the $T_{1\rho}^H$ values were extracted from contact data by two procedures:

- Method A: Least-squares analysis of a semilog plot of spectral intensity versus time (e.g., eq 5) of contact data greater than ~3 ms provided $T_{1\rho}^H$ and $M(0)$.
- Method B: The contact curve was edited to remove points between 150 μs and 3 ms. A four-parameter simplex program was used to fit the contact data to eq 1 using $T_{1\rho}^H$ fixed at the value found in method A and $T_{1\rho}^C$ fixed at 10 s. Values of T_{CH} and $M(0)$ are obtained from the simplex treatment. Examples of simplex fits using method B are shown in Figure 2 (dotted lines). Values of $M(0)$ from sp^2 (aromatic

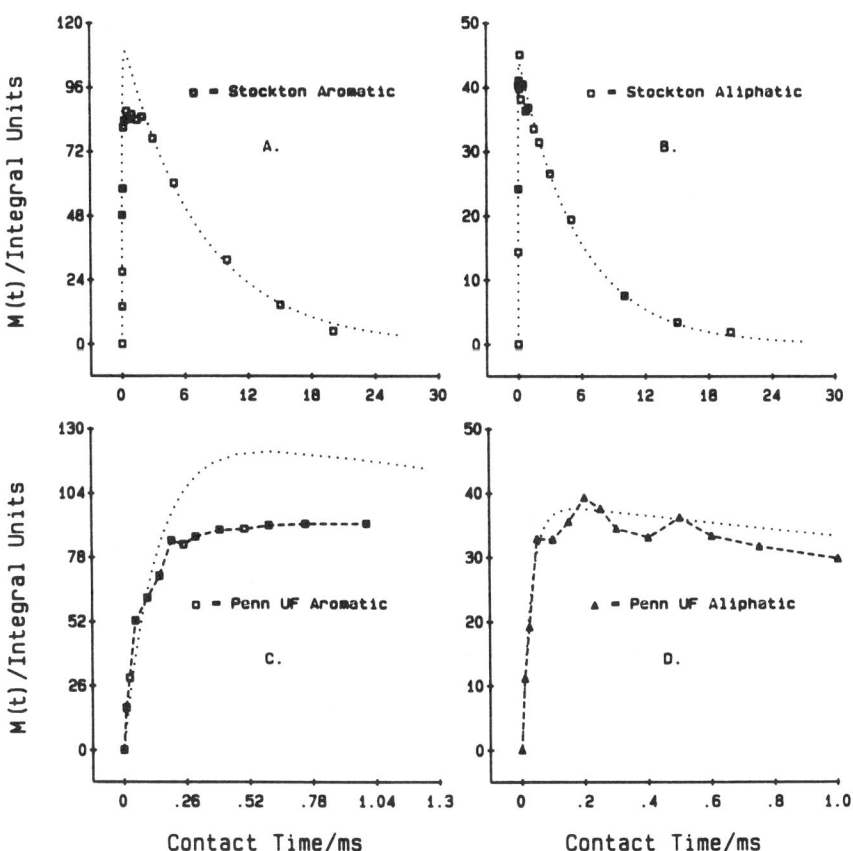

Figure 2. Contact curves for Lewiston–Stockton and Upper Freeport coals. Dotted lines are simplex fits of the data to eq 1, excluding data between 200 μs and 3 ms (method B).

plus carbonyl) and aliphatic spectral regions were combined to estimate aromaticities, $f_a = M(0)_{ar}/(M(0)_{ar} + M(0)_{al})$, where ar and al denote aromatic and aliphatic portions, respectively.

Table I presents $T_{1\rho}{}^H$ values determined by method A and T_{CH} values obtained from method B. Errors in T_{CH} and $T_{1\rho}{}^H$ are ±30 µs and ±0.4 ms (1σ), respectively. Table II provides chemical-shift distributions from the integration of Bloch-decay spectra. Table III presents chemical-shift distributions from CP–MAS data generated by integrating the spectral region at the maximum of the aromatic or aliphatic contact array, respectively. The aromatic and aliphatic integral distributions were then scaled to agree with method-A aromaticities. Because different functional groups evolve at different rates across the spectrum, this method, which was also used in the study by Solum et al. (35), introduces error into the distributions. However, the experimental time to acquire sufficient signal-to-noise ratios for measurement of the evolution of the smaller regional integrals, such as those of the carboxyl and aryl ether carbon regions, was prohibitive, being longer than the time required for the preferred Bloch-decay experiment.

Discussion

Time Dependence of Magnetization Growth and Decay.
Representative contact time curves are shown for Lewiston–Stockton and Pennsylvania Upper Freeport coals in Figure 2. The decay portions of the contact curves gave excellent fits to simple exponential decays for all of the Argonne Premium coals. The calculated curves for the aromatic regions (dotted lines, Figures 2A and 2C) result in values of $M(0)$ about 20% greater than the maximum observed experimental magnetization. Aromatic and aliphatic $T_{1\rho}{}^H$ values were consistently lower than values reported by Solum et al. (35). The lower $T_{1\rho}{}^H$ values observed here were suspected to be in part due to the relatively low proton field (21 kHz, corresponding to the 13-kHz mismatch dictated by the then-available carbon field of 34 kHz) compared to the 40-kHz matched carbon and proton fields employed by Solum et al. (35). However, a proton contact field of 47 kHz (34-kHz carbon field) gave similar aromaticities (see Figures 3 and 4) for the Argonne Premium coals. The degree to which the extrapolated values of $M(0)$ for the aromatic carbons from method A exceed the observed spectral intensities may reflect (1) the delayed growth of more slowly polarizing carbons and (2) excursions from the optimal Hartmann–Hahn match due to observe amplifier pulse droop. For example, the amplitude of the 75-MHz carbon (observe) pulse produced by the Varian 100-W amplifier was found to decrease ("droop") by approximately

Table I. Relaxation Parameters and Aromaticities for Argonne Coals: Bloch-Decay and CP–MAS Results

Coal	T_{CH} (μs)		$T_{1\rho}^H$ (ms)		f_a, Bloch Decay	f_a, CP–MAS			f_a, Previous Work[c]
	Ar	Al	Ar	Al		Method A[a]	Method B[b]	Average	
Beulah–Zap	41	18	6.3	5.5	0.77	0.67	0.63	0.65	0.66
Wyodak–Anderson	35	20	6.9	6.0	0.76	0.54	0.51	0.53	0.63
Blind Canyon	120	27	6.5	6.5	0.67	0.67	0.65	0.66	0.65
Illinois No. 6	104	20	6.9	5.9	0.70	0.73	0.70	0.72	0.72
Pittsburgh No. 8	144	30	7.2	7.4	0.72	0.79	0.78	0.78	0.72
Lewiston–Stockton	104	25	6.2	6.3	0.77	0.77	0.77	0.77	0.75
Upper Freeport	123	31	8.3	6.6	0.80	0.78	0.77	0.78	0.81
Pocahontas No. 3	85	22	6.9	5.7	0.89	0.86	0.84	0.85	0.86

NOTE: Our f_a values were obtained with the following experimental parameters: 75-MHz magnetic field and 13-kHz MAS spinning rate. Errors are ca. ±3% for f_a values, ca. ±0.3 ms (1σ) for $T_{1\rho}^H$, and ca. ±15 μs for T_{CH}. Abbreviations: Ar, aromatic; and Al, aliphatic.
[a]Method A determines f_a from least-squares treatment of contact data >3 ms using eq 5.
[b]Method B determines f_a from simplex fitting of data to eq 1 using fixed $T_{1\rho}^H$ from method A with exclusion of contact data between 200 μs and 3 ms.
[c]For comparison purposes, the f_a values obtained by Solum et al. (34) are included here.

Table II. 75-MHz–13-kHz MAS Bloch-Decay Results

Coal ppm:	>190	190–165	165–145	145–130	130–106	106–81	81–50	50–22	22–0	f_a
Beulah–Zap	0.7	6.6	13.2	18.1	30.8	7.6	5.3	13.9	3.8	77.1
Wyodak–Anderson	3.9	7.1	15.1	19.7	28.0	2.1	0.6	16.6	6.8	75.9
Blind Canyon	0.0	3.3	12.3	19.7	29.0	2.2	2.6	25.2	5.7	66.5
Illinois No. 6	0.0	1.1	10.8	22.4	32.3	3.8	3.9	20.4	5.3	70.4
Pittsburgh No. 8	0.0	1.7	10.7	23.9	34.4	0.9	1.6	18.9	7.9	71.5
Lewiston–Stockton	0.0	2.9	10.0	21.6	38.9	3.3	1.4	15.4	6.4	76.8
Upper Freeport	0.0	2.0	8.5	25.8	42.6	1.4	0.4	14.3	5.1	80.2
Pocahontas No. 3	0.0	1.8	6.7	25.5	51.1	3.4	0.0	7.6	3.9	88.5

NOTE: Except for f_a, column headings are shifts in parts per million, and values in the table are the fraction of carbons at those shifts. Errors in f_a are approximately ±2%. Absolute errors in regional integrals are approximately ±2–3%.

Table III. 75-MHz–13-kHz CP–MAS Results

Coal ppm:	>190	190–165	165–145	145–130	130–106	106–81	81–50	50–22	22–0	f_a
Beulah–Zap	0.5	4.5	11.5	25.5	25.5	6.5	5.9	20.1	7.6	67
Wyodak–Anderson	1.0	4.0	5.5	12.5	23.5	7.5	5.3	28.6	12.1	54
Blind Canyon	0.0	1.6	8.3	16.7	34.3	5.8	2.8	20.6	9.9	67
Illinois No. 6	0.9	1.6	9.7	20.9	37.4	2.4	0.3	17.6	9.2	73
Pittsburgh No. 8	1.0	4.0	9.0	20.2	40.7	3.5	2.8	14.5	4.1	79
Lewiston–Stockton	0.0	0.4	9.0	24.3	38.8	4.4	1.5	15.2	6.3	77
Upper Freeport	0.0	2.0	5.0	25.0	43.0	3.0	2.0	14.9	4.9	78
Pocahontas No. 3	0.0	2.0	5.0	28.0	47.0	4.0	1.4	8.0	4.5	86

NOTE: The results were obtained with the following experimental parameters: $\omega(^{13}C)/2\pi = 34$ kHz and $\omega(^{1}H)/2\pi = 21$ kHz at 13 kHz MAS. Column headings, except for f_a, are shifts in parts per million, and values in the table are the fraction of carbons at those shifts. Absolute errors in regional integrals are approximately ±2%. Errors in f_a are approximately ±2–3%.

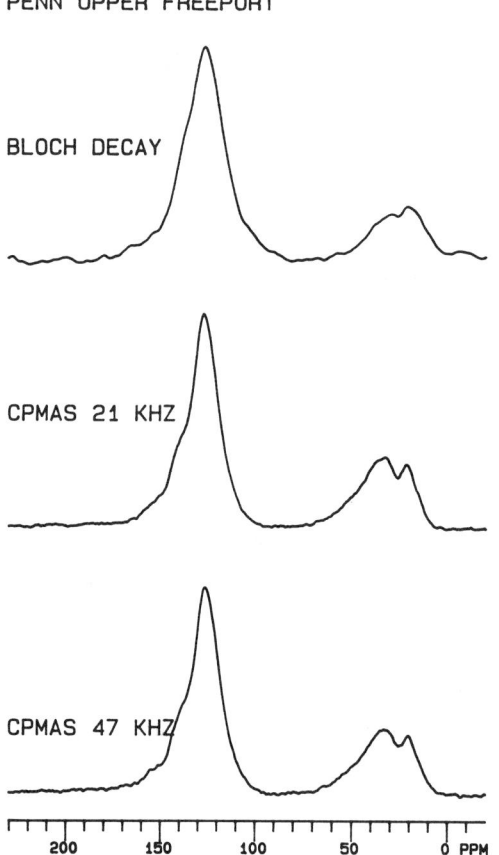

Figure 3. Bloch-decay spectrum ($f_a = 80.2\%$) and CP–MAS spectra ($f_a = 78\%$) for Upper Freeport coal at 21- and 47-kHz proton contact fields. Aromaticities agree to within 1% at the two mismatch maxima.

1% per millisecond. Pulse droop has the effect of decreasing the spectral intensity at contact times appreciably longer than the contact time used in measurement of the Hartmann–Hahn match maximum. Thus, excursion from the Hartmann–Hahn match will result in an apparent shorter $T_{1\rho}^H$ value.

To the extent that the first situation holds and $T_{1\rho}^H$ values are uniform and sufficiently long, the exponential decay will allow satisfactory determination of $M(0)$. The effects of higher contact fields (and improved pulse shape) were examined in more detail for the Lewiston–Stockton and Wyodak–Anderson coals. When contact fields (carbon/proton) were in-

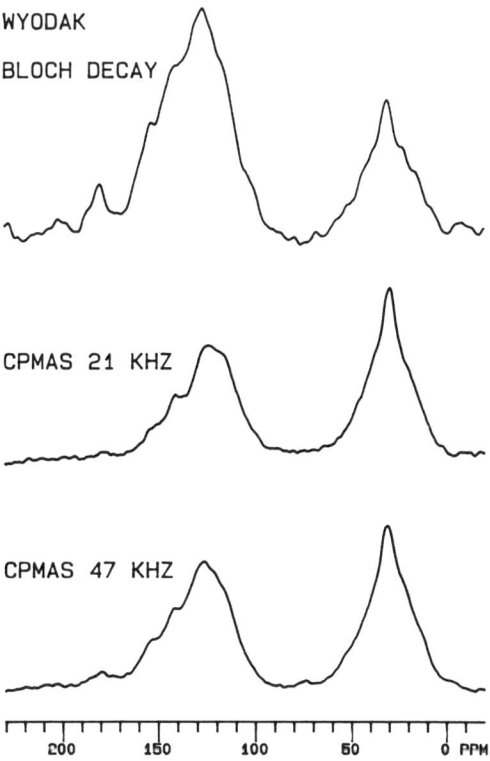

Figure 4. Bloch-decay spectrum ($f_a = 76\%$) and CP–MAS spectra ($f_a = 53\%$) of Wyodak–Anderson coal at 21- and 47-kHz proton contact fields. The growth of the carboxyl region around 177 ppm between 21 and 47 kHz is evident.

creased from 34/21 kHz (100-W observe amplifier) to 67/54 kHz (300-W observe amplifier), $T_{1\rho}^{H,ar}$ values of the Lewiston–Stockton coal increased only slightly, from 6.2 to 6.8 ms, and $T_{1\rho}^{H,al}$ values increased from 6.3 to 8.6 ms. The aromaticity decreased from 77 to 72% for the Lewiston–Stockton coal. The variable-contact-time data gave a better fit to eqs 1 and 2, and the value of $M(0)$ only slightly exceeded (by less than 5%) the maximum value of the aromatic contact curve. This result suggests that the higher aromaticity at the lower power level may be partially due to small excursions from the optimal Hartmann–Hahn match as a function of contact time. For the Wyodak–Anderson coal, however, identical aromaticity and $T_{1\rho}^{H}$ values were observed by using either the 100-W pulse amplifier or the 300-W amplifier, the latter providing higher contact power levels and an improved pulse shape. A rather intense dipolar oscillation (36) was observed for directly protonated carbons of coals and individual compounds (Figure 2D).

Comparison of Bloch-Decay, 75-MHz–13-kHz, and 25-MHz–4-kHz Results.

Table I provides a summary of aromaticities determined by Bloch decay, by 75-MHz–13-kHz variable-contact CP–MAS data analyzed by methods A and B and the average of A and B, and, for comparison, the 25-MHz–4-kHz CP–MAS results of Solum et al. (35). (The quantities in megahertz and kilohertz are the magnetic field and spinning rate, respectively.)

Table I reveals that the Bloch-decay aromaticities of the six bituminous coals agree within experimental error with the 4-kHz CP–MAS aromaticities determined by Solum et al. (35). The 13-kHz MAS results determined at 21-kHz proton and 34-kHz carbon fields agree within experimental error with Bloch-decay results for five of the six bituminous coals. The 13-kHz MAS results are 6% higher than the Bloch-decay results for Pittsburgh No. 8 coal, a result somewhat beyond the range of experimental error. The error may be due in part to a low $T_{1\rho}^H$ value caused by excursions from the Hartmann–Hahn match, which leads to an overestimate of $M(0)_{ar}$ in method A. Our $T_{1\rho}^H$ values for the Argonne Premium coals are uniformly lower than those of Solum et al. (35) and were not appreciably changed by operating at significantly higher proton and carbon contact fields. By contrast, $T_{1\rho}^H$ values were a strong function of the strength of the proton matching field for model compounds (see acenaphthene and 4,4'-dimethoxybibenzyl results in the following section). Although CP–MAS aromaticities at both low-speed and high-speed MAS for the majority of bituminous coals are in satisfactory agreement with Bloch-decay results, the 13-kHz MAS results for lignite and subbituminous coals show significant disagreement. The Bloch-decay results determined in this laboratory also suggest that the CP–MAS (with 4-kHz MAS) results of Solum et al. (35) for Beulah–Zap and Wyodak–Anderson coals are also low.

Figure 3 compares Bloch-decay and CP–MAS results at proton contact fields of 21 and 47 kHz for Pennsylvania Upper Freeport coal. The two CP–MAS spectra shown were collected at a contact time of 0.4 ms, which is typically the point of maximum intensity on the aromatic contact curves for the coals. The two CP–MAS aromaticities are identical. Careful examination of the Bloch-decay spectrum reveals higher substituted aromatic structures around 140 ppm balanced by an increase in methyl carbons relative to the methylenes compared to the CP–MAS spectra. The Upper Freeport coal serves as an example of a relatively well-behaved medium-rank coal showing nearly identical aromaticities and very similar structural distributions. Regional structural distributions from CP–MAS and Bloch-decay measurements agree within 2–3% for Illinois No. 6, Lewiston–Stockton, Pennsylvania Upper Freeport, and Pocahontas No. 3 coals.

The Bloch-decay results from the lignite and subbituminous coals stand in stark contrast to CP–MAS results. The aromaticity of subbituminous Wyodak–Anderson coal is 24% lower by CP–MAS at 75 MHz–13 kHz than by Bloch decay. The 25-MHz–4-kHz value reported by Solum et al. (35) is 13% lower than the Bloch-decay result. Figure 4 shows CP–MAS and Bloch-decay results at 21- and 41-kHz proton fields for the Wyodak–Anderson coal. The much higher aromaticity and significantly greater carboxyl content obtained by Bloch decay is readily apparent.

Careful examination of the carboxyl region at 177 ppm shows an increase in the carboxyl band on going from 21- to 47-kHz proton fields (Figure 4). Detectable increases in aromatic C–O, bridgehead, and alkylated aromatic regions around 155, 145, and >130 ppm are also apparent. The observation of identical aromaticities at the different power levels, in spite of the appreciably greater amounts of additional oxygenated and quaternary aromatic structures, illustrates the insensitivity of the aromaticity parameter to structural detail. The 13-kHz CP–MAS aromaticity of the Beulah–Zap lignite is in agreement with the results of Solum et al. (35); however, the Bloch-decay result shows 13% higher aromaticity. For both Wyodak–Anderson and Beulah–Zap coals, about 50% more carboxyl carbon is detected by Bloch decay than in the CP–MAS results (Tables II and III). Although carboxylic acid groups are readily detected in crystalline organic compounds by ^{13}C CP–MAS NMR spectroscopy at high-field and high-speed MAS, a substantial fraction of carboxyl groups and phenol–aryl ether carbons are not detected in coal by CP–MAS NMR spectroscopy with high-speed MAS. In Wyodak–Anderson coal, all categories of aromatic carbons are higher in the Bloch-decay spectrum. Oxygen-substituted and aromatic carbons (165–145 ppm) are present to fully threefold greater extent in the Bloch-decay spectrum.

Several causes of the large CP–MAS aromaticity errors for Wyodak–Anderson and Beulah–Zap coals are plausible. First, the presence of aqueous or mineral domains could lead to local radiofrequency (rf) inhomogeneity. This inhomogeneity would tend to reduce the intensity of nonprotonated-carbon signal because the widths of nonprotonated carbon matching peaks are narrower than those of protonated carbons. This effect would be much more evident with high-speed MAS than with conventional MAS speeds because the Hartmann–Hahn match is much more sensitive at high MAS speeds. Arguing against this view is the observation that extensive drying led to no differences in aromaticity for Wyodak–Anderson and Beulah–Zap coals.

Second, selective rotational motion of aromatic rings may be a feature of low-rank coals. If the low-rank coals are viewed as a predominately aliphatic network substituted with small pendant aromatic groups, greater rotational mobility would select against polarization transfer to aromatic carbons. Third, the packing characteristics of low-rank coals may simply lead to quaternary carbons, which are more isolated from neighboring C–H bonds, a feature leading to much longer T_{CH} values for nonprotonated structures. This effect would be exacerbated by the lower $T_{1\rho}^{H}$ values for the low-rank coals. Comparing the chemical-shift distributions in Table II with those in Table III for the Wyodak–Anderson coal shows that all classes of aromatic structures increase with Bloch decay, but the largest relative increases are in the carboxyl and quaternary phenolic regions rather than in the protonated aromatic regions.

Finally, the unique effect of high-speed spinning may be to reduce the intensity of the protonated aromatic carbons relative to nonprotonated aromatic carbons and protonated aliphatic carbons at the $f - \omega_r$ match. This effect is apparent for 4,4'-dimethoxybibenzyl (see Figure 1B) and is the effect reported for Torlon by Wind (32). However, this pattern was not observed for acenaphthene (see next section), and further study of model compounds is required. Again, Wyodak–Anderson coal displays the largest relative changes in quaternary aromatic carbons, a result that seems to indicate that the high-speed-spinning effects in suppression of protonated aromatic carbons at the $f - \omega_r$ match are small.

The reasonably satisfactory agreement between high-speed and low-speed MAS aromaticities for the majority of medium- to high-rank coals suggests that the increase in distortion of protonated versus nonprotonated spectral intensities introduced by high-speed MAS compared to conventional (4 kHz) MAS is low, if aromaticity is used as the (somewhat crude) measure of statistical structure. The sources of errors for low-rank coals remain the subject of further study.

Comparison of CP–MAS and Bloch-Decay Results for Acenaphthene and CP–MAS Results for 4,4'-Dimethoxybibenzyl.

To determine if the distortions in carbon intensities reported by Wind (32) for a single contact time persist in studies of the evolution of carbon intensities in a variable-contact-time study, we examined several organic compounds. Acenaphthene was somewhat difficult to examine because of long $T_1(H)$ values, which dictated a 60-s recycle time in the CP–MAS experiment. However, acenaphthene provides examples of aliphatic protonated and nonprotonated aromatic carbons and a quaternary carbon removed by two bonds from the nearest C–H bond (see structure 2). As shown in Table IV, T_1 values range from about 30 s for protonated

Table IV. Bloch-Decay, $T_1(C)$ and CP–MAS Results for Acenaphthene

Carbon	Chemical Shift	$T_1(C)$ (s)	T_{CH} (μs) (34/21)[a]	$T_{1\rho}^H$ (ms) (34/21)[a]	$T_{1\rho}^H$ (ms) (54/41)[a]	Bloch Decay	Carbon Intensity CP–MAS (34/21)[a]	CP–MAS (54/41)[a]	Structure
A[b]	128.5	48 ± 5	59	10.9	20.0	1.90	1.67	1.52	2
B[b]	123.9	44 ± 2	61	10.7	20.4	2.24	2.20	1.83	2
C[b]	119.8	43 ± 4	30	11.8	19.8	1.78	1.64	1.51	2
D	146.6	80 ± 10	287	15.9	23.9	(2)	(2)	(2)	(2)
E	140.3	105 ± 15	519	11.7	20.0	1.05	0.95	1.06	1
F	132.7	59 ± 9	285	11.7	20.8	1.09	0.99	1.03	1
G[b]	30.9	27 ± 6	24	10.8	19.8	1.98	2.12	2.09	2

NOTE: Results were obtained with the following experimental parameters: 75-MHz ^{13}C resonance frequency and 13-kHz MAS spinning rate. The values for acenaphthene aromaticity, with an error of ±2%, are as follows: theory, 83.3%; CP–MAS, 81.7%; and Bloch decay, 83.6%.
[a] Ratios indicate ^{13}C/^{1}H contact power levels $(f - \omega_r \text{ match})$.
[b] Protonated carbons.

aliphatic carbons to greater than 100 s for the buried quaternary carbon (carbon E). The methylene carbon (G) polarizes at about twice the rate of the aromatic methines, as expected. Buried quaternary carbon E exhibits the slowest polarization rate. The standard relative errors (percent standard relative error = $100 \times [(I^{\text{experimental}} - I^{\text{theory}})/n]^{0.5}$; I is the absorption intensity; $n = 12$ atoms) in integrated intensities compared to theory are as follows: Bloch decay, 5.8%; and CP–MAS, 7.9%. The standard relative errors in carbon intensities for this compound are comparable to errors reported by Alemany et al. (23, 24) for crystalline organic molecules at lower field, low-speed MAS.

Figure 5 shows the pronounced dipolar oscillation of methylene G, and the well-behaved evolution of the signal of buried carbon E. The scatter in intensities of protonated carbons is somewhat larger than unpro-

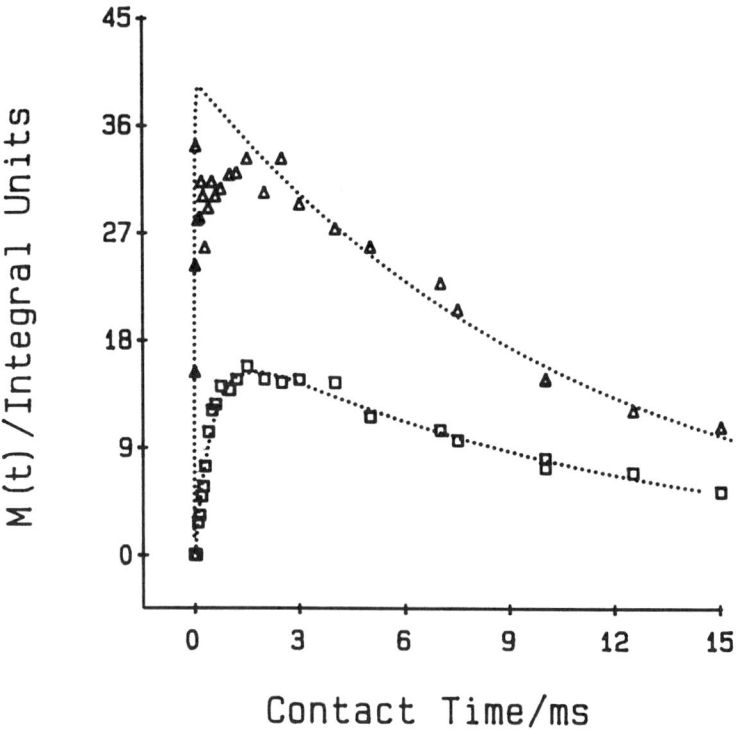

Figure 5. Contact curves for carbons E (one carbon) (□) and G (two carbons) (△) of acenaphthene determined with a proton contact field of 21 kHz. The ratio of G:E is 2.12:1 (6% error). See Table IV.

Table V. Contact-Power Dependence of Carbon Intensities and $T_{1\rho}^H$ Relaxation Times for 4,4'-Dimethoxybibenzyl

Carbon	Chemical Shift	Contact Power 54 kHz (^{13}C), 41 kHz (^1H)		Contact Power 67 kHz (^{13}C), 54 kHz (^1H)	
		M(0)	$T_{1\rho}^H$ (ms)	M(0)	$T_{1\rho}^H$ (ms)
A	158	1.0	19.7	1.0	27.2
B[a]	117	0.75	23.4	0.89	24.6
C[a]	110	0.73	21.6	0.81	26.9
D	42	0.89	22.5	1.07	24.9
E	56	0.90	21.2	0.99	28.5

NOTE: Carbons F and G (130 ppm) were unresolved and partially overlapped the broad, low-intensity signal of H centered at 137 ppm, a result precluding $T_{1\rho}^H$ measurements.

[a] Assignments may be reversed. A single peak at 113 ppm is observed in solution.

tonated carbons, but shows no consistent trend by comparison with nonprotonated carbons. The variations in the Bloch-decay carbon intensities in Table IV and their similarity in magnitude to CP–MAS errors is a very interesting, and reproducible, result. This result is not caused by violation of T_1, but the cause of the deviations is unknown.

The variation in carbon intensities and rotating-frame relaxation rates are shown for 4,4'-dimethoxybibenzyl in Table V. Values of $T_{1\rho}^H$ increase with increased contact power levels. The lower value of the protonated aromatic carbon intensity (carbon B) compared to the quaternary aromatic carbon intensity (carbon A) is similar to the single-contact-time result at the single contact time $f - \omega_r$ maximum of the Hartmann–Hahn curve (Figure 1B). Thus, although acenaphthene does not show a consistent pattern of reduced protonated aromatic intensity relative to the quaternary aromatic peaks, 4,4'-dimethoxybibenzyl shows a pattern of intensities similar to Wind's results (32) for the Torlon polymer. As shown in Table V, the errors in carbon-intensity ratios appear to decrease at higher power levels. In Table V, the aromatic carbons ortho to the alkyl linkage (carbons F and G) were not included because F and G were insufficiently resolved. Carbon H, which polarizes to only a small extent, was also broadened by motional effects, a result causing it to exhibit inadequate signal-to-noise ratio for $T_{1\rho}^H$ determination. Carbon H, which appears to suffer from a long T_{CH}, is a typical structural type that is lost in high-speed CP–MAS NMR of low-rank coals.

Conclusions

Variable-contact-time CP–MAS ^{13}C NMR spectroscopy at very high speed MAS can be used to characterize medium- to high-rank coals with aromaticities within a few percent of Bloch decay. The agreement in aromaticity between Bloch-decay and some of the CP–MAS results is somewhat fortuitous because an agreement in aromaticity may occur in spite of appreciable differences in structure. Low-rank, high-oxygen-containing coals, however, show significantly diminished aromaticities. Results from acenaphthene suggest that distortions caused by high-speed spinning on polarization transfer of the ratios of protonated to nonprotonated carbons tend to average out, yielding ratios of protonated to nonprotonated carbons that are in agreement with the structure. The magnitude of CP–MAS errors observed with acenaphthene are not large enough to account for errors in aromaticity observed for low-rank coals.

Other possible causes of low CP–MAS aromaticities for Wyodak–Anderson and Beulah–Zap coals include the attenuation of polarization transfer due to structural features leading to selective motion of aromatic rings, to the isolation of quaternary carbons from protons, or to severe dielectric effects of mineral or aqueous regions. As to the latter effect, residual water is not believed to be a cause of reduced aromaticities for Beulah–Zap and Wyodak–Anderson coals; the coals were dried for various lengths of time up to 24 h and exhibited reproducibly low aromaticities. Finally, although the enhanced sensitivity of the CP–MAS method will continue to make it important for semiquantitative measurements of relative changes in coal structure, neither high-speed-spinning CP–MAS nor conventional-spinning-speed CP–MAS appears to be useful as a quantitative tool for low-rank coals and humic materials. For these materials, Bloch decay is the method of choice.

Acknowledgement

We gratefully acknowledge the support of this work by the Director, Chemical Sciences Division, Office of Energy Research, U.S. Department of Energy, under Contract No. DE–AC06–1830–RLO.

References

1. Pines, A.; Gibby, M. G.; Waugh, J. S. *J. Chem. Phys.* **1973**, *59*, 569–573.
2. Hartmann, S. R.; Hahn, E. L. *Phys. Rev.* **1962**, *128*, 2042.
3. Pines, A.; Gibby, M. G.; Waugh, J. S. *J. Chem. Phys.* **1972**, *56*, 1776.

4. Andrew, E. R.; Bradbury, A.; Eades, R. G. *Nature (London)* **1959**, *183*, 1802.
5. Lowe, I. J. *Phys. Rev. Lett.* **1959**, *2*, 285.
6. Schaefer, J.; Stejskal, E. O.; Buchdahl, R. *Macromolecules* **1975**, *8*, 291.
7. Schaefer, J.; Stejskal, E. O. In *Topics in Carbon-13 NMR Spectroscopy;* Levy, G. C., Ed.; Wiley Interscience: New York, 1979; Vol. 3, p 283.
8. Garroway, A. N.; Moniz, W. B.; Resing, H. A. *Prepr. Div. Org. Coatings Plast. Chem.* **1976**, *36*, 133–138.
9. Lippmaa, E.; Alla, M.; Tuherm, T. *Proc. 19th Congr. Ampere (Heidelberg),* **1976**, 113–118.
10. VanderHart, D. L.; Retcofsky, H. L. *Fuel* **1976**, *55*, 202.
11. Zilm, K. W.; Pugmire, R. J.; Grant, D. M.; Wood, R. E.; Wiser, W. H. *Fuel* **1979**, *58*, 11–16.
12. Vasallo, A. M.; Wilson, M. A.; Edwards, J. H. *Fuel* **1987**, *66*, 622.
13. Russel, J. J.; Wilson, M. A.; Pugmire, R. J.; Grant, D. M. *Fuel* **1982**, *62*, 601–605.
14. Maciel, G. E.; Sullivan, J. J.; Petrakis, L.; Grandy, D. W. *Fuel* **1982**, *61*, 411–414.
15. Pugmire, R. J.; Zilm, K. W.; Woolfenden, W. R.; Grant, D. M.; Dyrkacz, G. R.; Bloomquist, A. A.; Horowitz, E. P. *Org. Geochem.* **1982**, *4*, 79–84.
16. Zilm, K. W.; Pugmire, R. J.; Larter, S. R.; Allan, J.; Grant, D. M. *Fuel* **1981**, *60*, 717–722.
17. Vassallo, A. M.; Wilson, M. A.; Collin, P.; Oades, J. M.; Water, A. G.; Malcolm, R. L. *Anal. Chem.* **1987**, *59*, 558.
18. Wilson, J. A.; Pugmire, R. J.; Karas, J.; Alemany, L. B.; Woolfenden, W. R.; Grant, D. M.; Given, P. *Anal. Chem.* **1984**, *56*, 933–943.
19. Davidson, R. M. *Nuclear Magnetic Resonance Studies of Coal;* IEA Coal Research: London, 1986.
20. Wershaw, R. L.; Mikita, M. A. *NMR of Humic Substances and Coal;* Lewis: Chelsea, MI, 1987.
21. Axelson, D. E. *Solid State Nuclear Magnetic Resonance of Fossil Fuels: An Experimental Approach;* Multiscience: Montreal, Canada, 1985.
22. Earl, W. L. In *NMR of Humic Substances and Coal;* Wershaw, R. L.; Mikita, M. A., Eds.; Lewis: Chelsea, MI, 1987; pp 167–187.
23. Alemany, L. B.; Grant, D. M.; Pugmire, R. J.; Alger, T. D.; Zilm, K. W. *J. Am. Chem. Soc.* **1983**, *105*, 2133–3141.
24. Alemany, L. B.; Grant, D. M.; Pugmire, R. J.; Alger, T. D.; Zilm, K. W. *J. Am. Chem. Soc.* **1983**, *105*, 2142–2147.
25. Botto, R. E.; Wilson, R.; Winans, R. E. *Energy Fuels* **1987**, *1*, 173–181.
26. Choi, C.; Muntean, J. V.; Thompson, A. R.; Botto, R. E. *Energy Fuels* **1989**, *3*, 528–533.
27. Hageman, E. W.; Chambers, R. R.; Woody, M. C. *Anal. Chem.* **1986**, *58*, 387.
28. Snape, C. E.; Axelson, D. E.; Botto, R. E.; Delpeuch, J. J.; Tetrely, P.; Gerstein, B. C.; Pruski, M.; Maciel, G. E.; Wilson, M. A. *Fuel* **1989**, *68*, 547–560.
29. Komorowski, R. A. In *High Resolution NMR Spectroscopy of Synthetic Polymers in Bulk;* Komorowski, R. A., Ed.; VCH: Deerfield Beach, FL, 1986; p 42.

30. Garroway, A. N.; Moniz, W. B.; Resing, H. A. *Carbon-13 NMR in Polymer Science;* Pasika, W. M., Ed.; ACS Symposium Series 103; American Chemical Society: Washington, DC, 1979; p 67.
31. Dec, S. F.; Wind, R. A.; Maciel, G. E.; Anthonio, F. E. *J. Magn. Reson.* **1986,** *70,* 355.
32. Wind, R. A.; Dec, S. F.; Lock, H.; Maciel, G. E. *J. Magn. Reson.* **1988,** *79,* 136–139.
33. Stejskal, E. O.; Schaefer, J.; Waugh, J. S. *J. Magn. Reson.* **1977,** *28,* 105–112.
34. Andrew, E. R. *Philos. Trans. R. Soc. London* **1981,** *A299,* 505.
35. Solum, M. S.; Pugmire, R. J.; Grant, D. M. *Energy Fuels* **1989,** *3,* 187–193.

RECEIVED for review June 8, 1990. ACCEPTED revised manuscript January 15, 1991.

High-Resolution ^{1}H NMR Studies of Argonne Premium Coals

Antoni Jurkiewicz[1], Charles E. Bronnimann, and Gary E. Maciel[2]

Department of Chemistry, Colorado State University, Fort Collins, CO 80523

> *Proton combined rotation and multiple-pulse spectroscopy (CRAMPS) studies were performed on untreated and pyridine-treated samples of seven Argonne Premium coals. The proton CRAMPS spectrum of an untreated sample usually consists of two substantially overlapping peaks, an aliphatic peak and an aromatic peak. In the spectrum of a sample saturated with perdeuteropyridine, the aliphatic and aromatic peaks are better resolved, and more spectral and structural details can be elucidated. The addition of pyridine can dramatically change the structural or dynamic characteristics of a coal. In a proton CRAMPS dipolar-dephasing experiment on an untreated coal, the protons of aliphatic moieties relax faster than those attached to aromatic rings. In pyridine-treated samples, the dynamics are far more complicated and vary within each sample. The spectral resolution increases with increasing dephasing interval. Peaks due to protons belonging to heterocyclic nitrogen and sulfur compounds were identified. A correlation between the chemical shift of the aliphatic peak and the hydrogen/carbon ratio was found for 19 coal samples.*

[1]On leave from the Institute of Coal Chemistry, Polish Academy of Sciences, 44–100 Gliwice, ul. 1 Maja 62, Poland
[2]Corresponding author

THE LINE-BROADENING EFFECT associated with spin–spin (T_2) relaxation due to magnetic dipole–dipole interactions has limited the use of solid-state ^{1}H NMR spectroscopy for coal-structure studies. The usual lack of resolution in the proton NMR spectra of coals has permitted the structural interpretation of the data only in a very gross sense, with relatively small emphasis on structural details. The proton combined rotation and multiple-pulse spectroscopy (CRAMPS) technique now provides the capability of obtaining high-resolution ^{1}H spectra of solids by means of multiple-pulse dipolar averaging combined with magic-angle spinning (MAS) (1–5). This technique typically permits the observation of two main peaks in the ^{1}H CRAMPS spectrum of an untreated coal, an aliphatic peak and an aromatic peak (6, 7).

Our previous proton CRAMPS studies of the Utah Blind Canyon Argonne Premium coal (No. 601) (2) and a Polish coal (WK22) (3) demonstrated the dramatic improvement in spectral resolution often achieved by saturating a coal sample with perdeuteropyridine. The resulting spectral resolution was not the same for these two coals, a result suggesting that spectral resolution could vary substantially with the identity and properties of a coal. The phenomena responsible for the enhancement of spectral resolution of a coal due to saturation with perdeuteropyridine are not well understood at this time. One of the reasons for enhancement may be the anisotropic bulk-susceptibility effect (5).

Pyridine has a great influence not only on the ^{1}H spectral resolution, but also on the dynamics in coal. This phenomenon has been observed previously by using wide-line ^{1}H NMR spectroscopy (8–12).

In our previous ^{1}H CRAMPS studies (8, 9), we found that a dipolar-dephasing experiment is extremely useful for investigating the dynamics of molecular and macromolecular parts of coal. In this experiment, a dephasing period is inserted between the initial 90° preparation pulse and the beginning of detection under multiple-pulse dipolar averaging. During the dephasing period, the dipolar interactions between protons, which largely govern the behavior of the proton magnetization, are modulated by molecular motion. Therefore, this method is useful for studying the nature of the ^{1}H–^{1}H dipolar interaction and the effects of molecular motion on this interaction. The molecular and macromolecular phases of coal manifest different molecular and segmental dynamics because of differences in the number and types of cross-links, size of molecules, local symmetry, steric conditions, etc. Therefore a variety of motional behaviors are observed for different chemical functionalities in coal.

Experimental Details

These studies were performed on seven Argonne Premium coal samples. The ultimate analyses (13) for all the Argonne samples are presented in Table I.

Table I. Ultimate Analyses of the Argonne Premium Coal Samples

Number	Coal	C	H	N	O	S
101	Upper Freeport	88.08	4.84	1.60	4.72	0.76
202	Wyodak–Anderson	76.04	5.42	1.13	16.90	0.48
301	Illinois No. 6	80.73	5.20	1.43	10.11	2.47
401	Pittsburgh No. 8	84.95	5.43	1.68	6.90	0.91
501	Pocahontas No. 3	91.81	4.48	1.34	1.66	0.51
701	Lewiston–Stockton	85.47	5.44	1.61	6.68	0.67
801	Beulah–Zap	74.05	4.90	1.17	19.13	0.71

NOTE: All values are given as percents on a dry, mineral-matter free basis.
SOURCE: Data are taken from reference 13.

Analogous studies on the Argonne Premium coal No. 601 (Blind Canyon) were reported previously (8). All coal samples were dried at ambient temperature overnight at a pressure of 3×10^{-3} torr (0.40 Pa). Pyridine-d_5 (99.5% perdeuteropyridine from MSD Isotopes) was added to each coal in a ratio of 2:1 (weight ratio of pyridine to coal). Each sample was then sealed under vacuum in a 5-mm glass tube, which forms the chamber of the MAS device. The CRAMPS experiments were performed at a proton Larmor frequency of 187 MHz on a modified NT-200 spectrometer (5). Spectra were obtained by using the BR-24 pulse sequence (14) with $\tau = 3$ μs. A single 180° pulse, placed midway in the dephasing interval, was used to suppress chemical-shift effects during the dephasing period. MAS speeds were 2.2–3.0 kHz.

Results

The proton CRAMPS spectra of the untreated premium coals, including the previously reported Blind Canyon coal, are shown in Figure 1. Generally, the spectrum of each coal consists of two peaks, an "aliphatic" peak (due to protons attached to sp^3 carbons) and an "aromatic" peak (due to protons attached to aromatic rings). The chemical shift of the aliphatic peak is centered around 1 ppm, and that for the aromatic peak is centered at about 6.8 ppm. However, for the Pocahontas No. 3 coal the aliphatic peak is shifted toward lower shielding, with a maximum at about 3.2 ppm. The line widths for coals with higher carbon contents (Pocahontas No. 3 and Upper Freeport) are larger than those for other coals. Small rotor lines (5) can be observed in the spectra of those two coals. The eight ^{1}H spectra in Figure 1 are fairly typical CRAMPS spectra of untreated coals, and these spectra yield the partial resolution of broad aromatic and aliphatic peaks.

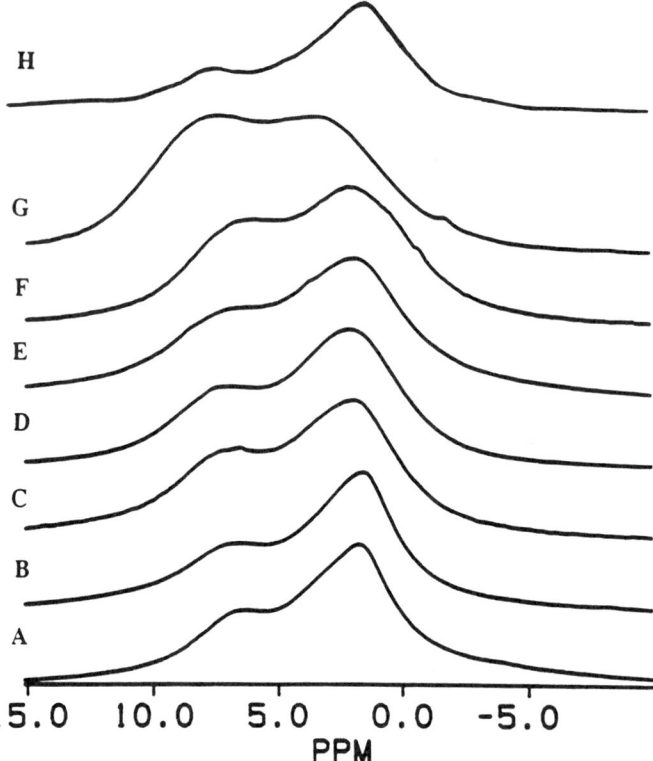

Figure 1. Proton CRAMPS spectra of untreated coals: (A) Beulah–Zap, (B) Wyodak–Anderson, (C) Illinois No. 6, (D) Pittsburgh No. 8, (E) Lewiston–Stockton, (F) Upper Freeport, (G) Pocahontas No. 3, and (H) Blind Canyon.

The effect that perdeuteropyridine saturation has on the CRAMPS spectra is shown for the same coals in Figure 2. The enhanced resolution in most cases makes it possible to observe far more spectral detail. The spectrum of Beulah–Zap coal (Figure 2A) displays three aliphatic peaks at 1.25, 1.8, and 3.4 ppm; the aromatic peak still shows only one maximum, which is centered at 6.6 ppm. The spectrum of the Wyodak–Anderson coal (Figure 2B) shows less resolution in the aliphatic band than that of the Beulah–Zap coal; nevertheless, two aliphatic peaks, at 1.1 and 1.9 ppm, are recognizable, and the peak at 1.9 ppm is broad. The aromatic pattern in the ^{1}H CRAMPS spectrum of the Wyodak–Anderson coal has only one maximum, at 6.7 ppm.

The saturation of Illinois No. 6 coal with perdeuteropyridine (Figure 2C) made it possible to observe six different peaks in the ^{1}H CRAMPS spectrum. Distinct aliphatic peaks are present at 1.1 and 2 ppm and aro-

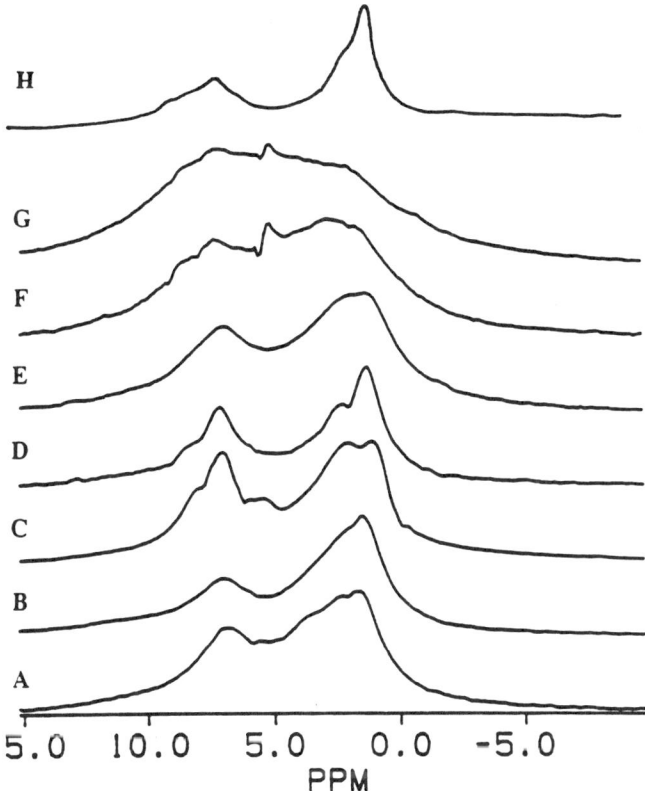

Figure 2. Proton CRAMPS spectra of pyridine-saturated coals: (A) Beulah–Zap, (B) Wyodak–Anderson, (C) Illinois No. 6, (D) Pittsburgh No. 8, (E) Lewiston–Stockton, (F) Upper Freeport, (G) Pocahontas No. 3, and (H) Blind Canyon.

matic peaks at 6.8 and 8.1 ppm. An aliphatic shoulder occurs at about 3.5 ppm; at 5.2 ppm is a residual water signal, which may be due partly to the added pyridine.

The ^{1}H CRAMPS spectrum of Pittsburgh No. 8 coal (Figure 2D) displays very clearly resolved aliphatic peaks at 0.9 and 1.8 ppm and two aromatic peaks at 6.8 and 8.2 ppm. A broad aliphatic shoulder at about 3.5 ppm can also be seen.

The ^{1}H CRAMPS spectrum of high-volatile bituminous (HVB) Lewiston–Stockton coal shows an aliphatic peak centered at 1.1 ppm and a second, broad aliphatic peak centered at about 1.7 ppm. The aromatic signal is centered at 6.6 ppm.

The ^{1}H CRAMPS spectra of the Upper Freeport and Pocahontas No. 3 coals (Figures 2F and 2G, respectively) both show broad, poorly resolved patterns. Nevertheless, it is possible to recognize in the spectrum of Upper Freeport coal aliphatic peaks at 1.4 and 2.5 ppm and aromatic peaks at 6.9 and 8.1 ppm; the water peak is evident at 4.9 ppm. The aromatic signal of the Pocahontas No. 3 coal spectrum is better resolved than the aliphatic pattern. Aromatic peaks occur at 6.9 and 8.2 ppm, but the aliphatic pattern shows only one broad peak. The water resonance at 4.9 ppm is also evident. The previously reported ^{1}H CRAMPS spectrum of the Utah Blind Canyon coal (Figure 2H) shows two aliphatic peaks at 0.9 and 1.7 ppm and three aromatic peaks centered at about 6.9, 8.1, and 9 ppm.

Dipolar-dephasing was performed on all of the untreated samples. The spectra for untreated Wyodak–Anderson subbituminous (SB) coal are presented in Figure 3 in terms of absolute intensities (Figure 3B) and in terms of an intensity scaled arbitrarily for convenient visual evaluation (Figure 3A). Figure 3B clearly shows that the spectra decay quickly under dephasing; after a 30-μs dephasing time the signal has almost disappeared. As is also obvious, the aliphatic peak relaxes faster than the aromatic peak. Dipolar-dephasing experiments on the other untreated premium coals yielded similar decay patterns.

The dipolar-dephasing experiment provides far more interesting results on the pyridine-saturated coals than on untreated coals. The results of this experiment for perdeuteropyridine-treated Beulah–Zap lignite (L) are presented in Figure 4. In contrast to the results obtained on the untreated coals, an aliphatic peak in the pyridine-treated sample survives

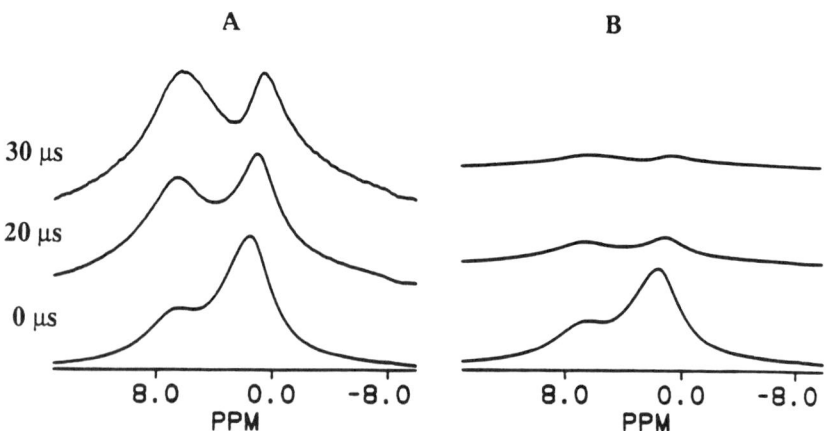

Figure 3. ^{1}H *CRAMPS dipolar-dephasing results for the untreated Wyodak–Anderson coal sample with dephasing periods of 0, 20, and 30 μs. A, arbitrarily scaled intensity; and B, absolute intensity.*

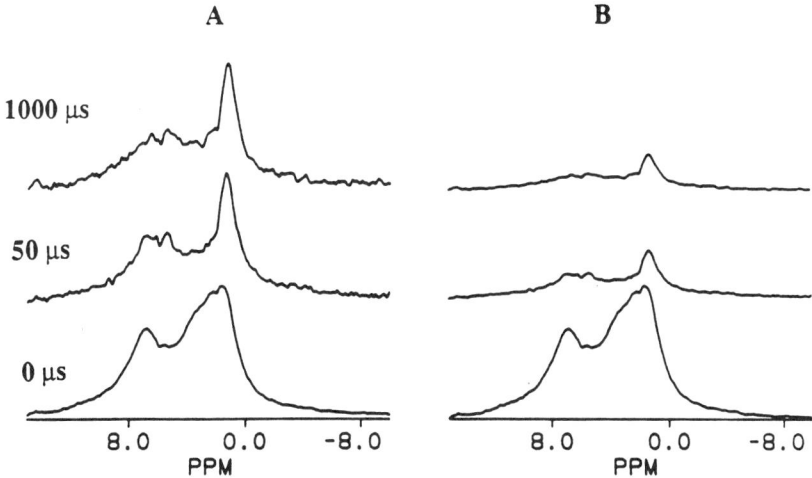

Figure 4. Proton CRAMPS dipolar-dephasing results for pyridine-saturated Beulah–Zap sample with dephasing periods of 0, 50, and 1000 μs. A, arbitrarily scaled intensity; and B, absolute intensity.

even after 1000 μs of dipolar-dephasing. A reasonable interpretation of this behavior is that this peak is due to a methyl group terminating an aliphatic chain. The other two aliphatic peaks are strongly attenuated after a 50-μs dephasing time. Similar decay behaviors are observed for Wyodak–Anderson SB coal (Figures 5A and 5B) and Lewiston–Stockton coal (Figures 5C and 5D). The sharp aliphatic peaks due to the methyl groups in these cases relax more slowly, and the second, broad aliphatic peak relaxes faster in each of these two coals.

The dipolar-dephasing behaviors of the Illinois No. 6 and Pittsburgh No. 8 coals are not very similar to the behavior of the other coals, as can be seen in Figures 6A and 6B and Figures 6C and 6D, respectively. Within the spectrum of each of these two coals, the various peaks decay with relatively similar rates; for example, the methyl-group magnetization dephases with roughly the same rate as does the magnetization of the other aliphatic structures. In both cases, the spectrum is essentially the same after 50- and 400-μs dephasing times. Dipolar-dephasing is slower in the Pittsburgh No. 8 coal than in the Illinois No. 6 coal.

The medium-volatile bituminous (MVB) Upper Freeport coal and the low-volatile bituminous (LVB) Pocahontas No. 3 coal display rapid dipolar-dephasing decays in spite of the presence of pyridine, as shown in Figures 7A and 7B and Figures 7C and 7D, respectively. After a 50-μs dephasing time, little intensity persists. In the Upper Freeport coal, both of the aliphatic peaks decay with approximately the same rate, and the ali-

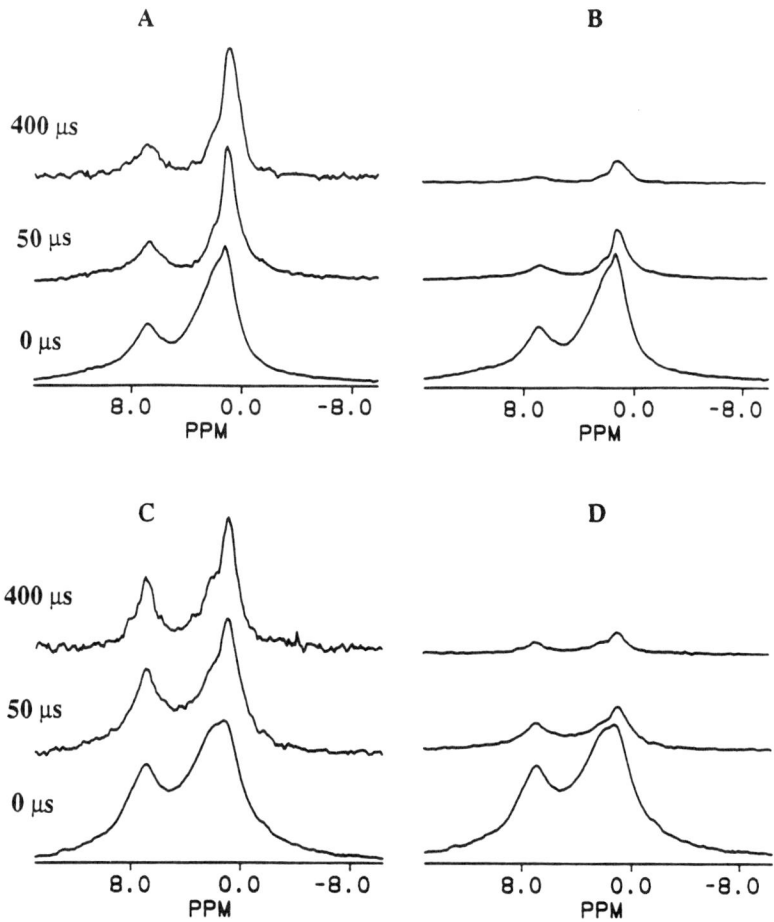

Figure 5. Proton CRAMPS dipolar-dephasing results for pyridine-saturated samples of Wyodak–Anderson SB coal (A, arbitrary scaling; and B, absolute intensity) and Lewiston–Stockton coal (C, arbitrary scaling; and D, absolute intensity). The dipolar-dephasing periods were 0, 50, and 400 μs.

phatic peaks of this spectrum decay faster than the aromatic peaks. The same behavior pattern is observed in the untreated coal. In the pyridine-saturated Pocahontas No. 3 coal, the aromatic and aliphatic resonances decay with approximately the same time constant. The slowest decaying magnetization for these coals is that of water (because it is a fluid). Even in these two coals with very broad aliphatic and aromatic bands, a significant improvement in spectral resolution with increasing dephasing time can be seen.

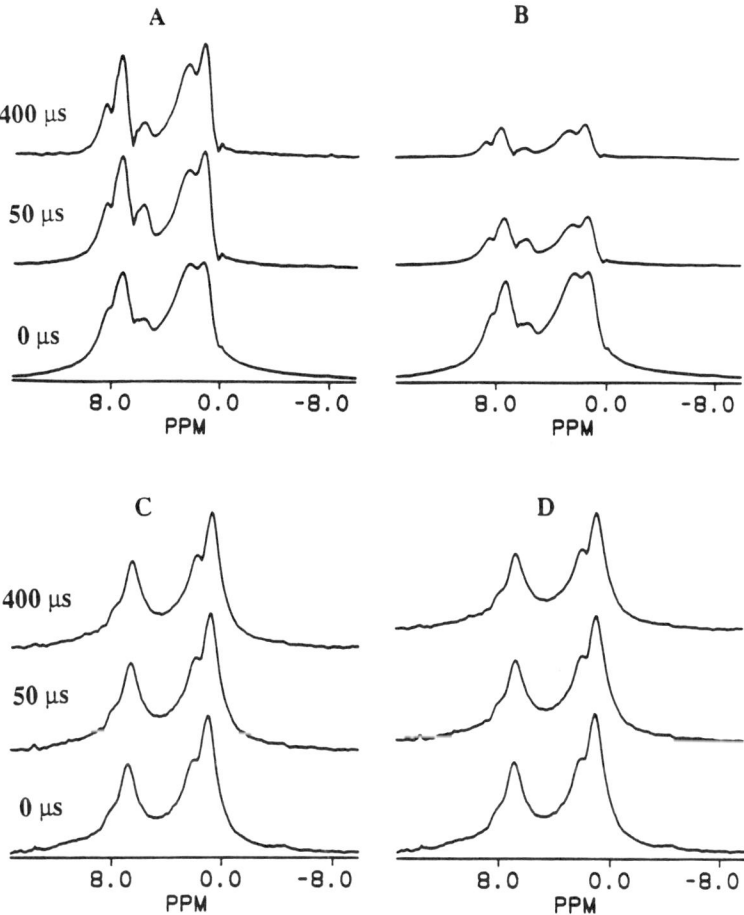

Figure 6. Proton CRAMPS dipolar-dephasing results for pyridine-saturated samples of Illinois No. 6 coal (A, arbitrary scaling; B, absolute intensity) and Pittsburgh No. 8 coal (C, arbitrary scaling; D, absolute intensity). The dipolar-dephasing periods were 0, 50, and 400 μs.

Discussion

^{1}H CRAMPS spectral resolution is generally enhanced for coals by treatment with perdeuteropyridine. The best resolution was obtained for Illinois No. 6 and Pittsburgh No. 8 coals. Both of these are HVB coals. Pyridine treatment is also very effective for Beulah–Zap lignite and Wyodak-Anderson SB coal. The improvement of spectral resolution is less dramatic for MVB Upper Freeport and LVB Pocahontas No. 3 samples.

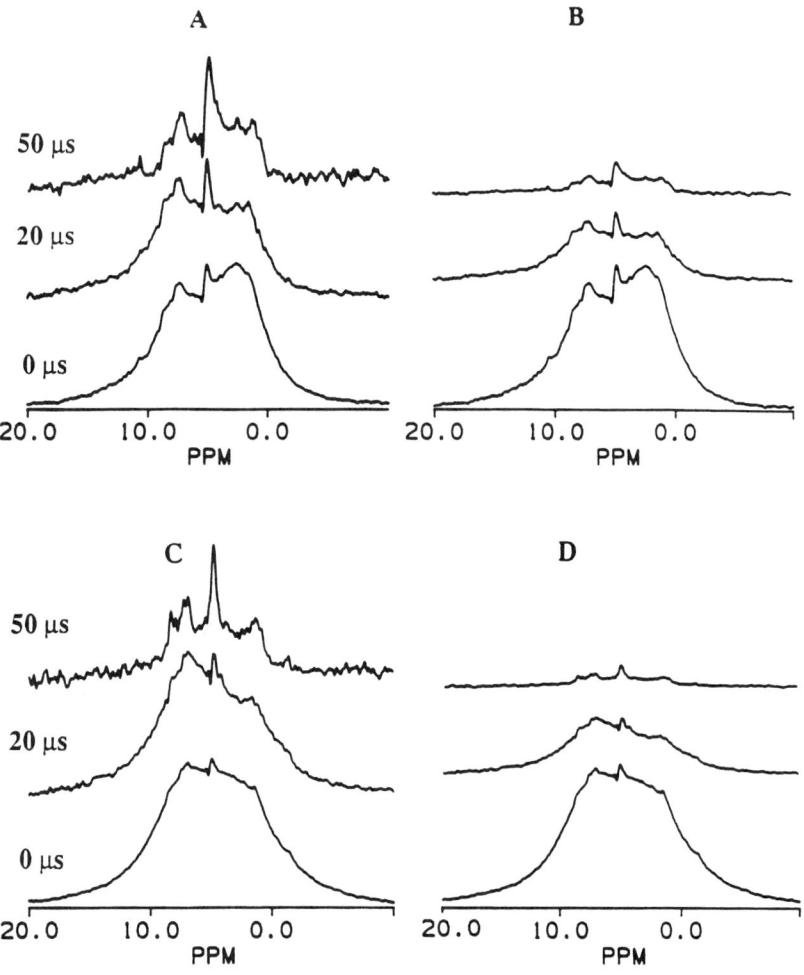

Figure 7. Proton CRAMPS dipolar-dephasing results for pyridine-saturated samples of Upper Freeport coal (A, arbitrary scaling; B, absolute intensity) and Pocahontas No. 3 coal (C, arbitrary scaling; D, absolute intensity). The dipolar-dephasing periods were 0, 20, and 50 μs.

Among the coals of this study, these last two coals have the highest carbon contents and largest aromaticities (15). However, the Pittsburgh No. 8 coal, which has an aromaticity that is similar to that of the Upper Freeport coal, showed much better spectral resolution enhancement. Therefore, the extent to which the spectral resolution characteristics depend on coal aromaticity is unclear.

We have not found a correlation between a coal's unpaired electron spin concentration and resolution enhancement by pyridine treatment. The available data on the maceral content and mineral matter content of the Premium coals (*13*) give no indication of a simple relationship between these properties and the spectral resolution observed in this study. However, the present study is based on too small a number of samples to warrant strong conclusions about these topics. Nevertheless, it is reasonable to suggest that LVB coals with high carbon contents will have poor CRAMPS spectral resolution, and pyridine saturation is not very effective in improving it. For other coals, pyridine treatment is more effective, and the resulting spectral resolution allows greater elucidation of structural detail.

For all samples, with the exception of Pocahontas No. 3, a relatively sharp aliphatic peak was observed. The chemical shift of this aliphatic peak varies from sample to sample within the range of 0.9–1.4 ppm. This range corresponds to methyl groups that are β, γ, or further removed from an aromatic ring (*16, 17*) and to paraffinic CH_3 groups. The second aliphatic peak is broader, and its chemical shift is more difficult to specify. Nevertheless, we can conclude that the chemical shift of this band varies from sample to sample over the range of 1.7–2.5 ppm. This chemical-shift range is characteristic of CH_3 groups that are α or β to an aromatic ring, of CH_2 groups β to aromatic rings, or of alicyclic or paraffinic CH_2 groups. For the Beulah–Zap lignite, a broad pattern centered around 3.4 ppm was observed. A shoulder with the same chemical shift is also noticeable in the spectra of Illinois No. 6 and Pittsburgh No. 8 HVB coals. This resonance can be associated with ring-joining methylene bridges; aliphatic O–CH_2 groups; indene CH_2 groups; and with CH_3, CH_2, and CH groups α to aromatic rings.

The broad resonance pattern with its center between 6.5 and 7.0 ppm in all the spectra of this study is characteristic of aromatic hydrogen. This spectral region may also include some phenolic OH hydrogens. The 8.2–8.4-ppm spectral region is in the chemical-shift range characteristic of heterocyclic aromatic structures. A distinct peak or shoulder is not observed at 8.2–8.4 ppm for all the samples of this study; such features appear most distinctly in the spectra of Illinois No. 6 and Pittsburgh No. 8 coals. The ^1H CRAMPS spectra of these two coals have particularly well-resolved aromatic peaks after pyridine treatment. The spectra of these two coals also show the highest degree of mobility (i.e., slowest dephasing) among the coals studied (Figure 6). Resolution of the peak at 8.2–8.4 ppm in these two coals may therefore simply be the result of the greater overall resolution in their CRAMPS spectra that is afforded by efficient swelling of these coals. Nevertheless, a correlation may exist between the intensity of the shoulder at 8.2–8.4 ppm and the sum A_T (%) of sulfur content (A_S, %) and nitrogen content (A_N, %) given in Table I. This parameter, A_T (%) = A_N (%) + A_S (%), has the following values for the samples of this study:

- Illinois No. 6, 3.9%
- Pocahontas No. 3, 1.85%
- Pittsburgh No. 8, 2.79%
- Upper Freeport, 2.36%
- Lewiston–Stockton, 2.28%
- Blind Canyon, 1.96%
- Wyodak–Anderson, 1.61%
- Beulah–Zap, 1.88%

Only for the two coals with the highest A_T values is a distinct 8.2–8.4-ppm peak observed, and for coals with smaller A_T values, a clearly defined feature at 8.2–8.4 ppm is not detected. (An exception may be Pocahontas No. 3.) This correlation supports the view that the 8.2–8.4-ppm spectral feature originates in the nitrogen and sulfur compounds in coal.

In the dipolar-dephasing experiment, the degree of spectral resolution often changes with the dephasing time. When the dephasing interval is increased, the line width typically becomes smaller, and resolution is thereby enhanced. The more mobile components of the coal structure are better resolved at longer dephasing times. The largest effect of this kind was observed for the Pocahontas No. 3, Upper Freeport, Beulah–Zap, Wyodak–Anderson, and Lewiston–Stockton coals. For the Illinois No. 6 and Pittsburgh No. 8 coals, the effect of increasing the dephasing time on spectral resolution is small. Of course, the structures in these two coals are mobilized to the greatest degree by pyridine treatment (vide supra), hence they should, and do, show the best resolution in the total spectrum, with or without dephasing. In the Lewiston–Stockton sample, the 8.2-ppm peak is observed in the spectrum obtained after a 400-μs dephasing time. However, there is no clear evidence of this peak in the dipolar-dephased spectra with the same dephasing time interval for the Wyodak–Anderson SB coal or the Beulah–Zap lignite, but these two samples have low nitrogen and sulfur contents.

As mentioned previously, the aliphatic-proton magnetization relaxes faster than the aromatic-proton magnetization in untreated coals. In most of the pyridine-saturated samples, the aliphatic magnetization relaxes more slowly than or approximately with the same rate as the aromatic magnetization. The dephasing time constants (inverse rate constants) depend on the distance between pairs of protons and their relative motions. The distances between aromatic protons are typically larger than those between aliphatic protons. Therefore, for relatively rigid structures the aliphatic protons would be expected to relax (dephase) faster than aromatic protons. This conclusion is consistent with the dipolar-dephasing results described in preceding sections for nontreated coals. After saturation of the sample with perdeuteropyridine, the coal becomes swollen, and the structure acquires much more freedom of motion; and as a result the proton magnetization accordingly decays more slowly. Hence, the dipolar-dephasing experiment shows that the influence of pyridine saturation on the motion of the aliphatic and aromatic protons is not the same. The aliphatic protons are in structures that are apparently less resistant to becoming mobilized by pyridine treatment than are the structures containing aromatic protons. This behavior raises the following question: Why are the structures containing aliphatic protons easier to mobilize than those containing aromatic protons?

To discuss this problem, we will consider and contrast the dephasing experiment, under conditions of perdeuteropyridine saturation, for the protons of a methyl group and those of other aliphatic structures. In the Beulah–Zap, Wyodak–Anderson, and Lewiston–Stockton coals, the methyl-group protons relax slower than protons of aromatic structures. The 1.8- or 3.3-ppm magnetization still dephases faster than the aromatic magnetization for these coals. Pyridine has a much greater effect on methyl group relaxation than on methylene group relaxation for all of the coals except Illinois No. 6 and Pittsburgh No. 8. In these two samples, all structures relax at similar rates. This more uniform behavior is the behavior that would be expected in the extensive rotational averaging of intramolecular dipolar couplings throughout the entire coal structure. In the limit of complete rotational averaging of dipolar interactions, as, for example, in the well-known case of adamantane, local features of the dipolar decay are removed, and all structures relax according to intermolecular dipolar interactions with the surrounding matrix. Another regime, in which none of the aliphatic components seem to be mobilized by pyridine saturation, is represented by the Upper Freeport coal, in which both aliphatic components relax faster than the aromatic protons.

Because of its size, shape, and symmetry, a methyl group needs only a small activation energy to rotate about its local C_3 axis (*18, 19*). Additionally, the C_3 axis can in many cases also rotate relative to the remainder of a large molecule. The effect of pyridine is most probably to make this last kind of motion more facile. The correlation time of this motion should depend on the location of the methyl group in the structures; for example, the C_3 axis of a methyl group terminating a long aliphatic chain will have more freedom of motion than if it were attached at the α position on a large polycondensed aromatic ring system.

The Beulah–Zap, Wyodak–Anderson, and Lewiston–Stockton coals have low aromaticities (*12*). Hence, the methyl groups in these coals are probably connected mainly to ring-attached aliphatic chains at positions relatively far from the aromatic ring. In this kind of structure, the CH_3-containing moiety should be easily mobilized by pyridine swelling. In coals with higher aromaticities, the methyl groups should terminate shorter aliphatic chains or be attached directly to aromatic rings; therefore they may be more difficult to mobilize with solvent addition. The other aliphatic protons (e.g., in paraffinic methylene groups or alicyclic moieties), because of their geometry, may have significantly less freedom of motion; they should require a higher activation energy for motion. Therefore, pyridine is less effective in mobilizing them than is the case for methyl protons (8, 9). Aromatic protons situated in aromatic clusters, which may be connected by methylene cross-links, have limited freedom of movement, and therefore they are also relatively difficult to mobilize by pyridine treatment.

As mentioned previously, the chemical-shift maximum of the aliphatic peak of untreated Pocahontas No. 3 coal has a very different value (3.2 ppm) than is observed for a coal with a low carbon content (1.5 ppm). This observation stimulated us to look for some correlation between this pattern and other coal properties, for example, carbon content, H/C ratio, aromaticity, or volatile matter. Unfortunately, on the basis of only eight coals (including the previously studied Blind Canyon coal), we did not find a reasonable correlation. On the other hand, it can be misleading to draw conclusions about correlations in coals on the basis of studying only a few samples. For that reason we extended our effort to a broader set of samples. Accordingly, potential correlations with the chemical-shift position of the maximum of the aliphatic peak were explored for a total of 19 coals by including 11 Polish coals. The relevant ^{1}H chemical-shift data for the additional 11 coals are included in Table II. The attempted correlations between the chemical-shift and other coal parameters are presented in Figure 8.

Very poor correlation was found between carbon content and the chemical-shift parameter (Figure 8A); nevertheless, the coals with the highest carbon contents have the largest chemical shifts, δ_A, and samples with the lowest carbon contents can be expected to have small aliphatic-proton chemical shifts. The correlation between coal aromaticity (f_a) and δ_A (Figure 8B) is also weak; however, for a high-aromaticity coal, the chemical shift of the aliphatic resonance is generally larger than for low-aromaticity coals. For coals with low-volatile matter, δ_A is large and typically increases with increasing volatile matter (Figure 8C). The best correlation was found between the aliphatic chemical shift and the H/C ratio (Figure 8D). For a small H/C ratio, the largest δ_A value is observed, and for a large H/C ratio, δ_A is smaller. This result is consistent with the view that, when the H/C ratio increases, protons of the aliphatic structures of coal are on the average closer to aromatic rings; that is, the aliphatic chains attached to aromatic rings are shorter. For example, the chemical shift of protons of a methylene group attached directly to an aromatic ring

Table II. The Chemical-Shift Position (δ_A) of the Maximum in the Aliphatic Peak of ^{1}H CRAMPS Spectra of Untreated Coals

Number	Coal	δ_A (ppm)
101	Upper Freeport	1.8
202	Wyodak–Anderson	1.5
301	Illinois No. 6	1.6
401	Pittsburgh No. 8	1.8
501	Pocahontas No. 3	3.2
601	Blind Canyon	1.4
701	Lewiston–Stockton	1.7
801	Beulah–Zap	1.5
WK13	Polish	3.0
WK14	Polish	1.8
WK22	Polish	1.7
WK25	Polish	2.1
WK26	Polish	1.9
WK30	Polish	2.5
WK31	Polish	2.3
WK32	Polish	1.7
WK35	Polish	1.9
WK36	Polish	1.8
WK37	Polish	2.0

NOTE: The analytical data for the Argonne Premium coals are listed in Table I. The data for the Polish coals are taken from reference 15.

is in the range of 3–5 ppm. Also the proton chemical shift of a methyl or methylene group in an aliphatic chain attached on an aromatic ring increases when the group is placed closer to the ring.

Conclusions

Saturation of the coal sample with perdeuteropyridine enhances the proton CRAMPS spectral resolution of coal. The magnitude of this effect is not the same for all Argonne Premium coal samples. Low-volatile coals with high carbon content have poor spectral resolution, and pyridine treatment is not very effective for resolution enhancement. For other coals,

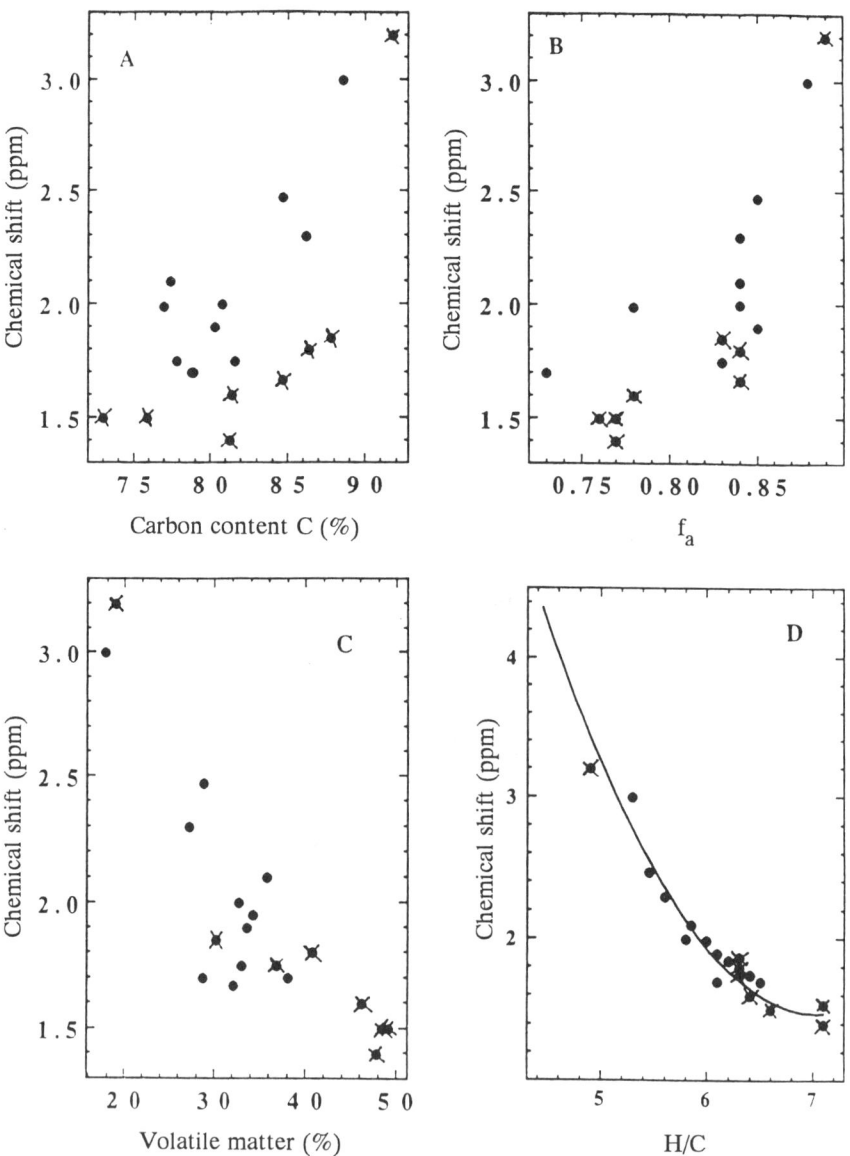

Figure 8. Correlation between the chemical-shift position of the maximum in the aliphatic peak for untreated coals and (A) carbon content, (B) carbon aromaticity, (C) volatile matter, (D) H/C ratio for both the Argonne Premium coals and 11 Polish coals. Points for the Argonne Premium coals are marked with an ×.

pyridine treatment is very effective, and the enhanced spectral resolution yields more structural detail.

In an untreated coal sample, the aliphatic protons decay in the dipolar-dephasing experiment significantly faster than aromatic protons. Pyridine saturation dramatically changes the dynamics of coal structure. The largest dynamic influence pyridine has is on methyl groups. In the pyridine-treated samples, the protons of methyl groups relax more slowly than those of aromatic moieties for the Wyodak–Anderson, Beulah–Zap, Upper Freeport, and Lewiston–Stockton coals. For the pyridine-saturated Illinois No. 6 and Pittsburgh No. 8 coals, all protons relax with approximately the same rate.

For samples with high sulfur and nitrogen contents, a peak or shoulder at 8.2 ppm is observed. A correlation was found between the chemical-shift position of the intensity maximum of the aliphatic peak and the H/C ratio. The chemical shift of the aliphatic peak tends to have a larger value for a sample with a larger H/C ratio.

Acknowledgments

We gratefully acknowledge use of the Colorado State University Regional NMR Center, funded by National Science Foundation Grant No. CHE–8616437 and partial support by U.S. Department of Energy Contract No. DE–AC22–88PC88813. Antoni Jurkiewicz is grateful for a Fulbright sponsorship.

References

1. Haeberlen, U. *High Resolution NMR in Solids: Selective Averaging;* Academic: New York, 1976.
2. Mehring, M. *Principles of High Resolution NMR in Solids*, 2nd ed.; Springer Verlag: New York, 1983.
3. Gerstein, B. C.; Chou, C.; Pembleton, R. G.; Wilson, R. C. *J. Phys. Chem.* **1977**, *87*, 565.
4. Schnabel, B.; Haubenreisser, U.; Scheler G.; Müller, R. *Proc. 19th Congr. Ampere (Heidelberg), 1976; p 441.*
5. Maciel, G. E.; Bronnimann, C. E.; Hawkins, B L. In *Advances in Magnetic Resonance: The Waugh Symposium;* Warren, W. S., Ed.; Academic: San Diego, CA, 1990; Vol. 14.
6. Rosenberger, H.; Scheler, G.; Kunstner, E. *Fuel* **1988**, *67,* 509.
7. Bronnimann, C. E.; Maciel, G. E. *Org. Geochem.* **1989**, *14*, 189.
8. Jurkiewicz, A.; Bronnimann, C. E.; Maciel, G. E. *Fuel* **1990**, *69,* 804.
9. Jurkiewicz, A.; Bronnimann, C. E.; Maciel, G. E. *Fuel* **1989**, *68*, 872.

10. Jurkiewicz, A.; Marzec, A.; Pislewski, N. *Fuel* **1982**, *61*, 647.
11. Barton, W. A.; Lynch, L. J.; Webster, D. S. *Fuel* **1984**, *63*, 1262.
12. Kamienski, B.; Pruski, M.; Gerstein, B. C.; Given, P. H. *Energy Fuels* **1987**, *1*, 45.
13. Vorres, K. S. *Energy Fuels* **1990**, *4*, 420.
14. Burum, D. P.; Rhim, W. K. *J. Chem. Phys.* **1979**, *71*, 944.
15. Jurkiewicz, A.; Wind, R. A.; Maciel, G. E. *Fuel* **1990**, *69*, 830.
16. Davidson, R. M. *Nuclear Magnetic Resonance Studies of Coal;* Report No. 1CT/S/TR32; ICA Coal Research: London, Jan. 1986.
17. Bhacca, N. S.; Johnson, L. F; Shoolery, J. N. *High Resolution NMR Spectra Catalog;* Varian Associates: Palo Alto, CA, 1962.
18. Jurkiewicz, A.; Pislewski, N.; Kunert, K. A. *Macromolecules Sci.-Chem.* **1982**, *18*(4), 511.
19. Jurkiewicz, A.; Tritt-Goc, J.; Pislewski, N; Kunert, K. A. *J. Polym. Sci. A* **1983**, *21*(4), 1195.

RECEIVED for review June 8, 1990. ACCEPTED revised manuscript June 17, 1991.

22

Measurement of ^{13}C Chemical-Shift Anisotropy in Coal

Anita M. Orendt[1], Mark S. Solum[1], Naresh K. Sethi[1,3], Craig D. Hughes[1], Ronald J. Pugmire[2,4], and David M. Grant[1]

[1]Department of Chemistry and [2]Department of Fuels Engineering, University of Utah, Salt Lake City, UT 84112

> *The methods available in NMR spectroscopy to obtain the principal values of the chemical-shift tensor are discussed. Applications to coal and to compounds with model structures that might be important in coal are presented. The composition of aromatic carbons in coal as determined by chemical-shift powder patterns is compared to results obtained by cross-polarization with magic-angle spinning and dipolar dephasing.*

A NUMBER OF TECHNIQUES in NMR spectroscopy can be used to obtain chemical-shift anisotropy (CSA) information from solid samples. In this chapter, the various methods will be described, along with the application of some of the techniques to both coal and substituted polycyclic aromatic compounds whose structures may be important in understanding the structure of coal. The various experimental methods will be discussed in terms of the ^{13}C chemical shift even though the techniques can be applied to any nuclei. The methods for obtaining the CSA include measurement and analysis of the static powder line shape (1), analysis of the variable-angle sample-spinning line shapes (2, 3), and analysis of the intensity pattern obtained from spinning at the magic angle at a rotation

---

[3]Current address: Amoco Corporation, P.O. Box 400, Naperville IL 60566
[4]Corresponding author

rate less than the width of the anisotropy (4, 5). Techniques that require more elaborate equipment and control over the orientation of the sample with respect to the magnetic field include the magic-angle hopping experiment (6) and the dynamic angle spinning technique (7, 8). In both of these two-dimensional (2-D) experiments, the orientation of the sample is changed during each scan. A related experiment is the stop-and-go technique (9) in which the sample rotation rate is rapidly changed from a stationary state to a high spinning rate during each scan. Another technique for studying the CSA involves spinning at the magic angle and using radio-frequency (rf) pulses to regain the CSA information that is averaged by the spinning (10). Finally, the application of a 2-D chemical-shift–chemical-shift correlation technique (11) to coal samples will be discussed.

Information Available from Chemical-Shift Anisotropy

The CSA is a reflection of the interaction of the electronic environment of the nuclei with the external magnetic field. Depending on the orientation of a given molecule in the field, a different value of the chemical shift is observed. If the sample simultaneously exists in all possible orientations (i.e., a random powder sample), a superposition of all the possible shift values for the nuclei in the molecule is observed. This pattern is the so-called *powder pattern*. The frequencies obtained from the break points or discontinuities of the powder pattern are the principal values of the chemical-shift tensor (1). For a nucleus (e.g., ^{13}C) that is in an anisotropic environment, three distinct values are observed, and these values correspond to the three elements (δ_{11}, δ_{22}, and δ_{33}) found when the chemical-shift tensor is diagonal. From the study of a single crystal, both the principal values of the shift tensor and their orientation in the molecular framework can be determined (1). Single-crystal studies of aromatic compounds show that δ_{33} is always perpendicular to the plane of the aromatic ring (12, 13), and its value in the aliphatic region reflects its independence of the π system. The other two tensor components, δ_{11} and δ_{22}, are oriented in the plane of the aromatic system. Theoretical studies have placed δ_{11} nearly perpendicular to the C–C bond with the largest π-bond character in a number of polycondensed aromatic hydrocarbons (14).

In most of the experimental methods to be described, only the aromatic region of the coal spectrum will be analyzed. For convenience, the aromatic carbons in coal can be grouped into four categories: protonated, substituted (having an alkyl carbon substituent), phenolic (having an oxygen substituent), and bridgehead or condensed. In Figure 1, ideal line shapes for the four types of carbons are shown. These ideal line shapes

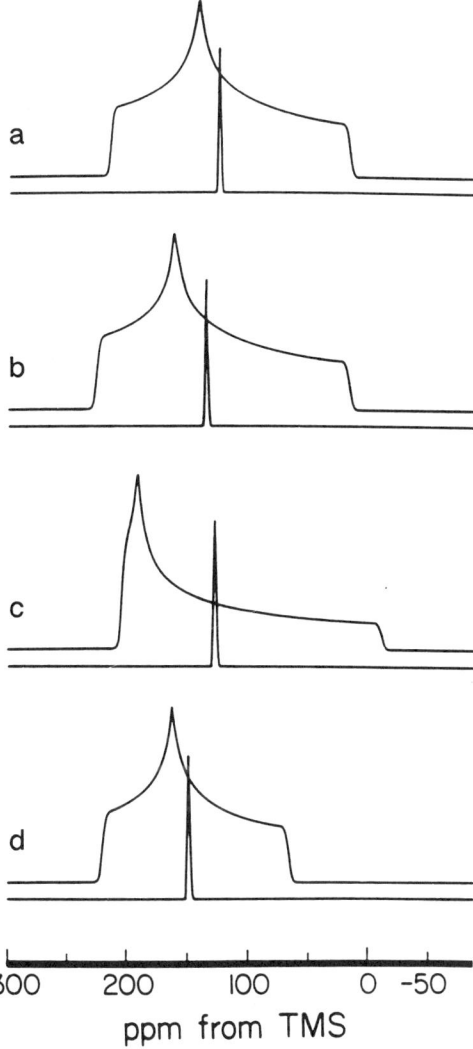

Figure 1. Ideal tensor patterns for the four types of aromatic carbons in coal: (a) protonated, (b) alkyl substituted, (c) condensed, and (d) phenolic.

were obtained by tabulating literature values for the shift tensor components for aromatic carbons in simple organic compounds (15–20) measured by a variety of methods (including the variable-angle sample spinning results from this laboratory). These results are presented in Table I.

Table I. Average Chemical-Shift Tensor Values for Aromatic Carbons

Carbon Type	δ_{11}	δ_{22}	δ_{33}	δ_{ave}
Protonated (73)	209 ± 21	138 ± 14	15 ± 10	121 ± 11
Substituted (32)	223 ± 8	160 ± 17	18 ± 8	134 ± 7
Phenolic (33)	219 ± 17	163 ± 9	69 ± 5	150 ± 7
Condensed (10)	206 ± 6	193 ± 7	−4 ± 11	131 ± 5

NOTE: The numbers in parentheses following the carbon type are the number of measurements considered in the analysis. All values are given in parts per million.

For protonated aromatic carbons, a highly asymmetric tensor with an isotropic chemical shift of 121 ppm relative to tetramethylsilane (TMS) is expected. The substituted aromatic carbons have an average isotropic chemical shift of 134 ppm, and the phenolic aromatic carbons have an average isotropic chemical shift of 150 ppm. The downfield isotropic shift observed due to the electronegativity of oxygen is almost entirely due to the shift in the value of δ_{33}. Furthermore, the principal values of both substituted and phenolic carbons are slightly less asymmetric than that of the protonated carbons (i.e., the difference between δ_{11} and δ_{22} is smaller in both cases than is observed for the protonated carbons). For the condensed aromatic carbons, an average isotropic chemical shift of 131 ppm is observed, with a shift tensor that is nearly axially symmetric ($\delta_{11} \approx \delta_{22}$) due to the local symmetry (approaching C_3) in the plane of the molecule at the bridgehead carbons.

Experimental Methods

Static Powder Samples. The most straightforward experimental method to obtain the chemical-shift information in a solid is to take the ^{13}C spectrum under static conditions (1). The powder sample contains all orientations with respect to the magnetic field and as such gives a broad line shape from which the principal values of the chemical-shift tensor can be obtained. Experimentally, either a simple Fourier transform (FT) experiment with high-power ^1H decoupling to remove the ^1H–^1H and the ^1H–^{13}C dipolar interactions or a cross-polarization (CP) experiment (21) with ^1H decoupling can be performed. The cross-polarization technique is generally used because of the fourfold enhancement in the ^{13}C signal that can be realized.

In Figure 2, a static powder pattern for 1,2,3,6,7,8-hexahydropyrene, along with its best simplex fit of the aromatic region using the POWDER method (22), is shown. The principal values of the shift tensor obtained from this spectrum

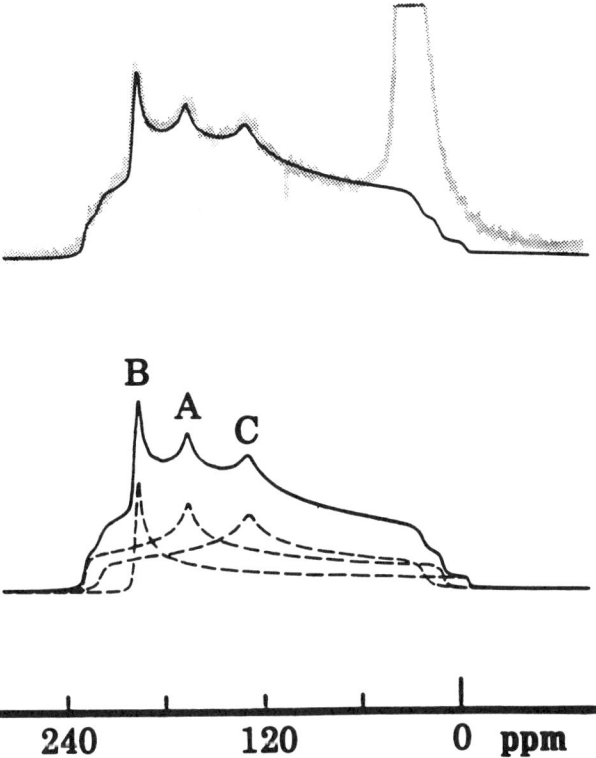

Figure 2. Static line shape with best fit of aromatic region for 1,2,3,6,7,8-hexahydropyrene. Part A indicates the pattern for the substituted aromatic carbons, B indicates the pattern for the bridgehead or condensed carbons, and C indicates the pattern for the protonated aromatic carbons. (Reproduced from reference 23. Copyright 1987 American Chemical Society.)

are given in Table II. This example clearly illustrates the limitations of the technique. The upfield components of the aromatic carbon shift tensors are often obscured by the signal from any aliphatic carbons in the sample. With samples that have many carbons with similar isotropic chemical shifts, determining the break points becomes very difficult. A practical limit for this experiment is two or three aromatic tensors.

Earlier work on the analysis of static powder patterns from coal samples by Wemmer et al. (15) used a set of data from aromatic carbons in a large number of compounds to derive ideal line shapes for aromatic, condensed aromatic, alkoxy, and aliphatic carbons. Static coal spectra were then fit to obtain the composition in terms of these four predetermined tensors.

In the work presented here, no assumptions were made about the value of the tensor components of the various types of aromatic carbons. Because of the problem of overlap of the aromatic σ_{33} component and the aliphatic carbon tensor patterns, this technique is useful only on coals whose aromaticities are high (i.e., greater than 95%). In these cases, the small amount of aliphatic signal, when spread over its CSA, will usually be lost in the base-line noise. A representative static spectrum of a coal sample, PSOC-1468, is shown along with the results of the simplex fit in Figure 3. The improvement in the fit in going from a single tensor to two tensors is evident from the comparison of Figures 3a and 3b. The sum of squares between the experimental and simulated spectrum, which is the measure of the goodness of the fit, improved from 0.200 to 0.062. The single-tensor fit gave components at 199, 175, and −9 ppm, for an average of 122 ppm; the results of the two-tensor fit are given in Table III. In the spectral-fitting routine, all of the tensor components, as well as the relative amounts of each type of carbon, were allowed to vary. The results obtained to date on anthracite coals studied in this laboratory are summarized in Table III. Given the inherent errors associated with the experiment, the correlation with the elemental analysis data (atomic H/C ratio) is remarkably good. The tensor components in these three coals are remarkably similar to those given in Table I for model compounds, even though no assumptions about the components were made in the fitting process.

Variable-Angle Sample Spinning. The technique of variable-angle sample spinning (VASS) (2, 3), spinning at an angle other than the magic angle, has been shown to be very useful in the study of model compounds and appears to be useful for the study of coal. The expression that governs the averaging of the CSA line shape during rapid spinning (spinning speeds greater than the CSA) is given in eq 1.

$$W(\theta, \alpha, \beta) = W_0 \left\{ \delta_i + \frac{1}{2} \left(3\cos^2\theta - 1 \right) \left[\frac{1}{2} \left(3\cos^2\beta - 1 \right) \left(\delta_{33} - \delta_i \right) \right. \right.$$
$$\left. \left. + \frac{1}{2} \left(\sin^2\beta \cos 2\alpha \right) \left(\delta_{11} - \delta_{22} \right) \right] \right\} \quad (1)$$

In this equation, α and β are two of the Euler angles relating the principal axis system (PAS) of the chemical-shift tensor to the spinner frame; δ_{11}, δ_{22}, and δ_{33} are the principal values of the chemical-shift tensor; δ_i is the average or isotropic chemical shift; and $W_0 = -\gamma B_0$, where B_0 is the magnetic field and γ is the gyromagnetic ratio. For θ, the angle the spinning axis makes with respect to the magnetic field, such that $(3\cos^2\theta - 1) = 0$ (i.e., at the magic angle), this expression reduces to the isotropic chemical shift, δ_i. For angles less than the magic angle, a partially averaged and scaled powder pattern is obtained, whereas for angles greater than the magic angle the pattern is inverted as well as scaled. Ex-

Table II. Tensor Components for Aromatic Carbons in 1,2,3,6,7,8-Hexahydropyrene

Carbon Type	δ_{11}	δ_{22}	δ_{33}	δ_{ave}	Aromatic Carbons (%)
Protonated	225	128	22	125	40
Substituted	231	168	9	136	40
Condensed	203	197	−6	131	20

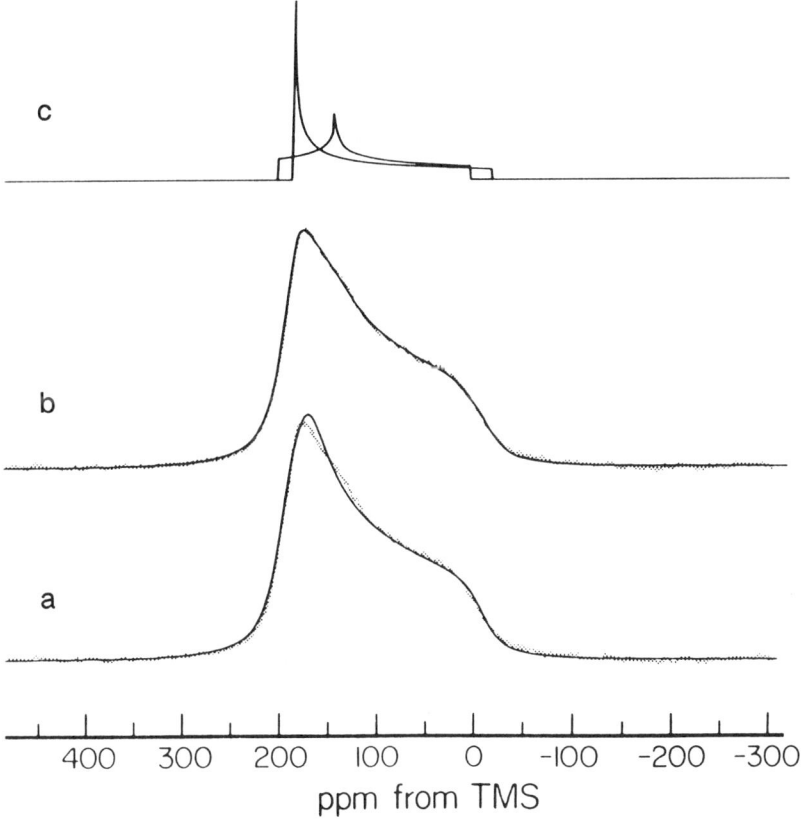

Figure 3. Static line shape with best fit for PSOC-1468 coal: (a) best single-tensor fit, (b) best two-tensor fit, and (c) individual unbroadened tensor patterns in b scaled by percentages obtained in the fit.

Table III. Tensor Components for Aromatic Carbons in Coals

Carbon Type	δ_{11}	δ_{22}	δ_{33}	δ_{ave}	Total Carbons (%)	Protonated Aromatic Carbons (%) ea[a]	dd[b]
PSOC-628[c] (96% aromatic)							
Protonated	210	145	9	121	34	43[d]	
Condensed	192	185	1	126	62		
PSOC-867[c] (100% aromatic)							
Protonated	215	138	21	125	13	12	15
Condensed	193	181	−8	122	87		
PSOC-1468[e] (95% aromatic)							
Protonated	205	148	7	120	27	24	26
Condensed	190	188	−15	121	73		

[a] Percentage of protonated aromatic carbons determined by elemental analysis (i.e., H/C ratio).
[b] Percentage of protonated aromatic carbons determined from dipolar-dephasing and CPMAS results.
[c] Data are from ref. 23. Unlike the other anthracite coals studied with aromaticities greater than 95%, the aliphatic contribution to the powder pattern of this coal could not be neglected, as it appeared in a fairly narrow chemical-shift range and contributed significantly to the line shape.
[d] Elemental analysis data was obtained from Galbraith Laboratories from the same sample that was used for the NMR study.
[e] The CPMAS spectrum of this coal has a rather sharp line centered in the 20–30-ppm region (24). The f_a value is 0.96 as determined by a variable-contact-time experiment (25). If this aliphatic peak is assumed to be due primarily to CH_3 groups, the renormalized aromatic H/C ratio from elemental analysis is 0.32. If the aliphatic carbons are assumed to be due to CH_2 groups, the normalized H/C ratio is 0.36. Either value is within the estimated error of the experimental data (i.e., 34%).

perimentally, five spectra are generally recorded: one at the magic angle, two at larger angles, and two at smaller angles. Each of the four spectra recorded off from the magic angle have their own distinctive line shape; they are not just scaled versions of the same line shape. The four off-angle spectra can then be fit simultaneously with the isotropic chemical shifts locked to the values obtained from the spectrum at the magic angle.

VASS removes some of the difficulties associated with the analysis of static powder patterns. By both the partial averaging of the CSA and the inversion at angles larger than the magic angle this technique removes the overlap observed between the aromatic and aliphatic powder patterns, thereby allowing the analysis of samples with substantial aliphatic components. Less time is required to obtain a spectrum with a satisfactory signal-to-noise ratio because of the scaling of the CSA. Because of the number of spectra that are being fit and the different line shapes, the limitation of the number of aromatic carbon tensors that can be analyzed (relative to a static powder pattern) increases to about four or five. This condition still restricts the study of model polycondensed aromatic hydrocarbons to those that are symmetrically substituted. The results of applying this method to several model compounds are given in Table IV. The set of VASS spectra for

Table IV. Tensor Components for Model Compounds

Carbon Type	δ_{11}	δ_{22}	δ_{33}	δ_{ave}	Aromatic Carbons (%)
1,2,3,6,7,8-Hexahydropyrene					
Protonated	225	128	22	125	40
Substituted	231	168	9	136	40
Condensed	203	197	−6	131	20
Pyracene					
Protonated	209	142	19	124	40
Substituted	226	166	36	142	40
Condensed	202	192	24	139	20
1,4,7,8-Tetramethylnaphthalene					
Protonated	235	123	32	130	40
Substituted	231	161	14	135	40
Condensed	204	202	1	135	20
9,9,10,10-Tetramethyl-9,10-dihydroanthracene					
Protonated	229	147	7	128	33
Substituted	224	143	17	128	33
Condensed	219	186	20	142	33

1,2,3,6,7,8-hexahydropyrene along with the simulations are shown in Figure 4. A comparison of Figures 2 and 4 shows the advantages of the VASS technique as compared to the analysis of static powder patterns.

This technique has also been applied to several coals, with the results shown in Table V. The aromaticities given in the table are from variable-contact-time cross-polarization–magic-angle spinning (CP–MAS) experiments. A representative set of two VASS spectra along with the best fit is shown in Figure 5. The fitting of the spectra is complicated because of the lack of clear break points. Hence, no unique solution to the simplex fitting routine is possible (i.e., many local minima exit or occur on the surface). Many of the solutions could be eliminated by the fact that the fitted values for the shift tensor components and iso-

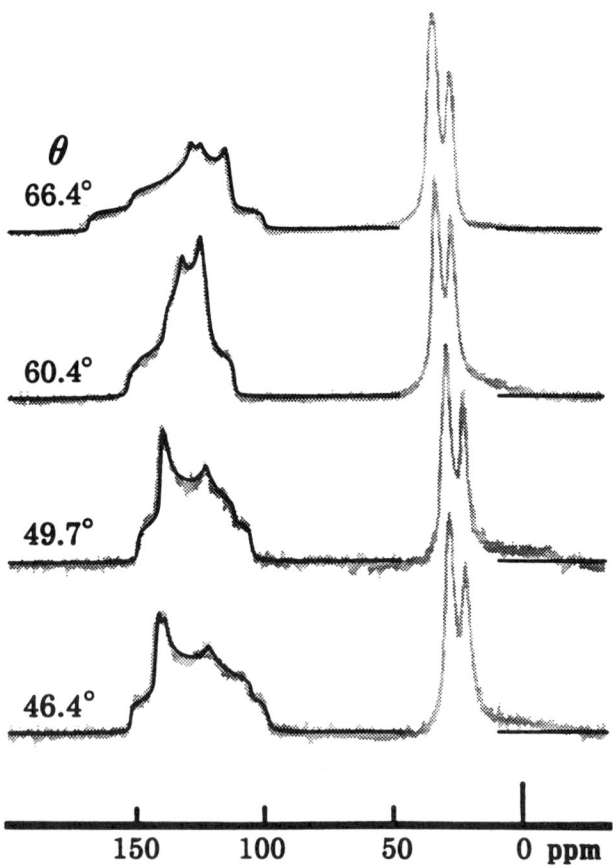

Figure 4. VASS spectra of 1,2,3,6,7,8-hexahydropyrene with best fits. (Reproduced from reference 23. Copyright 1987 American Chemical Society.)

Table V. Tensor Components and Population Factors for Coals Obtained by VASS

Carbon Type	δ_{11}	δ_{22}	δ_{33}	δ_{ave}	Total Carbons (%) VASS	dd[a]
Illinois No. 6 (72% aromatic)[b]						
Protonated	204	142	32	126	27	26
Substituted	231	161	46	146	18	18
Condensed	212	184	−9	129	22	22
Phenolic	215	155	80	150	5	6
Upper Freeport (81% aromatic)[b]						
Protonated	208	141	20	123	28	28
Substituted	217	158	33	136	21	20
Condensed	203	189	−11	127	30	29
Phenolic	229	159	71	153	2	2
Pocahontas No. 3 (86% aromatic)[b]						
Protonated	215	145	20	127	33	33
Substituted	220	162	31	138	17	17
Condensed	203	191	−10	128	34	34
Phenolic	225	158	73	152	2	2
Aldwarke Silkstone Fusinite (88% aromatic)[c]						
Protonated	223	149	17	130	34	34
Substituted	231	161	46	146	6	
Condensed	204	192	−30	122	48	
Phenolic	—	—	—	—	0	

[a] Percentages were obtained from dipolar-dephasing experiments.
[b] Percent aromatic and percent dd values were taken from ref. 25.
[c] Percent aromatic carbon and percent protonated aromatic carbon values were taken from ref. 26.

tropic shift values were quite different from those observed in model compounds and in the anthracite coals. Other solutions could be eliminated because the percentages of the protonated and nonprotonated shielding tensors were obviously wrong, the tensor components were not in the expected range, or the sum of squares was lowered by adding extensive broadening in the simulation. By using information from CP–MAS and dipolar-dephasing studies on the percentage of the aromatic carbons that are protonated as a starting point (but not a locking point) in the fitting process, a reasonable solution was obtained for each of the coals. Therefore, by using information obtained from CP–MAS and dipolar-dephasing experiments, the VASS technique yields principal values of the shift

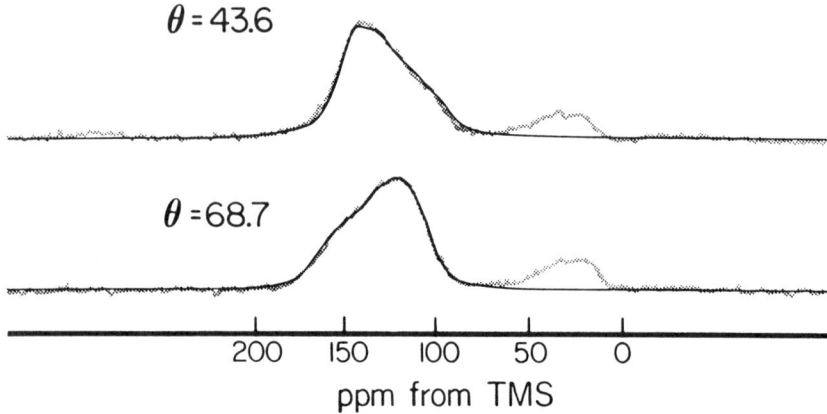

Figure 5. VASS spectra of Pocahontas No. 3 coal with best fits.

tensor for the aromatic carbons in coal that are consistent with those observed in model compounds.

Slow Magic-Angle Spinning. If the rate of spinning at the magic angle is much less than the CSA, the resulting spectrum will be a narrow line at the isotropic chemical shift with sidebands spaced at integer numbers of the spinning frequency for the width of the CSA (4, 5). Analysis of the peak intensities of this sideband pattern yields the principal values of the shielding tensor. The analysis procedure depends on a careful measurement of the sideband intensities and the use of a reasonably complex mathematical analysis; thus this procedure has been used only in relatively few cases. In addition, this procedure requires a constant spinning speed (i.e., ±10 Hz) during the entire experiment; if the spinning speed varies, the line widths of the sidebands are increased, and the peak intensities become less reliable. If a compound contains more than about three inequivalent carbons with isotropic shifts in the same region, the complexity of the spectrum increases, and it becomes difficult to determine the connectivity between the central peak and its associated sidebands. Morgan and Newman (20) used this approach to obtain the tensor components of a series of 12 benzene compounds with oxygenated substituents. One group has extended this technique to a 2-D method (27) in which the various sideband patterns are separated, as discussed in the next section.

The slow-spinning method was applied to the study of several Australian coals by Burgar and co-workers (28, 29). In this work, average tensor components for three types of aromatic carbons (protonated, substituted, and bridgehead) taken from the work of Wemmer et al. (15) were used, and the percentage composition was varied to obtain the best fit. The percentage of aliphatic and alkoxy carbons were also determined with a reported accuracy of 5%.

Chemical-Shift—Chemical-Shift Correlation Spectroscopy. A new technique for measuring ^{13}C chemical-shift tensors in powdered solid samples has great potential for providing new information about coals and similar materials. This technique is a modification of the well-known 2-D exchange NMR experiment (30, 31) in which the sample is rapidly reoriented by 90° along an axis perpendicular to the magnetic field, B_0, during the mixing time of a 2-D experiment. The spectra have projections onto the two chemical-shift axes, which are the conventional powder ^{13}C solid-state chemical-shift spectra, in each of the two orientations of the sample relative to the external field. The complete spectrum, however, contains more detailed, correlative information about the orientational distribution of ^{13}C atoms in the sample, and as such, offers the capability to characterize order in partially ordered samples (32), a task that is difficult to accomplish from the measurement of a normal powder pattern. Another great potential of this technique is the ability to sort overlapping chemical-shift patterns and to make unequivocal connections between the principal values of a given chemical-shift tensor in powdered polycrystalline samples.

As mentioned previously, a powder has randomly distributed ^{13}C atoms, so the chemical-shift patterns in any orientation with the external field are identical. This random orientation gives a 2-D spectrum that is symmetrical across its diagonal. The unique representation of chemical-shift tensors in this experiment allows all of the principal values of one tensor to be measured by connecting peaks in the spectrum in triangular (for axially symmetric shift tensors) or hexagonal (for general shift tensors) patterns, as shown in Figures 6a and 6b. A molecule in the first orientation, $F1$, that gives rise to a signal at $\delta_{\|}$, must after a 90° sample rotation, give a signal at $\delta_\perp$ in $F2$. A 90° rotation likewise can take a molecule that gives a signal at $\delta_\perp$ in $F1$ to an orientation that gives a signal at either $\delta_{\|}$ or $\delta_\perp$. These relationships lead to the triangular pattern observed for an axially symmetric tensor as shown in Figure 6a. For an asymmetric tensor, a 90° sample rotation must take one principal value into the remaining two components, leading to the hexagonal pattern shown in Figure 6b. In both cases, molecules oriented such that their signals are between the principal values give rise to the intensity in the center of the triangle or hexagon. Each unique chemical-shift tensor in the sample has its own unique hexagonal or triangular pattern, and the added spectral space and inherent redundancy of the second frequency dimension permits good resolution of different tensors in cases where conventional one-dimensional (1-D) spectroscopy might fail.

The ability of this technique to sort overlapping chemical-shift patterns in coal samples is demonstrated in Figures 7a and 7b for Upper Freeport and Pocahontas No. 3 coals (Argonne Premium coal samples), respectively. In a conventional 1-D static solid-state spectrum of coal, a large degree of overlap between aromatic and aliphatic signals occurs, whereas in the 2-D chemical-shift–chemical-shift correlation spectrum, the two regions are completely separated. This separation allows for an accurate determination of the aromaticity from a non-MAS spectrum. The calculated aromaticities from the 2-D spectra are 0.78

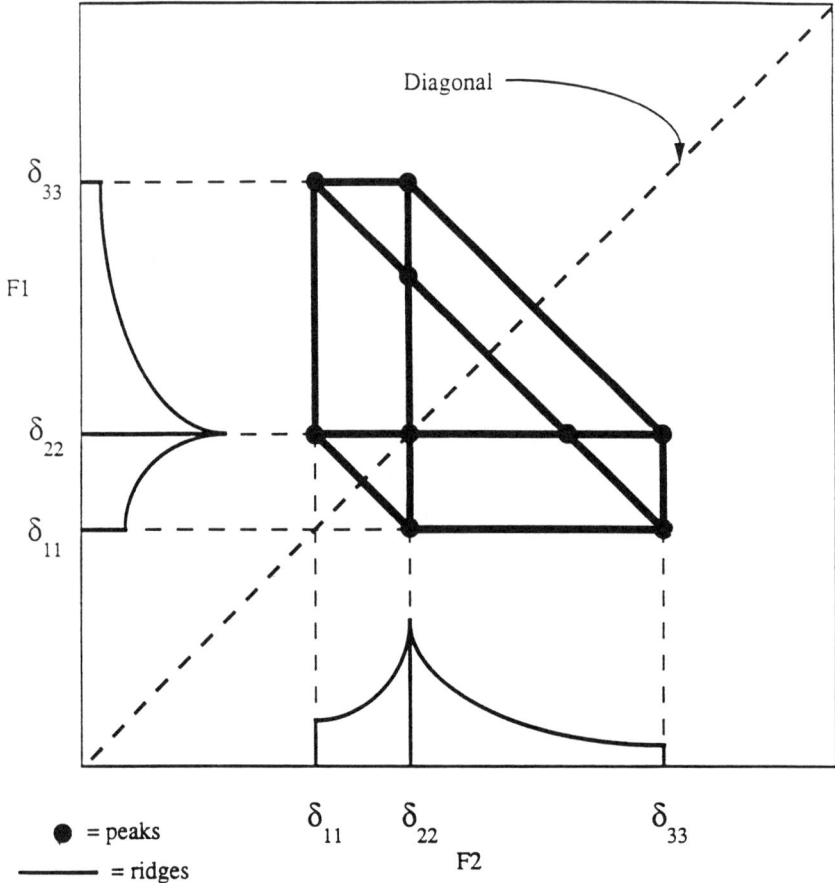

Figure 6a. Ideal shielding patterns obtained in the 2-D chemical-shift–chemical-shift experiment for the asymmetric tensor. Here and in Figure 6b, the peaks, marked by ●, are the principal values of the chemical-shielding tensor.

for Upper Freeport and 0.87 for Pocahontas No. 3 coals, results that are in good agreement with the CP–MAS results of 0.81 and 0.86, respectively (25). In addition, the line shapes of the aromatic and aliphatic regions show a distribution of chemical-shift patterns, and a model-dependent estimate of this distribution should be possible. Efforts are also underway to determine if any ordering information can be obtained from spectra recorded on samples that have been preserved in their original bedded form.

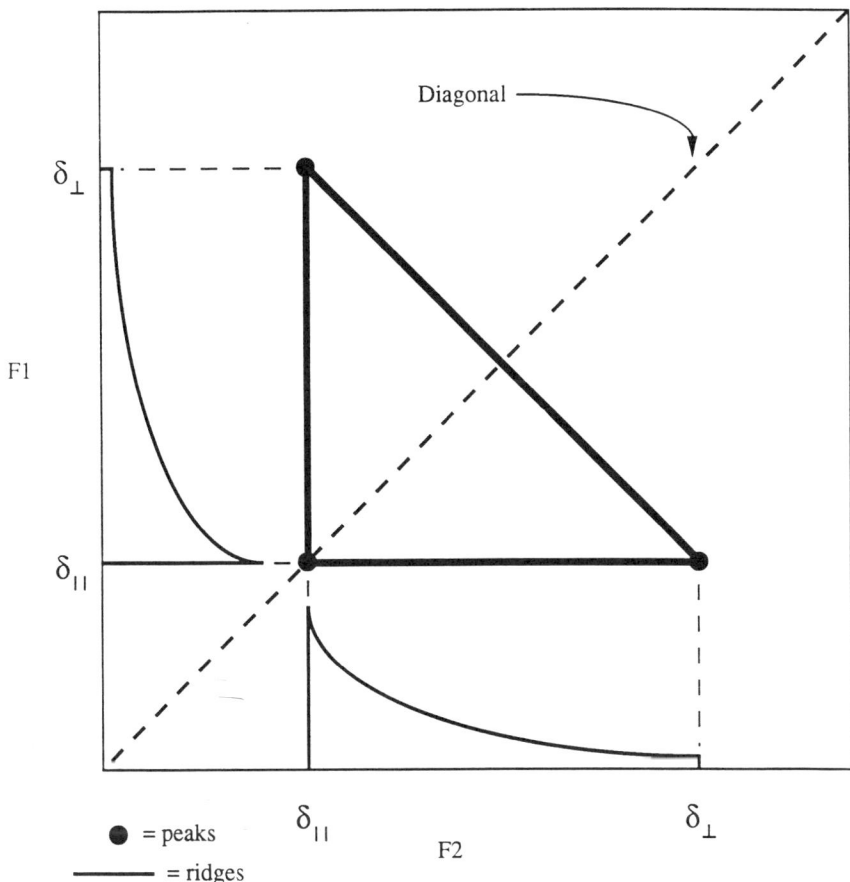

Figure 6b. Ideal shielding patterns obtained in the 2-D chemical-shift–chemical-shift experiment for the axially symmetric tensor.

Magic-Angle Hopping, Stop-and-Go, and Dynamic Angle Spinning. These three techniques all share the feature of separating the CSA information of the individual inequivalent carbons. In addition, the mechanical aspects of the experiments are very demanding. The methods of magic-angle hopping (MAH) (6) and dynamic angle spinning (DAS) (7, 8) both require that the sample be reoriented with respect to the magnetic field. The stop-and-go (STAG) method (9) requires that the sample be taken from a high spinning speed to rest during each acquisition. In the MAH experiment, discrete jumps are made between three orientations, which are related to each other by 120° rotations about the magic angle, with equal time being spent at each orientation. In the DAS experiment, the rapidly spinning sample is reoriented from the magic angle

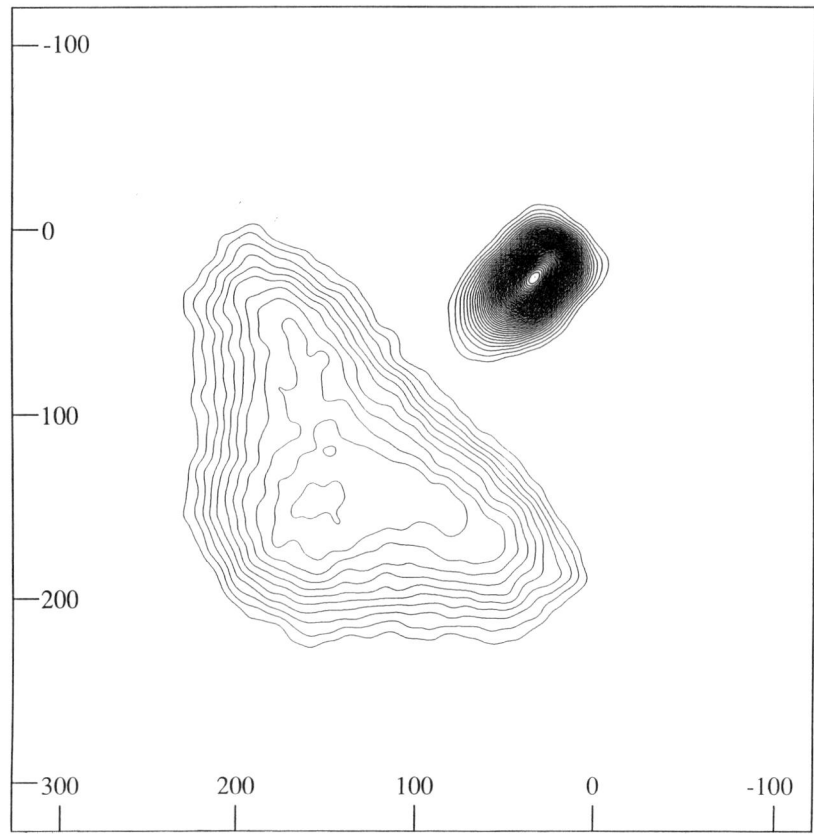

Figure 7a. Equal-intensity contour plot of the 2-D chemical-shift–chemical-shift spectrum of the Upper Freeport coal sample. Here and in Figure 7b, chemical-shift scales are in parts per million (ppm) from TMS.

to any other angle; the choice of the second angle determines the degree of scaling observed in the tensor patterns obtained in the second dimension. Finally, in the STAG sequence the experiment is done under stationary conditions except during the detection period. The major problem with all three of the techniques is the mechanical difficulties encountered in moving the sample (either stop-and-start or reorienting a spinning sample) in a very short period of time. In addition, all three methods require much more time for acquisition than any of the 1-D methods in order to get reasonable signal-to-noise ratios in the slices of the 2-D plot. DAS probes that promise reorientation times in the neighborhood of 20 ms are now commercially available.

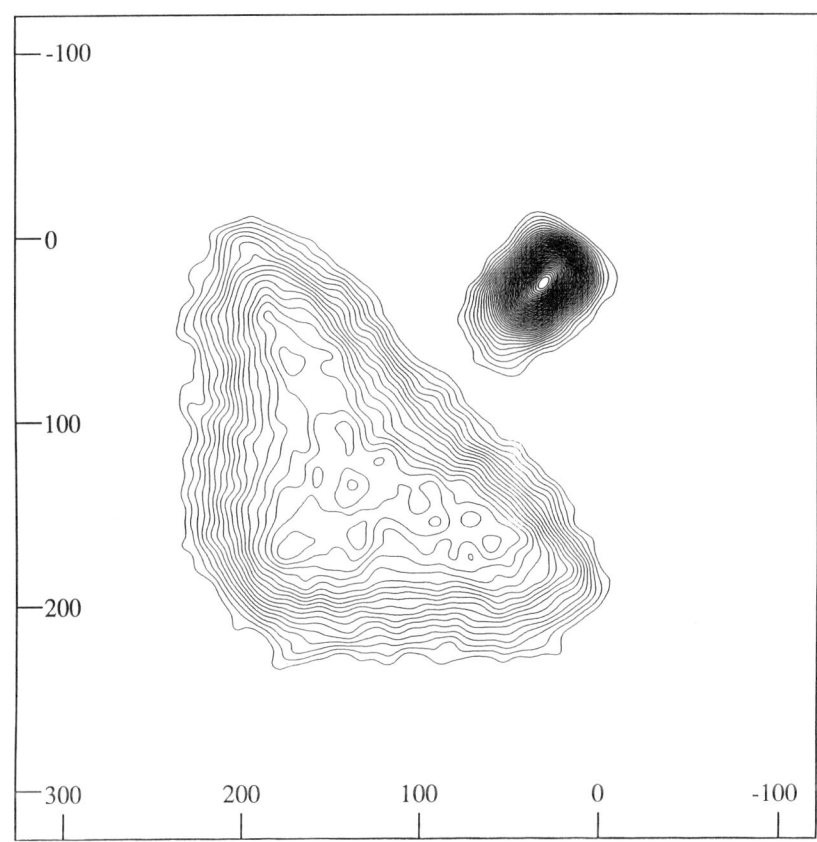

Figure 7b. Equal-intensity contour plot of the 2-D chemical-shift–chemical-shift spectrum of the Pocahontas No. 3 Argonne Premium coal sample.

All of these experiments give the same result in the final spectrum. The 2-D spectra have the MAS spectrum of the isotropic chemical shift in one dimension, and the second dimension contains all the anisotropic information in the form of separated powder patterns, which may or may not be scaled. This separation in the second dimension removes the limitation of the number of inequivalent carbons that can be studied. If the carbon has an isotropic chemical shift that can be resolved from all the other isotropic shifts, the powder pattern in the second dimension will be of only one carbon. Therefore, these techniques will definitely be powerful tools in the study of more complicated and less symmetrical model compounds. An example of the power of the DAS technique in resolving a large number of tensor patterns is shown in Figure 8 (*8*). In this spectrum of cholic acid, 14 carbons have isotropic chemical shifts between 10 and 50 ppm, and all are completely separated in the slices of the 2-D plot.

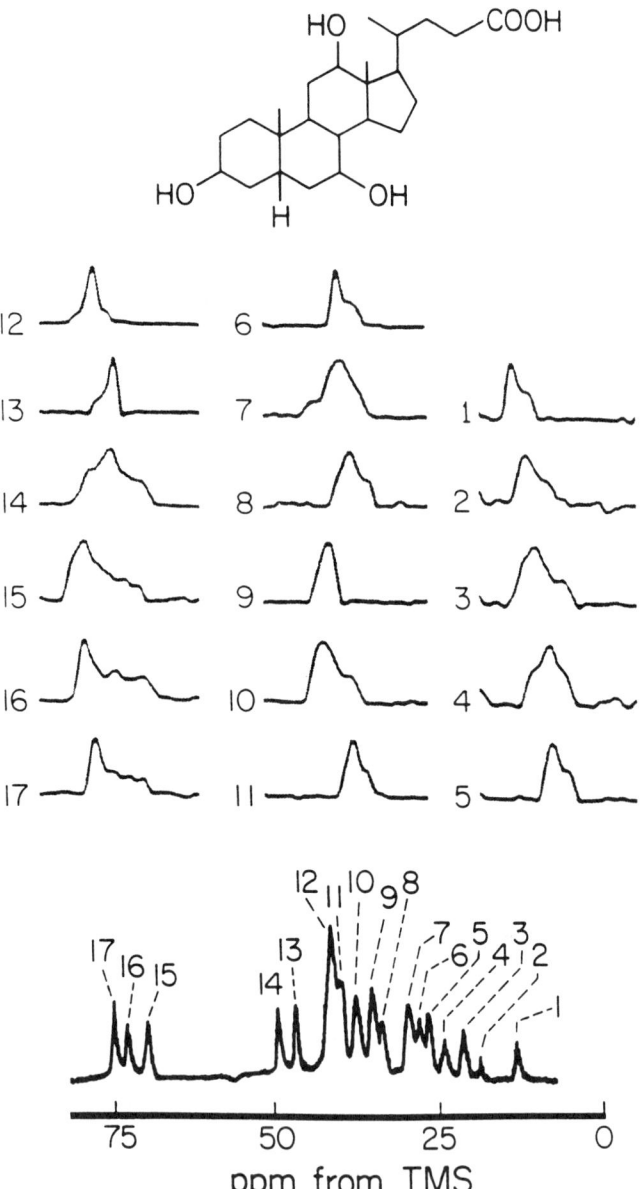

Figure 8. Results of a DAS experiment on cholic acid. The structure of cholic acid is shown at the top of the figure, the slices of the 2-D plot of each carbon are shown in the middle of the figure, and the MAS spectrum is shown at the bottom. (Reproduced with permission from reference 8. Copyright 1985.)

The value of these techniques for the study of the coal has yet to be seen. The short carbon T_1 (spin–lattice relaxation time) reported in coals (33) make a short reorientation time a necessity. In addition, because the inequivalent carbons are not resolved, what information can be obtained from the overlapping carbons contained in the slices in the second dimension and how difficult the analysis of these slices may be remain to be seen.

Magic-Angle Spinning with Synchronous Application of Radio-frequency Pulses. This method retrieves the CSA information from a MAS experiment by the application of a train of π pulses that are synchronized with the sample rotation, a technique first introduced by Lippmaa and co-workers (10). This type of experiment is very sensitive to pulse imperfections. Since its introduction, several groups have made modifications to minimize the effects of pulse imperfections (34) and to extend the method into a 2-D technique (35). Tycko et al. (36) modified the basic method to render the intensities of the CSA patterns obtained in the second dimension undistorted, making it easier to obtain reliable tensor components.

To date, no work on coal has been reported. This method does not suffer from the problem of relaxation as do the methods discussed in the previous section. However, the questions concerning the time required for acquisition and the interpretation of the slices in the second dimension still remain.

Summary and Conclusions

A number of techniques exist for the determination of the CSA in solids. The choice of a method is often limited by the availability of the necessary equipment and by the difficulty of the experiment in terms of the required pulse sequences and the mechanical aspects of sample spinning and reorientation. Furthermore, the complexity of the compound can rule out some experiments because of the complexity of the resultant spectrum.

From the limited number of coal samples on which the tensor information has been obtained, several conclusions can be reached. The shift values obtained (both tensor components and isotropic chemical shifts) are consistent with those obtained in the substituted polycyclic aromatic compounds that are often used as model compounds. The majority of the results are within one standard deviation of the average taken from aromatic carbons in a large number of compounds. In addition, the breakdown of the aromatic carbons into the four classes outlined in the first section as determined by both static or VASS line-shape analysis is consistent with the results of both dipolar-dephasing experiments and elemental analysis.

The study of the CSA in coal has several advantages over doing only the more common CP–MAS experiments. Most obviously, the three principal values that provide additional information on the electronic environment of the nucleus are determined instead of only the average of these three values, that is, the isotropic chemical shift. The information on the percentages of bridgehead and phenolic aromatic carbons obtained from the fitting of the chemical-shift pattern may be more reliable than those obtained from CP–MAS and dipolar-dephasing experiments in which all of the intensity in a specific frequency range is assumed to be due to a certain type of aromatic carbon.

Finally, the 2-D techniques that separate the individual static or scaled line shapes are very powerful tools and have the potential of being extremely useful. The usefulness of DAS for the study of complex compounds has already been demonstrated (8). The full potential of this method, as well as the other 2-D techniques, in the study of coal still remains to be explored. The 2-D chemical-shift–chemical-shift correlation method may also prove to be a valuable tool in the study of coal. Refinements of the experimental procedure are currently under study.

Acknowledgments

This work is supported by the National Science Foundation through the Advanced Combustion Engineering Research Center (Contract No. CDR–8522618), the Pittsburgh Energy Technology Center through the Consortium for Fossil Fuel Liquefaction Science (Contract No. DE–FC22–89PC89852), and the U.S. Department of Energy Office of Basic Energy Sciences (Grant No. DE–FG02–86ER13510).

References

1. Mehring, M. *Principles of High Resolution NMR in Solids;* Springer-Verlag: New York, 1983.
2. Stejskal, E. O.; Schaefer, J.; McKay, R. A. *J. Magn. Reson.* **1977**, *25*, 569.
3. Sethi, N. K.; Grant, D. M.; Pugmire, R. J. *J. Magn. Reson.* **1987**, *71*, 476.
4. Maricq, M. M.; Waugh, J. S. *J. Chem. Phys.* **1979**, *70*, 3300.
5. Herzfeld, J.; Berger, A. E. *J. Chem. Phys.* **1980**, *73*, 6021.
6. Bax, A.; Szeverenyi, N. M.; Maciel, G. E. *J. Magn. Reson.* **1983**, *52*, 147.
7. Bax, A.; Szeverenyi, N. M.; Maciel, G. E. *J. Magn. Reson.* **1983**, *55*, 494.
8. Maciel, G. E.; Szeverenyi, N. M.; Sardashti, M. *J. Magn. Reson.* **1985**, *64*, 365.
9. Zeigler, R. C.; Wind, R. A.; Maciel, G. E. *J. Magn. Reson.* **1988**, *79*, 299.

10. Lippmaa, E.; Alla, M.; Tuherm, T. *Proc. 19th Congr. Ampere* (Heidelberg), 1976; p 113.
11. Hughes, C. D; Sherwood, M. H.; Alderman, D. W.; Grant, D. M. *J. Magn. Reson.* **1992**, accepted for publication.
12. Veeman, W. S. *Prog. NMR Spectrosc.* **1984**, *16*, 193.
13. Facelli, J. C.; Grant, D. M.; Michl, J. *Acc. Chem. Res.* **1987**, *20*, 152, and references therein.
14. Facelli, J. C.; Grant, D. M. *Theor. Chim. Acta* **1987**, *71*, 277.
15. Wemmer, D. E.; Pines, A.; Whitehurst, D. D. *Philos. Trans. R. Soc. London, A* **1981**, *300*, 15.
16. Duncan, T. M. *J. Phys. Chem. Ref. Data* **1987**, *16*, 125.
17. Carter, C. M.; Alderman, D. W.; Facelli, J. C.; Grant, D. M. *J. Am. Chem. Soc.* **1987**, *109*, 2639.
18. Carter, C. M.; Facelli, J. C.; Alderman, D. W.; Grant, D. M.; Dalley, N. K.; Wilson, B. E. *J. Chem. Soc. Faraday Trans. 1* **1988**, *84*, 3673.
19. Facelli, J. C.; Grant, D. M. *Molecular Structure and Carbon-13 Chemical Shielding Tensors Obtained from Nuclear Magnetic Resonance: Topics in Stereochemistry;* John Wiley: New York, 1989; Vol. 19, p 1.
20. Morgan, K. R.; Newman, R. H. *J. Am. Chem. Soc.* **1990**, *112*, 4.
21. Pines, A.; Gibby, M. G.; Waugh, J. S. *J. Chem. Phys.* **1973**, *59*, 569.
22. Alderman, D. W.; Solum, M. S.; Grant, D. M. *J. Chem. Phys.* **1986**, *84*, 3717.
23. Prepr. Pap. Am. Chem. Soc. Div. Fuel Chem. **1987**, *32*, 155.
24. Sethi, N. K.; Pugmire, R. J.; Facelli, J. C.; Grant, D. M. *Anal. Chem.* **1988**, *60*, 1574.
25. Solum, M. S.; Pugmire, R. J.; Grant, D. M. *Energy Fuels* **1989**, *3*, 187.
26. Soderquist, A.; Burton, D. J.; Pugmire, R. J.; Beeler, A. J.; Grant, D. M.; Durand, B.; Huk, A. Y. *Energy Fuels* **1987**, *1*, 50.
27. Aue, W. P.; Ruben, D. J.; Griffin, R. G. *J. Magn. Reson.* **1981**, *43*, 472.
28. Burgar, M. I. *Fuel* **1984**, *63*, 1621.
29. Burgar, M. I.; Kalman, J. R.; Stephens, J. F. *Proc. 1985 Int. Conf. Coal Sci.* (Sydney, Australia) **1985**, 780.
30. Jeener, J.; Meier, B. H.; Bachmann, P.; Ernst, R. R. *J. Chem. Phys.* **1979**, *71*, 4546.
31. Carter, C. M.; Alderman, D. W.; Grant, D. M. *J. Magn. Reson.* **1985**, *65*, 183.
32. Henrichs, P. M. *Macromolecules* **1987**, *20*, 2099.
33. Botto, R. E.; Axelson, D. E. *Prepr. Pap. Am. Chem. Soc. Div. Fuel Chem.* **1988**, *33*, 50.
34. Bax, A.; Szeverenyi, N. M.; Maciel, G. E. *J. Magn. Reson.* **1983**, *51*, 400.
35. Yarim-Agaev, Y.; Tutunjian, P. N.; Waugh, J. S. *J. Magn. Reson.* **1982**, *47*, 51.
36. Tycko, R.; Dabbagh, G.; Mirau, P. A. *J. Magn. Reson.* **1989**, *85*, 265.

RECEIVED for review June 8, 1990. ACCEPTED revised manuscript June 13, 1991.

ELECTRON PARAMAGNETIC RESONANCE SPECTROSCOPY

23

2-mm Band and X-Band Electron Spin Resonance and Electron Spin-Echo Investigations of Some Carbonaceous Materials

Yuri D. Tsvetkov, Sergei A. Dzuba[1], and Victor I. Gulin

Institute of Chemical Kinetics and Combustion, Russian Academy of Science, Siberian Branch, Novosibirsk, 630090, Russia

Argonne Premium coal samples were studied by using 2-mm band and X-band continuous-wave electron spin resonance (CW ESR) and X-band electron spin-echo (ESE) spectroscopy. The line widths and g factors (Landè g factor, spectroscopic splitting factor) were determined. The correlation between $\Delta g = g_\parallel - g_\perp$ and the carbon content in coal samples was established. Paramagnetic centers in coals could be attributed to radicals with partial redistribution of spin density from polycyclic π-system to peroxide-type structures. The degree of this redistribution depends on the degree of carbonization. Phase relaxation times, T_2, for these coals were determined by using ESE spectroscopy.

THE ELECTRON SPIN RESONANCE (ESR) SPECTRA of natural coals consist of singlet lines with widths from 0.1 to 10 G and with a *g* factor (Landè *g* factor, spectroscopic splitting factor) close to that of a free electron. High-resolution ESR spectroscopic methods are very important in interpreting these spectra because the latter have no hyperfine structure or other characteristic properties.

[1] Corresponding author

Experimental Details

We studied the natural coals with a 2-mm band continuous-wave (CW) spectrometer (1) and an X-band Bruker ESP-300 spectrometer. Electron spin-echo experiments were made with the spectrometer described by Salikhov et al. (2).

The g factors were measured by using standard samples containing Mn^{2+} ions in MgO and the 1,1-diphenyl-2-picrylhydrazyl (DPPH) samples. The concentration of the paramagnetic particles was determined by using single crystals of $CuCl_2 \cdot 2H_2O$ as a reference. The coal sample measurements that were obtained with the X-band ESP-300 spectrometer were compared with standard samples by using a double resonator. For the 2-mm band measurements, the MgO powder was placed in the resonator together with a sample. ESR spectra were obtained at room temperature (X-band) and at 200 K (2-mm band). The internal diameter of the sample tube was 3 mm for the X-band measurements and 0.4 mm for the 2-mm band measurements.

The coals were provided by K. S. Vorres (Argonne National Laboratory) within the framework of the Argonne Premium Coal Sample Program. Eight samples were supplied by different mines, and these samples included lignite and bituminous coals with different degrees of volatility and with other well-known properties described by Vorres (3) (Table I). In Table I, the samples are listed in order of increasing carbon content (degree of carbonization). The order in which the samples were supplied by the Argonne Program (3) is also indicated.

The coal samples were finely dispersed powders held in an inert nitrogen atmosphere in sealed glass tubes. The coals were studied in open air both immediately after (type A samples) and more than an hour after (type B samples) opening the tube.

Results and Discussion

The ESR spectra of the A samples at both bands are represented by two overlapping lines, one narrow and one broad (Figure 1). The narrow line vanishes with time after exposure to air. This phenomenon is known as the *oxygen effect* (4). The loss of the narrow line is accompanied by a broadening of the wide line in the B samples. For example, the width of the broad line of sample 7, on transition from A to B type, changes from 3.5 to 5.8 G at the X-band and from 5.3 to 10.8 G at the 2-mm band. The line widths for the B samples are listed in Table II. The g factors of the broad line measured at the X-band at room temperature are shown in Table III for the B samples. These results indicate that the g factor increases with the decreasing degree of carbonization of the coal.

The shape of the broad lines at the X-band was nearly symmetric for all of the B samples (Figure 2a). Transition to the 2-mm band changes

Table I. Argonne Premium Coal Samples and Some Characteristics

Sample No.	Argonne No.	Seam	State	Rank	C	H	O	S	Ash
1	8	Beulah–Zap	ND	L	73	4.8	20	0.8	10
2	2	Wyodak–Anderson	WY	S	75	5.4	18	0.6	9
3	3	Illinois No. 6	IL	HVB	78	5.0	14	4.8	15
4	6	Blind Canyon	UT	HVB	81	5.8	12	0.6	5
5	4	Pittsburgh No. 8	PA	HVB	83	5.3	9	2.2	9
6	7	Lewiston–Stockton	WV	HVB	83	5.3	10	0.7	20
7	1	Upper Freeport	PA	MVB	86	4.7	8	2.3	13
8	5	Pocahontas No. 3	VA	LVB	91	4.4	2	0.7	5

ABBREVIATIONS: LVB, low-volatile bitumirous; MVB, medium-volatile bituminous; HVB, high-volatile bituminous; SB, subbituminous; and L, lignite.

SOURCE: Adapted from reference 3.

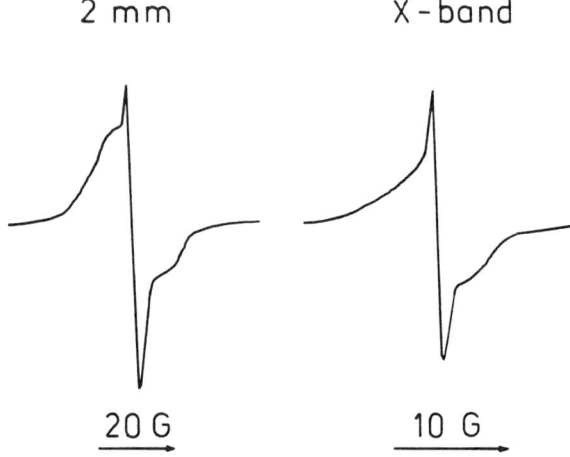

Figure 1. ESR spectra of sample 7 at 2-mm band and X-band obtained just after exposure to air.

Table II. Paramagnetic Center Concentrations, Line Widths, and Phase Relaxation Times for B Samples

Sample No.	$[R]$ $\times 10^{-18}$ (g^{-1})	X-band ΔH (G)	2-mm ΔH (G)	T_2 (μs)
1	3.4	6.4	—[a]	1.3
2	3.9	6.6	—	1.0
3	4.2	6.8	—	0.69
4	8.5	7.3	—	0.60
5	8.3	$3.5^b, 6.3$	5.3^b	0.66
6	8.2	6.1	14	0.73
7	8.4	$3.5^b, 5.8$	$5.3^b, 10.8$	0.69
8	11.0	5.2	14.4	0.64

[a] Dash indicates asymmetrical lines.
[b] Results obtained just after exposure to air (A samples).

the situation completely (Figures 2b, 2c, and 2d). B samples with a low degree of carbonization display a noticeable axial anisotropy of their g factors in the 2-mm band ESR spectra. The experimental values for $g_\parallel$ and $g_\perp$ are given in Table III.

The following correlation may be made from the data in Table III: the smaller the degree of carbonization, the greater the $g_\parallel$ value. The value of $g_\perp$ is also dependent on the degree of carbonization, but to a les-

Table III. g Factor Values for B Samples

Sample No.	X-band g	2-mm $g_\perp$	2-mm $g_\parallel$	$g_{iso} = 1/3(g_\parallel + 2g_\perp)$	$(g_\parallel - g_\perp) \times 10^4$
1	2.0035	2.00290	2.00503	2.00361	21.2
2	2.0035	2.00290	2.00490	2.00357	20
3	2.0030	2.00260	2.00400	2.00307	14
4	2.0031	2.00261	2.00371	2.00298	11
5	2.0028	2.00261	—[a]	—	<5
6	2.0027	2.00261	—	—	<5
7	2.00285	2.00259	—	—	<5
8	2.0028	2.00267	—	—	<5

[a] Dash indicates that measurements are impossible (*see* Figure 2d).

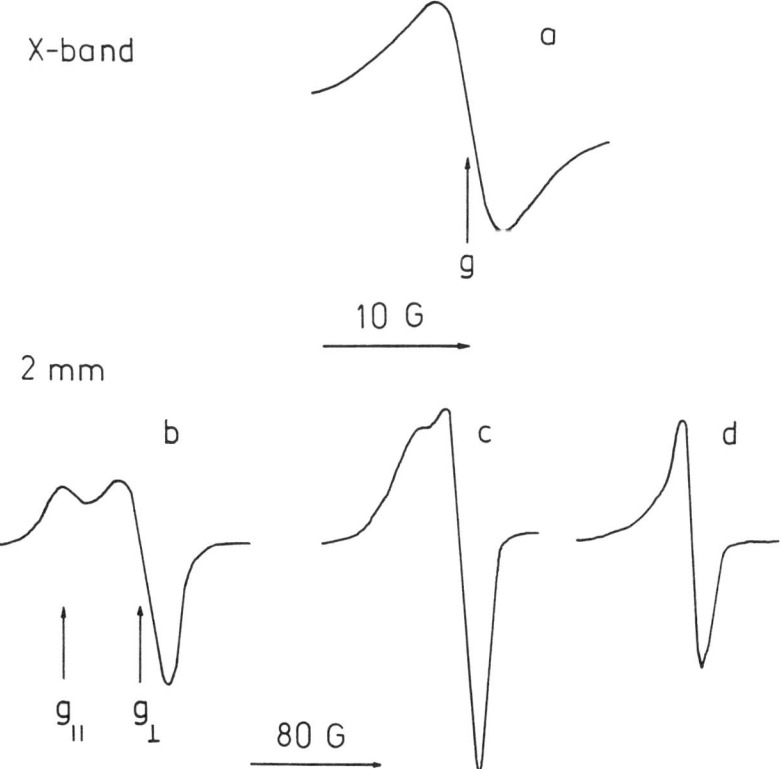

Figure 2. ESR spectra of sample 2 at X-band (a) and of samples 2 (b), 4 (c) and 6 (d) at 2-mm band. The spectra were obtained from samples that had been exposed to air for more than an hour.

ser extent. The 2-mm and X-band ESR data are in fair agreement if the g-factor isotropic values in both cases are compared:

$$g_{iso} = \frac{1}{3}(g_\parallel + 2g_\perp) \tag{1}$$

Thus, in the X-band ESR spectra the increase in the g factor of poorly carbonized coals is related to the increase of the $g_\parallel$ value.

These data are in good agreement with the hypothesis that the paramagnetic centers in highly carbonized coals are practically fully localized in a carbon-containing polycyclic π system. The g factors of such centers (denoted here by $\dot{\pi}$) are 2.0027 ± 0.0001 (5). This polycyclic π system occurs in the highly carbonized coals (samples 5–8). The increase in g-factor values for coals with a low degree of carbonization (samples 1–4) could result from the presence of oxygen (see Table I) in our system. Indeed, from Table III it follows that $g_\parallel > g_\perp$. This relation holds for radicals of a peroxide type $R\dot{O}_2$ (5), for which $g_\perp = 2.002$–2.010 and $g_\parallel = 2.03$–2.04. We think that the g factors determined in this work may be attributed to a partial redistribution of spin density from $\dot{\pi}$ to a peroxide structure $\pi\dot{O}_2$. Assuming x to be the spin density of $R\dot{O}_2$, and $1 - x$ that of $\dot{\pi}$ in the $\pi\dot{O}_2$ structure, the g-tensor in this redistribution changes in proportion to spin density fractions on the structures of both types:

$$\hat{g}_{\dot{\pi}O_2} = \hat{g}_{\dot{\pi}}(1 - x) + x\hat{g}_{R\dot{O}_2} \tag{2}$$

For sample 1, the x value will be approximately 0.1.

Table II also provides the values of the phase relaxation times, T_2, of the paramagnetic centers. The T_2 values were determined at room temperature by analyzing the decay of the two-pulse ESE signal. The kinetics of this decay can be described by an exponential dependence, $\exp(-2\tau/T_2)$, where τ is the time between two pulses. A lower degree of carbonization results in higher T_2 values. In this case, the effect may be assigned to the influence of the concentration of the paramagnetic centers on T_2 (Table II) via dipole–dipole interactions between the different spins (2). As shown in Table II, the concentration of paramagnetic centers increases with the degree of carbonization.

We have also tried to measure the spin–lattice relaxation times (T_1) in these samples by using the saturating pulse followed by observation of the ESE signal recovery (2). However, the recovery times appeared to be comparable with the time resolution of this technique (on the order of a microsecond). It seems that in this case T_1 and T_2 are almost the same.

References

1. Grinberg, O, Ya.; Dubinsky, A. A.; Lebedev, Ya. S. *Uspechi Khimii* (in Russian) **1983**, *52*, 1490.
2. Salikhov, K. M.; Semenov, A. G.; Tsvetkov, Yu. D. *Electron Spin Echo and Its Applications* (in Russian); Nauka: Novosibirsk, Russia, 1976.
3. Vorres, K. S. *Users Handbook for the Argonne Premium Coal Sample Program;* Argonne National Laboratory: Argonne, IL, 1989.
4. Ingram, D. *Biological and Biochemical Applications of Electron Spin Resonance;* Adam Hilger: London, 1969.
5. *Landolt-Bornstein Zahlen-Werte und Funktionen aus Naturwissenschaften und Technik;* Hellwege, K.-H., Ed.; Springer-Verlag: Berlin, Germany, 1977; Vol. 9, part b; 1979; Vol. 9, part c.

RECEIVED for review June 8, 1990. ACCEPTED revised manuscript December 26, 1990.

24

Electron Spin Resonance, Electron Nuclear Double Resonance, and Electron Spin-Echo Spectroscopic Studies of Argonne Premium Coals

Xinhua Chen, Hugh McManus, and Larry Kevan[1]

Department of Chemistry, University of Houston, Houston, TX 77204-5641

Six coals provided by the Argonne Premium Coal Sample Program were studied by using electron spin resonance (ESR) and electron spin-echo (ESE) spectroscopy, and five of these coals were studied by using electron nuclear double resonance (ENDOR) spectroscopy. Measurements were conducted at various temperatures. Sample pretreatment was done by exposing to air, evacuating at room temperature, or evacuating at 673 K. Spin concentrations, g values, and relaxation parameters were all affected by the pretreatments. The average radius of the unpaired-electron wave function was determined from proton-matrix ENDOR results. ESE spectra show that ^{13}C and ^1H nuclei have significant interactions with the unpaired electrons in some coals.

ELECTRON MAGNETIC RESONANCE SPECTROSCOPY has proven to be a powerful means of studying paramagnetic species in coal (1–10). Conventional continuous-wave (CW) electron spin resonance (ESR) can

---

[1] Corresponding author

be used to determine the geometrical and electronic structure and the concentration of the paramagnetic species in coal. An ESR spectrum of coal shows a convolution of a number of unresolved hyperfine structures. The insufficient resolving power of ESR spectroscopy does not permit a determination of the identity of the interacting nuclei or of the hyperfine coupling. Resolution enhancement can be achieved by electron nuclear double resonance (ENDOR) spectroscopy (11). Analysis of matrix ENDOR line widths in solids gives the distance between the paramagnetic species and the interacting nuclei. The pulsed-electron-magnetic-resonance method, electron spin-echo (ESE) spectroscopy, also provides high sensitivity for probing weak hyperfine interactions. Analysis of the echo decay provides a direct determination of electron-spin-relaxation parameters (12), and simulation of the echo-modulation pattern yields the identity and number of interacting nuclei as well as the interaction distances (13). In the present study, ESR and ESE spectroscopy were employed in investigating six selected coals from the Argonne Premium Coal Sample Program, and ENDOR spectroscopy was employed in studying five of these coals.

Experimental Details

Coal samples were provided by the Argonne Premium Coal Sample Program. These coals were collected from freshly exposed seams and were stored under nitrogen under carefully controlled conditions with minimum exposure to air. The coals as received had been sieved to 100 mesh and sealed under nitrogen. The coal names and the abbreviation codes are listed in the following table in order of decreasing coal rank.

Name	Rank	Code
Pocahontas No. 3 seam from Virginia	LVB	POC
Blind Canyon seam from Utah	HVB	UT
Pittsburgh No. 8 seam from Pennsylvania	HVB	PITT
Illinois No. 6 seam from Illinois	HVB	IL
Wyodak–Anderson seam from Wyoming	SB	WY
Beulah–Zap seam from North Dakota	L	ND

NOTE: The order of decreasing carbon content is POC > UT ~ PITT > IL ~ ND > WY > ND.

ABBREVIATIONS: LVB, low-volatile bituminous; HVB, high-volatile bituminous; SB, subbituminous; and L, lignite.

Approximately 5 mg of coal was transferred to 2-mm i.d. × 3-mm o.d. Suprasil quartz tubes in a glove box under a nitrogen atmosphere. The three pretreatments used to prepare individual samples from each coal were (1) exposing the coal to air overnight and sealing in air; (2) evacuating the coal at room temperature (RT) overnight with a final residual pressure of approximately 1×10^{-5} torr (1.33 mPa) and sealing under vacuum; and (3) evacuating at RT, heating under vacuum to 673 K for ~2 h, and sealing under vacuum.

ESR spectra were recorded at RT, 77 K, and 4 K with a Bruker ESP 300 ESR spectrometer operating at X-band (9 GHz) with 100-kHz magnetic-field modulation. The microwave frequency and the magnetic field were monitored with a Hewlett-Packard 5350B frequency counter and a Bruker ER035 gaussmeter, respectively. The g values of the samples were measured by using 1,1-diphenyl-2-picrylhydrazyl (DPPH) as a standard with $g = 2.0036$, and the spin concentrations were determined by comparing the sample results with a Bruker strong-pitch reference containing 3×10^{15} spins/cm.

ENDOR spectra were recorded at RT and at 150 K with the ESP 300 ESR spectrometer interfaced with a Bruker ENDOR unit. Frequency modulation was performed without magnetic-field modulation, and the result was a first-derivative presentation of the ENDOR spectra. Typical experimental conditions were as follows: microwave power, 2 mW; microwave frequency, 9.46 GHz; radio-frequency power, 100 W; frequency modulation, 12.5 kHz; and modulation depth, 158 kHz. For low-temperature ENDOR experiments, a Bruker ER411 variable-temperature unit was used.

ESE spectra were recorded at RT, 77 K, and 4 K with a home-built X-band ESE spectrometer (14). Both two- and three-pulse experiments were performed. In the two-pulse experiments, pulse widths of 26 and 52 ns were used to generate $90°-\tau-180°$ pulse sequences where τ is the interpulse time. In the three-pulse experiments, three 26-ns pulses were used to generate $90°-\tau-90°-T-90°$ pulse sequences, and the phase sequence $\{ (0\ 0\ 0) + (\pi\ \pi\ 0) \} - \{ (\pi\ \pi\ \pi) + (0\ 0\ \pi) \}$ was employed for successive pulse sequences to eliminate two-pulse interferences (15).

Results and Discussion

ESR Spectroscopy. The ESR spectra of the studied coals can be categorized into two groups: type-1 spectra, in which both narrow and broad components can be seen; and type-2 spectra, in which only a broad signal can be seen. The narrow component and the broad signal have peak-to-peak line widths of ~1 G and ~3–11 G, respectively.

The ESR results of all coal samples are sumarized in Table I, and an example of the ESR spectra from PITT coal is shown in Figure 1. The

Table I. Summary of ESR Parameters of Argonne Premium Coals

Sample Code	Measurement Temperature	Pretreatment	g Value	Line Width (G)	Line-Shape Parameter	Spin Concentration $\times 10^{19}$ spins/g
POC	RT	In air	2.0039	3.1	13	3.9
		RT evac	2.0030	5.3	5	3.2
		673-K evac	2.0029	5.7	6	5.5
	77 K	In air	2.0028	1.3	98	
		RT evac	2.0029	5.5	5	
		673-K evac	2.0030	5.9	5	
	4 K	In air	2.0028	1.3	11	
		RT evac	2.0029	5.9	5	
		673-K evac	2.0035	6.0	5	
UT	RT	In air	2.0028	5.5	22	1.0
		RT evac	2.0029	4.9	20	0.8
		673-K evac	2.0029	5.0	7	6.4
	77 K	In air	2.0028	5.6	17	
		RT evac	2.0023	5.6	14	
		673-K evac	2.0028	5.4	6	
	4 K	In air	2.0027	6.9	13	
		RT evac	2.0028	6.9	8	
		673-K evac	2.0027	5.3	7	

PITT	RT	In air	2.0029	0.9	8	1.1
		RT evac	2.0029	1.0	9	0.9
		673-K evac	2.0032	6.3	4	5.6
	77 K	In air	2.0029	0.9	5	
		RT evac	2.0029	0.8	5	
		673-K evac	2.0030	6.7	5	
	4 K	In air	2.0040	0.8	8	
		RT evac	2.0028	0.7	8	
		673-K evac	2.0031	6.5	4	
IL	RT	In air	2.0030	1.3	10	1.4
		RT evac	2.0029	0.9	9	0.7
		673-K evac	2.0027	1.1	17	2.5
	77 K	In air	2.0031	1.0	8	
		RT evac	2.0029	0.8	8	
		673-K evac	2.0029	1.0	10	
	4 K	In air	2.0030	1.2	5	
		RT evac	2.0029	0.6	13	
		673-K evac	2.0029	1.0	10	
WY	RT	RT evac	2.0034	7.5	3	0.2
		673-K evac	2.0027	7.6	5	0.6
	77 K	RT evac	2.0040	8.6	3	
		673-K evac	2.0034	8.0	4	
	4 K	RT evac	2.0033	8.6	3	
		673-K evac	2.0024	11.1	5	

Continued on next page

Table I. Summary of ESR Parameters of Argonne Premium Coals—Continued

Sample Code	Measurement Temperature	Pretreatment	g Value	Line Width (G)	Line-Shape Parameter	Spin Concentration $\times 10^{19}$ spins/g
ND	RT	In air	2.0037	6.1	3	0.9
		RT evac	2.0040	7.6	3	1.4
		673-K evac	2.0030	5.4	7	14.2
	77 K	In air	2.0035	7.7	4	
		RT evac	2.0036	8.6	3	
		673-K evac	2.0031	5.2	6	
	4 K	In air	2.0046	7.5	3	
		RT evac	2.0030	11	5	
		673-K evac	2.0035	8.3	6	

NOTE: The estimated errors are ± 0.0001 in g, ± 0.1 G in the line width, and $\pm 0.1 \times 10^{19}$ spins/g in the spin concentration.

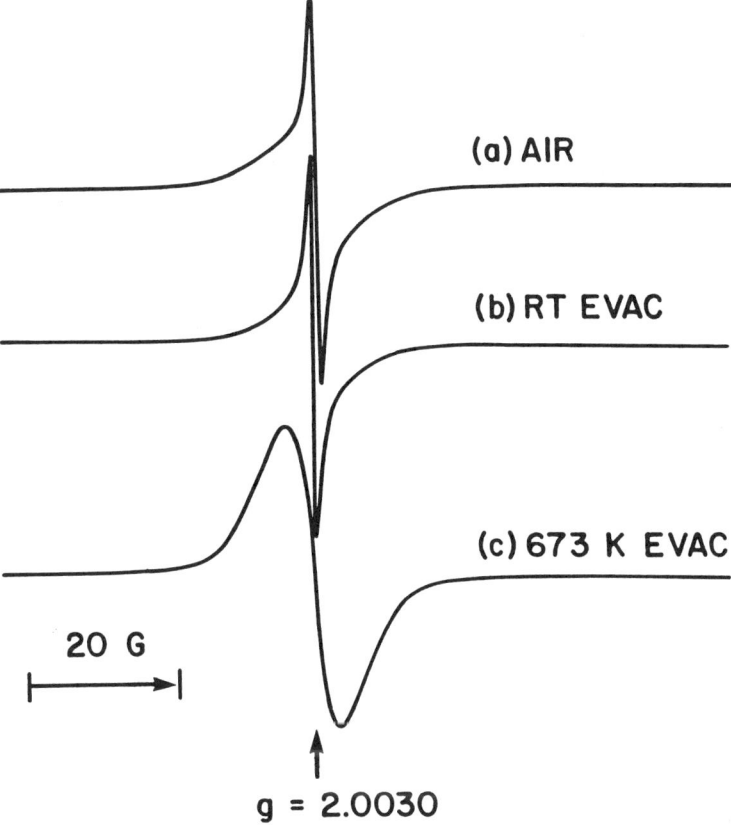

Figure 1. ESR spectra recorded at 77 K for PITT coal after different sample treatments. The sharp component disappears in c.

g values of all samples range from 2.0023 to 2.0046. ND coal samples have higher g values. Generally, the g value increases with decreasing rank and carbon content and increasing heteroatom (mainly oxygen) content, a result that was reported previously (3). PITT (in air and RT evacuation), POC (in air, measured at 77 and 4 K), and IL coals have type-1 ESR spectra as described previously, and the other coals have type-2 ESR spectra. This observation is confirmed by the line widths listed in Table I. Generally, a specific coal gave only one type of spectrum, independent of sample preparation and measuring temperature, with the following two exceptions: (1) Evacuation at 673 K made the IL coal spectrum change from type-1 to type-2; and (2) POC coal (in air) gives a type-2 spectrum when measured at RT, but it gives type-1 spectra when measured at 4 and 77 K.

We characterized the line shapes by using the ratio of negative to positive slopes of the derivative line shapes with positive phase at low field as a parameter; see Table I. The slope ratio is 2.2 for Gaussian lines and 4.0 for Lorentzian lines (*16*). However, for most samples the determined ratios were even greater than 4.0. This situation resulted from the overlap of two or more components in the spectra. Even when only one broad signal occurred in type-2 spectra, the slope ratios for most spectra were still much greater than 4.0, a result suggesting that the type-2 spectra also consisted of two or more components.

Spin concentrations increased by 2 to 10 times after the samples were evacuated at 673 K. The mechanism for this increase has not been determined.

ENDOR Spectroscopy. The ENDOR response was recorded for five coals (excluding WY) with the three different pretreatments and with the magnetic field fixed at the center of the ESR line. In prior work, the strongest ENDOR signal from IL coal occurred at 150 K (*7, 8*). In all cases, the line shape was predominantly Gaussian. The full width at half height (FWHH) of the absorption line was determined from the peak-to-peak width of the first-derivative line assuming a Gaussian line shape. The signal intensity was 20–30% stronger at 150 K than at RT. The ND sample gave no ENDOR response. For the IL, PITT, and UT samples, the ENDOR signal observed after pretreatment evacuation at RT was about an order of magnitude stronger than the ENDOR signals of samples prepared in air. Of the samples evacuated at 673 K, only the IL and POC coals gave ENDOR signals, and these signals were observed only at 150 K. Thus the ENDOR intensity is very sensitive to the pretreatment conditions. Table II gives a summary of the ENDOR results for all the coal samples.

The matrix ENDOR line shapes were analyzed by using a model developed by Hyde et al. (*17*) and the procedures of Helbert et al. (*18*). The expression for the line shape can be written as:

$$f(\nu) = N \int_0^\pi \int_\delta^\infty \frac{\cos^2\theta \sin^3\theta}{r^4} \left(\frac{1}{\alpha^2 + \left[\nu - (q/r^3)\right]^2} + \frac{1}{\alpha^2 + \left[\nu + (q/r^3)\right]^2} \right) d\theta\, dr \quad (1)$$

where α is the half width at half height of the nuclear-spin-packet line shape. The lower limit for the integration over r, δ, represents the dis-

Table II. Summary of ENDOR Parameters of Argonne Premium Coals

Sample Code	Measurement Temperature	Pretreatment	ΔH_{pp} (kHz)	$FWHH^a$ (kHz)	S/N^b	δ (Å)
POC	RT	In air	445 ± 40	535 ± 50	8:1	3.8 ± 0.4
	150 K	In air	450 ± 35	540 ± 35	13:1	3.7 ± 0.6
		RT evac	500 ± 50	650 ± 175	<2:1	4.2 ± 0.2
		673-K evac	460 ± 100	460 ± 100	<2:1	3.7 ± 1.0
UT	RT	In air	405 ± 40	475 ± 50	2:1	4.2 ± 0.4
		RT evac	405 ± 20	475 ± 25	53:1	4.1 ± 0.2
	150 K	In air	385 ± 50	450 ± 60	15:1	4.2 ± 0.4
		RT evac	385 ± 50	450 ± 60	47:1	4.2 ± 0.2
PITT	RT	In air	385 ± 40	455 ± 50	4:1	4.2 ± 0.5
		RT evac	365 ± 10	430 ± 18	41:1	4.1 ± 0.1
	150 K	In air	385 ± 40	455 ± 35	12:1	4.2 ± 0.3
		RT evac	385 ± 10	455 ± 10	97:1	4.2 ± 0.1
IL	RT	In air	480 ± 50	565 ± 60	5:1	3.6 ± 0.4
		RT evac	405 ± 10	475 ± 12	59:1	4.1 ± 0.1
	150 K	In air	445 ± 40	525 ± 35	21:1	3.8 ± 0.2
		RT evac	405 ± 10	475 ± 10	114:1	4.1 ± 0.1
		673-K evac	430 ± 100	505 ± 120	<2:1	3.0 ± 0.9

NOTE: ENDOR signals were not observed for some of the samples that were pretreated by evacuation at room temperature or 673 K. In these cases, no listing is included in the table. No ENDOR signals were observed from the ND coal sample. The WY sample was not investigated by ENDOR.

[a]The Gaussian absorption line width as the full width at half height (FWHH) is determined from the peak-to-peak derivative line width (ΔH_{pp}) of the ENDOR signal by using the formula: FWHH = $(2 \ln 2)^{1/2} \Delta H_{pp}$.
[b]Signal-to-noise ratio.

Table III. Summary of ESE Parameters of Argonne Premium Coals

Sample Code	Measurement Temperature	Pretreatment	Two-Pulse		Three-Pulse	
			$T_m(\mu s)$[a]	Nuclear Mod.	$T_1(\mu s)$[a]	Nuclear Mod.
POC	RT	In air	0.17	^{1}H	1.6	^{1}H
		RT evac	0.10	—[b]	1.3	^{1}H
		673-K evac	0.08	—	0.9	—
	77 K	In air	0.22	^{1}H	3.6	—
		RT evac	0.15	^{1}H	>5	—
		673-K evac	0.15	^{1}H	>5	^{1}H
	4 K	In air	0.19	^{1}H	4.7	^{1}H
		RT evac	0.15	^{1}H	2.1	^{1}H, ^{13}C
		673-K evac	0.22	—	0.8	^{1}H, ^{13}C
UT	RT	In air	0.21	^{1}H	>5	^{1}H, ^{13}C
		RT evac	0.26	^{1}H	>5	^{1}H
	77 K	In air	0.19	^{1}H	>5	^{1}H, ^{13}C
		RT evac	0.24	—	>5	^{1}H, ^{13}C
	4 K	In air	0.19	^{1}H	4.3	^{1}H, ^{13}C
		RT evac	0.17	^{1}H	>5	^{1}H, ^{13}C
		673-K evac	0.08	^{1}H	>5	^{1}H, ^{13}C
PITT	RT	In air	0.20	—	>5	^{1}H
		RT evac	0.33	^{1}H	>5	^{1}H
	77 K	In air	0.19	^{1}H	2.9	^{1}H
		RT evac	0.32	^{1}H	>5	^{1}H, ^{13}C
		673-K evac	0.08	—		

Sample	Temp	Condition	T_m (μs)[a]	Nucleus	T_1 (μs)[a]	Nucleus
IL	4 K	In air	0.22	^{1}H	>5	^{1}H
		RT evac	0.23	^{1}H	>5	^{1}H, ^{13}C
		673-K evac	0.14	^{1}H	1.2	^{1}H, ^{13}C
	RT	In air	0.15	^{1}H	>5	^{1}H, ^{13}C
		RT evac	0.28	^{1}H	>5	^{1}H, ^{13}C
		673-K evac	0.10	^{1}H	—	—
	77 K	In air	0.20	^{1}H	4.7	^{1}H, ^{13}C
		RT evac	0.33	^{1}H	>5	^{1}H
		673-K evac	0.09	^{1}H	>5	—
	4 K	In air	0.16	^{1}H	3.7	^{1}H
		RT evac	0.32	^{1}H	—	—
		673-K evac	0.20	^{1}H	1.2	^{1}H
WY	4 K	RT evac	0.08	^{1}H	—	—
ND	RT	In air	0.10	—	>5	—
		RT evac	0.18	—	3.4	—
	4 K	In air	0.17	^{1}H	0.5	^{1}H
		RT evac	0.47	—	>5	—

NOTE: In some cases, no ESE signals were observed. In these cases, either the appropriate column is blank or the entire category was omitted from the table.

[a] Estimated errors are ±0.04 μs in T_m and ±0.1 μs in T_1.

[b] Dashes indicate that there was no nuclear modulation or that the spectrum was too noisy to see nuclear modulations.

tance from the unpaired electron that the "matrix" nuclei begin. The angle θ is between the applied magnetic field and the radial vector, r, between the electron and the nucleus; ν is the frequency offset from the center of the ENDOR line; N is a constant that affects the magnitude, but not the shape, of the ENDOR line; and $q = (8\pi^2)^{-1} \gamma_e \gamma_N h (3 \cos \theta - 1)$. The magnetogyric ratios of the electron and nucleus are γ_e and γ_N, respectively, and h is Planck's constant. Equation 1 was numerically integrated on a DEC VAX computer. The equation contains two parameters: δ and α. The parameter α was chosen to be 100 kHz (17, 18), and δ varied from 3 to 10 Å. For each value of δ, a simulated ENDOR line, $f(\nu)$, was obtained. From this line, a FWHH was determined, and the FWHH was then plotted as a function of δ. This plot was used to determine a value of δ for each of the experimental ENDOR lines obtained from the coal samples. The results of this analysis are given in Table II. The unpaired-electron radius is approximately 4 Å. As the carbon content decreases, this radius may decrease slightly. The results for the POC coal indicate a radius smaller than 4 Å.

ESE Spectroscopy. Two- and three-pulse ESE experiments were carried out at RT, 77 K, and 4 K on the six coal samples. ESE spectra were obtained from most samples. The ESE parameters are summarized in Table III. Some examples of ESE spectra are shown in Figures 2–4. The measured phase-memory times, T_m, from two-pulse ESE experiments range from 0.08 to 0.47 μs. The T_m values reported previously (6) were 0.23 μs for natural PITT coal and 0.25 μs for heat-treated PITT coal; those results compare with 0.22 and 0.14 μs, respectively, in this study. The small discrepancy may be attributed to differences in the original sample preparation, because the coals used in this study were collected under controlled conditions that were established for the Argonne Premium Coal Program. From the three-pulse ESE spectra, T_1, the time required for the echo to decay to e^{-1} of its original intensity, was measured. Values of T_1 range from 0.5 μs to greater than 5 μs, the upper limit of the time sweep used in the ESE experiments. The echo-decay time, T_1, can be identified approximately as the spin–lattice relaxation time. The phase-memory time, T_m, is related to the spin–spin relaxation time T_2, but in solids T_m is longer than the true T_2. In general, both T_m and T_1 decrease after 673-K evacuation. Apparently, T_m is longer for lower rank coals. This result may reflect an instantaneous-diffusion contribution to T_m in higher rank coals.

Nuclear-modulation patterns in the ESE signals provide information on the interacting nuclei. Modulations from ^1H were observed in most two-pulse ESE signals, and both ^1H and ^{13}C modulations were observed in some three-pulse ESE signals (Figure 4). Modulations from ^{13}C in natural abundance were observed in IL coal in this study, as previously re-

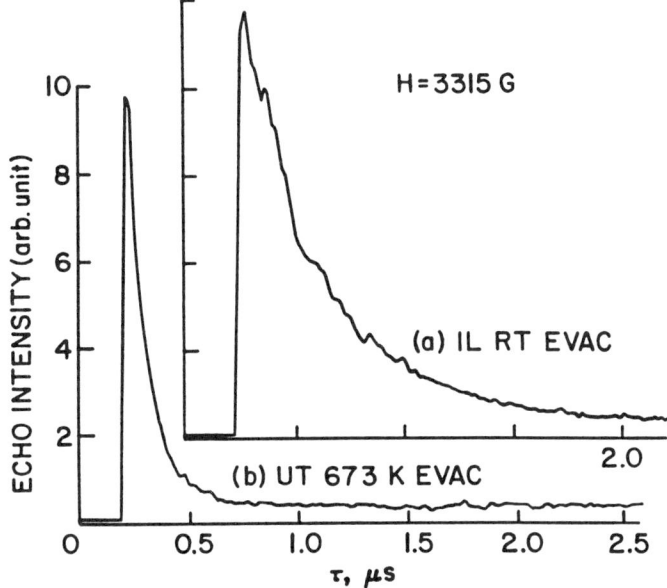

Figure 2. Two-pulse ESE spectra recorded at 4 K showing slow decay ($T_m = 0.32$ μs) for IL coal (a) and fast decay ($T_m = 0.08$ μs) for UT coal (b).

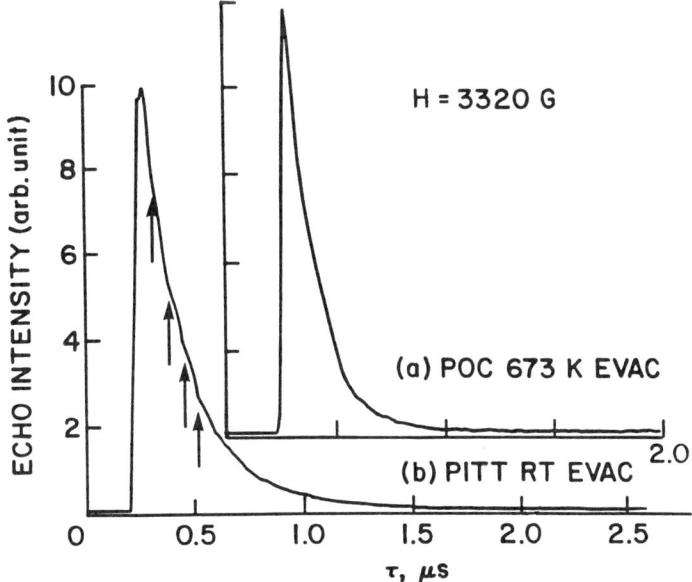

Figure 3. Two-pulse ESE spectra at 4 K showing differences in modulation depth. No modulation is visible in a, and ^1H modulation is indicated by arrows in b.

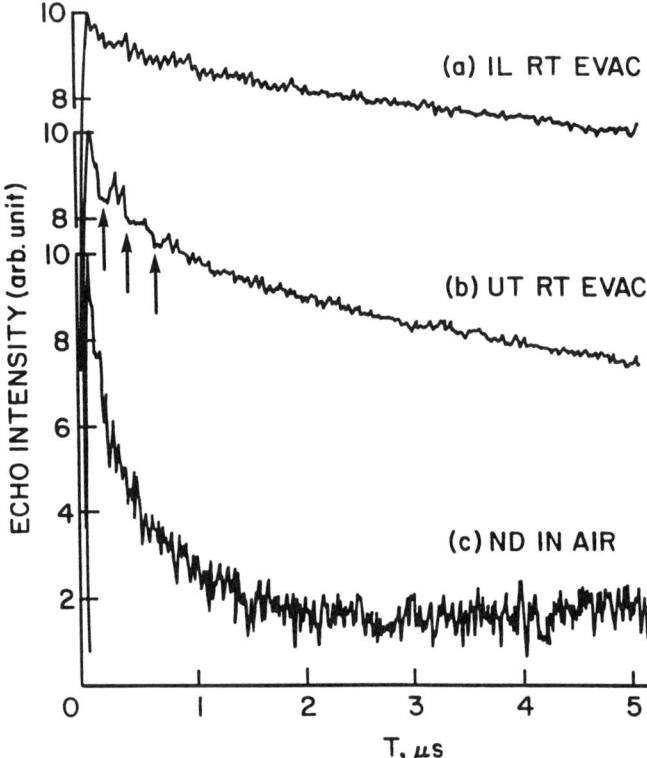

Figure 4. Three-pulse ESE spectra. (a) IL coal evacuated at RT and measured at 77 K. The peaks show ^1H modulation. (b) UT coal evacuated at RT and measured at 4 K. The peaks show ^1H modulation, and the arrows indicate superimposed ^{13}C modulation. (c) ND coal exposed to air and measured at 4 K shows rapid decay ($T_1 = 0.17$ μs).

ported (19). However, the ^{13}C modulation depth was greater in UT coal than in IL coal. The ^{13}C modulation must also be present in the two-pulse ESE signals, but apparently the modulation is not clearly observable with the available signal-to-noise ratio. The low signal-to-noise ratio also precluded meaningful Fourier transformation and computer simulation. Although the number of nuclei interacting with the paramagnetic centers and the interaction distance have not been quantified into a possible molecular environment for the paramagnetic centers, the observed modulation patterns do indicate the presence of ^1H and ^{13}C nuclei within 5–6 Å of the paramagnetic centers in these coals.

Summary

A major goal of this study was to provide comparative electron-magnetic-resonance data for a variety of magnetic parameters of the well-characterized coals in the Argonne Premium Coal Program. These parameters are reported with reference to well-defined measurement conditions and a minimum of experimental data post-treatment. These are the most comprehensive ENDOR and ESE data on a variety of coals to date. The summary of magnetic-resonance parameters should be useful for evaluating the effects of the controlled-collection and -storage conditions of coals under the Argonne Premium Coal Program. These data can serve as benchmarks for other laboratories. To try to delineate any unique molecular structure and molecular environment for the paramagnetic centers in such coals would be premature at this stage.

Acknowledgment

This research was partially supported by the Division of Chemical Sciences, Office of Basic Energy Sciences, Office of Energy Research, U.S. Department of Energy and the National Science Foundation.

References

1. Uebersfeld, J.; Etinenne, A.; Combrisson, J. *Nature (London)* **1954**, *174*, 614.
2. Ingram, D. J. E.; Tapley, J. G.; Jackson, R.; Bond, R. L.; Murnahgan, A. R. *Nature (London)* **1954**, *174*, 797.
3. Retcofsky, H. L.; Stark, J. M.; Friedel, R. A. *Anal. Chem.* **1968**, *40*, 1699.
4. Retcofsky, H. L. In *Coal Science;* Gorbaty, M. L.; Larsen, J. L.; Wendler, I., Eds.; Academic: New York, 1982; Vol. 1.
5. Schlick, S.; Narayana, M.; Kevan, L. *Fuel* **1983**, *62*, 1251.
6. Schlick, S.; Narayana, M.; Kevan, L. *Fuel* **1986**, *64*, 873.
7. Schlick, S.; Narayana, P. A.; Kevan, L. *J. Am. Chem. Soc.* **1978**, *100*, 3322.
8. Schlick, S.; Kevan, L. In *Magnetic Resonance. Introduction, Advanced Topics and Applications to Fossil Energy;* Petrakis, L.; Fraissard, J. R., Eds.; Reidel: Dordrecht, Netherlands, 1984; p 655.
9. Wiser, W. H. In *Magnetic Resonance. Introduction, Advanced Topics and Applications to Fossil Energy;* Petrakis, L.; Fraissard, J. R., Eds.; Reidel: Dordrecht, Netherlands, 1984; p 325.
10. Schlick, S.; Kevan, L. In *Spectroscopic Analysis of Coal Liquids;* Kershaw, J. R., Ed.; Elsevier: New York, 1989; Chapter 9.
11. Kevan, L.; Kispert, L. D. *Electron Spin Double Resonance Spectroscopy;* Wiley-Interscience: New York, 1976.

12. Mims, W. B. In *Electron Paramagnetic Resonance;* Geschwind, S., Ed.; Plenum: New York, 1972; Chapter 4.
13. Kevan, L. In *Time Domain Electron Spin Resonance;* Kevan, L.; Schwartz, R. N., Eds.; Wiley-Interscience: New York, 1979; Chapter 8.
14. Ichikawa, T.; Kevan, L.; Narayana, P. A. *J. Phys. Chem.* **1979,** *83,* 3378.
15. Fauth, J.-M.; Schweiger, A.; Braunschweiler, L.; Forrer, J.; Ernst, R. R. *J. Magn. Reson.* **1986,** *66,* 74.
16. Alger, R. S. *Electron Paramagnetic Resonance: Techniques and Applications;* Interscience: New York, 1968; p 42.
17. Hyde, J. S.; Rist, G. H.; Eriksson, L. E. G. *J. Phys. Chem.* **1968,** *72,* 4269.
18. Helbert, J.; Kevan, L.; Bales, B. L. *J. Chem. Phys.* **1972,** *57,* 723.
19. Snetsinger, P. A.; Cornelius, J. B.; Clarkson, R. B.; Bowman, M. K.; Belford, R. L. *J. Phys. Chem.* **1988,** *92,* 3696.

RECEIVED for review June 8, 1990. ACCEPTED revised manuscript December 18, 1990.

25

Electron Paramagnetic Resonance Spin-Probe Studies of Porosity in Solvent-Swelled Coal

Ross Spears[1], Janina Goslar[2], and Lowell D. Kispert[3]

Chemistry Department, University of Alabama, Tuscaloosa, AL 35487–0336

> *An electron paramagnetic resonance (EPR) spin-probe method was used to examine the changes in the size and number distribution of the accessible regions in five Argonne Premium coal samples as a function of rank upon swelling with the solvents toluene, nitrobenzene, and pyridine. As the basicity of the solvent increased (from that of toluene to that of pyridine), the number and length of the cylindrical pores increased with decreasing coal rank. The number of cylindrical pores also increased with oxygen content (with decreasing rank), a result suggesting a destruction of the hydrogen-bond network upon swelling with pyridine.*

THE PORE STRUCTURE AND SWELLING PROPERTIES OF COALS have been studied in order to gain a better understanding of coal structure. Studies of surface area, pore-size distribution, pore volume, and even pore shape have been performed in order to understand the surface

---

[1]Current address: U.S. Bureau of Mines, P.O. Box L, Tuscaloosa, AL 35486–9777
[2]Current address: Institute of Molecular Physics, Polish Academy of Sciences, 60–179 Poznan, Poland
[3]Corresponding author

properties of coals (*1*). Additionally, studies of coal–solvent interactions and the resulting changes in the pore structure of coals have been performed (*2–6*). As a complement to solvent-swelling studies, experiments have been carried out to determine how guest molecules diffuse into coal as it swells (*7, 8*). These studies not only lead to a better understanding of coal structure, but they are also important for understanding coal liquefaction and how liquefaction catalysts gain access to coal.

Coal consists of cross-linked macromolecular subunits. Cross-links consist of aliphatic bridges as well as hydrogen bonds. Cross-link density appears to increase with coal rank. Although not completely understood, the solvent swelling of coal apparently occurs via two mechanisms. Nonpolar solvents such as toluene appear to swell coal by solvating molecules trapped within the pore structure. However, nonpolar solvents are not capable of disrupting the cross-links in coal and consequently cannot gain access to molecules trapped deep inside the coal particle. Thus, nonpolar solvents are not good swelling solvents for coal.

Hydrogen bonds are very important in cross-linking coal macromolecules (*7*). Polar, hydrogen-bonding solvents such as pyridine appear to replace the coal–coal hydrogen bonds with coal–solvent hydrogen bonds. This replacement decreases the cross-link density of the coal, a phenomenon that allows the coal to swell (*8*). Furthermore, the ability of a solvent to swell coal is related to its pK_b (*7*).

Hydrogen bonding in coal is most likely due to oxygen functionalities within the coal (*9*). Because oxygen content decreases with rank and cross-link density increases with rank, the cross-links in higher ranked coals must be primarily aliphatic bridges. Nonpolar swelling solvents have very limited access to the interior of the coal because of the high cross-link density, and hydrogen-bonding solvents have no means of reducing the cross-link density. Consequently, solvents are less able to swell high-rank coals.

Until the early part of the last decade, electron paramagnetic resonance (EPR) spectroscopy was used primarily to study the concentration of free radicals in coal and to determine *g* values. However, in 1981 Silbernagel et al. (*10*) showed the possibility of diffusing nitroxide spin probes into the coal structure and observing the probes by EPR spectroscopy. 4-Hydroxy-2,2,6,6-tetramethylpiperidinoxyl (TEMPOL, structure 1) was placed in a hexane solution and diffused into either Wyodak or Illinois No. 6 coal. The broadening and reduction of the nitroxide spin-probe EPR signal was associated with the diffusion of the TEMPOL molecules into the coal matrix. The results of this experiment suggested that surface absorption and diffusion of spin probes into coal was possible.

The aforementioned method was expanded in our laboratory (*11*), and EPR studies of doped Alabama (*12*), Pennsylvania State University

OH
|
(structure)
|
N
|
O•

1

Coal Research Section (PSOC) (*4*), and Argonne (*2, 3*) Premium coals were performed. From these EPR studies, the relative accessibility of different-size spin probes to the pore structure of coals in the presence of a swelling solvent can be estimated. Nitroxide spin probes with different sizes and shapes were diffused into various coal samples that were swelled with toluene (an intermediate swelling solvent), benzene, and pyridine. Untrapped probe molecules were removed with a nonswelling wash solvent such as cyclohexane. Relative pore shapes and size distributions were then determined from the EPR spin probes trapped within the coal matrix.

More recently, the results of an EPR spin-probe study (*2*) showed that the relative number distribution of acidic functionalities measured by the spin-probe method is linearly related to the ratio of phenolic to alkyl −OH groups determined by diffuse reflectance infrared (DRIFT) measurements (*13*). Thus the relative number distribution of acidic functionalities can be measured by the spin-probe method. In addition, the predicted increase in elongated voids in Pittsburgh No. 8 (Argonne Premium coal sample [APCS] No. 4) upon swelling with pyridine was confirmed (*2*). This confirmation showed that pore-size distribution can be estimated by using the spin-probe method. An electron nuclear double resonance (ENDOR) study (*14*) of five nitroxide spin-probe-doped Argonne Premium coal samples was used to identify matrix protons in the coal, the protons of the nitroxide spin probe, and the interaction of the surrounding matrix-coal protons with the nitroxide spin probes.

In this study, the variation in the size and number of cylindrical pores was measured as a function of both rank and the degree of swelling for selected Argonne Premium coal samples in which the percent carbon (dry, mineral-matter-free [dmmf]) values vary from 74.05% (Beulah–Zap) to 91.81% (Pocahontas No. 3). Toluene, nitrobenzene, and pyridine were used as swelling solvents, and cyclohexane was used as the wash solvent. Our results extend earlier work (*2*) and were confirmed by independent small-angle neutron scattering (SANS) (*5, 6*) and DRIFT studies (*13*).

Experimental Details

Five Argonne Premium coal samples (100 mesh) were used:

APCS Number	Name	Rank
3	Illinois No. 6	HVB
4	Pittsburgh No. 8	MVB
5	Pocahontas No. 3	LVB
6	Utah Blind Canyon	HVB
8	Beulah–Zap	L

ABBREVIATIONS: LVB, low-volatile bituminous; MVB, mid-volatile bituminous; HVB, high-volatile bituminous; and L, lignite.

The coal samples were doped with nitroxide spin probes by using a procedure that was found to be optimal and that was reported previously (2). A 30-mg sample of coal under an argon atmosphere was mixed with 2 mL of a 10^{-3} M toluene, nitrobenzene, or pyridine spin-probe solution. The mixture was heated in an oil bath at 50–60 °C and was stirred for 18 h. The mixture was then filtered, and the solid residue was dried for 2 h at room temperature and 10^{-1} torr (13.3 Pa). The dried coal was washed with 3 mL of dry cyclohexane for 3 min to remove any spin probe that was attached to the exterior of the coal or that was in pores or accessible regions too large to trap the probe. The mixture was filtered, washed with 5 mL of cyclohexane, and vacuum dried at room temperature for 0.5 h. After drying, 10 mg of the sample was placed in an EPR tube and evacuated for 0.5 h at 5×10^{-3} torr (0.67 Pa) and sealed. During the entire procedure, the doped coal was exposed to air for only a few minutes.

A swelling temperature of 50–60 °C was optimal, because in that temperature range the swelling probes did not decay and a swelling equilibrium was reached after 12 h (2). The spin concentration of the undoped coal was not studied as a function of solvent. The samples were stored under liquid nitrogen until the EPR spectra were recorded.

To determine the nitroxide-radical spin concentration for a particular sample, the area under only the low-field peak (z component of the nitrogen hyperfine coupling) of the spin-probe-doped-coal EPR spectrum was measured. The nitroxide-radical concentration could not be determined by integrating the entire EPR spectrum of the doped-coal sample because the radical concentration of the undoped coal is typically 10–100 times greater than the nitroxide radical concentration. The typical error in measuring radical concentration by EPR spectroscopy is 10–20% under the most ideal conditions (the system here is far from ideal), and the error in determining the radical concentration in the undoped coal sample would exceed (up to a factor of 10) the spin-probe concentration. Furthermore, determining a base line for the low-field peak is difficult, so typical EPR integration programs fail. This difficulty was solved by using the Kurta tablet digitizer interfaced to an IBM PC AT microcomputer and Sigma Scan software (Jandel Scientific Corp.) that was designed for determining irregular areas.

The total area of the spin-probe EPR spectrum was compared to the area of the EPR spectrum of the Cr(III) intensity standard (SRM-2601) from the National Bureau of Standards, and an absolute spin concentration could then be determined by using methods described previously (2). The fraction of the total nitroxide spectral area measured (typically 10%) was established by determining the area under the low-field EPR peak of a frozen toluene or pyridine solution containing a 1 mM concentration of the same spin probe. To reduce integration errors, the spectra were integrated three times and then averaged.

The compositions of the undoped Argonne Premium coals are given in Table I. The following spin probes were used in this study:

- V, 4-octadecanoylamino-2,2,6,6-tetramethylpiperidinoxyl
- X, 4-amino-2,2,6,6-tetramethylpiperidinoxyl-4-pyridinecarboxaldimine
- XII, 4-hexylamino-2,2,6,6-tetramethylpiperidinoxyl
- XIII, 4-nonylamino-2,2,6,6-tetramethylpiperidinoxyl

Spin probes X, XII, XIII, and V are made up of 4-, 6-, 9-, and 15-carbon chains, respectively, attached to the same nitroxyl-radical substituent (*see* structures on page 473). Spin probes I–IX were defined previously (3), and the numbering system for spin probes I–XIII used in our previous papers is maintained here for consistency.

EPR powder spectra were recorded on a Varian E-12 continuous-wave (CW) EPR spectrometer at room temperature. The magnetic fields were calibrated by using a Bruker ER035M NMR gaussmeter, and the microwave frequency was measured with a Hewlett–Packard 5246L frequency counter.

Results and Discussion

In Figure 1, the radical concentration of coals swelled with toluene solutions of spin probes V, X, XII, and XIII is shown as a function of carbon content. The radical concentration of doped coals swelled with toluene showed a similar dependence with carbon content for probes I, II, and V (2). Thus, although there are differences in spin-probe radical concentration as a function of percent carbon, the radical concentration decreases with rank until a level of 85% carbon and then levels off or increases. For spin probe X, the smallest probe for which results are reported here, the radical concentration decreased from 15.4×10^{17} spins/g at 74.05% carbon

Table I. Analytical Data for the Argonne Premium Coal Samples Studied

APCS Number	Coal	C	H	N	S_{org}	Cl	F	O
8	Beulah–Zap	74.05	4.90	1.17	0.71	0.04	0.00	19.13
3	Illinois No. 6	80.73	5.20	1.40	2.47	0.06	0.00	10.11
6	Blind Canyon	81.32	5.81	1.59	0.37	0.03	0.00	10.88
4	Pittsburgh No. 8	84.95	5.43	1.68	0.91	0.12	0.00	6.9
5	Pocahontas No. 3	91.81	4.48	1.34	0.51	0.20	0.00	1.66

NOTE: Dry, mineral-matter-free (dmmf) values are given in percent, except for sulfur values, which are given in dry percent of the coal sample.

(dmmf) (APCS No. 8) to 1.0×10^{17} spins/g at 84.95% carbon (APCS No. 4) and then increased to 3.0×10^{17} spins/g at 91.81% carbon (APCS No. 5). Similar behavior was noticed by Goslar for a toluene solution of spin probe II, which is a probe very similar in structure to X.

Spin probes XII, XIII, and V have similar shapes and differ from each other primarily in length. Spin probe XII is slightly longer than spin probe X. For coals doped with XII, the radical concentration decreased rapidly from 16.5×10^{17} spins/g at 74.05% carbon (APCS No. 8) to 3.2×10^{17} spins/g at 81.3% carbon (APCS No. 6), then decreased much more slowly to 0.5×10^{17} spins/g at 91.81% carbon (APCS No. 5).

For coals doped with the two largest spin probes studied, XIII and V, the radical concentration decreased with percent carbon. For example, the concentration of XIII decreased linearly from 11.1×10^{17} to 1.4×10^{17} spins/g, and the radical concentration of V decreased from 1.5×10^{17} to

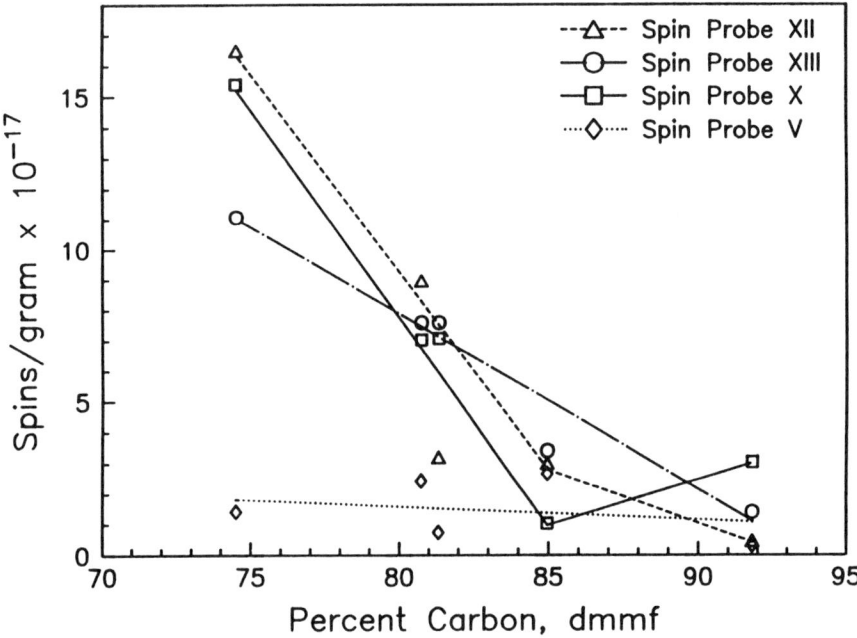

Figure 1. Nitroxide radical concentration vs. percent carbon for APCS Nos. 3, 4, 5, 6, and 8 swelled with a toluene solution of spin probes X, XII, XIII, and V.

0.3×10^{17} spins/g over the range of coals studied. Upon swelling with toluene, apparently fewer pores can accommodate spin probe V than can accommodate spin probe XIII.

In general, the carbon content of coal increases with rank (15). Figure 1 shows that for all spin probes studied, the nitroxide-radical concentration decreased with carbon content and, consequently, with rank. This result indicates that as the rank of the coal samples increases, the degree of swelling will be lower, and fewer areas will be accessible to the spin probes.

In Figure 2, the concentration of nitroxide radicals within the pores of the coal swelled with nitrobenzene solutions of different spin probes is shown as a function of oxygen content. As noted earlier, the nitroxide-radical concentrations of doped coals that were swelled with polar solvents seems to relate to percent oxygen rather than to percent carbon. Nitrobenzene is a stronger swelling solvent than toluene and thus causes a larger change in the pore structure. Consequently, the spin-probe concentrations are different for coals swelled in nitrobenzene. The nitroxide-radical concentration of coals doped with spin probe X increased from 0.10×10^{17} spins/g at 1.66% oxygen (dmmf) (APCS No. 5) to 3.0×10^{17}

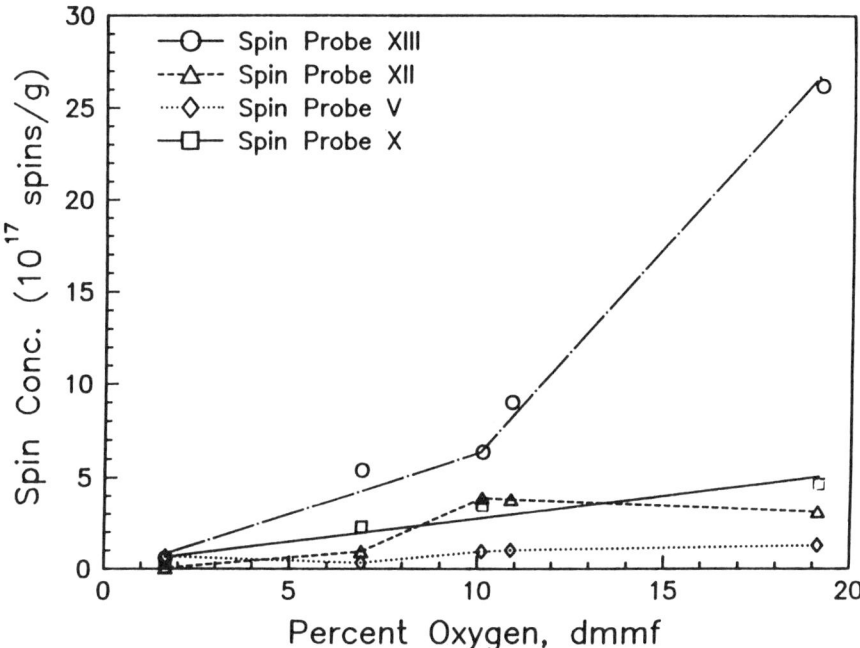

Figure 2. Nitroxide radical concentration vs. percent oxygen for APCS Nos. 3, 4, 5, 6, and 8 swelled with a nitrobenzene solution of spin probes X, XII, XIII, and V.

spins/g at 10.88% oxygen (APCS No. 6) and then levels off. Coals doped with spin probe XII behave in a similar manner. The concentration of nitroxide radicals within the pores of the coal increased from 0.1×10^{17} spins/g at 1.66% oxygen (APCS No. 5) to 3.9×10^{17} spins/g at 10.11% oxygen (APCS No. 3) and then decreased slightly to 3.1×10^{17} spins/g at 19.13% oxygen (APCS No. 8).

Of the two largest spin probes studied with nitrobenzene, the coals doped with XIII had the highest nitroxide-radical concentrations, and coals doped with V had the lowest nitroxide-radical concentrations. For coals doped with XIII, the radical concentration increased from 0.6×10^{17} spins/g at 1.66% oxygen (APCS No. 5) to 9.0×10^{17} spins/g at 10.88% oxygen (APCS No. 6) then increased to 26×10^{17} spins/g at 19.13% oxygen (APCS No. 8). The nitroxide-radical concentration in coals doped with V increased from 0.3×10^{17} to only 1.3×10^{17} spins/g over the range of coals studied. Like toluene, nitrobenzene did not swell the coals sufficiently to accommodate spin probe V.

In Figure 3, the nitroxide-radical concentration in coals swelled with pyridine solutions of different spin probes is shown as a function of oxygen content. Pyridine is a much stronger swelling solvent than either tol-

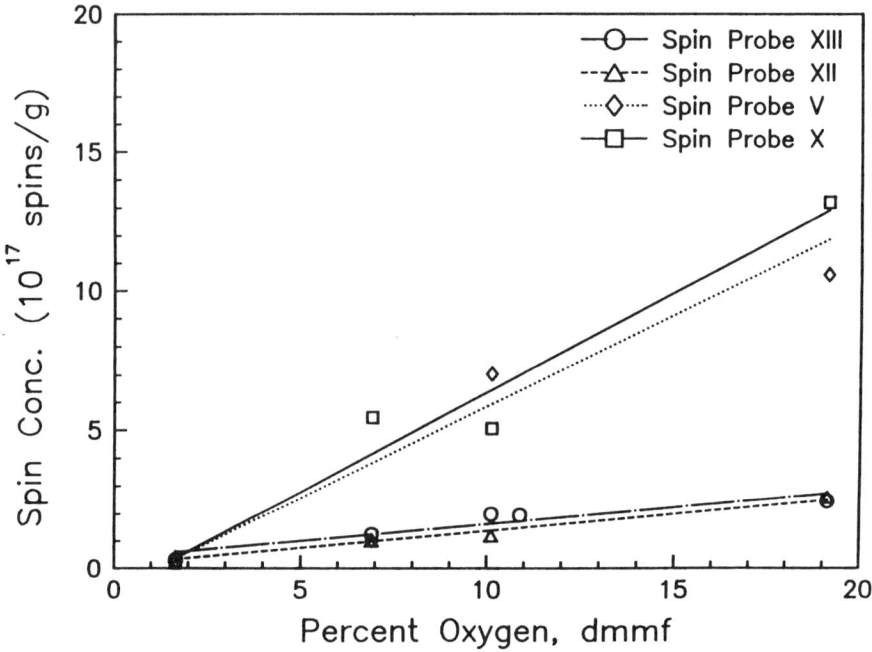

Figure 3. Nitroxide radical concentration vs. percent oxygen for APCS Nos. 3, 4, 5, 6, and 8 swelled with a pyridine solution of spin probes X, XII, XIII, and V.

uene or nitrobenzene, and the behavior of the spin probes is considerably different than in coals swelled by either of the two latter solvents. Of the spin probes studied, coals doped with either the largest (V) or the smallest (X) spin probe showed the highest nitroxide-radical concentrations. For coals doped with X, the nitroxide-radical concentration linearly increased from 0.2×10^{17} to 13×10^{17} spins/g over the range studied. For coals doped with V, the nitroxide radical concentration increased from 0.3×10^{17} to 11×10^{17} spins/g over the range of coals examined. Pyridine opens up the pore structure large enough that even the large probe (V) can be trapped before pyridine is removed. The reasoning behind the seemingly unusual behavior of spin probe X will be discussed in more detail later. For coals doped with XII and XIII, the nitroxide radical concentration increased from 0.3 and 0.4×10^{17} to 2.5 and 2.6×10^{17} spins/g, respectively, over the range of coals studied.

Figures 2 and 3 show the spin-probe content of coals swelled with nitrobenzene and pyridine, respectively, as a function of oxygen content. Oxygen content decreases with rank. For all solvent systems and all spin probes studied, the nitroxide radical concentration decreased with decreasing oxygen content and, consequently, with increasing rank. Again, this indicates that the ability of the solvent to swell the coal and open areas of accessibility for the spin probes decreases with rank.

One further interesting observation from Figures 1–3 is that for the highest ranked coals the nitroxide radical concentration changes very little regardless of probe or swelling solvent. This result confirms that in higher ranked coals cross-linking consists of covalent bonds rather than hydrogen bonds.

The ability of solvents to swell coals and open areas of accessibility is strongly dependent upon coal rank. This observation leads to the question of how the pore structure changes with swelling within a given coal rank. By using SANS measurements of Pittsburgh No. 8 coal, Winans and Thiyagarajan (6) determined that pores in unswelled coal are spherical, and that pores in coal swelled by pyridine are elongated, needlelike voids. Goslar and Kispert (2) confirmed their result by using a spin-probe study. The five Argonne coals used in the study by Goslar and Kispert (2) were swelled in toluene or pyridine using either a spherical or cylindrical probe. The results are shown in Table 3 of reference 2. Toluene mildly swells the coal and should change the pore structure only minimally. Mildly swelled coals doped with the spherical probe (I) contained significant radical concentrations, a result indicating that a significant number of spherically shaped pores were present (2). However, for severely swelled coals (i.e., those swelled with pyridine), the radical concentration of coals doped with I was very small, a result indicating that spherically shaped pores had all but disappeared.

On the other hand, coals doped with the cylindrically shaped probe (II) and swelled with toluene had significant radical concentrations. Swelling with pyridine slightly decreased radical concentrations in APCS Nos. 8 and 5, and in the three high-volatile bituminous (HVB) coals (APCS Nos. 3, 4, and 6) radical concentration was significantly increased. This result shows that in severely swelled coals the pores become cylindrical.

Several questions immediately arise:

- Upon learning the shape of pores in swelled coals, can the size of these pores be further defined?
- How does the pore structure of coal change when solvents with intermediate swelling ability are used?
- How does the size distribution of the pores change as the coal is swelled?
- How does the shape of the pores change with swelling?

To answer these questions, the five Argonne Premium Coal samples described in the "Experimental Details" section were swelled with either toluene, nitrobenzene, or pyridine and were labeled with three spin probes

(XII, XIII, and V) that are similar in shape and width but that differ in length. Toluene has an electron donor number of 0.1, the electron donor number of nitrobenzene is 4.4, and that of pyridine is 33.1. The electron donor number has been used (15) as a measure of coal swelling ability, although the variation in size and shape among various solvents makes the correlation a poor one (7). A better measure of swelling ability is the basicity of the solvent and the solvent's ability to break hydrogen bonds.

In Figure 4, the radical concentration in the labeled coals is shown as a function of spin probe length for coals swelled with toluene. For the toluene system, spin probe XII results in the highest radical concentrations of the five Argonne coals. As the probe size increased, the probe concentration decreased. Toluene, a poor swelling solvent, has a limited capacity to open areas of accessibility to spin probes. The pores are elongated only slightly, and larger spin probes had difficulty entering the coals.

In Figures 5 and 6, the radical concentration in the labeled coals is shown as a function of spin probe length for coals swelled with nitrobenzene and pyridine, respectively. In Figure 5, the maximum spin concentration at the probe length corresponds to that of spin probe XIII. This

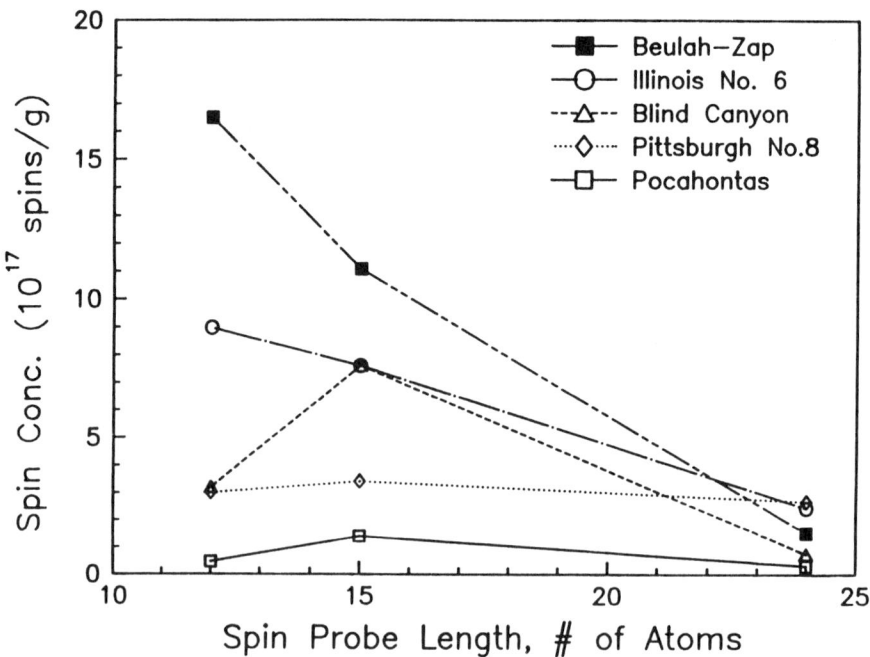

Figure 4. Nitroxide radical concentration for five coals swelled with a toluene–spin probe solution vs. spin probe length.

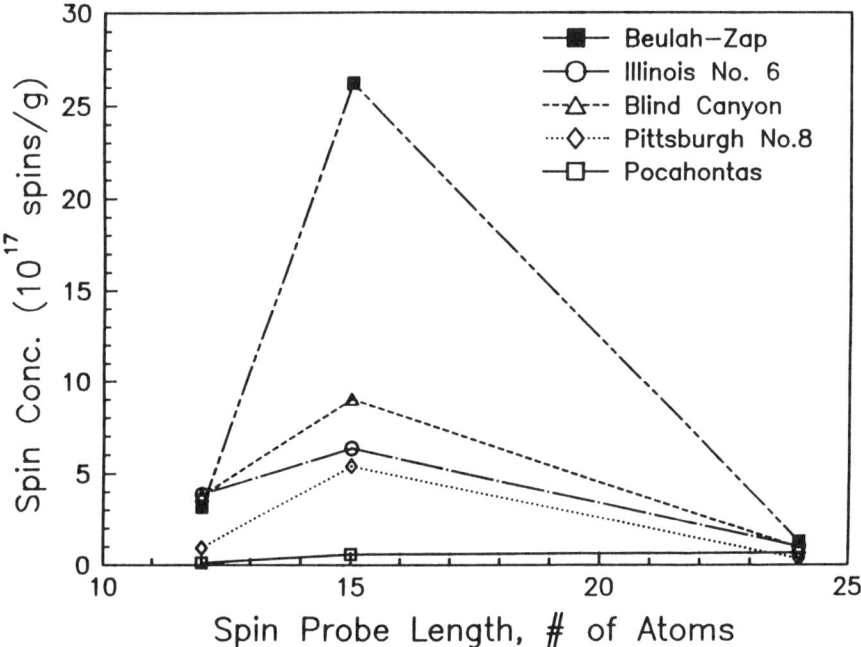

Figure 5. Nitroxide radical concentration for five coals swelled with a nitrobenzene–spin probe solution vs. spin probe length.

probe was the right size to be trapped within the pore structure of the swelled coal without being removed by the cyclohexane wash, as apparently happened to the smaller probes. However, nitrobenzene is only a moderate swelling solvent, and larger probes such as V have difficulty gaining access to the interior of the coal. On the other hand, there is no apparent maximum in Figure 6; the largest spin probe studied (V) had the highest concentration of all the spin probes. Pyridine, the best swelling solvent studied, opens up the pore structure of the coals to the greatest extent. Large spin probes such as V have easy access to the interior of the coal, and small spin probes are easily washed out. These results lead to two additional questions:

- How large a spin probe can be readily introduced into pyridine-swelled coals?
- What effects do solvents with even larger electron donor numbers have on the pore structure of coals?

Our hope is that these questions will be answered in a later study.

Spin probe X has a different structure than the other three spin probes studied. Spin probe X is shorter in length and, because of its steric

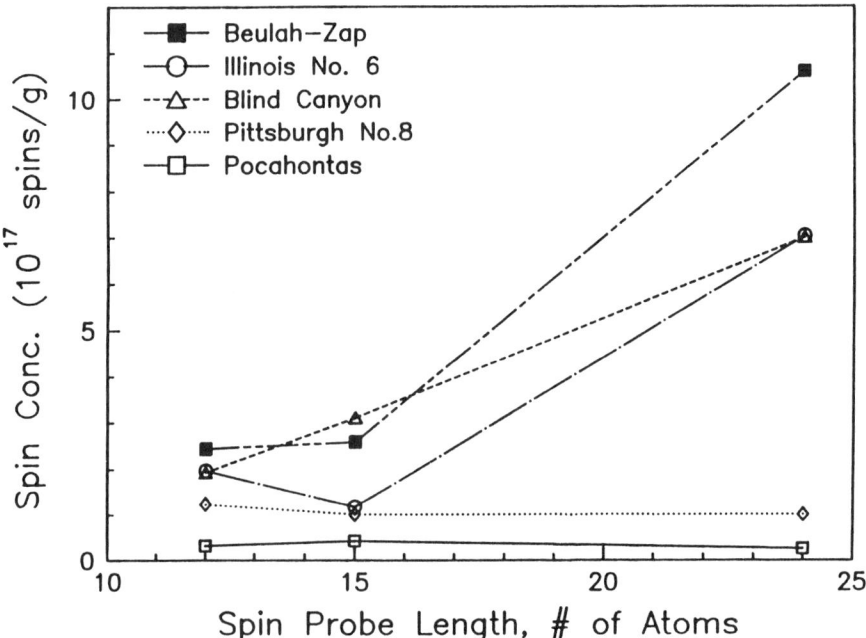

Figure 6. Nitroxide radical concentration for five coals swelled with a pyridine–spin probe solution vs. spin probe length.

rigidity, less bulky than spin probe XII, the smallest of the other three probes studied. In Table II, the nitroxide radical concentration is shown as a function of the swelling solvent for Argonne coals labeled with spin probe X. The radical concentration passes through a minimum when nitrobenzene is used as the swelling solvent. Both toluene- and pyridine-swelled coals retain significant amounts of probe X within the pore structure. This phenomenon is supported by data from earlier work (2). In toluene-swelled coals, the pore structure is opened up only slightly, and the result is that small cylindrical probes are trapped. Coal also contains a micropore system that is too small to be accessed by these probes when the coal is only mildly swelled. Nitrobenzene opens up areas of accessibility large enough to prevent retention of X and XII (Figure 4), but does not significantly open the micropore system. Pyridine, however, is a much stronger swelling solvent than either toluene or nitrobenzene. Pyridine apparently not only opens areas of accessibility large enough for spin probes like V, but also opens the micropore system enough to allow access by X. Thus the coal swelling process undergoes several stages.

Table II. Nitroxide-Radical Concentration for Coals Swelled with Solutions of Spin Probe X and Various Swelling Solvents

APCS Number	Coal	Toluene	Nitrobenzene	Pyridine
8	Beulah–Zap	15.4	4.6	13.2
3	Illinois No. 6	7.0	3.4	5.0
6	Blind Canyon	7.1	—	9.4
4	Pittsburgh No. 8	1.0	2.6	5.4
5	Pocahontas No. 3	3.0	0.1	0.20

NOTE: Concentrations are given in spins per gram $\times 10^{17}$.

Conclusions

In conclusion, the use of spin probes with swelling solvents is a convenient method with which to increase understanding of the pore structure of coals and how the pore structure changes in the presence of swelling solvents. As coal swells, the pores become cylindrical and elongated. As the severity of the swelling solvent is increased, the pores become larger until even the micropore system opens up to allow access to fairly large molecules. Regardless of the severity of the swelling solvent, the pore structure of high rank coals changes very little. Finally, a spin-probe EPR spectroscopic study can be used to observe changes in coal structure as solvents of different donor or swelling strength are used.

Acknowledgment

We gratefully acknowledge the U.S. Department of Energy, Pittsburgh Energy, Technology Center, University Coal Research Program, for financial support under Grant No. DE–FG 22–86 PC90502. We gratefully acknowledge Xinahi Chen for technical support.

References

1. Bartholomew, C. H.; White, W. E.; Thornock, D.; Wells, W. F.; Hecker, W. C.; Smoot, L. D.; Williams, F. F. *Prepr. Am. Chem. Soc. Div. Fuel Chem.* **1988**, *33*, 24–31 and references cited therein.
2. Goslar, J.; Kispert, L. D. *Energy Fuels* **1989**, *3*, 589–594.
3. Goslar, J.; Cooray, L. S.; Kispert, L. D. *Fuel* **1989**, *68*, 1402–1407.

4. Cooray, L. S.; Kispert, L. D.; Wuu, S. K. *Prepr. Am. Chem. Soc. Div. Fuel Chem.* **1988**, *33*, 32–37.
5. Gethner, J. S. *J. Appl. Phys.* **1986**, *59*, 1068–1084.
6. Winans, R. E.; Thiyagarajan, P. *Prepr. Am. Chem. Soc. Div. Fuel Chem.* **1987**, *32*, 227–231.
7. Hall, P. J.; Marsh, H.; Tomas, K. M. *Fuel* **1988**, *67*, 863–866.
8. Larsen, J. W., Green, T. K.; Kovac, J. *J. Org. Chem.* **1985**, *50*, 4729–4735.
9. Painter, P. C.; Sobkowiak, M.; Youtcheff, J. *Fuel* **1987**, *66*, 973–978.
10. Silbernagel, B. G.; Ebert, L. B.; Schlosberg, R. H.; Long, R. B. *Coal Structure;* Gorbaty, M. L.; Ouchi, K., Eds.; Advances in Chemistry 192; American Chemical Society: Washington, DC, 1981; pp 23–35.
11. Wuu, S. K.; Kispert, L. D. *Fuel,* **1981**, *64*, 1681–1686.
12. Kispert, L. D.; Cooray, L. S.; Wuu, S. K. *Prepr. Am. Chem. Soc. Div. Fuel Chem.* **1987**, *32*, 286–292.
13. Martin, K. A.; Chao, S. S. *Prepr. Am. Chem. Soc. Div Fuel Chem.* **1988**, *33*, 17–23.
14. Goslar, J.; Kispert, L. D. *Fuel* **1990**, *69*, 564–569.
15. Szeliga, J.; Marzec, A. *Fuel* **1983**, *62*, 1229.

RECEIVED for review June 7, 1990. ACCEPTED revised manuscript December 5, 1990.

26

Dynamic In Situ 9-GHz Electron Paramagnetic Resonance Studies of Argonne Premium Coal Samples

H. A. Buckmaster and Jadwiga Kudynska

Department of Physics and Astronomy, The University of Calgary, Calgary, Alberta, T2N 1N4, Canada

> *This chapter describes the preliminary determination of the spectroscopic parameters and $a_0 d_0$ Argand diagrams at 20 °C for as-received, dried, and moisture-saturated samples of the eight Argonne Premium coals using 9-GHz continuous-wave electron paramagnetic resonance (CW EPR) spectroscopy. The Argand line-shape diagrams characterize these coals according to rank. The low-temperature oxidation of as-received, dried, and moisture-saturated samples of the Wyodak–Anderson subbituminous and Blind Canyon high-volatile bituminous coals were studied by using dynamic, in situ 9-GHz CW EPR spectroscopy. The spectroscopic parameter changes below 100 °C are due primarily to one free radical species, and those above 100 °C are due to the other species and attain broad maxima near 120 °C. Exposure to nitrogen causes the maximum spin concentration to occur at a lower temperature in the dry Wyodak–Anderson coal but nitrogen exposure has the opposite effect in the Blind Canyon coal, a result suggesting that storage of coals in nitrogen may have altered the chemical properties of the Wyodak–Anderson coal.*

MOST CONTINUOUS-WAVE electron paramagnetic resonance (CW EPR) measurements have been performed on coal samples that have been contained in evacuated and sealed containers. The objective was to remove the air and water from the sample to avoid contamination effects, with particular emphasis on the presence of oxygen (*1–5*). Recently, Buckmaster, Kudynska, and co-workers (*6–9*) have performed exploratory studies of selected Alberta coals; these studies were directed to an assessment of the feasibility of using CW EPR spectroscopy to determine the susceptibility of Alberta coals to spontaneous combustion. A methodology was developed to study the low-temperature oxidation processes in these coals at temperatures below 250 °C to determine those factors that might play an important role in the sequence of events leading to spontaneous combustion.

The industrial importance of this knowledge is self-evident. For controlled low-temperature oxidation, the relative total spin concentration increases by a factor of 6 between 25 and 100 °C for a sample of a subbituminous (SB) coal but by a factor of only 2 for a sample of a high-volatile bituminous (HVB) coal, and the presence of air is essential for these changes to occur (*6*). Demineralization of the HVB coal sample reduced the relative total spin concentration increase in this temperature interval by 25% (*7*). A study of the exinite-depleted and -enhanced as well as the intact samples of an HVB coal revealed that the low-temperature process occurred in the vitrinite maceral, if the contribution of the inertinite maceral to the change in the relative total spin concentration was assumed to be either negligible or temperature independent (*8*).

The role of moisture content in the HVB sample was also investigated. The moisture content is the most important factor determining the maximum percentage increase in the relative total spin concentration (*9*).

As a result of this experience, we decided to repeat the same type of measurement methodology on samples of the eight Argonne Premium coals. Preliminary measurements were made on all eight samples. In addition, comprehensive measurements have been completed on samples of SB (Wyodak–Anderson) and HVB (Blind Canyon) coals. These two coals were selected because their physical and chemical properties were similar to those of the Alberta coals that have been studied previously (*6–9*). This chapter describes the results of those measurements.

Experimental Details

Sample Preparation. The sealed glass vials containing the samples were opened in a small, dry-argon-atmosphere glove box. Each vial was divided into three equal parts, which were then placed in a miniature desiccator jar. The sample tube assembly described in the next section was loaded with 50 mg of the as-received sample. This assembly was then removed from the glove box and inserted into the tapered ground-glass joint in the high-sample-temperature resonant cavity glassware. The glove box was prepared for another sample transfer when the first EPR measurement sequence was completed. Samples of the Wyodak–Anderson, Pocahontas No. 3, and Blind Canyon Argonne Premium coals were measured on the day that the vials were opened. All eight coals were then measured 4 days later. All of the coals had dehydrated slightly when these latter measurements were made because the glove box contained dry argon.

The glove box also contained two larger desiccators. One, which contained phosphorus pentoxide, was used to dry one fraction of each sample, and the other, which contained a saturated potassium sulfate solution, was used to moisture-saturate another fraction of each sample. Experience indicated that 3 days was sufficient time for the samples to attain their new states of moisture equilibria.

The physical and chemical characterization of the eight Argonne Premium coals samples have been studied by many researchers and have been summarized elsewhere and in the references therein (10).

CW EPR Measurements. The 9-GHz CW EPR spectrometer used synchronous demodulation at the microwave and magnetic-field modulation frequencies (10 kHz). A Stanford Research SR 530 lock-in amplifier was used at 10 kHz because it could be operated under computer control via an IEEE-488 bus. The EPR data acquisition was controlled by a Zenith Z-158 computer, and the data sets were transferred to a SUN 3/160 workstation for analysis. Software was developed to facilitate the calculation of the CW EPR spectral parameters and for the creation and plotting of the radial-difference $a_0 d_0$ Argand diagrams (RD$a_0 d_0$AD). (The Fourier absorption component is a_0; the Fourier dispersion component is d_0.)

The relative total spin concentration, $N(T)$, at temperature T was calculated by double integration of the first Fourier absorption coefficient, a_1. The value of a_0 was calculated by single integration of a_1, and d_0 was obtained from a_0 by using a Hilbert transform (6). No attempt was made to determine the absolute spin concentration because this study was concerned with only relative changes in the total spin concentration. The values and their errors in the spectral parameters were determined by repeating the computer spectral analysis 10 times. The errors were confirmed independently by repeating the low-temperature oxidation experiment with new samples of the same coal.

The following values were obtained: the spectroscopic splitting error, Δg, = ±0.00002; the error in effective line width at half maximum amplitude, $\Delta(\Delta B_{1/2})$, = ±10 μT; and the error in relative spin concentration, $\Delta N_{rel}/N_{rel}$, = 0.05.

The cylindrical TE$_{011}$-mode Bruker ER 4114 MT high-sample-temperature 9-GHz resonant cavity was used with a new, laboratory-designed sample-temperature controller that incorporated the Bruker B-TC 80115 power supply used previously. This controller features direct sample-temperature measurement, 1-K accuracy, long-term temperature stability of ∼0.1 K, temperature-set-point repeatability of ∼0.1 K, very fast temperature-response time (a few seconds for a 5-K set-point change), and near-critical damping response.

The EPR measurements were made in constant temperature–time increments to ensure that all measurement sequences had the same thermal "history" and to enable those measurement sequences on the same sample source to be repeated and compared reliably (11). A special sample holder that was designed permitted various gases to flow through the sample at a rate of 200 mL/min and enabled the temperature of the sample to regulate the temperature controller feedback loop. The total elapsed time between setting the temperature and the end of recording the EPR spectrum at each temperature was 15 min. Measurements were made in 5 °C steps from 25 to 150 °C and in 20 °C steps from 150 to 250 °C. At each temperature, the 50-mg sample, of which 35 mg was inside the resonant cavity, was exposed to either 15 min of dry air flow or 5 min of dry air flow followed by 10 min of dry nitrogen flow. The EPR spectrum was recorded during the last 5 min of either sequence.

Radial-Difference a_0d_0 Argand Diagrams.

The measurement of RDa_0d_0AD is a very sensitive diagrammatic technique that can be used to determine the nature of the magnetic resonance line-shape distortion (12). This technique is based on the fact that the graph of a_0 as a function of d_0 is a circle for a Lorentzian line-shaped resonance. This graph is called an a_0d_0 Argand diagram (a_0d_0AD). The presence of a distortion mechanism will cause the graph to deviate from a circle, and the details of this deviation are characteristic of the mechanism. Buckmaster and Duczmal (12) modeled the effect of two overlapping resonances under two conditions, and Shams Esfandabadi (11) has studied experimentally the effect of modulation broadening. A method of increasing the sensitivity of this diagrammatic technique is to modify the a_0d_0AD obtained from experimental data by subtracting the a_0d_0AD for a Lorentzian line shape (i.e., a circle) normalized to the maximum amplitude. The new graph is referred to as a radial-difference a_0d_0AD (RDa_0d_0AD). The changes in these RDa_0d_0AD graphs with temperature can be used to determine the portion of the resonance that has changed. This information can be interpreted in terms of a two-resonance model, and this interpretation is known as the Larsen–Marzec model for a coal (13–15).

In an independent study (11), the RDa_0d_0AD was reproduced with an accuracy better than $\sim 5\%$, but only when the output noise from the microwave synchronous demodulator was accurately minimized by extremely careful adjustment of the microwave reference power phase into this demodulator when the sample resonant cavity was critically coupled and precisely tuned. The necessary use of magnetic-field modulation to display the Fourier coefficients (usually a_1) of the resonance also modulation-broadens this resonance. Consequently, the RDa_0d_0AD for a modulation-broadened Lorentzian line-shape resonance exhibits a characteristic symmetric shape. This component can be eliminated by subtracting the RDa_0d_0AD obtained at one temperature from that obtained at the next higher temperature. This approach assumes that the line-width change is sufficiently small to be neglected in this temperature interval. The data analysis in this chapter makes use of these incremental RDa_0d_0ADs.

Results

20 °C Measurements of the Eight Argonne Premium Coal Samples.

Table I gives the 20 °C, 9-GHz CW EPR effective g-factor, effective peak-to-peak line width ΔB_{pp}, effective line width at half maximum amplitude $\Delta B_{1/2}$, relative peak-to-peak amplitude A_{pp}, and relative total spin concentration N_{rel} for samples of the Wyodak–Anderson SB, Blind Canyon HVB and Pocahontas No. 3 low-volatile bituminous (LVB) Argonne Premium coals (202, 601, and 501), respectively, that were measured within 10 min of their sealed glass vials being opened (AR–F). It also gives the values of these parameters when these samples were remeasured 4 days later (AR–S). The samples were at a higher relative humidity when stored in the flame-sealed vials than after they were opened and stored in dry argon. The effective g-factors all increased by 0.02% in the 4 days, ΔB_{pp} and $\Delta B_{1/2}$ both increased in Pocahontas No. 3 (LVB) and Blind Canyon (HVB) but decreased in the Wyodak–Anderson (SB) coal. The relative peak-to-peak amplitude, A_{pp}, increased in the Wyodak–Anderson and Blind Canyon coals but decreased in the Pocahontas No. 3 coal. The relative total spin concentration, N_{rel}, increased in all three of these coals. The ratios of the values of $\Delta B_{1/2}$ to those for ΔB_{pp} are significantly larger than those that would have been obtained for a pure Lorentzian line shape (1.15), and the ratio increases with the rank of the coal. These measurements support the hypothesis that the resonances observed for coals are the composite of two, and in some cases three, overlapping resonances (13–15) and that the overlap decreases with rank.

Table I. CW EPR Spectral Parameters at 20 °C for Samples of Three Argonne Premium Coals

Sample[a]	Seam[b]	Rank	g-factor	ΔB_{pp} (mT)	$\Delta B_{1/2}$ (mT)	A_{pp}	N_{rel}
202 F	WA	SB	2.00298	0.78	1.22	0.03	0.214
202 S			2.00351	0.76	1.19	0.04	0.263
601 F	BC	HVB	2.00269	0.58	1.03	0.03	0.127
601 S			2.00316	0.74	1.17	0.05	0.341
501 F	POC	LVB	2.00244	0.31	0.65	0.14	0.192
501 S			2.00284	0.41	0.85	0.09	0.264

[a] The sample designation refers to the Argonne Premium coal sample number, and F and S indicate as-received samples and as-received samples that have been stored 4 days under argon, respectively.

[b] ABBREVIATIONS: WA, Wyodak–Anderson; BC, Blind Canyon; and POC, Pocahontas No. 3.

Figures 1–3 show graphs of the CW EPR a_1 spectra for AR–F and AR–S samples of these three Argonne coals (a) and the corresponding RDa_0d_0ADs (b) for these spectra. The changes in the spectral parameters given in Table I can be correlated with the changes shown in Figure 3a. The line shape of the Wyodak–Anderson SB (Argonne 202) coal was not effected by storage, but it has significant symmetric distortion. This result shows that the relative number of free radicals in each species was not affected by drying due to storage. The line shape of the Blind Canyon HVB (Argonne 601) coal becomes more symmetric upon storage, and the low-field half becomes less distorted. The number of free radicals in the high-field (lower g-factor) species decreased as a result of drying due to storage. The line shape of the Pocahontas No. 3 LVB (Argonne 501) coal also becomes more symmetric upon storage. The distortion is probably due to three rather than two free radical species, and the result of drying due to storage is a decrease in the relative number of free radicals in the intermediate g-factor species.

Table II lists the effective g-factor, $\Delta B_{1/2}$, and N_{rel} at 20 °C for samples of the eight Argonne Premium coals (in increasing rank order) as received stored 4 days (AR–S), moisture saturated (MS), and dried (D). The values of these parameters are in good agreement with those reported previously (16, 17). The measurements reported in this chapter were obtained with the samples exposed to dry air and in various states of hydration. Both $\Delta B_{1/2}$ and N_{rel} increase dramatically in the Beulah–Zap lignite (L) (Argonne 801) and Wyodak–Anderson SB (Argonne 202) samples when they are dried, and these parameters change the least in the Pocahontas No. 3 LVB (Argonne 501) sample. The intermediate-rank samples are intermediate in this effect. Moisture-saturation caused the inverse behavior: N_{rel} increases as the rank increases, and $\Delta B_{1/2}$ decreases as the rank increases, with the exception of the HVB (Argonne 601) sample. The effective g-factor is a linear function of the carbon content of the coal sample and is in agreement with the well-known decreasing relationship (16, 17). The effects of drying and moisture-saturation do not change this relationship, but the slope is less for the latter than for the former.

Figures 4–7 show the RDa_0d_0ADs for as-received (AR–S), moisture-saturated (MS), and dried (D) samples of the eight Argonne Premium coals. The RDa_0d_0ADs for the lignite and subbituminous coals given in Figure 4 are similar because the effect of drying and moisture saturation is very small and the distortion is greater on the low-magnetic-field side of the resonance, which implies that the spin concentration in the high-field component is less than the spin concentration in the low-field component, that is, $N(g_{HF}) < N(g_{LF})$.

The diagrams for the four HVB coals can be divided into two groups of two. The diagrams in Figure 5 for the Illinois No. 6 and Pittsburgh No. 8 HVB coals are similar although the low-magnetic-field distortion is great-

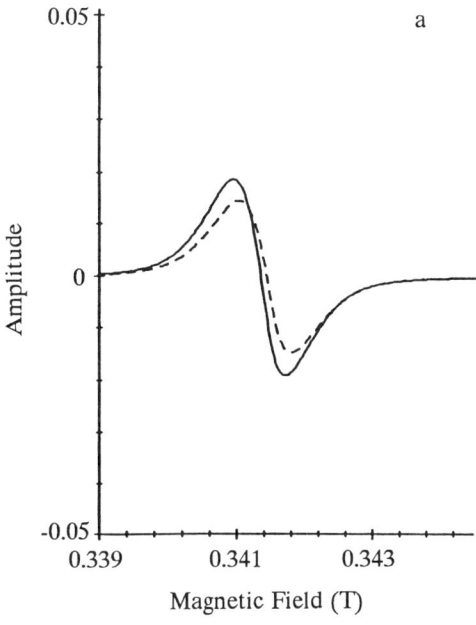

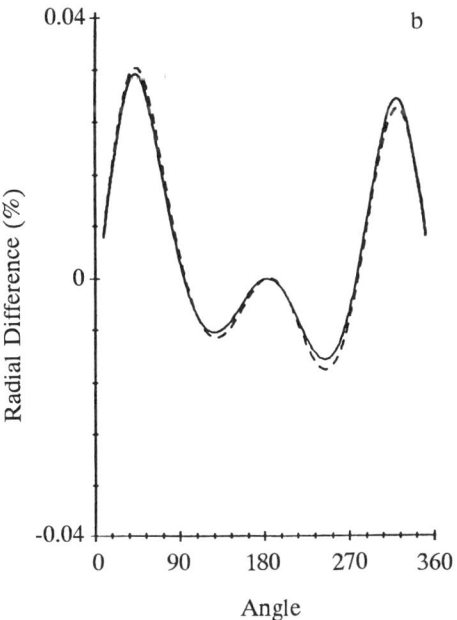

Figure 1. Graphs of the 20 °C CW EPR a_1 spectra (a) and RDa_0d_0AD (b) for as-received (dashed line) and as-received stored 4 days under argon (solid line) samples of Wyodak–Anderson SB coal.

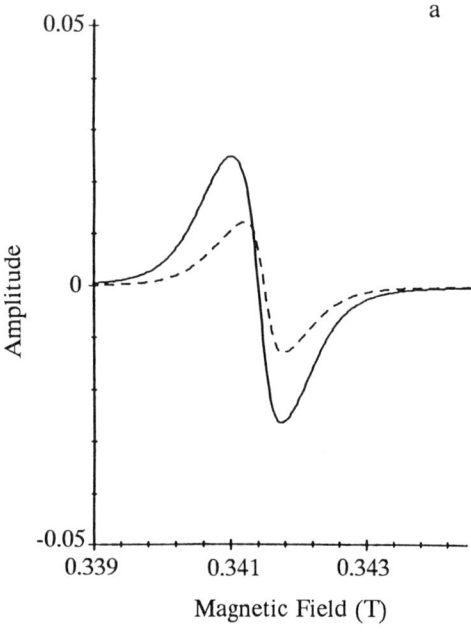

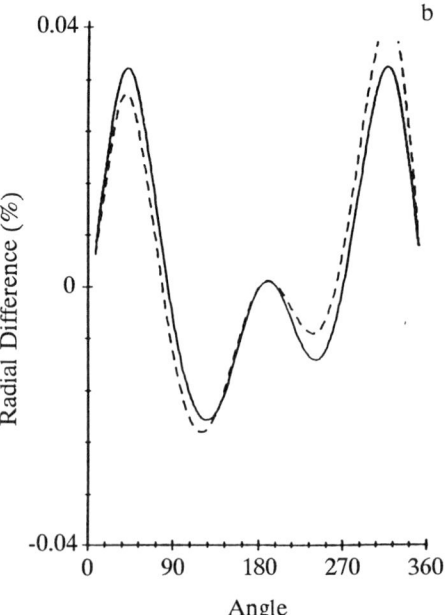

Figure 2. Graphs of the 20 °C CW EPR a_1 spectra (a) and RDa_0d_0AD (b) for as-received (dashed line) and as-received stored 4 days under argon (solid line) samples of Blind Canyon HVB coal.

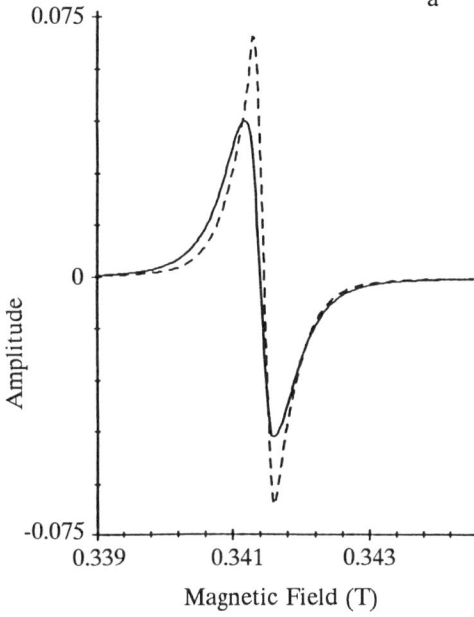

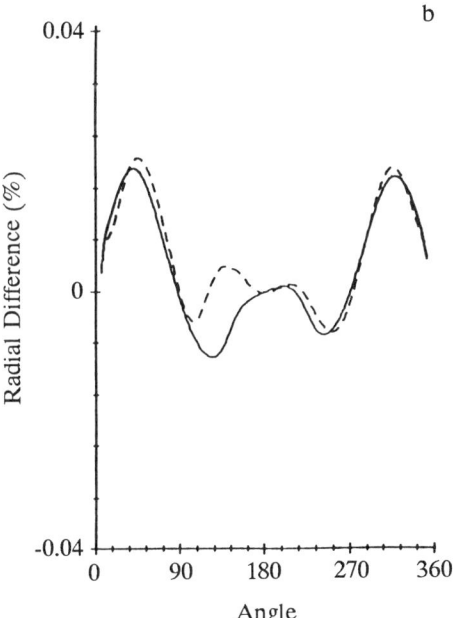

Figure 3. Graphs of the 20 °C CW EPR a_1 spectra (a) and RDa_0d_0AD (b) for as-received (dashed line) and as-received stored 4 days under argon (solid line) samples of Pocahontas No. 3 coal.

Table II. CW EPR Spectral Parameters at 20 °C for Various Samples of Eight Argonne Premium Coals

Sample	Seam	Rank[a]	Effective g-factor			$\Delta B_{1/2}$ (mT)			N_{rel}		
			AR–S[b]	MS[c]	D[d]	AR–S	MS	D	AR–S	MS	D
801	Beulah–Zap	L	2.00354	2.00316	2.00344	1.15	1.07	1.38	0.212	0.113	0.717
202	Wyodak–Anderson	SB	2.00351	2.00314	2.00345	1.19	1.11	1.43	0.263	0.134	1.000
301	Illinois No. 6	HVB	2.00316	2.00272	2.00299	1.08	1.04	1.26	0.154	0.131	0.318
401	Pittsburgh No. 8	HVB	2.00293	2.00263	2.00265	0.91	0.96	1.01	0.153	0.151	0.234
601	Blind Canyon	HVB	2.00316	2.00282	2.00296	1.17	1.16	1.23	0.341	0.209	0.514
701	Lewiston–Stockton	HVB	2.00298	2.00264	2.00278	1.02	1.01	1.08	0.383	0.256	0.430
101	Upper Freeport	MVB	2.00281	2.00253	2.00257	0.88	0.94	0.98	0.150	0.194	0.228
501	Pocahontas No. 3	LVB	2.00284	2.00251	2.00256	0.85	0.86	0.91	0.264	0.434	0.439

[a] ABBREVIATIONS: LVB, low-volatile bituminous; MVB, mid-volatile bituminous; HVB, high-volatile bituminous; SB, subbituminous; and L, lignite.
[b] As-received samples that were stored 4 days under argon.
[c] Moisture-saturated samples.
[d] Dried samples.

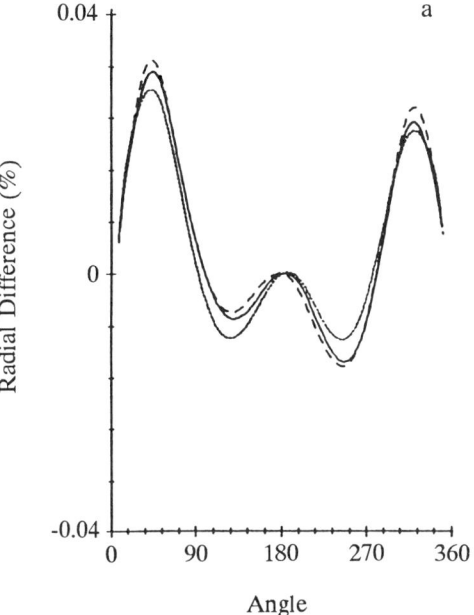

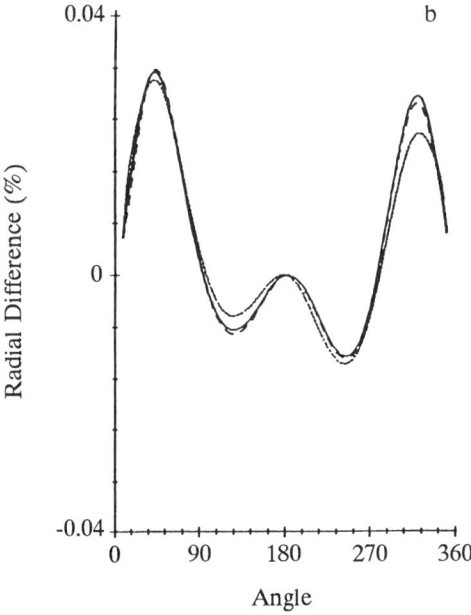

Figure 4. Graphs of the 20 °C 9-GHz RDa_0d_0ADs for as-received (solid line), moisture-saturated (dashed line), and dried (dotted line) samples of Beulah–Zap lignite (a) and Wyodak–Anderson SB (b) coals.

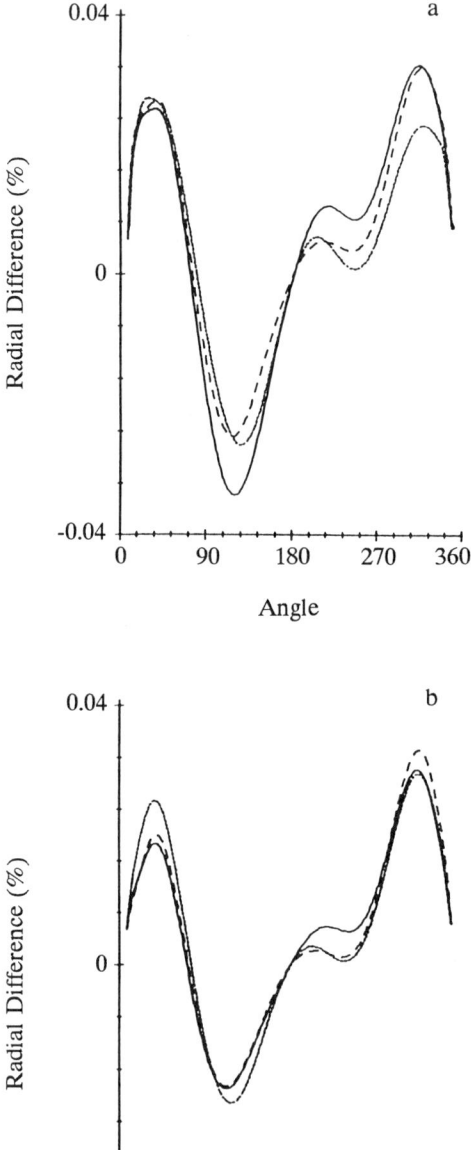

Figure 5. Graphs of the 20 °C 9-GHz RDa_0d_0ADs for as-received (solid line), moisture-saturated (dashed line), and dried (dotted line) samples of Illinois No. 6 HVB (a) and Pittsburgh No. 8 HVB (b) coals.

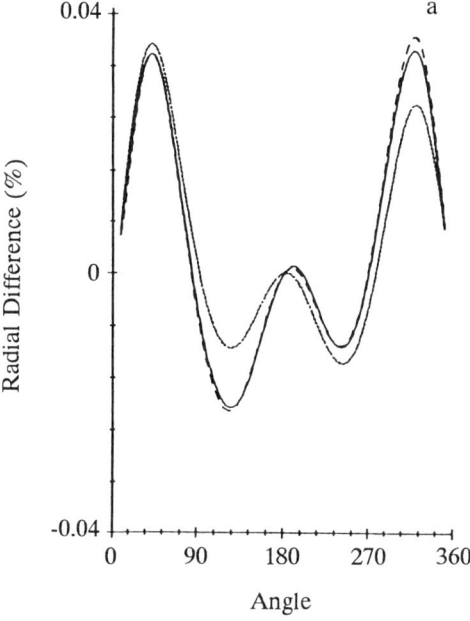

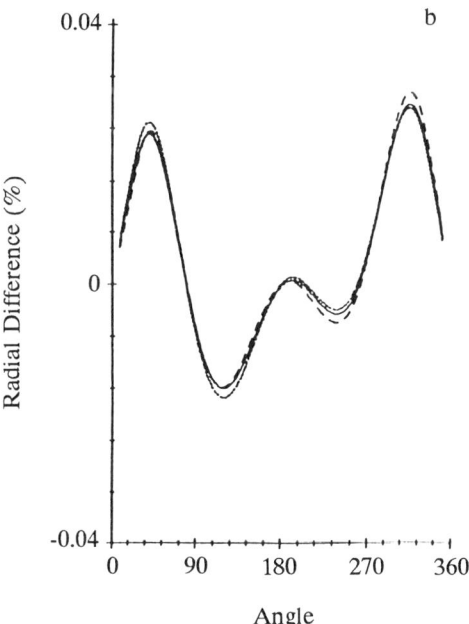

Figure 6. Graphs of the 20 °C 9-GHz RDa_0d_0ADs for as-received (solid line), moisture-saturated (dashed line), and dried (dotted line) samples of Blind Canyon HVB (a) and Lewiston–Stockton HVB (b) coals.

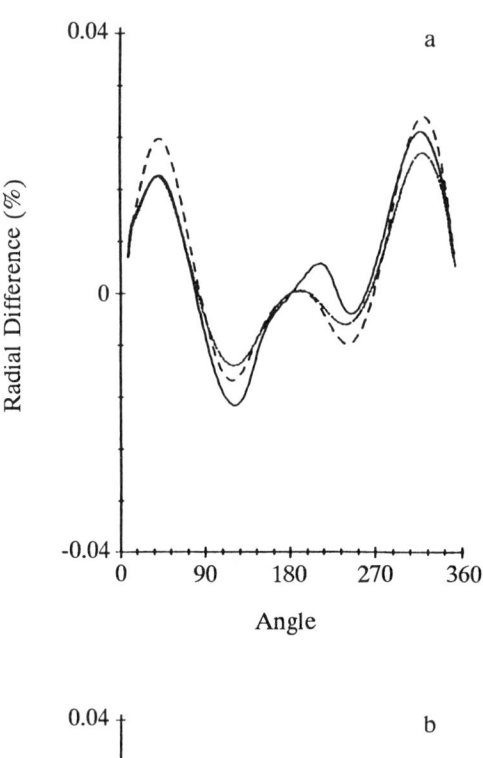
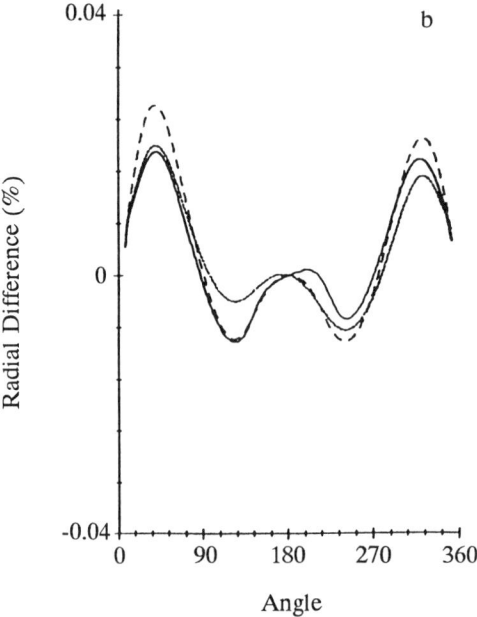

Figure 7. Graphs of the 20 °C 9-GHz RDa_0d_0ADs for as-received (solid line), moisture-saturated (dashed line), and dried (dotted line) samples of Upper Freeport MVB (a) and Pocahontas No. 3 LVB (b) coals.

er in the former, particularly in the as-received sample. In this case three rather than two free radical species may be present. The diagrams in Figure 6 for the Blind Canyon and Lewiston–Stockton HVB coals are also very similar, with the distortion slightly greater on the low-magnetic-field side of the resonance except when the Blind Canyon sample was dried, a result that implies that $N(g_{LF}) < N(g_{HF})$. The diagrams in Figure 7 for the Upper-Freeport MVB and Pocahontas No. 3 LVB coals are very similar, and the distortions may also indicate the existence of three rather than two free radical species (6). The rank of coal correlates with the shape of the RDa_0d_0AD and hence with the relative proportions and number of different free radical species present. The measured line shapes from which the RDa_0d_0ADs are derived are not very sensitive indicators of this structure because the deviations from a pure Lorentzian line shape are generally very small.

Low-Temperature Oxidation Studies. Samples of both the Wyodak–Anderson SB (Argonne 202) and Blind Canyon HVB (Argonne 601) coals were studied as received with air and air–nitrogen flow as well as dried and moisture-saturated with air–nitrogen flow.

Figures 8a and 8b are graphs of the effective g-factor as a function of the temperature from 20 to 250 °C for the SB (Argonne 202) and HVB (Argonne 601) coal samples, respectively. The effective g-factor for the dried samples of both coals exhibits a monotonic decrease with increasing temperature. This behavior was also observed in the Alberta SB and HVB coal samples studied previously under the same experimental regime (6, 9). As expected, the values of the effective g-factor for the SB coal (Argonne 202) samples exceed those for the HVB coal samples under the same conditions.

Figures 9a and 9b are graphs of the effective line width ($\Delta B_{1/2}$) as a function of the temperature from 20 to 250 °C for these same two SB and HVB coal samples, respectively. The behavior of the SB coal samples is distinctly different from that of the HVB coal samples. Initially the line widths for the SB coal samples that are subjected to various treatments decrease at about the same rate until a critical temperature is reached. The critical temperature is 70 °C for the as-received sample with air–nitrogen flow, 85 °C for the as-received sample with air flow, and 90 °C for the moisture-saturated sample, and the critical temperature decreased monotonically up to 250 °C for the dried sample. The line widths for the HVB coal samples subjected to various treatments increased to attain a broad maxima centered at 120 °C before decreasing, except for the dried sample. This line-width behavior is similar to that observed for Alberta HVB coal samples (6, 9). The line-width behavior of the Alberta SB samples mirrors that for the Alberta HVB samples except

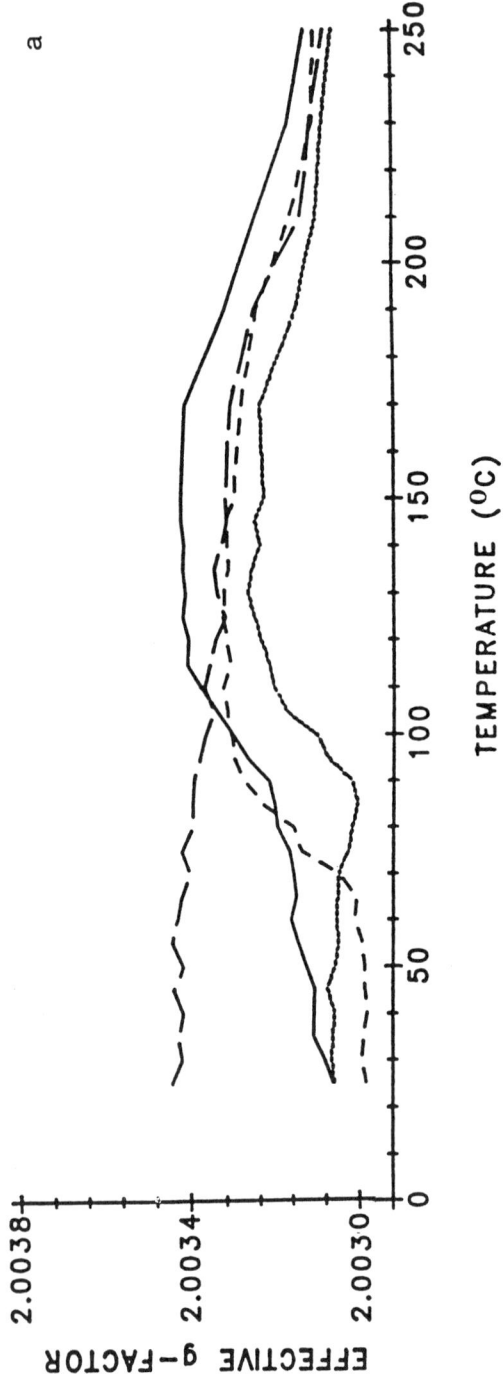

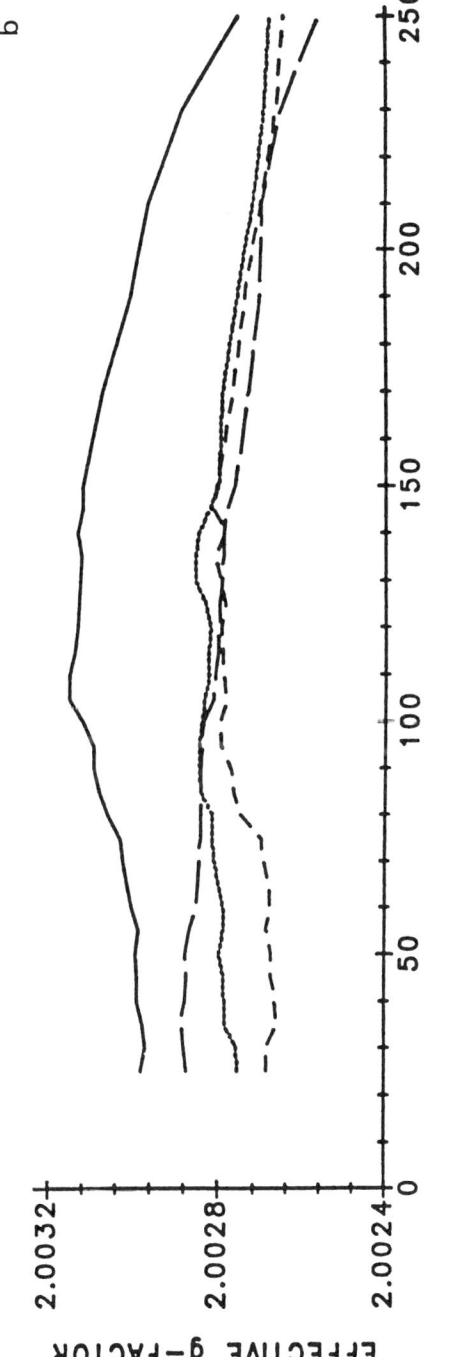

Figure 8. Graph of the temperature dependence of the effective g-factor for Wyodak–Anderson SB (a) and Blind Canyon HVB (b) coal samples as received, stored 1 day under argon, and exposed to air flow (solid line); as-received fresh (short dashes), moisture-saturated (dotted line), or dried (long dashes), and exposed to air–nitrogen flow.

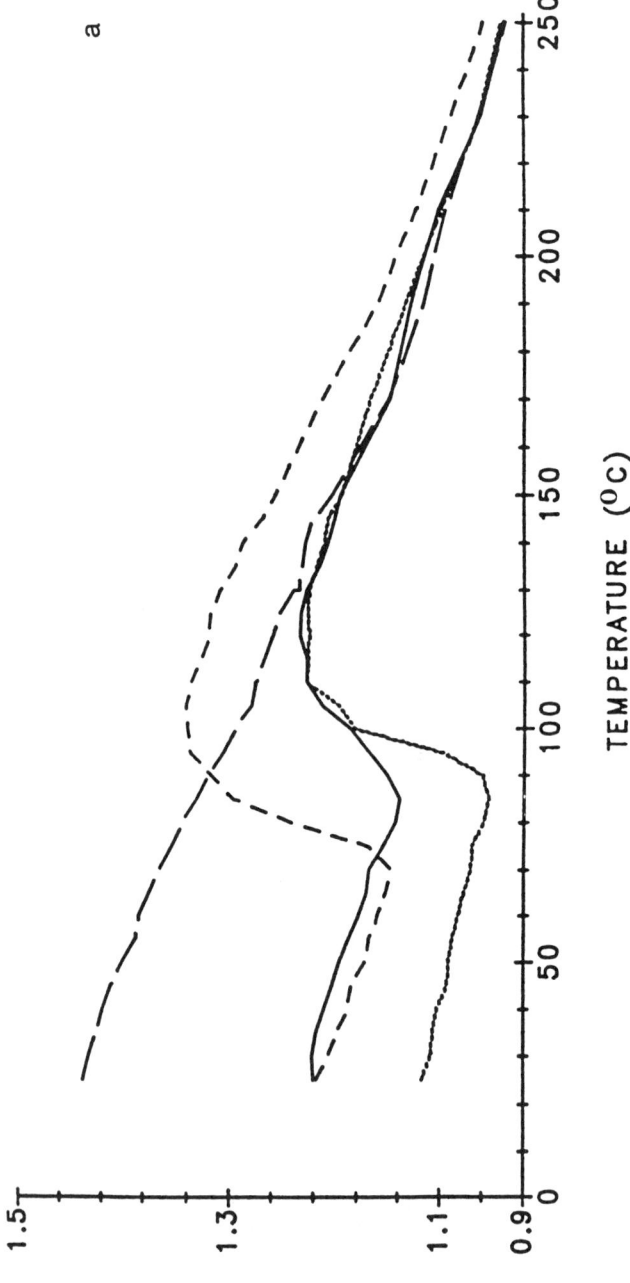

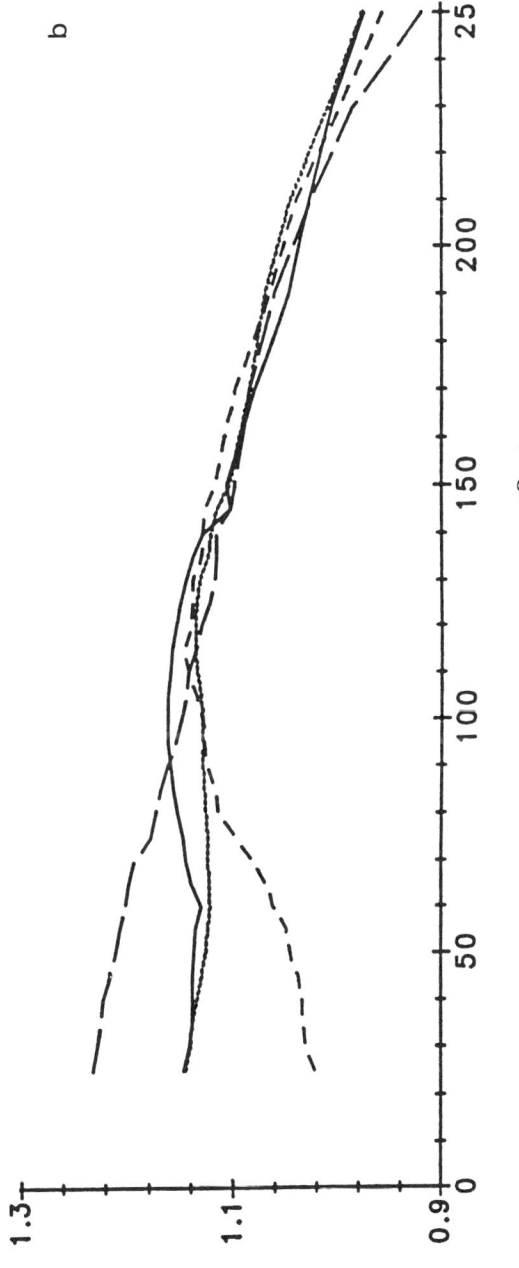

Figure 9. Graph of the temperature dependence of the effective line width ($\Delta B_{1/2}$) for Wyodak–Anderson SB (a) and Blind Canyon HVB (b) coal samples as received, stored 1 day under argon, and exposed to air flow (solid line); as-received fresh (short dashes), moisture-saturated (dotted line), or dried (long dashes), and exposed to air–nitrogen flow.

that the 20 °C value of the SB coal sample was 10% less than that for the HVB coal sample. This behavioral difference is an indicator that the lower rank Argonne coals were affected by their storage in a nitrogen atmosphere.

Figures 10a and 10b are graphs of the relative total spin concentration (N_{rel}) as a function of the temperature for these same two SB and HVB coal samples, respectively. The behavior is, in general, similar to that for the effective line width. The as-received SB coal sample that was subjected to air–nitrogen treatment exhibited the largest percent increase in N_{rel}, and the increase started at the lowest temperature (70 °C). Comparison of the N_{rel} values for the as-received SB coal samples that were subjected to air and air–nitrogen treatment leads to the conclusion that the storage of the Argonne Premium coal samples under nitrogen appears to have played a significant role in determining the low-temperature oxidation properties of this lower rank coal.

Storage of the Blind Canyon HVB coal sample in nitrogen appears to have suppressed the number of free radical spins created when the as-received sample was treated with air–nitrogen rather than air by a factor of greater than 2 at all temperatures up to 150 °C and shifted the maximum from 120 to 135 °C. The opposite effect was observed in the Wyodak–Anderson SB coal sample for which the suppression below 70 °C is replaced by a rapid creation of free radical spins that attains a maximum at 100 °C of the same value as that attained with air treatment at 120 °C. The susceptibility of this SB coal to combust spontaneously may have been altered by nitrogen storage.

These results should be compared with those that were obtained with comparable SB and HVB Alberta coals (6, 9), which were prepared and stored under argon rather than nitrogen. These samples always had lower N_{rel} when treated with nitrogen flow after air than those treated with only air flow. This result implied that some of the oxygen absorbed on the surface of the coal samples was flushed out by the nitrogen flow treatment. The interpretation of the changes in the effective g-factor, effective line width, and relative total spin concentration with temperature is facilitated by the use of incremental RDa_0d_0ADs (11) because these diagrams reveal the relative changes in the line shape and hence of the two or more free radical species involved. Further discussion of the application of this line-shape analysis technique is beyond the scope of this chapter.

Conclusions

The 20 °C CW EPR spectral parameters for the eight Argonne Premium coals exhibit properties that are similar to those reported in the literature

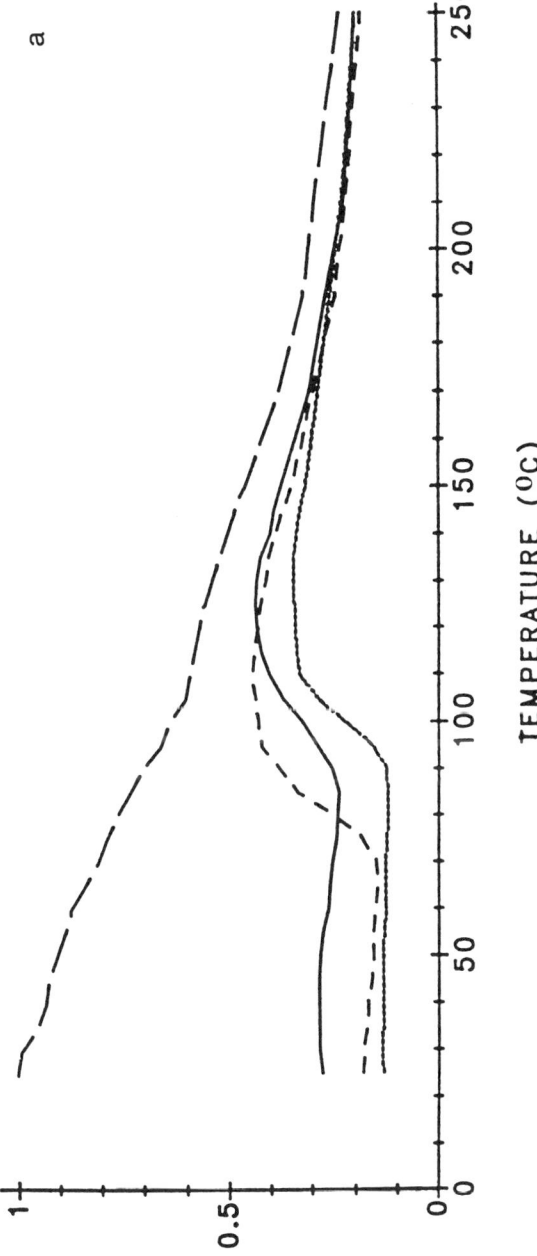

Figure 10a. Graph of the temperature dependence of the relative total spin concentration for Wyodak–Anderson SB coal samples as received, stored 1 day under argon, and exposed to air flow (solid line); as-received fresh (short dashes), moisture-saturated (dotted line), or dried (long dashes), and exposed to air–nitrogen flow.

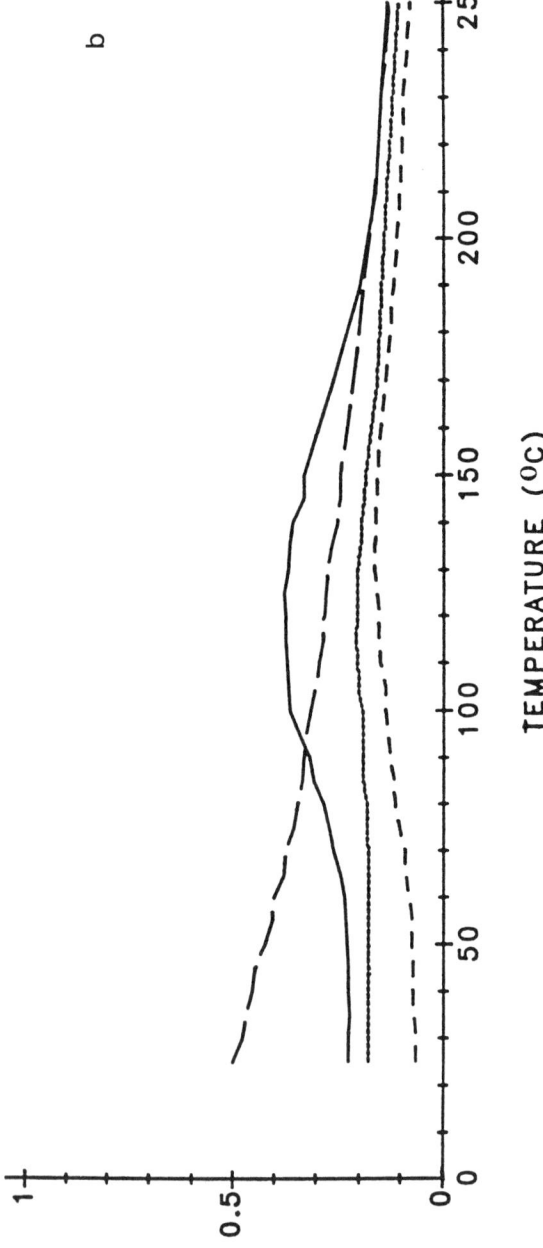

Figure 10b. Graph of the temperature dependence of the relative total spin concentration for Blind Canyon HVB coal samples as received, stored 1 day under argon, and exposed to air flow (solid line); as-received fresh (short dashes), moisture-saturated (dotted line), or dried (long dashes), and exposed to air–nitrogen flow.

for coals of comparable rank, carbon content, and atomic H–C ratio (*2, 16–18*). The RD$a_0 d_0$ADs reveal information concerning the line shape that has not been reported previously. Other workers (*19, 20*) have attempted to correlate various spectral parameters with the carbon content; however, their approach assumed that the line shape was symmetric. The measurements reported in this chapter demonstrate clearly that this assumption is incorrect. Coals of similar rank exhibit similar line shapes as analyzed by using their RD$a_0 d_0$ADs. The sensitivity to moisture has been found to vary with rank in a complex fashion, and the relative contributions of the free radical species present in different ranks varies both in character and the number of species present. A more precise analysis must await the development of a methodology for the quantitative analysis of RD$a_0 d_0$ADs.

Definitive conclusions cannot easily be drawn from the preliminary low-temperature oxidation studies of the samples of the Wyodak–Anderson SB and Blind Canyon HVB Argonne Premium coals reported in this chapter because the behavior of the effective g-factor, $\Delta B_{1/2}$, and N_{rel} as a function of temperature for the various sample states is distinctly different from that observed in comparable Alberta coals (*6, 9*). Consequently, further measurement sequences are required to determine the reasons for these differences. The method of sample preparation and storage may be the basic cause for these differences. Apparently, the storage of the Argonne Premium coal samples in a nitrogen atmosphere caused at least the lower rank samples to change their free radical content and nature. This result implies that there must exist reactions involving nitrogen that occur at 20 °C in moist, lower rank coals.

The CW EPR measurements and RD$a_0 d_0$AD analysis reported in this chapter indicate that the eight Argonne Premium coals were well chosen and have extremely interesting chemical properties. The decision to prepare and store the samples of these coals in a nitrogen atmosphere may have had unexpected consequences that are revealed only when low-temperature oxidation studies, such as those reported in this chapter, are attempted. The value of using RD$a_0 d_0$ADs for line-shape analysis has also been demonstrated.

Acknowledgments

This research was funded by Natural Sciences and Engineering Research Council of Canada operating grant (A0716) and the University of Calgary to H. A. Buckmaster. We acknowledge the assistance and advice of T. H. T. van Kalleveen concerning the development and refinement of the EPR spectral analysis and the RD$a_0 d_0$AD programs and V. Shams Esfandabadi concerning the application of the concept of an incremental RD$a_0 d_0$AD.

The development and maintenance of the computer-controlled 9-GHz CW EPR spectrometer was due to C. H. Hansen. This project could not have been completed without his encouragement and advice. We are indebted to M. Bowman and R. Botto for inviting us to participate in this study of the Argonne Premium coals and to report our results at PACIFICHEM '89 held in Honolulu, Hawaii, December 17–22, 1989.

References

1. Austen, D. E. G.; Ingram, D. J. E.; Given, P. H.; Binder, C. R.; Hill, L. W. In *Coal Science;* Given, P. H., Ed.; American Chemical Society: Washington, DC, 1966; pp 344–59.
2. Retcofsky, H. L.; Stark, J. M.; Friedel, R. A. *Anal. Chem.* **1968,** *40,* 1699–1704.
3. Petrakis, L.; Grandy, D. W. *Fuel* **1981,** *60,* 115–19.
4. Cole, D. A.; Herman, R. G.; Simmons, G. W.; Klier, K. *Fuel* **1985,** *64,* 303–6.
5. Schlick, S.; Narayana, M.; Kevan, L. *Fuel* **1986,** *65,* 873–76.
6. Kudynska, J.; Buckmaster, H. A.; Duczmal, T.; Bachelor, F. W.; Majumdar, A. *Fuel* **1992,** *71,* in press.
7. Buckmaster, H. A.; Kudynska, J. *Fuel* **1992,** *71,* in press.
8. Kudynska, J.; Buckmaster, H. A. *Fuel* **1992,** *71,* in press.
9. Buckmaster, H. A.; Kudynska, J. *Fuel* **1992,** *71,* in press.
10. Vorres, K. S. *Energy Fuels* **1990,** *4,* 420–426.
11. Shams Esfandabadi, V. unpublished M. Sc. Thesis, University of Calgary, Calgary, Alberta, Canada, 1990.
12. Buckmaster, H. A.; Duczmal, T. In *Electromagnetic Resonance of the Solid State;* Weil, J. A., Ed.; Chemical Society of Canada: Ottawa, Canada, 1987; pp 57–68.
13. Marzec, A.; Juzwa, M.; Betlej, K.; Sobkowiak, M. *Fuel Proc. Tech.* 19 **1979,** *2,* 35–44.
14. Larsen, J. W.; Kovac, J. In *Organic Chemistry of Coal;* Larsen, J. W., Ed.; ACS Symposium Series 71; American Chemical Society: Washington, DC, 1978; pp 36–49.
15. Duber, S.; Wiekowski, A. B. *Fuel* **1984,** *63,* 1641–43.
16. Retcofsky, H. L. In *Coal Science;* Gorbaty, M. L.; Larsen, J. W.; Wender, I., Eds.; Academic: New York, 1982; Vol. 1, pp 43–82.
17. Petrakis, L.; Grandy, D. W. In *Free Radicals in Coals and Synthetic Fuels;* Coal Science and Technology Series No. 5; Elsevier: Amsterdam, Netherlands, 1983; pp 71–136.
18. Petrakis, L.; Grandy, D. W. *Anal. Chem.* **1978,** *50,* 303–8.
19. Kwan, C. L.; Yen, T. F. *Anal. Chem.* **1979,** *51,* 1225–29.
20. Retcofsky, H. L.; Thompson, G. P.; Raymond, R.; Friedel, R. A. *Fuel* **1975,** *54,* 126–28.

RECEIVED for review June 8, 1990. ACCEPTED revised manuscript June 7, 1991.

27

Electron Magnetic Resonance of Standard Coal Samples at Multiple Microwave Frequencies

R. B. Clarkson, Wei Wang, D. R. Brown, H. C. Crookham, and R. L. Belford

Department of Chemistry and Illinois EPR Research Center, University of Illinois, Urbana, IL 61801

The naturally occurring unpaired electrons in coal provide a unique route for the nondestructive study of coal structure. Electron paramagnetic resonance (EPR), electron–nuclear double resonance (ENDOR), dynamic nuclear polarization (DNP), and electron spin-echo (ESE) techniques use the paramagnetic spins as probes of their environment, and the spins provide information about their immediate chemical surroundings as well as about the nature of more distant regions in the coal. In the context of an ongoing program in our laboratory to better determine the structure and bonding in the organic (maceral) components of whole coals (including Argonne Premium coal samples), coal extracts, and model coal systems by electron magnetic resonance methods (EPR, ENDOR, ESE), we have begun a multifrequency EMR study of coal, with the frequency range spanning nearly 2 orders of magnitude (2–250 GHz), to improve our understanding of the complex magnetic interactions present in the coal material and the origin of these effects in molecular structure and bonding. With this approach, we are developing methods for the elucidation of atomic and molecular structure in coal and other complex, disordered materials.

MULTIFREQUENCY ELECTRON MAGNETIC RESONANCE (EMR) spectroscopy is the method of making EMR measurements [e.g., electron paramagnetic resonance (EPR), electron–nuclear double resonance (ENDOR), and electron spin-echo (ESE)] on the same material at several substantially different microwave frequencies. Previous applications of this approach to EPR spectroscopy of coal have demonstrated its effectiveness, even over a modest two-frequency range (9.5 and 35 GHz) (*1–2*), and the entire subject was reviewed in 1987 by Belford et al. (*3*). Multifrequency studies of complex systems often can facilitate reliable interpretations of structure, bonding, and magnetic interactions that would otherwise remain uncertain or impossible to analyze in single-frequency experiments. An examination of the principal terms of the spin Hamiltonian provides a useful starting point for understanding the utility of multifrequency experiments.

The spin Hamiltonian, H_s, can be written as the sum of electronic (H_e) and nuclear (H_n) spin operators,

$$H_s = H_e + H_n \qquad (1)$$

in which

$$H_e = +|\mu_B| \, \mathbf{B}_o \cdot \mathbf{g}_e \cdot \mathbf{S} \qquad (2)$$

where the spin angular momentum $S = 1/2$, μ_B is the Bohr magneton, $\mathbf{B}_o$ is the external magnetic field vector, $\mathbf{g}$ is the matrix of g factors, $\mathbf{S}$ is the electronic spin angular momentum vector (matrix), and

$$H_n = -g_n |\mu_n| \, \mathbf{B}_o \cdot \mathbf{I} + \mathbf{S} \cdot \mathbf{A} \cdot \mathbf{I} \qquad (3)$$

where the nuclear spin quantum number $I = 1/2$, μ_n is the nuclear magneton, $\mathbf{I}$ is the nuclear spin angular momentum vector (matrix), and $\mathbf{A}$ represents the hyperfine interaction matrix. The first term in each part of H_s describes the Zeeman interaction (electronic or nuclear) and is characterized by a dependence on $\mathbf{B}_o$ and expressions containing g-factors. The second term in H_n describes the hyperfine interactions between electrons and nuclei.

Information on structure and bonding is contained in both the Zeeman and hyperfine interactions, which together form the basis for an experimentally observed EMR spectrum. Because of the complexity of the spectra of disordered, heterogeneous systems such as coal, analyzing or unambiguously interpreting the data can be very difficult. For example, the EPR line widths of coal spectra can include contributions from g-anisotropy and hyperfine interactions, and determining the relative importance of each contribution is usually difficult.

In this case, one would like to "switch off" the hyperfine interaction for one observation and then compare the resulting spectrum with a conventional one containing both Zeeman and hyperfine terms in order to separate the two contributions. Magic-angle spinning (MAS) attempts to do just this, for example, by averaging away the dipolar portion of $I_1 \cdot I_2$ interactions in NMR spectroscopy, and other NMR techniques address different portions of H_s to simplify spectra.

Unfortunately, the much more rapid electronic relaxation rates exhibited by paramagnetic systems have thus far made it impossible to apply techniques such as MAS to EMR spectroscopy. What is needed for EMR spectroscopy is a method for varying the importance of the terms in the spin Hamiltonian. This method must not depend critically on relaxation rates—a goal that the multifrequency approach accomplishes.

By applying EPR, ENDOR, or ESE techniques at different microwave frequencies (and hence in different B_0 ranges), Zeeman interactions can be emphasized or de-emphasized relative to hyperfine terms, allowing a more critical evaluation of many spectral effects, including g-dispersion (EPR) (4); nuclear Larmor frequencies and orientation selection (ENDOR) (5); echo-envelope modulation depth (6); orientation selection (7–8); exact cancellation (ESE) (9); and relaxation rates (seen in all methods). For example, Figure 1 shows EPR spectra taken at three different frequencies of an Illinois No. 6 coal (4). As g-dispersion increases with B_0, a low-field shoulder appears. In this example, important chemical information contained in the electronic Zeeman term could be revealed only at higher frequencies. Figure 2 also demonstrates the potential utility of the multifrequency approach. Here, the electron spin-echo envelope modulation (ESEEM) from the same Illinois coal is shown at two frequencies. The depth of the ESEEM pattern is greater at lower frequencies because of a reduction in the energy separation between electronic spin states and a corresponding increase in the transition probability of the nearly forbidden (electron–nuclear spin-flip) transitions.

In this chapter, we review our very high-frequency (VHF) (96 and 250 GHz) EPR work (4, 10) and report new low-frequency (3 GHz) pulsed EPR results on Argonne Premium coal samples. Related ongoing work in our laboratory includes multifrequency ENDOR spectroscopy of coal.

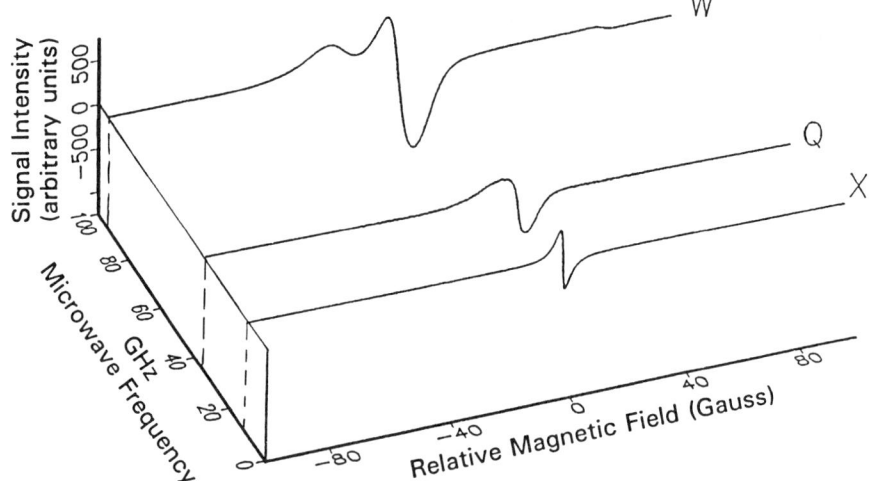

Figure 1. Dependence of EPR spectra of a powdered Illinois coal sample upon microwave frequency of the spectrometer. The magnetic field scale is centered at a value corresponding to g = 2.003 for each frequency. The frequencies are designated as microwave bands: X = 9.5 GHz, Q = 35 GHz, and W = 94 GHz.

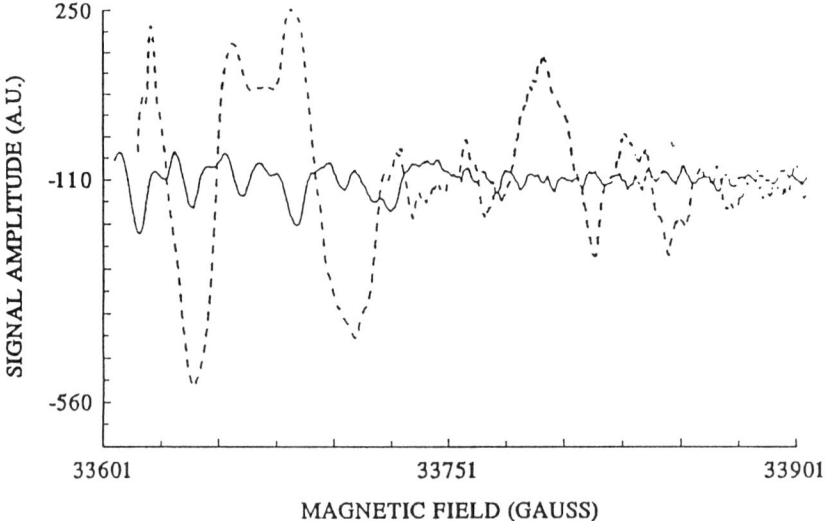

Figure 2. ESEEM time-domain patterns from a powdered Illinois coal sample taken at 3-GHz (- - -), and 9.4-GHz (—) microwave frequencies. (Reproduced from reference 6. Copyright 1989 American Chemical Society.)

Experimental Details

Samples of coal from the Argonne Premium Coal Sample Program were received in sealed glass ampules under nitrogen. The compositions of these eight samples have been reported (*11*). For EPR spectroscopy at 96 GHz (VHF EPR), the ampules were opened in air just prior to taking the spectra. Samples were loaded into 0.5-mm i.d. capillary tubes and placed in the cavity of the spectrometer. The W-band (96 GHz) instrument has been described elsewhere (*4*). Field strengths were measured with an NMR gaussmeter (Metrolab model 2025), and frequencies were monitored with a digital frequency counter (EIP model 578). A variety of sharp, solution-phase samples were used as g-markers.

A sample of Herrin No. 6 (essentially the same as Illinois No. 6) coal from the Illinois Coal Basin Sample Program (ICBSP No. 1, Southern Illinois University Registry No. 1822) was separated by density-gradient centrifugation prior to study, and vitrinite, sporinite, and fusinite maceral fractions were analyzed before taking EPR spectra. The analysis data for these samples are given in Table I.

The maceral composition of the Herrin No. 6 coal was determined to be 87% vitrinite, 4.4% liptinite, and 8.6% inertinite. The separated vitrinite from this coal was also studied on a 250-GHz spectrometer built by Lynch et al. at Cornell University (*12*).

Model compounds (perylene, dibenzothiophene, and dibenzofuran) were prepared as cation radicals either by adsorbing the materials as gases onto an activated silica alumina catalyst (Houdry M-46) (*13*) or by UV-irradiating boric acid glasses containing the compounds in 10–50-mM concentrations. Lower concentrations would, in some cases, be required to obtain the highest spectral resolution.

Samples for the ESE study were evacuated for 24 h at room temperature and at pressures of $<10^{-4}$ torr (13.33 mPa) in 5-mm thin-walled NMR tubes. The

Table I. Elemental and Proximate Data for Herrin No. 6 Coal

Sample	C	H	N	S	O
Whole coal	76.86	4.86	1.29	3.27	13.70
Vitrinite	75.68	5.25	1.41	2.59	11.57
Sporinite	76.18	5.90	1.11	3.87	12.94
Fusinite	79.61	4.24	1.35	2.03	12.77

NOTE: All values are given as percents on a dry, ash-free basis.

samples then were sealed in the glass tubes. The S-band (2–4 GHz) ESE spectrometer used in the study has been described previously (6).

Results and Discussion

Very High-Frequency EPR Spectroscopy. Figures 3a–3h show room-temperature W-band spectra of all eight Argonne Premium coal samples (APCS). The samples show a considerable range of line shapes; in some cases, these line shapes have features not seen at lower field strengths. Because of our special interest in high-sulfur Illinois coals, we begin the analysis of these spectra with an Illinois No. 6 coal (APCS 3). Many of the points that are considered in this analysis should be applicable to the interpretation of the VHF EPR spectra of other coals.

The most prominent features of the W-band spectrum of the Illinois No. 6 coal, in addition to the largest peak, include a low-field shoulder and a broad, weak, high-field wing.

To determine whether these features are characteristic of a particular maceral, W-band spectra were taken of vitrinite, sporinite, and fusinite components separated from the Herrin No. 6 coal sample (SIU No. 1822, which is analogous to the Argonne Premium Illinois No. 6 coal sample) by the density-gradient centrifugation method. Spectra of the three separated macerals are shown in Figures 4a–4c. These spectra clearly show substantial spectral variation with maceral type, and we will consider each maceral in turn.

The fusinite signal in Figure 4c is quite different from the other two macerals' spectra, and thus it must reflect a dominant interaction that is not as important in the other components. This interaction is usually assumed to be electron spin exchange, which results in a narrowing of the resonance line (1, 14). We have tested this hypothesis by performing S-band pulsed-EPR measurements on the fusinite. The result was a free induction decay (FID), as shown in Figure 5. The observation of an FID, combined with the absence of echoes, lends strong support for the spin-exchange model, particularly because the maceral contains almost as much hydrogen (4.24%) as the vitrinite and sporinite, which give strong spin echoes and proton ESEEM. Fourier transformation of the FID yields a frequency-domain spectrum with a line width nearly identical to that observed in the continuous-wave (CW) EPR spectrum at 3, 9, and 96 GHz, a result that is again in agreement with the prediction of a line width independent of B_o for exchange narrowing in the limit $B_e >> B_o$ or $B_e << B_o$, where B_e is the exchange field (1).

The simplest assumption is that the vitrinite spectrum in Figure 4a is due to a single paramagnetic species and that the spectrum can be simu-

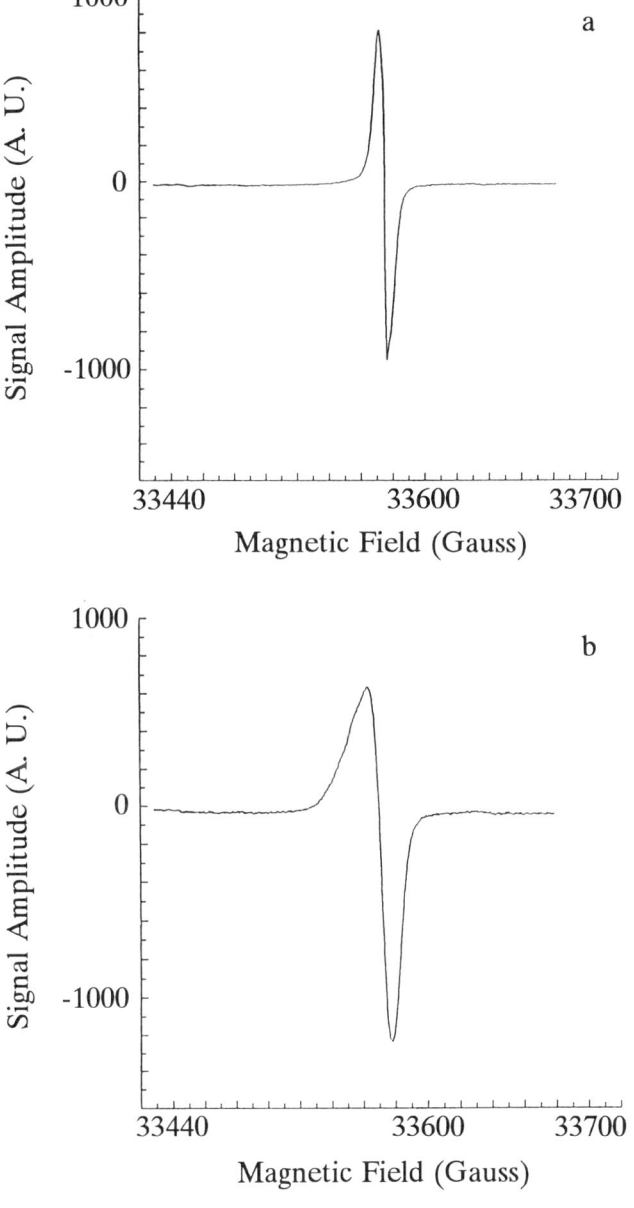

Figure 3. W-band (94-GHz) EPR spectra of eight Argonne Premium coal samples (APCS). First-derivative presentation. The field span (abscissa) for each spectrum is 200 G (= 0.02 T) centered at 33,600 G (3.36 T). Part a: APCS 1, Upper Freeport, Pennsylvania, coal. Part b: APCS 2, Wyodak–Anderson, Wyoming, coal. Continued on next page.

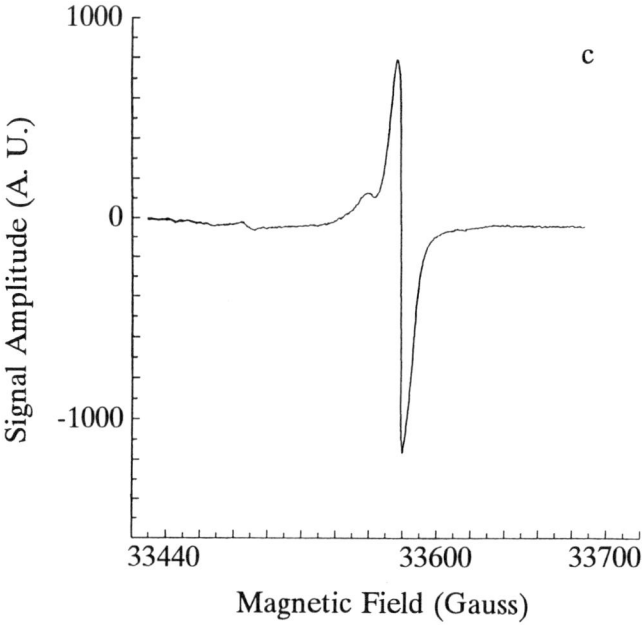

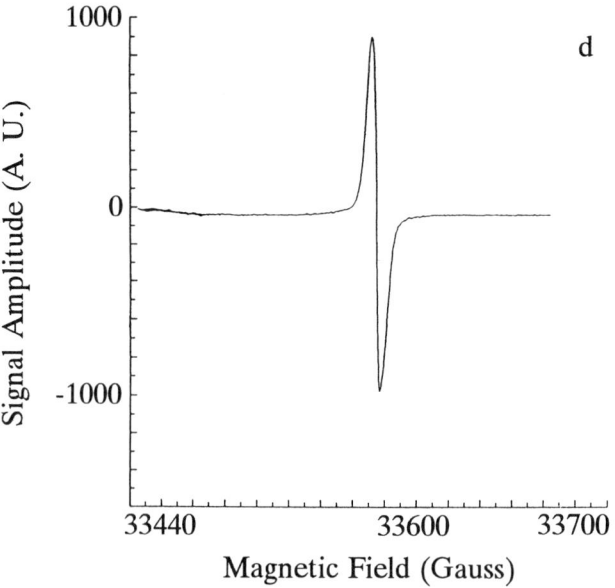

Figure 3. Continued. *Part c: APCS 3, Illinois No. 6 coal. Part d: APCS 4, Pittsburgh (No. 8), Pennsylvania, coal.*

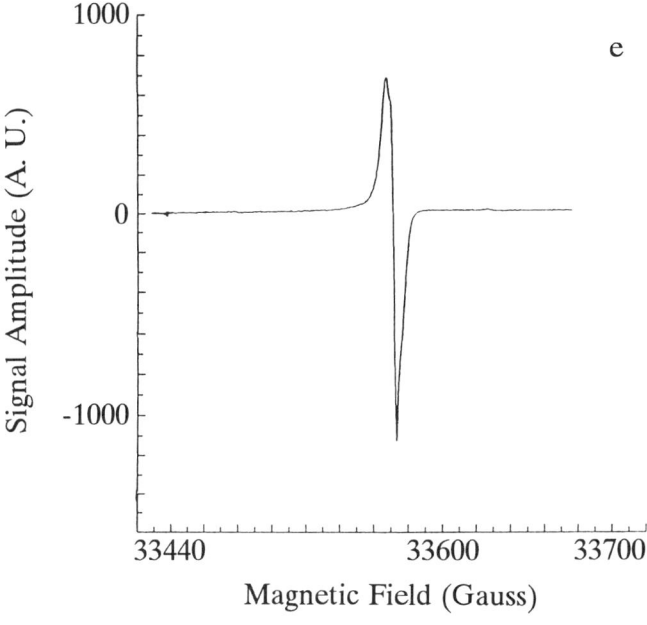

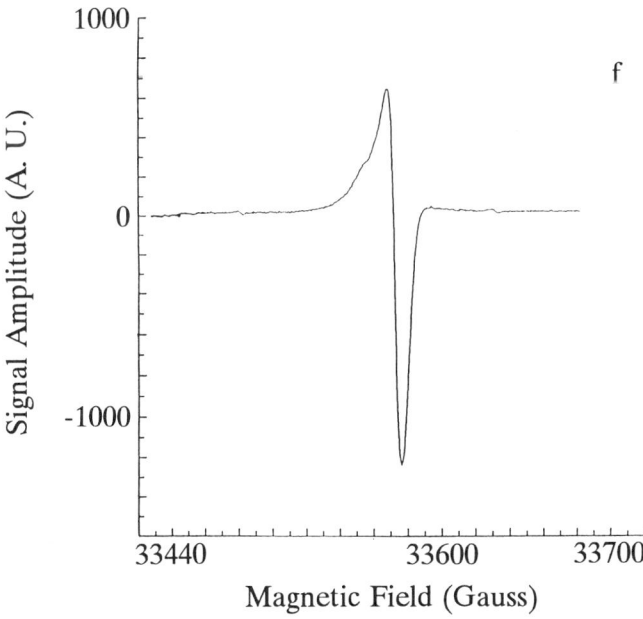

Figure 3. Continued. *Part e: APCS 5, Pocahontas No. 3, Virginia coal. Part f: APCS 6, Blind Canyon, Utah, coal.* Continued on next page.

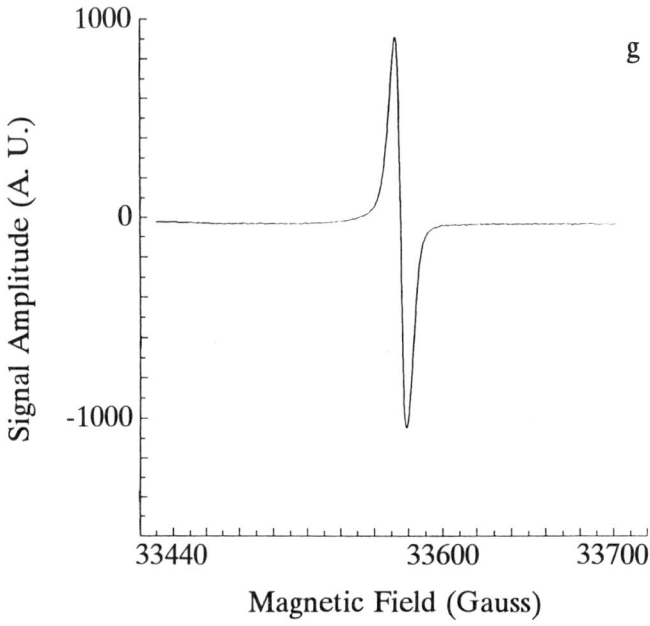

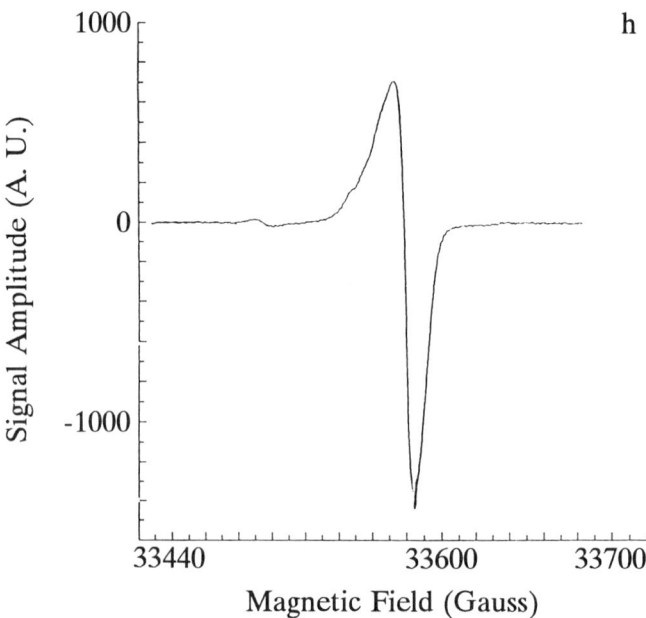

Figure 3. Continued. *Part g: APCS 7, Lewiston–Stockton, West Virginia, coal. Part h: APCS 8, Beulah–Zap, North Dakota, coal.*

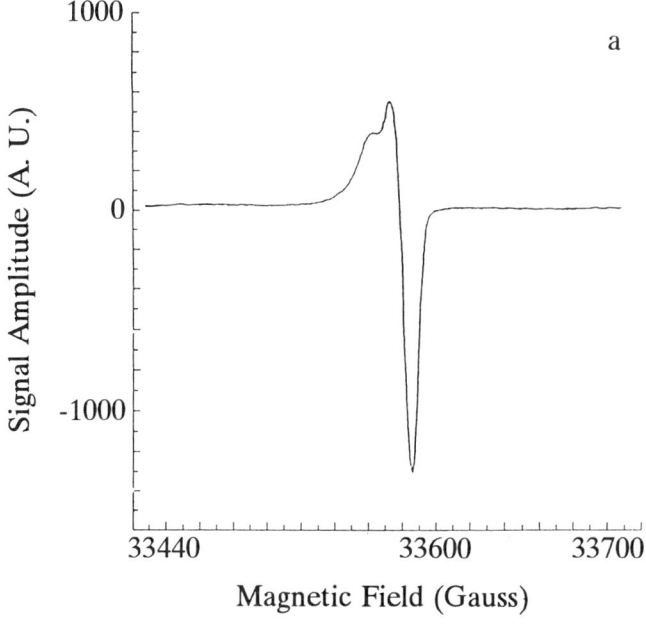

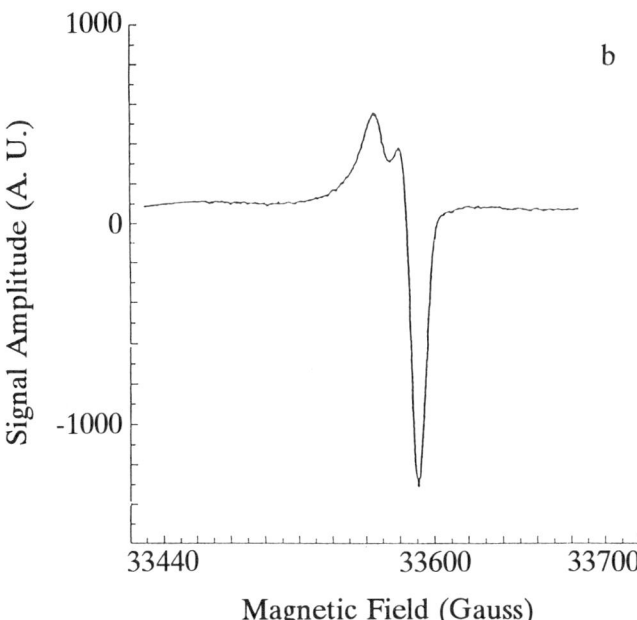

Figure 4. W-band (94-GHz) EPR spectra of powders of three macerals separated from an Illinois No. 6 coal (SIU No. 1822). Part a: vitrinite. Part b: sporinite. Continued on next page.

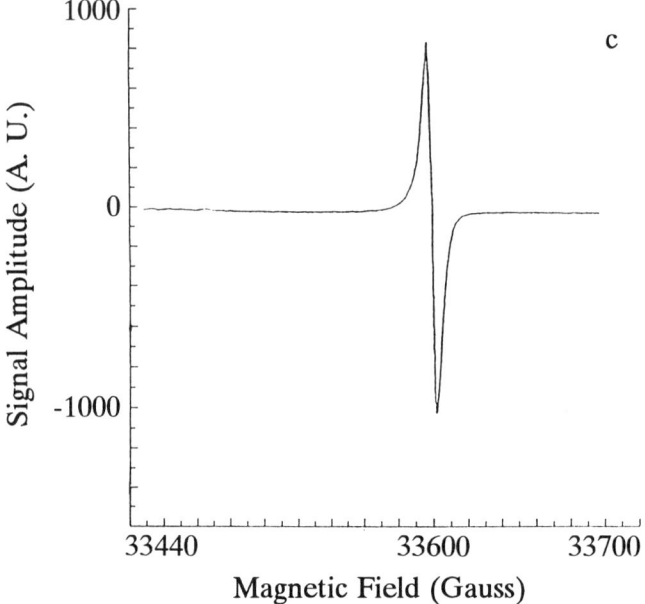

Figure 4. Continued. Part c: fusinite.

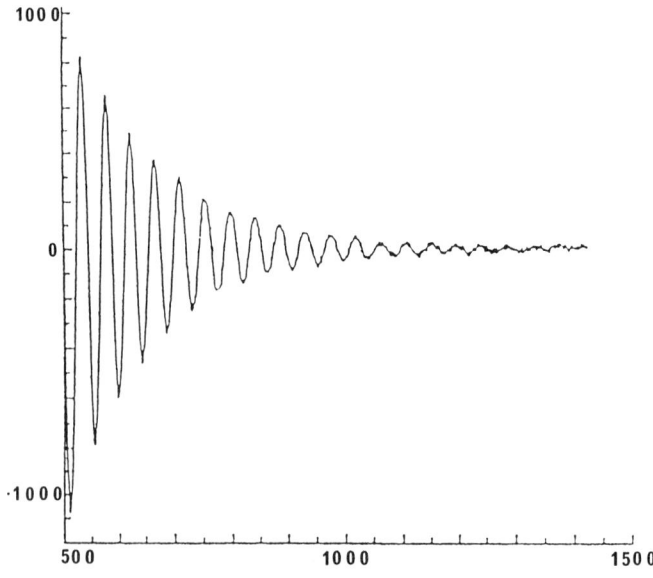

Figure 5. Free-induction decay (FID) after a single $\pi/2$ S-band microwave pulse applied to fusinite separated from Illinois (Herrin) No. 6 (SIU No. 1822). (Reproduced with permission from reference 10. Copyright 1990 Butterworth.)

lated with an anisotropic g-matrix and suitable line-width or line-shape parameters.

This simulation has been done, and the quite reasonable result is shown in Figure 6 for $g_1 = 2.0023$, $g_2 = 2.00274$, and $g_3 = 2.0042$. Other chemical and spectroscopic evidence strongly suggests that this EPR spectrum is the result of contributions from two or more radical species with different heteroatom compositions, so this simulation is useful for identifying the g-factors of the main features of the spectrum, but it should not be viewed as implying that the single-species model is correct. To confirm that the low-field peak in this spectrum is a part of an anisotropic line shape (as opposed to a partially resolved symmetric peak), a spectrum of the separated vitrinite was obtained at 250 GHz in the laboratory of Jack Freed at Cornell University. Figure 7 shows a comparison of the 96- and 250-GHz spectra and confirms that the low-field peak is indeed asymmetric and thus part of one or more overlapping anisotropic spectra.

Comparison of vitrinite and sporinite spectra (Figures 4a and 4b) illustrates the difficulties of the simple model because features with identical g-factors appear with different intensities in the two systems, a result suggesting the unequal contribution of two or more spectral components.

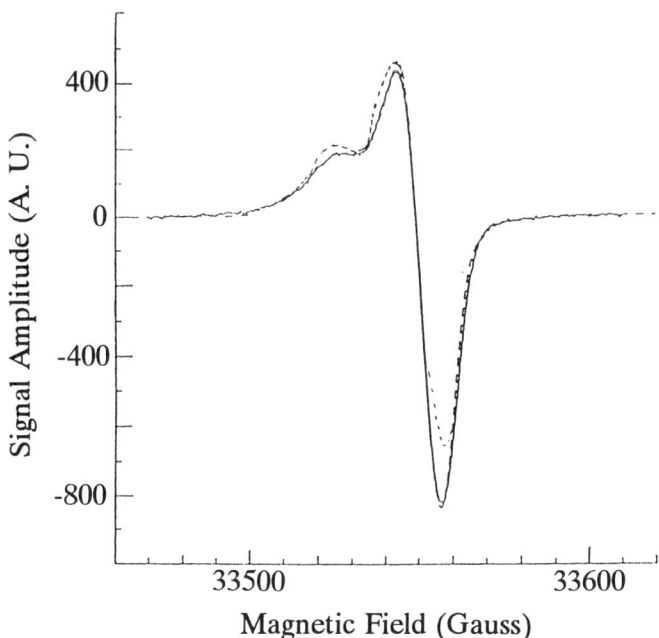

Figure 6. Experimental (—) compared with simulated (- - -) W-band (94-GHz) EPR spectrum of separated vitrinite maceral. (Reproduced with permission from reference 10. Copyright 1990 Butterworth.)

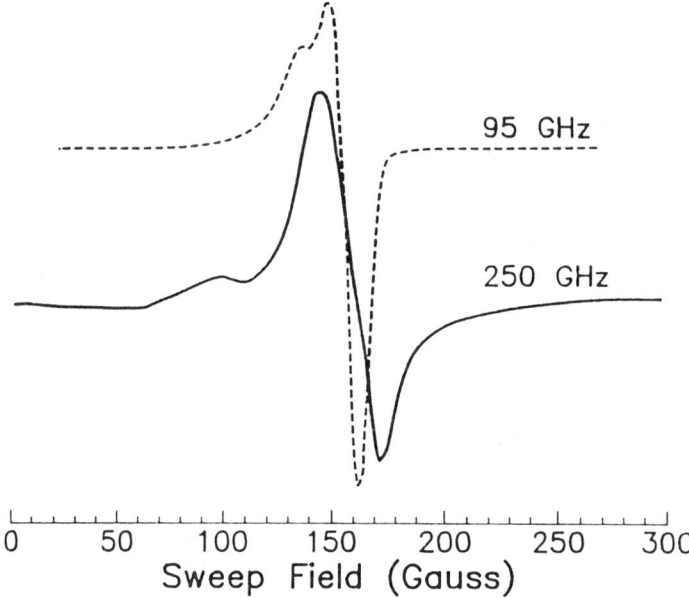

Figure 7. EPR spectra at two high microwave frequencies of a vitrinite maceral separated from Illinois No. 6 coal sample. (Reproduced with permission from reference 10. Copyright 1990.)

The lower two g-factors (g_1 and g_2) of the vitrinite are, in fact, the same as those observed for many different conjugated aromatic radicals, and which were predicted theoretically by Stone (15). It is tempting to see the spectrum as a composite formed from pure hydrocarbon and heteroatomic radical contributions. In all probability, the low-field shoulder in this spectrum is associated with heteroatomic radicals because spin-orbit (SO) coupling to sulfur (or oxygen) is the most likely mechanism to account for the higher value of g_3. The rather poor low-field fit of the single-species simulation then could be a reflection of this composite character. Clearly, this issue of spectral and sample heterogeneity is of utmost importance for the correct interpretation of EPR spectra of fossil fuels, even when the fuels are well-separated and purified, and we presently are working on several approaches designed to shed more light on the problem.

Deviations of g-factors (g-shifts) from the free-electron value (2.0023) and direction-dependence in g-factors (g-anisotropies) result from admixture of the orbital angular momentum (and therefore the magnetic moment) into the spin angular momentum to augment (or diminish) the spin-only Zeeman splitting.

An unpaired electron in an orbital that includes some *p*, *d*, or *f* character on atom A is subjected to a SO coupling interaction characteristic of that particular atomic orbital, and that interaction mixes excited-state spin orbitals into the ground-state spin orbitals Φ_α and Φ_β. As a rule, organic radicals that contain sulfur atoms display particularly strong g-shifts and g-anisotropies for the following reasons:

- Generally, $2p$ oxygen-atom orbitals are much more effective at SO coupling than carbon-atom orbitals, and $3p$ (or $3p-3d$ hybrid) sulfur-atom orbitals are a great deal more effective than oxygen. Moreover, sulfur heteroatoms in an aromatic π-system radical tend to trap the electron spin density more than oxygen heteroatoms, thus enhancing the intrinsic relative effectiveness of sulfur atoms in causing g-shifts.

- The g-anisotropies arise because the excited-state orbitals that are mixed into the ground-state semi-occupied molecular orbitals have directional properties.

- The g-shifts may be positive or negative depending on the type of excited state that SO coupling mixes into the ground state; $\chi^2\Phi^1-\chi^1\Phi^2$ and $\chi^0\Phi^1-\chi^1\Phi^0$ transitions cause g-shifts of opposite sign.

One promising avenue of investigation involves comparing W-band EPR spectra from coal with spectra obtained from model compounds believed to typify organic structures in coal. Figure 8a shows a spectrum from perylene cation radicals together with a preliminary theoretical simulation. The simulation used $g_1 = 2.0024$, $g_2 = 2.0030$, and $g_3 = 2.0032$; proton hyperfine values that were obtained experimentally by ENDOR spectroscopy (13); and spin-packet line widths that were determined by electron spin-echo measurements. Figure 8b shows the perylene radical cation spectrum superimposed on a vitrinite W-band spectrum to illustrate how a portion of the coal resonance line shape might originate from conjugated aromatic hydrocarbon radicals containing no heteroatoms.

Figure 9a shows a W-band spectrum of dibenzothiophene cation radicals together with a preliminary theoretical simulation ($g_1 = 2.0016$, $g_2 = 2.0054$, and $g_3 = 2.0106$). The effect of SO coupling with sulfur is seen in the higher g-factors and much more asymmetric line shapes. Attar and Dupuis (16) reported that thiophenes are the most abundant form of organic sulfur in Illinois coals (thiophenic sulfur, 58%; Ar–S–Ar, 20%; R–S–R, 18%; Ar–SH, 15%; R–SH, 7%); thus this model compound should give an indication of the sulfur-related effects that can be expected.

Because conjugated aromatic compounds are the predominant structural types for organic sulfur in the Illinois coal and because unpaired electrons are expected to be most abundant (and stable) in such chemical environments, it seems likely that EPR spectra of coal will strongly reflect the delocalization of electrons. This condition implies that a very small number of sulfur (or oxygen) atoms can exert a large effect on spectral

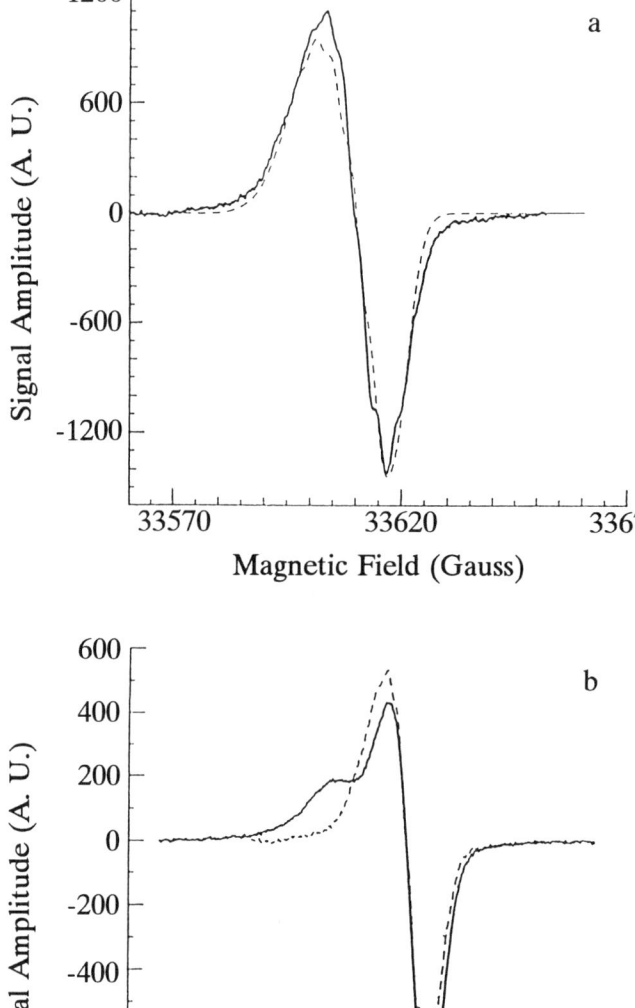

Figure 8. Part a: W-band (94-GHz) EPR spectrum (—) of perylene radical cation compared with simulation (- - -) using anisotropic g_e matrix and proton hyperfine matrices. Part b: W-band (94-GHz) EPR spectrum of perylene radical cation (- - -) compared with that of vitrinite (—). (Reproduced with permission from reference 10. Copyright 1990.)

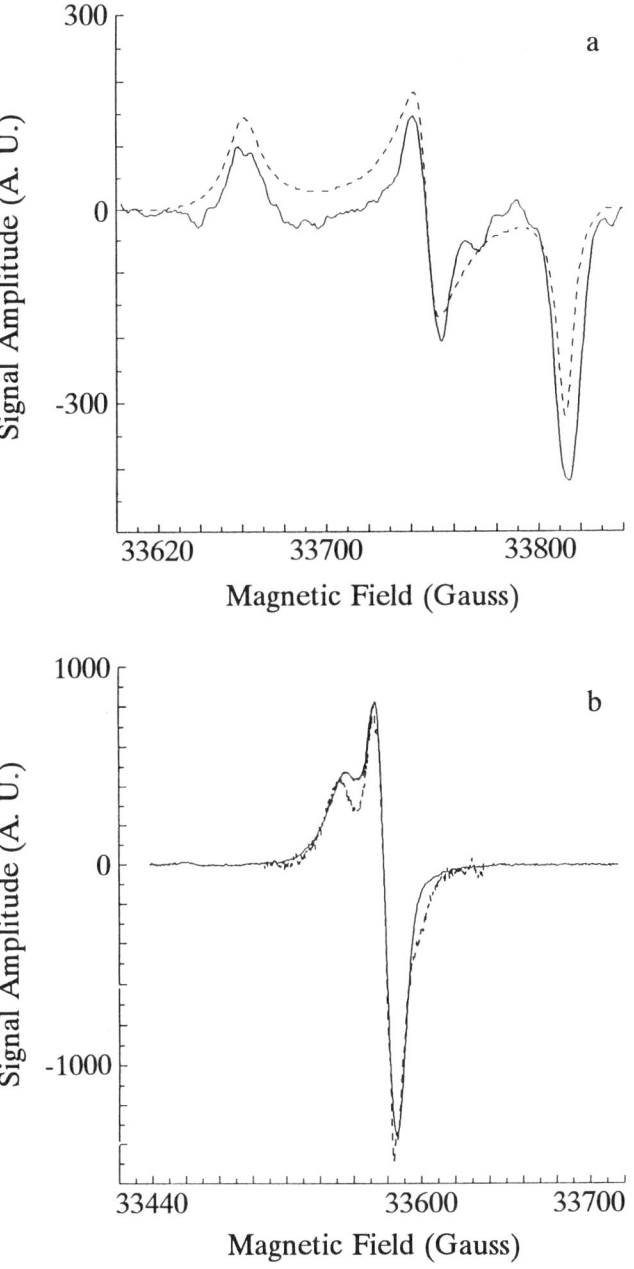

Figure 9. Part a: W-band (94-GHz) EPR spectrum (—) of dibenzothiophene radical cation compared with simulation (- - -) using anisotropic g_e matrix. Part b: W-band (94-GHz) EPR spectrum (—) of vitrinite compared with composite spectrum of perylene and dibenzothiophene radical cations (- - -). (Reproduced with permission from reference 10. Copyright 1990.)

line shapes, a phenomenon making this technique uniquely sensitive to heteroatoms with larger SO coupling constants (λ). Because the effect of a sulfur atom on g-factors decreases as an electron is delocalized over a larger number of carbon atoms (and the S–C ratio falls), the precise line shapes from sulfur in coal will depend on the size of the aromatic structures as well as on the type of bonding. This fact represents another opportunity for the nondestructive analysis of high-sulfur coals by W-band EPR spectroscopy, because it may be possible to model the spectra in order to get more detailed chemical information about the forms of organic sulfur that are present. To illustrate this concept, Figure 9b shows a W-band spectrum of vitrinite. Superimposed on it is a composite spectrum constructed by adding the spectrum from perylene radical cation and a portion of that from dibenzothiophene radical cation.

Although the agreement between the data and the construct is not perfect, a comparison does suggest that molecular forms like those of perylene and dibenzothiophene are contributing to the experimental spectrum, even though it is not likely that either of these species, as isolated structures, is a major component of coal. In the future, we hope to use spectral-addition methods in a more quantitative way to gain a better understanding of the molecular structure of the organic components of coal, including organic sulfur.

Low-Frequency Pulsed EPR Spectroscopy. Pulsed S-band (2–4 GHz) EPR spectroscopy was performed on all eight Argonne Premium coals. By means of a two-pulse ($\theta/2 - \theta$) sequence, the phase-memory time T_M of the spin echo from each sample was studied at 100 K as a function of magnetic field (B_o), pulse duration, and turning angle (θ). Figure 10 shows the two-pulse echo envelope from the Upper Freeport coal, together with a best fit assuming a single, exponential decay of the form

$$I(2\tau) = I_o \, e^{-2\tau/T} M \qquad (4)$$

Generally, there is in these data considerable nonexponential character to the first part of the decay curves, a factor that reduces the reliability of exponential fits employing a single time constant T_M. When the microwave pulse power is reduced so that the length of a π pulse is about 170 ns, the ESEEM effect disappears in two-pulse decay curves, and in most samples an almost perfect exponential decay is obtained with a time constant that is somewhat longer than the T_M observed for shorter, more powerful pulses (when strong ESEEM is observed). Also, when θ is reduced to values well below π, the apparent T_M values in many samples

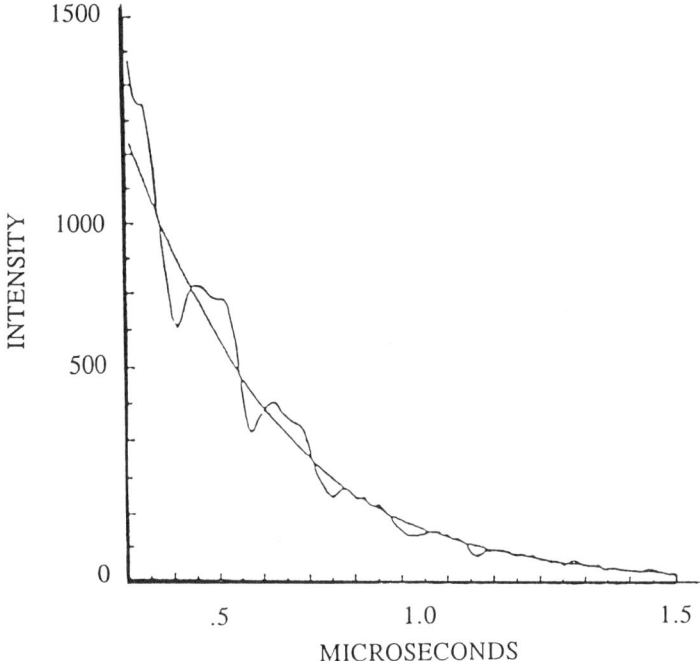

Figure 10. Two-pulse S-band electron spin-echo envelope (with ESEEM) of the Upper Freeport coal sample (APCS 1), together with a best-fit single exponential curve. $T_M = 626$ ns, $B_o = 1436$ G, and $\theta = \pi$, 40 ns.

(particularly in APCS 1, 3, 4, and 5) increase. This latter effect is much more pronounced when short, strong pulses are used.

Thomann et al. (17) have published pulsed EPR data taken at X-band (9.5 GHz) on isolated vitrinite macerals from a variety of Pennsylvania State University (PSOC) coals. They point out that T_M in coal can be a function of several contributing mechanisms, including spin–lattice relaxation, hyperfine interactions, and local spin dipole effects (e.g., instantaneous diffusion). At 9.5 GHz and with the duration of a pulse set at about 40 ns (and then reducing θ in successive experiments), they found that instantaneous diffusion makes a very significant contribution to T_M. Our data also demonstrate a θ dependence to T_M, a result indicating the presence of instantaneous diffusion as a contributing mechanism at lower frequencies in Argonne Premium coals. This effect is most pronounced in the highest rank Argonne Premium coals, in which higher spin densities are found and in which local spin dipole effects should be strongest.

Performing pulsed EPR measurements at 3 GHz instead of 9.5 GHz increases the importance of hyperfine interactions relative to electronic

Zeeman terms. This phenomenon was clearly seen in Figure 2, where a comparison was made between ESEEM patterns obtained at 3 and 9.5 GHz. The enhanced probability of the "forbidden" spin transitions at 3 GHz causes hyperfine interactions to play a much more significant role as a contributing mechanism for phase-coherence decay at this lower field. This enhanced probability is seen in the nonexponential character of initial echo-decay curves when short, hard microwave pulses are used to excite a wide band of branching transitions. Increasing pulse lengths (and reducing microwave power) narrows the bandwidth of the experiment until no branching transitions are excited and ESEEM disappears. Then the echo-decay curves are, in most cases, well fitted by single exponential functions. Thus, it seems likely that low-frequency (generally, multifrequency) ESE experiments will prove very useful for characterizing weak average hyperfine and local spin dipole fields in whole coals.

One should keep the previously mentioned concerns about the nonexponential character of echo-decay curves firmly in mind when examining Table II, which reports T_M values for six of the eight Argonne Premium coal samples, measured at S-band (4 GHz, in this instance). Significant magnetic-field (B_o) dependence was observed in many samples, so

Table II. Electron Spin-Echo Decay Times, T_M, for Argonne Premium Coals

Argonne Sample	Field (G)	T_M (ns)
1	1436	626
1	1449	821
1	1434	681
3	1434	534
3	1440	563
4	1440	525
4	1450	665
5	1442	493
5	1452	394
6	1432	378
6	1437	402
6	1442	379
6	1447	352
8	1398	341
8	1416	297
8	1436	281

data at several field points are given for each sample. The fact that T_M varies with B_o, and not in the same way for all samples, may arise from structure-related phenomena, such as field selection of subspecies in these heterogeneous whole coals. We observe a less consistent correlation of phase-memory time with rank of coal than the X-band results of Thomann et al. (*17*) revealed. However, their work was performed on carefully separated vitrinite macerals (in air), and the measurements reported here pertain to evacuated samples of Argonne Premium whole coals. The same general range of T_M values was observed in both studies. Clearly, considerably more information on separated macerals (from these coals) and model compound data would be useful.

Acknowledgments

Partial support for this work was provided by the U.S. Department of Energy (University Coal Research Program, PETC), the Illinois Department of Energy and Natural Resources (through the Center for Research on Sulfur in Coal), the Donors of the Petroleum Research Fund administered by the American Chemical Society, and the National Institutes of Health (RR01811). We gratefully thank Edwin Hippo and John Crelling (Southern Illinois University, Carbondale) for the separated macerals used in this study, Jack Freed and his students (Cornell University) for the 250-GHz EPR spectrum of a separated vitrinite, Mark J. Nilges for his development work on the 96-GHz spectrometer, and Kathleen M. E. J. Motsegood for extensive work in data presentation.

References

1. Schlick, S.; Kevan, L. In *Magnetic Resonance. Introduction, Advanced Topics, and Applications to Fossil Energy;* Petrakis, L.; Fraissard, J. P., Eds.; D. Reidel: Dordrecht, Netherlands, 1984; p 655.
2. Malhotra, V. M.; Buckmaster, H. A. *Org. Geochem.* **1985,** *8,* 235.
3. Belford, R. L.; Clarkson, R. B.; Cornelius, J. B.; Rothenberger, K. S.; Nilges, M. J.; Timken, M. D. In *Electron Magnetic Resonance of the Solid State;* Weil, J. A., Ed.; Chemical Institute of Canada: Ottawa, Ontario, Canada, 1987; p 21.
4. Clarkson, R. B.; Wang, W.; Nilges, M. J.; Belford, R. L. In *Processing and Utilization of High-Sulfur Coals III;* Markuszewski, R.; Wheelock, T. D., Eds.; Elsevier: New York, 1990; pp 67–79.
5. Hurst, G. C.; Henderson, T. A.; and Kreilick, R. W. *J. Am. Chem. Soc.* **1985,** *107,* 7294.
6. Clarkson, R. B.; Timken, M. D.; Brown, D. R.; Crookham, H. C.; Belford, R. L. *Chem. Phys. Lett.* **1989,** *163,* 277.

7. Flanagan, H. L.; Singel, D. J. *J. Chem. Phys.* **1987**, *87*, 5606.
8. Cornelius, J. B.; McCracken, J.; Clarkson, R. B.; Belford, R. L.; Peisach, J. *J. Phys. Chem.* **1990**, *94*, 6977.
9. Lai, A.; Flanagan, H. L.; Singel, D. J. *J. Chem. Phys.* **1988**, *89*, 7161.
10. Clarkson, R. B.; Wang, W.; Brown, D.R.; Crookham, H. C.; Belford, R. L. *Fuel* **1990**, *69*, 1405.
11. Vorres, K. S. *Users Handbook for the Argonne Premium Coal Sample Program;* Argonne National Laboratory: Argonne, IL, 1989.
12. Lynch, W. B.; Earle, K. A.; Freed, J. H. *Rev. Sci. Instrum.* **1988**, *59*, 1345.
13. Clarkson, R. B.; Belford, R. L.; Rothenberger, K. S.; Crookham, H. C. *J. Catal.* **1987**, *106*, 500.
14. Silbernagel, B. G.; Gebhard, L. A.; Dyrkacz G. R. In *Magnetic Resonance. 14.22 Introduction. Advanced Topics and Applications to Fossil Energy*; Petrakis, L.; Fraissard, J. P., Eds.; D. Reidel: Dordrecht, Netherlands, 1984; p 645.
15. Stone, A. J. *Mol. Phys.* **1964**, *7*, 311.
16. Attar, A.; Dupuis, F. *Am. Chem. Soc. Div. Fuel Chem. Prepr.* **1979**, *24*, 166.
17. Thomann, H.; Silbernagel, B. G.; Jin, H.; Gebhard, L. A.; Tindall, P. *Energy Fuels* **1988**, *2*, 333.

RECEIVED for review June 8, 1990. ACCEPTED revised manuscript October 16, 1991.

28

Novel Characterization of Argonne Premium Coals with Electron Donors or Acceptors by Electron Paramagnetic Resonance Spectroscopy

T. Keneko, M. Sasaki, T. Yokono, and Yuzo Sanada[1]

Faculty of Engineering, Hokkaido University, Sapporo, 060, Japan

Noncovalent interaction in coal molecular structure has been studied with respect to radical concentration of charge-transfer complexes between coal and electron donor or acceptor molecules. The forms of the charge-transfer complexes of coal–iodine and coal–7,7,8,8-tetracyanoquinodimethane (TCNQ) systems are quite different. TCNQ interacts with oxygen-containing functional groups, and iodine associates with aromatic rings.

COALS ARE COVALENTLY CROSS-LINKED macromolecular networks. In addition to the covalent bonds, noncovalent interactions play important structural roles. The role of noncovalent-bond interactions in coal structure has been discussed by Larsen et al. (*1*). The noncovalent interactions serve as virtual cross-links that help to hold the network together. They also bind potentially soluble fragments to the network and thus inhibit dissolution of the fragments.

Several types of noncovalent interactions occur, and their relative population and importance changes with coal rank. Three different types of specific noncovalent interactions are important to coal structure. These

---

[1] Corresponding author

are, in order of decreasing magnitude, ionic bonds, hydrogen bonds, and aromatic–aromatic interactions.

The interactions of electron acceptors and donors (Lewis acids and bases) form 1:1 or n:1 molecular compounds ranging from loose complexes to stable compounds. The theory was proposed originally by Mulliken and Person (2) and involves resonance between nonbond structures (A–B) and dative structures (A$^+$–B$^-$), where A is an acceptor atom, molecule, or ion, and B is a donor atom, molecule, or ion. Extensive investigations have been done by Akamatsu et al. (3) on complexes between polycyclic aromatic hydrocarbons and halogens. They found that molecular complexes of polycyclic aromatic hydrocarbons with bromine or iodine behave as typical semiconductors (3). A chemical fractionation method for pitches, in terms of charge-transfer fractionation, was developed by Zander et al. (4–6). Aromatic molecules in pitch form charge-transfer complexes with electron acceptors such as picric acid, iodine, or hydrogen chloride. The complexes are sufficiently stable to allow isolation.

The amount of iodine taken up from aqueous solution by coal varies with rank as expressed by carbon content (dry, ash-free basis) (7). The fact that the amounts of iodine taken up exceed accepted surface-area values suggests that strong interactions, such as the formation of charge-transfer complexes, occur between iodine and coal molecules. We have established (8) a good correlation between the radical concentration of polynuclear aromatic hydrocarbon–iodine complexes and the ring-compactness factor. The larger the ring condensation of the aromatic hydrocarbon is, the higher becomes the spin-concentration value in the complex.

The purpose of this study is to clarify the noncovalent interaction in coal molecular structure with respect to radical concentrations of charge-transfer complexes between coal and election donor or acceptor molecules.

Experimental Details

For the most part, the coals tested were Argonne Premium Coal Samples kindly supplied by K. S. Vorres of the Argonne National Laboratory. The analytical data of the samples are shown elsewhere (9). Iodine and 7,7,8,8-tetracyanoquinodimethane (TCNQ) were selected as electron acceptors. These chemicals were supplied by Wako Chemical Co. (purity >95%) and used without further purification. Triethylphosphine oxide (TEPO), which was selected as an electron donor, was supplied by ICN Co. and was used without further purification.

A known amount of guest substances dissolved in chloroform or benzene was added to the coal. The suspensions were placed in an ultrasonic bath for 10 min and stirred for 1 h at 20 °C. The solvent was then evaporated away, and the coals were stored in a vacuum oven at 20 °C overnight.

The solvents used to dissolve and penetrate the guest molecules into coal particles were chloroform and benzene. These solvents are not absolutely neutral in terms of electron donor–acceptor interaction. Despite the remarkable differences in the extent of electron-donating and -accepting abilities of the solvents, the solvents had no effect on the spin concentration of the Argonne coal samples, and therefore, no special attention was paid to the solvent effects.

Samples were degassed at 1.33×10^{-2} Pa and sealed in electron paramagnetic resonance (EPR) glass tubes. EPR spectra were obtained with a Varian model E109 spectrometer. Measurements of the electron-spin concentration of the reference standard, 1,1-diphenyl-2-picrylhydrazyl (DPPH), were found to be reproducible to within 2%.

Results and Discussion

EPR Spectral Shape. Figure 1 illustrates spectra obtained for Beulah–Zap lignite and Blind Canyon high-volatile bituminous coals and their complexes. The spectra of the Beulah–Zap lignite complexes with iodine or TEPO consist of one broad line. The spectrum of the complex between Beulah–Zap lignite and TCNQ shows two distinct lines, one narrow and the other broad. The spectra recorded for Blind Canyon and its complexes each consist of one broad line. The resolved spectra that result from interactions between guest electron-acceptor or guest donor molecules and special sites on aromatic host molecules, as illustrated in Figure 1, give some insight into the molecular structure of coal.

Retcofsky et al. (*10*) have studied the signals from two fundamental macerals, vitrinite and fusinite, separated from various coals. The EPR line obtained for vitrinite was broader than that for fusinite. On the other hand, the structural model of bituminous coal recently proposed by Kovac and Larsen (*11, 12*) is very well adapted to the interpretation of the EPR signals. Their model involves a small molecular (M) phase and a macromolecular (MM) phase. The MM phase forms a three-dimensional skeleton that consists of macromolecular fragments connected by cross-bonds. The M phase is distributed in the pores or on the edges of the MM phase. Analysis of the relaxation times indicated that the paramagnetic centers related to the narrow line are attached to the latticed macromolecular phase MM, and the paramagnetic centers related to the broad line are distributed throughout the molecular phase M (*13*).

The peak-to-peak line width, ΔH_{pp}, of coal increases with addition of electron-acceptor molecules. A broad line is due to an increase in the M

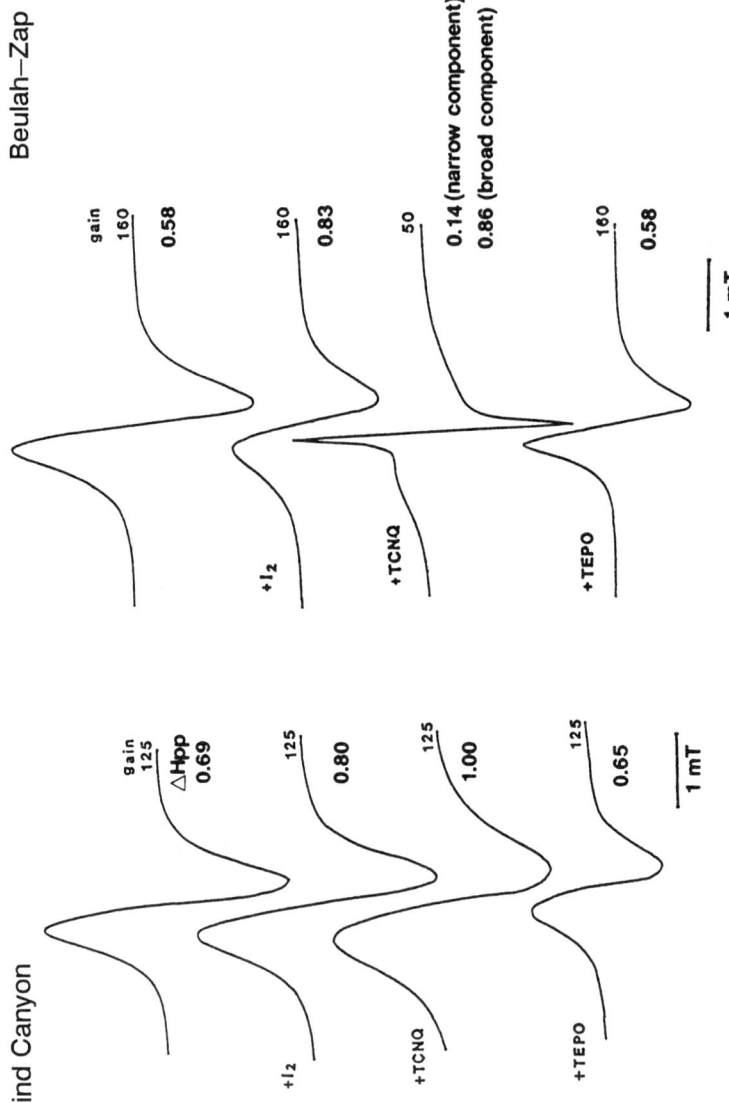

Figure 1. EPR spectra of Blind Canyon and Beulah–Zap coals with and without guest molecules.

phase in coal. The electron-acceptor molecule, such as iodine or TCNQ, attacks noncovalent bonds in coal. The intermolecular association of noncovalent bonds is broken by electron-acceptor molecules. The interaction between the coal molecule and the electron-acceptor molecule probably results in the increase in the M phase of coal. The appearance of a narrow component for the Beulah–Zap–TCNQ system might be due to different interactions such as a single electron transfer from coal to the TCNQ molecule. This question must be clarified with further measurements of g values and relaxation times.

Spin Concentration in Coal and Coal Complexes.

In Figure 2a, the absolute values of spin concentration, I_0, for the coal and for the coal complexes, I, as a function of carbon content (rank) of coal are given. Good correlations were obtained between I and rank. Although the values of I_0 of pristine coals were kept almost constant over the whole range of rank, a remarkable increase in the I values was recognized for high-rank coals with the addition of iodine as a dopant. On the other hand, the I values of the TEPO–coal-complex series were slightly decreased, as was expected. The values of I for the TCNQ–coal-complex series also increased with decreasing rank, as shown in Figure 2b. TCNQ as a dopant plays a particularly important role for low-rank coals.

The results shown in Figure 2 can be explained by taking into account the fact that aromatic molecules are electron donors and aromaticity and oxygen-containing functional groups change with coal rank. Figure 3 shows the relationship between the spin-concentration difference before and after the addition of guests, $I - I_0$, and coal rank. For iodine as a guest, the difference in spin concentration is mostly positive. On the other hand, the spin concentration of coal decreases with the addition of TEPO because the TEPO molecule has typical electron-donor characteristics.

The relationship between the spin-concentration difference before and after the addition of iodine, $I - I_0$, and H/C (atomic ratio) of coal can be depicted as in Figure 4. The value of $I - I_0$ is closely related to the H/C value of the parent coal. For TCNQ as a guest, no relationship was found between the value of $I - I_0$ and the H/C value of the parent coal.

Zander (6) reported that the equilibrium [aromatic donor + acceptor ⇌ charge-transfer complex] is governed by the first ionization potential, I_p, of the aromatic donor. The free energy of formation of the complexes as a measure of their thermodynamic stability (14) increases with decreasing ionization potential of the aromatic donors. The electron acceptor in this case was tetracyanoethylene. If the value of $I - I_0$ is a function of the thermodynamic stability of the complexes with iodine or TEPO (7), then

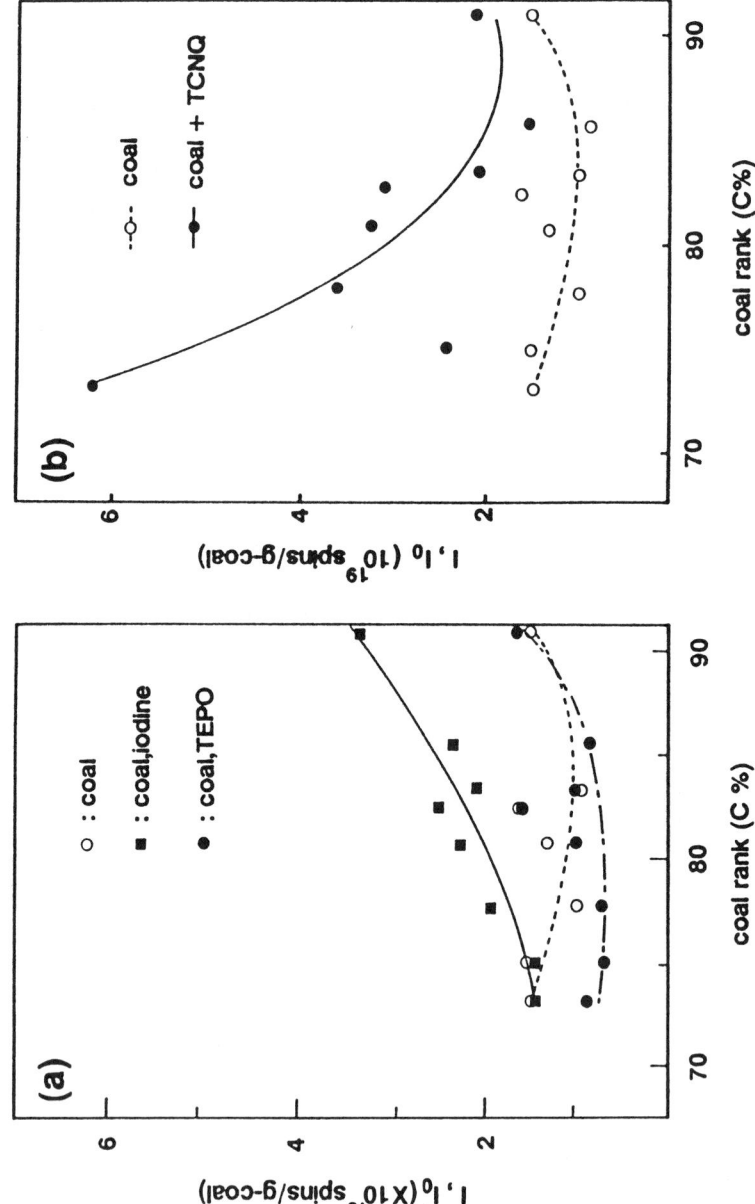

Figure 2. Spin concentration of original coal and coal–guest complexes vs. coal rank.

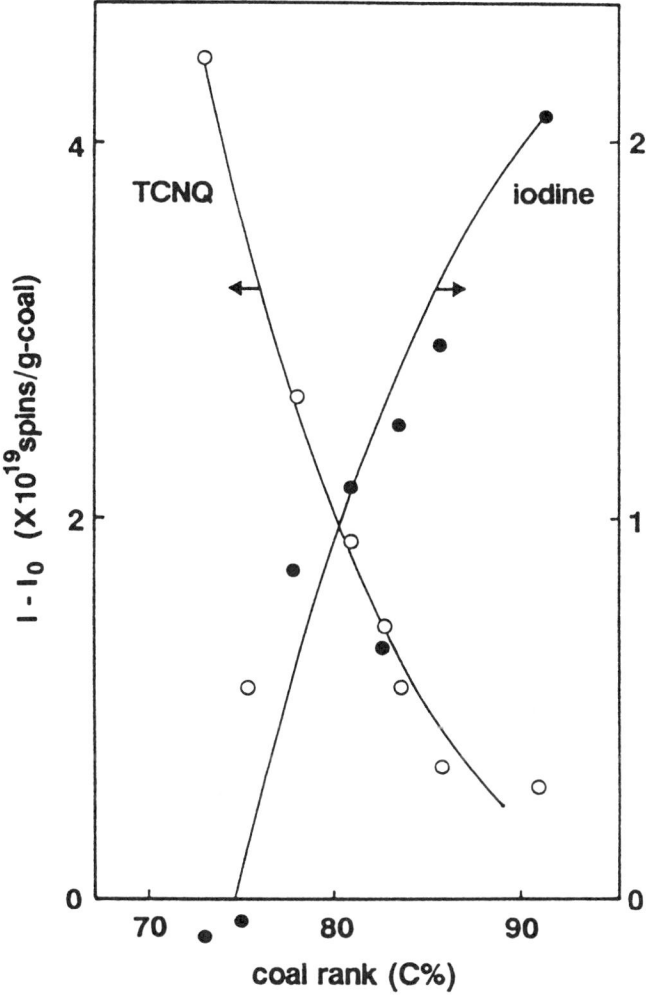

Figure 3. A plot of the difference in spin concentration of coal with and without guest molecules vs. rank.

the higher the value of $I - I_o$, the greater the magnitude of the aromatic–aromatic interaction in the coal molecular structure.

In TCNQ, the value of $I - I_o$ decreases monotonically with the increase in rank, although TCNQ is a molecule with behavior like an electron acceptor, such as iodine. As shown in Figure 5, an approximately linear relationship exists between the value of $I - I_o$ and the O/C atomic ratio of coal. Our interpretation is that the TCNQ molecule goes to the

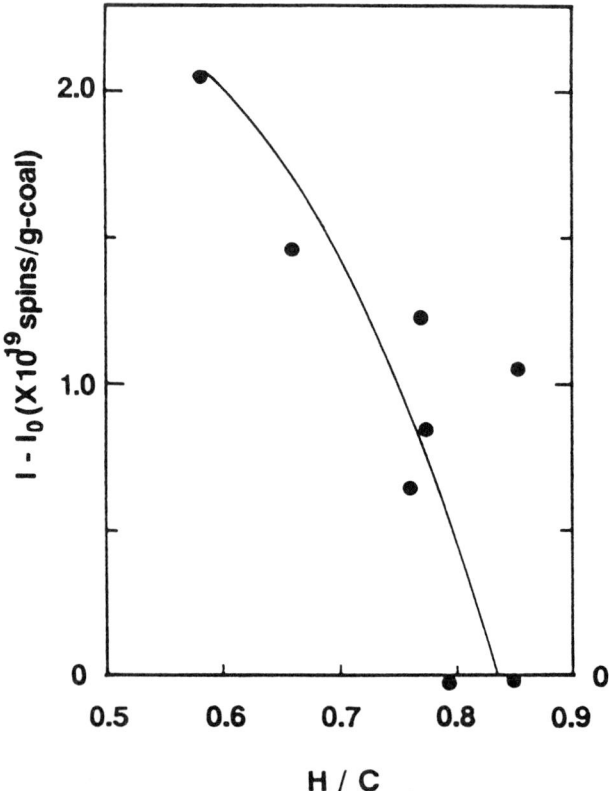

Figure 4. Relationship between the value of $I - I_o$ and the H/C ratio for the coal–iodine system.

sites associated with oxygen-containing functional groups that are able to form hydrogen bonds and interacts with these particular sites. The result of this interaction is the change in spin concentration of coal molecules. The different mechanisms of interaction between coal and TCNQ or iodine may be supported by the fact that the Beulah–Zap lignite–TCNQ system shows a spectrum consisting of two distinct lines, but the iodine system has a single broad line.

The importance of hydrogen bonding in coal structure is one of the first insights to result from the consideration of the macrostructure of coals (15). However, almost nothing is known about the intrinsic nature of hydrogen bonds in coal structure. When the mechanisms of interaction between coal molecules and TCNQ are elucidated, the strength and population of the hydrogen bonding will be solved. By using a Fourier trans-

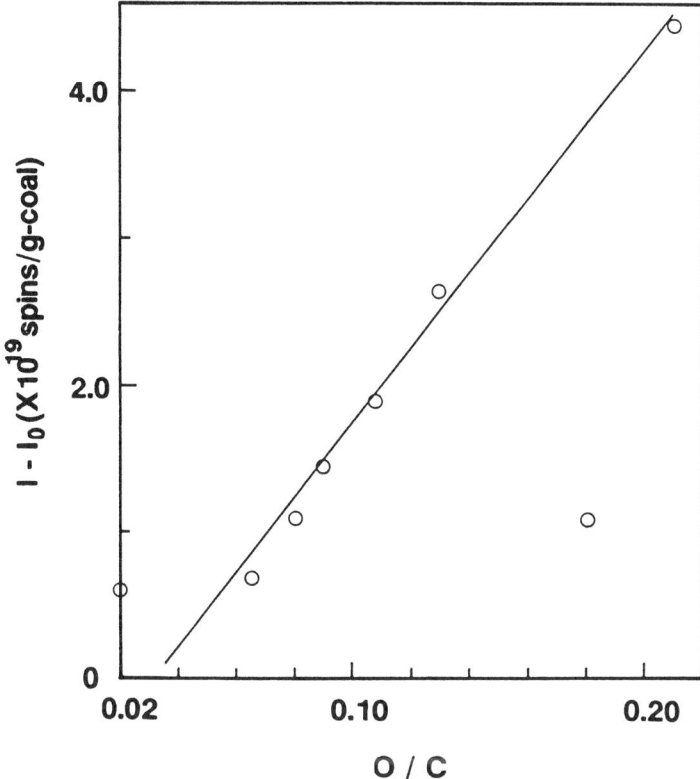

Figure 5. Relationship between the value of $I - I_o$ and the O/C ratio for the coal–TCNQ system.

form infrared (FTIR) technique to observe the shift of the vibration frequencies, we hope to elucidate the energetics of interactions between several polynuclear aromatic hydrocarbons and guests.

Conclusions

The forms of charge-transfer complexes of coal–TCNQ and coal–iodine are quite different. TCNQ interacts with oxygen-containing functional groups, and iodine interacts with aromatic rings. The study of guest and host interactions using EPR techniques will provide insight into the noncovalent interactions in coal molecular structure.

References

1. Larsen, J. W.; Nishioka, M.; Pum, C. S. *Proc. Int. Conf. Coal Sci.* **1989**, *1*, 9.
2. Mulliken, R. S.; Person, W. B. *Molecular Complexes;* John Wiley and Sons: New York, 1969; pp 1–8.
3. Akamatsu, H.; Inokuchi, H.; Matsunaga, Y. *Bull. Chem. Soc. Jpn.* **1956**, *29*, 213.
4. Zander, M. *Erdoel Kohle Erdgas Petrochem.* **1982**, *35*, 259.
5. Blumeer, G. P.; Kleffner, H. W.; Palm, J.; Zander, M. *Erdoel Kohle Erdgas Petrochem.* **1982**, *38*, 259.
6. Zander, M. *Fuel,* **1987**, *66*, 1459.
7. Rodriguez, N.; Yokono, T.; Takahashi, N.; Sanada, Y.; Marsh, H. *Fuel,* **1987**, *66*, 1743.
8. Yokono, T.; Takahashi, N.; Sanada, Y. *Energy Fuels* **1987**, *1*, 227.
9. Vorres, K. S. *User Handbook for the Argonne Premium Coal Sample Program,* Sept. 1, 1989; p 39.
10. Retcofsky, H. L.; Thompson, G. P.; Hough, M. R.; Friedel, R. A. *Organic Chemistry of Coal;* Larsen, J. W., Ed.; ACS Symposium Series 71; American Chemical Society: Washington, DC, 1978; p 142.
11. Kovac, J.; Larsen, J. W. *Am. Chem. Soc. Div. Fuel Chem.* **1977**, *22*, 181.
12. Larsen, J. W.; Kovac, J. *Organic Chemistry of Coal;* Larsen, J. W., Ed.; ACS Symposium Series 71; American Chemical Society: Washington, DC, 1978; pp 1, 142.
13. Duber, S.; Wieckowski, A. B. *Fuel,* **1982**, *61*, 431.
14. Merrifield, R. E.; Phillips, W. D. *J. Am. Chem. Soc.* **1958**, *80*, 2278.
15. Larsen, J. W.; Green, T. K.; Kovac, J. *J. Org. Chem.* **1985**, *50*, 4729.

RECEIVED for review June 7, 1990. ACCEPTED revised manuscript December 28, 1990.

29

Electron Paramagnetic Resonance and Electron Spin-Echo Spectroscopy of Argonne Premium Coals

Bernard G. Silbernagel, L. A. Gebhard, Marcelino Bernardo, and H. Thomann

Corporate Research, Exxon Research and Engineering Company, Route 22 East, Annandale, NJ 08801

> *The series of Argonne Premium coals was examined by using electron paramagnetic resonance (EPR) and electron spin-echo (ESE) techniques. EPR measurements indicate the presence of both carbon radical species and transition metal ions, principally iron and manganese, in the samples. The carbon radical densities and the ease of saturation of the radical signal do not increase with increasing coal rank, as was observed in previous studies of demineralized, isolated coal macerals. ESE techniques discriminate between the narrow (inertinitic) and broad (vitrinitic) components of the carbon radical signal, and the line widths and relaxation rates have been determined for both components. Instantaneous diffusion was observed for all bituminous coals but not for the lignite coal. The data suggest that transition metal species significantly affect the carbon radical resonance properties of these coals.*

CAREFUL SELECTION AND EXTENSIVE DOCUMENTATION (*1*) have made the Argonne Premium coals a benchmark for the future study of coals. This chapter presents a comprehensive survey of electron paramagnetic resonance (EPR) properties of the Argonne Premium coals. The results discussed here pertain to the "as received" coals, that is, before any other treatment. The properties of these materials are compared with previous results obtained on isolated coal macerals that were demineralized and separated by density-gradient centrifugation techniques (2). EPR studies of those isolated vitrinite macerals revealed a systematic variation of the resonance properties with increasing coal rank (3). Subsequent electron spin-echo (ESE) studies of the same samples demonstrated that these variations were associated with an increase in the carbon radical density and, for vitrinite macerals with carbon contents exceeding 82%, the onset of local order of the aromatic molecules serving as hosts for the carbon radicals (4). One goal of this study was to determine if a similar systematic behavior occurs for the Argonne Premium coals. A second goal was to identify the resonance properties of individual maceral types in these coal samples in which the macerals occur as a mixture.

Electron Magnetic Resonance Properties

Although EPR spectroscopy has a venerable history in coal science (5), ESE spectroscopy is a much more recent technique (6). This section describes the parameters of the paramagnetic species that can be determined by the two techniques and indicates their interrelationship.

EPR Spectroscopy. In EPR spectroscopy, the sample is exposed to a continuous, relatively weak microwave field, and the absorption of microwave energy is observed as the strength of an external magnetic field is varied. The resonance position, the magnetic field strength required to produce absorption of microwave energy at a given frequency, is usually defined in terms of the g-value: $g = h\nu_o/\mu_B H_o$, where h is Planck's constant, ν_o is the Larmor frequency, μ_B is the Bohr magneton, and H_o is the external magnetic field.

The width of the line, expressed in terms of the full width of the microwave absorption at half maximum, $\Delta H_{1/2}$, or the splitting between the maxima of the absorption derivative, ΔH_{pp}, and the shape of the line, which can be described in terms of the ratio, $\Delta H_{1/2}/\Delta H_{pp}$, are important parameters that are determined by the chemical and physical homogeneity, the physical state, and the molecular and spin dynamics. A general discussion of this topic is given in Chapter 30. In addition, the number of paramagnetic species in the sample can be determined by integrating the microwave absorption. The carbon radical density is defined as the number of radicals per gram of carbon in the sample. In most coals, these radical densities are $\sim 10^{18}$–10^{19} spins per gram of carbon. The response of the paramagnetic species to the applied microwave field can be determined by measuring the strength of the microwave absorption as a function of applied microwave field strength (7). At low microwave power, the absorption is proportional to the field strength, which by definition is proportional to the square root of the applied microwave power. At higher powers, this response becomes sublinear as the microwave field equilibrates populations in the spin states of the paramagnetic species, a phenomenon known as saturation.

One means of quantifying the saturation process is to plot the absorption intensity, I, divided by the square root of the power: $I/(P)^{1/2}$. In the linear region this ratio is constant, and it falls off at higher P values. The power needed to accomplish significant saturation is defined by the quantity $P_{1/2}$, the value of the microwave power for which $I/(P)^{1/2} = 0.5$.

ESE Spectroscopy. In ESE spectroscopy, the transient response to a short, intense pulse of microwave power is observed. This pulse prepares a predefined microscopic magnetization and imposes a phase coherence on the ensemble of carbon radical spins. As described in detail in the classic paper on the topic (8), the Fourier transform of the evolving magnetization after the application of a $\pi/2$ pulse (which tips the magnetization by 90° from the direction of the applied field) is equivalent to the microwave absorption obtained by EPR techniques. The return of the applied magnetization to equilibrium is called *electron spin–lattice relaxation* and occurs at a rate defined as $1/T_{1E}$. Loss of phase memory in the system occurs at a rate $1/T_M$, which is related to spin–spin interactions (described by a rate of $1/T_2$) and instantaneous-diffusion effects (9), the latter arising from changes in the local field resulting from tipping of the magnetization during the application of the microwave pulse. (T_M is the phase memory decay time, and T_2 is the spin–spin relaxation time.) As demonstrated in ref. 9, these processes can be described as

$$\frac{1}{T_M} = \frac{1}{T_2} + A <\sin^2 \frac{\theta}{2}> \qquad (1)$$

where θ is the angle through which the magnetization is tipped by the microwave pulse, and A is a measure of the local dipolar field strength. The ability to determine the relaxation parameters $1/T_M$, $1/T_2$, and $1/T_{1E}$ is a key strength of the ESE technique. These data are related to the EPR microwave saturation experiments because

$$\frac{I}{(P)^{1/2}} \propto \left[1 + \alpha P T_{1E} T_2\right]^{-1} \qquad (2)$$

Preparation and Properties

Sealed ampules of each of the Argonne Premium coals were kept in a nitrogen dry box. Portions of each sample were transferred to 4-mm quartz EPR tubes while in the box, and a pump-out valve was attached to the EPR tube before removing it from the dry box to ensure no contact with the atmosphere. The sample-filled EPR tubes were then transferred to a vacuum line and were subjected to a series of evacuation and backfilling steps before being sealed in a partial pressure of helium gas. The details of the evacuation process are important because the as-received coal samples have an appreciable amount of water in the coal pores. We have used proton NMR spectroscopy to estimate the amount of residual water in the pores after this evacuation process. In the most extreme case, the Beulah–Zap lignite, the water content was ~5 wt%. For most of the other samples, the water content is ≤ 1 wt%.

EPR Observations

Carbon Radical EPR Spectroscopy. The carbon radical signal for five of the eight samples consists of a superposition of a broad (~6–7 G) component and a narrow (<1 G) one, which have previously been associated with vitrinite and inertinite macerals, respectively (3, 4). An example, the Upper Freeport coal, is shown in Figure 1. In each case, the broader signal is the major component; the relative intensities of the two signals are shown in Table I. The g-values for the narrow components are ~2.0027, and the g-values for the broader components vary from 2.0031 to 2.0033 for the low-rank coals to 2.0028 for coals of higher rank. This behavior is consistent with the variations previously observed for isolated macerals (3). As seen in Table I, the radical densities vary widely from 10^{18} to 10^{19} in a manner that does not track with increasing coal rank and that is even more pronounced than in the isolated maceral studies (3). The microwave-saturation response for both the narrow and broad components is shown in Figures 2a and 2b. In these figures, $I/(P)^{1/2}$ is plotted as a function of the reduced parameter, $P/P_{1/2}$, so that the form of the variation can be compared for each of the samples. The saturation-response curves are significantly broader for the broad signal than for the narrow ones, a result suggesting a distribution of radical types in the vitrinitic macerals. In fact, the response for the narrow component falls quite close to the idealized Bloch model of eq 2 (7).

Two ways in which the present saturation data differ dramatically from previous observations on the isolated macerals are (1) there is little correlation between $P_{1/2}$ and coal rank for the broad component (in con-

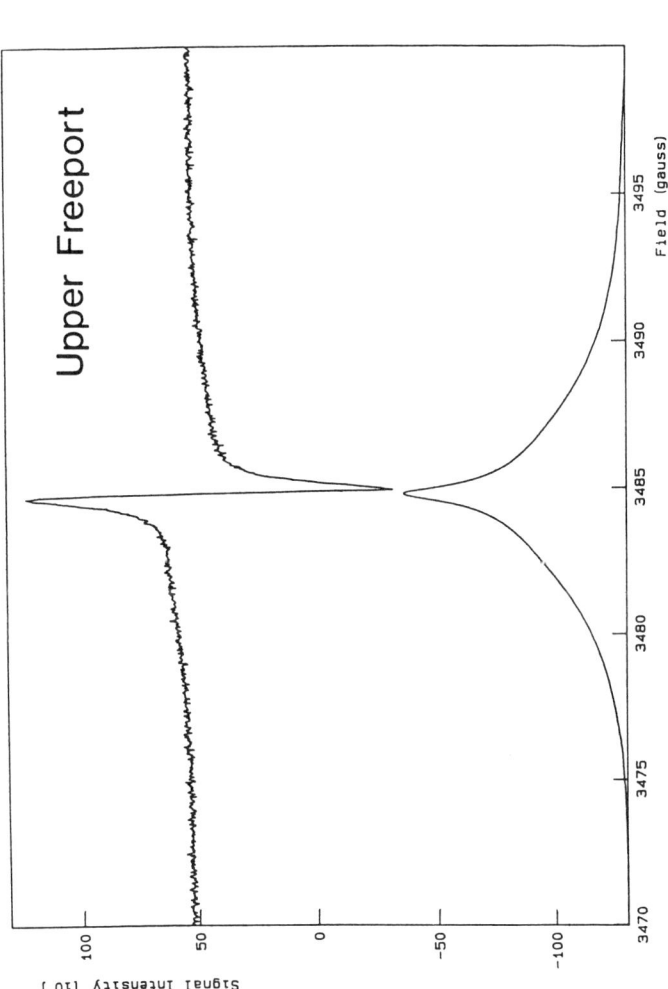

Figure 1. The two-component carbon radical signal observed in most Argonne Premium coals. Displayed on this trace are the first derivative of the microwave absorption, the absorption itself, and the integral of the absorption, which is proportional to the number of spins in the sample.

Table I. EPR Properties of Argonne Premium Coals

Coal	Rank	% C	Total Radical Density ($\times 10^{18}$/g C)	I_{inert}/I_{vit}	$P_{1/2}$ (mW) Vitrinite	$P_{1/2}$ (mW) Inertinite
Pocahontas No. 3	LVB	91.0	2.31	0.15	3.0	19
Upper Freeport	MVB	85.5	8.62	0.10	4.0	21
Pittsburgh No. 8	HVB	83.2	6.35	0.05	7.2	71
Lewiston–Stockton	HVB	82.6	10.61	—	6.5	48
Blind Canyon	HVB	80.7	7.02	—	2.0	—
Illinois No. 6	HVB	77.7	2.44	0.05	8.1	44
Wyodak–Anderson	SB	75.0	10.16	—	10.0	—
Beulah–Zap	L	72.9	5.29	—	19.0	—

ABBREVIATIONS: LVB, low-volatile bituminous; MVB, medium-volatile bituminous; HVB, high-volatile bituminous; SB, subbituminous; L, lignite; inert, inertinite; and vit, vitrinite.

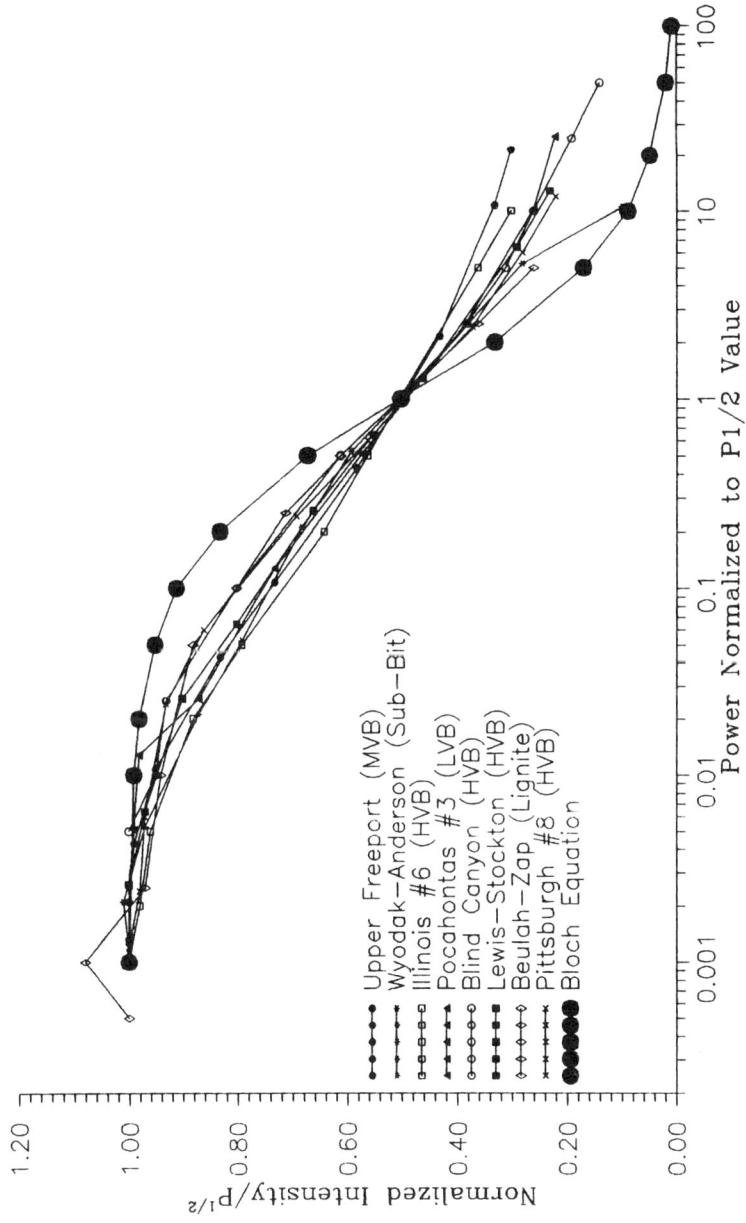

Figure 2a. *Saturation response for the broad component in each of the Argonne Premium coals.*

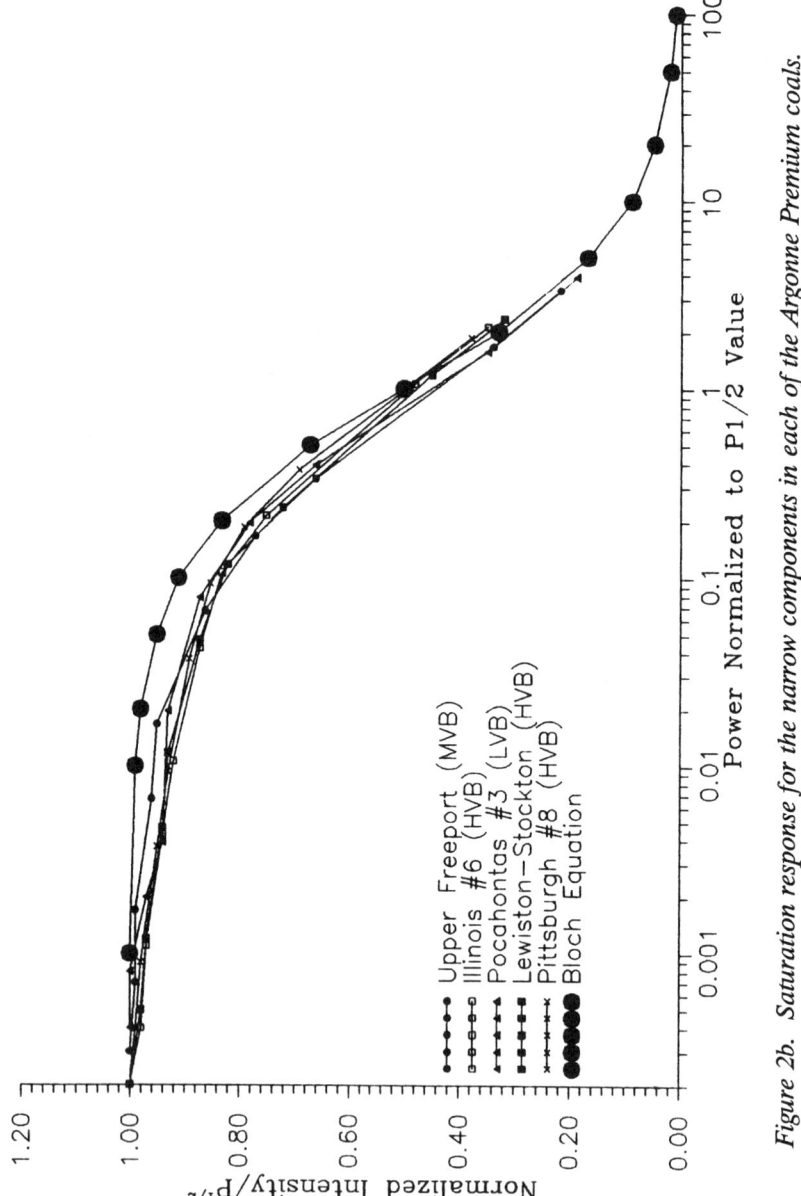

Figure 2b. Saturation response for the narrow components in each of the Argonne Premium coals.

trast to the isolated vitrinite case where $P_{1/2}$ increased nearly monotonically with coal rank), and (2) the magnitudes of $P_{1/2}$ differ substantially from the isolated maceral cases. For isolated vitrinite macerals, practically all subbituminous coal samples had $P_{1/2} \sim 1$ mW, with increases to levels of 50–70 mW for the highest rank macerals (low-volatile bituminous) in the survey. Here the lignite and subbituminous samples have $P_{1/2} \sim$ 10–20 mW, a result suggesting the presence of additional relaxation mechanisms. The situation is more perplexing for the high-rank coals, especially the low-volatile bituminous sample (Pocahontas No. 3), for which $P_{1/2} = 3$ mW. This anomalously low value, coupled with the low radical density of the sample, suggests that not all of the carbon radicals are observed in the ESR spectrum.

Transition Metal EPR Spectroscopy. In addition to the carbon radical signals, prominent signals from transition metal species were observed. Examples for three of the coals are shown in Figures 3a–3c. Although the signals are all observable at room temperature, the data in Figure 3 were taken at 4.8 K, where relaxation effects are less and the lines are significantly narrower. Three species are obvious. A six-fold signal in the vicinity of $g = 2$ can be attributed to Mn^{2+}, most likely as a substitutional impurity in carbonate minerals. In fact, the Mn^{2+} EPR intensity roughly correlates with the level of calcite in the samples as given in ref. 1. Two additional signals occur at lower fields, which by the Larmor equation $g = h\nu_o/\mu_B H_o$, would correspond to g-values of 4.2 and 5.7 and are typical of a number of iron ion species commonly encountered in minerals. Although g-values and line shapes can provide information about the valence and site symmetry of the ion, they can not provide unambiguous information about the chemical compound in which the ion occurs. The relative peak intensities, inferred from peak heights of the 4.8-K spectra, are presented in Table II. Wide variations occur in the three types of transition metal peaks from sample to sample.

Especially striking are the large amounts of the $g = 4.2$ iron species in the Beulah–Zap lignite, a very likely cause for the high $P_{1/2}$ values in that sample. As seen in Table II, the saturation properties do not obviously correlate with a specific paramagnetic type.

Electron Spin-Echo Properties

As mentioned previously, ESE experiments can distinguish between the sources of broadening and relaxation for carbon radicals. In addition, ESE can discriminate between the radicals in the narrow and broad EPR

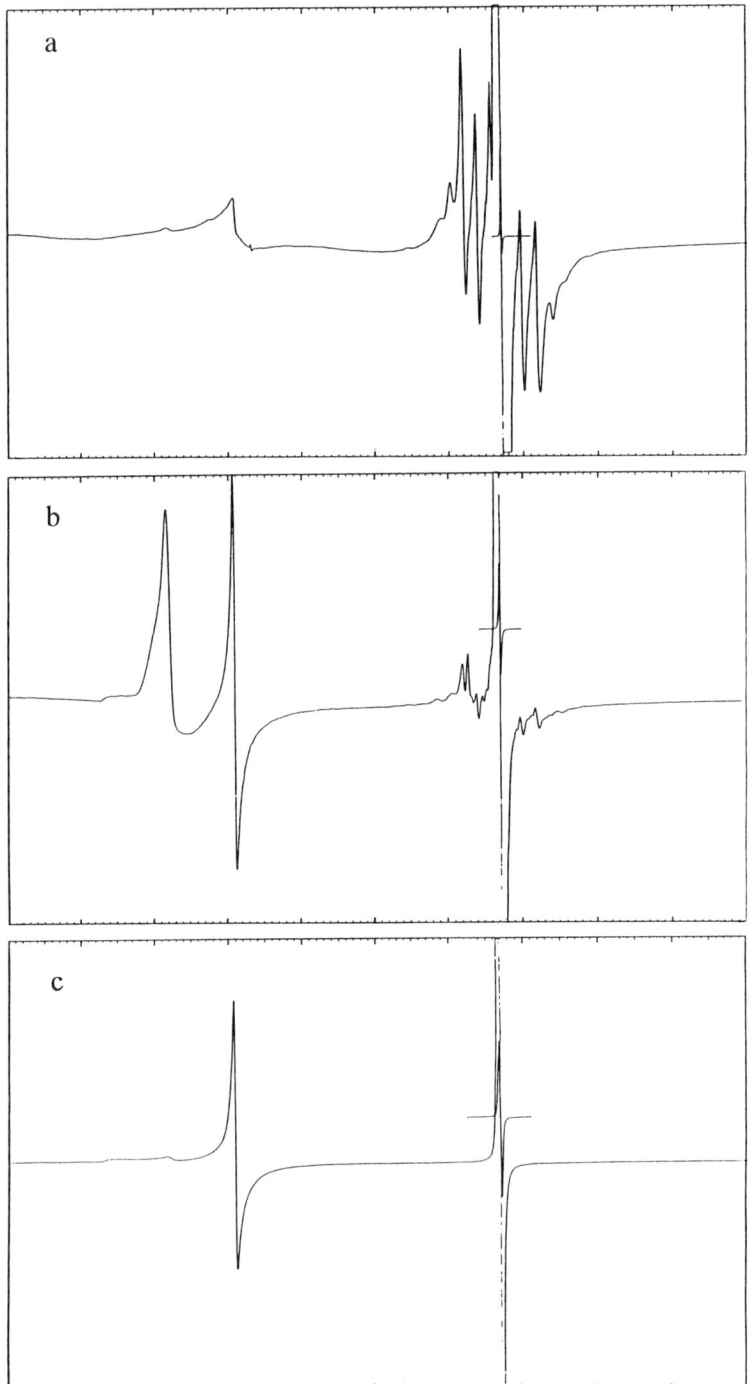

Figure 3. The transition metal EPR signals seen in (a) Pocahontas No. 3, (b) Blind Canyon, and (c) Wyodak–Anderson coals.

Table II. Transition Metal Species and Carbon-Radical Saturation

Coal	Rank	Fe (g = 5.7)	Fe (g = 4.2)	Mn^{2+}	$P_{1/2}$ (mW) Vitrinite	$P_{1/2}$ (mW) Inertinite
Pocahontas No. 3	LVB	0.12	1.44	9.44	3.0	19
Upper Freeport	MVB	0.20	1.84	5.64	4.0	21
Pittsburgh No. 8	HVB	2.87	0.58	0.79	7.2	71
Lewiston–Stockton	HVB	0.75	7.25	0.88	6.5	48
Blind Canyon	HVB	6.16	10.0	2.08	2.0	—
Illinois No. 6	HVB	1.30	10.2	33.6	8.1	44
Wyodak–Anderson	SB	0.60	36.8	n/a	10.0	—
Beulah–Zap	L	3.00	134.0	12.0	19.0	—

NOTE: The values listed are relative intensities from peak heights at 4.8 K.

ABBREVIATIONS: LVB, low-volatile bituminous; MVB, medium-volatile bituminous; HVB, high-volatile bituminous; SB, subbituminous; and L, lignite.

lines because of their different resonance properties. The broad EPR line exhibits strong inhomogeneous broadening, and the free induction decay (FID) of this line falls rapidly to zero. By contrast, the narrow line is nearly homogeneously broadened, and its FID decays much more slowly. Therefore the line-shape and relaxation properties of the narrow line can be probed by observing the FID at times long enough that the the contribution from the broad signal has already decayed. Conversely, the strong inhomogeneous broadening found in the broad line can be capitalized upon by forming a spin echo (10) and determining $1/T_M$ and $1/T_{1E}$ for that component from the echo measurements because the nearly homogeneous line will not form a spin echo.

Narrow-Line Results. As shown in Figure 4, taking the complex Fourier transform of the narrow-component FID yields the absorption spectrum. The derivative is obtained by digitally differentiating the absorption. Resulting values of the half-width $\Delta\omega_{1/2}$ and $\Delta\omega_{pp}$ for the five samples containing a narrow component are shown in Table III. The magnitudes of the widths are in good agreement with the corresponding EPR values. The form of the line shape can be inferred by examining the ratio $\Delta\omega_1/\Delta\omega_{pp}$. For a Gaussian line shape this ratio should be 1.18 (2 ln 2), and for a Lorentzian line shape it would be $\sqrt{3}$ (11). For the higher rank coals, the ratio lies quite close to the Lorentzian limit, and the ratio increases in the lower rank coals. This increased ratio reflects a distribution of radical types. Similar results were obtained from previous analyses of the EPR spectra of inertinites in isolated macerals (3).

As shown in Figure 5, the spin–lattice relaxation can be determined by measuring the inversion recovery of the FID. The resulting data are fit to a two-exponential recovery model; the data and the residuals from the fitting process are shown in the figure. The longer relaxation times are the true spin–lattice relaxation times, T_{1E}. The shorter relaxation times may indicate the presence of additional radical types, but are more likely the result of cross-relaxation phenomena. These relaxation values are summarized in Table IV. Qualitatively, the short T_{1E} values are consistent with the higher $P_{1/2}$ values observed for the narrow-line component as compared to the broad-line component.

Broad-Line Results. Decay of the spin echo as a function of echo-pulse spacing is used to trace the loss of phase memory. As mentioned in the preceding section, this phase-memory loss can occur for two reasons: loss of phase memory from spin–spin coupling (the usual T_2 process) and instantaneous-diffusion effects associated with the application of the refocusing pulse that forms the spin echo. These two processes

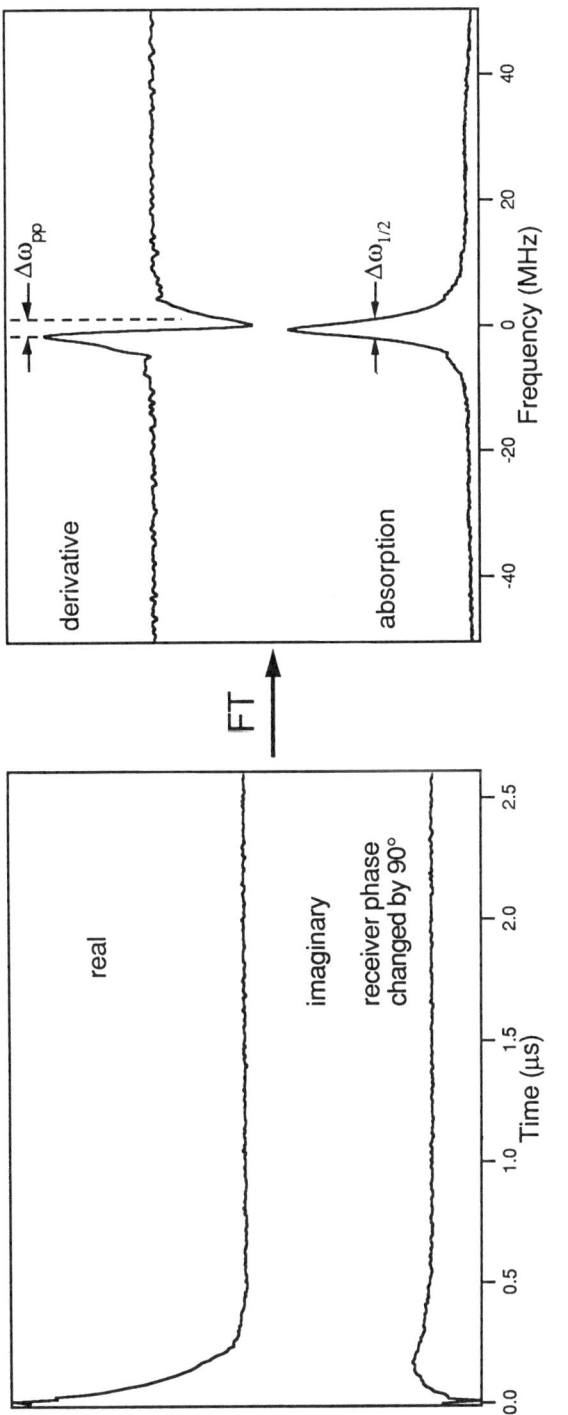

Figure 4. Selective FID detection of the narrow line. The Fourier transform of the FID is the EPR absorption.

Table III. Fourier Transform Line Widths for the Narrow Carbon-Radical Component

Coal	Rank	$\Delta\omega_{1/2}$ (MHz)	$\Delta\omega_{pp}$ (MHz)	$\Delta\omega_{1/2}/\Delta\omega_{pp}$
Pocahontas No. 3	LVB	2.2	1.4	1.6
Upper Freeport	MVB	2.2	1.2	1.8
Pittsburgh No. 8	HVB	3.5	1.8	1.9
Lewiston–Stockton	HVB	3.1	1.5	2.1
Blind Canyon	HVB	—	—	—
Illinois No. 6	HVB	3.2	1.2	2.7
Wyodak–Anderson	SB	—	—	—
Beulah–Zap	L	—	—	—

ABBREVIATIONS: LVB, low-volatile bituminous; MVB, medium-volatile bituminous; HVB, high-volatile bituminous; SB, subbituminous; and L, lignite.

Table IV. Spin–Lattice Relaxation of the Narrow Carbon-Radical Component

Coal	Rank	T_{1e}^{A} (μs)	T_{1e}^{B} (μs)	I
Pocahontas No. 3	LVB	0.93	6.54	0.09
Upper Freeport	MVB	0.50	1.92	0.50
Pittsburgh No. 8	HVB	0.59	1.93	0.56
Lewiston–Stockton	HVB	0.52	2.12	0.55
Blind Canyon	HVB	—	—	—
Illinois No. 6	HVB	0.56	7.17	0.11
Wyodak–Anderson	SB	—	—	—
Beulah–Zap	L	—	—	—

NOTE: A two-component recovery is observed, with characteristic times T_{1e}^{A} and T_{1e}^{B}, and an intensity ratio (I) of the two components.

ABBREVIATIONS: LVB, low-volatile bituminous; MVB, medium-volatile bituminous; HVB, high-volatile bituminous; SB, subbituminous; and L, lignite.

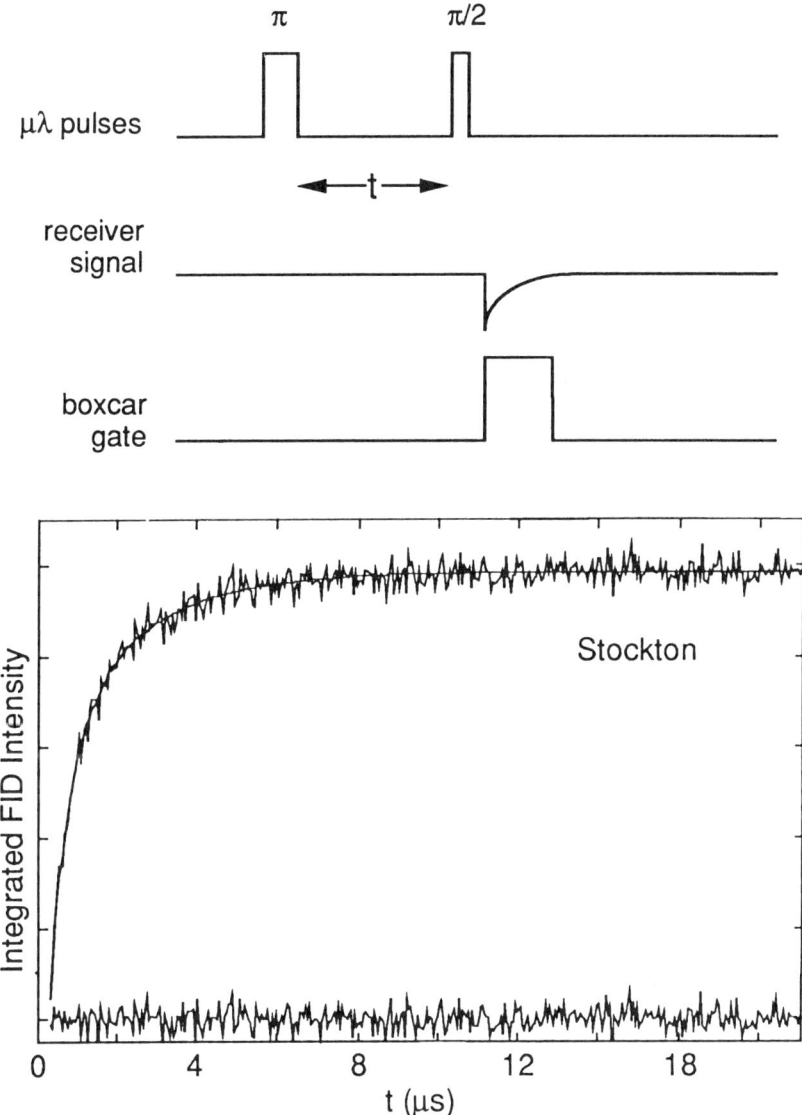

Figure 5. The measurements for the narrow component using FID inversion recovery.

can be differentiated by examining the variation of the echo decay for pulses of different tipping angles, $\theta_1 = \gamma H_1 t_w^i$, where γ is the gyromagnetic factor of the spin, H_1 is the microwave field strength, and t_w^i is the width of the ith pulse. The instantaneous-diffusion effect is a maximum for $\theta = 180°$.

Figure 6 shows the results for the high-rank Pocahontas No. 3 low-volatile bituminous coal and for the low-rank Beulah–Zap lignite. The instantaneous-diffusion effect is clearly evident for the high-rank coal and not observable for the lignite. A plot of the ratio of $1/T_M$ [$\theta = 90°$]/$(1/T_M)$ [$\theta = 29°$] is shown in Figure 7. With the exception of the

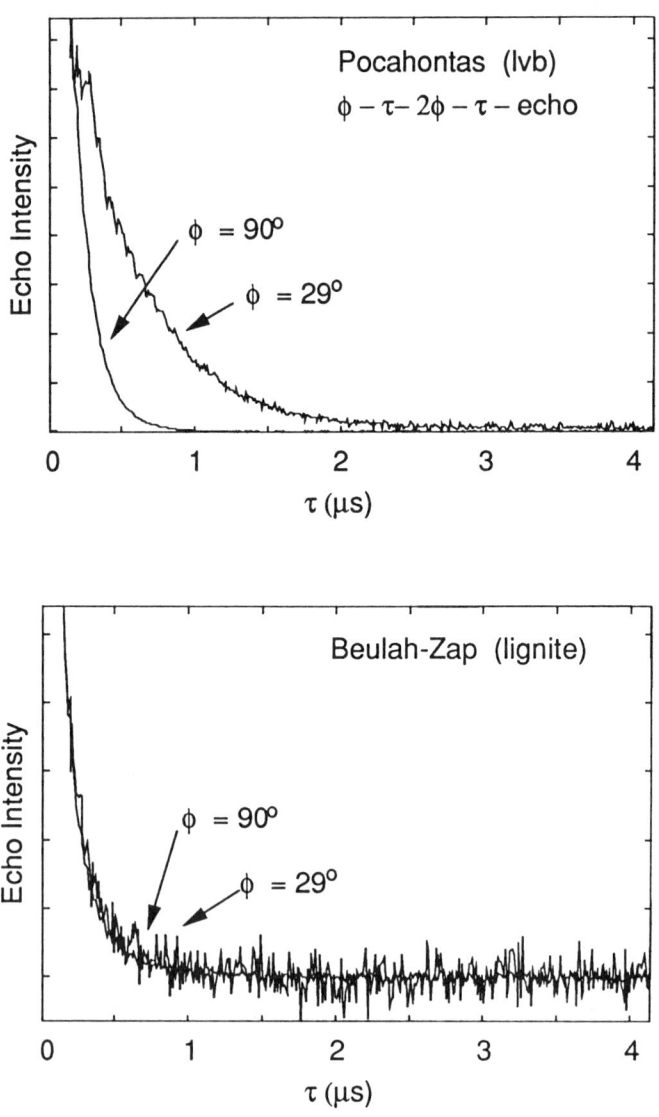

Figure 6. Probing instantaneous diffusion by using microwave pulses of different tipping angles.

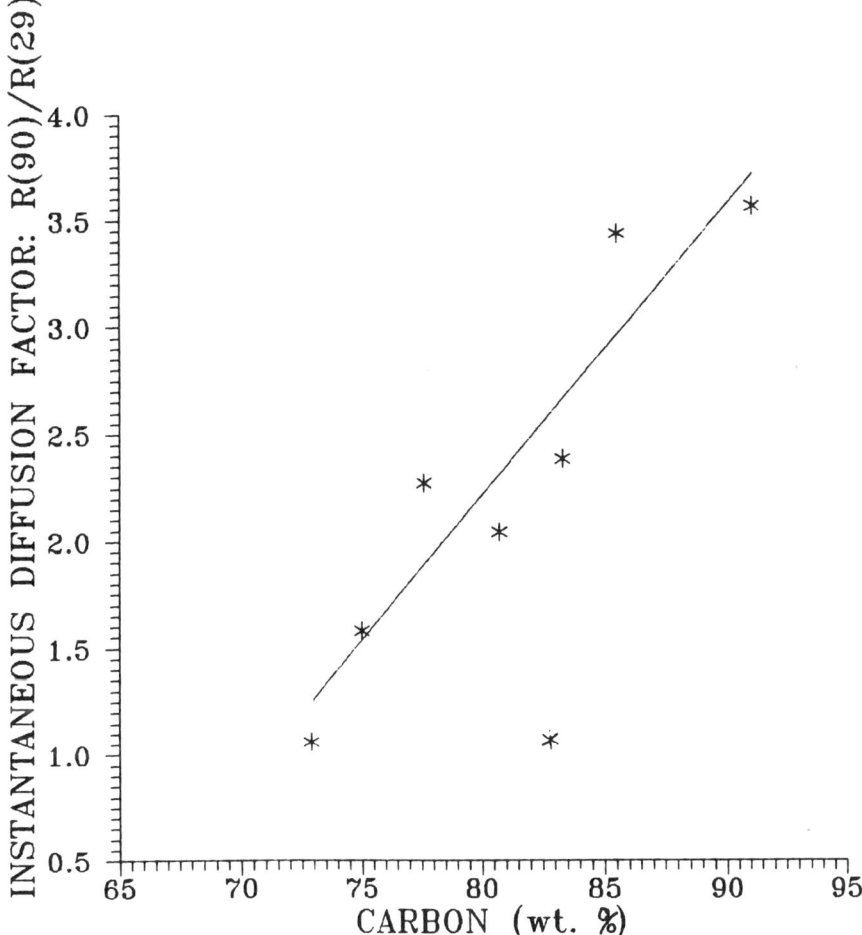

Figure 7. *Magnitude of instantaneous diffusion for coals of different ranks.*

Lewiston–Stockton coal, which has a peculiar maceral mixture, the other coals group into three categories: (1) low- and medium-volatile bituminous coals, with strong instantaneous diffusion; (2) high-volatile bituminous coals, with intermediate instantaneous diffusion; and (3) the low-rank coals, in which the instantaneous-diffusion effect is either small or absent. This trend is qualitatively consistent with the trend previously observed for isolated vitrinite coal macerals: The magnitude of the instantaneous-diffusion effect decreases with increasing rank for bituminous vitrains. However, the magnitude of the intrinsic phase-memory loss, $1/T_{ML}$, defined as the limit $1/T_M(\theta)$ as $\theta \to 0$, does not scale with the spin density, in contrast to the result seen for the isolated macerals.

Spin–lattice relaxation can be measured by using the three-pulse sequence shown in Figure 8. The results are again fit with a two-exponential model (the residuals from the fit are also seen in Figure 8), and the values for $1/T_{1E}$ are presented in Table V. Clearly, the magnitude of $1/T_{1E}$ is much smaller than for the narrow line, and the components

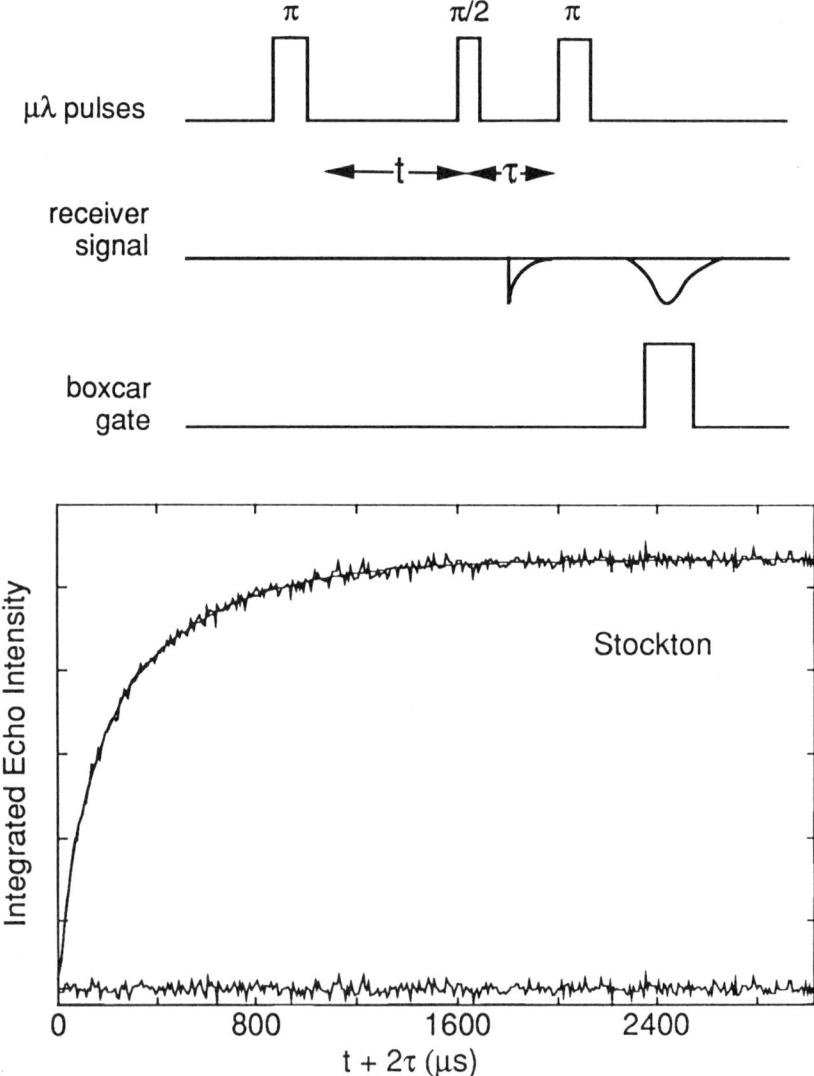

Figure 8. Spin-echo inversion-recovery technique to measure T_{1E} for the broad carbon radical component.

Table V. Spin–Lattice Relaxation of the Broad Carbon-Radical Component

Coal	Rank	T_{1e}^A (μs)	T_{1e}^B (μs)	I
Pocahontas No. 3	LVB	70	304	0.87
Upper Freeport	MVB	116	517	2.13
Pittsburgh No. 8	HVB	47	136	0.53
Lewiston–Stockton	HVB	94	423	1.36
Blind Canyon	HVB	102	354	0.49
Illinois No. 6	HVB	66	271	1.46
Wyodak–Anderson	SB	33	156	0.53
Beulah–Zap	L	28	182	0.29

NOTE: A two-component recovery is observed, with characteristic times T_{1e}^A and T_{1e}^B, and an intensity ratio (I) of the two components.

ABBREVIATIONS: LVB, low-volatile bituminous; MVB, medium-volatile bituminous; HVB, high-volatile bituminous; SB, subbituminous; and L, lignite.

are comparably represented. On the basis of the magnitudes, we propose that relaxation occurs in two stages: a cross-relaxation step, which is responsible for the short-term recovery, followed by an approach to equilibrium with the spins at a common spin temperature (11). The short relaxation times for the low-rank coals are particularly apparent and are consistent with the high values of $P_{1/2}$ measured by saturation EPR spectroscopy.

Conclusions

These EPR and ESE measurements provide a consistent picture of the Argonne Premium coals. EPR properties such as g-values, line widths, and line shapes are consistent with the changes in chemistry expected for coals of varying rank and are in agreement with previous observations on isolated coal macerals (3). Carbon radical densities vary widely and do not track rank, but previous isolated maceral observations suggest that such variability might be expected. The major departures from earlier work are in the behavior of the relaxation-related phenomena: microwave saturation and transient relaxation measurements.

For the isolated macerals, both the phase-memory decay and the spin–lattice relaxation were well fit by single exponential functions. The necessity for two exponential fits in the present case suggests a greater chemical and physical heterogeneity of the samples. Second, the magnitudes of the relaxation rates and $P_{1/2}$ values are significantly different than they were for the isolated macerals. Although the values of $1/T_{ML}$ are comparable to those observed previously, no values of $1/T_{1E}$ comparable to the rapid rates observed for the high-rank macerals are seen here. Conversely, the low-rank coals have significantly higher rates than previously observed. Both of these results and the comparatively low value of the measured paramagnetic spin density could be explained by invoking enhancement of the relaxation process by paramagnetic impurities. For the low-rank coals, this paramagnetic impurity effect is manifested as an increase in $1/T_{1E}$. For the high-rank coals, the relaxation can be so strong that the absorptions are no longer observable. This result would imply that we are seeing some nontypical components of the coal. The test of this hypothesis will rest in examination of demineralized and ion-exchanged samples of the Argonne Premium coals. [Note added in proof: This speculation has, in fact, been confirmed (12)].

Acknowledgments

We thank A. R. Garcia for preparation of the samples and for the proton NMR determination of the residual water content in the coals after sample sealing.

References

1. Vorres, K. S. *Energy Fuels* **1990**, *4*, 420–426.
2. Dyrkacz, G. R.; Horwitz, E. P. *Fuel* **1982**, *61*, 3–12.
3. Silbernagel, B. G.; Gebhard, L. A.; Dyrkacz, G. R.; Bloomquist, C. A. A. *Fuel*, **1986**, *65*, 558–65.
4. Thomann, H.; Silbernagel, B. G.; Jin, H.; Gebhard, L. A.; Tindall, P.; Dyrkacz, G. R. *Energy Fuels*, **1988**, *2*, 333–9.
5. See (e.g.) Retcofsky, H. R. In *Coal Structure;* Gorbaty, M. L.; Ouchi, K., Eds.; Advances in Chemistry 192; American Chemical Society: Washington, DC, 1981; pp 37–58.
6. See (e.g.) *Pulsed EPR: A New Field of Applications;* Keijers, C. P.; Reijerse, E. J.; Smidt, J., Eds; North Holland: Amsterdam, Netherlands, 1989.
7. Bloch, F. *Phys. Rev.* **1946**, *70*, 1–15.

8. Lowe, I. J.; Norberg, R. E. *Phys. Rev.* **1957**, *107*, 46–61.
9. Salikov, K. M.; Tsvetkov, Yu. D. In *Time Domain Electron Spin Resonance;* Kevan, L.; Schwartz, R. N., Eds.; John Wiley and Sons: New York, 1979; pp 232–77.
10. Hahn, E. L. *Phys. Rev* **1950**, *80*, 580–94.
11. Abragam, A. *The Principles of Nuclear Magnetism;* Clarendon: Oxford, England, 1961; p 107.
12. Silbernagel, B. G.; Gebhard, L. A.; Flowers, R. A.; Larsen, J. W. *Energy Fuels,* **1991**, *5*, 561–568.

RECEIVED for review June 8, 1990. ACCEPTED revised manuscript December 17, 1990.

30

Pulsed Electron Nuclear Double Resonance Spectroscopy of Argonne Premium Coals

Hans Thomann, Marcelino Bernardo, and Bernard G. Silbernagel

Corporate Research Laboratory, Exxon Research and Engineering Company, Route 22 East, Annandale, NJ 08801

> *Pulsed electron nuclear double resonance (ENDOR) studies on eight Argonne Premium coals are reported in this chapter. The coal samples were selected from the ranks of lignite to low-volatile bituminous. Two types of proton ENDOR signals were observed for all eight coal samples: a matrix line and local ENDOR lines. This result is in contrast to previous studies using conventional continuous-wave (CW) ENDOR techniques in which the matrix proton line dominated the spectra. Experimental conditions for recording of pulsed ENDOR spectra of coals are discussed. Local ENDOR proton hyperfine couplings of up to 20 MHz were observed, but no resolved structure was apparent in the ENDOR spectra. The magnitude of these couplings is consistent with that expected for aromatic protons on polynuclear aromatic (PNA) carbon radicals. The larger hyperfine couplings are enhanced in ENDOR spectra recorded at higher g-values as expected for PNA radicals with heteroatoms, most likely oxygen or sulfur.*

ALL COALS CONTAIN a substantial concentration of carbon-based free radicals ([1, 2]). Radical densities measured by electron paramagnetic resonance (EPR) spectroscopy are typically between 10^{18} and 10^{20} spins per

0065–2393/93/0229–0561$06.00/0
© 1993 American Chemical Society

gram of carbon for bituminous coals. These radicals are believed to be a direct consequence of the coalification process. One plausible explanation for their stability over geologic time is that the unpaired electron is resonance-stablized in polynuclear aromatic (PNA) molecules. PNA radicals are relatively inert toward chemical transformation. This resonance stabilization could also account for the stability of these PNA radicals during high-temperature conversion processing.

The PNA radical type and the relative distribution of types are believed to be a function of coal rank and maceral type (3). PNA radicals therefore provide a convenient method for the molecular classification of coals. Pulsed EPR evidence that has been obtained suggests that the PNA radical molecules are randomly distributed throughout the macromolecular structure of coal (4). Correlations between EPR parameters and conversion yields have also been established in studies of coal liquefaction (2). These combined results support the thesis that the study of the structural types and the relative distribution of these PNA radical molecules provides a natural probe of the complex three-dimensional macromolecular structure of aromatic and aliphatic hydrocarbons that together comprise the heterogeneous material known as coal.

In principle, it should be possible to identify the chemical structure of the PNA radical molecules from the analysis of the EPR spectrum. In practice, only a single inhomogeneously broadened EPR line with no hyperfine structure is observed for most coals. The overlap of the EPR spectra from the many PNA radicals in a whole coal, each with slightly different g-values and hyperfine splittings, results in a net broadening of the EPR spectrum. Additional line broadening may arise from the anisotropic hyperfine interactions expected from protons on the PNA radical molecules. The situation is improved in studies of isolated coal macerals, for which the absorption line width, line shape, g-value, and microwave-power-saturation properties are all correlated with coal rank and maceral type (3). However, no hyperfine structure is observed even for the isolated coal macerals.

Electron nuclear double resonance (ENDOR) spectroscopy is a well-established method for obtaining high-resolution spectra of nuclear hyperfine transitions (5). The ENDOR spectrum is a spectrum of NMR transitions detected via an EPR transition. NMR transitions from all nuclei that are magnetically coupled to the unpaired electron can contribute to the ENDOR spectrum. ENDOR experiments are particularly informative in cases where no hyperfine structure is apparent in the EPR spectrum.

For a two-spin system composed of an $S = 1/2$ electron coupled to an $I = 1/2$ nucleus with hyperfine coupling A, the allowed EPR transition frequencies are $\nu_{\pm}^{EPR} = \nu_e \pm A/2$, and the allowed NMR transition frequencies are $\nu_{\pm}^{NMR} = \nu_n \pm A/2$ (6). For carbon radicals typically encountered in coals, $\nu_n > A/2$, where $\nu_n = g_n \beta_n B_0/h$ is the nuclear Larmor frequency, g_n is the nuclear g-value, β_n is the nuclear Bohr magneton, B_0 is the applied magnetic field strength, and h is Planck's constant. In this case, second-order hyperfine perturbation corrections (on the order of $h^2A^2/g_e\beta_e B_0$) to the energy levels can be neglected so that the two ENDOR transitions are observed at $\nu_\pm = \nu_n \pm A/2$ for each inequivalent nucleus. In contrast to EPR spectroscopy, additional splittings or additional ENDOR lines are not observed for multiple sets of equivalent nuclei (7). This condition results in a greatly simplified spectrum because there is a one-to-one correspondence between each inequivalent nucleus and a pair of ENDOR lines.

Several previous ENDOR studies of coals have been reported (2, 8–11). The pulsed ENDOR studies of the Argonne Premium coals (APC) reported here were motivated by two considerations. First, the APC samples have a known and controlled sample history (12). The samples were selected to encompass a wide spectrum of coal types and properties. Many basic sample properties, densities, chemical composition, etc., have been documented (12). An aliquot of each coal sample was distributed to a participating research group for subsequent spectroscopic study. This protocol is in sharp contrast to most previous studies of coals in which the sample history was usually unknown. Comparison and synthesis of data from different laboratories was therefore usually not possible.

This study also was motivated by the recent developments in pulsed ENDOR techniques (13, 14, see also Chapter 4). Pulsed ENDOR techniques offer significant performance enhancements compared to the conventional, continuous-wave (CW) ENDOR experiment. These enhancements include greater sensitivity, higher spectral resolution, and the elimination of the dependence of the ENDOR signal on the detailed balance of the electron and nuclear spin relaxation rates. The latter is particularly significant because transitions in the CW ENDOR experiments are often not observed because of unfavorable relaxation rates.

Experimental Details

The pulsed ENDOR techniques used in the study of the APC samples are discussed in detail in Chapter 4. The APC samples were transferred from the sealed vial in which they were received to 4-mm EPR tubes in a dry box under an argon atmosphere. The tubes were evacuated to approximately 10^{-3} torr (0.13 Pa) and sealed. Samples were vacuum-pumped only briefly to minimize the potential removal of adsorbed fluids or gases from the coal matrix. Samples were stored in the sealed EPR tubes at room temperature.

The pulsed analog of the CW ENDOR experiment (15) is the Davies ENDOR experiment (16). As discussed in more detail in Chapter 4, the Davies EN-

DOR experiment consists of a pulse sequence in which a microwave pulse is applied to an EPR transition followed by a radiofrequency (rf) pulse applied to an NMR transition. The effect of the rf pulse (i.e., the ENDOR signal) is then detected as a change in the EPR transition intensity. It is convenient to monitor the EPR transition intensity by using an electron spin echo formed by two microwave pulses. The Davies ENDOR spectrum is then recorded by plotting the intensity of the electron spin-echo signal created by the final two microwave pulses as a function of the rf frequency. The rf frequency is stepped on successive pulse-sequence iterations.

Results

Davies ENDOR spectra of Illinois No. 6 coal, recorded at 298 and 85 K, are shown in Figures 1a and 1b, respectively. The spectra shown are the

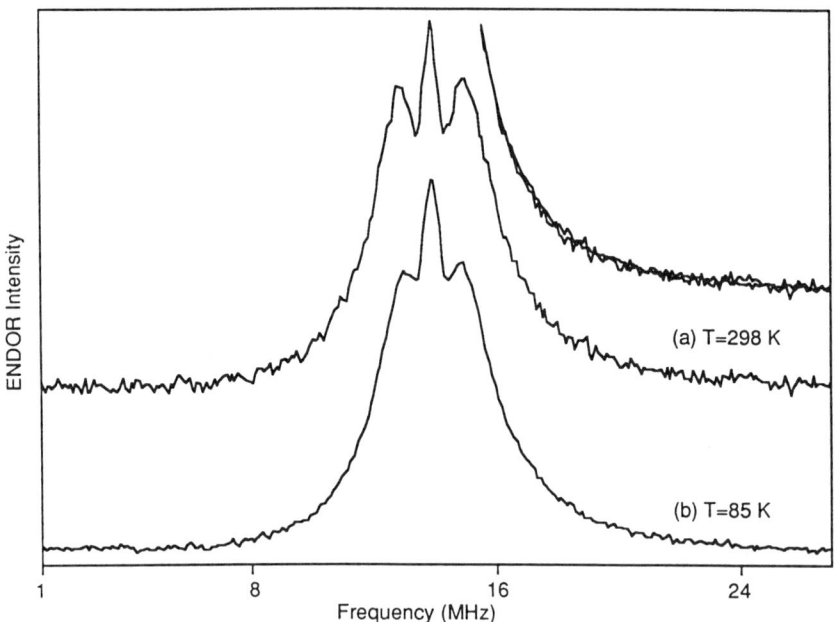

Figure 1. Davies ENDOR spectra for Illinois No. 6 (HVB) coal recorded at (a) T = 298 K and (b) T = 85 K. Other conditions are as follows: microwave frequency, 9.160 GHz; magnetic field strength, 3269.2 G; microwave preparation (π) pulse width, 0.40 μs; rf pulse width, 17.0 μs; resolution, 0.1 MHz/point; 30 echoes sampled per point; and 50 sweeps. The inset shows the overlay of the two spectra for the expanded region of the spectrum.

electron spin-echo (ESE) intensity plotted as a function of the frequency of the rf pulse. The most intense peak in Figure 1 is centered at 13.89 MHz, which is close to the proton nuclear Larmor frequency, $\nu_n = g_n \beta_n B_0 / h = 13.918$ MHz, calculated for a field of 3269.2 G. The 200-ppm difference between these values arises from the difference between the gaussmeter reading of the magnetic field and the true magnetic field at the sample. The peaks at approximately 1 MHz above and below the proton matrix line are a consequence of the pulse widths used to record these spectra. This feature is discussed further in following sections (*see* Figure 2).

The spectra are composed of two components, a superposition of matrix and local ENDOR signals. The matrix proton ENDOR signal arises from protons that sustain only a weak dipolar (hyperfine) interaction with the paramagnetic electron. Such interactions arise predominantly from protons on molecules in the host lattice other than the molecule supporting the paramagnetic electron. The local proton ENDOR signals arise from those protons that have both a Fermi contact as well as an anisotropic hyperfine interaction. These interactions arise from protons, most likely on the molecule hosting the paramagnetic electron, having a finite unpaired electron spin density. In the solid state, the local ENDOR line may arise from both isotropic and anisotropic hyperfine interactions.

For $\nu_{rf} > |\nu_n \pm 1|$ MHz, the local ENDOR signal smoothly and continuously decreases in intensity with increasing and decreasing frequency. On the high-frequency side, the ENDOR signal is still above the noise level at 24 MHz (i.e., at least 10 MHz above the proton Larmor frequency). This ENDOR signal intensity means that protons with hyperfine couplings of 20 MHz are observed in the present spectra. For a 20-MHz coupling, a finite ENDOR signal intensity should be observed at 4 MHz (i.e., 10 MHz below the proton Larmor frequency). However, the ENDOR signal appears to approach the noise level at roughly 7 MHz under the experimental conditions used to record the spectra in Figure 1. Careful examination also reveals that the ENDOR line shapes are not completely symmetric about the proton Larmor frequency. This asymmetry is a function of the experimental rf conditions. The intensity of the ENDOR signal at low rf frequency can be increased by using longer rf pulses. However, this increase has the effect of suppressing the ENDOR signal at higher rf frequency. This effect arises from the frequency-dependent nutation angle of the sublevel magnetization (*see* Chapter 4).

Careful consideration of the experimental conditions is required for a quantitative interpretation of the pulsed ENDOR spectra. The experimental conditions used for recording each ENDOR spectrum are given in the figure captions. We will demonstrate the effects of some important experimental parameters on the ENDOR spectrum observed. We will dis-

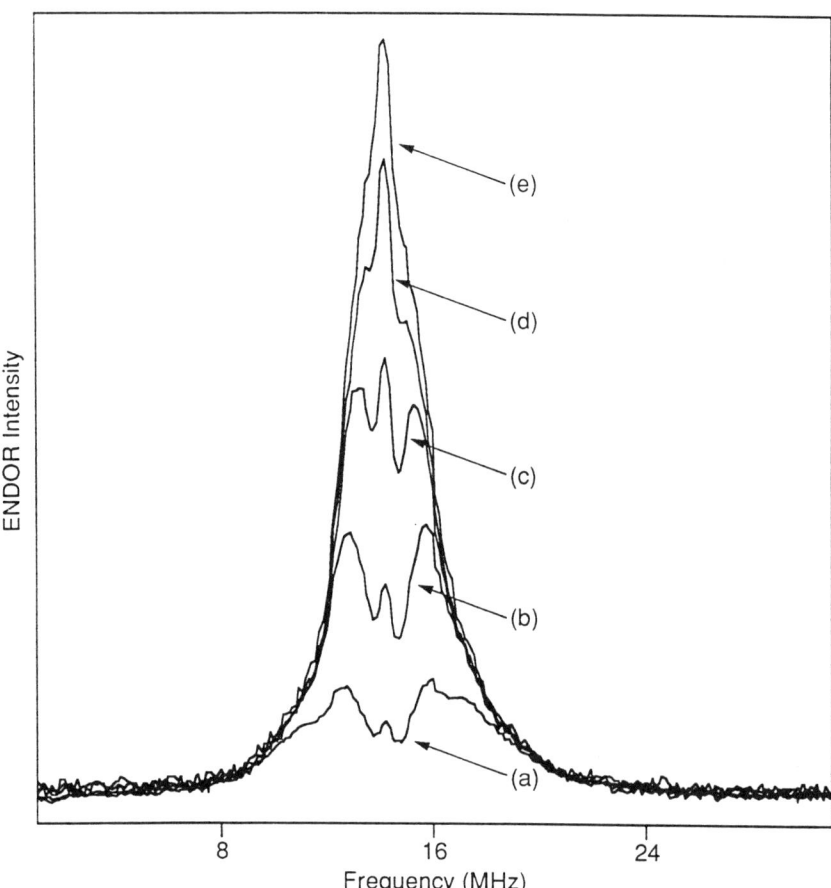

Figure 2. Davies ENDOR spectra for Illinois No. 6 (HVB) coal recorded for variable preparation-pulse bandwidths. Inversion pulse widths and relative transmitter attenuation levels are (a) 0.10 µs, 4.5 dB; (b) 0.20 µs, 16.0 dB; (c) 0.40 µs, 23.0 dB; (d) 0.60 µs, 25.0 dB; and (e) 0.80 µs, 27.0 dB. Other conditions were T = 100 K; microwave frequency, 9.335 GHz; magnetic field strength, 3335.8 G; rf mixing-pulse width, 6.75 µs; resolution: 0.117 MHz/point; 10 echoes sampled per point; and four sweeps.

cuss in detail the significance of the temperature at which the spectra are recorded, the effect of the microwave pulse width used to excite the EPR spectrum, and the dependence of the ENDOR spectra on the Zeeman magnetic field value selected. We will also discuss the importance of the delay time, τ, between the two microwave pulses used to detect the EPR signal; the effect of the pulse sequence repetition rate, τ_R^{-1}; and the effect of the width of the rf pulse.

The temperature is often a critical parameter in CW ENDOR experiments. In the CW ENDOR experiment, the ENDOR signal is detected as the change in the saturated EPR signal intensity as a function of applied rf magnetic field. For an ENDOR effect to be observed, the rate of induced NMR transitions must therefore compete with the rate of induced EPR transitions. This "rate-equality" condition requires large rf field intensities. If the rate of rf-induced NMR transitions is not comparable to the rate of induced EPR transitions, no CW ENDOR signal is observed; and this is in fact the reason CW ENDOR experiments often fail: The required rf magnetic field intensities in the CW ENDOR experiment are difficult to achieve. Often the temperature is used as an experimental parameter. Generally, the nuclear and electron spin–lattice relaxation rates do not have the same temperature dependence. The temperature is empirically adjusted so that the ratios of EPR and NMR transition rates are optimized to obtain the maximum CW ENDOR signal.

Perhaps one of the most important aspects of the Davies ENDOR experiment is the decoupling of the ENDOR signal from the ratio of the electron and nuclear spin-relaxation rates. This decoupling has the consequence that temperature is not a critical parameter in pulsed ENDOR experiment. In order to observe a Davies ENDOR signal, the electron spin phase-memory and electron spin–lattice relaxation times must only be sufficiently long that an electron spin echo can be observed. This condition is in sharp contrast to the CW ENDOR experiment in which it is not uncommon for ENDOR signals to be observed over only a narrow temperature range.

This decoupling of the Davies ENDOR signal from the spin dynamics has the important consequence that, in the absence of mechanisms that dynamically average hyperfine couplings, the ENDOR line shape remains independent of temperature. This temperature independence is illustrated in Figure 1 for a sample of Illinois No. 6 coal. The Davies ENDOR spectrum in Figure 1a was recorded at 298 K, and the spectrum in Figure 1b was recorded at 85 K. The inset shows an overlap for the portion of the spectra above 16 MHz with the vertical axis expanded. The line shapes for the two spectra are identical. The only direct manifestation of the effect of the temperature is in the magnitude of the observed ENDOR enhancement, $\Delta\chi/\chi$, where χ is the EPR susceptibility. This result is primarily a consequence of the temperature dependence of the electronic spin Boltzmann factor. The nuclear spin Boltzmann factor makes a negligible contribution to the ENDOR enhancement. In order to compare the enhancement factors for the local ENDOR signals, the enhancements are measured at $\nu_n \pm 1$ MHz. At 298 K, $\Delta\chi/\chi = 0.10$, and at 80 K, $\Delta\chi/\chi = 0.23$. The latter value is within experimental error equal to the factor of 2 increase expected from the Boltzmann factor. This result can be contrasted with the CW ENDOR results on bituminous coals for which no local ENDOR signals are observed at either temperature (8–11).

The optimum time delay between the Davies ENDOR pulse sequence iterations, τ_R, is a complex function of the spin-relaxation mechanisms that restore thermodynamic spin equilibrium. To avoid possible complications arising from rapid passage effects, $\tau_R \gg T_{1,\text{eff}}$, where $T_{1,\text{eff}}$ is the characteristic time for the dissipation of absorbed microwave energy. Rapid-passage effects (17), which can occur when $\tau_R < T_{1,\text{eff}}$, can result in distortions of spectral line shapes and even in shifts of the position of the ENDOR line. Passage effects are particularly pronounced if the rf sweep rate exceeds the spin–lattice relaxation rates. For data reported in this chapter, $\tau_R = 10$ ms, which is at least 20 times longer than the longest spin–lattice relaxation time (measured by spin-echo detection of the EPR signal following spin population inversion induced by a π pulse).

The delay time between the two microwave pulses that form the electron spin echo in the detection period, τ_{ESE}, was not a significant experimental variable. In principle, ENDOR spectra recorded at variable τ_{ESE} values could provide the opportunity for discrimination among paramagnetic radical types characterized by different phase-memory decay times, T_M. T_M-selective ENDOR spectroscopy may be particularly useful if the EPR spectra from individual radical types overlap (i.e., have the same g-value). However, in the present study of the APC whole coal samples, no T_M selectivity was observed.

The important effect of the width of the microwave pulse used to excite the EPR spectrum is illustrated in Figure 2. All of the spectra in Figure 2 were recorded under identical conditions, as described in the figure caption, except for the variable widths and relative power levels of the microwave pulses. The microwave power levels were adjusted to maintain the nominal π pulse condition for varying pulse widths. The Davies ENDOR spectra in Figures 2a–2e correspond to a progressively smaller fraction of the spins excited in the EPR spectrum. This property has an effect on the structure in the ENDOR spectrum near the proton Larmor frequency. This effect can be understood from the following discussion.

In an inhomogeneously broadened EPR absorption line, the microwave preparation pulse burns a hole with a width of $\Delta\omega_1 \simeq t_p^{-1}(\pi)$. This hole is sometimes referred to as a "spin-alignment hole" because no differential nuclear sublevel spin polarization is produced for sublevels with hyperfine couplings $A_j < \Delta\omega_1$. The spin-alignment hole is related to the phenomenon of microwave-power suppression of weak hyperfine couplings observed in CW ENDOR spectroscopy (5). The polarization transfer caused by the rf pulse in the Davies ENDOR experiment corresponds to the filling in of the spin-alignment hole by transfer of magnetization from outside the saturated region in the EPR spectrum. To observe weak hyperfine couplings a very narrow saturation hole must be created in the EPR spectrum. Thus a low-power, wide pulse favors the observation of weak hyperfine couplings. Conversely, high-power, short-pulse conditions optimize the observation of large hyperfine couplings, albeit at the expense of sacrificing sensitivity toward weak couplings. In fact, this phenomenon can be used to discriminate between overlapping ENDOR signals if the overlapping signals arise from nuclei sustaining weak and strong hyperfine interactions (18).

A compromise must be made between experimental conditions that optimize the sensitivity for larger hyperfine couplings while minimizing the suppression of weak couplings. Fortunately, the microwave-pulse excitation bandwidth effects are well defined and can be readily selected so that $A_j > \Delta\omega_1$, where A_j is the smallest hyperfine coupling of interest. For comparative ENDOR results on the Argonne Premium coal samples described in this chapter, a preparation pulse width of 0.40 μs was chosen as a compromise between optimizing the sensitivity to larger hyperfine couplings and minimizing the suppression of weak couplings.

The microwave pulse width effect complicates the identification of the matrix and local ENDOR lines for small values of the hyperfine coupling. Evidence that these two types of signals are in fact observed is derived from the following. First, the intensity of the narrow ENDOR line at ν_n attributed to the matrix proton line exhibits a different power dependence than the local ENDOR signal. The matrix line is observed even when microwave preparation pulse conditions are used that correspond to the largest $\Delta\omega_1$ bandwidths. Second, the intensity of the matrix ENDOR signal exhibits a different temperature dependence than the local ENDOR signal. As shown in Figure 1, the relative intensity of the peak at 13.89 MHz (ν_n) compared to the peaks at 12.90 MHz (ν_-) or 14.90 MHz (ν_+) is larger at 80 K than at 298 K.

These results suggest that the matrix and local ENDOR signals may be governed by different spin dynamics mechanisms. This difference is presumably due to the different spin relaxation mechanisms of the protons associated with the matrix and local proton ENDOR couplings. Thus the matrix and local ENDOR enhancements arising from weakly coupled nuclei may be spectroscopically distinguished by their different spin-relaxation mechanisms.

As a working hypothesis, we adopt the model that proton-matrix ENDOR lines arise from weak dipolar hyperfine interactions with no Fermi contact coupling (7), and local ENDOR lines have finite Fermi contact couplings. The interactions of protons sustaining no or only a small resonance frequency shift from the hyperfine interaction should be dominated by the dipolar magnetic field arising from the abundance of protons in typical bituminous coal samples. Protons sustaining progressively larger resonance frequency offsets should be progressively detuned from other protons. The spin dynamics of these frequency-shifted protons is expected to be dominated by the spin dynamics of the paramagnetic electron.

One method to test these anticipated trends in the NMR spin dynamics of protons sustaining strong and weak hyperfine interactions is to directly measure the NMR relaxation rates. Nuclear sublevel (i.e., NMR) proton spin–spin relaxation times, T_{2n}, values were measured at $\nu_{rf} = \nu_n \pm 1$ MHz. The pulse sequence for measuring the nuclear sublevel proton T_{2n} values has been discussed in greater detail in Chapter 4. Briefly, the NMR spin echo is created by using a Hahn spin-echo pulse sequence, $\pi/2 - \tau_n - \pi - \tau_n - S(2\tau_n)$, with selective NMR pulses determined by the rf frequency, ν_{rf}. The enhanced sensitivity necessary to observe the limited number of nuclei that may contribute at a selected rf frequency is achieved by the non-Boltzmann nuclear-spin polarization created by the EPR preparation (π) pulse. The nuclear coherence created by the NMR $\pi/2$ pulse freely precesses during time τ_n and is stationary in the reference frame precessing at the angular frequency ω_{rf}. The nuclear coherence is refocused at time $2\tau_n$ and converted to a sublevel population difference by a third rf pulse applied at time $2\tau_n$. This population difference is then observed as an amplitude modulation of the electron spin echo in the detection period.

Nuclear sublevel spin echoes can be observed by sweeping the evolution time, t_1, through τ_n starting from $t_1 < \tau_n$, as shown in Figure 3a. Phase-cycling the third rf pulse by π cancels any electron-magnetization recovery that may occur during the t_1 period. A representative nuclear spin-echo decay curve, recorded by sweeping τ_n with $t_1 = \tau_n$, is shown in Figure 3b. The least-squares fit to a double exponential and the residual of the fit are also shown. All NMR sublevel spin-echo decay curves fit well to double exponentials, as indicated from the residuals in the representative data in Figure 3b. The T_{2n} values are the time constants for the more slowly relaxing component. The faster relaxing components, which had time constants in the range of 1.6–3.3 μs, are probably due to incomplete cancellation of the nuclear free induction decays (FIDs).

Nuclear sublevel T_{2n} values of 32.9, 13.3, and 29.3 μs were obtained with $\nu_{rf} = 13.00, 13.89$, and 14.78 MHz, respectively, for the Pocahontas No. 3 (LVB) coal. The T_{2n} value of 13.3 μs obtained with $\nu_{rf} = \nu_n$ (the proton Larmor frequency) is very close to spin–spin dephasing values (T_2^*) measured directly by NMR spectroscopy. Gerstein et al. (19) reported $T_2^* = 10$ μs measured from the decay of the FID by proton NMR spectroscopy at 56.5 MHz for the Pocahontas coal. Similar dephasing times for bituminous coals have been reported by Barton and Lynch using two-pulse NMR techniques (20). Carbon–proton dipolar-dephasing times of 16 and 20 μs have been reported for the Pocahontas coal sample (21). These values were measured from the decay of the carbon magneti-

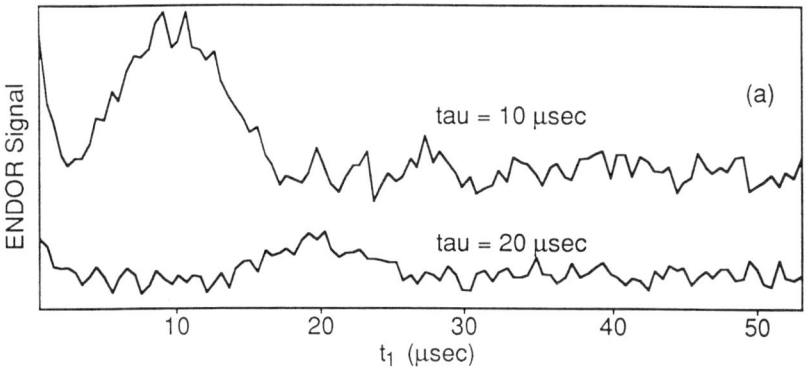

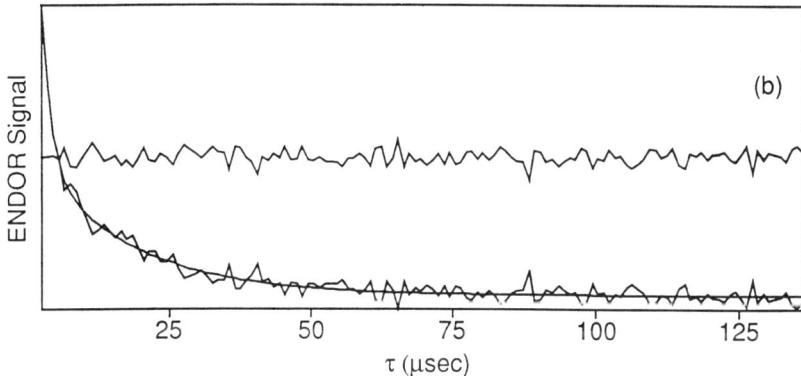

Figure 3. EPR detection of NMR spin echo on nuclear sublevels. Part a: Traces of NMR sublevel spin echoes for Pocahontas No. 3 (LVB) coal for two values of the refocusing delay time (NMR $t_p(\pi/2) = 3.58\ \mu s$, $\nu_{rf} = 14.78$ MHz, other experimental conditions were the same as in Figure 2). Part b: Nuclear sublevel spin-echo amplitude-decay waveform obtained with $\nu_{rf} = 13.00$ MHz. Least-squares double exponential fit and residuals of the fit are also shown.

zation and are therefore an upper limit for the proton dipolar-dephasing times. In these measurements, the proton dephasing time is attributed to the weighted average of the proton homonuclear dipolar interactions, which are the dominant mechanism for proton spin–spin relaxation. This interaction is proportional to the inverse cube of the sum of interproton distances. For naphthalene, the second moment in the rigid-lattice limit calculated by taking into account all protons within 6 Å of a single proton on a given naphthalene molecule leads to a damping constant, inferred from the second moment, of 12.3 μs *(19)*.

The close agreement between the nuclear sublevel T_{2n} value measured by pulsed ENDOR spectroscopy with $\nu_{rf} = \nu_n$ and the proton dephasing times measured directly by NMR spectroscopy suggests that the dominant mechanism for proton dephasing for protons that are dipole-coupled to the electron is still the proton dipolar interaction. Local-field fluctuations associated with the hyperfine interaction apparently make a negligible contribution to the proton T_{2n}, a conclusion that is in agreement with the conclusions of Barton and Lynch (20) and Gerstein et al. (19). The T_{2n} values at $\nu_{rf} = \nu_n \pm 1$ MHz are almost 3 times longer than at $\nu_{rf} = \nu_n$. This result is expected if the proton dipole interaction is the dominant mechanism for dephasing. The spectral density of spin states arising from protons decreases as $|\nu_{rf} - \nu_n| > 0$, so that a smaller number of protons is available to contribute to the proton dipolar interaction. On the basis of these results, the proton T_{2n} values can be used to distinguish between protons that contribute to matrix and local ENDOR lines. Protons that contribute to the matrix ENDOR line are characterized by nuclear sublevel T_{2n} values that are comparable to the proton spin–spin relaxation times measured directly by NMR spectroscopy.

Comparison of the Lorentzian line width of 23.9 kHz calculated from T_{2n} with the measured line width of 230 kHz (see following section) indicates that the ENDOR line is inhomogeneously broadened. One source of line broadening in the Davies ENDOR experiment is the finite bandwidth of the rf pulse. In the experiments reported here, this resolution limit is 0.12 MHz. If the ENDOR lines are inhomogeneously broadened, higher resolution may be obtained by using a more narrow excitation pulse bandwidth. However, longer rf pulses could introduce complications arising from spin dynamics similar to those encountered in CW ENDOR experiments.

The spectral resolution limit imposed by the bandwidth of the rf pulse can be obviated entirely by direct detection of the nuclear sublevel FID. This detection can be accomplished via a pulse sequence proposed by Mehring et al. (22). This pulse sequence is discussed in detail in Chapter 4. In the preparation period, an FID from a nuclear sublevel is created by a $\pi/2$ pulse whose frequency is resonant with the selected NMR transition. This nuclear coherence is detected point by point during the evolution period, t_1, by conversion to a population difference by the action of a second $\pi/2$ pulse. The population difference is detected as an amplitude-modulation of the electron spin echo in the detection period.

The line widths observed at 13.89 and 14.90 MHz are $\Delta\omega_{1/2} = 230$ and 620 kHz, respectively. Significantly, no additional structure is discernible in the NMR sublevel complex Fourier transform (FT) spectra (Figure 4). These line widths are significantly broader than the line widths (≤ 40 kHz) expected from the proton homonuclear dipole coupling. The larger line width observed at 14.90 MHz is consistent with the increased broad-

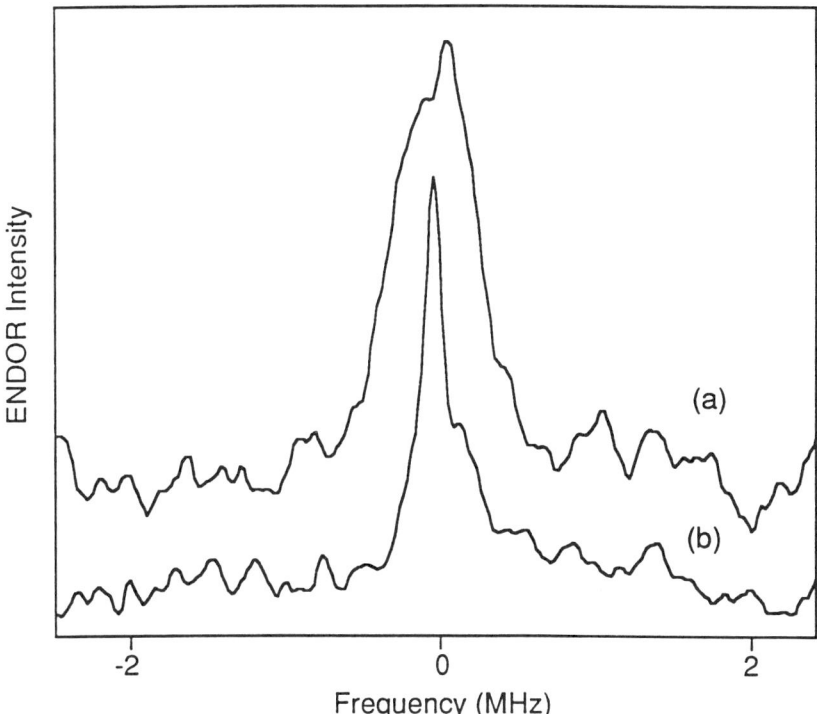

Figure 4. Complex FT sublevel spectra measured by electron spin-echo detection of the FID using rf frequencies of (a) 13.89 and (b) 14.90 MHz. For a, $t_{rf}(\pi/2) = 4.30$ μs, and for b, $t_{rf}(\pi/2) = 3.50$ μs. Increment in $t_1 = 0.10$ μs, and 261 data points were collected for in-phase and phase-quadrature FIDs. Figure 1 caption gives for other experimental conditions.

ening expected for protons that are more strongly shifted from ν_n by the larger hyperfine interaction. In fact, the line width in the spectrum observed at $\nu_{rf} = 14.90$ MHz is close to the excitation bandwidth limit of the rf pulse. The proton T_{2n} values, 13.3 and 29.3 μs measured at 13.89 and 14.90 MHz, respectively, indicate that the NMR lines are inhomogeneously broadened, a result that is consistent with the distribution of NMR resonance frequencies expected from the hyperfine interaction.

Having established the methodology for pulsed ENDOR experiments on carbonaceous samples, we can now explore the manner in which the spectra of the coal samples depend on the rank of the coal. Representative Davies ENDOR spectra for several APC samples are shown in Figure 5. All spectra were recorded at $g = 2.0023$ and $T = 80$ K. The only signif-

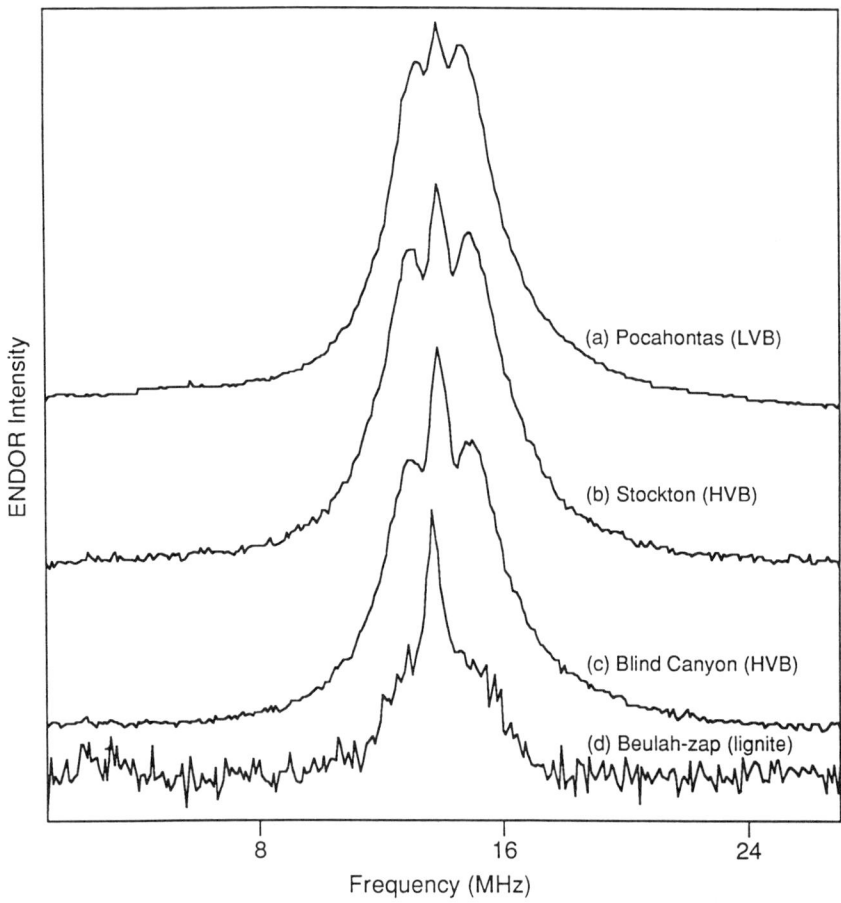

Figure 5. Comparative Davies ENDOR spectra for selected Argonne Premium coals. Experimental conditions: microwave frequency, 9.160 GHz for a–c and 9.050 GHz for d; magnetic field, 3268.7 G (a–c) and 3229.3 G (d); microwave preparation (π) pulse, 0.40 µs (a–d); rf mixing pulse, 17.0 µs (a–d); resolution, 0.1 MHz/point; temperature, 100 K (a–c) and 1.68 K (d); Number of echoes sampled per point, 30 (a–c) and 2 (d); and number of sweeps: (a) 40, (b) 14, (c) 26, and (d) 16. For spectrum d, a very broad signal component was digitally removed using a polynomial fit.

icant change in spectra for the bituminous, subbituminous (not shown), and lignite coals is the relative intensity of the matrix line relative to the local ENDOR enhancements. The relative intensity of the matrix proton line increases with decreasing rank. This result can be rationalized if we assume that the matrix line arises predominantly from the weak dipole coupling to protons on aliphatic portions of the hydrocarbon structures. This assumption is consistent with the widely held view that the aromaticity of coal increases with rank (1).

The magnitude of the ENDOR enhancement varied widely among the samples. Values of $\Delta\chi/\chi$ as low as 0.04 were observed for Pittsburgh No. 8 (HVB or high-volatile bituminous) and as high as 0.23 were observed for Illinois No. 6 (HVB) and Upper Freeport (MVB or medium-volatile bituminous) coals. No simple correlation of coal rank with EPR parameters (e.g., radical density) or ESE parameters (T_{1e} or T_{2e}) could be identified. The magnitude of the enhancements is probably strongly dependent on the mineral content, especially organically complexed transition metals. The largest enhancements were observed for those samples in which the relative intensities of the Fe peaks at $g = 4.2$ and 5.7 in the EPR spectrum were lowest (see Chapter 29).

An expanded region of the ENDOR spectra obtained for the bituminous coals is shown in Figure 6. These spectra were recorded under identi-

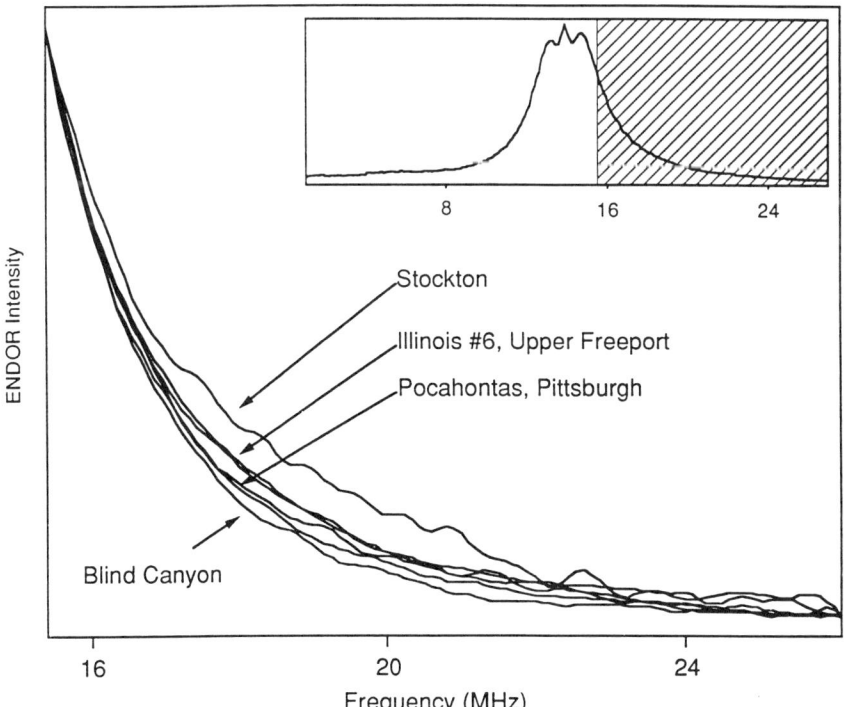

Figure 6. Expanded region of the Davies ENDOR spectrum for several bituminous Argonne Premium coals. The captions to Figures 2 and 4 give experimental conditions.

cal experimental conditions. With the exception of the Lewiston–Stockton (HVB) coal, there is no measurable change in the ENDOR line shapes in the higher frequency range of the ENDOR spectra. The larger enhancements between 17 and 21 MHz observed for the Lewiston–Stockton coal are most likely associated with the significantly higher exinite maceral content of this coal. The Wyodak–Anderson subbituminous and Beulah–Zap lignite coals also appear to have a different ENDOR lineshape profile at higher rf frequency, but a more quantitative comparison is difficult to make at this time. ENDOR spectra were difficult to obtain for these samples because of extremely small enhancement factors. This problem is likely due to the high concentration of minerals, which is consistent with the strong iron signals in the EPR spectra.

Discussion

This chapter reports results of the first pulsed ENDOR study of the carbon radicals in coals. Two overlapping carbon radical lines are observed at $g = 2$ in most of the APC samples. A narrow line, attributed to inertinitic fractions, is observed in all but the Blind Canyon (HVB), Wyodak–Anderson (SB), and Beulah–Zap (L) coals. This narrow line is characterized by comparatively short electron T_{1e} values ($T_{1e} \leq 7.2$ s), and homogeneous narrow line widths ($\Delta\omega_{1/2} \leq 3.5$ MHz, determined from the FID-detected complex FT EPR spectrum). No nuclear sublevel polarization can be created in a homogeneous EPR line or when $\Delta\omega_{1/2} < \Delta\omega_1$. Nuclear sublevel polarization is also difficult to detect if $T_{1e} < t_p(\nu_{rf})$. Under the present experimental conditions, the inertinitic component of the coal is not likely to contribute to the ENDOR spectra.

The ENDOR data presented in this chapter most likely arise from the broad resonance line centered at $g = 2$ in the EPR spectrum. This resonance arises from the carbon radicals in the vitrain fraction of the coal. The most significant findings of this preliminary study are

1. Both matrix and local ENDOR enhancements are observed.
2. Large proton hyperfine couplings, at least up to 20 MHz are observed.
3. Well-resolved ENDOR peaks are not observed.

As discussed in Chapter 29, resonance absorption lines from several types of paramagnetic species are observed in the EPR spectra of the APC.

The correlations between the carbon radical EPR parameters and coal rank and maceral types found in previous EPR studies of demineralized and isolated coal macerals were not observed in EPR studies of the APC samples. Magnetic transition metals can increase the electron spin relaxation rates of the carbon radicals; the result is smaller ENDOR enhancement factors in the pulsed ENDOR experiment. In CW ENDOR spectroscopy, this reduction in the ENDOR enhancement factors is more severe and may partly account for the vanishingly small enhancement factors for strongly coupled protons.

In most previously reported CW ENDOR studies of whole coals, a strong proton matrix line dominated the spectrum. However, in a CW ENDOR study reported by Retcofsky et al. (2), well-resolved hyperfine lines were reported for Adaville (subbituminous, type A, 76.3% C), Pittsburgh (HVB, type A, 82.6% C), and Pocahontas No. 4 (LVB or low-volatile bituminous, 90.4% C) coals. From four to seven discrete hyperfine coupling constants between the range of 1.1 and 27.4 MHz were reported. The results of the present study are consistent with the magnitudes of the coupling constants reported by Retcofsky et al. (2). However, no discrete ENDOR lines were observed in the present study. In the pulsed ENDOR spectra, the proton enhancement is maximum at the proton Larmor frequency and monotonically decreases with both increasing or decreasing rf frequency. Contamination by oxygen has been proposed as a possible mechanism for broadening of lines. Oxygen contamination is known to enhance both the electron and nuclear relaxation rates (10). Oxygen contamination is extremely unlikely to account for the absence of resolved ENDOR lines in the present study. The coal samples were maintained under an argon atmosphere, and the electron spin relaxation rates are much longer than expected for oxygen-exposed samples.

In CW ENDOR spectroscopy, discrete ENDOR lines corresponding to the principal values of the hyperfine tensor can be observed in powdered samples of organic π radicals if $T_D^{-1} << T_{1e}^{-1}$, where T_D^{-1} is the spectral diffusion rate (23, 24). If $T_D^{-1} >> T_{1e}^{-1}$, powder-type ENDOR spectra are observed with no resolved structure. In the pulsed ENDOR experiment, discrete ENDOR lines can be observed if $T_D^{-1} << t_m << T_{1e}^{-1}$, where t_m is the sublevel mixing time during which sublevel polarization transfer occurs (see Chapter 4). Discrete ENDOR lines are probably not observed in the present spectra of coals because each spectrum arises from a large distribution of molecular types with a wide range of hyperfine couplings. Given the chemical and physical heterogeneity of whole coals, this conclusion is certainly not unreasonable. Comparison of the powder ENDOR line shapes (see Figure 6) indicates either that the average size of the core of the host aromatic molecules does not change over the rank range studied (carbon content from 78 to 91%) or that the present level of sensitivity and resolution is not sufficient to detect the shifts in hyperfine couplings associated with the redistribution of spin densities resulting from changes in average aromatic core size. Future experiments on demineralized or isolated macerals should provide insight on whether these results are due to the intrinsic complexity of the coal or whether the spectral resolution is presently limited by some other mechanism such as the broadening and relaxation effects due to minerals. However, the sensitivity of the ENDOR line shape to the presence of a different maceral type observed for the Lewiston–Stockton coal suggests that the former interpretation is correct.

The ENDOR data presented here provide direct evidence that the unpaired electrons in coals are stabilized on polynuclear aromatic (PNA) structures as originally postulated more than 30 years ago (1). This postulation is also consistent with the conclusions from a large number of EPR (2, 3, 25) and NMR (19, 26) studies and with the chemical and physical trends observed with coal rank (1). The magnitudes of the hyperfine coupling constants observed in the present study are consistent with the couplings expected for aromatic protons. Isotropic hyperfine couplings of 17.4 MHz have been observed in solution ENDOR studies of the aromatic protons in phenylalenyl radicals (27). Assuming the usual α-proton hyperfine anisotropy (6), this could give rise to anisotropic couplings of up to 26 MHz. Thus, these data provide direct evidence for paramagnetic PNA radicals, a conclusion that is consistent with the conclusions of Retcofsky et al. (2). However, we also find evidence that PNA radicals with heteroatoms, most likely oxygen or sulfur, contribute to the spectrum. This evidence is derived from the increased ENDOR enhancements observed for the larger hyperfine couplings (see Figure 7) obtained when recording the ENDOR spectrum at higher g values.

ENDOR spectra of Pocahontas No. 3 (LVB) coal, recorded at $g = 2.0026$ and 2.0091, but with otherwise identical conditions to the other coals studied, are shown in Figures 7a and 7b, respectively. An expanded region of the high-frequency part of the spectrum is shown as an inset. Significantly stronger enhancements are observed at higher frequencies in the ENDOR spectrum recorded at $g = 2.0091$. The stronger enhancements are consistent with the g-value selectivity of aromatic free radicals with oxygen and sulfur heteroatoms in the polyaromatic ring system. This result is consistent with the EPR g-value shifts that have been correlated with the oxygen and sulfur content of vitrains. Additional evidence for paramagnetic PNAs with heteroatoms is derived from the g-shifted absorption observed by very high frequency EPR spectroscopy (28).

The present results provide insight into the type of chemistry associated with changes in coal rank. The decrease in the relative intensity of the matrix ENDOR line with increasing coal rank (from 78 to 91% C) is consistent with the expected progressive loss of aliphatic protons. Combined with the observation that the powder ENDOR line shapes do not change with rank over this same rank range, these data suggest that dealkylation and aromatic ring association, but not ring fusion, occur. More quantitative analysis of these chemical effects and trends will be deferred to studies of demineralized and ion-exchanged samples.

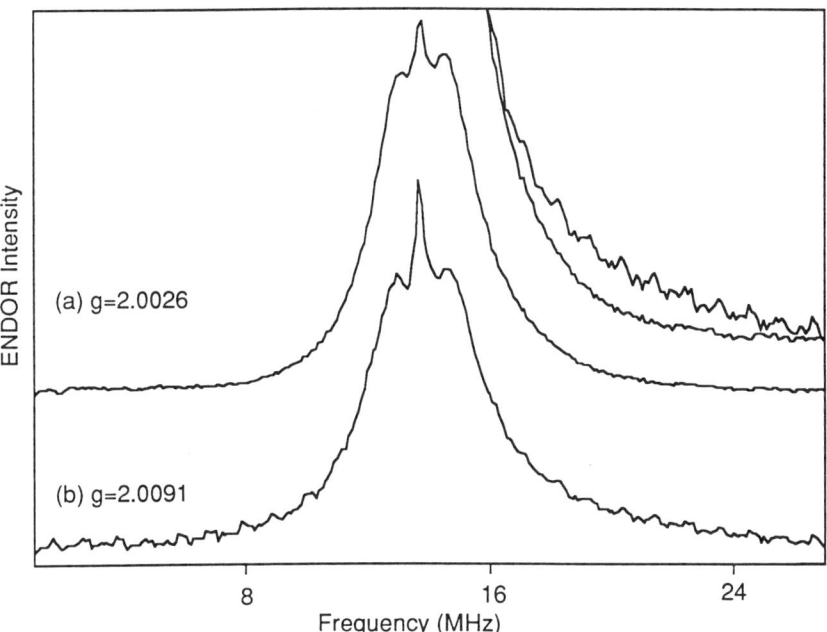

Figure 7. Davies ENDOR spectra for Pocahontas No. 3 (LVB) coal at two g values. Experimental conditions: microwave frequency, 9.075 GHz; 10 echoes sampled per point; T = 4.20 K; and number of sweeps, 4 at g = 2.0026, 6 at g = 2.0091. Other experimental conditions were the same as in Figure 1.

Acknowledgments

We thank M. T. Melchior for many stimulating discussions.

References

1. Van Krevelen, D. W. *Coal;* Elsevier: Amsterdam, Netherlands, 1961; p 393.
2. Retcofsky, H. L.; Hough, M. R.; Maguire, M. M.; Clarkson, R. B. In *Coal Structure;* Gorbaty, M. L.; Ouchi, K., Eds.; Advances in Chemistry 192; American Chemical Society: Washington, DC, 1981; pp 37–58.
3. Silbernagel, B. G.; Gebhard, L. A.; Dyrkacz, G. R.; Bloomquist, C. A. A. *Fuel* **1986**, *65,* 558–565.
4. Thomann, H.; Silbernagel, B. G.; Jin, H.; Gebhard, L. A.; Tindall, P. *Energy Fuels* **1988**, *2,* 333.
5. Kispert, L. D.; Kevan, L. *Electron Spin Double Resonance;* Wiley Interscience: New York, 1979.

6. Wertz, J. E.; Bolton, J. R. *Electron Spin Resonance;* Chapman and Hill: London, 1986.
7. Hyde, J. S.; Rist, G. H.; Eriksson, L. E. G. *J. Chem. Phys.* **1968,** *72,* 4269.
8. Decaillot, M.; Ubersfeld, J. *C. R. Seances Acad. Sci. Ser. B (Compt. Rend.* **1967,** *265,* 155.
9. Schlick, S.; Narayana, P. A.; Kevan, L. *J. Am. Chem. Soc.* **1978,** *100,* 3322.
10. Thomann, H.; Chui, C. W.; Goldberg, I. A.; Dalton, L. R. In *Magnetic Resonance: Introduction, Advanced Topics and Applications to Fossil Energy;* Petrakis, L.; Fraissard, J. P., Eds.; NATO ASI Series, Vol. 124; Reidel: Dordrecht, Netherlands, 1984.
11. Miyagawa, I.; Shibata, K.; Alexander, C., Jr. *Extended Abstracts* (15th Biennial Conference on Carbon, Philadelphia, PA, June 22–26, 1981); pp 464–465.
12. Vorres, K. S. *Energy Fuels* **1990,** *4,* 420–426.
13. Grupp, A.; Mehring, M. In *Modern Pulsed and Continuous Wave ESR;* Kevan, L.; Bowman, M. K., Eds.; Wiley: New York, 1990.
14. Schweiger, A. In *Modern Pulsed and Continuous Wave ESR;* Kevan, L.; Bowman, M. K., Eds.; Wiley: New York, 1990.
15. Feher, G. *Phys. Rev.* **1956,** *103,* 834.
16. Davies, E. R. *Phys. Lett.* **1974,** *47A,* 1.
17. Weger, M. *Bell System Tech. J.* **1960,** *39,* 1013.
18. Bernardo, M.; Thomann, H. *Chem. Phys. Lett.,* accepted, 1992.
19. Gerstein, B. C.; Chow, C.; Pembleton, R. G.; Wilson, R. C. *J. Phys. Chem.* **1977,** *81,* 565–570.
20. Barton, W. A.; Lynch, L. J. *J. Magn. Reson.* **1988,** *77,* 439–459.
21. Solum, M. S.; Pugmire, R. J.; Grant, D. M. *Energy Fuels* **1989,** *3,* 187–193.
22. Mehring, M.; Hofer, P.; Grupp, A. *Phys. Rev.* **1986,** *A33,* 3523.
23. Dalton, L. R.; Kwiram, A. L. *J. Chem. Phys.* **1972,** *57,* 1132.
24. Colligiani, A.; Pinzino, C.; Maniero, A. L.; Brustolon, M.; Corvaja, C. *J. Magn. Reson.* **1980,** *39,* 55–64.
25. Singer, L. S.; Lewis, I. C. *Chem. Phys. Carbon* **1981,** *17,* 1–88.
26. VanderHart, D. L.; Retcofsky, H. L. *Fuel* **1977,** *55,* 202–204.
27. Broser, W; Kurreck, H.; Obstreich-Janzen, S.; Schlomp, G.; Fey, H.-J.; Kirste, B. *Tetrahedron* **1979,** *35,* 1159–1166.
28. Clarkson, R. B.; Wang, W.; Nilges, M. J.; Belford, R. L. In *Processing and Utilization of High Sulfur Coal;* Markuszewski, R.; Wheelock, T. D., Eds.; Elsevier: New York, 1989.

RECEIVED for review June 8, 1990. ACCEPTED revised manuscript September 5, 1991.

31

Temperature Dependence of the Electron Paramagnetic Resonance Intensity of Whole Coals

The Search for Triplet States

Kurt S. Rothenberger, Richard F. Sprecher, Salvatore M. Castellano[1], and Herbert L. Retcofsky[2]

U.S. Department of Energy, Pittsburgh Energy Technology Center, P.O. Box 10940, Pittsburgh, PA 15236

> *The electron paramagnetic resonance intensity of coal as a function of temperature was investigated to determine the contribution of triplet-state species. The importance of considering the intensity behavior over an extended range of temperature without regard to a particular choice of energy separation, J, between the singlet and triplet states, is emphasized. Data collected on a wide variety of whole coals with a nitrogen gas-flow variable-temperature apparatus show a highly linear intensity versus reciprocal temperature relationship, in agreement with the Curie law, but with a nonzero intercept. To remove any ambiguity from these results, a closed-cycle refrigeration system was employed to collect data down to approximately 10 K. Non-Curie law be-*

---

[1]Current address: Istituto di Chimica Fisica, Università di Parma, Viale delle Scienze, 53100 Parma, Italy
[2]Current address: Burns and Roe Services Corporation, P.O. Box 18288, Pittsburgh, PA 15236

havior results at the very lowest temperatures have been attributed to experimental problems involving thermal contact, not to the presence of triplet states. Even if the reported effects were actually a consequence of triplet states caused by charge-transfer interactions, these interactions would have to be so small ($J < 10$ cm^{-1}) as to be insignificant to the chemistry of the system.

DESPITE EXTENSIVE SCIENTIFIC STUDY arising from its obvious economic importance, coal remains a very difficult material to understand. The origin of the electron paramagnetic resonance (EPR) signal in coal is no exception to this statement. Early EPR studies interpreted the signal, at least in most coals, to be due to organic free radicals (1–3). However, soon afterward, it was proposed (4) that donor–acceptor complexes were present in coals and could be responsible for this signal. Studies of petroleum-derived products by Yen (5, 6) lent support to this view. However, Retcofsky et al. (7, 8) subsequently pointed out that no evidence existed for such interactions in coals and argued that isolated doublet-state radicals ($S = 1/2$), primarily aromatic in nature, were responsible for the signal. Duber and Wieckowski (9) announced that triplet states ($S = 1$) were responsible for the broad component of their EPR signal, and that this conclusion was consistent with a model of donor–acceptor complexes in coal put forth in a solvent-extraction study by Marzec et al. (10).

On the other hand, an electron spin-echo study by Doetschman and Mustafi (11) indicated that, on the basis of the pulse duration required to generate a maximum spin-echo signal, the radical involved was likely to have doublet spin multiplicity. Castellano et al. (12) have shown, via a detailed statistical analysis, that within reasonable error limits the EPR signals of fractionated coal-derived residues obey the Curie law exactly and can be accounted for by purely doublet-state radicals.

The existence or absence of donor–acceptor complexes in coal is a matter of fundamental importance to questions involving coal structure. Unfortunately, selected papers from these conflicting EPR studies are often called upon as evidence to support one view or the other. These selective arguments only spread the confusion. Therefore, the purpose of the study reported here was to make a careful examination of the EPR evidence regarding this question by using a wide variety of commonly available coal samples and including data collected over an extended temperature range. A particular emphasis is placed on discussion of limitations of the experiment and the methods of interpreting the data.

The vast majority of previous work involved studies of EPR intensity, I, versus temperature, T. The conclusions as to the origin of the signal were largely based on whether this relationship obeyed the Curie law, which governs spin systems in the doublet state:

$$I = \frac{C}{T} \tag{1}$$

where C is a collection of fundamental constants, and it is assumed that the temperature is not too close to absolute zero (easily satisfied above helium temperatures for EPR spectroscopy). For the doublet-state spin system, a plot of I versus $1/T$ would be described by a linear fit through the origin, a plot of I versus T would be described by a hyperbola, and a plot of IT versus T would be described by a constant.

If the EPR signal were instead due to donor–acceptor interactions between nonradical constituents, such interactions would result in the formation of a coupled system containing a singlet state and a triplet state. For ground singlet-state systems having accessible triplet states and the same spin density as in the aforementioned doublet case (same concentration of spin-producing moieties), the EPR intensity is given by (13, 14)

$$I = \left[\frac{C}{T}\right]\left[\frac{4}{\exp(J/kT) + 3}\right] \tag{2}$$

where J is the energy separation between singlet and triplet states, and k is Boltzmann's constant. In eq 2, J can be either positive or negative. When J is positive, the singlet state lies below the triplet state; when J is negative, the triplet state lies below the singlet state.

To demonstrate the salient characteristics of eq 2 without recourse to a particular choice of the I parameter, we introduce a new variable, α,

$$\alpha = \frac{J}{kT} \tag{3}$$

and we develop the following new functions, $f(\alpha)$ and $g(\alpha)$, which behave analogously to the intensity, I, and the intensity–temperature product, IT, respectively:

$$f(\alpha) = \frac{\alpha}{\exp(\alpha) + 3} \tag{4}$$

$$g(\alpha) = \frac{1}{\exp(\alpha) + 3} \tag{5}$$

These new "intensities" are reported in arbitrary units, so constants of proportionality are neglected. The functions on the right of eqs 4 and 5 are shown in Figure 1. When plotted versus $1/\alpha$ (Figure 1a), they depict

a

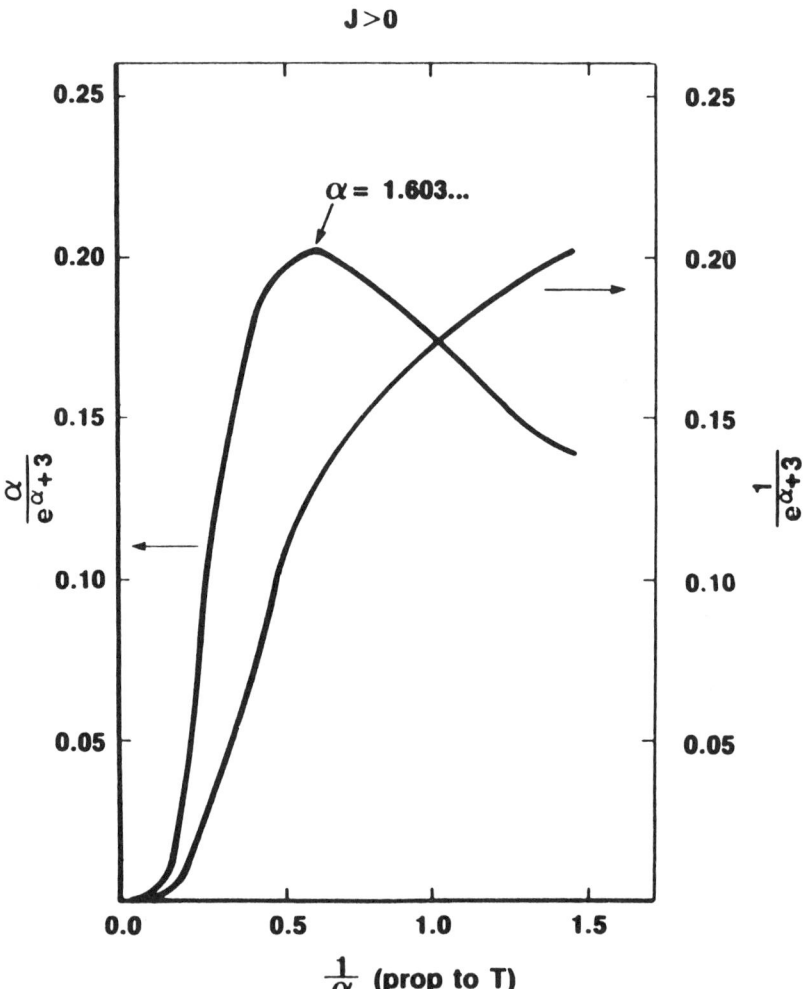

Figure 1. Theoretical plots depicting the arbitrary intensity function $\alpha/[\exp(\alpha) + 3]$ and the arbitrary intensity–temperature product function $1/[\exp(\alpha) + 3]$ vs. reduced temperature ($1/\alpha = kT/J$) for an interacting singlet–triplet system with the energy splitting J. Plot a is for the case where J > 0. The intensity function in a always goes through a maximum at the designated point $\alpha = 1.603....$

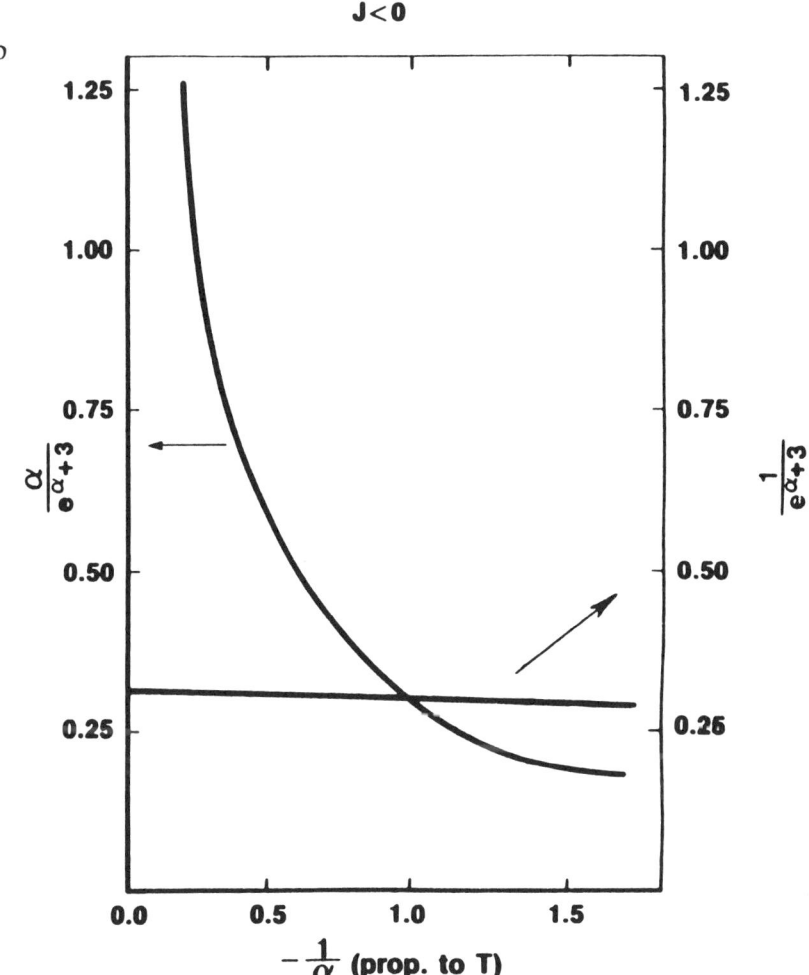

Figure 1. Continued. Plot b is for the case where J < 0.

the case in which $J > 0$, and when plotted versus $-1/\alpha$ (Figure 1b), they depict the case in which $J < 0$. The abscissa represents a reduced temperature; that is, specifying a particular value of I also specifies a linear temperature scale.

When $J < 0$, eq 2 and the plots of Figure 1 are very similar to the trends defined by the Curie law. This result is not unexpected, for at low temperatures the only difference is the presence of an additional degenerate transition. As the temperature is increased, the intensity decreases in both cases as the spin populations equalize. Therefore a ground triplet state would be difficult to distinguish from a doublet state solely by observing I versus T behavior. Fortunately, this possibility has already been addressed and ruled unlikely by electron spin-echo spectroscopy (11). We shall therefore confine further discussions to the existence of $J > 0$ interactions.

Because Figure 1 is plotted against the function, it is useful to gauge some sense of scale for the possible interactions described. By differentiating $f(\alpha)$ with respect to α in eq 4 and setting to zero, the extremum in Figure 1a can be located:

$$(\exp(\alpha) + 3) - \alpha \exp(\alpha) = 0$$

$$\alpha = 1.603\ldots \quad (6)$$

The second derivative reveals that this point always represents a maximum of the function. Therefore, a charge-transfer complex will have a maximum EPR intensity at the temperature

$$T = \frac{J}{1.603k} \quad (7)$$

In the classic model of a charge-transfer interaction, the lowest energy state of a molecular complex consisting of a neutral donor, D, and a neutral acceptor, A, is a singlet state [DA]. The first excited state is a "dative" state $[D^+A^-]$ formed by the transfer of an electron from D to A. The "free" ions D^+ and A^- would each then have one unpaired electron. These electrons could then interact to form a diamagnetic singlet state ($S = 0$) and a nearby paramagnetic triplet state ($S = 1$).

The diagram in Figure 1 is based on an analysis of spin population. Therefore, the energy splitting in question would be that between the ground state and the first available triplet state. These charge-transfer interactions are commonly observed in the UV–visible region from 15,000 to 30,000 cm^{-1}. Here, the maximum in Figure 1 defines a temperature in the range of tens of thousands of degrees. Room temperature would be represented by $1/\alpha$ values much less than 0.1, at which the intensity is negligible. Although such interactions represent valid examples of charge-transfer interactions, these triplet states are clearly not populated to any extent at normal temperatures. This study was not intended to investigate the existence of charge-transfer interactions in coal, but rather their contribution to the EPR signal. Charge-transfer interactions of this type, whether they exist in coal or not, would have no effect on the EPR signal.

For charge-transfer interactions to play an important role in the EPR spectrum of coal at normal temperatures, they would have to be of a type in which the dative, or ionic state [D$^+$A$^-$], formed the ground state of the complex. Complexes of this type would result from the interaction of strong donors and strong acceptors, and were first investigated by Kainer and co-workers (13, 14).

In this case, the J splitting in Figure 1 is that between the singlet and triplet states formed by the interaction of the two ionic species. To provide a maximum contribution to the EPR signal near ambient temperatures, J would be in the range of a few hundred cm^{-1} ($J = 334$ cm^{-1} for a maximum at 300 K). An example of such a complex and its magnetic behavior is given in the references (15, 16).

Experimental Details

Preparation of Samples for Nitrogen-Flow Intensity versus Temperature Studies. Samples were obtained from the Argonne Premium Coal Sample Program (17). The sample ampules, which had been sealed under nitrogen, were opened in a nitrogen-filled glove bag. A small, approximately 10-mm, plug of each sample was placed in a 4-mm o.d. quartz EPR tube that had been fitted with a standard taper joint. Sample plugs were kept short to minimize the development of a temperature gradient across the length of the sample during the variable-temperature experiment. The remainder of the coal sample was transferred to a glass vial that was tightly sealed with a Teflon-lined cap. These vials were then stored together in a desiccator under nitrogen for future sampling.

Upon opening the dry bag, the EPR tube with sample was immediately transferred to a vacuum line. The time from opening the bag to evacuation of the sample was less than 30 s. The samples were evacuated until an ionization gauge downstream of the sample read less than 2×10^{-7} torr (2.7×10^{-5} Pa). The samples were usually left on the line for several days, although the minimum pressure reading could be reached in 1–2 days. While on the vacuum line, sample tubes were flame-sealed with a torch.

Preparation of Samples for Spin-Concentration Measurements. Samples were drawn from the vials containing the unused portions of the Argonne coals from the I versus T experiment. Again, all sample handling was done inside a nitrogen-filled glove bag, except for the final movement to the vacuum line. Samples were degassed, and the tubes were flame sealed as before. The mass of sample removed from the vial was recorded along with the change in mass of the tube (parts) before and after sample preparation to obtain the mass of sample lost during evacuation. The sample plugs were made long enough (30 mm) to completely fill the sample cavity.

Mounting of Samples on Closed-Cycle Refrigerator Unit. 1,1-Diphenyl-2-picrylhydrazyl (DPPH) was obtained from Eastman Chemicals and purified by recrystallization from a chloroform–cyclohexane solution. A fresh vial of an Argonne Premium coal, analytical sample 1 (Pennsylvania Upper Freeport Coal), was opened for this experiment. To provide adequate thermal contact to the extension tip of the refrigeration unit (vide infra), the DPPH was dissolved in Duco cement and a drop of this mixture was glued to the tip. The coal powder was also attached to the tip using the same cement. In this case, the coal did not dissolve, but rather formed a pasty mixture, which was then applied to the tip.

Instrumentation and Software. A Varian model E-112 spectrometer equipped with a field-frequency lock, a model E-102 (X-band) microwave bridge, and a model E-232 dual sample cavity were used in this study. The microwave frequency was measured with an EiP model 350D high-frequency counter. The magnetic field strength and the field scan were calibrated with potassium nitrosodisulfonate (Fremy's salt) in an aqueous sodium carbonate solution on the day of each run. A g-value of 2.00550 for the Fremy's salt was assumed for the field calibration, and the hyperfine splitting was taken to be 1.3 mT for the scan calibration.

In all measurements, the sample was subjected to a 100-kHz magnetic field modulation and the reference, usually another coal sample, was modulated at 100 kHz. Saturation behavior was apparent in all spectra in which both broad- and narrow-line components were observed. All samples, except the Wyodak–Anderson coal, were run at the 0.01-mW microwave power setting (at the bridge), the lowest reliable power available with our bridge. Overmodulation (modulation amplitude of 2–3 G peak-to-peak [pp]) was employed to improve the signal-to-noise (S/N) ratio and to deemphasize the sharp component in many of the sample spectra. Although this overmodulation resulted in distortion of the spectral line shape, the measured intensities were unaffected. All instrumental parameters were held constant during the course of the run, except for sample gain, which was changed once (at the midpoint temperature) to make better use of the dynamic range of the digitizer.

The spectra were recorded in 100-G sweeps as a series of 4096 data points by a Compaq Deskpro 386 computer equipped with a Metrabyte AIO20 data-acquisition board, with data acquisition and processing software developed in this laboratory (by Sprecher and Rothenberger). Intensities were obtained by a double integration of the first-derivative EPR signal. The intensity obtained for each sample spectrum was compared against the intensity obtained for the reference spectrum run simultaneously with the sample. The intensity of the reference was calibrated daily in a separate measurement. In that calibration, the reference intensity was compared against the intensity of a standard sample of DPPH in silica, known to contain 6.31×10^{-6} mol of DPPH per gram of silica when prepared. The intensities were reported by the software in units of spins per gram. This unit is inappropriate for variable-temperature measurements, because it implies equal temperature and common origin between sample and reference. For this reason, the intensity data in the I versus T results are reported in arbitrary intensity units.

Nitrogen-Flow Variable-Temperature Measurements.

Both halves of the cavity were subjected to a stream of nitrogen gas. The reference-half stream was used merely to ensure a constant-temperature environment for the reference sample throughout the I versus T run. The sample-half gas stream was controlled by a Union Carbide FM-4550 mass flowmeter, cooled in liquid nitrogen, and then reheated by a Varian variable-temperature controller. The set points were calibrated with a copper–constantin thermocouple embedded in an EPR tube filled with alumina. In a separate calibration run, this tube was put through the same sequence of set points in the same order as the I versus T run for the coal samples.

The thermocouple voltages (FLUKE model 893A differential voltmeter) were converted to temperatures by using values from National Bureau of Standards (NBS) tables. The thermocouple itself was checked against the liquid-nitrogen boiling point, the carbon dioxide sublimation point, and the ice–water point, and the agreement was within $2°$ in each case. The temperature values corresponding to the set points agreed within $5°$ upon a second calibration check several weeks later. The temperature values presented in this chapter represent an average of the two calibrations.

Spectra were generally signal-averaged from 5 to 25 scans. Replicate intensity measurements done on the same sample consecutively usually agreed within $\pm 0.3\%$. Replicate intensity measurements done on the same sample on different days usually agreed within $\pm 3\%$. Reported intensities represent an average of the intensities over the number of measurements taken (minimum two) at a given temperature.

Closed-Cycle Variable-Temperature Refrigeration Equipment.

To collect data at temperatures below those accessible to the nitrogen-flow unit, an Air Products model DE-202 Displex closed-cycle refrigeration system was used. In this apparatus, cooling is obtained upon expansion of helium gas by a two-stage piston arrangement with maximum cooling at a "cold head" located at the end of the second piston. The temperature at the cold head was monitored with a gold (0.07% iron) versus Chromel thermocouple and adjusted via a 20-W resistive heater. The temperature was monitored at and controlled with an Air Products model 3700 digital temperature indicator–controller.

The sample was mounted at the end of an extension tip of either copper or sapphire that was screwed into threads in the cold head. Various greases were used to assist in making a good thermal contact. The whole piston–cold-head arrangement was covered with an APD Cryogenics model DMX–15 stainless steel vacuum shroud. The shroud was fitted with a 10.5-mm (o.d.) quartz cuvette to cover the extension tip and sample. The area contained by the shroud and cuvette was evacuated with a Leybold–Heraeus model TMV-10000 turbomolecular pumping system or sometimes simply with the rough pump from that system.

The refrigeration unit was suspended, tip down, on a harness from the bridge table. The extension tip with sample, covered by the cuvette, was inserted into the waveguide side of the E-232 cavity. The reference sample was again jacketed by a quartz insert with an ambient-temperature nitrogen-gas flow to maintain constant temperature. The presence of the extension tip in the cavity, especially the copper tip, and the mechanical vibration of the refrigeration unit resulted in a somewhat degraded S/N ratio. For the coal sample, five time-averaged scans constituted a measurement. The refrigeration system was cycled up and down in temperature, and the measurements, four in all, were made in both directions. Any intensity differences observed were within random error of the measurement, about 5%. Reported intensities represent an average of the four measurements taken.

Results

Spin-Concentration Measurements. EPR spectra of the eight Argonne Premium coal samples are shown in Figure 2. Intensity data for these samples is reported in Table I. Final intensities are reported on samples as prepared (after evacuation) and are adjusted for the presence of mineral matter in the sample. Therefore tabulated data represent the spin concentration in the organic part of the coal. These spin concentrations were used to scale the I versus T results for the nitrogen gas-flow variable-temperature runs presented in subsequent figures.

Variable-Temperature Measurements with Nitrogen Gas-Flow System. Curie law plots of intensity versus reciprocal temperature are shown for the eight Argonne Premium coal samples in Figure 3. All samples have an EPR intensity span of approximately 2.7-fold over the temperature range from 105 to 303 K. Each set of 12 data points has been fit by a linear least-squares regression. When these lines are extrapolated to infinite temperature ($1/T = 0$), they all show small positive intercepts.

To determine if this behavior was characteristic of the coal or of the experiment, the standard sample of DPPH in silica was subjected to the

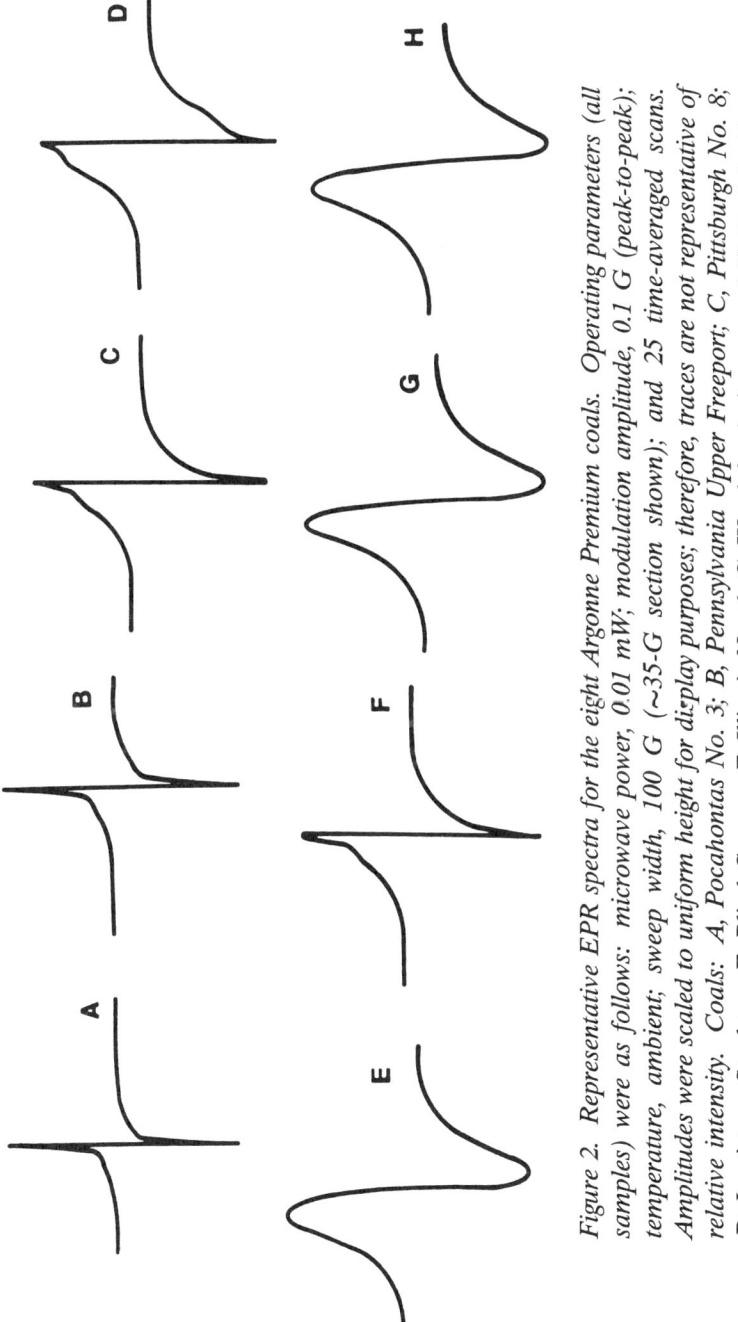

Figure 2. Representative EPR spectra for the eight Argonne Premium coals. Operating parameters (all samples) were as follows: microwave power, 0.01 mW; modulation amplitude, 0.1 G (peak-to-peak); temperature, ambient; sweep width, 100 G (~35-G section shown); and 25 time-averaged scans. Amplitudes were scaled to uniform height for display purposes; therefore, traces are not representative of relative intensity. Coals: A, Pocahontas No. 3; B, Pennsylvania Upper Freeport; C, Pittsburgh No. 8; D, Lewiston–Stockton; E, Blind Canyon; F, Illinois No. 6; G, Wyodak–Anderson, and H, Beulah–Zap.

Table I. Spin Concentration Values for Argonne Premium Coal Samples

Sample	Mineral Matter[a] (% LTA)	Intensity (10^{19} spins/g)	
		As Prepared	Adjusted
Pocohontas No. 3	5.5	2.932	3.102
Upper Freeport	15.3	1.893	2.171
Pittsburgh No. 8	10.9	1.701	1.909
Lewiston–Stockton	21.6	2.961	3.777
Blind Canyon	5.3	2.858	3.018
Illinois No. 6	18.1	1.169	1.427
Wyodak–Anderson	8.7	2.200	2.409
Beulah–Zap	8.7	3.059	3.350

[a] Mineral matter data were taken from reference 18. LTA means low-temperature ash.

same I versus T experiment as the coals. The DPPH exhibited the same behavior as the coals. In Figure 4, the ratio of the coal intensity to the DPPH intensity is plotted against temperature for the eight Argonne Premium coals. The result is a series of nearly horizontal lines (some slopes are slightly positive, and others are slightly negative), showing the variation within each sample set is small compared to the differences in intensities between samples.

Variable-Temperature Experiments with Closed-Cycle Refrigerator. The effect of different extension tips and attachment methods to the cold head is demonstrated in Figure 5. In this figure, the I versus T trace is shown for a sample of DPPH mounted at the end of a sapphire tip, a copper tip, and a copper tip attached to the cold head with thermally conductive copper-containing grease. Because the DPPH is expected to follow Curie law behavior to sub-liquid-helium temperatures, the signal intensity provides a method of measuring the actual temperature at the sample. As the cold head was controlled from room temperature to 10 K, the signal intensities varied approximately 3-fold, 9-fold, and 15-fold, for samples mounted at the end of the sapphire tip, the copper tip, and the copper tip attached to the cold-head with metal-containing conductive grease, respectively.

A copper tip with the conductive grease was used to mount the coal sample for the I versus $1/T$ data shown in Figure 6. The nitrogen-gas flow I versus $1/T$ data are also shown for comparison. Instead of scaling intensities to spin concentrations, the numbers from all runs were normalized to an intensity of 1.0 at 300 K so all data could be displayed on the same plot.

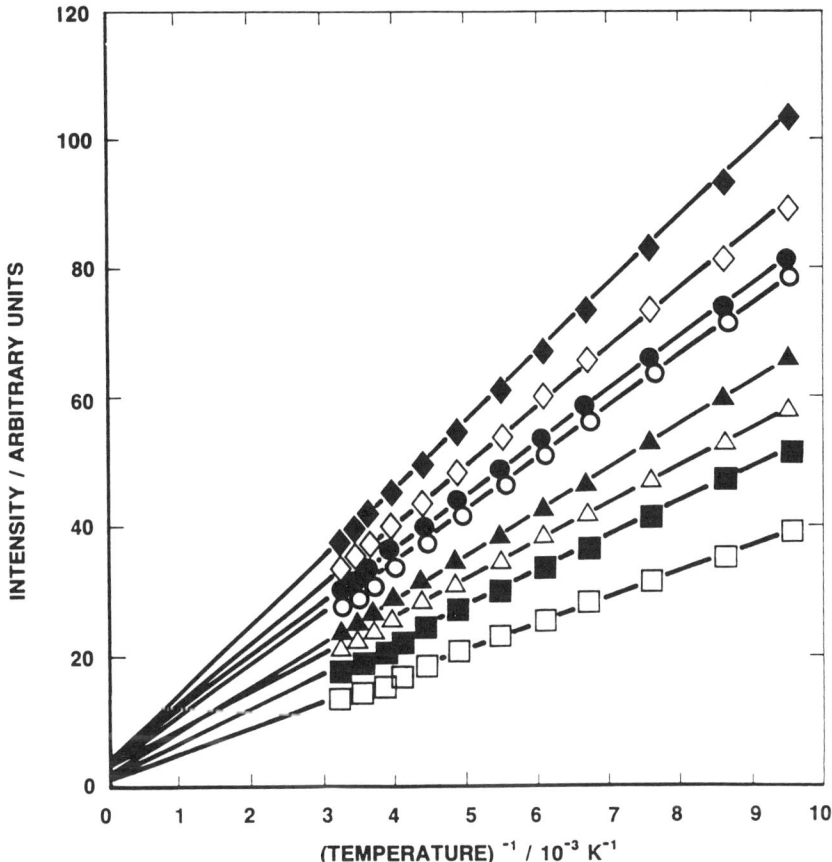

Figure 3. Variable-temperature EPR data for the Argonne Premium coals: ◇, *Lewiston–Stockton;* ◆, *Beulah–Zap;* ●, *Pocahontas No. 3;* ○, *Blind Canyon;* △, *Wyodak–Anderson;* ▲, *Pennsylvania Upper Freeport;* ■, *Pittsburgh No. 8; and* □, *Illinois No. 6. The linear fit was extrapolated to show the sign of the intensity intercept.*

Discussion

The I versus $1/T$ plots in Figure 3 are very similar to those published by Retcofsky et al. almost 10 years ago (8). Repeating the experiment with more data points per coal, better preserved samples, and instrumental advantages such as computer-controlled data acquisition and processing resulted in surprisingly little change in the character of the data. Excellent linear fits, a requirement of the Curie law, were obtained for all eight samples over the range from room temperature to approximately 100 K. A zero intercept, a more stringent requirement of the Curie law, is not sat-

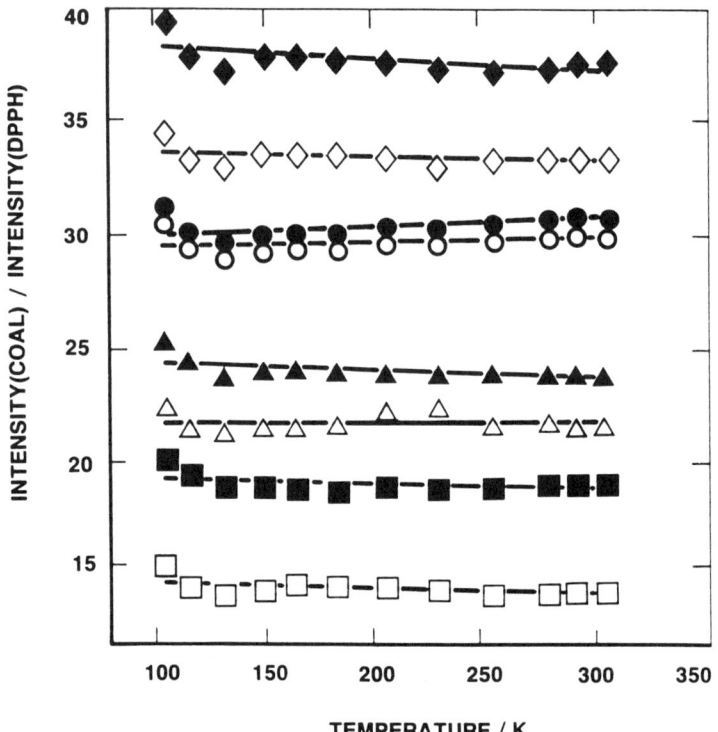

Figure 4. Intensity data from Figure 2 divided by intensity of the standard DPPH sample at same temperature as the coal. The resulting data were rescaled for the plot. The Argonne Premium coals are as follows: ◇, *Lewiston–Stockton;* ◆, *Beulah–Zap;* ●, *Pocahontas No. 3;* ○, *Blind Canyon;* △, *Wyodak–Anderson;* ▲, *Pennsylvania Upper Freeport;* ■, *Pittsburgh No. 8; and* □, *Illinois No. 6.*

isfied by the data. The intercepts obtained, however, are small and tend to be slightly positive. Retcofsky et al. (8) described this as "Curie law-type behavior" and used it as an argument to support the free radical hypothesis.

A small positive intercept in a plot of I versus $1/T$ would result in a small positive slope in a plot of IT versus T. Duber and Wieckowski (9), also with similar data, preferred the IT versus T representation and used the small slope present in the plot as evidence for donor–acceptor interactions. They reported a J value of 379 cm^{-1} via an analysis similar to that of Yen and Young (6). In a detailed error analysis, Castellano et al. (12) argued that the intercept of an I versus $1/T$ plot (or the slope of an IT

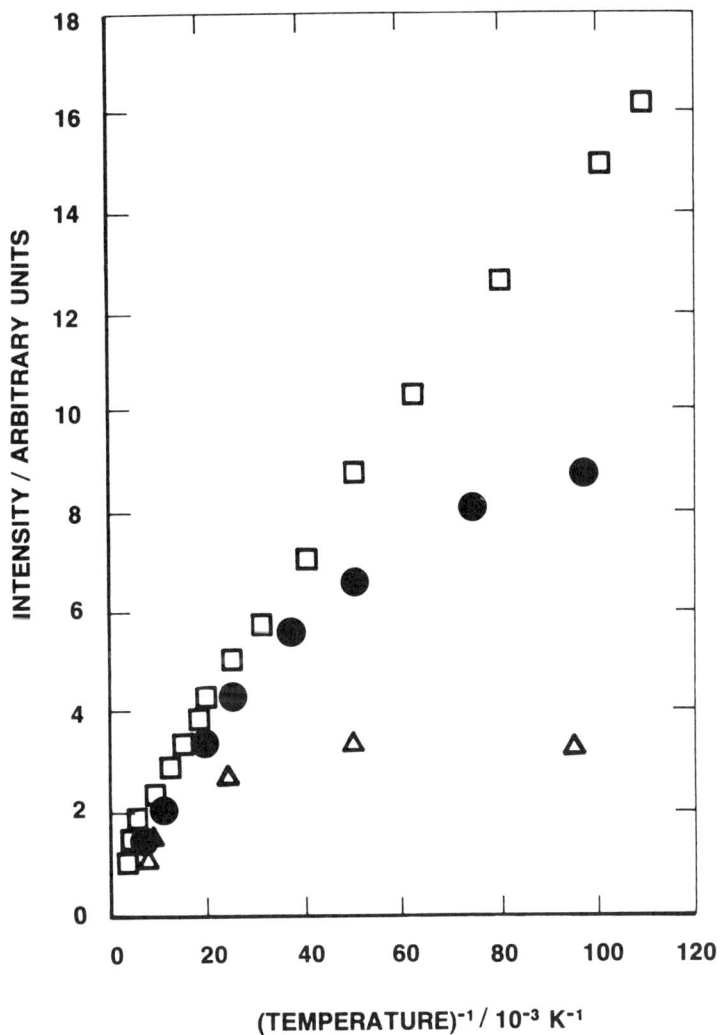

Figure 5. Refrigeration system sample thermal contact. EPR intensity data for DPPH are plotted vs. apparent temperature (read at the cold head) for the following sample attachment methods: △, sapphire extension tip; ●, copper extension tip; and □, copper extension tip with conductive metallic grease aiding contact.

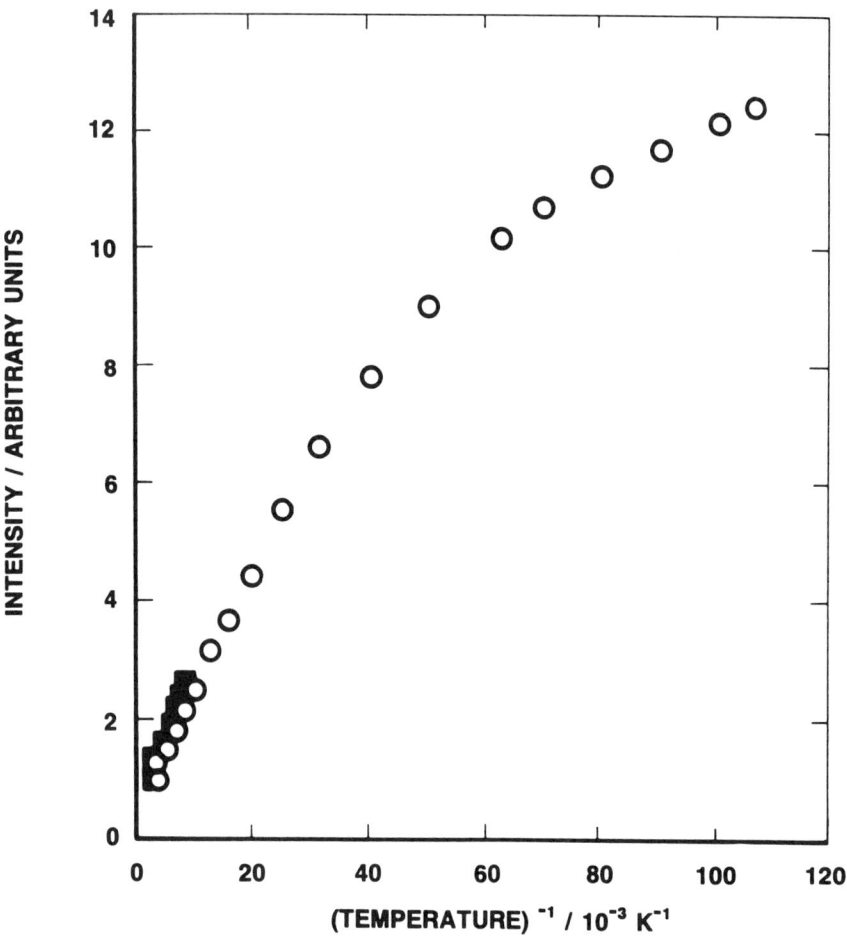

Figure 6. Variable-temperature EPR intensity data for Argonne coal sample from 10 to 300 K on closed-cycle refrigerator (○). Nitrogen gas-flow data (■) are shown on same scale for comparison. All intensities are normalized to 1.0 at 300 K.

versus T plot) plays the role of a sink for the cumulative effects of all systematic errors in the measurement and that failure to recognize it as such may lead to erroneous conclusions.

Much of the data interpretation hinges on whether the small intercept in an I versus $1/T$ plot (or the slope in an IT versus T plot) is meaningful. Perhaps some perspective can be brought to the situation by comparing the I versus T behavior of coal with that of DPPH physically diluted in silica. DPPH is a stable, relatively noninteracting free radical that we expect (and will later show) to obey the Curie law over a wide temperature range. When used as the object of study, the DPPH sample exhibits virtually the same I versus T behavior as the coal. In the plots of Figure 4, where the coal intensity has been divided by the DPPH intensity at each temperature, the resulting fits are a series of straight, nearly horizontal lines. The small but random slopes indicate that the DPPH too possesses a positive intercept that is intermediate in magnitude to that of the coals. The fact that eight coal samples of widely varying rank give nearly identical experimental trends and that the magnitude of differences within data sets are small compared to those between data sets should serve as a warning against attaching too much significance to small deviations from ideal behavior.

Although the DPPH data are illustrative, they are by no means conclusive, and a more rigorous examination of the question is in order. The difficulty in determining the presence (or absence) of triplet-state contribution to the EPR signal is that often data are collected over only a small portion of the reduced temperature $(1/\alpha)$ axis in Figure 1a. Over a small range, a linear IT versus T plot with a small positive slope could easily be interpreted as falling on a linear portion of the $1/[\exp(\alpha) + 3]$ versus $1/\alpha$ curve (see Figure 1a). Yen and Young's (6) calibration curves, used to fit their petroleum asphaltene data, simply represent sections of this plot with the abscissa (temperature) either expanded or compressed by a particular choice of J.

The problem of pursuing this type of analysis over a limited range of data can be illustrated as follows. Suppose that a set of EPR intensity data is fitted by the correlation

$$I = b + \frac{a}{T} \tag{8a}$$

or

$$IT = a + bT \tag{8b}$$

The plot of IT versus T is a straight line. If the Curie law is obeyed exactly, then $b = C$, the Curie constant from eq 1, and $a = 0$; that is, the line is parallel to the temperature axis. Otherwise, the line has a small slope, a.

By contrast, a system consisting of fractional contributions from both doublet-state and singlet-state–triplet-state species can be described by the following nonlinear equation:

$$IT = A + \frac{4B}{\exp(J/kT) + 3} \qquad (9)$$

where A and B are directly proportional to the fraction of doublet-state and singlet-state–triplet-state species, respectively.

Fitting the linear doublet-state function in eq 8 to the nonlinear triplet-containing function represented by eq 9 yields an illustrative result. This fit can be done, as even a nonlinear function is highly linear over a limited range about its flex points. These flex points can be found by taking the first and second derivatives of the function on the right side of eq 9 with respect to T. Setting the second derivative to zero and using eq 3 to express results in terms of α gives

$$2\alpha \exp(\alpha) - (\exp(\alpha) + 3)(\alpha + 2) = 0 \qquad (10)$$

Solving numerically yields the values of α at the flex points: $\alpha = 2.845...$ and $\alpha = -2.160....$

By forcing linear data to conform to this analysis, the best fit will result in the region near the flex points of the function. If T_M represents the median temperature of the set of measurement, then to have physical meaning, fractional contributions A and B must both be positive. Therefore, we can set

$$J = 2.845\, kT_M \qquad (11)$$

for positive intercepts, and

$$J = -2.160\, kT_M \qquad (12)$$

for negative intercepts.

For a typical nitrogen gas-flow experiment with data over the range 100–300 K, $T_M = 200$ K. Therefore, $J = 395$ cm^{-1} for positive intercepts, and $J = -300$ cm^{-1} for negative intercepts.

Quite often, when a fit of experimental data is made by this method, the validity of the results is judged not so much by the goodness of the fit, but rather by the physical plausibility of the value of the J parameter. Because charge-transfer interactions would contribute most strongly to the ambient-temperature EPR spectrum if the J value was approximately 340 cm^{-1} (as explained in the introductory comments), the positive intercept value of 395 cm^{-1} would seem to be perfectly reasonable. However, this value is clearly a consequence of the fitting procedure, and all sets of EPR intensity data gathered in the same temperature range with a positive intercept will yield approximately the same value of J. The fact that these values are physically plausible is purely accidental and does not by itself warrant an interpretation based on the existence of triplet states.

Another scenario from Figure 1a should also be considered. Perhaps all of the data were collected in a region of higher temperature than the α = 1.603 maximum. In this region, charge-transfer behavior is difficult to distinguish from Curie law behavior. The intensity function decreases with temperature in both cases, and the IT product function is constant for the Curie case, but is slowly increasing in the singlet–triplet model. This latter behavior is precisely that expected if a small positive intercept is present in otherwise linear data. Therefore, the possibility of a singlet–triplet model cannot be ruled out solely on the basis of data from this limited temperature range. Other reasons exist for increasing the temperature range. Although we have previously shown how small deviations from Curie law behavior may be misinterpreted, we have not shown conclusively that they are being misinterpreted. Clearly, a better experiment is needed.

However, extending the range of temperature below the nitrogen boiling point is not easy, as demonstrated by the results shown in Figure 5. The temperature measurement and heater control on the refrigeration unit are attached to the cold head, but the sample is located some 3 inches away at the end of the extension tip. Without proper thermal contact, the temperature at the two places may be vastly different.

The DPPH sample mounted at the end of the extension tip of the refrigeration system would be expected to closely follow Curie law behavior, or at the very least to provide a reproducible I versus $1/T$ trace. Therefore, a measurement of the temperature at the end of the extension tip can be made by projecting the actual intensity value for the DPPH onto a straight-line fit of the I versus $1/T$ data (or onto the very nearly straight-line fit obtained with the copper tip and conductive grease). Sapphire is a good conductor, but the temperature at the end of the sapphire extension tip apparently never got much below 70 K. The copper tip did better, but was still about 10° off the trace defined by using the same tip with conductive metallic grease to aid thermal contact. Even this best experimental arrangement gave intensity data that corresponded to a minimum temperature about 7° above what it would have been if the data were to follow an idealized Curie law line defined by the highest temperature points. There is no way of knowing for sure if the curvature on this trace is real or a result of experimental error. It is easy to see, however, that failure to consider the implications of poor thermal contact, heat leakage, or any other phenomena that would result in incorrect temperature readings may lead to drastic misconceptions about sample I versus T behavior.

With these caveats in mind, we consider the I versus $1/T$ data taken for the coal sample on the Displex refrigerator and plotted in Figure 6. For comparison, the nitrogen gas-flow temperature data sets were plotted on the same graph. The inverse temperature scale has been extended by a factor of 10 by using this refrigeration system. Although the plot has noticeable curvature, the I versus $1/T$ trace is surprisingly linear down to below 20 K ($10^3/T = 50$).

Obviously, the source of the curvature must be addressed. If the curvature is real and indicative of a nearby intensity maximum (i.e., $\alpha = 1.603$ in Figure 1a), this maximum would have to occur at a temperature of less than 10 K. This temperature would force J to be less than 11 cm^{-1}. In the search for important intermolecular forces in coal, the significance of this intensity maximum would be dubious at best. In a heterogeneous, chemically bonded "framework" model of coal, with radicals randomly distributed throughout the structure, weak exchange interactions may possibly take place between radicals as a result of their chance juxtaposition in the solid matrix. But this observation would clearly seem to be a poor reason to invoke a structural model that is heavily dependent on donor–acceptor interactions.

We would prefer to take a different view of the results of Figure 6. On the basis of the data presented in Figure 5, it is unrealistic to expect the trace of coal to follow an idealized I versus $1/T$ curve as well as does the DPPH sample, for reasons of thermal contact alone. The DPPH was actually solubilized by the cement—it was a frozen solution with thermal contact on the molecular scale. The coal sample, however, did not dissolve and was merely a physical admixture of powder and glue. Thus, we are left to rely on grain-to-cement or grain-to-grain thermal communication which, considering the extreme temperatures involved and the possibility of differential shrinkage, would certainly be less efficient. We must also consider the possibility of radiative heat-transfer from the room-temperature environment outside the quartz cuvette. For these reasons, we believe the curvature exhibited in the very low temperature region of Figure 6 is primarily of thermal origin and is not indicative of magnetic interactions between spins.

Finally, whether the possibility of fractional contributions to the EPR signal from both doublet and triplet species changes the analysis has not been considered. Although it may seem surprising, the argument is no different than that applied to the case of pure states. In an effort to address this question, Figure 7 was constructed. The six traces on each graph describe the I versus $1/T$ behavior [$A\alpha + (4B\alpha/[\exp(\alpha) + 3])$ versus α], and the IT versus T behavior [$A + (4B/[\exp(\alpha) + 3])$ versus $1/\alpha$], for various mixed systems. The fractional contributions of doublet and triplet states, respectively, range from pure doublet (top trace) to pure triplet (bottom trace) in 20% increments. Once again, the choice of J expands or

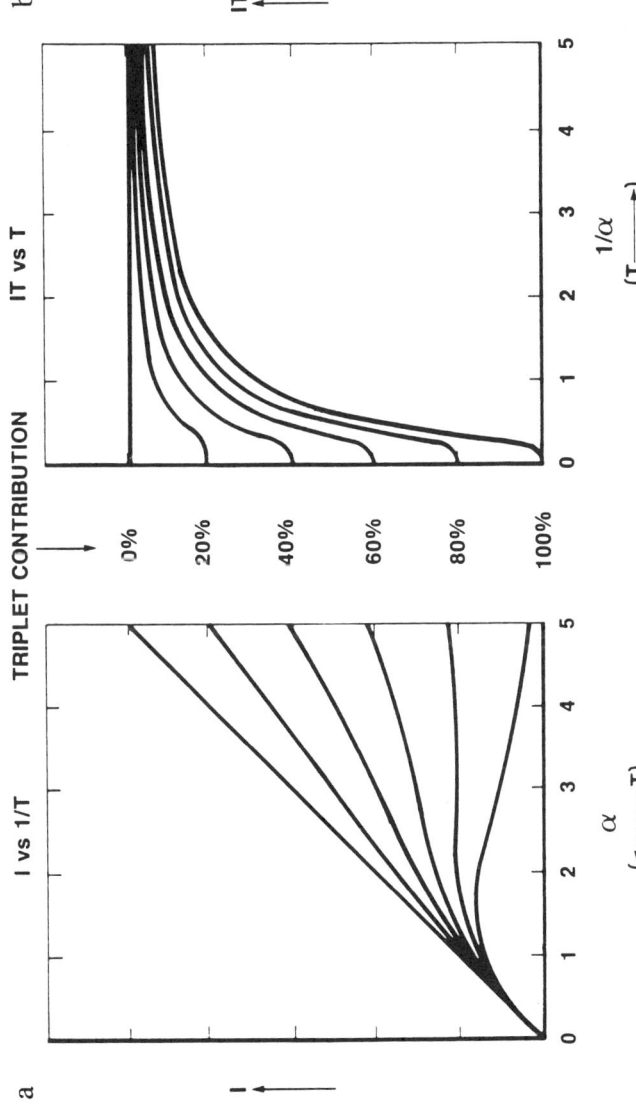

Figure 7. Theoretical plots depicting (a) arbitrary intensity function, $A\alpha + 4B\alpha/[exp(\alpha) + 3]$, vs. inverse reduced temperature ($\alpha = J/kT$) and (b) the intensity-temperature product function, $A + 4B/[exp(\alpha) + 3]$, vs. reduced temperature ($1/\alpha = kT/J$), where A and B represent the fractional contributions of doublet and singlet-triplet species, respectively. All plots were done for $J > 0$. The numbers between the two plots label the percentage contribution of triplet state for each trace from top to bottom.

compresses the temperature scale of the abscissa. The $1/[\exp(\alpha) + 3]$ trace of Figure 1a is the same as the pure triplet (100%) trace of Figure 7b.

A maximum does not occur within the range of our intensity data. For purposes of discussion, we first assume that all the data were taken in the low-temperature range and that any maximum still lies at a much higher temperature. With coal, the option to extend the range of study to higher temperature does not exist, as irreversible changes in the spin concentration will begin to take place relatively quickly upon heating (19). For large values of J (>1000 cm^{-1}), the triplet state is simply not populated, and doublet-state contributions as low as 0.03% will dominate the EPR spectrum at room temperature and below (20). This situation is similar to the case of the singlet ground-state system already discussed. Such interactions may exist, but they contribute nothing to the EPR intensity.

The second assumption for discussion is that the triplet contribution is substantial but not dominating enough to cause an outright maximum in the I versus $1/T$ plot. The doublet contribution is strong enough that a very nearly straight-line I versus $1/T$ plot results with triplet contributions up to almost 60% (see Figure 7a). However, here it is important to consider the magnitude of the intensity intercept resulting from extrapolation of the linear portion of the data to the infinite-temperature axis.

The magnitude of this extrapolated intercept is proportional to the fraction of triplet state in the system (Figure 7a). Even though the intensity units on the intercept of Figure 7a cannot be quantified in absolute terms, they can still be measured relative to the intensity values elsewhere in the plot. For example, a J-splitting value of 400 cm^{-1}, a singlet–triplet contribution of 40%, and data taken up to room temperature ($\alpha \approx 2$) would yield an extrapolated intercept of about one-third that of the room-temperature value. The extrapolated intercepts obtained in this study are at most 1/10 of the room-temperature values. Even if these intercepts were not in any way the result of systematic errors or experimental uncertainties, they would only represent very minor contributions to the EPR signal from singlet–triplet species.

The last assumption for discussion is that all the data were taken in the high-temperature region, above which the intensity contribution of the triplet state has reached a maximum. Here, all fractional possibilities quickly merge into a single line, and distinguishing doublet from triplet becomes impossible. But the situation is not lost, because in principle, the low-temperature option always exists. The importance of this low-temperature option is most easily seen on the plots of Figure 7b (analogous to IT versus T). Regardless of the fractional contributions, the triplet state will eventually depopulate, and if there is any amount of doublet state present, it will dominate the EPR spectrum. The temperature at which the triplet state depopulates is not dependent on the fraction present, but rather on the value of J.

An analogous argument can be put forth to address the possibility of multiple contributions from a series of different charge-transfer species with differing values of J. Eventually all of the triplet states will depopulate, leaving the doublet-state fraction to dominate or showing a decreasing intensity with decreasing temperature if no such doublet state exists. In practice, it may not always be possible to reach such low temperatures. However, it is possible, as we have done, to place an upper limit on the interaction, in effect "pushing" the value of J into a range that is not realistic or would have little bearing on the chemistry of the system. If the nonlinearity in the plot of Figure 6 is representative of triplet-state contributions, then such states have still not depopulated at a temperature of 10 K, leading to the conclusion that $J < 10$ cm^{-1}, and such species are essentially noninteracting doublet radicals.

Conclusion

The spins observed by EPR spectroscopy are not indicative of charge-transfer interactions in any of the eight coal samples studied. Interactions in which the "no-bond" structure [DA] lies lower than the dative structure [D$^+$A$^-$] will not have thermally accessible paramagnetic states and will not contribute to the EPR intensity. In strong donor and acceptor species, the dative structure may lie at a lower energy corresponding to a situation where complete electron transfer may occur. Previously diamagnetic donor and acceptor species will now exhibit paramagnetism. If the interaction is very strong, the triplet state will again be inaccessible and will not contribute to the EPR intensity. The case of moderate-to-weakly interacting species is the only one that could give rise to the effects observable by EPR spectroscopy. We have seen no evidence of such contributions and have shown that if they do exist, they would be the result of interactions so weak as to exhibit essentially independent doublet-state radical species.

Acknowledgments

This publication is based on work performed in the Postgraduate Research Training Program under Contract No. DE-AC05-76OR00033 between the U.S. Department of Energy and Oak Ridge Associated Universities.

References

1. Van Krevelen, D. W. *Coal;* Elsevier: Amsterdam, Netherlands, 1961; pp 394–399.
2. Tschamler, H.; De Ruiter, E. In *Chemistry of Coal Utilization;* Lowery, H. H., Ed; Wiley: New York, 1963; Suppl. Vol.; pp 35–118.
3. Ladner, W. R.; Wheatley, R. *Brit. Coal Util. Res. Assoc. Mon. Bull.* **1965,** *29,* 201–231.
4. Elofson, R. M.; Schultz, K. F. *Prepr. ACS Div. Fuel Chem.* **1967,** *11,* 513–520.
5. Yen, T. F. *Fuel,* **1973,** *52,* 93–98.
6. Yen, T. F.; Young, D. K. *Carbon,* **1973,** *11,* 33–41.
7. Retcofsky, H. L.; Thompson, G. P.; Hough, M.; Friedel, R. A. In *Organic Chemistry of Coal;* Larsen, J. W., Ed.; ACS Symposium Series 71; American Chemical Society: Washington, DC, 1978; pp 142–155.
8. Retcofsky, H. L.; Hough, M. R.; Maguire, M. M.; Clarkson, R. B. In *Coal Structure;* Gorbaty, M. L.; Ouchi, K., Eds.; Advances in Chemistry 192; American Chemical Society: Washington, DC, 1981; pp 37–58.
9. Duber, S.; Wieckowski, A. B. *Fuel* **1984,** *63,* 1474–1475.
10. Marzec, A.; Juzwa, M.; Betlej, K.; Sobkowiak, M. *Fuel Proc. Tech.* **1979,** *2,* 35–44.
11. Doetschman, D. C.; Mustafi, D. *Fuel,* **1986,** *65,* 684–693.
12. Castellano, S. M.; Chisolm, W. P.; Sprecher, R. F.; Retcofsky, H. L. *Anal. Chem.* **1987,** *59,* 1726–1731.
13. Kainer, H.; Bijl, D.; Rose-Innes, A. C. *Naturwissenschaften* **1954,** *13,* 303–304.
14. Bijl, D.; Kainer, H.; Rose-Innes, A. C. *J. Chem. Phys.* **1959,** *30,* 765–770.
15. Scott, J. C.; Garito, A. F.; Heeger, A. J. *Phys. Rev. B* **1974,** *10,* 3131–3139.
16. Tomkiewicz, Y.; Taranko, A. R.; Torrance, J. B. *Phys. Rev. B* **1977,** *15,* 1017–1023.
17. Vorres, K. S. *Int. Conf. Coal Sci.* **1987,** 937–939.
18. Harvey, R. In *User's Handbook for the Argonne Premium Coal Sample Program;* Vorres, K., Ed.; Argonne National Laboratory: Argonne, IL, 1989; Vol. 2, p 13.
19. Sprecher, R. F.; Retcofsky, H. L. *Fuel* **1983,** *62,* 473–476.

RECEIVED for review June 8, 1990. ACCEPTED revised manuscript December 20, 1991.

32

Measurement of Electron Dipolar Fields and Dynamics in Solids

Michael K. Bowman

Chemistry Division, Argonne National Laboratory, Argonne, IL 60439

> *The spin-temperature treatment used in solid-state NMR spectroscopy and the statistical treatment used in electron spin-echo (ESE) spectroscopy are compared. The statistical treatment is shown to be an exact solution of a multispin Hamiltonian with certain bilinear terms treated as relaxation processes. The magnitude of the dipolar field and its autocorrelation function have been identified in both treatments. This allows the transfer of experimental data on electron dipolar fields to nuclear spin systems in solids containing large numbers of paramagnetic centers. Measurements of the dipolar field and its relaxation time are demonstrated on samples from the Argonne Premium Coal Sample program. Even brief exposure to air has significant effects on the electron-spin dynamics. Sample preparation is an important aspect of magnetic resonance experiments on coal.*

MAGNETIC RESONANCE IN SOLIDS provides two quite different ways to characterize a spin system: by its spectrum and by its dynamics. The spectral features of a spin system are most useful when the spins interact weakly with each other. The spectra of weakly interacting systems are characteristic of the individual molecules in the system, and these spectra provide a powerful means of identification. The spin dynamics, in contrast, are largely determined by the environment of the individual electron spins through spin–lattice and spin–spin interactions that control relaxation rates. Those rates provide a unique probe of the organization of the solid and also determine the optimal conditions for obtaining magnetic resonance spectra.

To date, most electron paramagnetic resonance (EPR) measurements of relaxation in solids have concentrated on spin–lattice relaxation and phase-memory (comparable to spin–spin) relaxation (*1, 2*). Although the dynamics of the dipolar field can control phase-memory relaxation, there is little direct information (*3–5*). Yet, as solid-state NMR spectroscopy and the technique of dynamic nuclear polarization (DNP) are applied to samples, such as coal, with high concentrations of paramagnetic centers, the dynamics of electron dipolar fields become increasingly important (*6, 7*). The very different formalisms employed, spin temperature for NMR spectroscopy (*8*) and a statistical approach of Klauder and Anderson (*9*) for EPR spectroscopy, hamper the transfer of the available information. This chapter attempts to partially bridge that gap. These two formalisms are briefly outlined in the remainder of this section. The "Background" section shows that the magnitude and the dynamics of the electron dipolar field appear and can be measured in the context of both approaches. The rest of this chapter describes some initial experiments using pulsed EPR spectroscopy to measure the dynamics of the electron dipolar field in samples of Argonne Premium coals.

Spin temperature is an extremely powerful concept in the study of magnetic relaxation in solids. This formalism explains many puzzling results in systems with high concentrations of spins. Goldman (*8*) summarized the "rules of the spin-temperature game" on the basis of a large body of theoretical and experimental work. This approach to spin dynamics in solids rests on two premises. The first is that a Hamiltonian consisting of the sum of commuting operators, H_i, can describe the evolution of the density matrix, σ, of the system. The H_i are constants of motion for σ. The Hamiltonian may be truncated, with small nonsecular terms deleted so that the operators for the constants of motion commute with each other. The second assumption is that σ has the form

$$\sigma = 1 - \sum_i \beta_i H_i \tag{1}$$

where β_i is the inverse spin temperature of the ith constant of motion. The σ usually meets this condition in a time comparable to T_2 (the spin–spin relaxation time). Equation 1 not only requires that off-diagonal elements of the spin density matrix vanish, but also places restrictions on the diagonal elements.

If these two conditions hold, the "hypothesis is that spin systems behave in the same way as the systems usually considered in thermodynamics" (8). Normally, the constants of motion are the Zeeman energy for the spins in the sample and the dipolar energy between spins. The constants of motion are treated as thermal reservoirs in poor contact with each other and with the lattice. Spin temperature provides a simple mechanism to calculate energy flow among these reservoirs.

The Hamiltonian explicitly includes the secular part of the spin–spin interactions. This inclusion makes spin temperature attractive for concentrated spin systems. NMR, EPR, and DNP NMR experiments have shown the value of the spin-temperature formalism.

Pulsed EPR spectroscopy has not made extensive use of the spin-temperature concept to describe electron spins. The samples are usually dilute in electron spins to minimize relaxation caused by spin–spin interactions and to allow detection of the signals. Usually, the spin–spin interactions are consciously minimized. Experiments measure the intrinsic dynamics of noninteracting electron spins, which are easier to calculate. Yet those spin–spin interactions can provide valuable information about electron-spin systems in solids. Some elegant electron spin-echo (ESE) (4, 5) measurements have used electron spin–spin interactions to probe spatial correlations and to find spatial distributions of free radicals (5, 10–12).

The usual approach in ESE spectroscopy is based on work by Klauder and Anderson (9) examining the behavior of a spin in a time-dependent dipolar field of other spins. The temporal properties of the dipolar field are treated statistically. This statistical treatment is less satisfying than the spin-temperature formalism, but it does remove the restrictions inherent in eq 1. Unlike the spin-temperature formalism, the statistical treatment applies at times that are short compared to T_2 and describes the behavior of off-diagonal elements of the density matrix as well as the diagonal elements. The original treatment by Klauder and Anderson (9) appears to make several gross simplifications that have made many researchers suspicious of its validity. This statistical approach can be quite rigorous, as will be shown in the following section.

Background

Spin Temperature. The spin-temperature formalism is most useful for describing a special class of states that are far from equilibrium. Those states have density matrices with no off-diagonal terms and with their diagonal elements constrained by eq 1. The preparation of these states usually involves the transfer of Zeeman order or magnetization to other reservoirs of the spin system. The evolution of the density matrix is treated as a flow of thermal energy between different reservoirs. Because the spin-temperature formalism describes a restrictive class of nonequilibrium spin states, the relation between its variables and those used to describe equilibrium states or pulsed EPR experiments is not always clear.

Consider a spin system that can serve as a model for many systems. We will examine it first in the spin-temperature formalism and then with the statistical approach. The system contains one type of spin, S, with dipolar interactions between spins and with some inhomogeneous broadening uncorrelated with spatial position. The spin Hamiltonian for the system at high field is

$$H = H_Z + H_D + H_S \tag{2}$$

where the Zeeman, dipolar, and inhomogeneous broadening Hamiltonians are, respectively,

$$H_Z = \hbar\omega \sum_i S_z^i \tag{3}$$

$$H_D = \sum_{ij} \frac{3}{2} a_{ij} \left[S_z^i S_z^j - \frac{1}{3} S^i S^j \right] \tag{4}$$

$$H_S = \sum_i \Delta_i S_z^i \tag{5}$$

S_z^i and S_z^j are the z components of the spin for the i or j radicals, with z taken to be along the applied Aeeman field; ω is the electron Zeeman frequency; a_{ij} is the dipolar interaction between spins i and j; and Δ_i is the inhomogeneous shift of spin i from ω. The units used are such that $\hbar = 1$. The spin Hamiltonian is truncated with only secular terms retained. For convenience, H_D and H_S can be truncated so that they commute with each other (13). This truncation is equivalent to the assumption that $|\Delta_i - \Delta_j|$ is almost never comparable to $|a_{ij}|$ and generally holds if there is a continuous distribution of $(\Delta_i - \Delta_j)$ or a_{ij} about a value of zero. The origin of the inhomogeneous broadening is unspecified and may be produced by hyperfine interactions, g-factor anisotropy, or random lattice strains.

The Zeeman interaction and the dipolar and the inhomogeneous broadening terms commute with each other and are independent of each other, at least in the high-temperature limit (14, 15). Consequently three limiting cases exist:

1. Both H_D and H_S are separate constants of the motion of the density matrix (16).
2. The sum $(H_S + H_D)$ is a combined constant of motion (16, 17).

3. Each different value of Δ defines a separate spin system with separate constants of motion (as in Section 7 of ref. 15).

In the first two cases, the spin-temperature formalism requires the density matrix to relax rapidly to the form

$$\sigma = 1 - \beta_Z H_Z - \beta_D H_D - \beta_S H_S \tag{6}$$

characterized by three temperatures (if $H_D + H_S$ is a constant of motion, then $\beta_D = \beta_S$). In the final case, each value of Δ defines a separate spin system with many temperatures, one for each unique value of Δ.

$$\sigma = 1 - \beta_D H_D - \sum_{kl} \beta_{Zk}\left(\omega + \Delta_k\right) S_z^l \tag{7}$$

where the summation over l includes all spins with the same value of Δ, and the summation over k covers all distinct values of Δ. Equation 7 has lost much of the simplicity of the spin-temperature formalism and resembles the noninteracting spin-packet model used by Portis (*18*) to describe inhomogeneous broadening. When eqs 6 or 7 do not hold, the spin system lies outside the domain of the simple spin-temperature formalism.

When eq 6 describes σ, the different spin temperatures can be measured in continuous-wave (CW) or pulsed magnetic resonance experiments. Because the H_i (i = Z, D, or S) terms are independent, the expectation value of H_i is given by

$$\langle H_i \rangle = \mathrm{Tr}\left[\sigma H_i\right] = \mathrm{Tr}\left[\beta_i H_i^2\right] = \beta_i \mathrm{Tr}\left[H_i^2\right] \tag{8}$$

where $\mathrm{Tr}(X)$ represents the normalized trace of X. Thus, β_D can be measured if the field, $\mathrm{Tr}(H_D^2)$, is known, or $\mathrm{Tr}(H_D^2)$ can be measured if β_D can be calculated. The measurement of β_D at various times after preparation of the initial state provides the dynamics of the dipolar field (*19*).

In most experiments measuring the inverse dipolar temperature, the spin state is prepared so that β_D is much greater than the inverse lattice temperature β_L (i.e., a very low dipolar temperature that warms back up to the lattice temperature). The autocorrelation function for the dipolar energy is

$$\beta_D(t) = \beta_D(0) \left[\frac{\langle H_D(0) H_D(t) \rangle}{\langle H_D^2 \rangle}\right] \tag{9}$$

$$\langle H_D(0)H_D(t)\rangle = \langle \sum_{ij} \frac{9}{4}a_{ij}^2 \left[S_z^i(0)S_z^j(0) - \frac{1}{3}\mathbf{S}^i(0)\cdot \mathbf{S}^j(0) \right]$$
$$\times \left[S_z^i(t)S_z^j(t) - \frac{1}{3}\mathbf{S}^i(t)\cdot \mathbf{S}^j(t) \right] \rangle \quad (10)$$

where the autocorrelation function is

$$\langle X(0)X(t)\rangle = \text{Tr}\left[\sigma X \exp(-iHt) X \exp(iHt)\right] \quad (11)$$

with H the total Hamiltonian. The autocorrelation function in eq 9 gives the dipolar spin–lattice relaxation time, T_{1D}, in the limit of no energy flow between the different reservoirs (8). Equations 9–11 connect β_D or T_{1D} with expectation values and autocorrelation functions of the dipolar interaction that pulsed EPR experiments can measure and that have meaning in an equilibrium system.

Statistical Approach. Klauder and Anderson (9) developed a statistical approach from ideas used previously to explain line widths (20). We follow a different development that arrives at the same results (12). The formal strategy rather than the computational details are emphasized. Several unnecessary approximations will be made in the process to keep the intermediate equations simple without affecting the final results.

In most pulsed EPR experiments, the ultimate observable is

$$\text{Tr}\left[\sigma S_+\right] = \sum_i \text{Tr}\left[\sigma S_+^i\right] \quad (12)$$

and it suggests a strategy of seeking a solution for σ with greatest accuracy for spin i. An exact solution for σ can be found if H_D in eq 4 is truncated further to

$$H_D^i = \sum_j \frac{3}{2}a_{ji}\left[S_z^j S_z^i - \frac{1}{3}\mathbf{S}^j\cdot \mathbf{S}^i\right] \quad (13)$$

The interactions between spin i and all the other electron spins are retained. The spin–spin interactions that do not involve the detected electron spin i are deleted. Those deleted terms are important nonetheless because they cause flip-flop transitions that can alter the dipolar field at spin i. At this point, a relaxation term is introduced to reproduce the statistical characteristics of the neighboring spins. The evolution of the density matrix becomes

$$\frac{\partial \sigma}{\partial t} = -i\left[\mathbf{H}^i, \sigma\right] + R\sigma \tag{14}$$

$$\mathbf{H}^i = \mathbf{H}_Z + \mathbf{H}_D^{\ i} + \mathbf{H}_S \tag{15}$$

where R, the relaxation operator from Redfield theory (21), can be treated phenomenologically. It is not necessary (but often possible) to calculate the elements of R. This Hamiltonian is solved easily if R does not transfer electron spin coherence between neighbors of spin i. The microwave pulses used in pulsed EPR measurements are incorporated in the strong pulse limit as simple rotations. Calculations yield very accurate solutions for spin i provided that R duplicates the statistics of the spins interacting with i.

The solution of eq 14 is simple because $\mathbf{H}_i$ contains no terms bilinear in electron spin except those involving spin i. Mims (22) showed that the density matrix for a many-spin system with an analogous Hamiltonian is exactly a tensor product of density matrices each involving spin i and another electron spin. In other words, the evolution of i under the influence of all the other spins is the product of its evolution under the individual influence of each spin. Thus, the many-spin problem reduces to a product of two-spin problems, each of which is readily solved (12). This solution is possible because the statistics of the spins surrounding i can be described by using R. The experimentally detected signal, $V(t)$ is given by

$$V(t) = \langle \prod_j V_j(t) \rangle = \exp \sum_j [1 - V_j(t)] \tag{16}$$

$$V_j(t) = \frac{\mathrm{Tr}\left[\sigma_{ij}(t)\mathbf{S}_+^{\ i}\right]}{\mathrm{Tr}\left[\sigma_{ij}(0)\mathbf{S}_+^{\ i}\right]} \tag{17}$$

after suitable averaging over an ensemble of spins i (5, 12), where σ_{ij} is the density matrix solution for two spins i and j. This is an extremely powerful and general result because it applies to any density matrix with this spin Hamiltonian. In the present case, eq 16 describes the signal obtained after a state has been prepared, allowed to evolve (relax) for time t, and then converted into an observable.

In contrast to the spin-temperature approach, it is not necessary to prepare a large, nonequilibrium dipolar field to measure the field or its relaxation. The statistical approach makes use of the thermal fluctuations about the average value. Thus, while the average $<H_D> \approx 0$ at high temperatures, the local fields are nonzero because $<H_D^2> \neq 0$. The measurement of the autocorrelation function gives the relaxation time of the dipolar field. Consequently, dipolar fields at equilibrium provide access to the dipolar spin–lattice relaxation time.

Pulsed EPR Spectroscopy. We now consider actual EPR pulse sequences and the dipolar parameters they can measure. The following pulse sequences will be considered:

1. the two-pulse, primary echo sequence and the instantaneous diffusion experiment
2. the "2 + 1" sequence for measuring dipolar interactions
3. a modification of the "2 + 1" sequence

These experiments deserve specific consideration because they measure dipolar parameters cleanly without contamination from inhomogeneous broadening, nuclear modulation effects, or spin–lattice relaxation.

Salikhov et al. (12) describe the evolution of σ in the statistical treatment outlined in the preceding sections for the primary echo sequence (4, 5). The experiment applies a strong pulse along the y axis in the rotating frame with a turning angle θ_1 followed by evolution for time τ, another pulse of θ_2 and evolution for τ, culminating in a spin echo. We will follow the experiment for a two-spin system that can then be generalized to the full many spin system. The first pulse converts σ from $1 - \beta_L H_Z$ to

$$\sigma = 1 - \beta_L \left[S_z^1 \cos\theta_1 + S_x^1 \sin\theta_1 + S_z^2 \cos\theta_1 + S_x^2 \sin\theta_1 \right] \quad (18)$$

which then evolves according to eq 14. If the elements of R are much smaller than the electron Zeeman frequencies, only a few elements of σ need to be followed, that is, σ_{13} and σ_{24}, where the basis set is

$$|1> = |\alpha_1\alpha_2> \quad (19a)$$
$$|2> = |\alpha_1\beta_2> \quad (19b)$$
$$|3> = |\beta_1\alpha_2> \quad (19c)$$
$$|4> = |\beta_1\beta_2> \quad (19d)$$

This basis set gives a pair of coupled differential equations in a frame rotating at the Zeeman frequency:

$$\frac{\partial \sigma_{13}}{\partial t} = -i\left[\Delta_{1'} + \frac{a_{12}}{2}\right]\sigma_{13} - \left[W_1 + W_2\right]\sigma_{13} + W_2\sigma_{24} \quad (20a)$$

$$\frac{\partial \sigma_{24}}{\partial t} = -i\left[\Delta_1 - \frac{a_{12}}{2}\right]\sigma_{24} - \left[W_1 + W_2\right]\sigma_{24} + W_2\sigma_{13} \quad (20b)$$

where W_1 and W_2 are the two independent elements of R in the two-spin problem. The solution for these two elements of σ gives

$$\sigma_{13}(t) = \exp\left[-i\Delta_1 t - \left[W_1 + W_2\right]t\right]$$

$$\times \left\{\sigma_{13}(0)\cosh R_2 t + \left[W_2\sigma_{24}(0) - i\frac{a_{12}}{2}\sigma_{13}(0)\right]\frac{\sinh R_2 t}{R_2}\right\} \quad (21)$$

$$\sigma_{24}(t) = \exp\left[-i\Delta_1 t - \left[W_1 + W_2\right]t\right]$$

$$\times \left\{\sigma_{24}(0)\cosh R_2 t + \left[W_2\sigma_{13}(0) - i\frac{a_{12}}{2}\sigma_{24}(0)\right]\frac{\sinh R_2 t}{R_2}\right\} \quad (22)$$

$$R_2^2 = W_2^2 - \frac{a_{12}^2}{4} \quad (23)$$

for $t < \tau$. The second microwave pulse produces more rotation about the microwave magnetic field. Evolution of the new σ continues until the echo appears at time 2τ. The echo amplitude, normalized to unity for $\tau = 0$ is

$$V_2(2\tau) = \left[\left(\cosh R_2\tau + \frac{W_2}{R_2}\sinh R_2\tau\right)^2 + \frac{a_{12}^2}{4R_2^2}\sinh^2 R_2\tau\right.$$

$$\left. - \frac{a_{12}^2}{2R_2^2}\sinh^2 R_2\tau \sin^2\frac{\theta_2}{2}\right]\exp\left\{-2\left[W_1 + W_2\right]\tau\right\} \quad (24)$$

for the interaction of the electron spin with only one other spin. This interaction has an explicit dependence on θ_2, the turning angle of the second pulse. Equation 24 successfully predicts the complete averaging of the dipolar interaction for large W_2, which becomes apparent in typical electron spin-echo experiments in the limit $W_2 > 10^5$ s^{-1} (11, 12). When $W_2 \ll |a_{12}|$, the echo decay reduces to

$$V_2(2\tau) = \left[1 + 2\sin^2\frac{a_{12}\tau}{2}\sin^2\frac{\theta_2}{2}\right]\exp\left[-2\left(W_1 + W_2\right)\tau\right] \quad (25)$$

In the limit that there are many electron spins interacting with each other, the experimentally measured echo decay is given by the ensemble average

$$\left\langle \prod_j V_j(2\tau) \right\rangle = \exp\left[-2\tau\sin^2\frac{\theta_2}{2}\left(\sum_j \frac{a_{ij}^2}{4}\right)^{1/2} - 2W_i\tau\right] \quad (26)$$

The decay of the echo with τ contains an exponential component independent of other echo-decay pathways whose rate is proportional to $\sin^2(\theta_2/2)$ times the sum of a_{ij}^2 (i.e., $\langle H_D^2 \rangle$!).

Thus, variation of the echo decay with the turning angle of the second microwave pulse allows the measurement of the dipolar field, eqs 9 and 10, that appears in the spin-temperature formalism. As W_2 becomes greater than approximately 10^5 s^{-1}, some averaging of the dipolar field takes place and, in addition, W_2 contributes directly to the echo decay. This experiment for measuring the dipolar field in the limit of small W_2 is called *instantaneous diffusion* because the second pulse changes the dipolar field instantly, producing spectral diffusion.

Kurshev et al. (23) described an ingenious generalization of the instantaneous-diffusion experiment where τ is fixed and a third pulse (the "+1" pulse) is swept between time zero and τ. This extra pulse causes instantaneous diffusion at various points during the evolution of σ. No background decay of the echo occurs because τ remains constant. They derived an echo decay (23) of

$$V(\tau,t) = \exp\left[-2\sin^2\frac{\theta_{+1}}{2}\cos\theta_2\left(\sum_j \frac{a_{ij}^2}{4}\right)^{1/2}t\right]V(2\tau) \quad (27)$$

where the "+1" pulse occurs at time $t < \tau$. If the microwave pulses are well characterized, this experiment provides a cleaner measurement of the dipolar field because nuclear modulation and phase-memory decay do not occur.

A new modification of this "2 + 1" pulse sequence permits measurement of the autocorrelation function of the dipolar field. The sequence resembles the three-pulse, stimulated-echo sequence with the addition of a "+1" pulse (Figure 1). The sequence starts with the same microwave pulse and evolution of σ (eqs 18 and 21–23), as do the primary echo or the "2 + 1" sequence. At some time before τ, the "+1" pulse is applied.

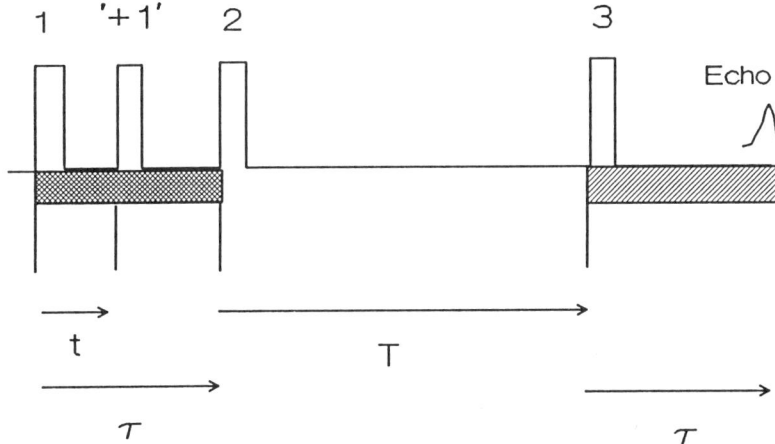

Figure 1. Schematic diagram of the "3 + 1" pulse sequence, showing the location of the the three microwave pulses and the signal echo. The echo is formed by the pulses labeled 1, 2, and 3, and the "+1" pulse alters the dipolar field. The shaded areas indicate the times during which the spins evolve under the influence of the local dipolar fields. The time periods for t, T, and τ are indicated below the pulses.

Because the echo is detected, only those i spins unaffected by the "+1" pulse contribute to the echo. However, the "+1" pulse does affect the spins interacting with the detected spin; consequently, the dipolar field experienced by i changes. Again σ_{13} and σ_{24} evolve, but in the new dipolar field. At time τ, the coherences that ultimately generate the stimulated echo are converted into polarizations, that is, diagonal elements of σ, by the "second" microwave pulse.

The dipolar field can change during the interval (Figure 1) between the pulses 2 and 3, T, but the diagonal elements of σ will not change, except for the Zeeman spin–lattice relaxation. The third pulse changes the polarization back to coherences that evolve in the dipolar field present following the third pulse. If the dipolar field remains correlated for T, the evolution of σ during the τ periods is correlated and depends on the action of the "+1" pulse. The echo decay is

$$V(\tau,t,T) = \exp\left[-\sin^2\frac{\theta_{+1}}{2}\cos\theta_2\cos\theta_3\langle H_D(0)H_D(T)\rangle^{1/2}t\right]V(\tau,0,T) \qquad (28)$$

which is valid for $<H_D(0)H_D(T)> \approx <H_D^2>$. A more general analysis will be given in a subsequent publication because the relaxation depends on the details of the autocorrelation function. This sequence provides direct experimental access to $<H_D(0)H_D(T)>$ and through it to T_{1D}.

As might be expected, the same dipolar interactions between electron spins appear in both the spin-temperature and the statistical treatments. Using this correspondence, the electron dipolar fields and relaxation times from pulsed EPR experiments, where the spin-temperature formalism might not be valid, can be transferred to DNP experiments where the statistical treatment is irrelevant.

Experimental Details

All experiments were performed on a home-made, Fourier transform EPR spectrometer (*24–26*) at room temperature using a $3\lambda/4$ slotted-tube resonator with an operating frequency of 9.152 GHz. The static magnetic field was set to the center of the EPR signal. Pulses were nominally 8 ns wide. A 350-W pulse gave a turning angle of $\pi/2$ radians as measured from the free induction decay (FID) of the samples themselves, a deuterated polyacetylene reference sample, and a stable solution of duroquinone radical anion. The turning angle in the experiments was varied by changing the pulse power. Phase-cycling suppressed FIDs and other unwanted signals.

Samples were carefully sealed in fused silica (Suprasil) tubes. Eight samples from four of the Argonne Premium coals were prepared. The intent was to obtain one set of samples as close to the material inside the sample vials as possible. The second set was used to test the effect of brief exposure to air. Samples were prepared from the following coals: Pocahontas No. 3 (ID-501, low-volatile bituminous), Blind Canyon No. 6 (ID-601, high-volatile bituminous), Beulah–Zap 3/4/87 (ID-801, lignite), and Pittsburgh No. 8 (ID-401, high-volatile bituminous) (*27*). All samples came from 5-g vials of 100 mesh. The sample vials were opened in a dry, oxygen-free nitrogen glove box after being mixed as recommended (*27*). A portion of each sample was transferred rapidly to a 4-mm o.d. tube so that the length of the sample was between 1.5 and 2.5 cm. An O-ring seal temporarily sealed the tubes, whose lower end was cooled in liquid nitrogen to reduce the pressure in the tube. The tube was then sealed with an oxygen–natural gas torch. The samples were exposed to the low temperature for less than 1 min to reduce cryopumping of volatile compounds from the coal onto the walls of the tube. The samples were kept in subdued light at room temperature. For preparation of the air-exposed samples, the opened vials were removed from the dry box and exposed to air for about 10 min. Sample tubes were filled in air and sealed as previously described.

Measurements made by using the echo amplitudes, the echo magnitudes,

Results

Zeeman Spin—Lattice Relaxation. The Zeeman spin–lattice relaxation rates were measured by using the saturation-recovery method, and the magnetization was detected by using a two-pulse echo sequence. The two pulses from the previous echo sequence saturated the spin system. The repetition rate of the experiment was varied to observe the recovery. A typical relaxation curve is shown in Figure 2. The recovery appears to consist of a single exponential starting from complete saturation, although the presence of a slow minor component is not excluded. Clearly the brief exposure to oxygen had a major effect on the spin–lattice relaxation rate. Table I gives the spin–lattice relaxation results for all

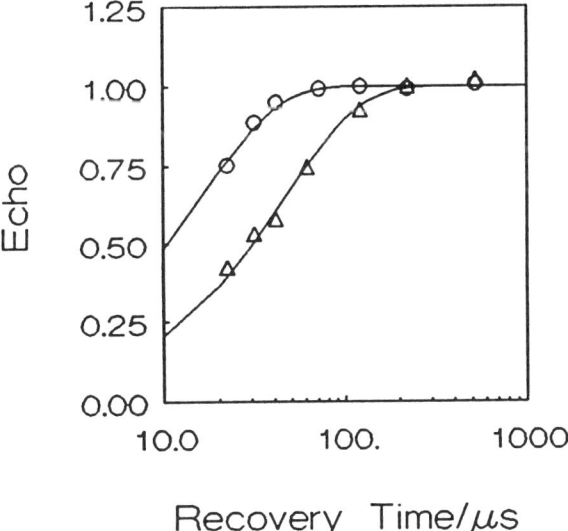

Figure 2. Recovery of the normalized electron spin-echo signal from saturation produced by the previous echo sequence. The recovery time is the pulse repetition period. The solid lines are least-squares fits of $[1 - \exp(-t/T_1)]$ to the data points. The samples were at room temperature: △, *Pocahontas No. 3 unexposed to air,* $T_1^{-1} = 22.8$ *kHz; and* ○, *Pocahontas No. 3 exposed to air as described in text,* $T_1^{-1} = 62.1$ *kHz.*

Table I. Spin–Lattice Relaxation Rates at 300 K

Coal	Nitrogen	Air
Pocahontas No. 3	22.7	66.2
Blind Canyon No. 6	12.4	29.4
Beulah–Zap	33.0	44.1
Pittsburgh No. 8	9.6	17.2

NOTE: All values are given in kilohertz.

eight samples. The spin–lattice relaxation rates were used to set the repetition rates for the other experiments and to verify that T_1^{-1} (i.e., W_2) < 10^5 s^{-1}.

Dipolar Fields. The dipolar field in the Pocahontas No. 3 sample that was not exposed to air was measured by using the "2 + 1" sequence. The measured dipolar field was 4.2 MHz (Figure 3). The "2 + 1" experiment was carried out on seven of the coal samples with $\tau = 1.0$ μs, a 2-kHz repetition rate, and $\pi/3$ turning angles for all three pulses. The results are given in Table II.

Dipolar Relaxation. The "3 + 1" experiment previously described was used to measure the dipolar relaxation in several samples. The decay of the echo was measured at fixed τ and T with t varied. The data, illustrated in Figure 4, are consistent with the exponential decay predicted by eq 28, but by no means do these results prove that the decay is exponential. The decay rate, equal to $\sin^2(\theta_{+1}/2) \cos \theta_2 \cos \theta_3$ $<H_D(0)H_D(T)>^{1/2}$, was measured at several different values of T (Figure 5) and shows a decrease in decay rate with increasing T, as expected. The change in decay rate with T was fit by an exponential in an attempt to estimate the dipolar spin–lattice relaxation time T_{1D}, although there is no reason, given the sample heterogeneity of coal, to expect an exponential form. The values obtained from the fits appear in Table III. The data were taken with τ in the range 0.6–1.0 μs. The large scatter in the data make it impossible to tell if the decay depended on τ.

Discussion

Applicability of the Spin-Temperature Formalism. The preceding discussion of the spin-temperature and statistical formalisms showed

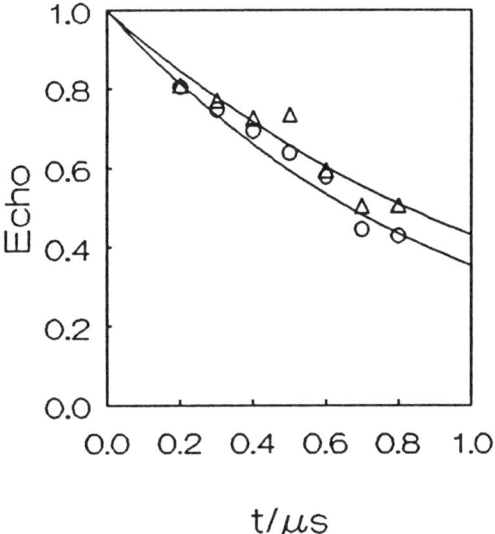

Figure 3. Normalized echo amplitude in the "2 + 1" experiment. Data were taken at room temperature with τ = 1.06 μs and a 2-kHz repetition rate. The solid lines are least-squares fits of exp (−2kt) to the data. Samples were Pocahontas No. 3 unexposed to air, k = 1.04 ± 0.08 MHz (Δ) and exposed to air, k = 0.84 ± 0.13 MHz (○). The values of k give the dipolar field after correction for the pulse turning angles.

Table II. Dipolar Fields Obtained from the "2 + 1" Experiment

Coal	Nitrogen	Air
Pocahontas No. 3	4.2 ± 0.3	3.3 ± 0.6
Blind Canyon No. 6	2.9 ± 0.7	2.9 ± 0.1
Beulah–Zap	—	3.0 ± 0.4
Pittsburgh No. 8	2.9 ± 0.6	3.0 ± 0.7

NOTE: All values are given in megahertz. The dash means data are not available.

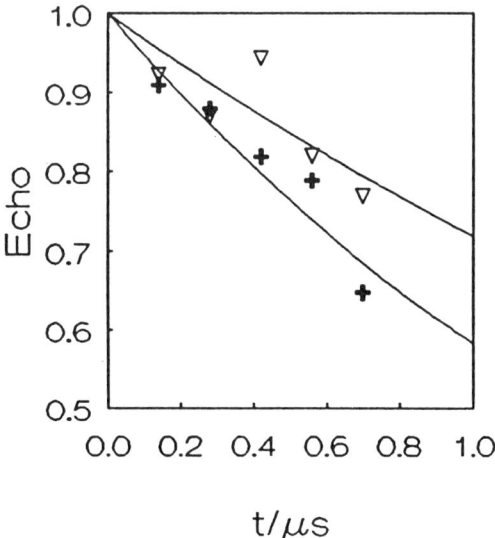

Figure 4. Normalized amplitude of the "3 + 1" echo with t for the Pittsburgh No. 8 sample that was not exposed to air. The data were taken at room temperature with $\tau = 0.86$ μs and a repetition rate of 4 kHz. The solid lines are least-squares fits of theoretical decays to the data. The time T was 5 μs, with a decay rate of 0.54 ± 0.09 MHz (+) or 30 μs with a decay rate of 0.33 ± 0.04 MHz (∇).

that both give access to β_D, $\langle H_D^2 \rangle$, and T_{1D}. This result raises the question of whether the spin-temperature formalism is sufficient for analysis of the pulsed EPR experiments described. Is the statistical treatment needed at all? Does the spin-temperature formalism apply in its simple form?

The answer is very simple for the primary, two-pulse spin-echo experiment. The first pulse converts σ from $1 - \beta_L H_z$ to $1 - \beta_L(S_z \cos \theta + S_x \sin \theta)$. The S_x term ultimately forms the echo signal and evolves as S_x and S_y for the entire time before the echo. The existence of the echo shows that eq 1 does not apply and that the simple spin-temperature formalism cannot be applied.

The spin echo depends on off-diagonal elements of the density matrix. When inhomogeneous broadening occurs, the intrinsic decay of the echo, not the FID time, is the relevant time constant. For the Pocahontas No. 3 coal sample that was not exposed to air, the limiting T_2 in the absence of instantaneous diffusion is 5.3 μs. The off-diagonal elements exist and can be "refocused" to produce another echo even under conditions of strong instantaneous diffusion. Thus, the two-pulse echo-type ex-

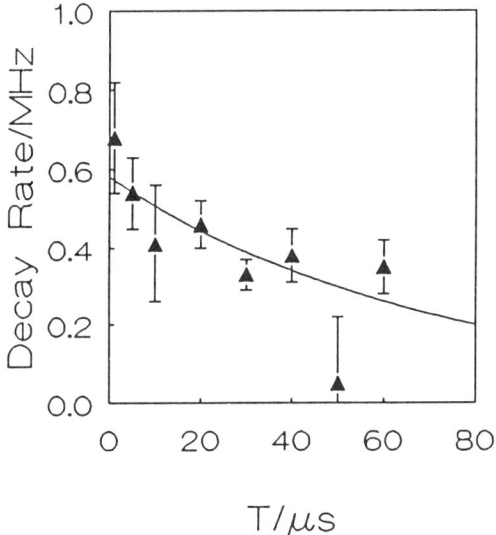

Figure 5. Measured decay rates for the "3 + 1" experiment as a function of the time T. Sample and other experimental conditions were the same as in Figure 4. The solid line is an exponential curve least-squares fitted to the data points with an autocorrelation time of 75 μs for the dipolar field.

Table III. Range of Dipolar Relaxation Rates Obtained from the "3 + 1" Experiment

Coal	Nitrogen	Air
Pocahontas No. 3	25–170	—
Blind Canyon No. 6	170	—
Beulah–Zap	200–1000	50–500
Pittsburgh No. 8	13–22	39

NOTE: All values are given in kilohertz. The dash means data are not available.

periments prepare σ in a state that does not fit eq 1, and the spin-temperature formalism does not apply until after an echo can no longer be observed.

Another class of echo experiments is based on the stimulated echo. In these experiments, magnetization persists in the diagonal elements as a polarization for times much longer than T_2 before generating an echo. During most of the time, there are no off-diagonal elements. In some pulsed experiments, notably the Jeener–Broekaert experiment (19), the spin-temperature formalism can be applied because the diagonal elements of σ conform to eq 1. However, this is not true of the stimulated-echo experiment or other experiments based on it, such as the "3 + 1" experiment. The reason lies in the form of σ. The stimulated echo comes from a polarization pattern (3–5) that oscillates as a function of the EPR frequency of the spins. Equation 1 does not describe σ under these conditions, but instead, σ contains very high-order polynomials in H_i. The decay of these higher order terms in σ produces echo decay. Again, the spin-temperature formalism does not become valid until the stimulated echo has decayed.

For both primary-echo and stimulated-echo experiments, the spin-temperature formalism applies only after measurements become impossible. The different values of the inhomogeneous broadening could be treated as defining separate spin systems. The sample would then have a single dipolar reservoir and many Zeeman reservoirs, a condition that could make stimulated-echo experiments applicable. However, this strategy lacks the simplicity and elegance of the original spin-temperature concept applied to a single-spin system.

In general, ESE experiments can measure the rate at which σ approaches a form describable by the spin-temperature formalism, but that process lies outside the spin-temperature description. A notable exception to this generalization occurs when σ is generated in the form of eq 1, as in the Jeener–Broekaert experiment (19). There is no fundamental reason why the spin-temperature formalism cannot be applied to other EPR experiments if eq 1 holds. This condition requires either sufficient time for relaxation of σ or careful preparation of σ.

Electron Spin Relaxation in Coal. Experiments on electron spin relaxation in coal face two problems. The first problem is that coal is a heterogeneous material on both a macroscopic and a microscopic scale. Coals also differ depending on their location of origin and their rank (27). Another type of heterogeneity results from the macerals that make up a piece of coal (27). All macerals contain free radicals with different properties and environments. A sample of coal contains a mixture of radicals differing in chemical identity and surroundings, and it is very easy to select a subset of those radicals when performing EPR experiments. For instance, EPR experiments may favor radicals with a longer or shorter T_1 or T_2, larger or smaller line width, or spectral shifts caused by g-factor or hyperfine interactions. The second difficulty with coal measurements lies in the easy mutability of coal. The samples used in this study have water contents ranging from 0.65 to 32.25% and volatile matter ranging between

18.48 and 43.72% (*27*). Thus, evacuation of samples causes some water and volatiles to be lost. Also, the high surface area of coals allows significant adsorption and exchange of gases and even oxidation of the coal.

The Argonne Premium Coal Sample Program addresses the problem of sample heterogeneity by ensuring that samples of a coal start out as similar as possible. We tried not to alter the coal during sample preparation. The samples never were evacuated, they were cooled only once, and exposure to air was carefully controlled. Thus, our results will be interesting to compare with those from samples prepared in other ways.

The spin–lattice relaxation rates for the eight different coal samples reported here vary by a factor of 7. The microwave pulses of the previous echo sequence produce saturation over a frequency range wider than the EPR spectrum, a condition that reduces problems caused by spectral diffusion (*28*). The recoveries are consistent with single exponentials, although systematic deviations would be difficult to see given the small number of points on the recovery curves. Nevertheless, the recoveries extrapolate back to a state of complete saturation, a result suggesting a single, major exponential component. This exponential recovery is surprising given the microscopic heterogeneity of coal. The different macerals would be expected to produce different relaxation rates and nonexponential relaxation. Spin diffusion, rapid on a 50-μs time scale, must occur to average relaxation of the different macerals to values similar to those reported for vitrinite macerals (*29*).

The dipolar fields measured in the "2 + 1" experiment cluster around 3–4 MHz, a result corresponding to local spin concentrations of about 4×10^{18} spins/cm^3. The spin–lattice relaxation in every sample is below the 100-kHz rate that marks the start of dipolar averaging in the instantaneous-diffusion experiment (*11, 12*) and probably in the "2 + 1" experiment as well. These values for the dipolar field are similar to those reported for vitrinite macerals (*29*).

Dipolar spin–lattice relaxation times showed a great variability for different experimental conditions, sometimes changing 10-fold. Part of that variation undoubtedly comes from unintentional selection of different populations of spins in the experiments. For instance, when τ is long, the echo emphasizes spins with a long T_2. These spins may well have a small dipolar field and a long dipolar relaxation time. A second problem is that if the relaxation is nonexponential, the fitted relaxation time depends on the exact values of T used in the measurements. The numerical values for the dipolar relaxation time should not be taken as exact values; their significance lies in the fact that it is possible to observe relaxation of the dipolar field in these experiments. It is difficult to tell from Figure 5 or any of the other curves whether the relaxation is exponential, which would be unlikely given the heterogeneity of coal.

The dipolar spin–lattice relaxation time refers to a condition of no energy flow between the dipolar reservoir and the other reservoirs of the system (8). In the spin-temperature formalism, dipolar relaxation may differ from the dipolar spin–lattice relaxation rate if energy from the Zeeman or other reservoirs relaxes through the dipolar reservoir. This point is difficult to consider because the reservoirs and the dipolar spin–lattice relaxation are concepts from the spin-temperature formalism, which does not apply here. Part of the variability in the observed dipolar relaxation may be due to differences in the initial σ producing different relaxation. These issues need to be investigated further both theoretically and by using a more homogeneous sample.

Effect of Sample Preparation. The brief exposure of the sample to air (probably the oxygen) produced a very noticeable effect on the spin–lattice relaxation rates. The relative change in the relaxation rates with oxygen exposure is smallest in the sample with the highest water content [Beulah–Zap, ~33% water by weight versus 1–5% for the others (27)]. Whether this correlation shows a cause-and-effect relationship or results from chance or correlation with another property is impossible to determine. The change in the spin–lattice relaxation rate is large enough that any magnetic resonance experiment that depends on electron spin–lattice relaxation will be sensitive to air exposure. Such experiments include EPR, NMR, ENDOR (electron nuclear double resonance), and DNP spectroscopy.

The spin–lattice relaxation rates and dipolar-field magnitudes also appear to depend on sample treatment. This dependence is evident when the present results are compared to similar measurements in this volume. The samples measured here were never evacuated or heated, and thus they are different from samples of the same coal that were evacuated or heated. Apparently, the removal of water and other volatiles from the coal changes the magnetic resonance properties of the coal.

The relaxation rates for the dipolar field also seem sensitive to air, although that effect is difficult to quantify at present. The air-exposed Pocahontas and Blind Canyon coal samples have no reported relaxation rates. Those relaxation rates seem to be too short to measure and much shorter than those of the Beulah–Zap and Pittsburgh No. 8 samples.

Conclusions

Both the spin-temperature formalism and the statistical treatment used to analyze pulsed EPR experiments measure some of the same properties of a spin system. The dipolar field from electron spins and its time dependence can be measured by using instantaneous-diffusion, the "2 + 1", and the "3 + 1" pulse sequences. This capability affords new methods for characterizing spin systems and promises better ways of optimizing magnetic resonance conditions.

The electron spin relaxation of samples of Argonne Premium coals show the effect of brief exposure to air. Even a brief exposure to air increases the spin–lattice relaxation rate. Relaxation in coal also appears sensitive to evacuation and heating of the sample. The sensitivity likely arises from removal of water and other volatiles.

Acknowledgments

I thank R. Wind for pressing me to explore the relation between spin temperature and the statistical treatment, J. R. Norris for insisting that the statistical treatment could be placed on a more rigorous basis, and A. M. Raitsimring for valuable discussions about the "3 + 1" sequence. This work was supported by the Office of Basic Energy Sciences, Division of Chemical Sciences, U.S. Department of Energy, under Contract No. W–31–109–Eng–38.

References

1. *Time Domain Electron Spin Resonance;* Kevan, L.; Schwartz, R. N., Eds.; Wiley: New York, 1979.
2. *Electronic Magnetic Resonance of the Solid State;* Weil, J. A.; Bowman, M. K.; Morton, J. R.; Preston, K. F., Eds.; Canadian Society for Chemistry: Ottawa, Ontario, Canada, 1987.
3. Mims, W. B.; Nassau, K.; McGee, J. D. *Phys. Rev.* **1961**, *23*, 2059–2069.
4. Mims, W. B. *Electron Paramagnetic Resonance;* Geschwind, S., Ed.; Plenum: New York, 1972; p 263–351.
5. Salikhov, K. M.; Semenov, A. G.; Tsvetkov, Yu. D. *Electron Spin Echoes and Their Applications;* Nauk: Novosibirsk, 1976.
6. Botto, R. E.; Axelson, D. E. *Am. Chem. Soc. Div. Fuel Chem.* **1988**, *33*, 50–57.
7. Wind, R. A.; Jurkiewicz, A.; Maciel, G. E. *Fuel* **1989**, *68*, 1189–1197.
8. Goldman, M. *Spin Temperature and Nuclear Magnetic Resonance in Solids;* Oxford University Press: London, 1970.
9. Klauder, J. R.; Anderson, P. W. *Phys. Rev.* **1962**, *125*, 912–932.
10. Brown, I. M. *Time Domain Electron Spin Resonance;* Kevan, L.; Schwartz, R. N., Eds.; Wiley: New York, 1979; pp 195–229.
11. Salikhov, K. M.; Tsvetkov, Yu. D. *Time Domain Electron Spin Resonance;* Kevan, L.; Schwartz, R. N., Eds.; Wiley: New York, 1979; pp 231–277.

12. Salikhov, K. M.; Dzuba, S. A.; Raitsimring, A. M. *J. Magn. Reson.* **1981**, *42*, 255–276.
13. Goldman, M. *Spin Temperature and Nuclear Magnetic Resonance in Solids;* Oxford University Press: London, 1970, Appendix.
14. Philippot, J. *Phys. Rev.* **1964**, *133*, A471–A477.
15. Jeener, J.; Eisendrath, H.; Van Steenwinkel, R. *Phys. Rev.* **1964**, *133*, A478–A490.
16. Buishvili, L. L.; Zviadadze, M. D.; Khutsishvili, G. R. *Zh. Eksp. Teor. Fiz.* **1969**, *56*, 290–298 (English translation: *Sov. Phys. JETP* **1969**, *29*, 159–163).
17. Clough, S.; Scott, C. A. *J. Phys. C* **1968**, *1*, 919–931.
18. Portis, A. M. *Phys. Rev.* **1953**, *91*, 1071–1078.
19. Jeener, J.; Broekaert, P. *Phys. Rev.* **1967**, *157*, 232–240.
20. Anderson, P. W.; Weiss, P. R. *Rev. Mod. Phys.* **1953**, *25*, 269–276.
21. Redfield, A. G. *Phys. Rev.* **1955**, *98*, 1787–1809.
22. Mims, W. B. *Phys. Rev. B* **1972**, *5*, 2409–2419.
23. Kurshev, V. V.; Raitsimring, A. M.; Tsvetkov, Yu. D. *J. Magn. Reson.* **1989**, *81*, 441–454.
24. Massoth, R. J. Ph.D. Dissertation, University of Kansas, 1987.
25. Angerhofer, A.; Massoth, R. J.; Bowman, M. K. *Isr. J. Chem.* **1988**, *28*, 227–238.
26. Bowman, M. K. In *Modern Pulsed and Continuous-Wave Electron Spin Resonance;* Kevan, L.; Bowman, M. K., Eds.; Wiley: New York, 1990; pp 1–42.
27. Vorres, K. S. *Users Handbook for the Argonne Premium Coal Sample Program;* ANL/PCSP–89/1; Argonne National Laboratory: Argonne, IL, 1989.
28. Bowman, M. K. *Time Domain Electron Spin Resonance;* Kevan, L.; Schwartz, R. N., Eds.; Wiley: New York, 1979; pp 67–105.
29. Thomann, H.; Silbernagel, B. G.; Jin, H.; Gebhard, L. A.; Tindall, P. *Energy Fuels* **1988**, *2*, 333–339.

RECEIVED for review June 8, 1990. ACCEPTED revised manuscript August 12, 1991.

CONCLUSION

33

Advanced Magnetic Resonance Techniques Applied to Argonne Premium Coals

Bernard G. Silbernagel[1] and Robert E. Botto[2]

[1]Corporate Research, Exxon Research and Engineering Company, Annandale, NJ 08801
[2]Chemistry Division, Argonne National Laboratory, Argonne, IL 60439

> *The application of a wide variety of magnetic resonance techniques to Argonne Premium coals provides a great deal of information about both the coal samples and the methodology being employed for their examination. Electron paramagnetic resonance (EPR) observations at high frequencies and as a function of temperature give a clearer picture of the different types of carbon radicals present in these coals. Electron spin-echo (ESE) measurements of the radical relaxation times reveal significant variations in coals of different ranks. Electron nuclear double resonance (EN-DOR) techniques, both continuous wave and pulsed, reveal details of the interaction of the unpaired electron of the radical with the nuclei on its host molecule and adjacent molecules. Various NMR techniques have been applied to the quantitation of the percentage of aromatic carbon in the various coals. These advanced techniques, coupled with studies of coal changes with chemical (oxidation, pyrolysis, and reactions with donor and acceptor molecules) and physical (physisorption and solvent swelling) changes expand our understanding of these coals.*

IN THE AREA OF MAGNETIC RESONANCE great progress has been made, and recent advances have had consequences in the study of a common series of coal samples: the Argonne Premium coals (1). Many of the techniques discussed in this book have been developed over the past few years, but their consistent application to a common series of fossil-fuel materials has not heretofore occurred. A discussion of these new techniques and the associated theory was presented in Chapters 1–7. Here we will focus on the specific applications to the Argonne Premium coals and describe the additional knowledge that is available in principle from such applications.

Magnetic resonance in coals is a rich scientific field with a long history (2). A variety of species can be observed: nuclei (particularly ^{13}C and ^{1}H), the unpaired electrons associated with carbon radicals in the system, and paramagnetic ions of transition metals that may be incorporated in minerals or occur as individual ions in the organic matrix of the coal. As indicated in Figure 1, NMR and EPR (electron paramagnetic resonance) studies have largely been mutually exclusive. Nuclei lying within ~10 Å of a carbon radical or paramagnetic impurity have been unobservable because of broadening and relaxation effects of the electron spin. Con-

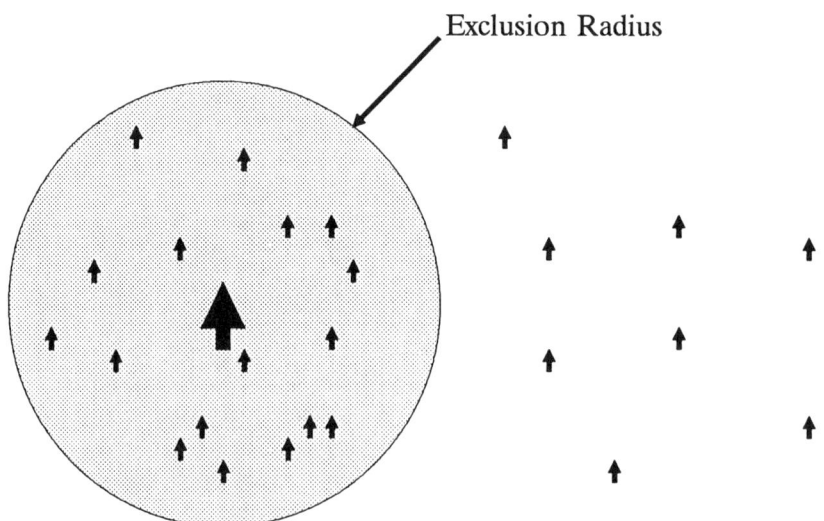

Figure 1. Schematic diagram of electron spin (large arrow) and nearby nuclear spins (small arrows) observed in coals. Nuclei lying within the exclusion radius (indicated by the circle) are not observable by NMR spectroscopy because of broadening effects from the electron spin.

versely, interactions of the electron spin with nuclei (mostly protons) in their environment are responsible for the width of the EPR absorption, but the EPR spectra observed in coals are usually featureless, and a quantitative conclusion about the radical environment cannot be drawn from the result. New techniques, particularly electron spin-echo (ESE) spectroscopy, electron spin-echo modulation, and pulsed electron nuclear double resonance (ENDOR) spectroscopy are providing new, quantitative information about these nearby nuclei.

Electron magnetic resonance techniques differ from those of NMR in another important way. The protons and ^{13}C nuclei are distributed throughout the organic matter of the coal sample, and NMR probes the molecules in the coal "democratically" when the experiments are properly performed. By contrast, only one unpaired electron is present for every 10^3 to 10^4 carbon atoms, and simple chemical considerations indicate that they will be associated with the aromatic molecules in the system.

Electron Magnetic Resonance Spectroscopy of Argonne Premium Coals

In any attempt to relate coal chemistry to EPR observations, important considerations are (1) how homogeneously the radicals may be distributed in the coal and (2) how representative these radicals are of the total coal. Such questions will benefit from the availability of the widest possible amount of information.

We will explore the following questions:

1. What does electron magnetic resonance say about the molecular chemistry and the molecular structure of the coals?

2. How do these chemical and structural properties vary with rank?

3. How is the coal changed by the treatments to which it may be subjected, such as oxidation, pyrolysis, chemical processing, swelling, or reactions with reagent molecules, and what role do the associated minerals play in this process?

Another implicit question concerns the effect that handling the samples before examining them may have on the observed properties.

Nature of Carbon Radicals in Coal. Regarding the nature of the paramagnetic centers in coal, are they molecules bearing a single unpaired electron, or does the EPR absorption arise from pairs of spins such as might be found in a donor–acceptor complex? In a donor–acceptor complex, a paramagnetic response would be expected only from complexes in which the spins were in a triplet state, because the singlet configuration is diamagnetic. As the temperature varies, the relative populations in the triplet and singlet states of the two spin systems would vary, and the variation of the intensity of the EPR signal would not follow Curie law behavior. Some workers have observed non-Curie law behavior and have postulated significant levels of donor–acceptor complexes in coals (3). However, such experiments must be done with great care because non-Curie law behavior can occur as an artifact of the experimental procedure in a number of ways. In Chapter 31, Rothenberger et al. report that the temperature dependence of the EPR intensity for the coals of all ranks, from lignite (Beulah–Zap) to low-volatile bituminous (Pocahontas No. 3) follows Curie law for temperatures from 100 to 300 K. A detailed analysis of lower temperature measurements (to ~10 K) suggests that the deviations from Curie law that do occur are associated with thermal effects such as heat transfer.

EPR Absorption Line Shapes. The EPR absorption has long been recognized to be complex. It often consists of a broad and a narrow component. The broad component is associated with macerals like vitrinite, the lignin-based component of the coal, and waxy materials like exinite or sporinite, which come from the leaves and spores of trees. The narrow component is attributed to highly metamorphosed, inertinitic materials in the coal. Because the EPR spectra accumulated at the conventional X-band microwave frequency (~10 GHz) do not usually have well-defined features, line-shape analyses are difficult to do with much precision. Measurements at 35 GHz, also quite commonly available, indicate significantly more asymmetry to the line. High-frequency EPR spectrometers operating at 94 and 240 GHz, as reported by Clarkson et al. in Chapter 27, and 2 mm (~150 GHz), as reported by Tsvetkov et al. in Chapter 23, dramatically illustrate the extent of anisotropy in these lines.

Such a variation can arise from two factors: (1) several lines of roughly similar shape but displaced in resonance frequency from one another (i.e., having different g values, where the g value is the EPR analog of a chemical shift of the absorption line) or (2) one or more lines that have a g-value anisotropy. One way of testing this factor is to do high-frequency measurements on individual maceral types of samples that have been isolated by density centrifugation techniques. For an Illinois No. 6 coal, examination of the isolated maceral samples demonstrates that both frequency shifts and g-value anisotropies are present for sporinite and vitrinite macerals. The data of both Chapters 23 and 27 indicate that this anisotropy is most pronounced for low-rank coals (up to a carbon content of 81%), a finding that suggests a combination of heterocyclic and aromatic radical species. Pulsed ENDOR experiments, as reported by Thomann et al. in Chapter 30, also provide strong evidence for the existence of significant levels of heteroatom radicals in the low-rank coals.

Discriminating Radical Types Using ESE Relaxation Techniques. Differences in the relaxation properties of different radical types allow these species to be studied individually. In parallel EPR and ESE examinations, Silbernagel et al. (Chapter 29) made a systematic attempt to establish whether the properties inferred from the ESE measurements of T_M (the phase memory decay time) and T_{1E} (the electron spin–lattice relaxation time) were consistent with the two component signals observed in the EPR. Five of the coal samples (Pocahontas No. 3, Upper Freeport, Pittsburgh No. 8, Lewiston–Stockton, and Illinois No. 6) showed both broad and narrow EPR absorptions, and spectral decomposition allows the determination of their widths and relative intensities. The broad signal was known to be inhomogeneously broadened, and spin echoes were easily observed, with the width of the spin echoes, $1/T_2^*$ (T_2^* is the spin dephasing time), in good agreement with the EPR line width.

By contrast, the narrow line is homogeneously broadened, and no echo was formed. However the rate of decay of the free induction decay (FID) gave a value of $1/T_2^*$ that was in excellent agreement with the observed EPR values. Detailed measurement of T_{1E} for these systems indicates that the relaxation processes are complex because the results are not described by a single exponential. A portion of this nonexponentiality is associated with spectral diffusion of the magnetization during the course of the relaxation process, which is particularly significant for the higher rank coals. However, some of the effect probably is associated with a distribution of T_{1E} values for different molecular types in the same coal.

A quantitative evaluation of this effects remains to be done, with an appropriately demineralized series of samples. A direct attempt to associate relaxation properties with maceral types (4) is a useful exercise, but may encounter the problem that there is not a strict division in relaxation properties among radicals in macerals of different types.

Electron–Nuclear Interactions Using ENDOR Spectroscopy. The broad signal seen in coals is the result of interactions between the electron spins and nuclei in the vicinity, but a profile of the magnitudes of the interactions encountered can be obtained only by using double-resonance techniques. These observations can be done either by conventional CW (continuous-wave) ENDOR techniques, as reported by Chen et al. in Chapter 24, or by pulsed ENDOR techniques, as reported by Thomann et al. in Chapter 30. The CW technique is dependent on the relative values of the electronic and nuclear relaxation processes and is difficult to observe in many cases. For each coal, a matrix signal, resulting from coupling of the electron spin to protons at considerable distance from the radical, was observed in CW ENDOR.

By contrast, the pulsed ENDOR sees both a matrix signal and a broader signal associated with the local hyperfine interactions from protons residing on the radical molecule itself; the result is a spectrum that extends over a frequency range of $\sim$20 MHz. This result is in contrast to the $\sim$0.5-MHz width associated with the matrix ENDOR signals.

The results of the pulsed ENDOR work are particularly revealing: The size of the matrix ENDOR contribution falls with increasing coal rank, a reflection of the decrease in proton density in these higher rank coals. Even more significant is the fact that the shape of the local ENDOR signal is nearly identical for all of the bituminous coals, a feature suggesting that major changes in the size of aromatic molecules (i.e., the number of aromatic rings) do not occur during this phase of the coalification process. This report is the first direct spectroscopic evidence of this important result.

Another opportunity afforded by the pulsed double-resonance techniques is the chance to actually study the dynamics of the nuclei that reside within the exclusion radius and are therefore inaccessible to conventional NMR spectroscopy. In this case the dynamics of the nuclei are detected by their perturbation of the electron spin properties that occurs as a result of their mutual hyperfine coupling.

Coal Structure and Swelling. In addition to the individual molecules, the properties of coal and its reactivity are strongly influenced by the extent to which reagents can be introduced into the solid coal. As a first consideration, access is afforded by the network of pores in the coal. However, the coal solid can also swell to accommodate molecules within the organic structure itself. The ability of the organic matter to swell will be determined by the nature and the strength of the interactions among the individual molecules. Several factors can influence this macrostructure in the coal: covalent bonds between units, charge-transfer interactions and π-bonding between aromatic units, and hydrogen bonding resulting from polar interactions between heteroatomic components of the coal. These latter polar interactions can be perturbed by the introduction of polar swelling solvents, such as pyridine, in the coal. The extent of the local swelling then can be probed by introducing paramagnetic molecules of different geometries into the swollen coal and determining the number of such molecules that can be incorporated. By using a variety of polar and nonpolar solvents and spin probes of different geometries, a rank-dependent swelling behavior is documented, as reported by Spears et al. in Chapter 25. In low-rank coals, pores become much more cylindrical when swelled by polar solvents. In high-rank coals, where hydrogen bonding is less likely to play a role, the pore geometry is not changed by the swelling process.

Rank-Dependent Effects in Argonne Premium Coals. Because the Argonne Premium coal series spans a very wide range of coal ranks, it is important to determine if the EPR and ESE parameters observed are consistent with our fundamental understanding of coal chemistry. In a series of observations on isolated macerals from Pennsylvania State University (PSOC) coals of comparable range in rank, systematic variations of EPR parameters were observed (5). For increasing coal rank, that is, increasing weight percent carbon of the organic matter in the coal, g values fell (to values expected for aromatic radicals), the density of radicals generally increased, the line width of the carbon radical signal increased (the line-width increase reflects this radical density increase), and the radicals became progressively less easy to saturate by the application of higher microwave powers. ESE measurements revealed increasing values of $1/T_M$ and $1/T_{1E}$ as well as increased spectral diffusion with increasing coal rank, all of which are consistent with a higher density of radicals and an increased aromaticity in the higher rank coals (6).

The present data do not always follow this trend: Wide variations are found in the measured radical density for a given coal sample, g values that are observed lie outside of the range expected for either aromatic or heterocyclic molecules, and a monotonic increase of the relaxation properties with increasing coal rank is not observed. Although some valid questions about the sample-handling history of some of the PSOC coal samples remain, these systematics suggest that another variable is at work in the Argonne samples that remains to be completely specified.

Electron Magnetic Resonance and the Chemical Transformation of Coals. In the previous section, we restricted our discussion to the examination of the coals in the "native" state, assuming that the effects of handling are negligible and that handling the coal samples has not resulted in a change of their resonance properties. Actually, the very process of opening the sample ampules will change the samples, presumably causing a loss of some of the water vapor that resides in the pores of the coals and by introducing molecular oxygen when the samples are exposed to air. Clarkson et al. report in Chapter 27 that the samples were opened and immediately run. Tsvetkov et al. report in Chapter 23 that the samples were run immediately after opening and 1 h later. The narrow line attributed to inertinites vanished, and the broader line broadened further after a 1-h exposure to air. These effects can be explained by the introduction of paramagnetic O_2 molecules into the coal with resulting magnetic broadening (7).

Silbernagel et al. (Chapter 29) prepared the samples in a dry box, which was then evacuated and back-filled with helium and sealed without any exposure to air. Rothenberger et al. (Chapter 31) transferred the samples in a nitrogen atmosphere and sealed them without evacuation; this approach provided the closest approximation to the "as-received" coals. Treatment can also be varied. Chen et al. (Chapter 24) immediately evacuated and sealed a portion of the samples while another portion was exposed to air and then sealed. Buckmaster and Kudynska (Chapter 26) report that their sample was prepared as-received, dried in an inert atmosphere, and saturated with moisture.

The ESE results for the samples that have been treated differently show different behavior. Parallel measurements on air-exposed and evacuated samples (Chapter 24) in most cases do show an increase in T_M at room temperature with evacuation of the sample, but T_{1E} actually falls with evacuation in two cases, from opposite ends of the rank spectrum (Pocahontas No. 3 low-volatile bituminous coal and Beulah–Zap lignite). The values of T_M and T_2 (the electron spin–spin relaxation time) vary considerably from one series of measurements to the next.

These ambient-temperature handling processes can have significant effects on the samples. The most detailed survey of these phenomena, reported in Chapter 26, is an EPR study where changes in the shape of the EPR line can be detected with great sensitivity by using an Argand diagram analysis. Even storing the samples in an argon atmosphere changes g values and radical densities because of water loss that occurred during storage. Oxidation occurs at temperatures as low as 50 °C, and its rate depends critically on the amount of water in the samples at the time. Some of these changes are the result of interactions with the organic matter, but some result from changes of the mineral matter and the organically coordinated transition metal ions present in the as-received coal.

Pyrolysis of the coals, another form of chemical reaction, is briefly surveyed in Chapter 24. Treatment at 673 K in an evacuated tube led to a substantial increase in the radical densities of the samples and a corresponding decrease in T_M. In several instances it also led to significant decreases in T_{1E}, which again might be expected for increasingly aromatic, ordered hydrocarbons (6).

Another series of experiments involves the reaction of the coal with donors (such as triethylphosphine oxide, TEPO) and acceptor molecules (such as iodine and tetracyanoquinone, TCNQ). A significant increase in the carbon radical intensity occurs at higher ranks with the addition of an I_2 acceptor. At low ranks, adding the TEPO donor decreases the radical intensity, but adding a TCNQ acceptor increases the radical intensity. These systematics suggest that the radical species in the coal are radical cations and that the electron-transfer chemistry depends on the chemical form (and perhaps the steric character) of the probe molecules.

Solid ^{13}C NMR Spectroscopy of Argonne Premium Coals: An Interlaboratory Comparison

The subject of quantitation in solid ^{13}C NMR analyses of coal and other insoluble carbonaceous materials has been the focus of intensive debate over the past decade. As in other disciplines of fuel science, the complex nature of carbonaceous fuels, together with the diversity of samples chosen for study, has made comparisons between published results difficult. With the inception of the Argonne Premium Coal Sample Program, it is now possible to conduct an interlaboratory study on a common suite of pristine coal samples spanning a diverse rank range. This interlaboratory study permits a valid comparison of NMR data measured with different experimental techniques and under a variety of different experimental conditions.

Table I is a compilation of carbon aromaticity values for the eight Argonne Premium coals reported in 13 independent studies. The NMR measurements were carried out at static field strengths ranging from 1.4 to 9.4 T (corresponding to carbon resonant frequencies of 15–100 MHz) and using a variety of methods, including cross-polarization–magic-angle spinning (CP–MAS), single-pulse excitation (SPE) with MAS, MAS with total suppression of spinning sidebands (TOSS), and dynamic nuclear polarization–cross-polarization (DNP–CP) without MAS. Carbon aromaticities were determined from CP experiments employing contact times generally between 1 and 2 ms, and in certain instances, by mathematically fitting signal intensities derived from variable contact-time experiments. Proton decoupling field strengths were typically between 40 and 82 kHz; MAS frequencies varied from 3 to 13 kHz. Recycle-delay times ranged from 1–3 s in CP experiments to 60 s in SPE experiments.

The mean values derived for carbon aromaticities of the Argonne Premium coals and standard deviations obtained from the data in Table I are presented in Table II. The data have been analyzed according to groups that correspond to three experimental categories: mean aromaticity values from all CP–MAS measurements, values determined by CP–MAS at a static field of 2.3 T only, and those values derived from SPE (90° pulse–acquire–recycle delay) measurements.

Table I. Survey of Carbon Aromaticity Values for Argonne Premium Coals

Laboratory	BZ	W	ILL	PITT	BC	LS	UF	POC
A[a]	0.69	0.61	0.67	0.70	0.61	0.72	0.77	0.83
B[b]	0.67	–	–	–	–	0.67	–	–
	0.65	–	–	0.63	–	–	–	0.76
C[c]	–	0.59	–	0.67	–	–	–	–
D[d]	–	0.69	0.72	0.75	0.63	0.76	0.78	0.86
	–	0.58	0.60	0.61	0.53	0.63	0.66	0.77
E[e]	0.71	0.59	0.66	0.70	0.59	0.71	0.78	0.84
F[f]	0.77	0.76	0.70	0.72	0.67	0.77	0.80	0.89
	0.65	0.53	0.72	0.78	0.66	0.77	0.78	0.86
	0.66	0.63	0.72	0.72	0.65	0.75	0.81	0.86
G[g]	0.61	0.63	0.72	0.72	0.65	0.75	0.81	0.86
H[h]	0.69	0.63	0.71	–	–	–	0.81	0.86
I[i]	0.76	0.77	0.78	0.84	0.77	0.83	0.83	0.89
J[j]	0.66	0.67	0.72	0.73	0.63	0.78	0.82	0.85
K[k]	0.72	–	–	–	0.64	–	–	0.81
	0.73	–	–	–	0.61	–	–	0.83
	0.83	–	–	–	0.56	–	–	0.87
L[l]	0.74	0.66	0.72	0.75	0.68	0.71	0.83	0.89
	0.70	0.65	0.72	0.74	0.67	0.75	0.82	0.86
M[m]	0.76	0.75	0.78	0.77	0.65	0.79	0.81	–
	0.65	0.63	0.70	0.72	0.63	0.69	0.77	–

ABBREVIATIONS: BZ, Beulah–Zap; W, Wyodak–Anderson; ILL, Illinois No. 6; PITT, Pittsburgh No. 8; BC, Blind Canyon; LS, Lewiston–Stockton; UF, Upper Freeport; and POC, Pocahontas.

[a] Chapter 13: 4.7-T CP–MAS; 90° pulse, 3.5 μs; t_{CP}, 2 ms; ν_r, 3 kHz.
[b] Reference 11: 7.05-T CP–MAS (TOSS); 90° pulse, 5 μs; t_{CP}, 2–3 ms; ν_r, 4 kHz.
[c] Chapter 14: 4.7-T CP–MAS; 90° pulse, 6 μs; t_{CP}, 1 ms; ν_r, 3.7–4.5 kHz.
[d] Chapter 16: 9.4-T CP–MAS (TOSS); 90° pulse, 4 μs; t_{CP}, 1 ms; ν_r, 4 kHz.
[e] Chapter 17: 7.05-T CP–MAS; 90°x pulse, 5 μs; t_{CP}, 2 ms; ν_r, 4 kHz.
[f] Chapter 20: 7.05-T SP and CP–MAS (2.3-T CP–MAS); 90° pulse, 5 μs; t_{CP} curve, ν_r, 14 kHz (4 Hz).
[g] Reference 12: 2.3-T CP–MAS; 90°x pulse, 6 μs; t_{CP} curve, ν_r, 4 kHz.
[h] Reference 9: 4.7-T TOSS; 90°x pulse, 5 μs; t_{CP}, 2 ms; ν_r, 4 kHz.
[i] Reference 13: 1.4-T DNP–CP–MAS; nonspinning.
[j] Reference 9: 2.3-T CP–MAS; 90°x pulse, 3.5 μs; t_{CP}, 1.5 ms; ν_r, 4 kHz.
[k] Private communication (V. Pang and G. E. Maciel): 2.3-T SP and CP–MAS; t_{CP} curve.
[l] Reference 14: 2.3-T SP and CP–MAS; 90° pulse, 4.5 μs; t_{CP} curve, ν_r, 4 kHz.
[m] Reference 15: 2.3-T SP and CP–MAS; t_{CP} curve, ν_r, 4.5 kHz.

Table II. Carbon Aromaticity for Argonne Premium Coals

Coal Type	CP–MAS	CP–MAS at 2.3 T	SPE
Pocahontas	0.85 ± 0.01 (12)	0.85 ± 0.01 (5)	0.86 ± 0.05 (3)
Upper Freeport	0.80 ± 0.02 (10)	0.82 ± 0.01 (4)	0.81 ± 0.02 (3)
Lewiston–Stockton	0.75 ± 0.02 (9)	0.76 ± 0.02 (4)	0.76 ± 0.04 (3)
Blind Canyon	0.63 ± 0.03 (11)	0.64 ± 0.02 (5)	0.66 ± 0.02 (4)
Pittsburgh No. 8	0.72 ± 0.03 (11)	0.73 ± 0.01 (4)	0.75 ± 0.03 (3)
Illinois No. 6	0.71 ± 0.02 (10)	0.72 ± 0.00 (4)	0.73 ± 0.04 (3)
Wyodak–Anderson	0.62 ± 0.04 (11)	0.65 ± 0.02 (4)	0.74 ± 0.03 (3)
Beulah–Zap	0.69 ± 0.05 (12)	0.67 ± 0.05 (5)	0.76 ± 0.03 (4)

NOTE: All entries are mean values and standard deviations; the number of measurements is in parentheses.

For Illinois No. 6 and coals higher in rank, the mean carbon aromaticity values that were determined for the three data sets can be considered the same within the limits of the standard deviations. This condition means that the variability in carbon aromaticities reported by different laboratories for these coals is greater than the differences observed between experimental techniques, that is, CP and SPE experiments. The mean values from SPE experiments, however, always tend to be on the high end of the range of CP values.

Differences in mean carbon aromaticity values for CP and SPE experiments are significant for the two lower rank coals, Wyodak–Anderson subbituminous and Beulah–Zap lignite samples. For these coals, the mean carbon aromaticities for SPE measurements are clearly higher, by approximately 10 and 20%, respectively. These differences may be accounted for, in part, by complications from paramagnetic transition metal ions that are intimately associated with the coal organic phase (8, 9). The presence of these paramagnetic metal species in Wyodak–Anderson and Beulah–Zap coals adversely affects the spin–lattice relaxation properties of these coals (9). Shortening of proton T_1 relaxation times via interactions with paramagnetic species clearly results in reduced cross-polarization efficiencies, and thus leads to lower derived CP–MAS aromaticity values (see Chapter 18).

Variances associated with the CP–MAS data of the two low-rank coals are the highest for the entire suite of Argonne coals. In fact, the variance in the CP–MAS data, for which there is a statistically relevant number of measurements, is generally seen to increase with decreasing coal rank. These larger variations in experimental data for the lower rank coals may be a result of different sample-handling procedures, which could affect the analytical outcome of the experiment because of the presence of different amounts of water in the samples or incidental oxidation during transfer and analysis. Clearly, the lower rank coals should be most susceptible to the effects of water removal and oxidation, because the low-rank coals contain the highest levels of both water and paramagnetic transition metal ions. Any changes in these properties due to different sample handling would have a more profound influence on the CP–MAS measurements for these samples.

Comparing mean CP and SPE aromaticity values for Argonne coals with aromaticities obtained by other pulse techniques shows the expected trends. Carbon aromaticities obtained using the TOSS pulse sequence by laboratory H (Table II) are in good agreement, but those reported by laboratories B and D are significantly lower than mean CP or SPE aromaticity values. This disparity in TOSS results illustrates the severe demands that must be placed on proper implementation of the TOSS pulse sequence for quantitative analysis. On the other hand, aromaticity values obtained via DNP–CP experiments (laboratory I) tend to be high, a result that is entirely consistent with a methodology involving polarization transfer of magnetization from free radicals that reside predominantly on aromatic structures in coals.

Finally, it is instructive to determine whether mean carbon aromaticities derived from this interlaboratory exercise show rank dependencies for the Argonne coals that are consistent with our fundamental view of coal structure. Previous observations on coals spanning comparable ranges in rank consistently have shown systematic trends between NMR carbon aromaticities and coal rank as defined by carbon content or hydrogen-to-carbon ratio of the coal. The data in Table II, however, do not show definitive trends with rank. In fact, within the error limits of the analyses, CP- and SPE-derived aromaticities seem to be independent of rank through the (Beulah–Zap) lignite to (Lewiston–Stockton) high-volatile bituminous (HVB) coal range. Moreover, data for the Blind Canyon coal is anomalous and appears to correlate better with hydrogen-to-carbon ratio rather than carbon content. Only for the higher rank coals are trends between carbon aromaticities and coal rank obvious. Perhaps changes in aromaticity with rank for low-rank coals are less pronounced than had been previously thought. However, the number of samples in the Argonne suite may be insufficient statistically to allow for such an analysis of the data.

Discussion

A wide variety of new magnetic resonance techniques can be successfully applied to coals. In some cases, like CW ENDOR, such a success is by no means a foregone conclusion. These experimental efforts are particularly important because they have been performed on the same series of coal

Table III. Survey of EPR Carbon Radical Intensities

Coal Type	Chapter 31	Chapter 23	Chapter 29	Ref. 4	Chapter 24	Chapter 28	Mean Values
Pocahontas	3.10	1.10	0.21	2.33	3.4	1.54	1.95 ± 1.22
Upper Freeport	2.17	0.84	0.74	0.74	–	0.87	1.07 ± 0.62
Pittsburgh No. 8	1.91	0.82	0.53	1.80	1.0	1.00	1.17 ± 0.55
Lewiston–Stockton	3.78	0.83	0.38	–	–	1.64	1.78 ± 1.38
Blind Canyon	3.02	0.85	0.56	–	0.8	1.36	1.32 ± 1.00
Illinois No. 6	1.43	0.42	0.19	–	0.9	1.00	0.79 ± 0.49
Wyodak–Anderson	2.41	0.39	0.76	3.29	–	1.54	1.68 ± 1.19
Beulah–Zap	3.35	0.34	0.39	2.52	1.5	1.51	1.60 ± 1.18

NOTE: All values are in units of spins per gram of organic material, except those of Chapter 23, which are in spins per cubic centimeter.

samples, which furthermore have been treated in a very uniform manner up to the time when they were opened in the individual laboratories. Once the samples were opened, the variety of different handling procedures that were employed undoubtedly resulted in alterations of the samples from their pristine state because of the loss or addition of water or such incidental oxidation as may have occurred.

Enough overlap exists in the experiments performed for comparisons to be made of what would nominally be the same measurements. Significant variations in the ESE relaxation properties are found for the same coals, some of which can be explained by the different sample-handling procedures. The same may be said of the variation in NMR carbon aromaticity values, particularly for the lower rank coals. Furthermore, variations in the EPR parameters (g value, line width, and radical density) are large enough to suggest that other factors may also be at play. As an example of this effect, the EPR radical densities reported by the various researchers are presented in Table III. Variations greater than an order of magnitude are observed, and no systematic trend is seen in the variation of the radical density in the results of a given researcher. We propose that the major reason for this variability is the presence of magnetic transition metal ions, principally iron, that are in intimate contact with the organic matter in the coal (9, 10).

The Argonne Premium coals are unique in the fact that they have not been subjected to any drying or oxidation processes during their collection and preparation. As a result, actual free transition metal ions, interacting with the functionalized surface of the coal pores, are still present in these samples. In particular, Fe^{3+} ions, which would be expected to be quite abundant in these materials (because the iron content in the coals is on the order of 0.5–5 wt% of the whole coal), should be present as single ions or multiple-ion complexes. In the presence of oxidants, these soluble complexes oligomerize and ultimately form hematite (Fe_2O_3) (10), but subsequent EPR examinations demonstrate that the ions can occur at high density ($\sim 10^{20}$ per gram of coal) in the as-received coals (9). Citric acid washing is adequate to remove these signals from the coal and to lead to carbon radical values that are much less variable and other EPR properties that are consistent with the known variations in coal chemistry.

It is perhaps ironic that the variations of EPR and NMR properties that have been seen in these observations on the Argonne coals are most likely the result of the great care with which the samples have been handled in the course of their preparation.

Acknowledgments

B. G. Silbernagel gratefully acknowledges stimulating discussions with the conference participants as well as with J. W. Larsen, R. A. Flowers, L. A.

Gebhard, G. N. George, H. Thomann, and M. L. Gorbaty. R. E. Botto gratefully acknowledges support from the Office of Basic Energy Sciences, Division of Chemical Sciences, U.S. Department of Energy, under Contract No. W-31-109-ENG-38, and many helpful discussions with R. E. Winans, G. R. Dyrkacz, R. Hayatsu, and K. S. Vorres, as well as associations with several colleagues over the years: C.-Y. Choi, D. C. French, J. V. Muntean, L. M. Stock, A. R. Thompson, and C.-J. Tsiao.

References

1. Vorres, K. S. *Energy Fuels* **1990**, *4*, 420–426.
2. For a detailed review of the methodology and applications of EPT techniques to coal, see Retcofsky, H. L. In *Coal Structure;* Gorbaty, M. L.; Ouchi, K., Eds.; Advances in Chemistry 192; American Chemical Society: Washington, DC, 1981; pp 37–58. For a detailed review on the application of NMR to coal aromaticity, see Gerstein, B. C.; Murphy, P. D.; Ryan, L. M. In *Coal Structure;* Meyers, R. A., Ed.; Academic: New York, 1982; pp 87–126.
3. Duber, S.; Wieckowski, A. B. *Fuel* **1984**, *63*, 1474–1475.
4. Doetschman, D. C.; Dwyer, D. W.; Mustafi, D. Presented at the International Chemical Congress of Pacific Basin Societies, Honolulu, HI, December 17–22, 1989.
5. Silbernagel, B. G.; Gebhard, L. A.; Dyrkacz, G. R.; Bloomquist, C. A. A. *Fuel* **1986**, *65*, 558–65.
6. Thomann, H.; Silbernagel, B. G.; Jin, H.; Gebhard, L. A.; Tindall, P.; Dyrkacz, G. R. *Energy Fuels* **1988**, *2*, 333–339.
7. Thomann, H.; Goldberg, I. B.; Chiu, C.; Dalton, L. R. In *Magnetic Resonance: Introduction, Advanced Topics, and Applications to Fossil Energy;* Petrakis, L.; Fraissard, J. P., Eds.; NATO ASI Series C124; D. Reidel: Dordrecht, Netherlands, 1984.
8. Silbernagel, B. G.; Gebhard, L. A.; Flowers, R. A.; Larsen, J. W. *Energy Fuels* **1991**, *5*, 561–568.
9. Axelson, D. E.; Botto, R. E. *Prepr. Div. Fuel Chem. Am. Chem. Soc.* **1988**, *33*, 50–57.
10. Cotton, F. A.; Wilkenson, G. In *Advanced Organic Chemistry: A Comprehensive Text;* Interscience-Wiley: New York, 1972; pp 858–64.
11. Shi, J.-F.; Wang, Y.-S.; Qiu, L.-S.; Chen, S.-B.; Wu, X.-L.; Wu, X.-W. Presented at the International Chemical Congress of Pacific Basin Societies, Honolulu, HI, December 17–22, 1989.
12. Solum, M. S.; Pugmire, R. J.; Grant, D. M. *Energy Fuels* **1990**, *3*, 187.
13. Jurkiewicz, A.; Wind, R. A.; Maciel, G. E. *Fuel* **1990**, *69*, 830.
14. Muntean, J. V. Ph.D. Thesis, University of Chicago, 1990.
15. Love, G. D.; Snape, C. E.; Carr, A. *Proc. Int. Conf. Coal Sci.* (Newcastle Upon Tyne, England), 1991; pp 64–67.

RECEIVED for review September 29, 1990. ACCEPTED revised manuscript November 4, 1991.

INDEXES

AUTHOR INDEX

Adachi, Yoshio, 269
Axelson, David E., 253
Barton, Wesley A., 175, 229
Belford, R. L., 107, 507
Bernardo, Marcelino, 65, 539, 561
Botto, Robert E., 3, 341, 629
Bowman, Michael K., 91, 605
Bronnimann, Charles E., 27, 401
Brown, D. R., 507
Buckmaster, H. A., 483
Castellano, Salvatore M., 581
Chen, Xinhua, 451
Clarkson, R. B., 107, 507
Crookham, H. C., 507
Cyr, Natsuko, 281
Dela Rosa, Luisita, 359
Dzuba, Sergei A., 443
Egiebor, Nosa O., 281
Franz, James A., 377
Gebhard, L. A., 539
Gerstein, Bernard, 359
Goslar, Janina, 467
Grant, David M., 419
Gulin, Victor I., 443
Hayamizu, Kikuko, 295
Hayashi, Shigenobu, 295
Hu, Jiangzhi, 311
Hughes, Craig D., 419
Jurkiewicz, Antoni, 401
Kamiya, Kunio, 295
Kawamura, Mitsutaka, 295
Kawashima, Hiroyuki, 323
Keneko, T., 529
Kevan, Larry, 451
Kispert, Lowell D., 467
Kudynska, Jadwiga, 483
Lewis, Russ, 45
Li, Liyun, 311
Linehan, John C., 377
Lock, Herman, 45
Lynch, Leo J., 139, 175, 229
MacPhee, J. Anthony, 323
Maciel, Gary E., 3, 27, 45, 401
McManus, Hugh, 451
Nakamizo, Minoru, 269
Orendt, Anita M., 419
Pruski, Marek, 359
Pugmire, Ronald J., 419
Retcofsky, Herbert L., 581
Ridenour, Cynthia F., 27
Rothenberger, Kurt S., 581
Sakurovs, Richard, 229
Sanada, Yuzo, 139, 529
Sasaki, M., 529
Sethi, Naresh K., 419
Silbernagel, Bernard G., 539, 561, 629
Simms, Greg, 175
Solum, Mark S., 419
Spears, Ross, 467
Sprecher, Richard F., 581
Thomann, Hans, 65, 539, 561
Tsiao, Chihji, 341
Tsvetkov, Yuri D., 443
Vassallo, Anthony M., 201
Wang, Wei, 507
Webster, David S., 175
Wind, Robert A., 3, 45, 217
Yamada, Yoshio, 323
Yamashita, Yasumasa, 323
Ye, Chaohui, 311
Yokono, T., 529

AFFILIATION INDEX

Academia Sinica, 311
Alberta Research Council, 281
Argonne National Laboratory, 3, 91, 341, 605, 629
Battelle Northwest, 377
CANMET Energy Research Laboratories, 323
Chemagnetics, Inc., 3, 45
Colorado State University, 3, 27, 45, 217, 401
Commonwealth Scientific and Industrial Research Organisation, 139, 175, 229, 201
Exxon Research and Engineering Company, 65, 539, 561, 629

Government Industrial Research
    Institute, 269
Hokkaido University, 139, 529
Iowa State University, 359
NMR Technologies, Inc., 253
National Chemical Laboratory, 295
National Institute for Resources and
    Environment, 323
Russian Academy of Science, 443

U.S. Department of Energy, 581
University of Alabama, 281, 467
University of Calgary, 483
University of Houston, 451
University of Illinois
    at Urbana–Champaign, 107
University of Illinois, 507
University of Utah, 419

SUBJECT INDEX

A

A strain, line width effect, 110
Acenaphthene
  characterization procedure, 384
  comparison of CP–MAS and Bloch-decay results, 394-397
Acid treatment related to ESR free radical spin density, 308f, 309
Alberta coals, argon storage effect, 502
Aliphatic/aromatic hydrocarbon ratio, values for Argonne Premium coals, 276–278
Alkylpyridines sorbed onto coal, ^{13}C NMR spectroscopy, 201–215
  bonding site, 207–208
  cross-polarization spectra, 210, 211–212f
  examples, 208, 210
  relaxation behavior, 210, 213t, 214f
Anthracite, ^{13}C spectra, 321f, 322
Anthracite coal, estimation of percentage of undetected carbons, 9–10
Apparent aromaticity, calculation, 313
Apparent relaxation, definition, 92
Argonne Premium coals
  2-mm band and X-band ESR and ESE spectroscopy, 443–448
  Bloch-decay and CP–MAS ^{13}C NMR spectroscopy, 377–397
  ^{13}C CSA measurement, 419–437
  ^{13}C NMR spectroscopy, 637–641
  ^{13}C NMR spectroscopy of sorbed pyridine and alkylpyridines, 201–215
  characterization with electron donors or acceptors by EPR spectroscopy, 529–537
  classifications based on relaxation times, 305
  CW ESR spectroscopy, 443–448
  dynamic in situ 9-GHz EPR spectroscopy, 483–504
  ENDOR spectroscopy, 451–464, 571–579
  EPR spectroscopy, 539–549, 631–637
  ESE spectroscopy, 462–464,
    539–542, 547, 550–557
  ESR spectroscopy, 451–464
  ^{1}H CRAMPS spectra, 33f, 34
  ^{1}H NMR spectroscopy and spin–lattice relaxation, 295–309
  ^{1}H NMR thermal analysis, 229–249
  high-field NMR studies, 311–322
  high-resolution ^{1}H NMR spectroscopy, 401–417
  high-temperature ESR methods, 140–150
  high-temperature NMR methods, 149–169
  multifrequency EMR spectroscopy, 508–527
  proton quantitation by solid-state ^{1}H NMR spectroscopy, 359–375
  pulsed ENDOR spectroscopy, 571–579
  relaxation times vs. percent carbon, 303–304, 305f
  solid-state ^{13}C NMR spectroscopy, 323–338
  spin–lattice relaxation measurement, 341–357
  structural parameters, 269, 278
  temperature dependence of EPR intensity, 582–603
  concentration of water and organic hydrogen, 367, 368t
  See also individual names of coals
Aromatic carbons in coal
  distribution, 288t
  ideal line shapes, 420–422
Aromatic compounds, dipolar dephasing, 335–338
Aromaticity
  Bloch-decay and CP–MAS values, 386, 387t
  cross-polarization related to time, 330–332f, 333t
  determination, 326, 327f
  H/C ratio, 330–331, 333f

INDEX

Australian bituminous coals, PMR thermal analysis, 246, 247–248f, 249

B

Beulah–Zap coal
  ^{1}H NMR spectra of fresh, degassed, and dried samples, 297–300
  CW EPR spectroscopy, 483–504
  dipolar-dephasing results, 406, 407f
  EPR spectra with and without guest molecules, 531, 532f, 533
  ^{1}H CRAMPS of untreated and pyridine-saturated coals, 403–406
  PMR thermal analysis, 243–244, 245f
  truncated second moment-temperature pyrograms, 242f
  W-band EPR spectrum, 512, 516f
Bicyclo[3.2.1]–4-pyrrolidino-N-methyl-8-octane triflate to determine undetected carbons in coal, 10
Bituminous coals
  ^{13}C magnetization vs. matching time, 16, 17f
  diffusion-coupled relaxation and submicroscopic structure, 175–197
  ^{1}H NMR relaxation behavior, 176
  low-volatile, *See* Low-volatile bituminous coal
  molecular structure, *See* Molecular structure of bituminous coals
  PMR thermal analysis, 246–249
Bituminous vitrinites, properties, 191t
Bivariate correlations, principal-component analysis, 256–259
Blind Canyon coal
  CW EPR spectroscopy, 483–504
  EPR spectra with and without guest molecules, 531, 532f, 533
  ^{1}H CRAMPS of untreated and pyridine-saturated coals, 404–406
  ^{1}H NMR spectra of fresh, degassed, and dried coal, 297–298, 299f, 300
  PMR thermal analysis, 246
  stacked plot of ^{1}H NMR time-domain signals, 234, 237f
  transition metal EPR signals, 547, 548f, 549t
  truncated second moment-temperature pyrograms, 242, 243f
  W-band EPR spectrum, 512, 515f
Bloch-decay ^{13}C NMR spectroscopy of Argonne Premium coals, 377–397
  acenaphthene characterization procedure, 384

Bloch-decay ^{13}C NMR spectroscopy of Argonne Premium coals—*Continued*
  aromaticities, 386, 387t
  carbon magnetization determination methods, 384–386
  chemical-shift distributions, 386, 388t
  comparison of 4, 4′-dimethoxybibenzyl results with those of CP–MAS, 397t
  comparison of acenaphthene results with those of CP–MAS, 394–397
  comparison of aromaticities with those of CP–MAS, 392, 394
  contact time curves, 385f, 386
  experimental procedure, 384
  ^{1}H rotating-frame spin–lattice relaxation time, 384, 385f, 386
  relaxation parameter values, 386, 387t
  sample preparation, 383
  time dependence of magnetization growth and decay, 386, 390–392
  Upper Freeport coal, 386, 390f
  Wyodak-Anderson coal, 386, 391f
Boxcar gate and integrator, 96
BR-24 pulse sequence, use in ^{1}H CRAMPS, 39
Brown coals, PMR thermal analysis, 243–245

C

^{13}C aromaticity, quantitative accuracy, 315–316
^{13}C CSA in coal, 419–437
  average chemical-shift tensor values for aromatic carbons, 421, 422t
  chemical-shift–chemical-shift correlation spectroscopy, 431, 432f, 434–435f
  dynamic angle spinning, 433–435, 436f, 437
  experimental procedure, 422–437
  information obtained, 420, 421f, 422t
  limitations, 423
  magic-angle hopping, 433–435, 437
  magic-angle spinning with synchronous application of rf pulses, 437
  slow magic-angle spinning, 430
  static line shape with best fit, 424, 425f
  static powder pattern for hexahydropyrene, 422, 423f
  stop-and-go method, 433–435, 437
  techniques, 419–420
  tensor components for aromatic carbons in coal, 424, 426t

^{13}C CSA in coal—*Continued*
  tensor components for aromatic carbons in hexahydropyrene, 422–423, 424t
  variable-angle sample spinning, 424–425, 427–430
^{13}C cross-polarization–MAS solid-state NMR spectroscopy, coal and coal oxidation studies, 323–338
^{13}C cross-polarization NMR spectroscopy, applications, 217–218
^{13}C line broadening, influencing factors, 7–8
^{13}C magnetization, time dependence, 17
^{13}C NMR spectroscopy, distortion-free, rotating-frame DNP, 217, 226
^{13}C NMR spectroscopy of carbonaceous solids, quantitation technique, 3–24
^{13}C NMR spectroscopy of pyridine and alkylpyridines sorbed onto coal, 201–215
  alkyl-substituted pyridines, 210–212
  coal rank effect, 207, 209f
  comparison of Bloch-decay and CP spectra, 204–206
  elemental analysis of coal, 202, 203t
  experimental procedure, 202, 204
  implications for structure of sorbed pyridine, 214–215
  magic-angle spectroscopy for removal of line broadening from CSA, 206
  relaxation behavior of sorbed pyridine, 210–214
  temperature effect, 207, 208f
  time vs. CP signal intensity, 206f, 207
^{13}C nuclei, interaction with unpaired electrons, 6–11
Carbon-based free radicals, densities, 561–562
Carbon radicals, EPR spectroscopy in coal, 631–632
Carbon rotating-frame spin–lattice relaxation time, 351–355
Carbon spin–lattice relaxation time
  acenaphthene characterization, 384
  aliphatic carbons vs. coal rank, 347, 349f
  aromatic carbons vs. coal rank, 347, 348f
  calculated values, 346, 347t
  correlation of those from maxima with other analytical methods, 349
  description, 344–345
  determination, 346
  multicomponent magnetization decay, 347
  validity, 348–349
  weighted-average relaxation times vs. coal rank, 347, 350f

Carbonaceous materials
  high-temperature NMR studies at atmospheric pressure, 156–170
  high-temperature NMR studies of thermal transformation, 152–156
Carbonaceous solids
  ^{1}H NMR spectroscopy, 27–42
  quantitation in ^{13}C NMR spectroscopy, 3–24
Carboxyl carbon/protonated aromatic carbon ratio, values for Argonne Premium coals, 276–278
Cavity, sample, requirement for high-frequency EPR spectroscopy, 133–134
Cellulosic char, calculation of Overhauser enhancement of DNP, 52–53
Ceramic fiber, rare-spin spectra, 58, 59f
Charge-transfer complex, temperature of maximum EPR intensity, 586–587
Chemical-shift–chemical-shift correlation spectroscopy, for measurement of ^{13}C
Chemical-shift anisotropy, 431–432
Chemical transformation of coals, EMR effects, 635–637
Chemical treatment, removal of effects of unpaired electrons, 10
Chemometrics and multivariate techniques for data analysis, 254
Citric acid, ^{1}H CRAMPS spectrum, 32f, 33
Coal(s)
  Argonne Premium, *See* Argonne Premium coals
  bituminous, *See* Bituminous coals
  carbon-based free radical concentration, 561
  complexity, 5–6
  cost of optimizing processes, 253–254
  cross-link composition, 468
  difficulty in understanding, 582
  distortion-free ^{13}C NMR spectroscopy, 217–226
  ^{1}H NMR spectroscopy, 27–42
  hydrogen bond importance in cross-linking macromolecules, 468
  multiple-phase structure model, 309
  origin of EPR signal, 582
  physical and chemical transformations related to utility, 229
  polynuclear aromatic radical type vs. coal rank, 562
  population factors using variable-angle sample spinning, 428, 429t
  pyridine bonding to surface, 202

INDEX

Coal(s)—*Continued*
  rank vs. amount taken up from aqueous solution, 530
  role of noncovalent bond interactions, 529
  solid-state NMR relaxation time, principal-component analysis, 253–266
  structural determination of organic radicals using high-frequency EPR spectroscopy, 125
  tensor components using variable-angle sample spinning, 428, 429t
  unpaired electrons, 6–11
Coal liquefaction process, conventional, 281–282
Coal liquefaction residues characterized by solid-state ^{13}C NMR spectroscopy, 281–292
Coal rank, cross-polarization signal effect for pyridine, 207, 209f
Coal–solvent interactions
  importance, 201
  solvent swelling effect, 201–202
Coal–tetrahydroquinoline, high-temperature ^{1}H NMR spectra, 156f
Combined rotation and multiple-pulse spectroscopy, *See* Proton CRAMPS
Computer simulation, low-frequency EPR spectroscopy, 117, 119
Conservative diffusion experiment, application to bituminous coals, 189–197
Continuous-wave ENDOR spectroscopy, advantages and disadvantages, 66–67
Continuous-wave EPR spectroscopy, Argonne Premium coals, 483–504
Continuous-wave ESR spectroscopy, Argonne Premium coals, 443–448
Conventional coal liquefaction processes, 281–282
Copper complexes, low-frequency EPR spectroscopy, 108–122
Cross-polarization experiments, influencing factors, 13–21
Cross-polarization–MAS ^{13}C NMR spectroscopy of Argonne Premium coals, 377–397
  acenaphthene characterization, 384
  aromaticities, 386, 387t
  carbon magnetization determination, 384–386
  chemical-shift distributions, 386, 389t
  comparison of 4, 4′-dimethoxybibenzyl results with those of Bloch-decay, 397t

Cross-polarization–MAS ^{13}C NMR spectroscopy of Argonne Premium coals, 377–397—*Continued*
  comparison of acenaphthene results with those of Bloch-decay, 394–397
  comparison of aromaticities with those of Bloch-decay, 392–394
  contact time curves, 385f, 386
  experimental procedure, 383–384
  high-speed spinning, 380–383
  ^{1}H rotating-frame spin–lattice relaxation time determination methods, 384–386
  relaxation parameter values, 386, 387t
  sample preparation, 383
  time dependence of magnetization growth and decay, 386, 390–392
  Upper Freeport coal spectra, 386, 390f
  Wyodak-Anderson coal spectra, 386, 391f
Cross-polarization–MAS spectroscopy, applications, 4
Cross-polarization NMR technique
  advantages and disadvantages for quantitation in ^{13}C NMR spectroscopy, 23–24
  rf pulse sequences, time constants, and time delays, 4, 5f
Cross-polarization solid-state NMR spectroscopy, probe for pyridine bonding to coal, 202
Cross-polarization spin dynamics, analysis techniques, 16–21
Cross-polarization times, 326–333
Cumulant expansion, example for ESR measurement, 99–100
Cytochrome c oxidase, transmission EPR spectrum, 130, 132f

D

Data analysis, ESR measurement, 99
Davies ENDOR spectroscopy, 74–76
Decoupling of electron spins, removal of effects of unpaired electrons, 11
Diamond, vapor-deposited, ^{13}C spectrum, 58–59, 60f
Diffusion-coupled relaxation of bituminous coal, 175–197
  concepts, 178, 179t
  conservative diffusion data analysis, 189–191f
  conservative diffusion experimental procedure, 184
  conservative diffusion regime, 181

Diffusion-coupled relaxation of bituminous coal, 175–197
  correspondence between magnetic and thermal parameters, 179t
  data requirements, 177
  dynamic diffusion approach, 194–196
  dynamic diffusion data analysis, 187, 188f, 189
  dynamic diffusion experiment, 183–184
  Edzes–Samulski approach, 191–195f
  Edzes–Samulski data analysis, 184–187
  Edzes–Samulski experiment, 182–183
  macroscopic two-phase lamellar system, 180f, 181
  nonconservative diffusion regime, 181
  reduced two-phase model, 182
  selective excitation techniques, 177–178
  theory, 178–182
4,4′-Dimethoxybibenzyl
  comparison of CP–MAS and Bloch-decay results, 397t
  Hartmann–Hahn curve, 381f, 382
Dipolar dephasing to discriminate between types of carbons, 334–337
  advantages for ^{1}H CRAMPS, 402
  Argonne Premium coals, 273, 276t
  fresh coals, 334, 335f
  oxidized coals, 334–335, 336f
  protonated aromatic carbon determination, 334–336, 337t
  signal intensities of aromatic carbons vs. delay time, 273, 275f
  spectra for Illinois No. 6 coal, 273, 274f
  validity for protonated to nonprotonated carbon determination, 335–338f
Dipolar fields, measurement, 618, 619
Dipolar relaxation, measurement, 618, 619f, 621
Dipolar spin–lattice relaxation time, effect of experimental conditions, 623–624
Direct thermal mixing effect mechanism, DNP, 47f
Disordered glasses, ESR measurement effect, 97–99
Distortion-free ^{13}C NMR spectroscopy in coal, rotating-frame DNP, 217–226
Double ENDOR, *See* Electron nuclear triple resonance
Dry coals, proton relaxation rates, 193, 194t
Durene, ^{1}H CRAMPS spectrum, 32f, 33
Dynamic-angle spinning for measurement of ^{13}C CSA, 433

Dynamic diffusion experiments
  application to bituminous coals, 194–196
  selective saturation method, 183–184
Dynamic nuclear polarization
  ^{13}C spectra of anthracite, 321f, 322
  direct thermal mixing effect mechanism, 47f
  enhancement of nuclear magnetization, 219
  nuclear polarization due to static electron–nuclear interactions, 48–50
  Overhauser enhancement, 47f, 51–53
  overview, 45–47
  types, 219–221
Dynamic nuclear polarization–CP spectroscopy, comparison to DNP–SP spectroscopy, 56–58
Dynamic nuclear polarization NMR spectroscopy for solid materials research, 45–61
  applications, 53–60
  sensitivity, 54, 55f
  separation of enhancement curves, 53, 54f
Dynamic nuclear polarization–single-pulse spectroscopy, 56–60
  advantages, 60–61
  ceramic fiber rare-spin spectra, 59
  electron density distribution, 56–58
  molecular structure analysis in vicinity of unpaired electrons, 56, 57f
  three-spin effect, 58
  vapor-deposited diamond ^{13}C spectrum, 58–60f
Dynamics of EPR system, 91

E

Echo decay, 613–616
Edzes and Samulski experiment, 182–183
Eigenvalues and eigenvectors, principal-component analysis, 260, 261t, 262
Elastic modulus of coal, calculation of average molecular weight between cross-links, 350–351
Electron dipolar field and dynamics measurement in solids, 605–624
  applicability of spin-temperature formalism, 618, 620–622
  dipolar fields, 618, 619f,t

Electron dipolar field and dynamics
  measurement in solids—
  Continued
  dipolar relaxation, 618, 620–621
  ESR in coal, 622–624
  experimental procedure, 616–617
  pulsed EPR spectroscopy, 612–616
  spin temperature theory, 607–610
  statistical approach theory, 610–612
  Zeeman spin–lattice relaxation, 617f, 618t
Electron donors or acceptors for EPR
  characterization of Argonne Premium
  coals, 529–537
  difference in spin concentration of coal
    with and without guest molecules vs.
    coal rank, 533, 535f
  difference in spin concentration of coal
    with and without iodine vs. H/C ratio
    of coal, 533, 535, 536f
  EPR spectral shape with and without guest
    molecules, 531, 532f, 533
  experimental materials, 530–531
  hydrogen bonding importance, 536–537
  spin concentration of coal–guest molecule
    complexes vs. coal rank, 533, 534f
  spin concentration of coal vs. coal rank,
    533–534f
Electron–electron interactions, 7–9
Electron flip-flop transition rate, 8–10
Electron magnetic resonance spectroscopy,
  applications, 451–452
Electron magnetic resonance spectroscopy
  of Argonne Premium coals, 631–636
  absorption line shapes, 632–633
  chemical transformation effects, 635–637
  coal structure and swelling, 634–635
  electron–nuclear interactions, 633–634
  influencing factors, 642
  nature of carbon radicals in coal, 631–632
  radical type determination, 633
  rank-dependent effects, 635
Electron nuclear double resonance
  (ENDOR) spectroscopy, 65–89
  Argonne Premium coals, 458–465
  electron–nuclear interactions, 633–634
  energy level diagram for four-level
    system, 69f, 70
  EPR-detected sublevel coherence, 81–88
  EPR-detected sublevel population transfer,
    71–82
  factors affecting enhancement, 70
  features of spectra, 66
  Hamiltonian, 66–67
  limitation, 86

Electron nuclear double resonance
  (ENDOR) spectroscopy—
  Continued
  pulse sequences, 70–71
  relative enhancement, 68
  spin eigenvalues, 69
  triphenylmethyl radical spectrum, 66
Electron nuclear double resonance (ENDOR)
  spectroscopy of Argonne Premium coals,
  pulsed, *See* Pulsed ENDOR
    spectroscopy of Argonne Premium coals
Electron–nuclear interactions
  role in DNP, 46–47
  studies using ENDOR spectroscopy,
    633–634
Electron nuclear triple resonance
  hyperfine selective ENDOR, 79, 80f
  pulse sequence, 78f, 79
  sublevel EPR spectroscopy, 80–81, 82f
Electron paramagnetic resonance (EPR)
  spectroscopy
  absorption line shapes, 632–633
  applications, 468
  Argonne Premium coals, 467–481,
    539–557
    characterized with electron donors or
      acceptors, 529–537
    CW, 483–504
    EPR properties, 542, 544t
    microwave-saturation responses,
      542, 545f, 546f
    process and properties, 539–542
    temperature related to intensity,
      582–603
    transition metal spectra,
      547, 548f, 549t
    two-component carbon radical signal,
      542, 543f
  comparison to NMR spectroscopy,
    630–631
  field-support system, factors affecting
    spectra, 91
  multifrequency, *See* Multifrequency EPR
    spectroscopy
  triphenylmethyl radical spectrum, 66
  spin-probe spectroscopy of porosity in
    solvent-swelled coal, 467–481
  sublevel coherence detected with
    multiple-quantum ENDOR, 81–88
  sublevel population transfer detected with
    Davies ENDOR, 71–82
  See also Electron spin resonance
    spectroscopy
Electron spin, schematic diagram, 630f

Electron spin-echo (ESE) spectroscopy of
 Argonne Premium coals, 462–464,
 539–557
 advantages, 547, 550
 instantaneous diffusion using microwave
  pulses of different tipping angles,
  550, 554f
 magnitude of instantaneous diffusion vs.
  coal rank, 554, 555f
 nuclear modulation patterns, 462
 process and properties, 541
 selective free induction decay detection
  of narrow line, 550, 551f, 552t
 spin–lattice relaxation for broad line,
  556f, 557t
 spin–lattice relaxation for narrow line,
  550, 552t, 553
 two- and three-pulse experiments, 462
 X band, See X-band ESE spectroscopy of
  Argonne Premium coals
Electron spin-echo (ESE) spectroscopy of
 irradiated malonic acid, 67f, 68
Electron spin-echo relaxation for
 radical type discrimination, 633
Electron spin decoupling, removal of effects
 of unpaired electrons, 11
Electron spin–lattice relaxation, 541
Electron spin relaxation measurement,
 91–103
 "2 + 1" experiment, 102–103
 data analysis, 99–100
 detection of signal vs. spectrum, 95–97
 experimental problems, 622–623
 function of boxcar integrator, 96
 inversion-recovery experiment, 100–101
 pulse sequence effect, 96–97
 relaxation times vs. experimental
  conditions, 623–624
 samples studied, 97–99
 two-pulse spin-echo experiment, 101–102
Electron spin resonance (ESR) spectroscopy
 2-mm band, See Continuous-wave ESR
  spectroscopy of Argonne Premium coals
 applications, 141
 Argonne Premium coals, 453–458
 description, 140–141
 free radical formation, 141
 free radical spin densities, 308f, 309
 high-temperature, 139–149
 natural coals, description, 443
 X band, See X-band CW ESR spectroscopy
  of Argonne Premium coals
 See also Electron paramagnetic resonance
  spectroscopy

Electronic spin Hamiltonian, definition, 508
Enhancement factor of laboratory-frame
 DNP, 219

F

Factor loadings, definition, 258
Feed coals, stacked plots of spectra,
 284, 285f
Field-dependent terms, definition, 114
Free-induction decay of Argonne Premium
 coals, 365f, 366t
Frequencies, high-frequency EPR
 spectroscopy, 122–123
Frequency-shift perturbation technique,
 analysis of low-frequency EPR
 spectroscopy, 121, 122f
Fulvic acid, ^{1}H CRAMPS spectrum, 34f, 35
Fusinite
 free-induction decay and W-band EPR
  spectrum, 512, 518f
 line width of EPR vs. frequency,
  111–112f, 114f

G

g factor, temperature dependence,
 497–499f
g strain, line width effect,
 110, 112f

H

^{1}H–^{1}H dipolar-dephasing ^{1}H CRAMPS,
 molecular-level mobility analysis,
 35–37f
^{1}H–^{1}H dipolar interactions, ^{1}H NMR
 spectroscopy, 27–28
^{1}H decoupling, interference from molecular
 motions with quantitation in ^{13}C NMR
 spectroscopy, 11
^{1}H magnetization, spin locking, 15f, 16
^{1}H NMR relaxation behavior, bituminous
 coals, 176
^{1}H NMR spectroscopy
 carbonaceous solids, 27–42
 for characterizing molecular structures
  and motions in organic solids,
  295–296
 water analysis in coals, 361

INDEX

^{1}H NMR spectroscopy of Argonne Premium coals, 295–309
  degassed coal spectra, 297–300
  dried coal spectra, 297–300
  experimental procedure, 297
  fresh coal spectra, 297, 298–299f, 300
  ^{1}H relaxation times, 301, 302–304f
  importance of moisture removal, 296
  line widths of dried coals vs. percent carbon, 301f
  relaxation times vs. percent carbon, 303–304, 305–307f
^{1}H NMR transverse-magnetization signals for bituminous coals, 176
Hamiltonian equation, 28
Hardgrove grindability index, second-moment parameter and temperature of maximum rate of increase, 167, 168f
Hartmann–Hahn matching condition, quantitation in CP experiment, 14
Heteroatom sensitivity, high-frequency EPR spectroscopy, 125–127f
Heterogeneity of sample, *See* Sample heterogeneity
1,2,3,6,7,8-Hexahydropyrene, variable-angle spinning spectra, 427, 428f
High-field NMR studies of Argonne Premium coals, 311–322
  apparent aromaticities, 313, 316t, 317
  apparent aromaticity calculation, 313
  aromaticity-extraction process, 313, 314f
  aromaticity vs. method, 317–318
  ^{13}C CP NMR spectra, 313, 315f
  carbon-structure distribution parameters, 313, 316t, 320
  CRAMPS, 313–314, 317f
  proton aromaticities, 313–315, 318t, 320
  requirements, 311, 312
  rotating-echo intensity ratio, 318–320
  sensitivity, 312
High-frequency EPR spectroscopy, 121–134
  advantages, 122–123
  applications, 123
  attenuation and insertion loss of waveguide components, 131, 133
  economics, 134, 136
  frequencies, 122–123
  heteroatom sensitivity, 125–126, 127f
  instrumentation, 130–131, 133–134, 135t
  molecular geometric sensitivity, 128
  nitroxide free radical in frozen toluene, 123, 124f
  nitroxide radical mixture, 125, 126f
  organic free radicals, 125

High-frequency EPR spectroscopy—*Continued*
  passage effects, 128f
  resolution, 123–129
  sample cavity requirement, 133–134
  sensitivity vs. frequency, 133–134, 135t
  structural determination of organic radicals in coal, 125
  waveguide component availability, 130–131
  zero-field splitting, 129–130, 131–132f
High-resolution ^{1}H NMR spectroscopy, Argonne Premium coals, 401–417
High-speed spinning, CP–MAS ^{13}C NMR spectroscopy, 380–381f, 382–383
High-temperature ESR spectroscopy, 139–169
  high-pressure methods, 142–146
  pyrolysis studies on pristine and oxidized coals, 146–151f
High-temperature–high-pressure ESR spectroscopy, 142–146
  advantages, 146
  apparatus assembly, 145f
  apparatus design, 142–146
  cavity, 143, 144f
  heating rates, 146
  spin concentration calculations, 146
  tube, 143f
High-temperature NMR spectroscopy
  applications, 149, 151
  carbonaceous material studies at CSIRO, 156–170
  carbonaceous material studies at Hokkaido University, 152–156f
  methods, 149–169
  temperature ranges, 151–152
Highvale coal
  aromatic carbon distribution, 288t
  hydrogen distribution, 288, 289t
  liquefaction data, 287t, 288
  stacked plots of spectra, 284–286f
Humic acid, ^{1}H CRAMPS spectrum, 34f, 35
Hydrated keratin, reduced proton-magnetization relaxation, 194, 195f
Hydrogen assay procedures, 360–361
Hydrogen bonding, importance in cross-linking macromolecules, 468
Hydrogen distribution in coals, 288, 289t
Hydrogen gas in coal liquefaction residues, 289, 290–292
Hydrogen-to-carbon ratio, aromaticity effect, 330–331, 333f
4-Hydroxy–2, 2, 6, 6-tetramethylpiperidinoxyl, ESR studies in coal, 468–469

Hyperfine selective ENDOR spectroscopy, irradiated malonic acid, 79, 80f

I

Illinois No. 6 coal
aromatic carbon distribution, 288t
^{13}C NMR spectra vs. spinning rate, 271f, 272
CW EPR spectroscopy, 483–504
Davies ENDOR spectra, 564–570
dipolar-dephasing spectra, 273, 274f, 407, 409f
^{1}H CRAMPS of untreated and pyridine-saturated coals, 403–406
hydrogen distribution, 288, 289t
liquefaction data, 287t, 288
stacked plot of ^{1}H NMR time-domain signals, 234, 236f
stacked plots of CP–MAS and DD NMR spectra, 284, 285f
temperature vs. truncated second moment, 241f
truncated second moment-temperature pyrograms, 242, 243f
W-band EPR spectrum, 512, 514f
In situ measurement technique, 140
Indirect thermal mixing effect, 47f
Inhomogeneously broadened lines, 66
Instantaneous diffusion, 614
Inverse electron–electron flip-flop time, 49
Inversion-recovery experiment, ESR measurement, 100–101
Iodine, coal rank vs. amount taken up from aqueous solution by coal, 530
Irradiated malonic acid
Davies ENDOR spectrum, 74, 75f
electron spin-echo spectrum, 67f, 68
hyperfine selective ENDOR, 79, 80f
Mims ENDOR spectrum, 77f, 78
multiple-quantum ENDOR spectrum, 87, 88f
sublevel EPR spectroscopy, 81, 82f
sublevel free induction decay, 83, 84f, 85
sublevel Rabi oscillations, 72, 73f
Isotope-dilution technique, 360

K

Kerogen concentrate, ^{1}H CRAMPS spectrum, 34f, 35

L

Laboratory-frame DNP signal enhancement vs. microwave offset frequency, 219, 220f
Larsen–Marzec model, description, 486
Lewiston–Stockton coal
^{1}H relaxation times, 301–302, 303f
CW EPR spectroscopy, 483–504
dipolar-dephasing results, 407, 408f
^{1}H CRAMPS of untreated and pyridine-saturated coals, 403–406
Hartmann–Hahn matching curve, 380f, 381
truncated second moment-temperature pyrograms, 242, 243f
W-band EPR spectrum, 512, 516f
Line shape
EPR absorption, influencing factors, 632–633
low-frequency effect, 109–111f
Line width
low-frequency effect, 110–113f
NMR line in absence of electrons, 49
temperature dependence, 497, 500–502
Loadings, plot for principal-component analysis, 262, 263f
Loop-gap resonator, schematic representation, 119, 120f
Low-frequency EPR spectroscopy, 108–121
data analysis, 120–121, 122f
economics, 134, 136
instrumentation, 119, 120f
line shape vs. frequency, 109–111f
line width vs. frequency, 110–113f
methodology, 119–120
signal-to-noise ratios, 119–120
spectral simulation, 117, 119
state mixing effect, 114–118
Low-frequency pulsed EPR spectroscopy, 524–527
factors affecting phase-memory time, 524–525
frequency effect, 525, 526
spin-echo decay times vs. magnetic field, 526t, 527
two-pulse echo envelope, 524, 525f
Low-volatile bituminous coal
^{13}C-NMR spectra, 55–57f
DNP enhancement curves, 53, 54f

M

Macerals, aromaticity, 330

INDEX

Macroscopic two-phase lamellar system, 180–181
Macroscopic Zeeman magnetization, 178
Magic-angle hopping, description, 12
Magic-angle hopping for measurement of ^{13}C CSA, 433
Magic-angle spinning
  ^1H NMR effects on monoethyl fumarate, 29, 31f, 32
  interference with molecular motion in ^{13}C NMR spectroscopy, 12
  modulation of polarization-transfer time, quantitation in CP experiment, 14–15
Magnetic parameters, correlation to thermal parameters, 179t
Magnetic resonance in coals, electron and nuclear spins, 630
Magnetic resonance in solids, methods to characterize spin system, 605
Magnetic resonance techniques applied to Argonne Premium coals, 629–642
Magnetization growth and decay, time dependence, 386, 390–391, 392
Malonic acid
  Davies ENDOR spectrum, 74, 75f
  electron spin-echo spectrum, 67f, 68
  hyperfine selective ENDOR, 79, 80f
  Mims ENDOR spectrum, 77f, 78
  multiple-quantum ENDOR spectrum, 87, 88f
  sublevel EPR spectroscopy, 81, 82f
  sublevel free induction decay, 83, 84f, 85
  sublevel Rabi oscillations, 72, 73f
Matching time, relationship to spin–lattice relaxation time in rotating reference frame and polarization-transfer time, 16–19f
Methane gas in coal liquefaction residues, 289–292
Microwave pulse, description, 70–71
Microwave resonator, function, 96
Millimeter waves, description, 122
Mims ENDOR spectroscopy
  irradiated malonic acid radical spectrum, 77f, 78
  magnetization pattern and pulse sequence, 76–77
Molecular dynamics in coals, 342
Molecular-level mobility, analysis using ^1H–^1H dipolar-dephasing ^1H CRAMPS, 35–37
Molecular motions, ^1H decoupling, interference with quantitation and MAS in ^{13}C NMR spectroscopy, 11–12

Molecular structure and properties of bituminous coals studied with ^1H NMR techniques, 175–176
Molecular weight between cross-links in coal calculated using elastic modulus of coal, 350–351
MREV–8 pulse sequence, use in ^1H CRAMPS, 39
Multicomponent magnetization decays in macromolecular systems, role of carbon spin–lattice relaxation time distributions, 347–348
Multifrequency EMR spectroscopy of Argonne Premium coals, 507–527
  applications, 508
  elemental and proximate data for Herrin No. 6 coal, 511t
  EPR spectra vs. microwave frequency, 509, 510f
  ESE envelope modulation time domain patterns vs. microwave frequency, 509, 510f
  experimental procedure, 511–512
  low-frequency pulsed EPR spectroscopy, 524, 525f, 526t, 527
  method for varying importance of spin Hamiltonian terms, 509
  very high frequency EPR spectroscopy, 512–524
Multifrequency EPR spectroscopy, 107–136
  advantages, 108
  frequency ranges, 107–108
  high frequency, 121–135
  low frequency, 108–122
  zero field, 121
Multiple-phase structure model, coal, 309
Multiple-pulse line-narrowing technique, ^1H NMR effects on monoethyl fumarate, 29–32
Multiple-quantum ENDOR spectroscopy, irradiated malonic acid, 87, 88f
Multivariate statistical analysis and pattern recognition, principal-component analysis, 253–266

N

Nitrogen gas in coal liquefaction residues, 285–289
Nitroxide free radical in frozen toluene, high-frequency EPR spectroscopy, 123, 124f
Nitroxide radical mixture, high-frequency EPR spectroscopy, 125, 126f

NMR relaxation rate measurement, 571
NMR spectroscopy
  comparison to EPR spectroscopy, 630–631
  for characterization of carbonaceous samples, 27
  high field, See High-field NMR studies of Argonne Premium coals
  high-temperature, 149–169
  solid materials research, 45–61
  techniques to probe for pyridine bonding to coal, 202
  thermal analysis, 140
NMR spin echo, EPR detection, 570, 571f
Nonconservative diffusion regimes, 181
Noncovalent interactions in coal structure, 529–530
Nonequilibrium spin states, spin-temperature formalism, 607–610
Nuclear polarization enhancement, 48–50
Nuclear spin Hamiltonian, 508
Nuclear spins, schematic diagram, 630f

O

Oil shale, ^{1}H CRAMPS spectrum, 34f, 35
Organic free radicals, high-frequency EPR spectroscopy, 125
Overhauser enhancement of DNP, average enhancement, 47, 51–53
Oxidized coal
  pyrolysis studies using high-temperature ESR spectroscopy, 148–151f
  solid-state ^{13}C NMR spectroscopy, 323–338
Oxygen, high-frequency EPR spectroscopic effect, 125–126

P

Perdeuteropyridine, saturation effect on ^{1}H CRAMPS, 35f
Phase cycling, function, 95
Pittsburgh No. 8 coal
  CW EPR spectroscopy, 483–504
  dipolar-dephasing results, 407, 409f
  ^{1}H CRAMPS of untreated and pyridine-saturated coals, 403–406
  truncated second moment-temperature pyrograms, 242, 244f
  W-band EPR spectrum, 512, 514f

Pocahontas No. 3 coal
  ^{1}H NMR spectra of fresh, degassed, and dried, 297–300
  chemical-shift–chemical-shift correlation spectra, 431–432, 434f
  CW EPR spectroscopy, 483–504
  dipolar-dephasing results, 407, 408, 410f
  ^{1}H CRAMPS of untreated and pyridine-saturated coals, 403–406
  stacked plot of ^{1}H NMR time-domain signals, 234, 238f
  transition metal EPR signals, 547–549t
  truncated second moment-temperature pyrograms, 242, 244f
  variable-angle spinning spectra, 428–430f
  W-band EPR spectrum, 512, 515f
Polarization transfer between electron and nuclear spin systems, 45
Polarization-transfer time
  determination, 328, 329t
  quantitation in CP experiment, 14
  relationship to spin–lattice relaxation time, 16–21
Polycrystalline powders, ESR measurement effect, 97–99
Polynuclear aromatic radicals, structure analysis techniques, 562–563
Pore structure of coals, 467–468
Porosity, EPR spin-probe spectroscopy in solvent-swelled coal, 467–481
Powder pattern, definition, 420
Primary echo pulse sequence, measurement of dipolar parameters, 612–614
Principal-component analysis of coal solid-state NMR relaxation times, 253–266
  advantages, 255–256
  bivariate correlations, 256, 257f
  characterization data, 258, 259t
  correlation-coefficient matrix, 260, 261t
  data form, 258
  eigenanalysis, 256, 258
  eigenvalues and eigenvectors, 260–262
  interpretation of principal components, 258
  loadings, 262, 263f
  multivariate statistical analysis and pattern recognition, 253–266
  process, 256
  Q-mode factor analysis, 264, 265f, 266
  scores, 262, 263–264f
Principal components, definition, 258

INDEX

Pristine coal, pyrolysis studies using high-temperature ESR spectroscopy, 146–149f
Proton aromaticities, calculation, 314
Proton CRAMPS, 29–42
  Argonne Premium coal spectra, 33f, 34
  citric acid spectrum, 32f, 33
  durene spectrum, 32f, 33
  fulvic acid spectrum, 34f, 35
  humic acid spectrum, 34f, 35
  kerogen concentrate spectrum, 34f, 35
  line-narrowing techniques, 29, 31f, 32
  molecular-level mobility analysis, 35–36, 37f
  oil shale spectrum, 34f, 35
  operating parameter selection, 40–41
  perdeuteropyridine saturation effect, 35f
  pulse sequence selection, 36, 38–39
  static magnetic field strength selection, 39–40
  strategy, 29, 30f
  water problem, 41–42
Proton CRAMPS of Argonne Premium coals, 401–417
  advantages of dipolar dephasing, 402
  aromatic hydrogen content, 411
  chemical shift of aliphatic group, 411
  chemical-shift position of aliphatic peak maximum, 414, 415t
  chemical-shift position of aliphatic peak maximum vs. coal parameters, 414, 416f
  dipolar dephasing for untreated and pyridine-saturated samples, 406, 407–409f
  experimental procedure, 402–403
  factors affecting resolution, 402–413
  proton type vs. mobilization, 413–414
  spectra of pyridine-saturated coals, 404, 405f, 406
  spectra of untreated coals, 403, 404f
  sulfur and nitrogen content, 411–412
  ultimate analyses of samples, 402, 403t
Proton laboratory-frame spin–lattice relaxation time, 342
Proton magnetic resonance thermal analysis, 159–169
  comparison to conventional analysis, 169t, 170
  correction for temperature-sensitive effects, 162–163
  description and applications, 230
  Hardgrove grindability index, 167, 168f
  hydrogen-loss pyrogram, 163–164, 165f
  instrumentation, 159, 160f, 162

Proton magnetic resonance thermal analysis—*Continued*
  operations in pyrolysis experiment, 159, 161f
  rigid-remaining hydrogen pyrogram, 164, 166f
  second-moment parameter vs. temperature, 166, 167f
  signal averaging, 162
  stacked plots of signals, 163–164f
  temperature of minimum second-moment parameter, 167, 168f
Proton magnetic resonance thermal analysis of Argonne Premium coals, 229–249
  analytical data, 231–233t
  bituminous coals, 246–249
  coal composition effect, 249
  experimental procedure, 231
  quantification of ^{1}H NMR signal, 234, 239–243
  shape of transverse-magnetization signal, 231, 234
  stacked plots of time-domain signals vs. coal type, 234–238f
  subbituminous coals, 243–245f
Proton quantitation in Argonne Premium coals by solid-state ^{1}H NMR spectroscopy, 359–375
  experimental procedure, 363
  free induction decay components, 365f, 366t
  Gaussian decay vs. water content, 371
  Lorentzian decay vs. water content, 371
  proximate and ultimate analyses of coals, 362t, 363
  quantitative measurements, 371–375f
  samples, 361, 362t
  solid-echo measurements, 364, 365
  spin counting, 367–375
  water content vs. ^{1}H NMR spectra of coals, 368, 369f
  water content vs. CRAMPS decoupled spectra, 369, 370f
Proton relaxation times, calculation, 297
Proton rotating-frame spin–lattice relaxation time
  determination, 328, 329t
  interpretation difficulties, 342–343
  values for aromatic and aliphatic resonances in Argonne Premium coals, 355t, 356
  weathering effect, 356f, 357
Proton spin–spin relaxation times, measurement, 570–571

Protonated aromatic carbons, determination using dipolar dephasing, 334–338
Pulse sequence
  "2 + 1", measurement of dipolar parameter, 614
  ENDOR spectroscopy, types, 70–71
  ESR effect, 94–95
  Mims ENDOR spectroscopy, 76f, 77
  multiple-quantum ENDOR spectroscopy, 87f
  selection for ^{1}H CRAMPS, 38–39
  sublevel free induction decay, 82, 83f
  sublevel spin echoes, 85f
Pulsed ENDOR spectroscopy, advantages and disadvantages, 66–67
Pulsed ENDOR spectroscopy of Argonne Premium coals, 561–579
  advantages, 563
  chemistry vs. coal rank, 578
  coal rank vs. spectra, 573–576
  evidence for paramagnetic polynuclear aromatic radicals, 578, 579f
  experimental condition effect, 563–567
  ^{1}H spin–spin relaxation time measurement, 570–572
  line broadening vs. bandwidth, 572, 573f
  NMR relaxation rate measurement, 570
  NMR spin-echo detection on nuclear sublevels, 570, 571f
  occurrence of discrete lines, 577–578
  oxygen contamination effect, 577
  preparation-pulse bandwidths vs. spectra, 565, 566f, 568–569
  spin dynamics mechanisms, 569–572
  stabilization of unpaired electrons on polynuclear aromatic structures, 578
  temperature vs. spectra, 564f, 565, 567–568
Pulsed EPR spectroscopy
  modification of "2 + 1" pulse sequence, 614–616
  primary echo pulse sequence, 612–614
Pulsed NMR spectroscopy, hydrogen assay, 360–361
Pyridine
  bonding to coal surface, 202
  ^{1}H spectral resolution and dynamics effects, 402
  sorbed onto coal, ^{13}C NMR spectroscopy, 201–215
Pyridine-swollen Borehole vitrinite, recovery of ^{1}H magnetization vs. evolution time, 196, 197f

Pyrolysis studies using high-temperature ESR spectroscopy, pristine and oxidized coals, 146–149

Q

Q-mode factor analysis, 264–266
Quantitation in ^{13}C NMR spectroscopy of carbonaceous solids, 3–24
  factors affecting CP experiments, 13–21
  influencing factors, 4–13
  interference from molecular motions, 5–6
  limiting factors and possible remedies, 21, 22t
  MAS, 11–12
  optimal experimental conditions, 21, 23
  pulse sequence selection, 23–24
  recycle delay, 12–13
  sample heterogeneity vs. quantitativeness, 5–6
  unpaired electrons vs. quantitativeness, 6–11
Quantitation in cross-polarization experiments
  advantages and disadvantages, 13–14
  CP spin dynamics vs. quantitativeness, 16–21
  Hartmann–Hahn matching condition vs. quantitativeness, 14
  MAS modulation of polarization transfer time, 14–15
  spin locking of ^{1}H magnetization vs. quantitativeness, 15f, 16
Quantitation of protons in Argonne Premium coals by solid-state ^{1}H NMR spectroscopy, 359–375
Quaternary/tertiary carbon ratio, values for Argonne Premium coals, 276–278

R

Rabi oscillations, pulse sequence for detection, 72, 73f
Radical-difference absorption–dispersion component Argand diagrams, 486–497
Radical types, discrimination using ESE relaxation techniques, 633
Radio frequency pulse, 70–71
Radius of nuclei not contributing to bulk NMR signal, 49–50
Rank, EMR parameter effects, 635
Reaction time methods, 140

Recycle delay, selection criteria and quantitation, 12–13
Relative ENDOR spectroscopy, definition, 68
Relaxation
  EPR system, view of spin system, 91
  mechanism, 93
  parameters, Bloch-decay and CP–MAS values, 386, 387t
  process, definition, 93
  spin system, 92–103
  time, determination, 297
Resolution, high-frequency EPR spectroscopy, 123–129
Resonance phenomenon, frequencies, 107
Rotating-echo intensity ratio, 318–320
Rotating-frame DNP
  advantages, 220–221
  enhancement factor, 219
  signal enhancement vs. microwave offset frequency, 219, 220f
Rotating-frame DNP–CP in coal
  advantages, 221
  ^{13}C spectra of LVB coal, 223f, 224
  experimental procedure, 221–222
  limitations, 225
  pulse sequence, 221, 222f

S

Sample cavity, requirement for high-frequency EPR spectroscopy, 133–134
Sample heterogeneity, complexity of coal, 5–6
Saturation-recovery experiment, *See* Inversion-recovery experiment
Score, plot for principal-component analysis, 262–264
Second moment of frequency-domain ^{1}H NMR spectrum, 234
Second-moment parameter in proton magnetic thermal analysis, 166–168
Selective saturation experiments
  application to bituminous coals, 190–197
  data analysis, 184–191
  experimental procedure, 182–184
Signal losses due to unpaired electrons, determination using NMR standard, 10
Signal processing, ESR measurement, 96
Signal separation methods, 94–95
Simulation, computer, low-frequency EPR spectroscopy, 117, 119

Single crystals, ESR measurement effect, 97–99
Single-pulse NMR technique
  advantages and disadvantages for quantitation, 23–24
  rf pulse sequences, time constants, and time delays, 4, 5f
Slow magic-angle spinning for measurement of ^{13}C CSA, 430
Solid echo, measurement for Argonne Premium coals, 364–365
Solid fossil fuels, study using solid-state NMR techniques, 311
Solid materials, research with NMR and DNP spectroscopy, 45–61
Solid-state ^{13}C NMR spectroscopy
  Argonne Premium coals, 269–278, 637–642
  potential applications, 3–4
  to characterize coal liquefaction residues, 281–292
  to study coal and coal oxidation, 323–338
Solid-state effect mechanism, dynamic nuclear polarization, 46, 47f
Solid-state ^{1}H NMR spectroscopy, proton quantitation in Argonne Premium coals, 359–375
Solid-state high-resolution NMR spectroscopy, use of CP and MAS, 269–270
Solid-state NMR relaxation times of coal, principal-component analysis, 253–266
Solid-state NMR techniques for study of solid fossil fuels, 311
Solids
  carbonaceous, quantitation in ^{13}C NMR spectroscopy, 3–24
  measurement of electron dipolar fields and dynamics, 605–624
Solvent-swelled coal, EPR spin-probe spectroscopy of porosity, 467, 481
Solvent swelling, 201
Sorbed pyridine, relaxation behavior, 210, 213–215
Spin-fluid phase magnetization, 179
Spin–lattice relaxation of Argonne Premium coals, 295–309, 341–357
  carbon rotating-frame spin–lattice relaxation time, 351–355
  carbon spin–lattice relaxation time, 344–351
  early studies, 342–343
  ESR free radical spin densities vs. acid treatment, 308f, 309

Spin–lattice relaxation of Argonne Premium coals—*Continued*
  experimental procedure, 297, 343–345f
  influencing factors, 309
  interpretation difficulties, 342–343
  ^{1}H relaxation times, 297, 301–304f
  ^{1}H rotating-frame spin–lattice relaxation time, 355–357
  pulse sequences for rotating-frame spin–lattice relaxation time measurements, 344, 345f
  relaxation time determination, 297
  sample composition, 343, 344t
Spin-lattice relaxation time
  in rotating reference frame, 16–21
  See also Carbon spin–lattice relaxation time
Spin-temperature formalism
  applicability, 618, 620–622
  states, 607–610
Spin concentration, total, temperature dependence, 502, 503–504f
Spin counting for Argonne Premium coals, values, 367t
Spin dynamics
  cross-polarization, analysis techniques, 16–21
  function, 605–606
Spin eigenvalues, ENDOR, 69
Spin Hamiltonian
  definition, 508
  for EPR active system, 114–115
Spin locking of ^{1}H magnetization, quantitation in CP experiment effect, 15f, 16
Spin systems
  importance of spectra of weakly interacting systems, 605
  views, 91
Spin temperature, theory, 606–610
Spinning sidebands, quantitation in ^{13}C NMR spectroscopy, 11–12
Sporinite, W-band EPR spectrum, 512, 517f
State mixing
  determination, 115–117
  electron spin-echo envelope modulation, 117, 118f
  electron Zeeman interaction, 114, 115t
  field-dependent spin Hamiltonian, 114
  low field-frequency, 115–117
  total intensity vs. forbidden intensity vs. frequency, 113–116f

Static electron–nuclear interactions, role in DNP, 48–50
Static magnetic field strength selection for ^{1}H CRAMPS, criteria, 39–40
Statistical approach to spin behavior, 607, 610–612
Stimulated-echo ENDOR spectroscopy, *See* Mims ENDOR spectroscopy
Stop-and-go method
  description, 15
  for measurement of ^{13}C chemical-shift anisotropy, 433
Structural parameters of Argonne Premium coal samples using CP–MAS–DD ^{13}C NMR spectroscopy
  determination, 275
  ^{13}C NMR spectra vs. spinning rate, 269–278
  aliphatic/aromatic hydrogen ratio, 276
  carbon distributions, 272t, 273
  carboxyl carbon/protonated aromatic carbon ratio, 276
  dipolar dephasing, 273–276t
  experimental procedure, 270–271
  listing, 277, 278t
  number of rings, 277
  quaternary/tertiary aromatic carbon ratio, 276
  substituted/protonated and bridgehead aromatic carbon ratio, 277
Structure related to swelling, 634–635
Subbituminous coals, PMR thermal analysis, 243–245f
Subbituminous Taiheiyo coal, high-temperature ^{1}H NMR spectra, 155f
Sublevel coherence, EPR-detected, multiple-quantum ENDOR, 81–88
Sublevel EPR spectroscopy
  irradiated malonic acid spectra, 81, 82f
  spectrum generation, 80–81
Sublevel free induction decay
  irradiated malonic acid spectra, 83–85
  pulse sequence, 82, 83f
Sublevel population transfer, EPR-detected, Davies ENDOR, 71–82
Sublevel spin echoes, envelope decay, 85, 86
Submicroscopic structure of bituminous coal
  conservative diffusion approach, 191
  data requirements, 177
  dynamic diffusion approach, 194, 195f, 196
  Edzes–Samulski approach, 184–187, 191–195
  selective excitation techniques, 177–178

INDEX 663

Substituted/protonated and bridgehead aromatic carbon ratio, values for Argonne Premium coals, 277, 278*t*
Substitution coefficient, values for Argonne Premium coals, 276–278
Sulfur, high-frequency EPR spectroscopic effect, 125–126, 127*f*
Swelling
  properties of coals, 467–468
  related to structure, 634–635

T

Temperature
  cross-polarization signal effect for pyridine, 207, 208*f*
  ENDOR effect, 567
Temperature dependence of EPR intensity of coals, 581–603
  analytical problems with limited data range, 597–599
  arbitrary intensity function plot, 583–586
  arbitrary intensity–temperature product function plot, 583–586
  closed-cycle variable-temperature refrigeration equipment, 589–590
  comparison of 1, 1-diphenyl-2-picrylhydrazyl intensity vs. temperature behavior to that of coal, 596–597
  experimental procedure, 587–590
  factors affecting temperature readings, 599
  fractional contributions of doublet and triplet states vs. EPR signal, 600–603
  instrumentation, 588–589
  intensity based on Curie law, 582–583
  intensity based on donor–acceptor interactions, 583
  nitrogen-flow variable-temperature measurements procedure, 589
  sample preparation, 587–588
  software, 588–589
  spin concentration measurements, 590–592*t*
  temperature of maximum EPR intensity for charge-transfer complex, 586–587
  theory, 582–587
  variable-temperature experiments, 592–597
Tetrakis(trimethylsilyl)silane to determine undetected carbons in coals, 10
Thermal analysis methods
  information obtained, 230
  examples, 140

Thermal equilibrium value of ^{13}C magnetization, calculation, 17
Thermal parameters, correlation to magnetic parameters, 179*t*
Thermally induced changes in coals, study methods, 139–140
Thermoplastic bituminous coal
  hydrogen-loss pyrogram, 163–165*f*
  rigid-remaining hydrogen pyrogram, 164, 166*f*
Three-spin effect, 58
Time, cross-polarization signal effect for pyridine, 206*f*, 207
Time-averaged field, determination, 49
Time-dependent carbon magnetization, 378–380
Total suppression of sidebands technique, spectral distortion, 317–320
Transition metal EPR spectroscopy, Argonne Premium coals, 547–549*t*
Transverse ^1H magnetization, 36
Triphenylmethyl radical, EPR vs. ENDOR spectra, 66
Truncated second moment of frequency-domain ^1H NMR spectrum, 234, 239–242
Two-pulse spin-echo experiment, ESR measurement, 101–102

U

Undetected carbons in coal
  determination, 10
  relaxation effects, 21
Unpaired electron–^{13}C nuclei interactions, 6–11
  estimation of percentage of undetected carbons, 7–11
  factors affecting ^{13}C line broadening, 7–8
  problem, 6–7
  undetected carbons vs. unpaired electron concentration, 8, 9*f*
Upper Freeport coal
  ^1H relaxation times, 303, 304*f*
  chemical-shift–chemical-shift correlation spectra, 431–432, 434*f*
  CW EPR spectroscopy, 483–504
  dipolar-dephasing results, 407, 408, 410*f*
  ^1H CRAMPS spectra of untreated and pyridine-saturated coals, 403–406
  S-band spin-echo envelope, 524, 525*f*
  time dependence of magnetization growth and decay, 386, 390–392

Upper Freeport coal—*Continued*
  truncated second moment-temperature pyrograms, 242, 244f
  W-band EPR spectrum, 512, 513f

V

Vapor-deposited diamond, ^{13}C spectrum, 58–59, 60f
Variable-angle sample spinning, 424–430
  advantages, 424, 427
  experimental procedure, 427
  line shape expression, 424–425
  population factors for coals, 428, 429t
  spectra of coal, 428–429, 430f
  spectra of hexahydropyrene, 427, 428f
  tensor components for coals, 428, 429t
  tensor components for model compounds, 427t
Very high frequency EPR spectroscopy of Argonne Premium coals, 512–524
  comparison of W-band spectra, 521–524
  experimental and simulated W-band spectra, 512–516
  maceral type vs. W-band spectra, 512, 517–519
  sensitivity to sulfur and oxygen, 521, 524
Vitrinite
  electron spin-echo envelope modulation vs. frequency, 118
  EPR spectra at high microwave frequencies, 519, 520f
  experimental and simulated W-band EPR spectrum, 512, 519f
  W-band EPR spectrum, 512, 517f, 519

W

Wagner far-IR spectrometer, schematic representation, 130, 131f
WAHUHA multiple-pulse homonuclear dipolar line-narrowing technique, ^1H NMR effects on monoethyl fumarate, 29, 31f, 32
WAHUHA pulse sequence, use in ^1H CRAMPS, 38–39
Weighted-average spin–lattice relaxation time, 346, 347, 350
Wyodak-Anderson coal
  ^1H relaxation times, 301, 302f
  aromatic carbon distribution, 288t
  CW EPR spectroscopy, 483–504
  dipolar-dephasing results, 406f, 407, 408f
  ^1H CRAMPS of untreated and pyridine-saturated coals, 403–406
  hydrogen distribution, 288, 289t
  liquefaction data, 287t, 288
  PMR thermal analysis, 243–245f
  stacked plots of spectra, 284, 285f
  time dependence of magnetization growth and decay, 386, 390–392
  transition metal EPR signals, 547–549t
  truncated second moment-temperature pyrograms, 242f
  W-band EPR spectrum, 512, 513f

X

X-band CW ESR spectroscopy of Argonne Premium coals, 443–448

Z

Zeeman magnetization density, 178
Zeeman spin–lattice relaxation measurement, 617f, 618t
Zero-field EPR, frequencies, 121
Zero-field splitting
  cytochrome c oxidase transmission EPR spectrum, 130, 132f
  high-frequency EPR spectroscopy, 129–132f
  instrumentation, 130, 131f
  parameter estimation, 129–130
Zero-field terms, definition, 114

Copy editing: Janet S. Dodd and Julie Poudrier Skinner
Indexing: Deborah H. Steiner
Production: Donna Lucas
Cover design: Ronna Hammer
Acquisition: Cheryl Shanks

Printed and bound by Maple Press, York, PA

Bestsellers from ACS Books

The ACS Style Guide: A Manual for Authors and Editors
Edited by Janet S. Dodd
264 pp; clothbound, ISBN 0–8412–0917–0; paperback, ISBN 0–8412–0943–X

Chemical Activities and Chemical Activities: Teacher Edition
By Christie L. Borgford and Lee R. Summerlin
330 pp; spiralbound, ISBN 0–8412–1417–4; teacher ed. ISBN 0–8412–1416–6

Chemical Demonstrations: A Sourcebook for Teachers,
Volumes 1 and 2, Second Edition
Volume 1 by Lee R. Summerlin and James L. Ealy, Jr.;
Vol. 1, 198 pp; spiralbound, ISBN 0–8412–1481–6;
Volume 2 by Lee R. Summerlin, Christie L. Borgford, and Julie B. Ealy
Vol. 2, 234 pp; spiralbound, ISBN 0–8412–1535–9

Writing the Laboratory Notebook
By Howard M. Kanare
145 pp; clothbound, ISBN 0–8412–0906–5; paperback, ISBN 0–8412–0933–2

Developing a Chemical Hygiene Plan
By Jay A. Young, Warren K. Kingsley, and George H. Wahl, Jr.
paperback, ISBN 0–8412–1876–5

Introduction to Microwave Sample Preparation: Theory and Practice
Edited by H. M. Kingston and Lois B. Jassie
263 pp; clothbound, ISBN 0–8412–1450–6

Principles of Environmental Sampling
Edited by Lawrence H. Keith
ACS Professional Reference Book; 458 pp;
clothbound; ISBN 0–8412–1173–6; paperback, ISBN 0–8412–1437–9

Biotechnology and Materials Science: Chemistry for the Future
Edited by Mary L. Good (Jacqueline K. Barton, Associate Editor)
135 pp; clothbound, ISBN 0–8412–1472–7; paperback, ISBN 0–8412–1473–5

Personal Computers for Scientists: A Byte at a Time
By Glenn I. Ouchi
276 pp; clothbound, ISBN 0–8412–1000–4; paperback, ISBN 0–8412–1001–2

Polymers in Aqueous Media: Performance Through Association
Edited by J. Edward Glass
Advances in Chemistry Series 223; 575 pp;
clothbound, ISBN 0–8412–1548–0

For further information and a free catalog of ACS books, contact:
American Chemical Society
Distribution Office, Department 225
1155 16th Street, NW, Washington, DC 20036
Telephone 800–227–5558